Fields Institute Communications

VOLUME 67

For further volumes:
http://www.springer.com/series/10503

Radu Laza • Matthias Schütt • Noriko Yui
Editors

Arithmetic and Geometry of K3 Surfaces and Calabi-Yau Threefolds

The Fields Institute for Research
in the Mathematical Sciences

Editors
Radu Laza
Mathematics Department
Stony Brook University
Stony Brook, NY
USA

Matthias Schütt
Institut für Algebraische Geometrie
Leibniz Universität Hannover
Hannover, Germany

Noriko Yui
Department of Mathematics and Statistics
Queen's University
Kingston, ON
Canada

ISSN 1069-5265   ISSN 2194-1564 (electronic)
ISBN 978-1-4614-6402-0   ISBN 978-1-4614-6403-7 (eBook)
DOI 10.1007/978-1-4614-6403-7
Springer New York Heidelberg Dordrecht London

Library of Congress Control Number: 2013932015

Mathematics Subject Classification (2010): 14Jxx, 11Gxx, 14Dxx, 14Cxx

Cover design: Drawing of J.C. Fields by Keith Yeomans

Printed on acid-free paper

Springer is part of Springer Science+Business Media (www.springer.com)

Preface

The workshop on *Arithmetic and geometry of K3 surfaces and Calabi–Yau threefolds* was held at the Fields Institute and University of Toronto from August 16 to 25, 2011. The workshop was organized by Charles F. Doran (Alberta), Shigeyuki Kondō (Nagoya), Radu Laza (Stony Brook), James D. Lewis (Alberta), Matthias Schütt (Hannover), and Noriko Yui (Kingston/Fields).

This proceedings volume for the 2011 Calabi–Yau workshop is edited by Radu Laza, Matthias Schütt, and Noriko Yui. The editors wish to express their appreciation to all the contributors for preparing their manuscripts for the Fields Communications Series, which required extra effort in presenting not only current developments but also some background material on the discussed topics. All papers in this volume were peer-reviewed. We are deeply grateful to all the referees for their efforts in evaluating the articles, in particular, in the limited time frame.

The workshop was financially supported by various organizations. In addition to the Fields Institute, the workshop was supported by NSF (grant no. 1100007), JSPS (Grant-in-Aid (S), No. 22224001), and DFG (GRK 1463 "Analysis, Geometry and String Theory"). Additionally, several participants have used their individual grants (e.g., NSF or NSERC) to cover their travel expenses. We are thankful to all these sponsor organizations: their support made possible the participation of a large number of junior participants and of a significant number of researchers from outside North America. This in turn led to a very dynamic and active workshop.

Some of the articles were copy-edited by Arthur Greenspoon of Mathematical Reviews. The editors are grateful for his help towards improving both the stylistic and mathematical presentations.

Last but not least, we thank Debbie Iscoe of the Fields Institute for her help in re-formatting articles in the Springer style and assembling this volume for publication.

Stony Brook, NY	Radu Laza
Hannover, Germany	Matthias Schütt
Kingston, ON	Noriko Yui

Introduction

In recent years, research in K3 surfaces and Calabi–Yau varieties has seen spectacular progress from both arithmetic and geometric points of view, which in turn continues to have a huge influence and impact on theoretical physics, in particular, on string theory. The workshop was designed to bring together experts and junior researchers who are aspiring to become experts for 10 days at the Fields Institute in August 2011 to review recent developments, inspire graduate and postdoctoral fellows and young researchers, and also explore future directions of the subjects. With 114 (officially registered) participants, there was a wide geographical representation, with a very significant presence of European and Japanese participants (in addition to US and Canadian participants).

The workshop started with a 3-day introductory session aimed at graduate students and postdoctoral fellows, followed by a 1 week research conference with Sunday off. The introductory lectures were intended to give some background and a brief overview of the vast topic of Calabi–Yau varieties and K3 surfaces. At the subsequent research conference, there were in total 35 research talks presented on wide ranges of topics around K3, Enriques and other surfaces, and Calabi–Yau threefolds and higher-dimensional varieties and manifolds, some of which can be found in this volume.

As a consequence of the significant interest in the subject, we are organizing a follow-up extended concentration period on Calabi–Yau varieties, in the form of a semester long thematic program *Calabi–Yau varieties: arithmetic, geometry and physics* at the Fields Institute (July to December 2013). This thematic program is devoted to the arithmetic and geometry of Calabi–Yau varieties and the connections to physics, especially string theory.

Scientific Focus of the Workshop

The workshop concentrated on topics (on the geometry and arithmetic of Calabi–Yau varieties) that have either seen great progress recently or shown a high potential for future inventions. Specifically, the major topics covered included:

1. Families and degenerations of Calabi–Yau varieties—moduli theoretic and arithmetic viewpoints.
2. Modularity: Galois representations of Calabi–Yau varieties and their connections to automorphic forms, in particular to classical, Hilbert, and Siegel modular forms.
3. Calabi–Yau varieties of CM type and with special automorphisms, especially K3 surfaces with symplectic and non-symplectic automorphisms.
4. Algebraic cycles and motives: divisors, CM cycles, and motives arising from K3 surfaces and Calabi–Yau threefolds.
5. Variations of mixed Hodge structures, periods, and Picard–Fuchs differential equations.

Overview of This Volume

In the following paragraphs we give a brief overview of the volume. There are in total 24 articles. Some of the articles are written-up versions of the talks presented at the workshop, while others report on subsequent developments on the subject matter of the workshop. Roughly the articles are divided into three categories, namely:

- *Introductory lectures*
- *Arithmetic and geometry of K3, Enriques and other surfaces*
- *Arithmetic and geometry of Calabi–Yau threefolds and higher dimensional varieties*

   The workshop's program contained several other talks on related topics. Additional documentation from the workshop is available on the homepage maintained by the Fields Institute.[1]

Introductory Lectures

There are four survey papers by Kondō, Lewis, Schütt, and Yui which comprise a selection of the lectures given by the organizers during the 3-day introductory period of the workshop. These lectures were mostly aimed at junior participants of the workshop and often geared specifically towards some of the talks to be given

---

[1] http://www.fields.utoronto.ca/programs/scientific/11-12/CalabiYau/index.html.

during the research conference. Similarly for this volume, the survey papers can be used as a starting point or as a guide to the subject.

The surveys by Kondō and Schütt review the geometry and arithmetic of K3 surfaces, including basics such as lattice theory. Yui's paper presents the current status on modularity of Calabi–Yau varieties in its different incarnations. The focus lies on Calabi–Yau varieties of dimension at most three. Lewis reviews transcendental aspects of algebraic cycles and, specializing to the Calabi–Yau situation, explains some recent developments in the field.

Arithmetic and Geometry of K3, Enriques, and Other Surfaces

A common theme of many papers in this section are elliptic fibrations. Most notably, Bertin and Lecacheux classify all elliptic fibrations on a specific K3 surface. From a similar elliptic modular surface, Anema and Top derive explicit algebraic coverings of a pointed torus. On the moduli side, Besser and Livné relate specific elliptic K3 surfaces to abelian surfaces with quaternionic multiplication; this produces explicit Shimura curves. Special cycles in moduli spaces of lattice polarized K3 surfaces are treated by Kudla. These higher Noether–Lefschetz loci are the input of certain generating series whose modularity is known.

Elliptic fibrations form also the key ingredient for Kerr's approaches to the computation of transcendental invariants of indecomposable algebraic K_1 classes. Moreover Kerr's work builds on toric geometry which features prominently in the paper of Whitcher et al. as well. Here three-dimensional reflexive polytopes with S_4 symmetry are related to natural one-parameter family of K3 surfaces with symplectic S_4 action. Picard–Fuchs equations are studied not only in this paper but also by Gährs for certain one-parameter families associated with invertible polynomials, using the GKZ system.

Oguiso's paper is concerned with a classical problem: it proves that there is a smooth quartic K3 surface automorphism that is not derived from a Cremona transformation. Almost as classical a problem for Enriques surfaces, Dolgachev extends results for cohomologically or numerically trivial automorphisms to arbitrary characteristics. Contrary to previous approaches, the key tool is Lefschetz' fixed point formula. Enriques surfaces of Hutchinson–Göpel type are investigated by Mukai and Ohashi. Starting from the projective geometry of Jacobian Kummer surfaces, they give a sextic presentation for these Enriques surfaces and then describe their intrinsic symmetries.

On the less classical side, Hulek and Ploog extend the theory of Fourier–Mukai partners to include the presence of polarizations, producing a counting formula for the number of partners. In a different direction, Schoen's paper computes invariants of the degeneration of a product of elliptic curves upon split-muliplicative reduction. He expatiates on divisor class group, (co)homology, and Picard group of the closed fibers. To close the circle on fibrations, Heijne and Kloosterman study a special class of surfaces, the so-called Delsarte surfaces. Singled out by their accessibility to explicit computations, a classification of some specific fibrations on these surfaces is given.

Arithmetic and Geometry of Calabi–Yau Threefolds and Higher Dimensional Varieties

For one-dimensional families of Calabi–Yau manifolds, Cynk and van Straten compute Picard–Fuchs operators based on the expansion of a period near a conifold point. The algorithm is explained in detail and illustrated by some concrete examples consisting in double octics. In contrast, the paper by Gouvêa, Kiming, and Yui considers rigid Calabi–Yau threefolds defined over **Q**. Motivated by the geometric realization problem they pose the question of whether the Calabi–Yau threefolds admit quadratic twists, giving answers for a number of examples. Automorphic forms are also the main players in the papers by Kondō and Movasati. Kondō uses Borcherds theory of automorphic forms on orthogonal groups to construct a rational map from the Segre cubic threefold to its dual, the Igusa quartic threefold. The major novelty of Movasati's paper consists in a modification of the interplay between moduli of polarized Hodge structures of a fixed type and Griffiths period domains. While in the classical case (for Hermitian symmetric domains) one obtains automorphic forms and algebraic structures on the mentioned moduli spaces, Movasati's formulation leads to a notion of quasi-automorphic forms.

The foundations for Donaldson–Thomas invariants for stable sheaves on algebraic threefolds with trivial canonical bundle are reviewed in Gulbrandsen's contribution. Special emphasis lies on abelian threefolds. Chen and Lewis prove density statements about the subgroup of invertible points on intermediate Jacobians. They focus on the points in the Abel–Jacobi image of nullhomologous algebraic cycles on projective algebraic manifolds. Last but not least, the paper by Pearlstein and Schnell concerns the infinitesimal invariant of a normal function on a complex manifold. When the manifold is quasi-projective and the function is admissible, they show that this zero locus is constructible in the Zariski topology.

Acknowledgment

Let us conclude by expressing our sincere thanks to all the contributors to this volume, the referees, the sponsoring institutions, and, most of all, all participants of the workshop for creating the stimulating atmosphere from which this volume arose.

List of Participants

Ahmadi, Ruhi (Seyedruhallah)	University of Regina
Bertin, Marie José	Université Paris 6
Cattaneo, Andrea	Università degli Studi di Milano
Caviedes Castro, Alexander	University of Toronto
Chen, Xi	University of Alberta
Chung, Min In	
Clingher, Adrian	University of Missouri–St. Louis
Colombo, Elisabetta	Università di Milano
Cynk, Slawomir	Jagiellonian University
Dolgachev, Igor	University of Michigan
Doran, Charles	University of Alberta
Dubey, Umesh Kumar Vanktesh	The Institute of Mathematical Sciences
Dumitrescu, Olivia	University of California, Davis
El basraoui, Abdelkrim	CRM-CICMA
Elizondo, E. Javier	UNAM
Elkies, Noam D.	Harvard University
Filippini, Sara Angela	Università degli Studi dell'Insubria
Gährs, Swantje	Leibniz Universität Hannover
Gallardo, Patricio	Stony Brook University
Goto, Yasuhiro	Hokkaido University of Education
Grieve, Nathan	Queen's University
Gualtieri, Marco	University of Toronto
Gulbrandsen, Martin	Stord/Haugesund University College
Halic, Mihai	King Fahd University of Petroleum and Minerals
Halle, Lars Halvard	University of Oslo
Harder, Andrew	Queen's University
Hashimoto, Kenji	Korea Institute for Advanced Study

Hayama, Tatsuki	National Taiwan University
Hoffman, Jerome	Louisiana State University
Hulek, Klaus	Leibniz Universität Hannover
Hwang, DongSeon	Korea Institute for Advanced Study
Ito, Hiroki	Nagoya University
Ito, Hiroyuki	Tokyo University of Science
Jang, Junmyeong	Korea Institute for Advanced Study
Jensen, David	SUNY Stony Brook
Kanazawa, Atsushi	University of British Columbia
Kelly, Tyler	University of Pennsylvania
Kerr, Matthew	Washington University in St. Louis
Keum, JongHae	Korea Institute for Advanced Study
Kiming, Ian	University of Copenhagen
Kloosterman, Remke	Humboldt Universitaet zu Berlin
Kondo, Shigeyuki	Nagoya University
Kooistra, Remkes	The King's University College
Kudla, Stephen	University of Toronto
Kumar, Abhinav	Massachusetts Institute of Technology
Kuwata, Masato	Chuo University
Lang, William	Brigham Young University
Laza, Radu	Stony Brook University
Lecacheux, Odile	Université de Paris 6
Lee, ChongGyu	University of Illinois at Chicago
Lee, Hwayoung	Korea Institute for Advanced Study
Lee, Jae Hyouk	Ewha Womans University
Lewis, Jacob	Universität Wien
Lewis, James D.	University of Alberta
Liedtke, Christian	Stanford University
Lyons, Christopher	University of Michigan
Mase, Makiko	Tokyo Metropolitan University
Matias, Rodrigo	Concordia University
Mayanskiy, Evgeny	Pennsylvania State University
McKay, John	Concordia University
Miura, Makoto	University of Tokyo
Mongardi, Giovanni	Università degli studi "RomaTRE"
Moraru, Ruxandra	University of Waterloo
Movasati, Hossein	IMPA
Mukai, Shigeru	Kyoto University
Nikulin, Viacheslav	University of Liverpool
Novoseltsev, Andrey	University of Alberta
O'Grady, Kieran	Sapienza Università di Roma
Oguiso, Keiji	Osaka University
Ohashi, Hisanori	Nagoya University
Okawa, Shinnosuke	University of Tokyo
Pasten, Hector	Queen's University

Pearlstein, Gregory	Michigan State University
Prez-Buenda, Rogelio	Concordia University
Rams, Slawomir	Jagiellonian University
Rana, Julie	UMass Amherst
Rohde, Jan Christian	Universitaet Hamburg
Saber, Hicham	University of Ottawa
Saito, Sachiko	Hokkaido University of Education, Asahikawa Campus
Sakovics, Dmitrijs	Universitry of Edinburgh
Sano, Taro	University of Warwick
Sarti, Alessandra	University of Poitiers
Schnell, Christian	University of Illinois
Schoen, Chad	Duke University
Schröer, Stefan	University Duesseldorf
Schulze, Frithjof	Leibniz Universität Hannover
Schütt, Matthias	Leibniz Universität Hannover
Sijsling, Jeroen	IMPA
Smirnov, Ilia	Queen's University
Solis, Pablo	University of California, Berkeley
Song, Ruifang	Harvard University
Takayama, Yukihide	Ritsumeikan University
Taki, Shingo	Tokyo Denki University
Talamanca, Valerio	Università Roma Tre
Tanimoto, Sho	Courant Institute of Mathematical Sciences, New York University
Thompson, Alan	University of Oxford
Tian, Zhiyu	Stony Brook University
Tuncer, Serhan	University of Alberta
Ufer, Dominik	University of Ulm
van den Dries, Bart	Utrecht University
van Straten, Duco	University of Johannes Gutenberg
Wandel, Malte	Leibniz Universität Hannover
Wen, Jun	Stony Brook University
Whang, Junho Peter	Queen's University
Whitcher, Ursula	Harvey Mudd College
Yang, Sen	Louisiana State University
Yoshikawa, Ken-Ichi	Kyoto University
Yu, Jeng-Daw	National Taiwan University
Yui, Noriko	Queen's University
Zhang, Letao	Rice University
Zhang, Ying	University of Pennsylvania
Zhang, Yongsheng	Stony Brook University
Zhang, Zheng	Stony Brook University

Contents

Part III Research Articles: Arithmetic and Geometry of Calabi-Yau Threefolds and Higher Dimentional Varieties

Part I
Introductory Lectures

K3 and Enriques Surfaces

Shigeyuki Kondō

Abstract This is a note on my introductory lectures on $K3$ and Enriques surfaces in the workshop "Arithmetic and Geometry of K3 surfaces and Calabi–Yau threefolds" held at the Fields Institute. No new results are included.

Key words: $K3$ surfaces, Enriques surfaces, Torelli type theorem, Automorphisms, Automorphic forms

Mathematics Subject Classifications (2010): Primary 14J28; Secondary 14C34, 14J50, 11F55, 32N10

1 Introduction

In this note we give a brief survey on the theory of $K3$ and Enriques surfaces. We consider only complex $K3$ and Enriques surfaces. The main topics in this note are moduli and automorphisms. For $K3$ and Enriques surfaces, an analogue of the Torelli theorem for compact Riemann surfaces holds. Let C be a compact Riemann surface of genus $g \geq 1$. Let $\omega_1, \ldots, \omega_g$ be a basis of $H^0(C, \Omega^1_C)$ and let $\alpha_1, \ldots, \alpha_g, \beta_1, \ldots, \beta_g$ be a symplectic basis of $H_1(C, \mathbf{Z})$. Then the $g \times 2g$ matrix

$$\left(\int_{\alpha_j} \omega_i, \ \int_{\beta_j} \omega_i \right)_{1 \leq i,j \leq g}$$

is called the period matrix of C. Roughly speaking, the Torelli theorem for compact Riemann surfaces states that the isomorphism class of C is determined by the period matrix.

S. Kondō (✉)
Graduate School of Mathematics, Nagoya University, Nagoya 464-8602, Japan
e-mail: kondo@math.nagoya-u.ac.jp

R. Laza et al. (eds.), *Arithmetic and Geometry of K3 Surfaces and Calabi–Yau Threefolds*,   3
Fields Institute Communications 67, DOI 10.1007/978-1-4614-6403-7_1,
© Springer Science+Business Media New York 2013

For $K3$ surfaces, their periods are defined by the integrals of a holomorphic 2-form over the second homology classes. Thanks to the Torelli type theorem for $K3$ surfaces, due to Piatetskii-Shapiro and Shafarevich [35] for the algebraic case and Burns and Rapoport [9] for the general case, we can reduce many geometric problems to those of lattices. We shall explain how to use the lattice theory. Recall that for a $K3$ surface X, $H^2(X, \mathbf{Z})$ together with the cup product is an even unimodular lattice of signature $(3, 19)$. Moreover, for an algebraic $K3$ surface, the Picard lattice S_X and its orthogonal complement T_X, called the transcendental lattice, are important primitive sublattices in $H^2(X, \mathbf{Z})$. The Picard lattice S_X determines the distribution of curves on X and the group $\mathrm{Aut}(X)$ of automorphisms of X is isomorphic, up to finite groups, to the group $\mathrm{O}(S_X)/W(X)$ where $\mathrm{O}(S_X)$ is the orthogonal group of S_X and $W(X)$ is the subgroup generated by reflections associated to (-2)-vectors in S_X. On the other hand, the transcendental lattice T_X describes the moduli. The associated period domain is a bounded symmetric domain of type IV. It would be interesting to study the moduli by using automorphic forms.

In Sect. 2 we recall the basic theory of lattices. We discuss classification of even unimodular lattices. Since the Euler number of a $K3$ surface is 24, even definite unimodular lattices of rank 24, called Niemeier lattices, will be important. We also introduce an important invariant, called the discriminant quadratic form, of even lattices. We explain the notion of overlattices making a new lattice from a given lattice and primitive embeddings of even lattices into even unimodular lattices in terms of the discriminant quadratic forms. In Sect. 3 we recall periods and period domains for $K3$ and Enriques surfaces. In Sect. 4 we state the Torelli type theorem for algebraic $K3$ surfaces (Theorem 4.2), and then we discuss the group of automorphisms of $K3$ and Enriques surfaces. Also, we mention what kind of finite groups can act on $K3$ surfaces as automorphisms. Here a sporadic finite simple group called the Mathieu group appears. Finally, in Sect. 5, we recall Borcherds' theory on automorphic forms on bounded symmetric domains of type IV associated to lattices of signature $(2, n)$. We mention some applications of Borcherds' theory to the moduli of $K3$ and Enriques surfaces.

The author thanks the referee for his careful reading of the manuscript and many suggestions.

2 Lattices

Main reference of this section is Nikulin [30].

2.1 Definition

A *lattice* $(L, \langle, \rangle)$ is a pair of a free $\mathbf{Z}$-module L of rank r and a non-degenerate symmetric integral bilinear form $\langle, \rangle : L \times L \to \mathbf{Z}$. For simplicity we omit $\langle, \rangle$ if there is no confusion. For a lattice L, we denote by L^* the dual $\text{Hom}(L, \mathbf{Z})$ of L, and by A_L the quotient group L^*/L which is a finite abelian group. Denote by $d(L)$ the order of A_L. A lattice L is called *unimodular* if $d(L) = 1$, and *even* if $\langle x, x \rangle \equiv 0$ mod 2 for any $x \in L$. The *signature* of a lattice L is that of the quadratic form $L \otimes \mathbf{R}$. Also, we call L *negative definite*, *positive definite* or *indefinite* if $L \otimes \mathbf{R}$ is so. We denote by $L \oplus M$ the orthogonal direct sum of lattices L and M. Also, we denote by $L^{\oplus m}$ the orthogonal direct sum of m copies of L. For simplicity and applications to algebraic geometry we focus on the case of even lattices.

2.2 Examples

We denote by U the *hyperbolic plane*, that is, an even unimodular lattice of signature $(1, 1)$. For a lattice $(L, \langle , \rangle)$ and an integer m, we denote by $L(m)$ the lattice $(L, m\langle , \rangle)$. For example, $U(m)$ is defined by the matrix $\begin{pmatrix} 0 & m \\ m & 0 \end{pmatrix}$. A *root lattice* is a negative definite lattice generated by (-2)-vectors. We denote by A_m, D_n or E_k the even negative definite lattice defined by the Cartan matrix of type A_m, D_n or E_k, respectively. Any root lattice is isomorphic to the orthogonal direct sum of A_m, D_n, E_k. Usually root lattices are assumed to be positive definite. Here we use the opposite sign because it is more relevant in algebraic geometry.

2.3 Unimodular Lattices

Among lattices, unimodular lattices are fundamental. In the following we introduce the classification problem of unimodular lattices.

2.4 Proposition

Let L be an even unimodular lattice of signature (p, q). Then $p - q \equiv 0$ mod 8.

In the case of indefinite unimodular lattices, its isomorphism class is determined by its signature.

2.5 Proposition

Let L be an even unimodular indefinite lattice. Then L is uniquely determined by its signature (p, q). If $p \geq q$, then $L \cong U^{\oplus q} \oplus E_8(-1)^{\oplus(p-q)/8}$. If $q \geq p$, then $L \cong U^{\oplus p} \oplus E_8^{\oplus(q-p)/8}$.

For the proofs of these propositions, we refer the reader to Serre [37]. On the other hand, it is a difficult problem to determine isomorphism classes of definite unimodular lattices. Up to rank 24, the classification is known. See the following Table 1.

Table 1: Classification of even definite unimodular lattices

r	8	16	24	32
n(r)	1	2	24	$\geq 8 \cdot 10^8$
L	E_8	$E_8^{\oplus 2}$, Γ_{16}	Niemeier lattices	?

Here r is the rank of L and $n(r)$ is the number of isomorphism classes of even negative definite unimodular lattices of rank r. Γ_{16} is an even negative definite unimodular lattice of rank 16 whose root sublattice is D_{16}. Later we will give a construction of Γ_{16} (see Example 2.9, (1)). We call an even negative definite unimodular lattice of rank 24 a *Niemeier lattice*. The isomorphism class of any Niemeier lattice N is determined by its root sublattice $R(N)$. The root sublattice $R(N)$ has remarkable properties, for example, if $R(N) \neq \emptyset$, then rank$(R(N)) = 24$. A Niemeier lattice without (-2)-vectors is called a *Leech lattice*. Later we will give an example of a Niemeier lattice with $R(N) = A_1^{\oplus 24}$ (see Example 2.9, (2)). For more details, we refer the reader to Conway and Sloane [10].

2.6 Discriminant Quadratic Form

Next we introduce the most important invariant for even lattices. Let L be an even lattice. Define the maps

$$q_L : A_L \to \mathbf{Q}/2\mathbf{Z}, \quad b_L : A_L \times A_L \to \mathbf{Q}/\mathbf{Z}$$

by $q_L(x + L) = \langle x, x \rangle$ mod $2\mathbf{Z}$ and $b_L(x + L, y + L) = \langle x, y \rangle$ mod $\mathbf{Z}$. We call q_L the *discriminant quadratic form* and b_L the *discriminant bilinear form*. Let O(L) be the orthogonal group of L, that is, the group of isomorphisms of L preserving the bilinear form. Similarly O(q_L) denotes the group of isomorphisms of A_L preserving q_L. Let $g \in$ O(L). Then g acts on L^* canonically and hence acts on $A_L = L^*/L$. Thus there is a natural map from O(L) to O(q_L).

2.7 Overlattices

Let L be an even lattice. For a given L, we introduce a simple method to get a new even lattice L'. An even lattice L' is called an *overlattice* of L if L is a sublattice of L' of finite index. Let $H \subset A_L$ be an isotropic subgroup with respect to q_L, that is, $q_L|H \equiv 0$. Then we can define a subgroup L_H of L^* by

$$L_H = \{x \in L^* : x \bmod L \in H\}.$$

The condition $q_L|H \equiv 0$ implies that L_H is an even lattice. Thus we have an over-lattice L_H of L. Note that $d(L_H) = d(L)/[L_H : L]^2$. Conversely, if L' is an overlattice of L, then we have an isotropic subspace L'/L of A_L. Thus we have the following proposition.

2.8 Proposition

The set of overlattices of L bijectively corresponds to the set of isotropic subgroups of A_L.

2.9 Examples

We give two examples of even unimodular lattices by using Proposition 2.8.

(1) Let $L = D_{16}$. Then we can easily see that $A_L \cong (\mathbf{Z}/2\mathbf{Z})^2$ and A_L consists of three isotropic vectors and one non-isotropic vector. Let $\alpha \in A_L$ be a non-zero isotropic vector and let $H \cong \mathbf{Z}/2\mathbf{Z}$ be the isotropic subgroup of A_L generated by α. Then we have an overlattice Γ_{16} which is an even unimodular lattice because

$$d(\Gamma_{16}) = \frac{d(L)}{[\Gamma_{16} : L]^2} = \frac{4}{2^2} = 1.$$

Since D_{16} is not contained in $E_8^{\oplus 2}$, Γ_{16} is not isomorphic to $E_8^{\oplus 2}$.

(2) Let $L = A_1^{\oplus 24}$. Then $A_L \cong \mathbf{F}_2^{24}$. Let $x = (x_1, \ldots, x_{24}) \in \mathbf{F}_2^{24}$ be the standard coordinate. Then

$$q_L(x) = -\frac{1}{2} \sum_{i=1}^{24} x_i.$$

There exists a subspace $\mathcal{G}$ of dimension 12 in A_L, called a *binary Golay code*, satisfying the conditions

(i) $(1, \ldots, 1) \in \mathcal{G}$,
(ii) For $x \in \mathcal{G}$, the number of non-zero entries of x is divisible by 4,
(iii) For $x \neq 0 \in \mathcal{G}$, the number of non-zero entries of $x \geq 8$.

The condition (ii) means that $\mathcal{G}$ is isotropic, and hence there exists an overlattice N of L with $[N : L] = 2^{12}$. Hence N is an even unimodular lattice. The third condition implies that the root sublattice $R(N)$ of N coincides with L. Thus we have a Niemeier lattice N with $R(N) = A_1^{24}$. Since any isometry of N preserves $R(N)$, $O(N)$ is a subgroup of $O(R(N))$. Obviously

$$O(R(N)) \cong (\mathbf{Z}/2\mathbf{Z})^{24} \cdot \mathfrak{S}_{24}$$

where $\mathfrak{S}_{24}$ is the symmetric group of degree 24 acting on the coordinates as permutations and $(\mathbf{Z}/2\mathbf{Z})^{24}$ is the Weyl group $W(R(N))$ of $R(N)$. Define

$$M_{24} = \{\sigma \in \mathfrak{S}_{24} \ : \ \sigma(\mathcal{G}) = \mathcal{G}\}.$$

Since $W(R(N))$ acts trivially on A_L, $W(R(N))$ is a normal subgroup of $O(N)$. The quotient $O(N)/W(R(N))$ is isomorphic to M_{24}. The group M_{24} is called the *Mathieu group* of degree 24, and is a finite sporadic simple group. For Mathieu groups, we refer the reader to Conway and Sloane [10].

2.10 Primitive Embeddings

In the following we discuss embeddings of an even lattice into an even unimodular lattice. Let L be an even unimodular lattice and let S be a sublattice of L. Here we assume that S is *primitive*, that is, L/S is torsion-free. Denote by T the orthogonal complement of S in L. Then T is a primitive sublattice of L. Obviously L is an overlattice of $S \oplus T$, and hence $H = L/(S \oplus T)$ is an isotropic subgroup of $A_S \oplus A_T$ with respect to $q_S \oplus q_T$. Let

$$p_S : A_S \oplus A_T \to A_S, \quad p_T : A_S \oplus A_T \to A_T$$

be the projections. Then we have the following proposition.

2.11 Proposition

The restrictions $p_S|H : H \to A_S$ and $p_T|H : H \to A_T$ are isomorphic and H is the graph of $\gamma = p_T \circ p_S^{-1} : A_S \to A_T$. Moreover $q_S(x) + q_T(\gamma(x)) \equiv 0 \bmod 2$.

Proof. First we show the injectivity of $p_S|H$. Let $x \in L$. Write $x = x_S + x_T$ where $x_S \in S^*$ and $x_T \in T^*$. Assume that $x_S = 0$ in A_S. Then $x_S \in S$ and hence $x_T = x - x_S \in L$. Since T is primitive in L, $x_T \in T$. Hence $x \in S \oplus T$. This implies the injectivity of $p_S|H$. Similarly $p_T|H$ is injective. Since $|H|^2 = |A_S| \cdot |A_T|$, $p_S|H$ and $p_T|H$ are isomorphic. The remaining assertions are obvious.

Conversely we have the following corollary.

2.12 Corollary

Let S, T be even lattices. Let $\gamma : A_S \to A_T$ be an isomorphism satisfying $q_T(\gamma(x)) \equiv -q_S(x) \bmod 2$ for any $x \in A_S$. Then there exist an even unimodular lattice L and *a primitive embedding of S into L such that T is isomorphic to the orthogonal complement of S in L.*

Proof. Consider the subgroup

$$H = \{(x, \gamma(x)) : x \in A_S\}$$

of $A_S \oplus A_T$. Then, by the assumption, H is isotropic with respect to $q_S \oplus q_T$, and hence, by Proposition 2.8, there exists an even unimodular lattice L. Since γ is isomorphic, S is primitive in L.

2.13 Example

We take E_8 to be an even unimodular lattice L. Let S be a root lattice primitively embedded in E_8 and let $T = S^\perp$ in L. Then T is given as in the following Table 2.

Table 2: Examples of primitive sublattices in E_8

S	A_1	A_2	A_3	A_4	D_4
T	E_7	E_6	D_5	A_4	D_4

The following corollaries will be used later.

2.14 Corollary

Let L be an even unimodular lattice and let S be a primitive sublattice of L. Let $T = S^\perp$ in L. Let $g \in O(S)$. Assume that g acts on A_S trivially. Then there exists an isometry $\tilde{g} \in O(L)$ of L such that $\tilde{g}|S = g$ and $\tilde{g}|T = 1_T$.

Proof. Put $\tilde{g} = (g, 1_T) \in O(S \oplus T)$. Then $\tilde{g}$ is an isometry of the dual $S^* \oplus T^*$. Since $\tilde{g}$ acts trivially on $A_S \oplus A_T$, it acts trivially on $L/(S \oplus T)$; that is, $\tilde{g}(L) = L$.

Similarly we can see the following.

2.15 Corollary

Let L be an even unimodular lattice and let S be a primitive sublattice of L. Let $T = S^\perp$ in L. Assume that the natural map

$$O(T) \to O(q_T)$$

is surjective. Then any $g \in O(S)$ can be extended to an isometry of L.

Finally let us consider the uniqueness of primitive embeddings of an even lattice S into an even unimodular lattice L. We need the uniqueness of isomorphism classes of its orthogonal complement $T = S^\perp$. Also, for the extension problem of isometries of S to those of L, we need the condition of T mentioned in Corollary 2.14. The following theorem gives a sufficient condition for these problems.

2.16 Theorem ([30])

Let T be an even indefinite lattice. Assume that

$$\mathrm{rank}(T) \geq l(T) + 2$$

where $l(T)$ is the number of minimal generators of the finite abelian group A_T. Then the isomorphism class of T is uniquely determined by its signature and q_T. Moreover, the natural map $\mathrm{O}(T) \to \mathrm{O}(q_T)$ is surjective.

The above theorem is due to Nikulin [30]. Nikulin proved a more general result. See [30], Theorem 1.14.2.

3 Periods of K3 and Enriques Surfaces

3.1 Periods of K3 Surfaces

A *K3 surface* is a compact complex surface with $H^1(X, \mathcal{O}_X) = 0$ and $K_X = 0$, where K_X is the canonical line bundle of X. The important fact is that $H^2(X, \mathbf{Z})$ together with the cup product is an even unimodular lattice of signature $(3, 19)$. This follows from Wu's formula, Poincaré duality, and Hirzebruch's index theorem. For more details, we refer the reader to Barth et al. [3].

Let L be an (abstract) even unimodular lattice of signature $(3, 19)$. By Proposition 2.5, there is an isomorphism

$$\alpha_X : H^2(X, \mathbf{Z}) \to L.$$

The pair (X, α_X) is called a *marked K3 surface*. By definition of a K3 surface, there exists a nowhere vanishing holomorphic 2-form ω_X on X which is unique up to constants. Recall that ω_X satisfies the Riemann condition

$$\langle \omega_X, \omega_X \rangle = 0, \quad \langle \omega_X, \bar{\omega}_X \rangle > 0.$$

Define

$$\Omega = \{[\omega] \in \mathbf{P}(L \otimes \mathbf{C}) : \langle \omega, \omega \rangle = 0, \quad \langle \omega, \bar{\omega} \rangle > 0\}.$$

Then Ω is a 20-dimensional complex manifold, which is called the *period domain* of marked $K3$ surfaces. For a marked $K3$ surface (X, α_X), we associate the point $\alpha_X(\omega_X)$ in Ω. The Torelli type theorem for $K3$ surfaces and the surjectivity of the period map imply that Ω is the set of isomorphism classes of marked $K3$ surfaces.

Next consider the case of polarized $K3$ surfaces. Let X be an algebraic $K3$ surface. A *polarization of degree* $2d$ on X is a primitive nef and big divisor H with $H^2 = 2d > 0$ (here a divisor D is called primitive if $D = mD'$ for some integer m and a divisor D' then $m = \pm 1$). We call the pair (X, H) a *polarized $K3$ surface of degree* $2d$. Let h be a primitive vector in L with $h^2 = \langle h, h \rangle = 2d$. Let L_{2d} be the orthogonal complement of h in L. Note that L_{2d} does not depend on the choice of h by Theorem 2.16. Moreover, for a polarized $K3$ surface (X, H), there exists an isomorphism

$$\alpha_X : H^2(X, \mathbf{Z}) \to L$$

which sends H to h. Define

$$\mathcal{D}_{2d} = \{ [\omega] \in \mathbf{P}(L_{2d} \otimes \mathbf{C}) \ : \ \langle \omega, \omega \rangle = 0, \quad \langle \omega, \bar{\omega} \rangle > 0 \}.$$

Then $\mathcal{D}_{2d}$ is a disjoint union of two copies of a bounded symmetric domain of type IV and of dimension 19. Let Γ_{2d} be the kernel of the natural map $O(L_{2d}) \to O(q_{L_{2d}})$ which acts on $\mathcal{D}_{2d}$ properly discontinuously. To a triple (X, H, α_X) we associate the point $\alpha_X(\omega_X)$ in $\mathcal{D}_{2d}$ because ω_X is perpendicular to algebraic classes. For a polarized $K3$ surface (X, H) we associate the point $\alpha_X(\omega_X)$ mod Γ_{2d} in $\mathcal{D}_{2d}/\Gamma_{2d}$ forgetting the marking α_X. The Torelli type theorem (see Theorem 4.2) tells us that $\mathcal{D}_{2d}/\Gamma_{2d}$ is the coarse moduli space of polarized $K3$ surfaces of degree $2d$. The Torelli type theorem for algebraic $K3$ surfaces was first given by Piatetskii-Shapiro and Shafarevich [35]. Later Burns and Rapoport [9] gave a Torelli type theorem for Kähler $K3$ surfaces.

3.2 Periods of Enriques Surfaces

An algebraic surface Y is called an *Enriques surface* if its geometric genus p_g and the irregularity q vanish, and $K_Y^{\otimes 2}$ is trivial. The 2-torsion K_Y defines an unramified double cover

$$\pi : X \to Y.$$

One can easily see that X is a $K3$ surface. Since $K3$ surfaces are simply connected, π is the universal cover of Y. Let σ be the covering transformation of π. The Enriques surface Y is completely determined by the pair of the $K3$ surface X and a fixed-point free involution σ of X.

Define

$$L_X^{\pm} = \{ x \in H^2(X, \mathbf{Z}) \ : \ \sigma^*(x) = \pm x \}.$$

Obviously $L_X^{\pm}$ are primitive sublattices of $H^2(X, \mathbf{Z})$. Moreover $L_X^+ \cong \pi^* H^2(Y, \mathbf{Z})$. By the definition of Enriques surfaces we can see that the second Betti number of Enriques surfaces is 10. The genus formula, Poincaré duality and the Hodge index theorem imply that the free part of $H^2(Y, \mathbf{Z})$ is an even unimodular lattice of signature $(1, 9)$ which is isomorphic to $U \oplus E_8$ by Proposition 2.5. Since π is an unramified double covering, $L_X \cong U(2) \oplus E_8(2)$. By Proposition 2.11, $q_{L_X^+} = -q_{L_X^-}$. In particular the isomorphism class of L_X^- is uniquely determined by its signature $(2, 10)$ and $q_{L_X^-}$ (Theorem 2.16). It is known that

$$L_X^- \cong U \oplus U(2) \oplus E_8(2).$$

We fix abstract lattices $L^+ = U(2) \oplus E_8(2)$ and $L^- = U \oplus U(2) \oplus E_8(2)$. It follows from Proposition 2.8 that there exists an even unimodular lattice L of signature $(3, 19)$ which is an overlattice of $L^+ \oplus L^-$. Let ι be an isometry $(1_{L^+}, -1_{L^-})$ of $L^+ \oplus L^-$. By Corollary 2.14, ι can be extended to an isometry of L, which is denoted by the same symbol ι. By using Theorem 2.16, we can see that for each Enriques surface Y or equivalently for each pair of a $K3$ surface and a fixed-point-free involution σ, there exists an isomorphism

$$\alpha_X : H^2(X, \mathbf{Z}) \to L$$

which sends L_X^+ to L^+. Now we define

$$\mathcal{D}(L^-) = \{[\omega] \in \mathbf{P}(L^- \otimes \mathbf{C}) : \langle \omega, \omega \rangle = 0, \quad \langle \omega, \bar{\omega} \rangle > 0\}.$$

Then $\mathcal{D}(L^-)$ is a disjoint union of two copies of a bounded symmetric domain of type IV and of dimension 10. Let Γ be the orthogonal group $O(L^-)$, which acts on $\mathcal{D}(L^-)$ properly discontinuously. For (X, σ, α_X) we associate the point $\alpha_X(\omega_X)$ in $\mathcal{D}$. Forgetting the marking α_X, we associate the point $\alpha_X(\omega_X)$ mod Γ in $\mathcal{D}(L^-)/\Gamma$.

3.3 Remark

Let $\omega \in \mathcal{D}(L^-)$. Then it follows from the surjectivity of the period map for $K3$ surfaces that there exists a marked $K3$ surface (X, α_X) satisfying $\alpha(\omega_X) = \omega$. However it may happen that X is *not* a covering of an Enriques surface, i.e., the isometry ι is not represented by an automorphism. The reason is as follows: assume that there is a (-2)-vector r in L^- with $\langle r, \omega \rangle = 0$. Then there exists a class $\delta \in L_X^-$ with $\delta^2 = -2$ and $\langle \delta, \omega_X \rangle = 0$. Then δ is algebraic and by the Riemann–Roch theorem we may assume that δ is effective. If ι is represented by an automorphism of X, then it preserves effective classes. This is impossible because $\delta \in L_X^-$.

For $r \in L^-$ with $r^2 = -2$, we define

$$r^{\perp} = \{[\omega] \in \mathcal{D}(L^-) : \langle \omega, r \rangle = 0\}, \quad \mathcal{H} = \bigcup_r r^{\perp}.$$

where r belongs to the set of vectors in L^- with $r^2 = -2$. The above Remark 3.3 implies that the periods of Enriques surfaces lie in $\mathcal{D}(L^-)\setminus\mathcal{H}$, and in fact, the moduli space of Enriques surfaces is given by

$$(\mathcal{D}(L^-)\setminus\mathcal{H})/\Gamma.$$

It is known that $\mathcal{H}/\Gamma$ is irreducible. $\mathcal{H}$ is called the *discriminant locus* of Enriques surfaces. The Torelli type theorem for Enriques surfaces was first given by Horikawa [18].

3.4 Remark

A generic Enriques surface does not contain a smooth rational curve. For example, assume that the Picard number of X is 10, that is, the Picard lattice S_X of X coincides with L_X^+. Then there are no smooth rational curves on Y. In fact, if C is a smooth rational curve on Y, then $\pi^*(C)$ is the disjoint union of two smooth rational curves $C^\pm$: $\pi^*(C) = C^+ + C^-$. The difference $C^+ - C^-$ is contained in L_X^- which contradicts the assumption $S_X = L_X^+$.

The above remark suggests to us the following important invariant for Enriques surfaces. Let Y be an Enriques surface and X the covering $K3$ surface. Let C be a smooth rational curve on Y. Then we have two disjoint smooth rational curves $C^\pm$ on X satisfying

$$C^+ + C^- \in L_X^+, \ (C^+ + C^-)^2 = -4, \ C^+ - C^- \in L_X^-,$$

$$(C^+ - C^-)^2 = -4, \ C^\pm = ((C^+ + C^-) \pm (C^+ - C^-))/2.$$

Let S_X be the Picard lattice of X and let $S_X^\pm$ be $S_X \cap L_X^\pm$. Now consider a vector $\delta_- \in S_X^-$ with $\delta_-^2 = -4$ such that there exists a vector $\delta_+ \in S_X^+$ with $\delta_+^2 = -4$ and $(\delta^+ + \delta^-)/2 \in S_X$. Denote by Δ_- be the set of all such vectors δ_- and by $[\Delta_-]$ the lattice generated by Δ_-. Let $K = [\Delta_-](1/2)$. Then K is an even negative definite lattice generated by (-2)-vectors, and hence is isomorphic to a root lattice. Finally define a homomorphism

$$\xi : K/2K \to S_X^+(1/2)$$

by sending δ_- to δ_+. Then the pair $(K, \mathrm{Ker}(\xi))$ of a root lattice and a finite abelian group is called the *root invariant* of the Enriques surface Y. It is important to study the distribution of smooth rational curves on Y and the automorphism group of Y. The root invariant was introduced by Nikulin [32].

3.5 Example

We give an example of Enriques surfaces which has a root invariant $(E_6, \{0\})$. Let S be a smooth cubic surface defined by a homogeneous polynomial $F(z_0, z_1, z_2, z_3)$ of degree 3. Then the Hessian polynomial of F, if it is not identically zero, defines a quartic surface H called the *Hessian quartic surface* of S. To study Hessian quartic surfaces, it is convenient to use the Sylvester form of S. It is classically known that a general cubic surfaces S can be written in the *Sylvester form*

$$\lambda_1 x_1^3 + \cdots + \lambda_5 x_5^3 = 0, \quad x_1 + \cdots + x_5 = 0$$

where $x_1, \ldots, x_5$ are linear forms in z_0, z_1, z_2, z_3 each four of which are linearly independent, and $\lambda_i \in \mathbf{C}^*$. The forms $x_1, \ldots, x_5$ are uniquely determined by F up to permutation and multiplication by a common non-zero scalar, and $\lambda_1, \ldots, \lambda_5$ are uniquely determine by F and x_i (Segre [36], Chap. IV). For a cubic surface defined by the Sylvester form, the corresponding Hessian quartic surface H is given by

$$\frac{1}{\lambda_1 x_1} + \cdots + \frac{1}{\lambda_5 x_5} = 0, \quad x_1 + \cdots + x_5 = 0.$$

The Hessian H has ten nodes p_{ijk} defined by $x_i = x_j = x_k = 0$, and contains ten lines l_{mn} defined by $x_m = x_n = 0$. For general F, H does not contain any other singular points. Let X be the minimal resolution of H. Then X is a $K3$ surface with 20 smooth rational curves, that is, exceptional curves E_{ijk} over 10 nodes p_{ijk} and strict transforms L_{mn} of 10 lines l_{mn}. The curve E_{ijk} meets exactly three curves L_{ij}, L_{ik} and L_{jk}, and conversely L_{ij} meets exactly three curves E_{ijk} $(k \neq i, j)$. Thus we have two sets $\{E_{ijk}\}$, $\{L_{mn}\}$ of smooth rational curves on X each of which consists of 10 disjoint curves, and each curve in one set meets exactly three curves in the other set.

The birational involution defined by

$$(x_1 : \cdots : x_5) \rightarrow (\frac{1}{\lambda_1 x_1} : \cdots : \frac{1}{\lambda_5 x_5})$$

induces a fixed-point free involution σ of X, and hence the quotient $Y = X/\langle \sigma \rangle$ is an Enriques surface. The involution σ switches two curves E_{ijk} and L_{mn} where $\{i, j, k, m, n\} = \{1, 2, 3, 4, 5\}$. We denote by $\bar{L}_{ij}$ the image of L_{ij} or E_{kmn} in Y. The curve $\bar{L}_{ij}$ meets exactly three curves $\bar{L}_{km}$, $\bar{L}_{kn}$ and $\bar{L}_{mn}$. We can easily see that the dual graph of ten smooth rational curves $\{\bar{L}_{ij}\}$ is isomorphic to the Petersen graph.

The 20 curves $\{E_{ijk}, L_{mn}\}$ generate a sublattice N of signature $(1, 15)$ in the Picard lattice S_X of X. Let M be the orthogonal complement of N in $L = H^2(X, \mathbf{Z})$. It is known that M is isomorphic to $U \oplus U(2) \oplus A_2(2)$ [14]. Let R be the orthogonal complement of M in L^-. Obviously R is a negative definite lattice of rank 6. We will show that R is isomorphic to $E_6(2)$. Consider the following classes,

$$E_{123} - L_{45}, \; E_{145} - L_{23}, \; E_{235} - L_{14}, \; E_{345} - L_{12}, \; E_{125} - L_{34}, \; E_{245} - L_{13},$$

which generate a lattice isomorphic to $E_6(2)$ in R. By comparing

$$A_{L^-} \cong (\mathbf{Z}/2\mathbf{Z})^{10} \text{ and } A_{E_6(2)\oplus M} \cong (\mathbf{Z}/2\mathbf{Z})^{10} \oplus (\mathbf{Z}/3\mathbf{Z})^2,$$

we can conclude that $E_6(2) \cong R$. For the geometry of Hessian quartic surfaces we refer the reader to Dolgachev and Keum [14], Dardanelli and van Geemen [12].

4 Automorphisms

In this section, we first mention the Torelli type theorem for algebraic $K3$ surfaces. Next we give its applications to a description of the group of automorphisms of an algebraic $K3$ surface, a relation between finite groups of symplectic automorphisms of $K3$ surfaces and the Mathieu group, and a description of the group of automorphisms of a generic Enriques surface.

4.1 Torelli Type Theorem and the Group of Automorphisms for an Algebraic K3 Surface

Let X be an algebraic $K3$ surface and let S_X be the Picard lattice. By the Hodge index theorem, S_X has signature $(1, \rho - 1)$. Let T_X be the orthogonal complement of S_X which is called the *transcendental lattice* of X. The signature of T_X is $(2, 20 - \rho)$. Put

$$\Delta(X) = \{\delta \in S_X : \delta^2 = -2\}.$$

For each $\delta \in \Delta(X)$, we can define an isometry s_δ of S_X by

$$s_\delta(x) = x + \langle \delta, x \rangle \delta.$$

Then s_δ is a reflection with respect to the hyperplane $\delta^\perp$. We denote by $W(X)$ the group generated by reflections s_δ, $\delta \in \Delta(X)$. Let

$$P(X) = \{x \in S_X \otimes \mathbf{R} : x^2 > 0\}.$$

We denote by $P(X)^+$ the connected component of $P(X)$ containing an ample class. Since the hyperplane $r^\perp$ has signature $(1, \rho - 2)$, $r^\perp \cap P(X)^+$ is non-empty. Hence s_δ preserves $P(X)^+$. Let $C(X)$ be the connected component of

$$P(X)^+ \setminus \bigcup_{r \in \Delta(X)} r^\perp$$

containing an ample class. It is known that any vector in $C(X) \cap S_X$ is represented by an ample divisor. Thus $C(X)$ is called the ample cone. The following is the Torelli type theorem for algebraic $K3$ surfaces due to Piatetskii-Shapiro and Shafarevich.

4.2 Theorem ([35])

Let X, X' be algebraic K3 surfaces. Let $\varphi : H^2(X, \mathbf{Z}) \to H^2(X', \mathbf{Z})$ be an isometry satisfying

(1) $\varphi(\omega_X) \in \mathbf{C} \cdot \omega_{X'}$,
(2) $\varphi(C(X)) \subset C(X')$.

Then there exists an isomorphism $g : X' \to X$ with $g^* = \varphi$.

Since any automorphism preserves ample classes, we have a natural map

$$\psi : \mathrm{Aut}(X) \to \mathrm{Aut}(C(X)) = \{\varphi \in O(S_X) : g(C(X)) = C(X)\}$$

where $\mathrm{Aut}(X)$ is the group of automorphisms of X. Also, the general theory of reflection groups implies that $C(X)$ is a fundamental domain of $W(X)$ with respect to the action on $P(X)^+$. Therefore we have an isomorphism

$$\mathrm{Aut}(C(X)) \cong O(S_X)/\{\pm 1\} \cdot W(X).$$

4.3 Theorem ([35])

The homomorphism ψ has finite kernel and finite cokernel.

Proof. We show that ψ has finite cokernel as an application of lattice theory. Consider the subgroup $\mathfrak{G}$ of $\mathrm{Aut}(C(X))$ given by

$$\mathfrak{G} = \mathrm{Ker}(\mathrm{Aut}(C(X)) \to O(q_{S_X})).$$

Since $O(q_{S_X})$ is a finite group, $\mathfrak{G}$ is of finite index in $\mathrm{Aut}(C(X))$. Let $\varphi \in \mathfrak{G}$. Then it follows from Corollary 2.14 that there exists an isometry $\tilde{\varphi}$ of $H^2(X, \mathbf{Z})$ with $\tilde{\varphi}|S_X = \varphi$ and $\tilde{\varphi}|T_X = 1$. Since $\omega_X \in T_X \otimes \mathbf{C}$, $\tilde{\varphi}$ preserves the period of X and, by definition of $\mathfrak{G}$, preserves the ample cone. It now follows from Theorem 4.2 that there exists an automorphism g of X with $g^* = \tilde{\varphi}$. Thus $\mathfrak{G}$ can be realized as automorphisms of X.

On the other hand, we show that the restriction of $\mathrm{Aut}(X)$ to T_X is a finite group. Consider the subspace W of $T_X \otimes \mathbf{R}$ generated by $\mathrm{Re}(\omega_X)$ and $\mathrm{Im}(\omega_X)$. The Riemann condition implies that W is positive definite. Since the signature of T_X is $(2, 20 - \rho)$, the orthogonal complement $W^\perp$ in $T_X \otimes \mathbf{R}$ is negative definite. Thus $\mathrm{Aut}(X)|T_X$ is a discrete subset of the compact set $O(W) \times O(W^\perp)$, and hence is finite. Hence ψ has finite kernel.

4.4 Corollary

The group $\mathrm{Aut}(X)$ *of automorphisms of a K3 surface is finite if and only if* $[\mathrm{O}(S_X) : W(X)] < \infty$.

The Picard lattices S_X with $[\mathrm{O}(S_X) : W(X)] < \infty$ were classified by Nikulin [31, 33] and Vinberg [39]. In general, it is difficult to calculate a fundamental domain of $W(X)$. However, for example, Vinberg [38] calculated $\mathrm{Aut}(X)$ for two algebraic $K3$ surfaces X with Picard number 20, and Kondō [23] gave generators of $\mathrm{Aut}(X)$ for a generic Jacobian Kummer surface X associated to a smooth curve of genus 2. In the following, we present an idea used in [23].

4.5 The Leech Lattice and the Group of Automorphisms of a Generic Jacobian Kummer Surface

Let L be an even unimodular lattice of signature $(1, 25)$ and let $W(L)$ be the group generated by all reflections s_δ associated to (-2)-vectors δ in L. Consider the action of $W(L)$ on a connected component, denoted by $P(L)^+$, of the set

$$P(L) = \{x \in L \otimes \mathbf{R} \; : \; x^2 > 0\}.$$

In this case, $[\mathrm{O}(L) : W(L)] = \infty$; however, Conway [10], Chap. 27, gave a concrete description of a fundamental domain of $W(L)$ as follows. Let Λ be the Leech lattice, that is, the even unimodular negative definite lattice of rank 24 without (-2)-vectors (see Sect. 2). It follows from Theorem 2.5 that L is isomorphic to $U \oplus \Lambda$. We fix an orthogonal decomposition

$$L = U \oplus \Lambda.$$

We write (m, n, λ) for a vector in L where m, n are integers, $\lambda \in \Lambda$ and the norm is given by $2mn + \lambda^2$. Let

$$\rho = (1, 0, 0) \in L.$$

Note that $\langle \rho, \delta \rangle \neq 0$ for any (-2)-vector $\delta \in L$ because Λ does not contain (-2)-vectors. A (-2)-vector $\delta \in L$ is called a *Leech root* if $\langle \delta, \rho \rangle = 1$. Denote by Δ the set of all Leech roots. Note that Λ bijectively corresponds to Δ by

$$\Lambda \ni \lambda \rightarrow (-1 - \lambda^2/2, 1, \lambda) \in \Delta.$$

Put

$$C = \{x \in P^+(L) \; : \; \langle x, \delta \rangle > 0, \delta \in \Delta\}.$$

Then

4.6 Theorem ([10], Chap. 27)

C is a fundamental domain of W(L).

Consider a root sublattice R of L generated by some Leech roots. Denote by S the orthogonal complement of R in L. Then S is an even lattice of signature $(1, 24 - \mathrm{rank}(R))$. Define

$$D(S) = C \cap P(S)^+,$$

where $P(S)^+ = P(L)^+ \cap S \otimes \mathbf{R}$. Let w be the projection of ρ into S^*. Then

4.7 Proposition ([4])

(1) *w is contained in $D(S)$. In particular $D(S)$ is non-empty.*
(2) *$D(S)$ is a finite polyhedron.*

Now assume that S is isomorphic to the Picard lattice S_X of a $K3$ surface X. Since any (-2)-vector in S_X is a (-2)-vector in L, we may assume that the ample cone $C(X)$ of X contains $D(S_X)$. Then w is the class of an ample divisor on X. Thus we have a finite polyhedron $D(S_X)$ in $C(X)$ and an ample class w which may help us to study the geometry of X. In the following we apply this to the case of a generic Jacobian Kummer surface.

Consider the root lattice $R = A_3 \oplus A_1^6$. We can embed R into L such that R is generated by Leech roots and $S = R^\perp$ is isomorphic to the Picard lattice S_X of the Kummer surface X associated to a generic smooth curve of genus 2. The faces of the finite polyhedron $D(S_X)$ consist of 316 $(= 32 + 32 + 60 + 192)$ hyperplanes perpendicular to (-2)-, (-4)-, (-4)- or (-12)-vectors in S_X respectively. These 32 (-2)-vectors correspond to 32 smooth rational curves on X forming the Kummer $(16)_6$-configuration. The ample class w defines an embedding of X into $\mathbf{P}^5$ whose image is the intersection of three quadrics. The group of symmetries of the finite polyhedron $D(S_X)$ is isomorphic to $(\mathbf{Z}/2\mathbf{Z})^5 \cdot \mathfrak{S}_6$ where $(\mathbf{Z}/2\mathbf{Z})^5$ acts on X as automorphisms (16 translations and 16 switches) and $\mathfrak{S}_6$ is the symmetry of the Weierstrass points on the curve of genus 2. Finally the remaining 32, 60, 192 hyperplanes of $D(S_X)$ correspond to classical automorphisms of X, that is, 16 projections and 16 correlations, 60 Cremona transformations and 192 Cremona transformations respectively. We can conclude that $\mathrm{Aut}(X)$ is generated by $(\mathbf{Z}/2\mathbf{Z})^5$ and these classical automorphisms. We should remark that Keum [19] found 192 automorphisms corresponding to 192 hyperplanes by the Torelli type theorem for $K3$ surfaces (Theorem 4.2). For more details, we refer the reader to Keum [19], Kondō [23], and Ohashi [34].

This method can be applicable to some other cases. For example, if we take the root lattice E_6 or D_6 as R, then we can obtain a fundamental domain of the *full* reflection group of the Picard lattice of the two most algebraic $K3$ surfaces given in

Vinberg [38]. Also Dolgachev and Keum [14] applied this method to calculate the group of automorphisms of a generic Hessian quartic surface.

4.8 Finite Groups of Automorphisms of K3 Surfaces

Let G be a finite group of automorphisms of a $K3$ surface X. Fix a nowhere vanishing holomorphic 2-form ω_X on X which is unique up to constants. For $g \in G$, one can define a non-zero constant $\alpha(g)$ by $g^*(\omega_X) = \alpha(g) \cdot \omega_X$. Thus we have a homomorphism $\alpha : G \to \mathbf{C}^*$. Since the image of α is a finite subgroup of the multiplicative group, it is cyclic. Hence we have an exact sequence

$$1 \to \mathrm{Ker}(\alpha) \to G \to \mathbf{Z}/m\mathbf{Z} \to 1.$$

An automorphism g is called *symplectic* if $\alpha(g) = 1$. Any finite group of automorphisms is an extension of a finite group of symplectic automorphisms by a cyclic group.

If X is algebraic, then any automorphism g of X acts on the transcendental lattice T_X as a finite cyclic group. More precisely, the following holds.

4.9 Proposition ([29])

Let X be an algebraic K3 surface and $g \in \mathrm{Aut}(X)$.

(1) *The restriction $g^*|T_X$ has finite order m.*
(2) *$\alpha(g) = 1$ if and only if $g^*|T_X = 1$.*
(3) *Assume $|g^*|T_X| = m > 1$. Then $\mathrm{rank}(T_X)$ is divisible by $\varphi(m)$, where φ is the Euler function.*

Proof. (1) The proof of the finiteness of $\langle g^*|T_X \rangle$ has been given in the proof of Theorem 4.3.
(2) Since $\omega_X \in T_X \otimes \mathbf{C}$, it suffices to show that if $\alpha(g) = 1$, then $g^*|T_X = 1$. Let $x \in T_X$. Then

$$\langle \omega_X, x \rangle = \langle g^*(\omega_X), g^*(x) \rangle = \langle \omega_X, g^*(x) \rangle$$

which implies that $\langle x - g^*(x), \omega_X \rangle = 0$. Hence $x - g^*(x) \in S_X \cap T_X = \{0\}$.
(3) By the assertion (2), $\alpha(g) = \zeta_m \neq 1$ is a primitive m-th root of unity. We show that g^* has no non-zero fixed vectors in $T_X \otimes \mathbf{Q}$. Let $x \in T_X \otimes \mathbf{Q}$ with $g^*(x) = x$. Then

$$\langle \omega_X, x \rangle = \langle g^*(\omega_X), g^*(x) \rangle = \langle \zeta_m \omega_X, x \rangle.$$

This implies that $\langle \omega_X, x \rangle = 0$ and hence $x \in (S_X \cap T_X) \otimes \mathbf{Q} = \{0\}$. The above argument shows that if $(g^*)^n \neq 1$, then $(g^*)^n$ has no non-zero fixed vectors. This implies that $g^*|T_X \otimes \mathbf{Q}$ is an irreducible representation of a cyclic group $\mathbf{Z}/m\mathbf{Z}$ of degree $\varphi(m)$ defined over $\mathbf{Q}$. Hence we have the last assertion.

Next we consider finite groups of symplectic automorphisms. We start with the following proposition.

4.10 Proposition ([29])

(1) *Let g be a finite symplectic automorphism of a K3 surface. Then $|g| \leq 8$.*
(2) *g has only isolated fixed points and the number of fixed points depends only on the order of g.*
(3) *Let g be of order m and let f_m be the number of fixed points of g. Then f_m is given in the following Table 3.*

Table 3: The number of fixed points of finite symplectic automorphisms of $K3$ surfaces

m	2	3	4	5	6	7	8
f_m	8	6	4	4	2	3	2

Recall that the Mathieu group M_{24} acts on the set $\Omega = \{1, \ldots, 24\}$ of 24 letters. Let M_{23} be the stabilizer subgroup of the letter 1. Then M_{23} is also a finite sporadic simple group, called the *Mathieu group* of degree 23. The conjugacy classes of M_{23} are determined by their orders and are given in the following Table 4.

Table 4: Conjugacy classes of M_{23}

$|\sigma|$	2	3	4	5	6	7	8
σ	$(2)^8$	$(3)^6$	$(4)^4(2)^2$	$(5)^4$	$(6)^2(3)^2(2)^2$	$(7)^3$	$(8)^2(4)(2)$

$|\sigma|$	11	14	15	23
σ	$(11)^2$	$(14)(7)(2)$	$(15)(5)(3)$	(23)

Denote by $\epsilon(|\sigma|)$ the number of fixed points of $\sigma \in M_{23}$ on Ω. Mukai [28] observed that $\epsilon(m) = f_m$ for $m \leq 8$. We have two representations of degree 24 defined over $\mathbf{Q}$: G acts on $H^*(X, \mathbf{Q}) \cong \mathbf{Q}^{24}$ and M_{23} acts on $\mathbf{Q}^\Omega \cong \mathbf{Q}^{24}$. Both actions have the same character for $m \leq 8$. In fact, Mukai proved a stronger assertion.

4.11 Theorem ([28])

Let G be a finite group. Then the followings are equivalent.

    (1) *G acts on a K3 surface as symplectic automorphisms.*

    (2) *G is a subgroup of M_{23} which has at least five orbits on Ω.*

The condition that G has at least five orbits is necessary because G fixes $H^0(X, \mathbf{Q})$, $H^4(X, \mathbf{Q})$, $\mathrm{Re}(\omega_X)$, $\mathrm{Im}(\omega_X)$ and a Kähler class. Mukai [28] determined maximal groups of symplectic automorphisms (11 types) and gave their explicit examples by equations. Later Kondō [22] and Mukai [22], Appendix gave a lattice theoretic proof of the above theorem by using the classification of Niemeier lattices.

4.12 Automorphisms of Enriques Surfaces

Let Y be an Enriques surface, X the covering $K3$ surface and σ the fixed-point free involution. We denote by S_X the Picard lattice of X and by ω_X a nowhere vanishing holomorphic 2-form on X. Let

$$\mathrm{Aut}(X, \sigma) = \{g \in \mathrm{Aut}(X) \ : \ g \circ \sigma = \sigma \circ g\}.$$

Since X is the universal cover of Y, we have

$$\mathrm{Aut}(Y) \cong \mathrm{Aut}(X, \sigma)/\langle \sigma \rangle.$$

An Enriques surface Y is called *generic* if X satisfies the following two conditions:

(1) $S_X = L_X^+$,
(2) ω_X does not lie in a proper subspace of $L_- \otimes \mathbf{C}$ defined over $\mathbf{Q}(\zeta_m)$, where ζ_m is a primitive m-th root of unity and $\varphi(m) \leq 12$ (see Proposition 4.9).

Now we can state the description for the group of automorphisms of a generic Enriques surface by Barth and Peters [2] and independently by Nikulin [31]. It follows from Theorem 4.3 that the group of automorphisms of an algebraic $K3$ surface with Picard number 1 is finite. Contrary to the case of $K3$ surfaces, a generic Enriques surface has an infinite group of automorphisms.

4.13 Theorem ([2, 31], Theorem 10.1.2)

Let Y be a generic Enriques surface. Then $\mathrm{Aut}(Y)$ *is isomorphic to*

$$\mathrm{Ker}(\mathrm{O}(L^+) \to \mathrm{O}(q_{L^+}))/\{\pm 1\}.$$

Proof. Note that the condition $S_X = L_X^+$ is equivalent to the condition $T_X = L_X^-$. Let g be an automorphism with $g \circ \sigma = \sigma \circ g$. Then $g^*|L_X^-$ has finite order and the second condition of the genericity implies that $g^*|L_X^- = 1$ (Proposition 4.9). Hence by Proposition 2.11 we see that g^* is contained in $\mathrm{Ker}(\mathrm{O}(L^+) \to \mathrm{O}(q_{L^+}))$. Obviously -1_{L^+} is not represented by any automorphism. Hence $\mathrm{Aut}(Y)$ is isomorphic to a subgroup of $\mathrm{Ker}(\mathrm{O}(L^+) \to \mathrm{O}(q_{L^+}))/\{\pm 1\}$. On the other hand, let $\varphi \in \mathrm{Ker}(\mathrm{O}(L^+) \to$

$O(q_{L^+})$). We may assume that φ preserves $P(S_X)^+$. Then there are no (-2)-vectors in $L_X^+ \cong U(2) \oplus E_8(2)$ and hence $C(X) = P(S_X^+)$ where $C(X)$ is the ample cone of X. By Corollary 2.14, there exists an isometry $\tilde{\varphi}$ of $H^2(X, \mathbf{Z})$ such that $\tilde{\varphi}|L_X^+ = \varphi$ and $\tilde{\varphi}|L_X^- = 1$. The isometry $\tilde{\varphi}$ preserves the period of X and the ample cone. It now follows from Theorem 4.2 that $\tilde{\varphi}$ is represented by an automorphism g of X. Obviously g commutes with σ and hence it induces an automorphism of Y.

Enriques surfaces with a finite group of automorphisms are very rare. Such Enriques surfaces were classified by Nikulin [32] and Kondō [20]. There are seven classes of such Enriques surfaces. Two of them consist of one-dimensional irreducible families and the others are unique. Moreover Nikulin [32] introduced the notion of the *root invariant* of an Enriques surface, which describes the group of automorphisms of the Enriques surface up to finite groups. For the root invariant, see Sect. 3.

We refer the reader to Dolgachev [13] for more examples of reflections and automorphisms.

5 Borcherds Products

Borcherds [7] gave a systematic method to construct an automorphic form on a bounded domain of type IV with known zeros and poles, called the *Borcherds product*. We give three examples of Borcherds products related to $K3$ and Enriques surfaces.

Let $T = \begin{pmatrix} 1 & 1 \\ 0 & 1 \end{pmatrix}$, $S = \begin{pmatrix} 0 & -1 \\ 1 & 0 \end{pmatrix}$ which are generators of $\mathrm{SL}(2, \mathbf{Z})$. Let L be an even lattice of signature $(2, n)$. For simplicity we assume that $n = 2b$ is even. Consider the group ring $\mathbf{C}[A_L]$ of $A_L = L^*/L$. Let e_α, $\alpha \in A_L$ be the standard generators. Let ρ_L be the Weil representation of $\mathrm{SL}(2, \mathbf{Z})$ on $\mathbf{C}[A_L]$ defined by:

$$\rho_L(T)(e_\alpha) = e^{\pi\sqrt{-1}q_L(\alpha)} e_\alpha, \quad \rho_L(S)(e_\alpha) = \frac{\sqrt{-1}^{b-1}}{\sqrt{|A_L|}} \sum_{\delta \in A_L} e^{-2\pi\sqrt{-1}b_L(\delta,\alpha)} e_\delta.$$

Let H^+ be the upper half-plane. A holomorphic map

$$f : H^+ \to \mathbf{C}[A_L]$$

is called a vector-valued modular form of weight k and type ρ_L if f satisfies the conditions (1) and (2):

(1) For any $\begin{pmatrix} a & b \\ c & d \end{pmatrix} \in \mathrm{SL}(2, \mathbf{Z})$ and $\tau \in H^+$,

$$f(\frac{a\tau + b}{c\tau + d}) = (c\tau + d)^k \rho_L \begin{pmatrix} a & b \\ c & d \end{pmatrix} \cdot f(\tau),$$

(2) f is meromorphic at cusps.

If f is holomorphic at cusps, then f is called a *holomorphic* vector-valued modular form.

Recall that

$$\mathcal{D}(L) = \{[\omega] \in \mathbf{P}(L \otimes \mathbf{C}) \ : \ \langle \omega, \omega \rangle = 0, \ \langle \omega, \bar{\omega} \rangle > 0\}$$

is a disjoint union of two copies of a bounded symmetric domain of type IV and dimension $2b$. Define

$$\tilde{\mathcal{D}}(L) = \{\omega \in L \otimes \mathbf{C} \ : \ \langle \omega, \omega \rangle = 0, \ \langle \omega, \bar{\omega} \rangle > 0\}.$$

Then the canonical map $\tilde{\mathcal{D}}(L) \to \mathcal{D}(L)$ is a $\mathbf{C}^*$-bundle. A meromorphic (holomorphic) function

$$\Phi : \tilde{\mathcal{D}}(L) \to \mathbf{C}$$

is called a *meromorphic (holomorphic) automorphic form of weight k with respect to Γ on $\mathcal{D}(L)$* if Φ is homogeneous of degree $-k$, that is, $\Phi(c \cdot \omega) = c^{-k} \Phi(\omega)$ for $c \in \mathbf{C}^*$, and is invariant under a subgroup Γ of $O(L)$ of finite index.

5.1 Theorem ([7])

Let f be a vector-valued modular form of weight $1 - b$ and type ρ_L. Let

$$f = \sum_{\alpha \in A_L} f_\alpha(\tau) \cdot e_\alpha = \sum_{\alpha \in A_L} \sum_{n \in \mathbf{Q}} c_\alpha(n) e^{2\pi \sqrt{-1} n \tau} \cdot e_\alpha$$

be the Fourier expansion. Assume that $c_\alpha(n) \in \mathbf{Z}$ for any $n \leq 0$. Then there exists a meromorphic automorphic form Ψ on $\mathcal{D}(L)$ of weight $c_0(0)/2$. The only zeros or poles of Ψ lie on rational quadratic divisors $\lambda^\perp$ where $\lambda \in L$ with $\lambda^2 < 0$. The order of zeros is given by

$$\sum_{0 < x \in \mathbf{R}, x\lambda \in L^*} c_{x\lambda}(x^2\lambda^2/2)$$

(or poles if this number is negative).

5.2 Example ([5])

Let L be an even unimodular lattice of signature $(2, 26)$. Then $A_L = \{0\}$. We take a modular form

$$f = 1/\Delta(\tau) = q^{-1} + 24 + \cdots$$

of weight -12, where

$$\Delta(\tau) = q \prod_{n>0}(1 - q^n)^{24}, \quad q = e^{2\pi\sqrt{-1}\tau}.$$

Then we have a holomorphic automorphic form Ψ_{12} on $\mathcal{D}(L)$ of weight $12 = 24/2$ with zeros along

$$\bigcup_{\lambda \in L, \; \lambda^2 = -2} \lambda^{\perp}.$$

In the following we discuss some applications of Ψ_{12} to the moduli spaces of polarized $K3$ surfaces. Recall that

$$L \cong U \oplus U \oplus E_8 \oplus E_8 \oplus E_8$$

(see Proposition 2.5). Consider the last component E_8 in the above decomposition of L, and let $x \in E_8$ be a primitive vector with $x^2 = -2d$. Let R be the orthogonal complement of x in E_8. Then

$$R^{\perp} = U \oplus U \oplus E_8 \oplus E_8 \oplus (-2d) \cong L_{2d}$$

where $(-2d)$ is the lattice generated by x. Thus we have a primitive embedding of L_{2d} into L which induces an embedding of the period domain $\mathcal{D}_{2d}$ of polarized $K3$ surfaces of degree $2d$ into $\mathcal{D}(L)$. Note that Ψ_{12} vanishes identically on $\mathcal{D}_{2d}$ because R contains (-2)-vectors. However Borcherds et al. [8] constructed an automorphic form on $\mathcal{D}_{2d}$ by using Ψ_{12} as follows: first divide Ψ_{12} by a product of linear forms each of which vanishes on the hyperplane perpendicular to a (-2)-vector in R, and then restrict it to $\mathcal{D}_{2d}$. In particular they showed that the moduli space of polarized $K3$ surfaces of degree 2 is isotrivial. By a similar method, Kondō [24] showed the existence of a cusp form of weight 19 for some d and proved that the Kodaira dimensions of the moduli spaces of polarized $K3$ surfaces of some degrees are nonnegative. Recently Gritsenko et al. [17] have determined the Kodaira dimensions of the moduli spaces of polarized $K3$ surfaces except for a finite number of d.

5.3 Example ([6, 7])

Let L be an even lattice $U \oplus U(2) \oplus E_8$ of signature $(2, 10)$. Then $A_L = (\mathbf{Z}/2\mathbf{Z})^2$. We denote the elements of $A_L = \mathbf{Z}/2\mathbf{Z} \times \mathbf{Z}/2\mathbf{Z}$ by $\{00, 01, 10, 11\}$ such that $00, 01, 10$ are isotropic with respect to q_L and 11 is not. Take a vector-valued modular form $f = \{f_\alpha(\tau)\}$ of weight -4 given by

$$f_{00}(\tau) = \frac{8\eta(2\tau)^8}{\eta(\tau)^{16}} = 8 + 128q + \cdots,$$

$$f_{01}(\tau) = f_{10}(\tau) = -\frac{8\eta(2\tau)^8}{\eta(\tau)^{16}} = -8 - 128q - \cdots,$$

$$f_{11}(\tau) = \frac{8\eta(2\tau)^8}{\eta(\tau)^{16}} + \frac{\eta(\tau/2)^8}{\eta(\tau)^{16}} = q^{-1/2} + 36q^{1/2} + \cdots,$$

where $\eta(\tau)$ is the Dedekind eta function. Then we have a holomorphic automorphic form Ψ_4 on $\mathcal{D}(L)$ of weight $4 = 8/2$ with zero along

$$\bigcup_{\lambda \in L,\ \lambda/2 \in L^*,\ \lambda^2 = -4} \lambda^\perp.$$

This automorphic form Ψ_4 can be considered as an automorphic form on the period domain of Enriques surfaces as follows. Recall that the period domain is associated to the lattice $L^- = U(2) \oplus U \oplus E_8(2)$. Note that $L^*(2)$ is isomorphic to L^-. Thus we have a natural isomorphism between $\mathcal{D}(L)$ and $\mathcal{D}(L^-)$. Moreover $O(L) \cong O(L^-)$ (see [21]). If we consider Ψ_4 as an automorphic form on $\mathcal{D}(L^-)$, its zero divisor is nothing but the discriminant locus $\mathcal{H}$ because $\lambda \in L, \lambda/2 \in L^*$ with $r^2 = -4$ correspond to $s \in L^-$ with $s^2 = -2$. Recall that the moduli space of Enriques surfaces is $(\mathcal{D}(L_-) \setminus \mathcal{H})/\Gamma$. Therefore the existence of an automorphic form vanishing exactly on the discriminant locus $\mathcal{H}$ implies that the moduli space of Enriques surfaces is quasi-affine [6].

We remark that the isomorphism $O(L) \cong O(L^-)$ induces an isomorphism between $\mathcal{D}(L^-)/O(L^-)$ and $\mathcal{D}(L)/O(L)$. The quotient $\mathcal{D}(L)/O(L)$ is birational to the moduli space of $K3$ surfaces X with a non-symplectic automorphism σ of order 2. It follows from Nikulin [31] that the fixed point set of σ is the disjoint union of a smooth curve of genus 5 and four smooth rational curves. Taking the quotient of X by σ and contracting exceptional curves, we get a plane quintic curve with a cusp. Therefore we have a birational map between the moduli space of Enriques surfaces and that of plane quintic curves with a cusp. It is known that the moduli space of plane quintic curves with a cusp is rational. Thus the moduli space of Enriques surfaces is rational, too. For more details, see [21].

5.4 *Example* ([15, 25])

Borcherds [7] gave another method, called *additive lifting*, to construct automorphic forms on a bounded symmetric domain of type IV. In the following we consider the case of Enriques surfaces. Let

$$\Gamma(2) = \mathrm{Ker}(O(L^-) \to O(q_{L^-})).$$

Then

$$O(L^-)/\Gamma(2) \cong O(q_{L^-}) \cong O^+(10, \mathbf{F}_2)$$

where $O^+(10, \mathbf{F}_2)$ is the group of isometries of the ten-dimensional quadratic form over $\mathbf{F}_2$ of even type (see [11], p. 146). The quotient $(\mathcal{D}(L^-) \setminus \mathcal{H})/\Gamma(2)$ is the moduli space of *marked* Enriques surfaces. For each holomorphic vector-valued modular form f of weight k and type ρ_{L^-}, there exists a holomorphic automorphic form F on $\mathcal{D}(L^-)$ of weight $k + \frac{10-2}{2} = k + 4$ with respect to $\Gamma(2)$ ([7], Theorem 14.3). This correspondence is called the additive lifting. We consider the simplest case, that is, holomorphic vector-valued modular forms of weight 0 and type ρ_{L^-}, which are nothing but elements in $\mathbf{C}[L^-]^{\mathrm{SL}(2,\mathbf{Z})}$. One can easily see that $\mathbf{C}[L^-]^{\mathrm{SL}(2,\mathbf{Z})} \cong \mathbf{C}^{187}$. By using the additive lifting, we can construct a 187-dimensional linear system of automorphic forms with respect to $\Gamma(2)$ on $\mathcal{D}(L^-)$ on which the finite group $O(q_{L^-})$ acts.

On the other hand, there are three types of vectors in $A_{L^-} \cong (\mathbf{Z}/2\mathbf{Z})^{10}$ denoted by type $00, 0, 1$ according to zero, non-zero isotropic, non-isotropic vectors. We consider a vector-valued modular form $h = \{h_\alpha(\tau)\}$ whose components h_α are given by functions h_{00}, h_0, h_1 depending only on the type α. Now we take a vector-valued modular form $\{h_\alpha(\tau)\}$ of weight -4 given by

$$h_{00}(\tau) = \frac{248\eta(2\tau)^8}{\eta(\tau)^{16}} = 248 + 3968q + \cdots,$$

$$h_0(\tau) = h_{10}(\tau) = -\frac{8\eta(2\tau)^8}{\eta(\tau)^{16}} = -8 - 128q - \cdots,$$

$$h_1(\tau) = \frac{8\eta(2\tau)^8}{\eta(\tau)^{16}} + \frac{\eta(\tau/2)^8}{\eta(\tau)^{16}} = q^{-1/2} + 36q^{1/2} + \cdots.$$

Then it follows from Theorem 5.1 that there exists a holomorphic automorphic form Φ on $\mathcal{D}(L^-)$ of weight 124 whose zero divisor is given by

$$\bigcup_{\lambda \in L^-, \, \lambda/2 \in (L^-)^*, \, \lambda^2 = -4} \lambda^\perp.$$

By comparing automorphic forms obtained by the additive lifting with Φ, we can see that the above 187-dimensional linear system is base-point free. This allows us to get an $O(q_{L^-})$-equivariant birational map from the moduli space of marked Enriques surfaces into $\mathbf{P}^{186}$. For more details, we refer the reader to Kondō [25] (the paper [25] contains a mistake which was pointed out and corrected by Freitag and Salvati-Manni [15]).

We remark that Allcock and Freitag [1] used the additive lifting first to get an $W(E_6)$-equivariant embedding of the moduli space of marked cubic surfaces into $\mathbf{P}^9$, where $W(E_6)$ is the Weyl group of type E_6. This embedding coincides with the one given by the *Cayley's cross ratios* of cubic surfaces. It would be interesting to study the geometric meaning of the linear system of automorphic forms in the case of Enriques surfaces. For more examples, see [16, 26, 27].

Acknowledgements The author was supported in part by JSPS Grant-in-Aid (S), No. 22224001, No. 19104001.

References

1. D. Allcock, E. Freitag, Cubic surfaces and Borcherds products. Comment. Math. Helv. **77**, 270–296 (2002)
2. W. Barth, C. Peters, Automorphisms of Enriques surfaces. Invent. Math. **73**, 383–411 (1983)
3. W. Barth, K. Hulek, C. Peters, A. Van de Ven, *Compact Complex Surfaces*, 2nd edn. (Springer, Berlin, 2003)
4. R. Borcherds, Automorphism groups of Lorentzian lattices. J. Algebra **111**, 133–153 (1987)
5. R. Borcherds, Automorphic forms on $O_{s+2,2}(\mathbf{R})$ and infinite products. Invent. Math. **120**, 161–213 (1995)
6. R. Borcherds, The moduli space of Enriques surfaces and the fake monster Lie superalgebra. Topology **35**, 699–710 (1996)
7. R. Borcherds, Automorphic forms with singularities on Grassmannians. Invent. Math. **132**, 491–562 (1998)
8. R. Borcherds, L. Katzarkov, T. Pantev, N.I. Shepherd-Barron, Families of $K3$ surfaces. J. Algebr. Geom. **7**, 183–193 (1998)
9. D. Burns, M. Rapoport, On the Torelli type problems for Kählerian $K3$ surfaces. Ann. Sci. Ec. Norm. Super. IV Ser. **8**, 235–274 (1975)
10. J.H. Conway, N.J.A. Sloane, *Sphere Packings, Lattices and Groups*, 3rd edn. Grundlehren Math. Wiss., Bd 290 (Springer, Berlin, 1999)
11. J.H. Conway, R.T. Curtis, S.P. Norton, R.A. Parker, R.A. Wilson, *Atlas of Finite Groups* (Oxford University Press, Oxford, 1985)
12. E. Dardanelli, B. van Geemen, Hessians and the moduli space of cubic surfaces. Contemp. Math. **422**, 17–36 (2007)
13. I. Dolgachev, Reflection groups in algebraic geometry. Bull. Am. Math. Soc. **45**, 1–60 (2009)
14. I. Dolgachev, J.H. Keum, Birational automorphisms of quartic Hessian surfaces. Trans. Am. Math. Soc. **354**, 3031–3057 (2002)
15. E. Freitag, R. Salvati-Manni, Modular forms for the even unimodular lattice of signature (2, 10). J. Algebr. Geom. **16**, 753–791 (2007)
16. E. Freitag, R. Salvati-Manni, The modular variety of hyperelliptic curves of genus three. Trans. Am. Math. Soc. **363**, 281–312 (2011)
17. V. Gritsenko, K. Hulek, G. Sankaran, The Kodaira dimension of the moduli of $K3$ surfaces. Invent. Math. **169**, 519–567 (2007)
18. E. Horikawa, On the periods of Enriques surfaces I, II. Math. Ann. **234**, 78–108 (1978); Math. Ann. **235**, 217–246 (1978)
19. J.H. Keum, Automorphisms of Jacobian Kummer surfaces. Comp. Math. **107**, 269–288 (1997)
20. S. Kondō, Enriques surfaces with finite automorphism groups. Jpn. J. Math. **12**, 191–282 (1986)
21. S. Kondō, The rationality of the moduli space of Enriques surfaces. Compos. Math. **91**, 159–173 (1994)
22. S. Kondō, Niemeier lattices, Mathieu groups, and finite groups of symplectic automorphisms of $K3$ surfaces (with an appendix by S. Mukai). Duke Math. J. **92**, 593–603 (1998)
23. S. Kondō, The automorphism group of a generic Jacobian Kummer surface. J. Algebr. Geom. **7**, 589–609 (1998)
24. S. Kondō, On the Kodaira dimension of the moduli space of $K3$ surfaces II. Comp. Math. **116**, 111–117 (1999)
25. S. Kondō, The moduli space of Enriques surfaces and Borcherds products. J. Algebr. Geom. **11**, 601–627 (2002)
26. S. Kondō, The moduli space of 8 points on $\mathbf{P}^1$ and automorphic forms. Contemp. Math. **422**, 89–106 (2007)
27. S. Kondō, Moduli of plane quartics, Göpel invariants and Borcherds products. Int. Math. Res. Notices **2011**, 2825–2860 (2011)
28. S. Mukai, Finite groups of automorphisms of $K3$ surfaces and the Mathieu group. Invent. Math. **94**, 183–221 (1988)

29. V.V. Nikulin, Finite groups of automorphisms of Kählerian surfaces of type $K3$. Moscow Math. Soc. **38**, 71–137 (1980)
30. V.V. Nikulin, Integral symmetric bilinear forms and its applications. Math. USSR Izv. **14**, 103–167 (1980)
31. V.V. Nikulin, Factor groups of groups of the automorphisms of hyperbolic forms with respect to subgroups generated by 2-reflections. J. Sov. Math. **22**, 1401–1475 (1983)
32. V.V. Nikulin, On a description of the automorphism groups of Enriques surfaces. Sov. Math. Dokl. **30**, 282–285 (1984)
33. V.V. Nikulin, Surfaces of type $K3$ with a finite automorphism group and a Picard number three. Proc. Steklov Inst. Math. **165**, 131–155 (1985)
34. H. Ohashi, Enriques surfaces covered by Jacobian Kummer surface. Nagoya Math. J. **195**, 165–186 (2009)
35. I. Piatetskii-Shapiro, I.R. Shafarevich, A Torelli theorem for algebraic surfaces of type $K3$. USSR Izv. **35**, 530–572 (1971)
36. B. Segre, *The Non-singular Cubic Surfaces* (Oxford, Oxford University Press 1942)
37. J.P. Serre, in *A Course in Arithmetic*, 3rd edn. Grundlehren Math. Wiss., Bd 290 (Springer, Berlin, 1999)
38. E.B. Vinberg, The two most algebraic $K3$ surfaces. Math. Ann. **265**, 1–21 (1983)
39. E.B. Vinberg, Classification of 2-reflective hyperbolic lattices of rank 4. Trans. Moscow Math. Soc. **68**, 39–66 (2007)

Transcendental Methods in the Study of Algebraic Cycles with a Special Emphasis on Calabi–Yau Varieties

James D. Lewis

Abstract We review the transcendental aspects of algebraic cycles, and explain how this relates to Calabi–Yau varieties. More precisely, after presenting a general overview, we begin with some rudimentary aspects of Hodge theory and algebraic cycles. We then introduce Deligne cohomology, as well as the generalized higher cycles due to Bloch that are connected to higher K-theory, and associated regulators. Finally, we specialize to the Calabi–Yau situation, and explain some recent developments in the field.

Key words: Calabi–Yau variety, Algebraic cycle, Abel–Jacobi map, Regulator, Deligne cohomology, Chow group

Mathematics Subject Classifications (2010): Primary 14C25; Secondary 14C30, 14C35

1 Introduction

These notes concern that part of Calabi–Yau geometry that involves algebraic cycles—typically built up from special subvarieties, such as rational points and rational curves. From these algebraic cycles, one forms various doubly indexed groups, called higher Chow groups, that mimic simplicial homology theory in algebraic topology. These Chow groups come equipped with various maps whose target space is a certain transcendental cohomology theory called Deligne cohomology.

More precisely these maps are called regulators, from the higher cycle groups of S. Bloch, denoted by $CH^k(X, m)$, of a projective algebraic manifold X, to Deligne cohomology, viz.:

J.D. Lewis (✉)
Department of Mathematical and Statistical Sciences, University of Alberta,
632 Central Academic Building, Edmonton, AB, Canada T6G 2G1
e-mail: lewisjd@ualberta.ca

R. Laza et al. (eds.), *Arithmetic and Geometry of K3 Surfaces and Calabi–Yau Threefolds*, 29
Fields Institute Communications 67, DOI 10.1007/978-1-4614-6403-7_2,
© Springer Science+Business Media New York 2013

$$\mathrm{cl}_{r,m} : \mathrm{CH}^r(X,m) \to H_{\mathscr{D}}^{2r-m}(X,\mathbf{A}(r)), \tag{1}$$

where $\mathbf{A} \subseteq \mathbf{R}$ is a subring, $\mathbf{A}(r) := \mathbf{A}(2\pi i)^r$ is called the "Tate twist", and as we will indicate below, some striking evidence that these regulator maps become highly interesting in the case where X is Calabi–Yau. As originally discussed in [39], we are interested in the following case scenarios, with the intention of also providing an update on new developments. For the moment we will consider $\mathbf{A} = \mathbf{Z}$; however we will also consider $\mathbf{A} = \mathbf{Q}$, $\mathbf{R}$ later on.

When $m = 0$, the objects of interest are the null homologous codimension 2 (= dimension 1) cycles $\mathrm{CH}^2_{\mathrm{hom}}(X) = \mathrm{CH}_{1,\mathrm{hom}}(X)$ on a projective threefold X, and where in this case, (1) becomes the Abel–Jacobi map:

$$\Phi_2 : \mathrm{CH}^2_{\mathrm{hom}}(X) \to J^2(X) = \frac{H^3(X,\mathbf{C})}{F^2 H^3(X,\mathbf{C}) + H^3(X,\mathbf{Z}(2))} \simeq \frac{\{H^{3,0}(X) \oplus H^{2,1}(X)\}^{\vee}}{H_3(X,\mathbf{Z}(1))}, \tag{2}$$

defined by a process of integration, $J^2(X)$ being the Griffiths jacobian of X. One of the reasons for introducing the Abel–Jacobi map is to study the Griffiths group $\mathrm{Griff}^2(X) \otimes \mathbf{Q}$. If we put $\mathrm{CH}^r_{\mathrm{alg}}(X)$ to be codimension r cycles algebraically equivalent to zero, then the Griffiths group is given by $\mathrm{Griff}^r(X) := \mathrm{CH}^r_{\mathrm{hom}}(X)/\mathrm{CH}^r_{\mathrm{alg}}(X)$.

When $m = 1$, the object of interest is the group

$$\mathrm{CH}^2(X,1) = \frac{\left\{\sum_{j,\mathrm{cd}_X Z_j = 1} (f_j, Z_j) \middle| \begin{array}{l} f_j \in \mathbf{C}(Z_j)^{\times} \\ \sum_j \mathrm{div}(f_j) = 0 \end{array}\right\}}{\mathrm{Image}(\mathrm{Tame~symbol})},$$

on a projective algebraic surface X. If we mod out by the subgroup of $\mathrm{CH}^2(X,1)$ where the f_j's $\in \mathbf{C}^{\times}$, then we arrive at the quotient group of indecomposables $\mathrm{CH}^2_{\mathrm{ind}}(X,1)$ which plays an analogous role to the Griffiths group above. Moreover if we assume that the torsion part of $H^3(X,\mathbf{Z})$ is zero, then in this case (1) becomes a map:

$$\underline{\mathrm{cl}}_{2,1} : \mathrm{CH}^2_{\mathrm{ind}}(X,1) \to \frac{[H^{2,0}(X) \oplus H^{1,1}_{\mathrm{tr}}(X)]^{\vee}}{H_2(X,\mathbf{Z})}, \tag{3}$$

where $H^{1,1}_{\mathrm{tr}}(X)$ is the transcendental part of $H^{1,1}(X)$, being the orthogonal complement to the subgroup of algebraic cocycles.

In the case $m = 2$, the objects of interest are the group of symbols:

$$\mathrm{CH}^2(X,2) = \left\{\xi := \prod_j \{f_j, g_j\} \middle| \begin{array}{l} f_j, g_j \in \mathbf{C}(X)^{\times} \\ \sum_{j,p \in X}\left((-1)^{v_p(f_j)v_p(g_j)}\left(\frac{f_j^{v_p(g_j)}}{g_j^{v_p(f_j)}}\right)(p), p\right) = 0 \end{array}\right\},$$

(v_p = order of vanishing at p), on a smooth projective curve X. In this case (1) becomes the regulator:

$$\mathrm{cl}_{2,2} : \mathrm{CH}^2(X,2) \to H^1(X,\mathbf{C}/\mathbf{Z}(2)). \tag{4}$$

As first pointed out in [39], if X is a smooth projective variety of dimension d, where $1 \leq d \leq 3$, then the maps and objects

- $\mathrm{cl}_{2,2}$ in (4) and $\mathrm{CH}^2(X, 2) \otimes \mathbf{Q}$ for $d = 1$
- $\underline{\mathrm{cl}}_{2,1}$ in (3) and $\mathrm{CH}^2_{\mathrm{ind}}(X, 1) \otimes \mathbf{Q}$ for $d = 2$
- Φ_2 in (2) and $\mathrm{Griff}^2(X) \otimes \mathbf{Q}$ for $d = 3$

become especially interesting and generally nontrivial in the case where X is a Calabi–Yau variety; moreover, in a sense that will be specified later, these maps are essentially "trivial" when restricted to indecomposables, for X either of "lower or higher order" to its Calabi–Yau counterpart. Cycle constructions involving nodal rational curves and torsion points, play a prominent role here.

Several recent developments in the context of algebraic cycles are included in these notes since the appearance of [39], which should be of interest to specialists. Having said this, these notes are prepared with the expressed interest in enticing a wider group of researchers into the subject.

We have benefited from conversations with Matt Kerr, Bruno Kahn and Xi Chen. We are also grateful to Bruno for sharing with us his preprint [29]. We owe the referee a debt of gratitude for doing a splendid job in recommending improvements and catching errors in an earlier version of this paper. We are also pleased that the referee made us aware of the interesting work of Friedman–Laza [20], and for raising the very interesting question of how to construct normal functions over the Calabi–Yau variations of Hodge structure that they construct.

2 Notation

Throughout these notes, and unless otherwise specified, $X = X/\mathbf{C}$ is a projective algebraic manifold, of dimension d. A projective algebraic manifold is the same thing as a smooth complex projective variety. If $V \subseteq X$ is an irreducible subvariety of X, then $\mathbf{C}(V)$ is the rational function field of V, with multiplicative group $\mathbf{C}(V)^\times$. Depending on the context (which will be made abundantly clear in the text), $\mathscr{O}_X$ will either be the sheaf of germs of holomorphic functions on X in the analytic topology, or the sheaf of germs of regular functions in the Zariski topology.

3 Some Hodge Theory

Some useful reference material for this section is [25, 36].

Let $E^k_X = \mathbf{C}$-valued C^∞ k-forms on X. (One could also use the common notation of $A^k(X)$ for C^∞ forms, but let's not.) We have the decomposition:

$$E^k_X = \bigoplus_{p+q=k} E^{p,q}_X, \quad \overline{E^{p,q}_X} = E^{q,p}_X,$$

where $E_X^{p,q}$ are the C^∞ (p,q)-forms which in local holomorphic coordinates $z = (z_1, \ldots, z_n) \in X$, are of the form:

$$\sum_{|I|=p,|J|=q} f_{IJ} dz_I \wedge d\bar{z}_J, \quad f_{IJ} \text{ are } \mathbf{C} - valued\ C^\infty functions,$$

$$I = 1 \le i_1 < \cdots < i_p \le d, \quad J = 1 \le j_1 < \cdots < j_q \le d,$$

$$dz_I = dz_{i_1} \wedge \cdots \wedge dz_{i_p}, \quad d\bar{z}_J = d\bar{z}_{j_1} \wedge \cdots \wedge d\bar{z}_{j_q}.$$

One has the differential $d : E_X^k \to E_X^{k+1}$, and we define

$$H_{\mathrm{DR}}^k(X, \mathbf{C}) = \frac{\ker d : E_X^k \to E_X^{k+1}}{dE_X^{k-1}}.$$

The operator d decomposes into $d = \partial + \bar{\partial}$, where $\partial : E_X^{p,q} \to E_X^{p+1,q}$ and $\bar{\partial} : E_X^{p,q} \to E_X^{p,q+1}$. Further $d^2 = 0 \Rightarrow \partial^2 = \bar{\partial}^2 = 0 = \partial\bar{\partial} + \bar{\partial}\partial$, by (p,q) type.

The above decomposition descends to the cohomological level, viz.,

Theorem 3.1 (Hodge decomposition).

$$H_{\mathrm{sing}}^k(X, \mathbf{Z}) \otimes_{\mathbf{Z}} \mathbf{C} \simeq H_{\mathrm{DR}}^k(X, \mathbf{C}) = \bigoplus_{p+q=k} H^{p,q}(X),$$

where $H^{p,q}(X) = d$-closed (p,q)-forms (modulo coboundaries), and

$$\overline{H^{p,q}(X)} = H^{q,p}(X).$$

Furthermore:

$$H^{p,q}(X) \simeq \frac{E_{X,d-\mathrm{closed}}^{p,q}}{\partial\bar{\partial}E_X^{p-1,q-1}}.$$

Some more terminology: *Hodge filtration.* Put

$$F^k H^i(X, \mathbf{C}) = \bigoplus_{p \ge k} H^{p,i-p}(X).$$

Now recall dim $X = d$.

Theorem 3.2 (Poincaré and Serre duality). *The following pairings induced by*

$$(w_1, w_2) \mapsto \int_X w_1 \wedge w_2,$$

are non-degenerate:

$$H_{\mathrm{DR}}^k(X, \mathbf{C}) \times H_{\mathrm{DR}}^{2d-k}(X, \mathbf{C}) \to \mathbf{C},$$

$$H^{p,q}(X) \times H^{d-p,d-q}(X) \to \mathbf{C}.$$

Therefore $H^k(X) \simeq H^{2d-k}(X)^\vee$, $H^{p,q}(X) \simeq H^{d-p,d-q}(X)^\vee$

Corollary 3.3.

$$\frac{H^i(X, \mathbf{C})}{F^r H^i(X, \mathbf{C})} \simeq F^{d-r+1} H^{2d-i}(X, \mathbf{C})^\vee.$$

3.4 Formalism of Mixed Hodge Structures

Definition 3.5. Let $\mathbf{A} \subset \mathbf{R}$ be a subring. An $\mathbf{A}$-Hodge structure (HS) of weight $N \in \mathbf{Z}$ is given by the following datum:

• A finitely generated $\mathbf{A}$-module V, and either of the two equivalent statements below:

•$_1$ A decomposition

$$V_{\mathbf{C}} = \bigoplus_{p+q=N} V^{p,q}, \quad \overline{V^{p,q}} = V^{q,p},$$

where $^-$ is complex conjugation induced from conjugation on the second factor $\mathbf{C}$ of $V_{\mathbf{C}} := V \otimes \mathbf{C}$.

•$_2$ A finite descending filtration

$$V_{\mathbf{C}} \supset \cdots \supset F^r \supset F^{r-1} \supset \cdots \supset \{0\},$$

satisfying

$$V_{\mathbf{C}} = F^r \bigoplus \overline{F^{N-r+1}}, \ \forall\, r \in \mathbf{Z}.$$

Remark 3.6. The equivalence of •$_1$ and •$_2$ can be seen as follows. Given the decomposition in •$_1$, put

$$F^r V_{\mathbf{C}} = \bigoplus_{p+q=N, p \geq r} V^{p,q}.$$

Conversely, given $\{F^r\}$ in •$_2$, put $V^{p,q} = F^p \cap \overline{F^q}$.

Example 3.7. $X/\mathbf{C}$ smooth projective. Then $H^i(X, \mathbf{Z})$ is a $\mathbf{Z}$-Hodge structure of weight i.

Example 3.8. $\mathbf{A}(k) := (2\pi i)^k \mathbf{A}$ is an $\mathbf{A}$-Hodge structure of weight $-2k$ and of pure Hodge type $(-k, -k)$, called the Tate twist.

Example 3.9. $X/\mathbf{C}$ smooth projective. Then $H^i(X, \mathbf{Q}(k)) := H^i(X, \mathbf{Q}) \otimes \mathbf{Q}(k)$ is a $\mathbf{Q}$-Hodge structure of weight $i - 2k$.

To extend these ideas to singular varieties, one requires the following terminology.

Definition 3.10. An **A**-mixed Hodge structure (**A**-MHS) is given by the following datum:

- A finitely generated **A**-module $V_{\mathbf{A}}$,
- A finite descending "Hodge" filtration on $V_{\mathbf{C}} := V_{\mathbf{A}} \otimes \mathbf{C}$,

$$V_{\mathbf{C}} \supset \cdots \supset F^r \supset F^{r-1} \supset \cdots \supset \{0\},$$

- An increasing "weight" filtration on $V_{\mathbf{A}} \otimes \mathbf{Q} := V_{\mathbf{A}} \otimes_{\mathbf{Z}} \mathbf{Q}$,

$$\{0\} \subset \cdots \subset W_{\ell-1} \subset W_{\ell} \subset \cdots \subset V_{\mathbf{A}} \otimes \mathbf{Q},$$

such that $\{F^r\}$ induces a (pure) HS of weight ℓ on $Gr_{\ell}^W := W_{\ell}/W_{\ell-1}$.

Theorem 3.11 (Deligne [16]). *Let Y be a complex variety. Then $H^i(Y, \mathbf{Z})$ has a canonical and functorial $\mathbf{Z}$-MHS.*

Remark 3.12. (i) A morphism $h : V_{1,\mathbf{A}} \to V_{2,\mathbf{A}}$ of **A**-MHS is an **A**-linear map satisfying:

- $h(W_{\ell}V_{1,\mathbf{A}\otimes\mathbf{Q}}) \subseteq W_{\ell}V_{2,\mathbf{A}\otimes\mathbf{Q}}, \quad \forall \ell,$
- $h(F^rV_{1,\mathbf{C}}) \subseteq F^rV_{2,\mathbf{C}}, \quad \forall r.$

Deligne ([16] (Theorem 2.3.5)) shows that the category of **A**-MHS is abelian; in particular if $h : V_{1,\mathbf{A}} \to V_{2,\mathbf{A}}$ is a morphism of **A**-MHS, then $\ker(h)$, $\mathrm{coker}(h)$ are endowed with the induced filtrations. Let us further assume that $\mathbf{A}\otimes\mathbf{Q}$ is a field. Then Deligne (*op. cit.*) shows that h is strictly compatible[1] with the filtrations $W_{\bullet}$ and $F^{\bullet}$, and that the functors $V \mapsto Gr_{\ell}^W V$, $V \mapsto Gr_F^r V$ are exact.

(ii) Roughly speaking, the functoriality of the MHS in Deligne's theorem translates to the following yoga: the "standard" exact sequences in singular (co)homology, together with push-forwards and pullbacks by morphisms (wherever permissible) respect MHS. In particular for a subvariety $Y \subset X$, the localization cohomology sequence associated to the pair (X, Y) is a long exact sequence of MHS. Here is where the Tate twist comes into play: Suppose that $Y \subset X$ is an inclusion of projective algebraic manifolds with $\mathrm{codim}_X Y = r \geq 1$. One has a Gysin map $H^{i-2r}(Y, \mathbf{Q}) \to H^i(X, \mathbf{Q})$ which involves Hodge structures of different weights. To remedy this, one considers the induced map $H^{i-2r}(Y, \mathbf{Q}(-r)) \to H^i(X, \mathbf{Q}(0)) = H^i(X, \mathbf{Q})$ via (twisted) Poincaré duality (see (5) below), which is a morphism of pure Hodge structures (hence of MHS). A simple proof of this fact can be found in Sect. 7 of [36]. Note that the morphism $H_Y^i(X, \mathbf{Q}) \to H^i(X, \mathbf{Q})$ is a morphism of MHS, and that accordingly $H_Y^i(X, \mathbf{Q}) \simeq H^{i-2r}(Y, \mathbf{Q}(-r))$ is an isomorphism of MHS (with Y still smooth).

---

[1] Strict compatibility means that $h(F^rV_{1,\mathbf{C}}) = h(V_{1,\mathbf{C}}) \cap F^rV_{2,\mathbf{C}}$ and $h(W_{\ell}V_{1,\mathbf{A}\otimes\mathbf{Q}}) = h(V_{1,\mathbf{A}\otimes\mathbf{Q}}) \cap W_{\ell}V_{2,\mathbf{A}\otimes\mathbf{Q}}$ for all r and ℓ. A nice explanation of Deligne's proof of this fact can be found in [44], where a quick summary goes as follows: For any **A**-MHS V, $V_{\mathbf{C}}$ has a **C**-splitting into a bigraded direct sum of complex vector spaces $I^{p,q} := F^p \cap W_{p+q} \cap [\overline{F^q} \cap W_{p+q} + \sum_{i\geq 2} \overline{F^{q-i+1}} \cap W_{p+q-i}]$, where one shows that $F^rV_{\mathbf{C}} = \oplus_{p\geq r} \oplus_q I^{p,q}$ and $W_{\ell}V_{\mathbf{C}} = \oplus_{p+q\leq \ell} I^{p,q}$. Then by construction of $I^{p,q}$, one has $h(I^{p,q}(V_{1,\mathbf{C}}) \subseteq I^{p,q}(V_{2,\mathbf{C}})$. Hence h preserves both the Hodge and complexified weight filtrations. Now use the fact that $\mathbf{A} \otimes \mathbf{Q}$ is a field to deduce that h preserves the weight filtration over $\mathbf{A} \otimes \mathbf{Q}$.

Example 3.13. Let $\overline{U}$ be a compact Riemann surface, $\Sigma \subset \overline{U}$ a finite set of points, and put $U := \overline{U} \backslash \Sigma$. According to Deligne, $H^1(U, \mathbf{Z}(1))$ carries a $\mathbf{Z}$-MHS. The Hodge filtration on $H^1(U, \mathbf{C})$ is defined in terms of a filtered complex of holomorphic differentials on U with logarithmic poles along Σ ([16], but also see (10) below). One can "observe" the MHS as follows. Poincaré duality gives us $H^1_\Sigma(\overline{U}, \mathbf{Z}) \simeq H_1(\Sigma, \mathbf{Z}) = 0$, and the localization sequence in cohomology below is a sequence of MHS:

$$0 \to H^1(\overline{U}, \mathbf{Z}(1)) \to H^1(U, \mathbf{Z}(1)) \to H^0(\Sigma, \mathbf{Z}(0))^\circ \to 0,$$

where

$$H^0(\Sigma, \mathbf{Z}(0))^\circ := \ker\left(H^2_\Sigma(\overline{U}, \mathbf{Z}(1)) \to H^2(\overline{U}, \mathbf{Z}(1))\right) \simeq \mathbf{Z}(0)^{|\Sigma|-1}.$$

Put $W_0 = H^1(U, \mathbf{Z}(1))$, $W_{-1} = \text{Im}(H^1(\overline{U}, \mathbf{Z}(1)) \to H^1(U, \mathbf{Z}(1)))$, $W_{-2} = 0$. Then $Gr^W_{-1} H^1(U, \mathbf{Z}(1)) \simeq H^1(\overline{U}, \mathbf{Z}(1))$ has pure weight -1 and $Gr^W_0 H^1(U, \mathbf{Z}(1)) \simeq \mathbf{Z}(0)^{|\Sigma|-1}$ has pure weight 0.

The following notation will be introduced:

Definition 3.14. Let V be an $\mathbf{A}$-MHS. We put

$$\Gamma_\mathbf{A} V := \text{hom}_{\mathbf{A}-\text{MHS}}(\mathbf{A}(0), V),$$

and

$$J_\mathbf{A}(V) = \text{Ext}^1_{\mathbf{A}-\text{MHS}}(\mathbf{A}(0), V).$$

In the case where $\mathbf{A} = \mathbf{Z}$ or $\mathbf{A} = \mathbf{Q}$, we simply put $\Gamma = \Gamma_\mathbf{A}$ and $J = J_\mathbf{A}$.

Example 3.15. Suppose that $V = V_\mathbf{Z}$ is a $\mathbf{Z}$ (pure) HS of weight $2r$. Then $V(r) := V \otimes \mathbf{Z}(r)$ is of weight 0, and (up to the twist) one can identify ΓV with $V_\mathbf{Z} \cap F^r V_\mathbf{C} = V_\mathbf{Z} \cap V^{r,r} := \epsilon^{-1}(V^{r,r})$, where $\epsilon : V \to V_\mathbf{C}$.

Example 3.16. Let V be a $\mathbf{Z}$-MHS. There is the identification due to J. Carlson (see [8, 28]),

$$J(V) \simeq \frac{W_0 V_\mathbf{C}}{F^0 W_0 V_\mathbf{C} + W_0 V},$$

where in the denominator term, $V := V_\mathbf{Z}$ is identified with its image $V_\mathbf{Z} \to V_\mathbf{C}$ (viz., quotienting out torsion). For example, if $\{E\} \in \text{Ext}^1_{\text{MHS}}(\mathbf{Z}(0), V)$ corresponds to the short exact sequence of MHS:

$$0 \to V \to E \xrightarrow{\alpha} \mathbf{Z}(0) \to 0,$$

then one can find $x \in W_0 E$ and $y \in F^0 W_0 E_\mathbf{C}$ such that $\alpha(x) = \alpha(y) = 1$. Then $x - y \in V_\mathbf{C}$ descends to a class in $W_0 V_\mathbf{C}/\{F^0 W_0 V_\mathbf{C} + W_0 V\}$, which defines the map from $\text{Ext}^1_{\text{MHS}}(\mathbf{Z}(0), V)$ to $W_0 V_\mathbf{C}/\{F^0 W_0 V_\mathbf{C} + W_0 V\}$.

4 Algebraic Cycles

For the next two sections, the reader may find it helpful to consult [37]. Recall $X/\mathbf{C}$ smooth projective, dim $X = d$. For $0 \leq r \leq d$, put $z^r(X)$ $(= z_{d-r}(X))$ = free abelian group generated by subvarieties of codim r $(= \dim \ d - r)$ in X.

Example 4.1. (i) $z^d(X) = z_0(X) = \{\sum_{j=1}^{M} n_j p_j \mid n_j \in \mathbf{Z}, \ p_j \in X\}$.

(ii) $z^0(X) = z_d(X) = \mathbf{Z}\{X\} \simeq \mathbf{Z}$.

(iii) Let $X_1 := V(z_2^2 z_0 - z_1^3 - z_0 z_1^2) \subset \mathbf{P}^2$, and $X_2 := V(z_2^2 z_0 - z_1^3 - z_1 z_0^2) \subset \mathbf{P}^2$. Then $3X_1 - 5X_2 \in z^1(\mathbf{P}^2) = z_1(\mathbf{P}^2)$.

(iv) $\mathrm{codim}_X V = r - 1$, $f \in \mathbf{C}(V)^\times$. $\mathrm{div}(f) := (f) := (f)_0 - (f)_\infty \in z^r(X)$ (principal divisor). (Note: $\mathrm{div}(f)$ is easy to define, by first passing to a normalization $\tilde{V}$ of V, then using the fact that the local ring $\mathscr{O}_{\tilde{V},\wp}$ of regular functions at $\wp$ is a discrete valuation ring for a codimension one "point" $\wp$ on $\tilde{V}$, together with the proper push-forward associated to $\tilde{V} \to V$.)

Divisors in (iv) generate a subgroup,

$$z^r_{\mathrm{rat}}(X) \subset z^r(X),$$

which defines the rational equivalence relation on $z^r(X)$.

Definition 4.2.

$$\mathrm{CH}^r(X) := z^r(X)/z^r_{\mathrm{rat}}(X),$$

is called the r-th Chow group of X.

Remark 4.3. On can show that $\xi \in z^r_{\mathrm{rat}}(X) \Leftrightarrow \exists \, w \in z^r(\mathbf{P}^1 \times X)$, each component of the support $|w|$ flat over $\mathbf{P}^1$, such that $\xi = w[0] - w[\infty]$. (Here $w[t] := \mathrm{pr}_{2,*}(\langle \mathrm{pr}_1^*(t) \bullet w\rangle_{\mathbf{P}^1 \times X})$.) If one replaces $\mathbf{P}^1$ by any choice of smooth connected curve Γ (not fixed!) and 0, ∞ by any 2 points P, $Q \in \Gamma$, then one obtains the subgroup $z^r_{\mathrm{alg}}(X) \subset z^r(X)$ of cycles that are algebraically equivalent to zero.[2] There is the fundamental class map (described later) $z^r(X) \to H^{2r}(X, \mathbf{Z})$ whose kernel is denoted by $z^r_{\mathrm{hom}}(X)$. More precisely, the target space and map requires some twisting, viz.,

$$z^r(X) \to H^{2r}(X, \mathbf{Z}(r)).$$

To explain the role of twisting here, we illustrate this with three case scenarios.

- Let $f : Y \to X$ be a morphism of smooth projective varieties, where dim $Y =$ dim $X - 1$. One has a commutative diagram of cycle class maps:

$$z^{r-1}(Y) \to H^{2(r-1)}(Y, \mathbf{Z}(r-1))$$

$$f_* \downarrow \qquad\qquad \downarrow f_*$$

$$z^r(X) \quad \to \quad H^{2r}(X, \mathbf{Z}(r))$$

---

[2] The fact that a smooth connected Γ will suffice (as opposed to a [connected] chain of curves) in the definition of algebraic equivalence follows from the transitive property of algebraic equivalence (see [36] (p. 180)).

Thus from the perspective of (mixed) Hodge theory, this diagram is "natural", as the right hand vertical arrow is a morphism of (M)HS.

• Let $U/\mathbf{C}$ be a smooth quasi-projective variety of dimension d, and $Y \subset U$ a closed algebraic subset. Using the twisted Poincaré duality theory formalism in this situation (see [28] (p. 82, p. 92)), Poincaré duality gives us an isomorphism of MHS:

$$H_Y^i(U, \mathbf{Z}(j)) \simeq H_{2d-i}(Y, \mathbf{Z}(d-j)) := H_{2d-i}(Y, \mathbf{Z})(j-d),$$

where $H_i(Y, \mathbf{Z}) := H_i^{BM}(Y, \mathbf{Z})$ is Borel–Moore homology.[3] For example if $U = Y = X$ is smooth projective, then $H^i(X, \mathbf{Z}(j))$ is a pure HS of weight $i - 2j$, and $H_a(X, \mathbf{Z}(b)) := H_a(X, \mathbf{Z})(-b)$ is known to be a pure HS of weight $2b - a$, hence $H_{2d-i}(Y, \mathbf{Z}(d-j))$ has weight $2(d-j) - (2d-i) = i - 2j$. Thus

$$H^i(X, \mathbf{Z}(j)) \simeq H_{2d-i}(X, \mathbf{Z}(d-j)), \tag{5}$$

is an isomorphism of HS.

Remark 4.4. Although tempting, from a "purist" point of view, it would be a mistake to interpret $H_a(X, \mathbf{Z}(b)) = H_a(X, \mathbf{Z})(b)$. This would imply that the Poincaré duality isomorphism in (5) would not preserve weights, and hence not an isomorphism of (M)HS in the sense given in Remark 3.12.

• Let $\mathscr{O}_X$ be the sheaf of analytic functions on X. Recall the exponential short exact sequence of sheaves

$$0 \to \mathbf{Z} \to \mathscr{O}_X \xrightarrow{\exp(2\pi i \cdot (-))} \mathscr{O}_X^\times \to 0,$$

where $\mathscr{O}_X^\times \subset \mathscr{O}_X$ is the sheaf of units. It is well-known that $H^1(X, \mathscr{O}_X^\times) \simeq \mathrm{CH}^1(X)$, and hence there is an induced Chern class map $\mathrm{CH}^1(X) \to H^2(X, \mathbf{Z})$. But this is not so natural as there is no canonical choice of i. Instead, one considers

$$0 \to \mathbf{Z}(1) \to \mathscr{O}_X \xrightarrow{\exp} \mathscr{O}_X^\times \to 0,$$

and accordingly the induced cycle class map $\mathrm{CH}^1(X) \to H^2(X, \mathbf{Z}(1))$.

One has inclusions:

$$z_{\mathrm{rat}}^r(X) \subseteq z_{\mathrm{alg}}^r(X) \subseteq z_{\mathrm{hom}}^r(X) \subset z^r(X).$$

Definition 4.5. Put

(i) $\mathrm{CH}_{\mathrm{alg}}^r(X) := z_{\mathrm{alg}}^r(X)/z_{\mathrm{rat}}^r(X),$

(ii) $\mathrm{CH}_{\mathrm{hom}}^r(X) := z_{\mathrm{hom}}^r(X)/z_{\mathrm{rat}}^r(X),$

(iii) $\mathrm{Griff}^r(X) := z_{\mathrm{hom}}^r(X)/z_{\mathrm{alg}}^r(X) = \mathrm{CH}_{\mathrm{hom}}^r(X)/\mathrm{CH}_{\mathrm{alg}}^r(X)$, called the Griffiths group.

The Griffiths group is known to be trivial in the cases $r = 0, 1, d$.

---

[3] We remind the reader that for singular homology $H_*^{sing}(U, \mathbf{Z})$ and ignoring twists, Poincaré duality gives the isomorphism $H_c^i(U, \mathbf{Z}) \simeq H_{2d-i}^{sing}(U, \mathbf{Z})$, where $H_c^i(U, \mathbf{Z})$ is cohomology with compact support; whereas $H^i(U, \mathbf{Z}) \simeq H_{2d-i}^{BM}(U, \mathbf{Z})$.

4.6 Generalized Cycles

The basic idea is this:

$$\mathrm{CH}^r(X) = \mathrm{Coker}\left(\bigoplus_{\mathrm{cd}_X V = r-1} \mathbf{C}(V)^\times \xrightarrow{\mathrm{div}} z^r(X) \right).$$

In the context of Milnor K-theory, this is just

$$\underbrace{\left(\to \cdots \bigoplus_{\mathrm{cd}_X V = r-2} K_2^M(\mathbf{C}(V)) \right) \xrightarrow{\mathrm{Tame}}}_{\text{building a complex on the left}} \bigoplus_{\mathrm{cd}_X V = r-1} K_1^M(\mathbf{C}(V)) \xrightarrow{\mathrm{div}} \bigoplus_{\mathrm{cd}_X V = r} K_0^M(\mathbf{C}(V)).$$

For a field $\mathbf{F}$, one has the Milnor K-groups $K_\bullet^M(\mathbf{F})$, where $K_0^M(\mathbf{F}) = \mathbf{Z}$, $K_1^M(\mathbf{F}) = \mathbf{F}^\times$ and

$$K_2^M(\mathbf{F}) = \left\{ \mathrm{Symbols}\ \{a,b\} \ \middle|\ a, b \in \mathbf{F}^\times \right\} \middle/ \left\{ \begin{array}{c} \text{Steinberg relations} \\ \{a_1 a_2, b\} = \{a_1, b\}\{a_2, b\} \\ \{a, b\} = \{b, a\}^{-1} \\ \{a, 1-a\} = 1, \ a \neq 1 \end{array} \right\}.$$

One has a Gersten–Milnor resolution of a sheaf of Milnor K-groups on X, which leads to a complex whose last three terms and corresponding homologies (indicated at $\updownarrow$) for $0 \leq m \leq 2$ are:

$$\bigoplus_{\mathrm{cd}_X Z = r-2} K_2^M(\mathbf{C}(Z)) \xrightarrow{T} \bigoplus_{\mathrm{cd}_X Z = r-1} \mathbf{C}(Z)^\times \xrightarrow{\mathrm{div}} \bigoplus_{\mathrm{cd}_X Z = r} \mathbf{Z}$$

$$\updownarrow \qquad\qquad\qquad \updownarrow \qquad\qquad\qquad \updownarrow \qquad\qquad\qquad (6)$$

$$\mathrm{CH}^r(X, 2) \qquad\qquad \mathrm{CH}^r(X, 1) \qquad\qquad \mathrm{CH}^r(X, 0)$$

where div is the divisor map of zeros minus poles of a rational function, and T is the Tame symbol map. The Tame symbol map

$$T: \bigoplus_{\mathrm{cd}_X Z = r-2} K_2^M(\mathbf{C}(Z)) \to \bigoplus_{\mathrm{cd}_X D = r-1} K_1^M(\mathbf{C}(D)),$$

is defined as follows. First $K_2^M(\mathbf{C}(Z))$ is generated by symbols $\{f, g\}$, $f, g \in \mathbf{C}(Z)^\times$.

For $f, g \in \mathbf{C}(Z)^\times$,

$$T(\{f, g\}) = \sum_D (-1)^{v_D(f) v_D(g)} \left(\frac{f^{v_D(g)}}{g^{v_D(f)}} \right)_D,$$

where $(\cdots)_D$ means restriction to the generic point of D, and v_D represents order of a zero or pole along an irreducible divisor $D \subset Z$.

Example 4.7. Taking cohomology of the complex in (6), we have:

(i) $\mathrm{CH}^r(X, 0) = z^r(X)/z^r_{\mathrm{rat}}(X) =: \mathrm{CH}^r(X)$.

(ii) $\mathrm{CH}^r(X, 1)$ is represented by classes of the form $\xi = \sum_j (f_j, D_j)$, where codim_X $D_j = r - 1$, $f_j \in \mathbf{C}(D_j)^\times$, and $\sum \mathrm{div}(f_j) = 0$; modulo the image of the Tame symbol.

(iii) $\mathrm{CH}^r(X, 2)$ is represented by classes in the kernel of the Tame symbol; modulo the image of a higher Tame symbol.

Example 4.8. (i) $X = \mathbf{P}^2$, with homogeneous coordinates $[z_0, z_1, z_2]$. $\mathbf{P}^1 = \ell_j :=$ $V(z_j)$, $j = 0, 1, 2$. Let $P = [0, 0, 1] = \ell_0 \cap \ell_1$, $Q = [1, 0, 0] = \ell_1 \cap \ell_2$, $R = [0, 1, 0] =$ $\ell_0 \cap \ell_2$. Introduce $f_j \in \mathbf{C}(\ell_j)^\times$, where $(f_0) = P - R$, $(f_1) = Q - P$, $(f_2) = R - Q$. Explicitly, $f_0 = z_1/z_2$, $f_1 = z_2/z_0$ and $f_2 = z_0/z_1$. Then $\xi := \sum_{j=0}^{2}(f_j, \ell_j) \in$ $\mathrm{CH}^2(\mathbf{P}^2, 1)$ represents a higher Chow cycle.

$$\begin{array}{c} \vee \\ \bullet P \\ \ell_1 / \backslash \ell_0 \\ --\ \bullet\ ---\ \bullet\ -- \\ Q/\quad \ell_2 \quad \backslash R \end{array}$$

This cycle turns out to be nonzero.[4] Consider the line $\mathbf{P}^1_0 := V(z_0 + z_1 + z_2) \subset \mathbf{P}^2$, and set $q_j = \mathbf{P}^1_0 \cap \ell_j$, $j = 0, 1, 2$. Then $q_0 = [0, 1, -1]$, $q_1 = [1, 0, -1]$, $q_2 = [1, -1, 0]$, and accordingly $f_j(q_j) = -1$. These Chow groups are known to satisfy a projective bundle formula (see [6], p. 269) which implies that

$$\mathrm{CH}^2(\mathbf{P}^2, 1) \simeq \{\mathbf{P}^1\} \otimes \mathrm{CH}^1(\mathrm{Spec}(\mathbf{C}), 1),$$

$$\mathrm{CH}^2(\mathbf{P}^1_0, 1) \simeq \{\mathbf{P}^1 \cap \mathbf{P}^1_0\}_{\mathbf{P}^2} \otimes \mathrm{CH}^1(\mathrm{Spec}(\mathbf{C}), 1),$$

where $\mathbf{P}^2 \to \mathrm{Spec}(\mathbf{C})$, and $\mathbf{P}^1_0 \to \mathrm{Spec}(\mathbf{C})$ are the structure maps, and $\mathbf{P}^1 \subset \mathbf{P}^2$ is a choice of line. It is well-known that $\mathrm{CH}^1(\mathrm{Spec}(\mathbf{C}), 1) = \mathbf{C}^\times$ ([6], see Example 5.3 below), and thus via restriction we have the isomorphisms:

$$\mathrm{CH}^2(\mathbf{P}^2, 1) \simeq \mathrm{CH}^2(\mathbf{P}^1_0, 1) \simeq \mathbf{C}^\times;$$

moreover under this isomorphism,

$$\xi \mapsto \prod_{j=0}^{2} f_j(q_j) = -1 \in \mathbf{C}^\times.$$

Hence $\xi \in \mathrm{CH}^2(\mathbf{P}^2, 1)$ is a nonzero 2-torsion class.[5]

---

[4] A special thanks to Rob de Jeu for supplying us this idea.

[5] Matt Kerr informed us of an alternate and slick approach to this example via the definition given in Example 4.7(ii). Namely one need only add $\mathrm{Tame}\{z_1/z_0, z_2/z_0\} = (-f_0^{-1}, \ell_0) + (f_1^{-1}, \ell_1) + (f_2^{-1}, \ell_2)$ to ξ to get the 2-torsion class $(-1, \ell_0)$, which is the same as ξ in $\mathrm{CH}^2(\mathbf{P}^2, 1)$.

(ii) Again $X = \mathbf{P}^2$. Let $C \subset X$ be the nodal rational curve given by $z_2^2 z_0 = z_1^3 + z_0 z_1^2$ (in affine coordinates $(x, y) = (z_1/z_0, z_2/z_0) \in \mathbf{C}^2$, C is given by $y^2 = x^3 + x^2$). Let $\tilde{C} \simeq \mathbf{P}^1$ be the normalization of C, with morphism $\pi : \tilde{C} \to C$. Put $P = (0, 0) \in C$ (node) and let $\{R, Q\} = \pi^{-1}(P)$. Choose $f \in \mathbf{C}(\tilde{C})^\times = \mathbf{C}(C)^\times$, such that $(f)_{\tilde{C}} = R - Q$. Then $(f)_C = 0$ and hence $(f, C) \in \mathrm{CH}^2(\mathbf{P}^2, 1)$ defines a higher Chow cycle.

Exercise 4.9. Show that $(f, C) = 0 \in \mathrm{CH}^2(\mathbf{P}^2, 1)$.

5 A Short Detour via Milnor K-Theory

This section provides some of the foundations for the previous section. In the first part of this section, we follow closely the treatment of Milnor K-theory provided in [2], which allows us to provide an abridged definition of the higher Chow groups $\mathrm{CH}^r(X, m)$, for $0 \leq m \leq 2$. The reader with pressing obligations who prefers to work with concrete examples may skip this section, without losing sight of the main ideas presented in this paper.

Let $\mathbf{F}$ be a field, with multiplicative group $\mathbf{F}^\times$, and put $T(\mathbf{F}^\times) = \bigoplus_{n \geq 0} T^n(\mathbf{F}^\times)$, the tensor product of the $\mathbf{Z}$-module $\mathbf{F}^\times$. Here $T^0(\mathbf{F}^\times) := \mathbf{Z}$, $\mathbf{F}^\times = T^1(\mathbf{F}^\times)$, $a \mapsto [a]$. If $a \neq 0, 1$, set $r_a = [a] \otimes [1 - a] \in T^2(\mathbf{F}^\times)$. The two-sided ideal R generated by the $\{r_a\}$'s is graded, and we put:

$$K_\bullet^M \mathbf{F} = \frac{T(\mathbf{F}^\times)}{R} = \bigoplus_{n \geq 0} K_n^M \mathbf{F}, \quad \text{(Milnor } K\text{-theory)}.$$

For example, $K_0(\mathbf{F}) = \mathbf{Z}$, $K_1(\mathbf{F}) = \mathbf{F}^\times$, and $K_2^M(\mathbf{F})$ is the abelian group generated by symbols $\{a, b\}$, subject to the Steinberg relations:

$$\{a_1 a_2, b\} = \{a_1, b\}\{a_2, b\}$$

$$\{a, 1 - a\} = 1, \text{ for } a \neq 0, 1$$
$$\{a, b\} = \{b, a\}^{-1}$$

Furthermore, one can also show that:

$$\{a, a\} = \{-1, a\} = \{a, a^{-1}\} = \{a^{-1}, a\}, \text{and } \{a, -a\} = 1. \qquad (7)$$

Quite generally, one can argue that $K_n^M(\mathbf{F})$ is generated $\{a_1, \ldots, a_n\}$, $a_1, \ldots, a_n \in \mathbf{F}^\times$, subject to:

$$(i) \qquad (a_1, \ldots, a_n) \mapsto \{a_1, \ldots, a_n\},$$

is a multilinear function from $\mathbf{F}^\times \times \cdots \times \mathbf{F}^\times \to K_n^M(\mathbf{F})$,

$$(ii) \qquad \{a_1, \ldots, a_n\} = 0,$$

if $a_i + a_{i+1} = 1$ for some $i < n$.

Next, let us assume given a field $\mathbf{F}$ with discrete valuation $v : \mathbf{F}^\times \to \mathbf{Z}$, with corresponding discrete valuation ring $O_\mathbf{F} := \{a \in \mathbf{F} \mid v(a) \geq 0\}$, and residue field $\mathbf{k}(v)$. One has maps $T : K_\bullet^M(\mathbf{F}) \to K_{\bullet-1}^M(\mathbf{k}(v))$. Choose $\pi \in \mathbf{F}^\times$ such that $v(\pi) = 1$, and note that $\mathbf{F}^\times = O_\mathbf{F}^\times \cdot \pi^\mathbf{Z}$. For example, if we write $a = a_0\pi^i$, $b = b_0\pi^j \in K_1^M(\mathbf{F})$, then $T(a) = i \in \mathbf{Z} = K_0^M(\mathbf{k}(v))$ and

$$T\{a, b\} = (-1)^{ij} \frac{a^j}{b^i} \in \mathbf{k}(v)^\times = K_1^M(\mathbf{k}(v)) \quad \text{(Tame symbol)}.$$

5.1 The Gersten–Milnor Complex

The reader may find [41] particularly useful regarding the discussion in this sub-section. Let $\mathscr{O}_X$ be the sheaf of regular functions on X, with sheaf of units $\mathscr{O}_X^\times$. As in [30], we put

$$\mathscr{K}_{r,X}^M := (\underbrace{\mathscr{O}_X^\times \otimes \cdots \otimes \mathscr{O}_X^\times}_{r \text{ times}})/\mathscr{J}, \quad \text{(Milnor sheaf)},$$

where $\mathscr{J}$ is the subsheaf of the tensor product generated by sections of the form:

$$\{\tau_1 \otimes \cdots \otimes \tau_r \mid \tau_i + \tau_j = 1, \quad \text{for some } i \text{ and } j, \ i \neq j\}.$$

For example, $\mathscr{K}_{1,X}^M = \mathscr{O}_X^\times$. Introduce the Gersten–Milnor complex (a flasque resolution of $\mathscr{K}_{r,X}^M$, see [17, 33]):

$$\mathscr{K}_{r,X}^M \to K_r^M(\mathbf{C}(X)) \to \bigoplus_{\mathrm{cd}_X Z=1} K_{k-1}^M(\mathbf{C}(Z)) \to \cdots$$

$$\to \bigoplus_{\mathrm{cd}_X Z=r-2} K_2^M(\mathbf{C}(Z)) \to \bigoplus_{\mathrm{cd}_X Z=r-1} K_1^M(\mathbf{C}(Z)) \to \bigoplus_{\mathrm{cd}_X Z=r} K_0^M(\mathbf{C}(Z)) \to 0.$$

We have

$$K_0^M(\mathbf{C}(Z)) = \mathbf{Z}, \qquad K_1^M(\mathbf{C}(Z)) = \mathbf{C}(Z)^\times,$$

$$K_2^M(\mathbf{C}(Z)) = \{\text{symbols } \{f, g\}/\text{Steinberg relations}\}.$$

The last three terms of this complex then are:

$$\bigoplus_{\mathrm{cd}_X Z = r-2} K_2^M(\mathbf{C}(Z)) \xrightarrow{T} \bigoplus_{\mathrm{cd}_X Z = r-1} \mathbf{C}(Z)^\times \xrightarrow{\mathrm{div}} \bigoplus_{\mathrm{cd}_X Z = r} \mathbf{Z} \to 0$$

where div is the divisor map of zeros minus poles of a rational function, and T is the Tame symbol map

$$T : \bigoplus_{\mathrm{codim}_X Z = r-2} K_2^M(\mathbf{C}(Z)) \to \bigoplus_{\mathrm{codim}_X D = r-1} K_1^M(\mathbf{C}(D)),$$

defined earlier.

Definition 5.2. For $0 \le m \le 2$,

$$\mathrm{CH}^r(X, m) = H_{\mathrm{Zar}}^{r-m}(X, \mathscr{K}_{r,X}^M).$$

Example 5.3.

$$\mathrm{CH}^1(X, 1) \simeq H_{\mathrm{Zar}}^0(X, \mathscr{K}_{1,X}^M) \simeq H_{\mathrm{Zar}}^0(X, \mathscr{O}_X^\times) \simeq \mathbf{C}^\times.$$

Remark 5.4. The higher Chow groups $\mathrm{CH}^r(W, m)$ were introduced in [6], and are defined for any non-negative integers r and m, and quasi-projective variety W over a field k. The formula in Definition 5.2 is only for smooth varieties X.

6 Hypercohomology

An excellent reference for this is the chapter on spectral sequences in [25].

The reader familiar with hypercohomology can obviously skip this section. Let $(\mathscr{S}^{\bullet \ge 0}, d)$ be a (bounded) complex of sheaves on X. One has a Cech double complex

$$(C^\bullet(\mathscr{U}, \mathscr{S}^\bullet), d, \delta),$$

where $\mathscr{U}$ is an open cover of X. The k-th hypercohomology is given by the k-th total cohomology of the associated single complex

$$(M^\bullet := \oplus_{i+j=\bullet} C^i(\mathscr{U}, \mathscr{S}^j), D = d \pm \delta),$$

viz.,

$$\mathbf{H}^k(\mathscr{S}^\bullet) := \varinjlim_{\mathscr{U}} H^k(M^\bullet).$$

Associated to the double complex are two filtered subcomplexes of the associated single complex, with two associated Grothendieck spectral sequences abutting to $\mathbf{H}^k(\mathscr{S}^\bullet)$ (where $p + q = k$):

$$'E_2^{p,q} := H_\delta^p(X, \mathscr{H}_d^q(\mathscr{S}^\bullet))$$

$$''E_2^{p,q} := H_d^p(H_\delta^q(X, \mathscr{S}^\bullet))$$

The first spectral sequence shows that quasi-isomorphic complexes yield the same hypercohomology:

Alternate take. Two complexes of sheaves $\mathscr{K}_1^\bullet$, $\mathscr{K}_2^\bullet$ are said to be quasi-isomorphic if there is a morphism $h : \mathscr{K}_1^\bullet \to \mathscr{K}_2^\bullet$ inducing an isomorphism on cohomology $h_* : \mathscr{H}^\bullet(\mathscr{K}_1^\bullet) \xrightarrow{\sim} \mathscr{H}^\bullet(\mathscr{K}_2^\bullet)$. Take a complex of acyclic sheaves $(\mathscr{K}^\bullet, d)$ (viz., $H^{i>0}(X, \mathscr{K}^j) = 0$ for all j) quasi-isomorphic to $\mathscr{S}^\bullet$. Then by the second spectral sequence:

$$\mathbf{H}^k(\mathscr{S}^\bullet) := H^i(H^0(X, \mathscr{K}^\bullet)).$$

For example if $\mathcal{L}^{\bullet,\bullet}$ is an acyclic resolution of $\mathscr{S}^\bullet$, then the associated single complex $\mathscr{K}^\bullet = \oplus_{i+j=\bullet}\mathcal{L}^{i,j}$ is acyclic and quasi-isomorphic to $\mathscr{S}^\bullet$.

Example 6.1. Let $(\Omega_X^\bullet, d)$, $(\mathscr{E}_X^\bullet, d)$ be the complexes of sheaves of holomorphic and $\mathbf{C}$-valued C^∞ forms respectively. By the holomorphic and C^∞ Poincaré lemmas, one has quasi-isomorphisms:

$$(\mathbf{C} \to 0 \to \cdots) \xrightarrow{\approx} (\Omega_X^\bullet, d) \xrightarrow{\approx} (\mathscr{E}_X^\bullet, d),$$

where the latter two are Hodge filtered, using an argument similar to that in (12) below. The first spectral sequence of hypercohomology shows that

$$H^k(X, \mathbf{C}) \simeq \mathbf{H}^k(\mathbf{C} \to 0 \to \cdots) \simeq \mathbf{H}^k((F^p)\Omega_X^\bullet) \simeq \mathbf{H}^k((F^p)\mathscr{E}_X^\bullet).$$

The second spectral sequence of hypercohomology applied to the latter term, using the known acyclicity of $\mathscr{E}_X^\bullet$, yields

$$\mathbf{H}^k(F^p\mathscr{E}_X^\bullet) \simeq \frac{\ker d : F^p E_X^k \to F^p E_X^k}{dF^p E_X^{k-1}} \simeq F^p H_{\mathrm{DR}}^k(X),$$

where the latter isomorphism is due to the Hodge to de Rham spectral sequence.

7 Deligne Cohomology

A standard reference for this section is [19] (also see [27]). For a subring $\mathbf{A} \subseteq \mathbf{R}$, we introduce the Deligne complex

$$\mathbf{A}_{\mathscr{D}}(r) : \quad \mathbf{A}(r) \to \underbrace{\mathscr{O}_X \to \Omega_X^1 \to \cdots \to \Omega_X^{r-1}}_{\text{call this } \Omega_X^{\bullet<r}}.$$

Definition 7.1. Deligne cohomology is given by the hypercohomology:

$$H^i_{\mathscr{D}}(X, \mathbf{A}(r)) = \mathbf{H}^i(\mathbf{A}_{\mathscr{D}}(r)).$$

Example 7.2. When $\mathbf{A} = \mathbf{Z}$, we have a quasi-isomorphism

$$\mathbf{Z}_{\mathscr{D}}(1) \approx \mathscr{O}_X^\times[-1],$$

hence

$$H^2_{\mathscr{D}}(X, \mathbf{Z}(1)) \simeq H^1(X, \mathscr{O}_X^\times) =: \mathrm{Pic}(X) \simeq \mathrm{CH}^1(X).$$

$$H^1_{\mathscr{D}}(X, \mathbf{Z}(1)) \simeq H^0(X, \mathscr{O}_X^\times) \simeq \mathbf{C}^\times \simeq \mathrm{CH}^1(X, 1).$$

Example 7.3. Alternate take on $H^1_{\mathscr{D}}(X, \mathbf{Z}(1))$. Look at the Cech double complex:

$$\begin{array}{ccc}
\mathscr{C}^0(\mathscr{U}, \mathbf{Z}(1)) & \to & \mathscr{C}^0(\mathscr{U}, \mathscr{O}_X) \\
\delta \downarrow & & \downarrow \delta \\
\mathscr{C}^1(\mathscr{U}, \mathbf{Z}(1)) & \to & \mathscr{C}^1(\mathscr{U}, \mathscr{O}_X)
\end{array}$$

So a class in $H^1_{\mathscr{D}}(X, \mathbf{Z}(1))$ is represented (after a suitable refinement) by $(\lambda := \{\lambda_{\alpha\beta}\}, f := \{f_\gamma\}) \in (\Gamma(U_\alpha \cap U_\beta, \mathbf{Z}(1)), \Gamma(U_\gamma, \mathscr{O}_X))$, with $f_\beta - f_\alpha =: \delta(f)_{\alpha\beta} = \lambda_{\alpha\beta}$. Note that $\exp(f) \in H^0(X, \mathscr{O}_X^\times)$ determines the isomorphism $H^1_{\mathscr{D}}(X, \mathbf{Z}(1)) \simeq H^0(X, \mathscr{O}_X^\times) \simeq \mathbf{C}^\times$.

Definition 7.4. The product structure on Deligne cohomology

$$H^k_{\mathscr{D}}(X, \mathbf{Z}(i)) \otimes H^l_{\mathscr{D}}(X, \mathbf{Z}(j)) \to H^{k+l}_{\mathscr{D}}(X, \mathbf{Z}(i+j)),$$

is induced by the multiplication of complexes $\mu : \mathbf{Z}_{\mathscr{D}}(i) \otimes \mathbf{Z}_{\mathscr{D}}(j) \to \mathbf{Z}_{\mathscr{D}}(i+j)$ defined by

$$\mu(x, y) := \begin{cases} x \cdot y, & \text{if } \deg x = 0, \\ x \wedge dy, & \text{if } \deg x > 0 \text{ and } \deg y = j > 0, \\ 0, & \text{otherwise.} \end{cases}$$

Example 7.5. For example, this product structure implies that

$$H^1_{\mathscr{D}}(X, \mathbf{Z}(1)) \cup H^1_{\mathscr{D}}(X, \mathbf{Z}(1)) = \{0\} \subset H^2_{\mathscr{D}}(X, \mathbf{Z}(2)).$$

Recall from Hodge theory, one has the isomorphism:

$$\mathbf{H}^i(\Omega_X^{\bullet \geq r}) \simeq F^r H^i(X, \mathbf{C}).$$

This together with the short exact sequence of complexes:

$$0 \to \Omega_X^{\bullet \geq r} \to \Omega_X^\bullet \to \Omega_X^{\bullet < r} \to 0,$$

implies that

$$\mathbf{H}^i(\Omega_X^{\bullet < r}) \simeq \frac{H^i(X, \mathbf{C})}{F^r H^i(X, \mathbf{C})}.$$

Thus applying $\mathbf{H}^\bullet(-)$ to the short exact sequence:

$$0 \to \Omega_X^{\bullet < r}[-1] \to \mathbf{A}_{\mathscr{D}}(r) \to \mathbf{A}(r) \to 0,$$

yields the short exact sequence:

$$0 \to \underbrace{\frac{H^{i-1}(X,\mathbf{C})}{H^{i-1}(X,\mathbf{A}(r)) + F^r H^{i-1}(X,\mathbf{C})}}_{=J_\mathbf{A}\left(H^{i-1}(X,\mathbf{A}(r))\right)} \to H^i_{\mathscr{D}}(X,\mathbf{A}(r)) \to \Gamma_\mathbf{A}(H^i(X,\mathbf{A}(r))) \to 0. \quad (8)$$

If we consider $\mathbf{A} = \mathbf{Z}$, and $i = 2r$, then (8) becomes:

$$0 \to J^r(X) \to H^{2r}_{\mathscr{D}}(X,\mathbf{Z}(r)) \to \mathrm{Hg}^r(X) \to 0,$$

where $J^r(X) = J(H^{2r-1}(X,\mathbf{Z}(r)))$ is the Griffiths jacobian, and where the Hodge group $\mathrm{Hg}^r(X)$ in untwisted form can be identified with:

$$\{w \in H^{2r}(X,\mathbf{Z}) \mid w \in H^{r,r}(X,\mathbf{C})\},$$

via $\ker(H^{2r}(X,\mathbf{Z}) \to H^{2r}(\Omega_X^{\bullet < r})))$. In particular, $\mathrm{Hg}^r(X)$ includes the torsion classes in $H^{2r}(X,\mathbf{Z}(r))$.

Next, if $\mathbf{A} = \mathbf{Z}$ and $i \leq 2r-1$, then from Hodge theory, $H^i(X,\mathbf{Z}(r)) \cap F^r H^i(X,\mathbf{C})$ is torsion. In particular, there is a short exact sequence:

$$0 \to \frac{H^{i-1}(X,\mathbf{C})}{F^r H^{i-1}(X,\mathbf{C}) + H^{i-1}(X,\mathbf{Z}(r))} \to H^i_{\mathscr{D}}(X,\mathbf{Z}(r)) \to H^i_{\mathrm{tor}}(X,\mathbf{Z}(r)) \to 0,$$

where $H^i_{\mathrm{tor}}(X,\mathbf{Z}(r))$ is the torsion subgroup of $H^i(X,\mathbf{Z}(r))$. The compatibility of Poincaré and Serre duality yields the isomorphism:

$$\frac{H^{i-1}(X,\mathbf{C})}{F^r H^{i-1}(X,\mathbf{C}) + H^{i-1}(X,\mathbf{Z}(r))} \simeq \frac{F^{d-r+1} H^{2d-i+1}(X,\mathbf{C})^\vee}{H_{2d-i+1}(X,\mathbf{Z}(d-r))}.$$

Next, if $\mathbf{A} = \mathbf{R}$ and $i = 2r-1$, then $H^i_{\mathrm{tor}}(X,\mathbf{R}(r)) = 0$; moreover if we set

$$\pi_{r-1} : \mathbf{C} = \mathbf{R}(r) \oplus \mathbf{R}(r-1) \to \mathbf{R}(r-1)$$

to be the projection, then we have the isomorphisms:

$$H^{2r-1}_{\mathscr{D}}(X,\mathbf{R}(r)) \simeq \frac{H^{2r-2}(X,\mathbf{C})}{F^r H^{2r-2}(X,\mathbf{C}) + H^{2r-2}(X,\mathbf{R}(r))}$$

$$\xrightarrow[\simeq]{\pi_{r-1}} \frac{H^{2r-2}(X,\mathbf{R}(r-1))}{\pi_{r-1}(F^r H^{2r-2}(X,\mathbf{C}))}$$

$$=: H^{r-1,r-1}(X,\mathbf{R}(r-1))$$

$$\simeq \{H^{d-r+1,d-r+1}(X,\mathbf{R}(d-r+1))\}^\vee.$$

7.6 Alternate Take on Deligne Cohomology

Let $h : (A^\bullet, d) \to (B^\bullet, d)$ be a morphism of complexes. We define

$$\mathrm{Cone}(A^\bullet \xrightarrow{h} B^\bullet)$$

by the formula

$$[\mathrm{Cone}(A^\bullet \xrightarrow{h} B^\bullet)]^q := A^{q+1} \oplus B^q, \quad \delta(a,b) = (-da, h(a) + db).$$

Example 7.7. $\mathrm{Cone}(\mathbf{A}(r) \oplus F^r \Omega_X^\bullet \xrightarrow{\epsilon - l} \Omega_X^\bullet)[-1]$ is given by:

$$\mathbf{A}(r) \to \mathscr{O}_X \xrightarrow{d} \Omega_X \xrightarrow{d} \cdots \xrightarrow{d} \Omega_X^{r-2} \xrightarrow{(0,d)} (\Omega_X^r \oplus \Omega_X^{r-1})$$

$$\xrightarrow{\delta} (\Omega_X^{r+1} \oplus \Omega_X^r) \xrightarrow{\delta} \cdots \xrightarrow{\delta} (\Omega_X^d \oplus \Omega_X^{d-1}) \to \Omega_X^d$$

Using the holomorphic Poincaré lemma, one can show that the natural map

$$\mathbf{A}_{\mathscr{D}}(r) \to \mathrm{Cone}(\mathbf{A}(r) \oplus F^r \Omega_X^\bullet \xrightarrow{\epsilon - l} \Omega_X^\bullet)[-1],$$

is a quasi-isomorphism.[6] Thus

$$H^k_{\mathscr{D}}(X, \mathbf{A}(r)) \simeq \mathbf{H}^r(\mathrm{Cone}(\mathbf{A}(r) \oplus F^r \Omega_X^\bullet \xrightarrow{\epsilon - l} \Omega_X^\bullet)[-1]).$$

Let $\mathscr{D}_X^\bullet$ be the sheaf of currents acting on C^∞ compactly supported $(2d - \bullet)$-forms. Further, let $\mathscr{D}_X^{p,q}$ be the sheaf of currents acting on C^∞ compactly supported $(d - p, d - q)$-forms. One has a decomposition

$$\mathscr{D}_X^\bullet = \bigoplus_{p+q=\bullet} \mathscr{D}_X^{p,q},$$

with a morphism of complexes $\mathscr{E}_X^\bullet \hookrightarrow \mathscr{D}_X^\bullet$ (induced by $\omega \mapsto (2\pi i)^{-d} \int_X \omega \wedge (-)$), and with $\mathscr{E}_X^{p,q} \hookrightarrow \mathscr{D}_X^{p,q}$, compatible with both ∂ and $\bar{\partial}$. Likewise, let $\mathscr{C}_X^\bullet = \mathscr{C}_{2d-\bullet,X}(\mathbf{A}(r))$ be the sheaf of (Borel–Moore) chains of real codimension $\bullet$. Identifying the constant sheaf $\mathbf{A}(r)$ with the complex $\mathbf{A}(r) \to 0 \to \cdots \to 0$, we have quasi-isomorphisms

$$\mathbf{A}(r) \xrightarrow{\approx} \mathscr{C}_X^\bullet(\mathbf{A}(r)), \quad \mathscr{E}_X^\bullet \xrightarrow{\approx} \mathscr{D}_X^\bullet$$

where the latter is (Hodge) filtered.

---

[6] Indeed first consider $(a, b) \in \Omega_X^r \oplus \Omega_X^{r-1} \xrightarrow{\delta} (-da, db - a) \in \Omega_X^{r+1} \oplus \Omega_X^r$. Then $\delta(a, b) = (0, 0) \Leftrightarrow da = 0$ & $a = db \Leftrightarrow a = db$. Therefore $\ker \delta / \mathrm{Im}(0, d) \simeq \Omega_X^{r-1}/d\Omega_X^{r-2} = \mathscr{H}^{r-1}(\mathbf{A}_{\mathscr{D}}(r))$. Next, for $j \geq 1$, $(a, b) \in \Omega_X^{r+j} \oplus \Omega_X^{r+j-1}$, $\delta(a, b) = 0 \Leftrightarrow (a, b) = \delta(-b, 0)$.

Observe that $\mathscr{D}_X^\bullet(X))[-1]$ is a subcomplex of $\mathrm{Cone}(\mathscr{C}_X^\bullet(X, \mathbf{A}(r)) \oplus F^r \mathscr{D}_X^\bullet(X) \xrightarrow{\epsilon - l}$ $\mathscr{D}_X^\bullet(X))[-1]$. Hence the cone complex description of:

$$H^i_{\mathscr{D}}(X, \mathbf{A}(r)) \simeq H^i(\mathrm{Cone}(\mathscr{C}_X^\bullet(X, \mathbf{A}(r)) \oplus F^r \mathscr{D}_X^\bullet(X) \xrightarrow{\epsilon - l} \mathscr{D}_X^\bullet(X))[-1]),$$

yields the exact sequence[7]:

$$\cdots \to H^{i-1}(X, \mathbf{A}(r)) \oplus F^r H^{i-1}(X, \mathbf{C}) \to H^{i-1}(X, \mathbf{C}) \tag{9}$$

$$\to H^i_{\mathscr{D}}(X, \mathbf{A}(r)) \to H^i(X, \mathbf{A}(r)) \oplus F^r H^i(X, \mathbf{C}) \to \cdots$$

7.8 Deligne–Beilinson Cohomology

The formulation of Deligne cohomology in Definition 7.1 above, which incidentally can be defined in the same way for any complex manifold (and is also called analytic Deligne cohomology), works well for projective algebraic manifolds X, but not so well for smooth open $U \subset X$. First of all, the naive Hodge filtration on U, viz., $\Omega_U^{\bullet \geq r}$ is the *wrong* choice. For example, if W is a Stein manifold, then $H^q(W, \Omega_W^i) = 0$ for all i and where $q \geq 1$. This tells us, via the Grothendieck spectral sequences associated to hypercohomology, that

$$H^j(W, \mathbf{C}) \simeq \frac{H^0(W, \Omega_W^j)_{d-\text{closed}}}{d H^0(W, \Omega_W^{j-1})}.$$

(Note: If W is a smooth affine variety, then by Grothendieck, one can use algebraic differential forms.) We hardly expect $H^j(W, \mathbf{C}) = F^j H^j(W, \mathbf{C})$ to be the case in general. Secondly, analytic Deligne cohomology fails to take into consideration the underlying algebraic structure of U. For instance $H^1_{\mathscr{D}}(U, \mathbf{Z}(1)) = H^0(U, \mathscr{O}_{U,an}^\times)$, i.e. the non-zero analytic functions on U. It would be preferable to recover the non-zero algebraic functions on U instead. Beilinson's remedy is to incorporate Deligne's logarithmic complex into the picture. By a standard reduction, we may assume that $j : U = X \backslash Y \hookrightarrow X$, where Y is a normal crossing divisor[8] with smooth components. We define $\Omega_X^\bullet \langle Y \rangle$ to be the de Rham complex of meromorphic forms on X, holomorphic on U, with at most logarithmic poles along Y. One has a filtered complex

$$F^r \Omega_X^\bullet \langle Y \rangle = \Omega_X^{\bullet \geq r} \langle Y \rangle,$$

with Hodge to de Rham spectral sequence degenerating at E_1. This gives

$$F^r H^i(U, \mathbf{C}) = \mathbf{H}^i(F^r \Omega_X^\bullet \langle Y \rangle) \subset \mathbf{H}^i(\Omega_X^\bullet \langle Y \rangle) = H^i(U, \mathbf{C}), \tag{10}$$

---

[7] The reader familiar with Deligne homology will see this definition as the same thing up to twist. Indeed this definition already incorporates Poincaré duality.

[8] Y is a normal crossing divisor, which in local analytic coordinates $(z_1, \ldots, z_d)$ on X, Y is given by $z_1 \cdots z_\ell = 0$, and so $\Omega_X^1 \langle Y \rangle$ has local frame $\{dz_1/z_1, \ldots, dz_\ell/z_\ell, dz_{\ell+1}, \ldots, dz_d\}$.

and defines the correct Hodge filtration. The weight filtration is characterized in terms of differentials with residues along $Y^{[\bullet]}$, where $Y^{[\bullet]}$ is the simplicial complex made up of the intersections of the irreducible components of Y.

Definition 7.9. Deligne–Beilinson cohomology is given by

$$H^i_{\mathscr{D}}(U, \mathbf{A}(r)) := \mathbf{H}^i(\mathbf{A}_{\mathscr{D}}(r)),$$

where

$$\mathbf{A}_{\mathscr{D}}(r) := \text{Cone}(Rj_*\mathbf{A}(r) \bigoplus F^r\Omega^\bullet_X\langle Y\rangle \xrightarrow{\epsilon-l} Rj_*\Omega^\bullet_U)[-1].$$

Here ϵ and l are the natural maps obtained after a choice of (the direct image of) injective resolutions of $\mathbf{A}(r)$ and $\Omega^\bullet_U$. One shows that this is independent of the good compactifications of U. One has a short exact sequence:

$$0 \to \frac{H^{i-1}(U, \mathbf{C})}{F^rH^{i-1}(U, \mathbf{C}) + H^{i-1}(U, \mathbf{A}(r))} \to H^i_{\mathscr{D}}(U, \mathbf{A}(r)) \to F^r \bigcap H^i(U, \mathbf{A}(r)) \to 0,$$

(11)

where

$$F^r \bigcap H^i(U, \mathbf{A}(r)) := \ker\left(F^rH^i(U, \mathbf{C}) \oplus H^i(U, \mathbf{A}(r)) \xrightarrow{\epsilon-l} H^i(U, \mathbf{C})\right).$$

We would like a more earthly description of $H^i_{\mathscr{D}}(U, \mathbf{A}(r))$. First observe that there are filtered quasi-isomorphisms

$$(F^r, \Omega^\bullet_X\langle Y\rangle) \hookrightarrow (F^r, \mathscr{E}^\bullet_X\langle Y\rangle) \hookrightarrow (F^r, \mathscr{D}^\bullet_X\langle Y\rangle),$$

(12)

where

$$F^r\mathscr{D}^\bullet_X\langle Y\rangle = \{F^r\Omega^\bullet_X\langle Y\rangle\} \otimes_{\Omega^\bullet_X} \mathscr{D}^\bullet_X.$$

To see this, one uses the argument in [27]. By a spectral sequence argument, it is enough to show that the associated graded pieces in (12) are quasi-isomorphic, viz.,

$$\Omega^r_X\langle Y\rangle \approx \Omega^r_X\langle Y\rangle \otimes_{\mathcal{O}_X} \mathscr{E}^{0,\bullet}_X \approx \Omega^r_X\langle Y\rangle \otimes_{\mathcal{O}_X} \mathscr{D}^{0,\bullet}_X,$$

where the differential is now $1 \otimes \bar{\partial}$. One now applies the $\bar{\partial}$ lemma together with the flatness of $\Omega^r_X\langle Y\rangle$ over $\mathcal{O}_X$, and using $\mathcal{O}_X$ as $\bar{\partial}$-linear. According to [34], $\mathscr{D}^\bullet_X\langle Y\rangle$ admits the interpretation of the space of currents acting on those (compactly supported) forms on X which "vanish holomorphically" on Y. Let $\mathscr{C}^i(X, \mathbf{A}(r))$ be the chains of real codimension i in X, and $\mathscr{C}^i_Y(X, \mathbf{A}(r))$ the subspace of chains supported on Y. Put

$$\mathscr{C}^i(X, Y, \mathbf{A}(r)) := \frac{\mathscr{C}^i(X, \mathbf{A}(r))}{\mathscr{C}^i_Y(X, \mathbf{A}(r))}.$$

One has a map of complexes:

$$(\mathscr{C}^\bullet(X, Y, \mathbf{A}(r)), d) \to (\mathscr{D}^\bullet_X\langle Y\rangle(X), d),$$

which induces a quasi-isomorphism

$$\mathscr{C}^\bullet(X, Y, \mathbf{A}(r)) \otimes \mathbf{C} \to \mathscr{D}_X^\bullet\langle Y\rangle(X).$$

Definition 7.10. Deligne–Beilinson cohomology is given by

$$H_{\mathscr{D}}^i(U, \mathbf{A}(r)) := H^i(\mathrm{Cone}(\mathscr{C}^\bullet(X, Y, \mathbf{A}(r)) \bigoplus F^r\mathscr{D}_X^\bullet\langle Y\rangle(X) \xrightarrow{\epsilon - l} (\mathscr{D}_X^\bullet\langle Y\rangle(X))[-1]).$$

Example 7.11. Let us compute $H_{\mathscr{D}}^1(U, \mathbf{Z}(1))$. First of all $\{\xi\} \in H_{\mathscr{D}}^1(U, \mathbf{Z}(1))$ is represented by a D-closed triple:

$$\xi = (a, b, c) \in (\mathscr{C}^1(X, Y, \mathbf{Z}(1)) \bigoplus F^1\mathscr{D}_X^1\langle Y\rangle(X) \bigoplus \mathscr{D}_X^0\langle Y\rangle(X)),$$

where $da = 0$, $db = 0$ and $a - b = dc$. Note that $\bar\partial$-regularity implies that $b \in \Omega_X^1\langle Y\rangle(X)_{d-\mathrm{closed}}$. Let $\hat\Omega_U^1$ be the sheaf of d-closed holomorphic 1-forms on U, and let's make the identification $\mathbf{C}^\times = \mathbf{C}/\mathbf{Z}(1)$. From the short exact sequence:

$$0 \to \mathbf{C}^\times \to \mathscr{O}_U^\times \xrightarrow{d\log} \hat\Omega_U^1 \to 0,$$

and the relation $a - b = dc$, it follows that

$$b \in \ker\,(H^0(U, \hat\Omega_U^1) \to H^1(U, \mathbf{C}^\times)),$$

and hence $b = d\log f$ for some $f \in \mathscr{O}_U^\times(U)$. Since $b \in \Omega_X^1\langle Y\rangle(X)$, it follows that $f \in \mathscr{O}_{U,\mathrm{alg}}^\times(U)$. Thus in Deligne cohomology[9]

$$\{\xi\} = (2\pi i T_\xi, \Omega_\xi, R_\xi),$$

where $T_\xi := \delta_{f^{-1}(\mathbf{R}^-)}$ is given by integration along $f^{-1}[-\infty, 0]$, and $\Omega_\xi = [d\log f]$, $R_\xi = [\log f]$ are the obvious defined currents.

Corollary 7.12.

$$\mathrm{cl}_{1,1} : \mathrm{CH}^1(U, 1) := \mathscr{O}_{U,\mathrm{alg}}^\times(U) \xrightarrow{\sim} H_{\mathscr{D}}^1(U, \mathbf{Z}(1)),$$

is an isomorphism.

Remark 7.13. We observe in passing the following. We deduce from (11) the short exact sequence:

$$0 \to \frac{H^0(U, \mathbf{C})}{F^1 H^0(U, \mathbf{C}) + H^0(U, \mathbf{Z}(1))} \to H_{\mathscr{D}}^1(U, \mathbf{Z}(1)) \to \Gamma(H^1(U, \mathbf{Z}(1))) \to 0,$$

---

[9] For compactly supported $\omega \in E_{U,c}^{2d-1}$, and $f \in \mathscr{O}_U^\times(U)$,

$$\int_U \frac{df}{f} \wedge \omega = \int_U d(\log f \wedge \omega) - \int_{U\setminus f^{-1}[-\infty,0]} \log f \wedge d\omega = 2\pi i \int_{f^{-1}[-\infty,0]} \omega + d[\log f](\omega),$$

where we use the principal branch of log.

which in turn from Corollary 7.12 yields the short exact sequence:

$$0 \to \mathbf{C}^\times \to \mathrm{CH}^1(U,1) \xrightarrow{d \log} \Gamma(H^1(U,\mathbf{Z}(1))) \to 0.$$

Remark 7.14. Let $U/\mathbf{C}$ be a smooth quasi-projective variety. If $H^\bullet_{\mathscr{D},an}(U,\mathbf{Z}(\bullet))$ denotes that analytic Deligne cohomology, then we know that $H^2_{\mathscr{D},an}(U,\mathbf{Z}(1)) \simeq H^1(U,\mathscr{O}^\times_{U,an})$, the holomorphic isomorphism classes of holomorphic line bundles over U. For Deligne–Beilinson cohomology, and using the fact that $H^1(U,\mathbf{Z}(1)) = W_0 H^1(U,\mathbf{Z}(1))$, it follows that there is a short exact sequence:

$$0 \to J(H^1(U,\mathbf{Z}(1))) \to H^2_{\mathscr{D}}(U,\mathbf{Z}(1)) \xrightarrow{\alpha} F^1 \cap H^2(U,\mathbf{Z}(1)) \to 0,$$

but in general

$$\Gamma(H^2(U,\mathbf{Z}(1))) = F^0 \cap W_0 H^2(U,\mathbf{Z}(1)) \subsetneq F^1 \cap H^2(U,\mathbf{Z}(1)),$$

where the shift $F^0 \mapsto F^1$ is really the same filtration, but the latter is in "untwisted" terminology. To remedy this, let us put $H^2_{\mathscr{H}}(U,\mathbf{Z}(1)) = \alpha^{-1}(\Gamma(H^2(U,\mathbf{Z}(1))))$. This turns out to be the same thing as the image $H^2_{\mathscr{D}}(\overline{U},\mathbf{Z}(1)) \to H^2_{\mathscr{D}}(U,\mathbf{Z}(1))$, where $\overline{U}$ is any smooth projective compactification of U. Then $H^2_{\mathscr{H}}(U,\mathbf{Z}(1))$ amounts to a special instance of Beilinson's absolute Hodge cohomology (see [3]). We then have the following:

Proposition 7.15. *Let $U/\mathbf{C}$ be a smooth quasi-projective variety. Then:*

$$H^2_{\mathscr{H}}(U,\mathbf{Z}(1)) \simeq H^1_{\mathrm{Zar}}(U,\mathscr{O}^\times_{U,\mathrm{alg}})$$

Proof. First recall that

$$H^1_{\mathrm{Zar}}(U,\mathscr{O}^\times_{U,\mathrm{alg}}) = H^1_{\mathrm{Zar}}(U,\mathscr{K}^M_{1,U}) = \mathrm{CH}^1(U).$$

There is a commutative diagram:

$$
\begin{array}{ccccccc}
0 \to & \mathrm{CH}^1_{\mathrm{hom}}(U) & \to & \mathrm{CH}^1(U) & \to & \frac{\mathrm{CH}^1(U)}{\mathrm{CH}^1_{\mathrm{hom}}(U)} & \to 0 \\[2mm]
 & \Phi_1 \downarrow & & \mathrm{cl}_1 \downarrow & & \wr \downarrow & \\[2mm]
0 \to & J(H^1(U,\mathbf{Z}(1))) & \to & H^2_{\mathscr{H}}(U,\mathbf{Z}(1)) & \to & \Gamma(H^2(U,\mathbf{Z}(1))) & \to 0
\end{array}
$$

It suffices to show that Φ_1 is an isomorphism. Let $\overline{U}$ be a smooth projective compactification of U. We may assume that $Y := \overline{U} \backslash U$ is a divisor. With regard to the short exact sequence:

$$0 \to H^1(\overline{U},\mathbf{Z}(1)) \to H^1(U,\mathbf{Z}(1)) \to H^2_Y(\overline{U},\mathbf{Z}(1))^\circ \to 0,$$

it is clear that $J(H^2_Y(\overline{U}, \mathbf{Z}(1))^\circ) = 0$, and hence the following diagram finishes the proof:

$$
\begin{array}{ccccccc}
\mathrm{CH}^1_Y(\overline{U})^\circ & \rightarrow & \mathrm{CH}^1_{\mathrm{hom}}(\overline{U}) & \rightarrow & \mathrm{CH}^1_{\mathrm{hom}}(U) & \rightarrow & 0 \\
\downarrow_? & & \downarrow_? & & \downarrow_{\Phi_1} & & \\
\varGamma H^2_Y(\overline{U}, \mathbf{Z}(1))^\circ & \rightarrow & J(H^1(\overline{U}, \mathbf{Z}(1))) & \rightarrow & J(H^1(U, \mathbf{Z}(1))) & \rightarrow & 0
\end{array}
$$

$\square$

8 Examples of $H^{r-m}_{\mathrm{Zar}}(X, \mathcal{K}^M_{r,X})$ and Corresponding Regulators

The reader is encouraged to consult for example [35, 43] (as well as works due to Bloch, Beilinson, Esnault and Goncharov), for various earlier incarnations of regulator type currents for higher Chow cycles. A complete description of the Beilinson/Bloch regulator in terms of polylogarithmic type currents for complex varieties can be found in [31, 32].

8.1 Case m = 0 and CY Threefolds

In this case we recall that

$$
H^r_{\mathrm{Zar}}(X, \mathcal{K}^M_{r,X}) = \mathrm{CH}^r(X).
$$

The fundamental class map:

$$
\mathrm{cl}_r : \mathrm{CH}^r(X) \rightarrow H^{2r}_{\mathrm{DR}}(X, \mathbf{C}) \simeq H^{2d-2r}_{\mathrm{DR}}(X, \mathbf{C})^\vee,
$$

can be defined in a number of equivalent ways:

(i) (See [18].) The $d\log$ map $\mathcal{K}^M_{r,X} \rightarrow \Omega^r_X$, $\{f_1, \ldots, f_r\} \mapsto \bigwedge_j d\log f_j$, induces a morphism of complexes in the Zariski topology $\{\mathcal{K}^M_{r,X} \rightarrow 0\} \rightarrow \Omega^{\bullet \geq r}_X[r]$, and thus using GAGA,

$$
\mathrm{CH}^r(X) = H^r_{\mathrm{Zar}}(X, \mathcal{K}^M_{r,X}) = \mathbf{H}^r(\{\mathcal{K}^M_{r,X} \rightarrow 0\}) \rightarrow \mathbf{H}^r(\Omega^{\bullet \geq r}_X[r])
$$

$$
= \mathbf{H}^{2r}(\Omega^{\bullet \geq r}_X) = F^r H^{2r}_{\mathrm{DR}}(X, \mathbf{C}).
$$

(ii) Let $V \subset X$ be a subvariety of codimension r in X, and $\{w\} \in H^{2d-2r}_{\mathrm{DR}}(X, \mathbf{C})$, (de Rham cohomology). Define

$$
\mathrm{cl}_r(V)(w) = \frac{1}{(2\pi \mathrm{i})^{d-r}} \delta_V := \frac{1}{(2\pi \mathrm{i})^{d-r}} \int_{V^*} w,
$$

and extend to $CH^r(X)$ by linearity, where $V^* = V\backslash V_{\text{sing}}$. Note that $\dim_{\mathbf{R}} V = 2d - 2r$. The easiest way to show that cl_r is well-defined (finite volume, closed current) is to first pass to a desingularization of V above, and apply a Stokes' theorem argument. The proof of a more direct approach can be found, for example, in [25].

(One way to connected (i) and (ii) is as follows. If we write Γ for $H^0(X, -)$, then there is a diagram that commutes up to sign:

$$\Gamma K_r^M(\mathbf{C}(X)) \to \Gamma \bigoplus_{\text{cd}_X Y=1} K_{r-1}^M(\mathbf{C}(Y)) \to \cdots \to \Gamma \bigoplus_{\text{cd}_X V=r} K_0^M(\mathbf{C}(X))$$

$$\int_X \frac{d\log_r}{(2\pi i)^d} \Big\downarrow \qquad \int_Y \frac{d\log_{r-1}}{(2\pi i)^{d-1}} \Big\downarrow \qquad \cdots \qquad \int_V \frac{d\log_0}{(2\pi i)^{d-r}} \Big\downarrow$$

$$\Gamma F^r \mathscr{D}_X^r \xrightarrow{d} \qquad \Gamma F^r \mathscr{D}_X^{r+1} \xrightarrow{d} \cdots \xrightarrow{d} \qquad \Gamma F^r \mathscr{D}_X^{2r}$$

where

$$d\log_r(\{f_1, \ldots, f_r\}) = \bigwedge_{j=1}^r d\log f_j, \qquad \int_V \frac{d\log_0}{(2\pi i)^{d-r}} = \frac{1}{(2\pi i)^{d-r}} \delta_V.$$

From the aforementioned filtered quasi-isomorphism $\Omega_X^{\bullet} \hookrightarrow \mathscr{D}_X^{\bullet}$, the prescriptions in (i) and (ii) can be seen as almost tautologies.)

(iii) Thirdly one has a fundamental class generator $\{V\} \in H_{2d-2r}(V, \mathbf{Z}(d-r)) \simeq H_V^{2r}(X, \mathbf{Z}(r)) \to H_{2d-2r}(X, \mathbf{Z}((d-r)) \simeq H^{2r}(X, \mathbf{Z}(r))$. In summary we have $\text{cl}_r : CH^r(X) \to \text{Hg}^r(X)$. This map fails to be surjective in general for $r > 1$ (see [36]).

Conjecture 8.2 (Hodge$_{\mathbf{Q}}$).

$$\text{cl}_r : CH^r(X) \otimes \mathbf{Q} \to \text{Hg}^r(X) \otimes \mathbf{Q}, \text{ is surjective.}$$

Next, the Abel–Jacobi map:

$$\Phi_r : CH_{\text{hom}}^r(X) \to J^r(X),$$

is defined as follows. Recall that

$$J^r(X) = \frac{H^{2r-1}(X, \mathbf{C})}{F^r H^{2r-1}(X, \mathbf{C}) + H^{2r-1}(X, \mathbf{Z}(r))} \simeq \frac{F^{d-r+1} H^{2d-2r+1}(X, \mathbf{C})^{\vee}}{H_{2d-2r+1}(X, \mathbf{Z}(d-r))},$$

is a compact complex torus, called the Griffiths jacobian.

Prescription for Φ_r: Let $\xi \in CH_{\text{hom}}^r(X)$. Then $\xi = \partial \zeta$ bounds a $2d - 2r + 1$ real dimensional chain ζ in X. Let $\{w\} \in F^{d-r+1} H^{2d-2r+1}(X, \mathbf{C})$. Define:

$$\Phi_r(\xi)(\{w\}) = \frac{1}{(2\pi i)^{d-r}} \int_\zeta w \quad \text{(modulo periods)}.$$

That Φ_r is well-defined follows from the fact that $F^\ell H^i(X, \mathbf{C})$ depends only on the complex structure of X, namely

$$F^\ell H^i(X, \mathbf{C}) \simeq \frac{F^\ell E^i_{X, d-\text{closed}}}{d(F^\ell E^{i-1}_X)},$$

where we recall that E^i_X are the C^∞ complex-valued i-forms on X.

Alternate take for Φ_r: Let $\xi \in \mathrm{CH}^r_{\text{hom}}(X)$. First observe that

$$H^{2r-1}_{|\xi|}(X, \mathbf{Z}) \simeq H_{2d-2r+1}(|\xi|, \mathbf{Z}) = 0,$$

as $\dim_{\mathbf{R}} |\xi| = 2d - 2r$. Secondly there is a fundamental class map $\xi \mapsto \{\xi\} \in H_{2d-2r}(|\xi|, \mathbf{Z}(d-r)) \simeq H^{2r}_{|\xi|}(X, \mathbf{Z}(r))$ (Poincaré duality). Further, since ξ is nulhomologous, we have by duality

$$[\xi] \in H^{2r}_{|\xi|}(X, \mathbf{Z}(r))^\circ := \ker(H^{2r}_{|\xi|}(X, \mathbf{Z}(r)) \to H^{2r}(X, \mathbf{Z}(r))).$$

Hence ξ determines a morphism of MHS, $\mathbf{Z}(0) \to H^{2r}_{|\xi|}(X, \mathbf{Z}(r))^\circ$. From the short exact sequence of MHS,

$$0 \to H^{2r-1}(X, \mathbf{Z}(r)) \to H^{2r-1}(X \backslash |\xi|, \mathbf{Z}(r)) \to H^{2r}_{|\xi|}(X, \mathbf{Z}(r))^\circ \to 0,$$

we can pullback via this morphism to obtain another short exact sequence of MHS,

$$0 \to H^{2r-1}(X, \mathbf{Z}(r)) \to E \to \mathbf{Z}(0) \to 0.$$

Then $\Phi_r(\xi) := \{E\} \in \mathrm{Ext}^1_{\mathrm{MHS}}(\mathbf{Z}(0), H^{2r-1}(X, \mathbf{Z}(r)))$. This class $\{E\}$ is easy to calculate in $J^r(X)$, in terms of a membrane integral. Note that via duality,

$$E \subset H^{2r-1}(X \backslash |\xi|, \mathbf{Z}(r)) \simeq H_{2d-2r+1}(X, |\xi|, \mathbf{Z}(d-r)),$$

and that if ζ is a real $2d - 2r + 1$ chain such that $\partial \zeta = \xi$ on X, then $\{\zeta\} \in H_{2d-2r+1}(X, |\xi|, \mathbf{Z})$. One can show that the class $x \in W_0 E$ corresponding to the current

$$\frac{1}{(2\pi i)^{d-r}} \int_\zeta,$$

maps to $1 \in \mathbf{Z}(0)$. Now choose $y \in F^0 W_0 E_{\mathbf{C}}$ also mapping to $1 \in \mathbf{Z}(0)$. By Hodge type alone, the current corresponding to $x - y$ in the Poincaré dual description of $J^r(X)$ is the same as for $x = \frac{1}{(2\pi i)^{d-r}} \int_\zeta$, which is precisely the classical description of the Griffiths Abel–Jacobi map. This next result is a consequence of the work of Griffiths (see [26], as well as Sect. 14 of [36]).

Theorem 8.3. *If* $F^{r-1} H^{2r-1}(X, \mathbf{C}) \cap H^{2r-1}(X, \mathbf{Q}(r)) = 0$, *then there is an induced map*

$$\underline{\Phi}_r : \mathrm{Griff}^r(X) \to J^r(X).$$

In particular $\Phi_r(\mathrm{CH}^r_{\mathrm{alg}}(X)) = 0 \in J^r(X)$. *This is the case for a general CY threefold with* $r = 2$.

Example 8.4. We define the cycle class map $\mathrm{cl}_r : \mathrm{CH}^r(X) \to H^{2r}_{\mathscr{D}}(X, \mathbf{Z}(r))$. Recall the short exact sequence:

$$0 \to J^r(X) \to H^{2r}_{\mathscr{D}}(X, \mathbf{Z}(r)) \to Hg^r(X) \to 0.$$

Let $\xi \in \mathrm{CH}^r(X)$ with support $|\xi|$. One has a similar LES as in (9):

$$\cdots \to H^{2r-1}_{|\xi|}(X, \mathbf{Z}(r)) \oplus F^r H^{2r-1}_{|\xi|}(X, \mathbf{C}) \to H^{2r-1}_{|\xi|}(X, \mathbf{C})$$

$$\to H^{2r}_{\mathscr{D},|\xi|}(X, \mathbf{Z}(r)) \to H^{2r}_{|\xi|}(X, \mathbf{Z}(r)) \oplus F^r H^{2r}_{|\xi|}(X, \mathbf{C}) \xrightarrow{x-y} H^{2r}_{|\xi|}(X, \mathbf{C}) \to \cdots$$

Via Poincaré duality, one has cycle class maps

$$\xi \mapsto [(2\pi i)^{r-d}(\{\xi\}, \delta_\xi)] \in \ker\left(H^{2r}_{|\xi|}(X, \mathbf{Z}(r)) \oplus F^r H^{2r}_{|\xi|}(X, \mathbf{C}) \to H^{2r}_{|\xi|}(X, \mathbf{C})\right);$$

moreover recall that $H^{2r-1}_{|\xi|}(X, \mathbf{C}) = 0$ (weak purity). Thus we have a class $[\xi] \in H^{2r}_{\mathscr{D},|\xi|}(X, \mathbf{Z}(r))$. Now use the forgetful map

$$H^{2r}_{\mathscr{D},|\xi|}(X, \mathbf{Z}(r)) \to H^{2r}_{\mathscr{D}}(X, \mathbf{Z}(r)),$$

to define $\mathrm{cl}_r(\xi) \in H^{2r}_{\mathscr{D}}(X, \mathbf{Z}(r))$. From the injection

$$H^{2r}_{\mathscr{D},|\xi|}(X, \mathbf{Z}(r)) \hookrightarrow H^{2r}_{|\xi|}(X, \mathbf{Z}(r)) \oplus F^r H^{2r}_{|\xi|}(X, \mathbf{C}),$$

and the aforementioned forgetful map, in terms of the cone complex, $\mathrm{cl}_r(\xi)$ is represented by $((2\pi i)^{r-d}\{\xi\}, (2\pi i)^{r-d}\delta_\xi, 0)$. If $\xi \sim_{\mathrm{hom}} 0$, then $\xi = \partial\zeta$, $(2\pi i)^{r-d}\delta_\xi = dS$ for some $S \in F^r \mathscr{D}^{2r-1}_X(X)$. So

$$D((2\pi i)^{r-d}\zeta, S, 0) + ((2\pi i)^{r-d}\{\xi\}, (2\pi i)^{r-d}\delta_\xi, 0) = \left(0, 0, (2\pi i)^{r-d}\int_\zeta - S\right).$$

For $\omega \in F^{d-r+1} H^{2d-2r+1}(X, \mathbf{C})$,

$$(2\pi i)^{r-d}\int_\zeta \omega - S(\omega) = \frac{1}{(2\pi i)^{d-r}}\int_\zeta \omega,$$

by Hodge type. This is the Griffiths Abel–Jacobi map.

Both maps (cl_r, Φ_r) can be combined to give

$$\mathrm{cl}_{r,0} : \mathrm{CH}^r(X) = \mathrm{CH}^r(X, 0) \to H^{2r}_{\mathscr{D}}(X, \mathbf{Z}(r)),$$

with commutative diagram:

$$0 \to CH^r_{hom}(X) \quad \to CH^r(X) \to \quad \frac{CH^r(X)}{CH^r_{hom}(X)} \to 0$$

$$\Phi_r \downarrow \qquad\qquad cl_{r,0} \downarrow \qquad\qquad cl_r \downarrow$$

$$0 \to \quad J^r(X) \quad \to H^{2r}_{\mathscr{D}}(X, \mathbf{Z}(r)) \to \quad Hg^r(X) \to 0.$$

8.5 Deligne Cohomology and Normal Functions

Suppose that $\xi \in CH^r(X)$ is given and that $Y \subset X$ is a smooth hypersurface. Then there is a commutative diagram

$$\begin{array}{ccc} CH^r(X) & \to & CH^r(Y) \\ \downarrow & & \downarrow \\ H^{2r}_{\mathscr{D}}(X, \mathbf{Z}(r)) & \to & H^{2r}_{\mathscr{D}}(Y, \mathbf{Z}(r)); \end{array}$$

Further, if we assume that the restriction $\xi_Y \in CH_{hom}(Y)$ is null-homologous, then $cl_{r,0}(\xi) \in H^{2r}_{\mathscr{D}}(X, \mathbf{Z}(r)) \mapsto J^r(Y) \subset H^{2r}_{\mathscr{D}}(Y, \mathbf{Z}(r))$. Next, if $Y = X_0 \in \{X_t\}_{t \in S}$ is a family of smooth hypersurfaces of X, then such a ξ determines a holomorphically varying map $v_\xi(t) \in J^r(X_t)$, called a normal function. The class $cl_r(\xi) = \delta(v_\xi) \in Hg^r(X)$ is called the topological invariant of v_ξ, i.e. v_ξ determines $cl_r(\xi)$. In [31], these ideas are extended in complete generality to the situation of the higher Chow groups, where the notion of "arithmetic normal functions" are introduced.

Example 8.6 (Griffiths' famous example ([26])). Let:

$$X = V(z_0^5 + z_1^5 + z_2^5 + z_3^5 + z_4^5 + z_5^5) \subset \mathbf{P}^5$$

be the Fermat quintic fourfold. Consider these three copies of $\mathbf{P}^2 \subset X$:

$$L_1 := V(z_0 + z_1, z_2 + z_3, z_4 + z_5),$$

$$L_2 := V(z_0 + \xi z_2, z_2 + \xi z_3, z_4 + z_5),$$

$$L_3 := V(z_0 + \xi z_1, z_2 + \xi z_3, z_4 + \xi z_5).$$

where ξ is a primitive 5-th root of unity. Then $L_1 \bullet (L_2 - L_3) = 1 \neq 0$, hence $\xi := [L_2 - L_3]$ is a non-zero class in $H^{2,2}(X, \mathbf{Z}(2))$. Further, if $\{X_t\}_{t \in U \subset \mathbf{P}^1}$ is a general pencil of smooth hyperplane sections of X, and if $t \in U$, then it is well known that $\xi_t \in CH^2_{hom}(X_t)$ by a theorem of Lefschetz. Since $\delta(v_\xi) = [L_2 - L_3] \neq 0$, it follows that $v_\xi(t)$ is non-zero for most $t \in U$. Therefore for general $t \in U$, $Griff^2(X_t)$ contains an infinite cyclic group by Theorem 8.3. The upshot is that if:

$$Y = V\left(z_0^5 + z_1^5 + z_2^5 + z_3^5 + z_4^5 + \left(\sum_{j=0}^{4} a_j z_j\right)^5\right) \subset \mathbf{P}^4,$$

for general $a_0, \ldots, a_4 \in \mathbf{C}$, then $\mathrm{Griff}^2(Y) \neq 0$ contains an infinite cyclic sub-group. H. Clemens was the first to show that the Griffiths group of a general quintic threefold in $\mathbf{P}^4$ is (countably) infinite dimensional, when tensored over $\mathbf{Q}$. Later it was shown by C. Voisin that the same holds for general CY threefolds. The idea is to make use of the rational curves on such threefolds.

Theorem 8.7 (See [7, 13, 22, 24, 26, 45]). *Let $X \subset \mathbf{P}^4$ be a (smooth) threefold of degree d. If $d \leq 4$, then $\Phi_2 : \mathrm{CH}^2_{\mathrm{hom}}(X) \xrightarrow{\sim} J^2(X)$ is an isomorphism. Now assume that X is general. If $d \geq 6$ then $\mathrm{Im}(\Phi_2)$ is torsion. If $d = 5$, then $\mathrm{Im}(\Phi_2) \otimes \mathbf{Q}$ is countably infinite dimensional.*

Theorem 8.8 ([45]). *If X is a general Calabi–Yau threefold, then $\mathrm{Im}(\Phi_2)$ is countably infinite dimensional, when tensored over $\mathbf{Q}$. In particular, since $\Phi_2(\mathrm{CH}^2_{\mathrm{alg}}(X)) = 0$, it follows that $\mathrm{Griff}^2(X; \mathbf{Q})$ is (countably) infinite dimensional over $\mathbf{Q}$.*

8.9 Case m = 1 and K3 Surfaces

Recall the Tame symbol map

$$T : \bigoplus_{\mathrm{codim}_X Z = r-2} K_2^M(\mathbf{C}(Z)) \to \bigoplus_{\mathrm{codim}_X D = r-1} K_1^M(\mathbf{C}(D)).$$

Then:

$$\mathrm{CH}^r(X, 1) = H_{\mathrm{Zar}}^{r-1}(X, \mathscr{K}_{r,X}^M) \simeq \left\{ \frac{\sum_j (f_j, D_j) \mid \sum_j \mathrm{div}(f_j) = 0}{T(\Gamma(\bigoplus_{\mathrm{codim}_X Z = r-2} K_2^M(\mathbf{C}(Z))))} \right\}.$$

We recall:

Definition 8.10. The subgroup of $\mathrm{CH}^r(X, 1)$ represented by $\mathbf{C}^\times \otimes \mathrm{CH}^{r-1}(X)$ is called the subgroup of decomposables $\mathrm{CH}^r_{\mathrm{dec}}(X, 1) \subset \mathrm{CH}^r(X, 1)$. The space of indecomposables is given by

$$\mathrm{CH}^r_{\mathrm{ind}}(X, 1) := \frac{\mathrm{CH}^r(X, 1)}{\mathrm{CH}^r_{\mathrm{dec}}(X, 1)}.$$

The map

$$\mathrm{cl}_{r,1} : \mathrm{CH}^r_{\mathrm{hom}}(X, 1) \to H_{\mathscr{D}}^{2r-1}(X, \mathbf{Z}(r)),$$

is given by a map

$$\mathrm{cl}_{r,1} : \mathrm{CH}^r_{\mathrm{hom}}(X, 1) \to \frac{F^{d-r+1} H^{2d-2r+2}(X, \mathbf{C})^\vee}{H_{2d-2r+2}(X, \mathbf{Z}(d-r))},$$

defined as follows. Assume given a higher Chow cycle $\xi = \sum_{i=1}^N (f_i, Z_i)$ representing a class in $\mathrm{CH}^r_{\mathrm{hom}}(X, 1)$. Then via a proper modification, we can view $f_i : Z_i \to \mathbf{P}^1$ as a morphism, and consider the $2d - 2r + 1$-chain $\gamma_i = f_i^{-1}([-\infty, 0])$. Then

$\sum_{i=1}^{N} \mathrm{div}(f_i) = 0$ implies that $\gamma := \sum_{i=1}^{N} \gamma_i$ defines a $2d - 2r + 1$-cycle. Since ξ is null-homologous, it is easy to show that γ bounds some real dimensional $2d - 2r + 2$-chain ζ in X, viz., $\partial \zeta = \gamma$. For $\omega \in F^{d-r+1}H^{2d-2r+2}(X, \mathbf{C})$, the current defining $\mathrm{cl}_{r,1}(\xi)$ is given by:

$$\mathrm{cl}_{r,1}(\xi)(\omega) = \frac{1}{(2\pi i)^{d-r+1}} \left[\sum_{i=1}^{N} \int_{Z_i \setminus \gamma_i} \omega \log f_i - 2\pi i \int_{\zeta} \omega \right],$$

where we choose the principal branch of the log function. (This is different branch from the one chosen in [35], for this regulator.) One can easily check that the current defined above is d-closed. Namely, if we write $\omega = d\eta$ for some $\eta \in F^{d-r+1}E_X^{2d-2r}$, then by a Stokes' theorem argument, both integrals above contribute to "periods" which cancel. The details of this argument can be found in [21], but quite generally can be found in [32].

Using the description of real Deligne cohomology given above, and the regulator formula, we arrive at the formula for the real regulator $r_{r,1} : \mathrm{CH}^r(X, 1) \to H_{\mathscr{D}}^{2r-1}(X, \mathbf{R}(r)) = H^{r-1,r-1}(X, \mathbf{R}((r-1)) \simeq H^{d-r+1,d-r+1}(X, \mathbf{R}(d-r+1))^\vee$. Namely:

$$r_{r,1}(\xi)(\omega) = \frac{1}{(2\pi i)^{d-r+1}} \sum_j \int_{Z_j} \omega \log |f_j|.$$

Example 8.11. Suppose that X is a surface. Then we have

$$\mathrm{cl}_{2,1} : \mathrm{CH}_{\mathrm{hom}}^2(X, 1) \to \frac{\{H^{2,0}(X) \oplus H^{1,1}(X)\}^\vee}{H_2(X, \mathbf{Z})}.$$

The corresponding transcendental regulator is defined to be

$$\Phi_{2,1} : \mathrm{CH}_{\mathrm{hom}}^2(X, 1) \to \frac{H^{2,0}(X)^\vee}{H_2(X, \mathbf{Z})},$$

$$\Phi_{2,1}(\xi)(\omega) = \int_{\zeta} \omega.$$

and real regulator

$$r_{2,1} : \mathrm{CH}^2(X, 1) \to H^{1,1}(X, \mathbf{R}(1))^\vee \simeq H^{1,1}(X, \mathbf{R}(1)),$$

$$r_{2,1}(\xi)(\omega) = \frac{1}{2\pi i} \sum_j \int_{Z_j} \log |f_j| \omega.$$

There is an induced map

$$\underline{r}_{2,1} : \mathrm{CH}_{\mathrm{ind}}^2(X, 1) \to H_{\mathrm{tr}}^{1,1}(X, \mathbf{R}(1)).$$

If X is a $K3$ surface, then $\mathrm{CH}_{\mathrm{hom}}^2(X, 1) = \mathrm{CH}^2(X, 1)$, hence there is an induced map

$$\underline{\Phi}_{2,1} : \mathrm{CH}_{\mathrm{ind}}^2(X, 1) \to \frac{H^{2,0}(X)^\vee}{H_2(X, \mathbf{Z})}.$$

Theorem 8.12. (i) *([40]) Let $X \subset \mathbf{P}^3$ be a smooth surface of degree d. If $d \leq 3$, then $r_{2,1} : CH^2(X, 1) \to H^{1,1}(X, \mathbf{R}(1))$ is surjective; moreover $CH^2_{\mathrm{ind}}(X, 1; \mathbf{Q}) = 0$. Now assume that X is general. If $d \geq 5$, then $Im(r_{2,1})$ is "trivial", i.e. its image in the transcendental part of $H^{1,1}(X, \mathbf{R}(1))$ is zero.*

(ii) *[Hodge-$\mathscr{D}$-conjecture for K3 surfaces ([10])] Let X be a general member of a universal family of projective K3 surfaces, in the sense of the real analytic topology. Then*

$$r_{2,1} : CH^2(X, 1) \otimes \mathbf{R} \to H^{1,1}(X, \mathbf{R}(1)),$$

is surjective.

(iii) *([12]) Let $X/\mathbf{C}$ be a general algebraic K3 surface. Then the transcendental regulator $\Phi_{2,1}$ is non-trivial. Quite generally, if X is a general member of a general subvariety of dimension $20 - \ell$, describing a family of K3 surfaces with general member of Picard rank ℓ, with $\ell < 20$, then $\Phi_{2,1}$ is non-trivial.*

Remark 8.13. (i) Regarding part (iii) of Theorem 8.12, one can ask whether $\Phi_{2,1}$ can be non-trivial for those K3 surfaces X with Picard rank 20, (which are rigid and therefore defined over $\overline{\mathbf{Q}}$)? In [12], some evidence is provided in support of this.

(ii) One of the key ingredients in the proof of the above theorem is the existence of plenty of nodal rational curves on a general K3 surface. Indeed, there is the following result:

Theorem 8.14 ([11]). *For a general K3 surface, the union of rational curves on X is a dense subset in the analytic topology.*

Remark 8.15. It is well known that for an elliptic curve E defined over an algebraically closed subfield $k \subset \mathbf{C}$, the torsion subgroup $E_{\mathrm{tor}}(\mathbf{C}) \subset E(k)$. An analogous result holds for rational curves on a K3 surface. Quite generally, the following result which may be common knowledge among experts, seems worthwhile mentioning:

Proposition 8.16. *Assume given $X/\mathbf{C}$ a smooth projective surface with $Pg(X) := \dim H^{2,0}(X) > 0$. If we write $X/\mathbf{C} = X_k \times_k \mathbf{C}$, viz., $X/\mathbf{C}$ obtained by base change from a smooth projective surface X_k defined over an algebraically closed subfield $k \subset \mathbf{C}$, and if $C \subset X/\mathbf{C}$ is a rational curve, then C is likewise defined over k.*

Proof. By a standard spread argument, there is a smooth projective variety S/k of dimension ≥ 0, and a k-family $\mathscr{C} \twoheadrightarrow S$ of rational curves containing C as a general member, with embedding h:

$$\mathscr{C} \xrightarrow{h} S \times_k X \xrightarrow{Pr_X} X$$
$$\underset{Pr_S}{\searrow} \qquad \swarrow$$
$$S$$

Since $Pg(X) > 0$, there are only at most a countable number of rational curves on $X/\mathbf{C}$, and hence $Pr_X(h(\mathscr{C})) = Pr_X(h(Pr_S^{-1}(t)))$ for any $t \in S(\mathbf{C})$. Now use the fact that $S(k) \neq \emptyset$. □

Now suppose that X is a $K3$ surface defined over $\overline{\mathbf{Q}}$. Let $\Sigma \subset X$ be the union of all rational curves on X. Then Σ is defined over $\overline{\mathbf{Q}}$. In discussions with Matt Kerr (personal communication), we naively raise the following:

Question 8.17. Is $X(\overline{\mathbf{Q}}) \subset \Sigma(\overline{\mathbf{Q}})$?

An affirmative answer to this question would not only imply that Σ is dense in $X(\mathbf{C})$ in the usual topology, but this would also provide a nontrivial instance of the Bloch–Beilinson conjecture on the injectivity of Abel–Jacobi maps for smooth projective varieties defined over $\overline{\mathbf{Q}}$. More specifically, by an application of the connectedness part of Bertini's theorem, Σ is connected, hence $\mathrm{CH}^2_{\mathrm{hom}}(X/\overline{\mathbf{Q}}) = 0$.

8.18 Torsion Indecomposables

The story about torsion indecomposable classes takes an interesting turn from the geometric story presented in Theorem 8.12(i). The situation is this, and for the moment let X be any projective algebraic manifold. An elementary consequence of the Merkurjev–Suslin theorem implies the following:

Theorem 8.19 (See [15]). *The kernel of the Abel–Jacobi map*

$$\underline{AJ}_X : \frac{\mathrm{CH}^2_{\mathrm{hom}}(X,1)}{\mathrm{CH}^2_{\mathrm{dec}}(X,1)} \to J\left(\frac{H^2(X,\mathbf{Z}(2))}{H^2_{\mathrm{alg}}(X,\mathbf{Z}(2))}\right),$$

is uniquely divisible. This implies that $\underline{AJ}_X$ is injective on torsion indecomposables $\left\{\frac{\mathrm{CH}^2_{\mathrm{hom}}(X,1)}{\mathrm{CH}^2_{\mathrm{dec}}(X,1)}\right\}_{\mathrm{tor}}$.

(Here we remind the reader that since we are working integrally, we have an inclusion that for torsion reasons, need not be an equality:

$$\frac{\mathrm{CH}^2_{\mathrm{hom}}(X,1)}{\mathrm{CH}^2_{\mathrm{dec}}(X,1)} \subseteq \frac{\mathrm{CH}^2(X,1)}{\mathrm{CH}^2_{\mathrm{dec}}(X,1)} =: \mathrm{CH}^2_{\mathrm{ind}}(X,1).)$$

On the other hand, one has the torsion subgroup $\{\mathrm{CH}^2_{\mathrm{ind}}(X,1)\}_{\mathrm{tor}}$. Put

$$H^2_{\mathrm{tr}}(X,\mathbf{Q}(2)/\mathbf{Z}(2)) = \mathrm{Cokernel}(H^2_{\mathrm{alg}}(X,\mathbf{Q}(2)/\mathbf{Z}(2)) \to H^2(X,\mathbf{Q}(2)/\mathbf{Z}(2))).$$

Theorem 8.20 ([29]). *There is an identification*

$$\{\mathrm{CH}^2_{\mathrm{ind}}(X,1)\}_{\mathrm{tor}} \xrightarrow{\sim} H^2_{\mathrm{tr}}(X,\mathbf{Q}(2)/\mathbf{Z}(2)).$$

In light of these two theorems, one expects that

$$\underline{AJ}_X : \left\{ \frac{\mathrm{CH}^2_{\mathrm{hom}}(X,1)}{\mathrm{CH}^2_{\mathrm{dec}}(X,1)} \right\}_{\mathrm{tor}} \xrightarrow{\sim} \left\{ J\left(\frac{H^2(X,\mathbf{Z}(2))}{H^2_{\mathrm{alg}}(X,\mathbf{Z}(2))} \right) \right\}_{\mathrm{tor}}.$$

For example, suppose that X is a $K3$ surface of Picard rank 20. Then $E :=$ $J\left(\frac{H^2(X,\mathbf{Z}(2))}{H^2_{\mathrm{alg}}(X,\mathbf{Z}(2))} \right)$ is an elliptic curve defined over a number field. In this case one expects the identification

$$\{\mathrm{CH}^2_{\mathrm{ind}}(X,1)\}_{\mathrm{tor}} \xrightarrow{\sim} \{E(\overline{\mathbf{Q}})\}_{\mathrm{tor}}.$$

8.21 Case m = 2 and Elliptic Curves

Regulator examples on $\mathrm{CH}^2(X,2)$. Let X be a compact Riemann surface. In [38] there is constructed a real regulator

$$r : \mathrm{CH}^2(X,2) \to H^1(X,\mathbf{R}(1)), \tag{13}$$

given by

$$\omega \in H^1(X,\mathbf{R}) \simeq H^1(X,\mathbf{R}(1))^{\vee} \mapsto \int_X \left[\log|f| d\log|g| - \log|g| d\log|f| \right] \wedge \omega$$
$$= 2 \int_X \log|f| d\log|g| \wedge \omega, \ (\text{by a Stokes' theorem argument}). \tag{14}$$

Alternatively, up to a twist, and real isomorphism, this is the same as the real part of the regulator $\mathrm{cl}_{2,2}$ in (4), viz.,

$$r_{2,2}(\omega) = \frac{1}{2\pi} \int_X [\log|f| d\arg g - \log|g| d\arg f] \wedge \omega, \tag{15}$$

where the formula for:

$$\mathrm{cl}_{2,2} : \mathrm{CH}^2(X,2) \to H^2_{\mathscr{D}}(X,\mathbf{Z}(2)) \simeq H^1(X,\mathbf{C}/\mathbf{Z}(2)) \simeq \frac{H^1(X,\mathbf{C})}{H_1(X,\mathbf{Z}(-1))} = \frac{H^1(X,\mathbf{C})}{H_1(X,\mathbf{Z})(1)}, \tag{16}$$

(for $\omega \in H^1(X,\mathbf{C})$), which can be found for example in [32], is induced, up to a factor[10] of $(2\pi i)^{-1}$, by:

$$\{f,g\} \quad \mapsto \tag{17}$$

---

[10] The decision to consider the factor $(2\pi i)^{-1}$ is somewhat "political", as reflected in the remark on page 2 of [32]. From a cohomological point of view, one works with $\mathbf{Z}(2)$ coefficient periods, whereas homologically, is it with $\mathbf{Z}(1)$ coefficients. This is neatly illustrated via the Poincaré duality isomorphism in (16).

$$\int_{X \setminus f^{-1}[-\infty,0]} \log f d \log g \wedge \omega - 2\pi i \int_{f^{-1}[-\infty,0] \setminus (f \times g)^{-1}[-\infty,0]^2} \log g \wedge \omega + (2\pi i)^2 \int_\zeta \omega,$$

where if we assume for the moment that $T\{f,g\}) = 0$, then ζ is a real membrane with $\partial \zeta = (f \times g)^{-1}[-\infty, 0]^2$. Otherwise if $T\{f,g\} \neq 0$, we are then dealing with a situation where $\{f,g\}$ is replaced by a given $\prod_\alpha \{f_\alpha, g_\alpha\}$, where

$$T\left(\prod_\alpha \{f_\alpha, g_\alpha\}\right) = \sum_\alpha T\{f_\alpha, g_\alpha\} = 0,$$

and accordingly arrive at a corresponding ζ. Note that (17) is really the current written in the slang form:

$$[\log f d \log g - 2\pi i \log g \delta_{f^{-1}(\mathbf{R}^-)} + (2\pi i)^2 \delta_\zeta] =: \tilde{R}.$$

To connect formulas (15) and (17), one takes the imaginary part of $\tilde{R}$ (consistent with $\mathbf{C}/\mathbf{Z}(2) \twoheadrightarrow \mathbf{C}/\mathbf{R}(2) \simeq \mathbf{R}(1)$). This gives us

$$\mathrm{Im}(\tilde{R}) = [\log |f| d \arg g + \arg f d \log |g| - 2\pi \log |g| \delta_{f^{-1}(\mathbf{R}^-)}].$$

Now add the coboundary current $d[\log|g| \arg f]$ and apply a Stokes' theorem argument.[11]

8.22 Constructing $K_2(X)$ Classes on Elliptic Curves X

We consider the following trick due to Bloch [5]. Let X be an elliptic curve and assume given $f, g \in \mathbf{C}(X)^\times$ such that $\Sigma := |\mathrm{div}(f)| \cup |\mathrm{div}(g)|$ are points of order N in $\mathrm{Pic}(X)$. Then

$$T(\{f,g\}^N) \in \bigoplus \mathbf{C}^\times \quad \text{and} \quad \mapsto 0 \in \mathrm{Pic}(X) \otimes \mathbf{C}^\times.$$

A clarification. This uses the Weil reciprocity theorem. Let X be a compact Riemann surface, $f, g \in \mathbf{C}(X)^\times$, and for $p \in X$, write

$$T_p\{f,g\} = (-1)^{v_p(g) v_p(f)} \left(\frac{f^{v_p(g)}}{g^{v_p(f)}}\right)\bigg|_p \in \mathbf{C}^\times.$$

Note that for $p \notin |\mathrm{div}(f)| \cup |\mathrm{div}(g)|$, we have $T_p\{f,g\} = 1$. Thus we can write $T\{f,g\} = \sum_{p \in X} T_p\{f,g\}$. Weil reciprocity says that $\prod_{p \in X} T_p\{f,g\} = 1$. Let us rewrite this as follows. If we write $T\{f,g\} = \sum_{j=1}^M (c_j, p_j)$, where $p_j \in X$ and $c_j \in \mathbf{C}^\times$, then $\prod_{j=1}^M c_j = 1$. Now fix $p \in X$ and let us suppose that $N p_j \sim_{\mathrm{rat}} N p$ for all j. Thus there exists $h_j \in \mathbf{C}(X)^\times$ such that $(h_j) = N p_j - N p$. Then $T\{h_j, c_j\} = (c_j^N, p) + (c_j^{-N}, p_j)$. The result is that

---

[11] Alternatively, taking $\mathrm{Re}((2\pi i)^{-1} \tilde{R})$ gives the formula in (15), viz., with the factor $(2\pi)^{-1}$, right on the nose.

$$T(\{f,g\}^N\{h_1,c_1\}\cdots\{h_M,c_M\}) = \prod_{j=1}^{M}(c_j^N,p) = (1,p) = 0.$$

Thus there exists $\{h_i\} \in \mathbf{C}(X)^\times$ and $\{c_i\} \in \mathbf{C}^\times$ such that $\{f,g\}^N \prod\{h_i,c_i\} \in$ $\mathrm{CH}^2(X,2)$. Note that the terms $\{h_i,c_i\}$ do not contribute to the regulator value by the formula in (14) above. Clearly this construction takes advantage of the existence of a dense subset of torsion points on X. Bloch (*op. cit.*) shows that the real regulator is nontrivial for general elliptic curves, and indeed A. Collino [14] shows that the regulator image of $\mathrm{CH}^2(X,2)$ for a general elliptic curve X is infinite dimensional (over $\mathbf{Q}$). Actually it is pretty easy to see why $r_{2,2}$ is non-trivial for a general elliptic curve:

Theorem 8.23 (Hodge-$\mathcal{D}$-conjecture for elliptic curves). *If X is a general elliptic curve in the real analytic Zariski topology, then $r_{2,2}$ is surjective.*

Proof. Let X be an elliptic curve given in affine coordinates by the equation $y^2 = h(x)$, where $h(x)$ is a cubic polynomial with distinct roots. A basis for $H^1(X,\mathbf{R})$ is given by

$$\omega_1 := \frac{dx}{y} + \frac{d\overline{x}}{\overline{y}} \quad ; \quad \omega_2 := i\left(\frac{dx}{y} - \frac{d\overline{x}}{\overline{y}}\right).$$

Next, we consider

$$f_1 := y + ix \quad ; \quad f_2 = y + x \quad ; \quad g_1 = g_2 = x.$$

We claim that for general X,

$$\det\begin{bmatrix} \int_X \log|f_1| d\log|g_1| \wedge \omega_1 & \int_X \log|f_1| d\log|g_1| \wedge \omega_2 \\ \int_X \log|f_2| d\log|g_2| \wedge \omega_1 & \int_X \log|f_2| d\log|g_2| \wedge \omega_2 \end{bmatrix} \neq 0. \quad (18)$$

Now let us first assume that X is given for which (18) holds, and note that the rational functions f_1, f_2, g_1, g_2 can each be expressed in the form L_1/L_2, where L_j are homogeneous linear polynomials in the homogeneous coordinates of $\mathbf{P}^2$ (and where $X \subset \mathbf{P}^2$). Since X has a dense subset of torsion points X_{tor}, and by Abel's theorem, one can find $\tilde{L}_j$ "close" to L_j, $j = 1,2$, such that $\tilde{L}_j \cap X \subset X_{\mathrm{tor}}$. Thus up to $\mathbf{C}^\times$ multiple, $\tilde{L}_1/\tilde{L}_2$ is "close" to L_1/L_2. Hence one can find $\tilde{f}_1, \tilde{f}_2, \tilde{g}_1, \tilde{g}_2$ for which

$$\left\{|\mathrm{div}(\tilde{f}_1)| \bigcup |\mathrm{div}(\tilde{f}_2)| \bigcup |\mathrm{div}(\tilde{g}_1)| \bigcup |\mathrm{div}(\tilde{g}_2)|\right\} \subset X_{\mathrm{tor}}, \quad (19)$$

and that by continuity considerations,

$$\det\begin{bmatrix} \int_X \log|\tilde{f}_1| d\log|\tilde{g}_1| \wedge \omega_1 & \int_X \log|\tilde{f}_1| d\log|\tilde{g}_1| \wedge \omega_2 \\ \int_X \log|\tilde{f}_2| d\log|\tilde{g}_2| \wedge \omega_1 & \int_X \log|\tilde{f}_2| d\log|\tilde{g}_2| \wedge \omega_2 \end{bmatrix} \neq 0. \quad (20)$$

Thus one can complete $\{\tilde{f}_1, \tilde{g}_1\}, \{\tilde{f}_2, \tilde{g}_2\}$ to classes $\xi_1, \xi_2 \in CH^2(X, 2)$, for which

$$\det \begin{bmatrix} r_{2,2}(\xi_1)(\omega_1) & r_{2,2}(\xi_1)(\omega_2) \\ r_{2,2}(\xi_2)(\omega_1) & r_{2,2}(\xi_2)(\omega_2) \end{bmatrix} \neq 0, \tag{21}$$

and so modulo the claim in (18), we are done. We sketch a proof of the claim. With regard to $dV = d\mathrm{Re}(x) \wedge d\mathrm{Im}(x)$:

$$\frac{d \log |x| \wedge \omega_1}{2} = \frac{1}{4}\left(\frac{1}{x\bar{y}} - \frac{1}{\bar{x}y}\right) dx \wedge d\bar{x} = \frac{\mathrm{Im}(\bar{x}y)}{|x|^2|y|^2} dV \tag{22}$$

$$\frac{d \log |x| \wedge \omega_2}{2} = -\frac{i}{4}\left(\frac{1}{x\bar{y}} + \frac{1}{\bar{x}y}\right) dx \wedge d\bar{x} = -\frac{\mathrm{Re}(\bar{x}y)}{|x|^2|y|^2} dV \tag{23}$$

Now let us degenerate X to the rational elliptic curve X_0 given by $y^2 = x^3$. Note that X_0 is given parametrically by $(x, y) = (z^2, z^3)$, $z \in \mathbf{C}$. Thus $\bar{x}y = |z|^4 z$, and up to a real positive multiplicative constant times the standard volume element on $\mathbf{C}$, which we will denote by dV_0, (22) and (23) become:

$$d \log |x| \wedge \omega_1 = \frac{\mathrm{Im}(z)}{|z|^4} dV_0 \quad ; \quad d \log |x| \wedge \omega_2 = -\frac{\mathrm{Re}(z)}{|z|^4} dV_0. \tag{24}$$

Let $\mathbf{H} = \{z \in \mathbf{C} \mid \mathrm{Im}(z) \geq 0\}$ be the upper half plane. Now one has the following *formal* calculations after degenerating to X_0, and using symmetry arguments:

$$\int_{X_0} \log |f_1| d \log |g_1| \wedge \omega_1 = \int_{\mathbf{C}} \log |z^3 + iz^2| \frac{\mathrm{Im}(z)}{|z|^4} dV_0 \tag{25}$$

$$= \int_{\mathbf{C}} \log |z + i| \frac{\mathrm{Im}(z)}{|z|^4} dV_0 = \int_{\mathbf{H}} \log \left|\frac{z+i}{\bar{z}+i}\right| \frac{\mathrm{Im}(z)}{|z|^4} dV_0 \mapsto +\infty,$$

using the fact

$$\left|\frac{z+i}{\bar{z}+i}\right| > 1 \Leftrightarrow \mathrm{Im}(z) > 0.$$

$$\int_{X_0} \log |f_2| d \log |g_2| \wedge \omega_1 = \int_{\mathbf{C}} \log |z + 1| \frac{\mathrm{Im}(z)}{|z|^4} dV_0 = 0. \tag{26}$$

For the remaining two formal calculations, put $w = iz$, and note that $\mathrm{Re}(z) = \mathrm{Im}(w)$, and that $|z + 1| = |w + i|$. Then

$$\int_{X_0} \log |f_2| d \log |g_2| \wedge \omega_2 = -\int_{\mathbf{C}} \log |z + 1| \frac{\mathrm{Re}(z)}{|z|^4} dV_0 \tag{27}$$

$$= -\int_{\mathbf{C}} \log |w + i| \frac{\mathrm{Im}(w)}{|w|^4} dV_0 = -\int_{\mathbf{H}} \log \left|\frac{z+i}{\bar{z}+i}\right| \frac{\mathrm{Im}(z)}{|z|^4} dV_0 \mapsto -\infty.$$

$$\int_{X_0} \log |f_1| d \log |g_1| \wedge \omega_2 = -\int_{\mathbf{C}} \log |z + i| \frac{\mathrm{Re}(z)}{|z|^4} dV_0 = 0. \tag{28}$$

Note that two of these integrals blow up over the singular point $z = 0$ of the singular curve X_0, as expected. By using the Lebesgue theory of integration, we can make the calculations in (25)–(28) more precise. First, by using the projection $(x, y) \mapsto x$, we have a double covering $X \rightarrow \mathbf{P}^1$. Thus for $f, g \in \mathbf{C}(X)$, and $\omega = \omega_1$ or $\omega = \omega_2$, we can express $\int_X \log |f| d \log |g| \wedge \omega$ as the integral of some Lebesgue integrable function $H(x)$ over $\mathbf{P}^1$. Next, by converting to polar coordinates, viz. $x = e^{it}$, we can Fubini integrate in $t \in [0, 2\pi]$ and $r \in [0, \infty]$. Let $h(r)$ be the result of integrating $H(x)$ with respect to t over $[0, 2\pi]$. As X degenerates to X_0, we can construct a sequence $\{h_n(r)\}$ which limits to $h_\infty(r)$ over X_0. In the cases of (25)–(28), we have that $h_\infty(r)$ is either zero, nonnegative, or nonpositive. By using the standard Lebesgue integral limit theorems, we arrive at the claim in (18), and hence the theorem. $\square$

For curves X of genus $g > 1$, the problem of constructing classes in $\mathrm{CH}^2(X, 2)$ seems to be related to the fact that under the Abel–Jacobi mapping $\Phi : X \rightarrow J^1(X)$, $p \mapsto \{p - p_0\}$, the inverse image of the torsion subgroup, $\Phi^{-1}(J^1(X)_{\mathrm{tor}})$, is finite, this being the import of the Mumford–Manin theorem (see [43] for a proof). Indeed as explained in [38] (as well as in [39]), one can prove a weak version of the Mumford–Manin theorem based on the fact that for a general curve X of genus $g > 1$, the image of the regulator map $\mathrm{cl}_{2,2} : \mathrm{CH}^2(X, 2) \rightarrow H^2_{\mathscr{D}}(X, \mathbf{Z}(2))$ is torsion (A. Collino [14]). Collino's approach (*op. cit.*) uses infinitesimal methods. The reader should also consult [23] for similar refined results in this direction. For the benefit of the reader, we will provide an ad hoc explanation as to why this is the case (in "Observation 1" below). In order to do so, we must first digress and consider the following setting.

Assume given a dominant morphism $\overline{\rho} : \overline{X} \rightarrow \overline{C}$ of smooth complex projective varieties, where $\overline{X}$ is a surface and $\overline{C}$ is a curve. Let $C \subset \overline{C}$ be an affine open subset over which $\overline{\rho}$ is smooth, and $\Sigma := \overline{C} \backslash C$, $X = \overline{\rho}^{-1}(C)$ and $\rho = \overline{\rho}|_X : X \rightarrow C$. For $t \in \Sigma$, we will assume that the singular set of X_t is a single node. Next, we will assume given a class $\{\xi\} \in \mathrm{CH}^2(X, 2)$. In particular $\partial \xi = 0$ on X (here ∂ is the same thing as the Tame symbol). Note that ξ is given by a product of symbols of the form $\{f, g\}$, where $f, g \in \mathbf{C}(X)^\times$. However, since $\mathbf{C}(X) = \mathbf{C}(\overline{X})$, one can also think of ξ as defined on $\overline{X}$ (call it $\overline{\xi}$) with $\partial \overline{\xi}$ supported on $\overline{X}_\Sigma := \overline{\rho}^{-1}(\Sigma)$. Now for $t \in \Sigma$, the contribution ("residue") of $\partial \overline{\xi}$ gives rise to a class in $\mathrm{CH}^1(X_t, 1)$. If X_t were smooth, then $\mathrm{CH}^1(X_t, 1) = \mathbf{C}^\times$; but here we are assuming that X_t has a single node $P \in X_t$ as singularity. Under the desingularization $\sigma : \tilde{X}_t \rightarrow X_t$, let $\{Q, R\} = \sigma^{-1}(P)$. Next if $Q - R \in \mathrm{CH}^1_{\mathrm{tor}}(\tilde{X}_t)$, then for some integer N, $N \cdot (Q - R) = \mathrm{div}(f)$ for some $f \in \mathbf{C}(\tilde{X}_t)^\times$. But on X_t, $\mathrm{div}(f) = 0$, and hence $\mathbf{C}^\times \subsetneq \mathrm{CH}^1(X_t, 1)$. The upshot is that if $\partial \overline{\xi}$ contributes to a nonzero element of $\mathrm{CH}^1(X_t, 1)/\mathbf{C}^\times$ for some $t \in \Sigma$, then via a residue calculation and a calculation of the MHS $H^2(X, \mathbf{Q}(2))$, the current $d \log \xi$ (induced by $\{f, g\} \mapsto d \log f \wedge d \log g$) will contribute to a nonzero class in $[d \log \xi] \in \Gamma H^2(X, \mathbf{Q}(2))$. *The converse statement also holds:* if $Q - R \notin \mathrm{CH}^1_{\mathrm{tor}}(\tilde{X}_t)$, for all such $t \in \Sigma$, then $[d \log \xi] = 0 \in \Gamma H^2(X, \mathbf{Q}(2))$. Next, the Leray spectral sequence associated to ρ (which by Deligne, degenerates at E_2, see [25] (p. 466)), together with the fact that since C is an affine curve (hence $H^2(C, R^0\rho_*\mathbf{Q}(2)) = 0$), yields the short exact sequence of MHS:

$$0 \to H^1(C, R^1\rho_*\mathbf{Q}(2)) \to H^2(X, \mathbf{Q}(2)) \to H^0(C, R^2\rho_*\mathbf{Q}(2)) \to 0.$$

Note that $\Gamma H^0(C, R^2\rho_*\mathbf{Q}(2)) = 0$ as $H^0(C, R^2\rho_*\mathbf{Q}(2))$ is of pure weight -2. Hence $\Gamma H^2(X, \mathbf{Q}(2)) = \Gamma H^1(C, R^1\rho_*\mathbf{Q}(2))$. On the other hand, for $t \in C$, ξ restricts to a class $\xi_t \in \mathrm{CH}^2(X_t, 2)$, and hence we have a normal function

$$\nu_\xi : C \to \bigcup_{t \in C} J(H^1(X_t, \mathbf{Z}(2))),$$

whose topological invariant is the aforementioned class $[d \log \xi] \in \Gamma H^1(C, R^1\rho_* \mathbf{Q}(2))$, and which we will now denote it by $\delta(\nu_\xi) := [d \log \xi]$. It is a general fact that there is a short exact sequence:

$$0 \to J(H^0(C, R^1\rho_*\mathbf{Q}(2))) \to \left\{ \begin{matrix} \text{Normal} \\ \text{functions} \end{matrix} \right\}_{\mathbf{Q}} \xrightarrow{\delta} \Gamma H^1(C, R^1\rho_*\mathbf{Q}(2)) \to 0, \quad (29)$$

where $\{\cdots\}_\mathbf{Q}$ means with respect to $\mathbf{Q}$-periods. We will explain this in more detail below, but comment in passing that the technical details can be found in [31]. If $\delta(\nu_\xi) = 0$, then $\nu_\xi \in J(H^0(C, R^1\rho_*\mathbf{Q}(2)))$, i.e. belongs to the fixed part of a corresponding variation of Hodge structure. The situation is not unlike what occurs in the short exact sequence involving Deligne cohomology in (8) above, and the nature of this argument is completely analogous to that in Example 8.6. We can frame this discussion in more precise terms. One has a cycle class map $\mathrm{cl}_{2,2} : \mathrm{CH}^2(X, 2) \to H^2_{\mathscr{D}}(X, \mathbf{Z}(2))$, (Deligne–Beilinson cohomology); moreover by a weight argument, there is a short exact sequence:

$$0 \to J(H^1(X, \mathbf{Z}(2))) \to H^2_{\mathscr{D}}(X, \mathbf{Z}(2)) \to \Gamma H^2(X, \mathbf{Z}(2)) \to 0. \quad (30)$$

For $t \in C$, X_t is a smooth curve. Then for such t, $H^2_{\mathscr{D}}(X_t, \mathbf{Z}(2)) = J(H^1(X_t, \mathbf{Z}(2)))$, and accordingly the map

$$t \in C \mapsto \mathrm{cl}_{2,2}(\xi_t) \in J(H^1(X_t, \mathbf{Z}(2))),$$

is our normal function ν_ξ; moreover the image of ξ via the composite

$$\mathrm{CH}^2(X, 2) \to H^2_{\mathscr{D}}(X, \mathbf{Q}(2)) \to \Gamma H^2(X, \mathbf{Q}(2)) = \Gamma H^1(C, R^1\rho_*\mathbf{Q}(2)),$$

is precisely $\delta(\nu_\xi)$. Finally to explain (29) more precisely, we observe that there is a short exact sequence:

$$0 \to H^1(C, R^0\rho_*\mathbf{Q}(2)) \to H^1(X, \mathbf{Q}(2)) \to H^0(C, R^1\rho_*\mathbf{Q}(2)) \to 0.$$

But $\Gamma H^0(C, R^1\rho_*\mathbf{Q}(2)) = 0$, hence we arrive at the short exact sequence:

$$0 \to J(H^1(C, R^0\rho_*\mathbf{Q}(2))) \to J(H^1(X, \mathbf{Q}(2))) \to J(H^0(C, R^1\rho_*\mathbf{Q}(2))) \to 0.$$

This together with (29) and (30)$_Q$ leads to the identification:

$$\left\{ \begin{array}{c} \text{Normal} \\ \text{functions} \end{array} \right\}_Q \simeq \frac{H^2_{\mathscr{D}}(X, \mathbf{Q}(2))}{J(H^1(C, R^0\rho_*\mathbf{Q}(2))))}.$$

Now having discussed the relationship between a cycle class $\xi \in \mathrm{CH}^2(X, 2)$, the associated normal function ν_ξ, and the topological invariant $\delta(\nu_\xi) \in \Gamma H^2(X, \mathbf{Q}(2))$ and how it is related to the "torsion" nature of the nodal singularities of the singular fibers $\{X_t\}_{t \in \Sigma}$, we are led to consider two divergent observations:

Observation 1. Suppose that X_0 is a general curve of genus $g > 1$. By general, we can assume that X_0 is a very general member of a pencil of curves $\{X_t\}_{t \in \mathbf{P}^1}$, defining a smooth surface $\overline{X}_{\mathbf{P}^1} := \bigcup_{t \in \mathbf{P}^1} X_t \to \mathbf{P}^1$, whose singular fibers are Lefschetz, i.e. admit a single ordinary node. Let $\xi_0 \in \mathrm{CH}^2(X_0, 2)$. After a suitable base extension $\overline{C} \to \mathbf{P}^1$, for some smooth projective curve $\overline{C}$, and corresponding $\overline{X} := \overline{C} \times_{\mathbf{P}^1} \overline{X}_{\mathbf{P}^1}$, with setting as in the above discussion, ξ_0 will then spread to a class $\xi \in \mathrm{CH}^2(X, 2)$, in a general family $\rho : X \to C$, where ρ is smooth and proper over an affine curve C. Granted that the singular fibers over $\Sigma \subset \overline{C}$ are not necessarily nodes (as $\overline{C} \to \mathbf{P}^1$ may ramify over the singular points), a similar line of reasoning as the nodal situation will occur, based on a parallel situation encountered in [9]. So for simplicity, let us assume that for each $t \in \Sigma$, that X_t is Lefschetz. Since $g(X_t) \geq 2$ for $t \in C$, it follows that for $t \in \Sigma$, $g(X_t) \geq 1$.

Proposition 8.24. *If X_0 is sufficiently general, then one can arrange for the following to hold:*

(i) $H^0(C, R^1\rho_*\mathbf{Q}(2)) = 0$.

(ii) *For every $t \in \Sigma$, the corresponding $Q - R$ is nontorsion in* $\mathrm{CH}^1(\tilde{X}_t)$.

Proof. Although we won't prove this, it goes without mentioning that (i) is a standard result in the deformation theory of curves and corresponding VHS. For (ii), one considers via deformation, a family of nodal curves of genus at least 1, together with an argument of Baire type using the fact that the torsion points on a curve of genus $g \geq 1$ is at most countable. $\square$

It follows that such a ξ would define a normal function for which $\delta(\nu_\xi) = 0 \in \Gamma H^1(C, R^1\rho_*\mathbf{Q}(2))$, and so $\nu_\xi \in J(H^0(C, R^1\rho_*\mathbf{Q}(2))) = 0$. This leads to $\mathrm{cl}_{2,2}(\xi_t) = 0 \in H^2_{\mathscr{D}}(X_t, \mathbf{Q}(2))$ for very general $t \in C$, and hence $\mathrm{cl}_{2,2}(\xi_0)$ is torsion as a class in $H^2_{\mathscr{D}}(X_0, \mathbf{Z}(2))$.

Observation 2. Consider an elliptic surface $\overline{\rho} : \overline{X} \to \overline{C}$. The singular fibers X_Σ are unions of rational curves. If for some $t \in \Sigma$, X_t is nodal with node $P \in X_t$, then on $\tilde{X}_t$, $Q - R \sim_{\mathrm{rat}} 0$, hence $\mathrm{CH}^1(X_t, 1)/\mathbf{C}^\times \neq 0$, and the possibility of a class $\xi \in \mathrm{CH}^2(X, 2)$, with nontrivial value $[d \log \xi] \in \Gamma H^2(X, \mathbf{Q}(2))$ arises. Assuming this is the case, then ν_ξ is nontrivial, and hence for general X_t, $\mathrm{cl}_{2,2}(\xi_t)$ is a nontorsion class (using a Baire category argument). This will be illustrated in Theorem 8.26 below, but as a preliminary warm-up, consider the nodal curve $D = \overline{V(y^2 - x^3 - x^2)} \subset \mathbf{P}^2$, with singular point $P = (0,0)$. By making the substitution $(x,y) = (x,ux)$, we end up with the desingularization $\sigma : \tilde{D} := \overline{V(u^2 = x + 1)} \to D$, and where $\sigma^{-1}(P) = \{Q = (0,1), R = (0,-1)\}$ in (x,u)-coordinates. Let

$$f = \frac{u+1}{u-1} = \frac{y+x}{y-x}.$$

Then viewing $f \in \mathbf{C}(\tilde{D})$, $\mathrm{div}_{\tilde{D}}(f) = R - Q$, and viewing $f \in \mathbf{C}(D) = \mathbf{C}(\tilde{D})$, $\mathrm{div}_D(f) = 0$. We apply this to the following.

Example 8.25. Let $\pi : X \to \mathbf{P}^1$ be the elliptic surface defined by

$$y^2 = x^3 + x^2 + t =: h(x),$$

and let $\Sigma \subset \mathbf{P}^1$ be the singular set of π. One shows that

$$\Sigma = \left\{0, \infty, \frac{-4}{27}\right\},$$

furthermore X_0, $X_{\frac{-4}{27}}$ are nodal curves, and X_∞ is a simply-connected tree of $\mathbf{P}^1$'s. We then have:

Theorem 8.26. *Let* $U = X\backslash\{X_0, X_{\frac{-4}{27}}, X_\infty\}$. *Then*

$$\Gamma(H^2(U, \mathbf{Q}(2))) \simeq \mathbf{Q}^2;$$

moreover it is generated by $[d\log(\xi_1)], [d\log(\xi_2)]$, *where*

$$\xi_1 = \left\{\frac{(y-x)^3}{8}, \frac{(y+x)^3}{8}\right\}\left\{\frac{y+x}{y-x}, t\right\}^3,$$

$$\xi_2 = \left\{\frac{(iy+x+2/3)^3}{8}, \frac{(iy-x-2/3)^3}{8}\right\}\left\{\frac{iy+x+2/3}{iy-x-2/3}, -t-4/27\right\}^3,$$

are classes in $\mathrm{CH}^2(U, 2; \mathbf{Q})$.[12]

Now choose a class $\xi \in \mathrm{CH}^2(U, 2)$ such that $[d\log\xi] \neq 0 \in \Gamma(H^2(U, \mathbf{Q}(2)))$. Thus for general $t \in \mathbf{P}^1$, $\mathrm{cl}_{2,2}(\xi_t)$ is nontorsion.

Remark 8.27. As pointed out by the referee, another class of examples pertaining to Observation 2 are the modular families of elliptic curves studied in [4], where every node P does give rise to such a class ξ, by Beilinson's Eisenstein symbol construction.

Acknowledgements Partially supported by a grant from the Natural Sciences and Engineering Research Council of Canada.

---

[12] M. Asakura informed me of his work in [1], which includes this theorem as a special case. Further he provides an upper bound for the rank of the $d\log$ image for variants of the family in Example 8.25.

References

1. M. Asakura, *On d log Image of K_2 of Elliptic Surface Minus Singular Fibers*, preprint (2006) [arXiv:math/0511190v4]
2. H. Bass, J. Tate, in *The Milnor Ring of a Global Field, in Algebraic K-Theory II*. Lecture Notes in Mathematics, vol. 342 (Springer, New York, 1972), pp. 349–446
3. A. Beilinson, Notes on absolute Hodge cohomology, in *Applications of Algebraic K-Theory to Algebraic Geometry and Number Theory*. Contemporary Mathematics, vol. 55, Part 1 (AMS, Providence 1986), pp. 35–68
4. A. Beilinson, Higher regulators of modular curves, in *Applications of K-Theory to Algebraic Geometry and Number Theory*, Boulder, CO, 1983. Contemporary Mathematics, vol. 55 (AMS, Providence, 1986), pp. 1–34
5. S. Bloch, *Lectures on Algebraic Cycles*. Duke University Mathematics Series, vol. IV (Duke University, Durham, 1980)
6. S. Bloch, Algebraic cycles and higher K-theory. Adv. Math. **61**, 267–304 (1986)
7. S. Bloch, V. Srinivas, Remarks on correspondences and algebraic cycles. Am. J. Math. **105**, 1235–1253 (1983)
8. J. Carlson, Extension of mixed Hodge structures, in *Journées de Géométrie Algébrique d'Angers 1979* (Sijthoff and Nordhoff, The Netherlands, 1980), pp. 107–127
9. X. Chen, J.D. Lewis, Noether-Lefschetz for K_1 of a certain class of surfaces. Bol. Soc. Mat. Mexicana (3) **10**(1), 29–41 (2004)
10. X. Chen, J.D. Lewis, The Hodge $-\mathscr{D}-$conjecture for $K3$ and Abelian surfaces. J. Algebr. Geom. **14**(2), 213–240 (2005)
11. X. Chen, J.D. Lewis, *Density of rational curves on $K3$ surfaces*. Math. Ann. [arXiv:1004.5167] (2011)
12. X. Chen, C. Doran, M. Kerr, J.D. Lewis, *Higher normal functions, derivatives of normal functions, and elliptic fibrations* (2011) (submitted) [arXiv:1108.2223]
13. C.H. Clemens, Homological equivalence, modulo algebraic equivalence, is not finitely generated. Publ. I.H.E.S. **58**, 19–38 (1983)
14. A. Collino, Griffiths' infinitesimal invariant and higher K-theory on hyperelliptic jacobians. J. Algebr. Geom. **6**, 393–415 (1997)
15. R. de Jeu, J.D. Lewis, *Beilinson's Hodge conjecture for smooth varieties*, J. of K-Theor. [arXiv:1104.4364] (2011)
16. P. Deligne, Théorie de Hodge, II, III. Inst. Hautes Études Sci. Publ. Math. **40**, 5–57 (1971); **44**, 5–77 (1974)
17. P. Elbaz-Vincent, S. Müller-Stach, Milnor K-theory of rings, higher Chow groups and applications. Invent. Math. **148**, 177–206 (2002)
18. H. Esnault, K.H. Paranjape, Remarks on absolute de Rham and absolute Hodge cycles. C. R. Acad. Sci. Paris t **319**, Serie I, 67–72 (1994)
19. H. Esnault, E. Viehweg, Deligne-Beilinson cohomology, in *Beilinson's Conjectures on Special Values of L-Functions*, ed. by Rapoport, Schappacher, Schneider. Perspectives in Mathematics, vol. 4 (Academic, New York, 1988), pp. 43–91
20. R. Friedman, R. Laza, *Semi-algebraic horizontal subvarieties of Calabi-Yau type*, preprint 2011 [arXive:1109.5632v1]
21. B.B. Gordon, J.D. Lewis, Indecomposable higher Chow cycles, in *The Arithmetic and Geometry of Algebraic Cycles*, Banff, AB, 1998. Nato Science Series C: Mathematical and Physical Sciences, vol. 548 (Kluwer, Dordrecht, 2000), pp. 193–224
22. M. Green, Griffiths' infinitesimal invariant and the Abel-Jacobi map. J. Differ. Geom. **29**, 545–555 (1989)
23. M. Green, P. Griffiths, The regulator map for a general curve, in *Symposium in Honor of C.H. Clemens*, Salt Lake City, UT, 2000. Contemporary Mathematics, vol. 312 (American Mathematical Society, Providence, 2002), pp. 117–127
24. M. Green, S. Müller-Stach, Algebraic cycles on a general complete intersection of high multi-degree of a smooth projective variety. Comp. Math. **100**(3), 305–309 (1996)

25. P. Griffiths, J. Harris, *Principles of Algebraic Geometry* (Wiley, New York, 1978)
26. P.A. Griffiths, On the periods of certain rational integrals: I and II. Ann. Math. **90**, 460–541 (1969)
27. U. Jannsen, Deligne cohomology, Hodge$-\mathscr{D}-$conjecture, and motives, in *Beilinson's Conjectures on Special Values of L-Functions*, ed. by Rapoport, Schappacher, Schneider. Perspectives in Mathematics, vol. 4 (Academic, New York, 1988), pp. 305–372
28. U. Jannsen, in *Mixed Motives and Algebraic K-Theory*. Lecture Notes in Mathematics, vol. 1000 (Springer, Berlin, 1990)
29. B. Kahn, Groupe de Brauer et (2, 1)-cycles indecomposables. Preprint (2011)
30. K. Kato, *Milnor K-theory and the Chow group of zero cycles*. Contemp. Math. Part I **55**, 241–253 (1986)
31. M. Kerr, J.D. Lewis, The Abel-Jacobi map for higher Chow groups, II. Invent. Math. **170**(2), 355–420 (2007)
32. M. Kerr, J.D. Lewis, S. Müller-Stach, The Abel-Jacobi map for higher Chow groups. Compos. Math. **142**(2), 374–396 (2006)
33. M. Kerz, The Gersten conjecture for Milnor K-theory. Invent. Math. **175**(1), 1–33 (2009)
34. J. King, Log complexes of currents and functorial properties of the Abel-Jacobi map. Duke Math. J. **50**(1), 1–53 (1983)
35. M. Levine, Localization on singular varieties. Invent. Math. **31**, 423–464 (1988)
36. J.D. Lewis, in *A Survey of the Hodge Conjecture*, 2nd edn. Appendix B by B. Brent Gordon. CRM Monograph Series, vol. 10 (American Mathematical Society, Providence, 1999), pp. xvi+368
37. J.D. Lewis, Lectures on algebraic cycles. Bol. Soc. Mat. Mexicana (3) **7**(2), 137–192 (2001)
38. J.D. Lewis, Real regulators on Milnor complexes. *K*-Theory **25**(3), 277–298 (2002)
39. J.D. Lewis, Regulators of Chow cycles on Calabi-Yau varieties, in *Calabi-Yau Varieties and Mirror Symmetry*, Toronto, ON, 2001. Fields Institute Communications, vol. 38 (American Mathematical Society, Providence, 2003), pp. 87–117
40. S. Müller-Stach, Constructing indecomposable motivic cohomology classes on algebraic surfaces. J. Algebr. Geom. **6**, 513–543 (1997)
41. S. Müller-Stach, Algebraic cycle complexes, in *Proceedings of the NATO Advanced Study Institute on the Arithmetic and Geometry of Algebraic Cycles*, vol. 548, ed. by J.D. Lewis, N. Yui, B. Gordon, S. Müller-Stach, S. Saito (Kluwer, Dordrecht, 2000), pp. 285–305
42. D. Ramakrishnan, Regulators, algebraic cycles, and values of *L*-functions, in *Contemporary Mathematics*, vol. 83 (American Mathematical Society, Providence, 1989), pp. 183–310
43. M. Raynaud, Courbes sur une variété abélienne et points de torsion. Invent. Math. **71**, 207–233 (1983)
44. J.H.M. Steenbrink, in *A Summary of Mixed Hodge Theory*, Motives, Seattle, WA, 1991. Proceedings of Symposia in Pure Mathematics, vol. 55, Part 1 (American Mathematical Society, Providence, 1994), pp. 31–41
45. C. Voisin, The Griffiths group of a general Calabi-Yau threefold is not finitely generated. Duke Math. J. **102**(1), 151–186 (2000)

Two Lectures on the Arithmetic of K3 Surfaces

Matthias Schütt

Abstract In these lecture notes we review different aspects of the arithmetic of K3 surfaces. Topics include rational points, Picard number and Tate conjecture, zeta functions and modularity.

Key words: K3 surface, Rational points, Elliptic fibration, Picard number, Singular K3 surface, Modular form, Class group

Mathematics Subject Classifications (2010): 14J28, 11F03, 11G05, 11G15, 11G25, 11G35, 14G05, 14G15, 14G25, 14J10, 14J27

1 Introduction

K3 surfaces are central objects of study in various areas of mathematics and physics such as algebraic, complex, and differential geometry, number theory and string theory. Naturally they featured prominently in the Fields workshop. These notes record two introductory lectures on the arithmetic of K3 surfaces with some bits of additional or supplementary material. Limitations of space and time do not possibly allow me to do justice to all important aspects of this area; I apologise for everything that may have been left out or not attributed correctly.

In brief these lecture notes aim to shed some light on the following three topics:

1. Rational points on K3 surfaces.
2. Picard numbers and the Tate conjecture for K3 surfaces.
3. Zeta functions and modularity for K3 surfaces.

M. Schütt (✉)
Institut für Algebraische Geometrie, Leibniz Universität Hannover,
Welfengarten 1, 30167 Hannover, Germany
e-mail: schuett@math.uni-hannover.de

R. Laza et al. (eds.), *Arithmetic and Geometry of K3 Surfaces and Calabi–Yau Threefolds*, Fields Institute Communications 67, DOI 10.1007/978-1-4614-6403-7_3,

Our survey is initiated by a brief motivation coming from algebraic curves which illustrates the thematic interplay between arithmetic and geometry.

2 Motivation: Rational Points on Algebraic Curves

Throughout this paper we are mostly concerned with varieties which are complex, smooth, and projective although many techniques that we discuss actually involve positive characteristic. On the level of curves, we can equivalently consider compact Riemann surfaces. Then there is a discrete invariant: the *genus g* of the compact Riemann surface, i.e. the number of holes or handles. It is a non-trivial fact that this purely topological invariant has an algebro-geometric counterpart: the geometric genus measuring the dimension of the space of regular 1-forms on the projective curve. It should come even more surprising how the genus governs the arithmetic of algebraic curves.

To see this, we assume that the algebraic curve C is defined over some number field K (which the reader may just as well assume to be $\mathbb{Q}$). Then it is a natural problem to investigate the set of K-rational points on C. It turns out that the cardinality of $C(K)$ falls into three cases according to the genus of C:

$g(C)$	$\#C(K)$	Comment
0	$0, \infty$	Rational if $C(K) \neq \emptyset$
1	$\leq \infty$	Elliptic if $C(K) \neq \emptyset$
≥ 2	$< \infty$	Faltings' Theorem [16]

The genus 1 case is particularly rich since the K-rational points form an abelian group (best visible on the model as a cubic in $\mathbb{P}^2$ with the group law that any three collinear points on the curve add up to zero). It goes back to Mordell and Weil that this group is finitely generated. Thanks to the group structure on elliptic curves, we deduce the following remarkable property of genus 1 curves C over a number field K: there exists some finite extension K'/K such that the K'-rational points are Zariski dense on C. To see this one first extends K to reach a rational point on C (such that C becomes elliptic) and then ensures, possibly by a further extension, that there is a rational point of infinite order. This concept is usually called potential density:

Definition 1. Let X denote a variety defined over some number field K. We say that X has or satisfies *potential density (of rational points)* if there exists some finite extension K'/K such that the K'-rational points lie dense on X.

Note that potential density unifies algebraic curves of genus 0 and 1. By Faltings' Theorem [16], however, curves of genus greater than 1 *never* satisfy potential density. It is a common belief that similar structures should hold in higher dimension, with the genus replaced by the *Kodaira dimension* κ. For instance, the Bombieri–Lang conjecture formulates that on varieties of maximal Kodaira dimension (so called *general type*: $\kappa(X) = \dim(X)$) the Zariski closure of the rational points over any number field forms a proper subvariety. It seems worthwhile noticing that potential density makes the arithmetic problem of rational points into a geometric notion which only depends on the $\bar{\mathbb{Q}}$-isomorphism class of the variety, but not on the precise model chosen.

In the next section, we discuss the problem of rational points on K3 surfaces which can be considered as a two-dimensional analogue of elliptic curves. Later we will see that also other concepts such as modularity carry over from elliptic curves to certain K3 surfaces in a decisive way.

3 K3 Surfaces and Rational Points

In essence there are two ways to extend the definition of elliptic curves to dimension 2. Requiring a group structure leads to abelian varieties which are fairly well understood in arithmetic and geometry. Almost automatically they come with potential density. On the other hand, we can impose the Calabi–Yau condition (in the strict sense); this leads to the notion of K3 surfaces:

Definition 2. A smooth projective surface X is called K3 if

$$\omega_X \cong \mathcal{O}_X \quad \text{and} \quad h^1(X, \mathcal{O}_X) = 0.$$

In fact, all K3 surfaces can be seen to be (algebraically) simply connected, and over $\mathbb{C}$, deformation equivalent (although the original argument for this went through non-algebraic K3 surfaces). For complex K3 surfaces we record the Hodge diamond which can be computed easily with Noether's formula:

$$
\begin{array}{ccccc}
 & & 1 & & \\
 & 0 & & 0 & \\
1 & & 20 & & 1 \\
 & 0 & & 0 & \\
 & & 1 & &
\end{array}
$$

The resulting Betti numbers also hold in positive characteristic (for ℓ-adic étale cohomology, say).

We give three examples. The first two mimic the definition of elliptic curves in two essentially different ways (which will surface again in Sect. 5) while the third relates to abelian surfaces. To ease the exposition, we limit ourselves to constructions outside characteristic 2.

Example 1. 1. Smooth quartics in $\mathbb{P}^3$.
  2. Double sextics, i.e. double coverings $X \to \mathbb{P}^2$ branched along a smooth sextic curve.
  3. Kummer surfaces Km(A) where A is an abelian surface and Km(A) is the minimal resolution of the quotient $A/\langle -1 \rangle$ with 16 rational double points.

In view of the deformation equivalence, we can also allow the quartic or sextic above to have isolated rational double points as singularities (ADE-type) and consider a minimal resolution which will then be K3.

In the following we will repeatedly consider Kummer surfaces of product type where the abelian surface is isomorphic to a product of elliptic curves $E \times E'$. Such Kummer surfaces come naturally with models as quartics or double sextics. To see this in an elementary way, write the elliptic curves in extended Weierstrass form

$$E : \quad y^2 = f(x), \quad E' : y'^2 = f'(x') \tag{1}$$

with cubic polynomials f, f' without multiple roots. Then a birational model of the Kummer surface $\mathrm{Km}(E \times E')$ is given by the double sextic

$$\mathrm{Km}(E \times E') : \quad w^2 = f(x)f'(x'). \tag{2}$$

Similarly quartic models are derived by bringing two linear factors from the RHS (over $\bar{K}$) to the LHS (multiply w by these factors). The fact that all these constructions produce indeed isomorphic K3 surfaces relies on general surface theory (birational maps between K3 surfaces are isomorphisms). There is one more incarnation of K3 surfaces that comes up handily on Kummer surfaces of product type: elliptic fibrations. Before sketching their theory in the next section, we indicate the relevance to the question of rational points:

Theorem 1 (Bogomolov, Tschinkel [5]). *Let X be a K3 surface over a number field. If X has an elliptic fibration or infinite automorphism group, then X satisfies potential density.*

The case of infinite automorphism group naturally implies potential density; indeed it suffices to exhibit a rational curve on X whose orbit under $\mathrm{Aut}(X)$ is infinite. Note that automorphisms of K3 surfaces are defined over number fields since the automorphism group is discrete and finitely generated (see [40, Sect. 7, Theorem 1], [57] and also [20, Proposition 2.1]). We will briefly explain the idea behind the elliptic fibrations after introducing the necessary background in the next section.

Apropos automorphisms, we mention as a sample of another yet completely different set of problems the question of the distribution of rational points, and in particular their periodicity under automorphisms. These issues lend K3 surfaces to the subject of dynamics. For instance, in [55] it is proved for certain K3 surfaces with infinite automorphism group (intersections of hypersurfaces of bidegree $(1, 1)$ and $(2, 2)$ in $\mathbb{P}^2 \times \mathbb{P}^2$) that the orbit of a rational point under the automorphism group is either finite or Zariski dense. The key ingredient here is a new notion of canonical height.

4 Elliptic K3 Surfaces

An elliptic surface is a smooth projective surface X together with a surjective morphism to a projective curve C,

$$X \to C,$$

such that almost all fibers are smooth curves of genus 1. Often one assumes the existence of a section (so that all fibers are in fact elliptic curves over the base field), but we will not restrict to these so-called *jacobian* fibrations here. In order to rule out products, for instance, one usually assumes that the fibration has a singular fiber. The possible singular fibers have been classified by Kodaira over $\mathbb{C}$ [24]; later Tate exhibited an algorithm for any perfect base field [60]. The reducible singular fibers consist solely of (-2)-curves (smooth rational curves) whose configuration corresponds to an extended Dynkin diagram (types $\tilde{A}_n, \tilde{D}_k, \tilde{E}_l$). The only irreducible singular fibers are the nodal and the cuspidal cubic.

Example 2 (Kummer surface of product type). The projections from $E \times E'$ to either factor induce elliptic fibrations on $\mathrm{Km}(E \times E')$. In the notation of (1), they could be given by twisted Weierstrass forms

$$f(x)y^2 = f'(x') \tag{3}$$

where x represents an affine parameter of the base curve $\mathbb{P}^1$. Visibly the fibration is isotrivial, with all smooth fibers quadratic twists of E' (in particular $\bar{K}$-isomorphic). The singular fibers (all of them reducible) are located at ∞ and the roots of $f(x)$, i.e. they correspond to the 2-torsion points of E. Each singular fiber has Kodaira type I_0^* (corresponding to $\tilde{D}_4$), a double $\mathbb{P}^1$ blown up in 4 A_1 singularities.

One advantage of elliptic surfaces is that they allow us to control the Néron–Severi group NS to some extent. Notably any two fibers are algebraically equivalent, but components of reducible fibers contribute non-trivially to NS. We call these curves *vertical*—as opposed to the multisections which are imagined in the horizontal direction. Clearly NS of an elliptic surface is generated by horizontal and vertical divisors; it is a non-trivial fact, however, that once there is a section, one can generate NS exclusively by fiber components and sections (see [50]). In particular, there is a closed expression (often referred to as the Shioda–Tate formula, cf. [50, Corollary 5.3]) for the Picard number

$$\rho(X) = \mathrm{rank}\, \mathrm{NS}(X),$$

involving only the reducible fibers (more precisely the number m_v of components of the fiber F_v) and the Mordell–Weil rank r:

$$\rho(X) = 2 + r + \sum_{v \in C} (m_v - 1). \tag{4}$$

Since the Picard number of an elliptic surface and its jacobian are the same, this formula indirectly also applies to any elliptic surface without section. Throughout the paper the Picard number should always be understood geometrically, i.e. over the algebraic closure of the base field (although we consider varieties over non-closed fields).

As an illustration, potential density holds for any jacobian elliptic (K3) surface with positive Mordell–Weil rank. To prove Theorem 1, one is thus led to consider elliptic K3 surface with MW-rank zero or no sections at all. In brief Bogomolov and Tschinkel show that any elliptic K3 surface with Picard number $\rho \leq 19$ admits infinitely many suitable multisections which are rational. Then they continue to prove that enough of these multisections are not related to torsion points. To finish the proof, they refer to a result by Shioda and Inose [53] that any K3 surface with $\rho = 20$ has infinite automorphism group (derived in the framework of Shioda–Inose structures, see Sect. 12).

We now turn to the problem how restrictive the assumptions in Theorem 1 are. There are at least two answers:

fairly restrictive   or   not terribly restrictive.

To justify the second answer, we mention a special feature of K3 surfaces: elliptic fibrations are completely governed by lattice theory. Namely, any divisor of self-intersection zero induces an elliptic fibration after [40, Sect. 3, Theorem 1]. Here one first applies reflections and inversions to D until it becomes effective by Riemann–Roch. Upon subtracting the base locus, the resulting linear system induces the elliptic fibration. One easily deduces:

Lemma 1. *Any K3 surface with Picard number $\rho \geq 5$ admits an elliptic fibration.*

Proof. The intersection pairing between curves equips NS with a non-degenerate quadratic form of signature $(1, \rho - 1)$ (compatible with cup-product on $H^2(X, \mathbb{Z})$). Since any such quadratic form of rank at least 5 represents zero, the claim follows from the discussion preceding the lemma.   □

Thus we find that the assumptions of Theorem 1 are not terribly restrictive in the following sense:

Corollary 1. *Any K3 surface of Picard number $\rho \geq 5$ over a number field satisfies potential density.*

In the opposite direction, it has to be noted that either assumption of Theorem 1 implies $\rho > 1$. A generic K3 surface, however, has $\rho = 1$. To argue that the assumptions are *fairly restrictive*, it therefore suffices to rule out that K3 surfaces over number fields somehow happen to lie on the countably many hypersurfaces in the moduli spaces comprising K3 surfaces with $\rho > 1$. This question will be discussed both from the theoretical and explicit view point in the next section.

5 Picard Number One

An elliptic curve always possesses a model as a plane cubic thanks to Riemann–Roch. Quite opposite to this, K3 surfaces have many different incarnations. We have seen two of them in Example 1: double sextics on the one hand and quartics in $\mathbb{P}^3$ on the other. While these cases certainly overlap, for instance on the Kummer surfaces (not only of product type), they differ in an essential way. This can be seen as follows.

By general moduli theory, both generic double sextic and generic quartic have Picard number $\rho = 1$ (for quartics, this result originally goes back to a conjecture of Noether, as proved by Tjurina). But then the hyperplane section H gives an ample divisor of self-intersection

$$H^2 = 2 \quad \text{resp.} \quad H^2 = 4.$$

In other words, NS as a lattice equals $\mathbb{Z}\langle 2 \rangle$ reps. $\mathbb{Z}\langle 4 \rangle$. As these two lattices are not isometric, generic double sextics and quartics cannot be isomorphic. In fact for any integer $d > 0$, there are K3 surfaces with a so-called polarization of degree $2d$ forming a 19-dimensional moduli space. The uniform approach would be to consider these as hypersurfaces in the 20-dimensional moduli space of all K3 surfaces (including non-algebraic ones). Similarly, K3 surfaces of Picard number $\rho \geq 2$ (such as in Theorem 1) lie on hypersurfaces in the moduli spaces of polarized K3 surfaces, described by lattice polarisations (cf. [33]). The solution whether all K3 surfaces over number fields might somehow happen to lie on these hypersurfaces was given by Terasoma and Ellenberg:

Theorem 2 (Terasoma [61], Ellenberg [14]). *For any integer $d > 0$, there exist $2d$-polarised K3 surfaces over $\bar{\mathbb{Q}}$ with $\rho = 1$.*

In this sense, the assumptions of Theorem 1 have to be considered as fairly restrictive. In particular, there was no K3 surface of $\rho = 1$ known to satisfy potential density until very recently Kharzemanov announced in [22] the existence of such K3 surfaces. In the sequel we shall discuss the first big obstacle to producing such examples: it is very hard to exhibit explicit K3 surfaces with $\rho = 1$!

6 Computation of Picard Numbers

The crux with the Picard number of an algebraic surface X is that it is in general very hard to compute. That is, unless the geometric genus of X vanishes—over $\mathbb{C}$ or if the surface lifts to such a surface in characteristic zero. In that case, we have $H^{1,1}(X) = H^2(X, \mathbb{C})$ so that Lefschetz' theorem returns

$$\text{NS}(X) = H^{1,1}(X) \cap H^2(X, \mathbb{Z}) = H^2(X, \mathbb{Z}). \tag{5}$$

Here we trivially deduce $\rho(X) = b_2(X)$. In case of non-zero geometric genus (or non-liftability), we would only be aware of the following procedure to compute the Picard number:

1. In the daytime, search systematically for (independent) curves on X giving lower bounds for $\rho(X)$.
2. In the nighttime, develop upper bounds for $\rho(X)$ until upper and lower bounds match.

We have to remark that it is unclear as of today whether either step can be implemented in an effective way. For instance, the daytime step requires to check on X for curves of increasing degrees which soon becomes computationally fairly expensive. There are, however, situations where we can do better. Notably the daytime step becomes vacuous if we aim to prove $\rho(X) = 1$. Also if X admits a jacobian elliptic fibration, then the shape of the curves to consider is very clear, as they can all be taken as sections where one raises the height successively (see the discussion at the end of Sect. 15 where this plays a crucial role).

The nighttime step concerning upper bounds for ρ is yet more delicate. A priori, one has only the following estimates

$$\rho(X) \leq \begin{cases} h^{1,1}(X) & \text{(over } \mathbb{C} \text{ by Lefschetz)} \\ b_2(X) & \text{(in any characteristic due to Igusa).} \end{cases}$$

Currently there are two approaches that have been worked out and tested in detail. The first requires the special situation where X admits non-trivial *automorphisms*. These give extra information about algebraic and non-algebraic classes in $H^2(X)$. The extreme situation consists of Fermat varieties where the big automorphism group gives complete control over all cohomology groups (see Example 4). More generally, as soon as an automorphism acts non-trivially on the regular 2-forms on the surface X, this can give upper bounds for $\rho(X)$. In his pioneering work [49] Shioda used this technique to exhibit an explicit quintic surface over $\mathbb{Q}$ with $\rho = 1$ (quite surprisingly one might want to add).

The second approach towards upper bounds for ρ consists in smooth *specialization*, mostly to characteristic $p > 0$. Here we consider a complex surface X with a model over some number field K (or its $\mathfrak{p}$-adic completion) with good reduction $X_{\mathfrak{p}}$ at some prime $\mathfrak{p}$. Since intersection numbers are preserved under specialization, we obtain an embedding of lattices

$$\text{NS}(X) \hookrightarrow \text{NS}(X_{\mathfrak{p}}). \tag{6}$$

Directly this gives the upper bound

$$\rho(X) \leq \rho(X_{\mathfrak{p}}).$$

Here the big advantage is that this upper bound is (theoretically) explicitly computable assuming the Tate conjecture (see Conjecture 1). Concretely $X_{\mathfrak{p}}$ is equipped with the Frobenius automorphism $\text{Frob}_{\mathfrak{p}}$ raising coordinates to their q-th powers where $q = \#\mathbb{F}_{\mathfrak{p}} = p^r$ is the norm of $\mathfrak{p}$. Then one is led to consider the induced action of $\text{Frob}_{\mathfrak{p}}^*$ on $H^2_{\text{ét}}(X_{\mathfrak{p}}, \mathbb{Q}_\ell)$. A crucial property of NS in this context is that it can always be generated by divisors defined over some finite extension of the base field. This implies that the absolute Galois group acts on NS through a finite group. Embedding

$$\text{NS}(X_{\mathfrak{p}}) \hookrightarrow H^2_{\text{ét}}(X_{\mathfrak{p}}, \mathbb{Q}_\ell) \tag{7}$$

via the cycle class map, we find that all eigenvalues of $\text{Frob}_{\mathfrak{p}}^*$ on the image of $\text{NS}(X_{\mathfrak{p}})$ take the shape ζq where ζ runs through roots of unity.

Conjecture 1 (Tate [58]). All eigenspaces of $\mathrm{Frob}_{\mathfrak{p}}^*$ in $H^2_{\text{ét}}(X_{\mathfrak{p}}, \mathbb{Q}_\ell)$ with eigenvalues as above are algebraic.

The Tate conjecture is known, for instance, for Fermat varieties and several kinds of K3 surfaces including elliptic ones [3] and those of finite height [38]. Recently, intriguing finiteness statements have been discovered to be equivalent to the Tate conjecture in [27].[1] At any rate, the above discussion gives an upper bound for $\rho(X_{\mathfrak{p}})$ in terms of the eigenvalues of $\mathrm{Frob}_{\mathfrak{p}}^*$ on $H^2_{\text{ét}}(X_{\mathfrak{p}}, \mathbb{Q}_\ell)$. We shall now indicate how to compute these eigenvalues.

The reciprocal characteristic polynomial $P_2(T)$ of $\mathrm{Frob}_{\mathfrak{p}}^*$ on $H^2_{\text{ét}}(X_{\mathfrak{p}}, \mathbb{Q}_\ell)$ appears as a factor of the zeta function of $X_{\mathfrak{p}}$ over $\mathbb{F}_{\mathfrak{p}}$. By the Weil conjecture, the zeta function can be computed by point counting over sufficiently many finite fields $\mathbb{F}_{q^r}$ through Lefschetz' fixed point formula. For instance, if $X_{\mathfrak{p}}$ is a regular surface over $\mathbb{F}_q$ (in the sense that $b_1(X_{\mathfrak{p}}) = 0$), then its zeta function takes the shape

$$\zeta(X, T) = \frac{1}{(1 - T)\, P_2(T)\, (1 - q^2 T)} \tag{8}$$

Hence point counting up to $r = \lceil (b_2(X_{\mathfrak{p}}) - 1)/2 \rceil$ will be sufficient to compute the zeta function thanks to the functional equation. Note that this will in practice still be fairly expensive, but there are improvements using p-adic cohomology [1].

With all these techniques at hand, here comes the major drawback of the specialization method: assuming the Tate conjecture, non-algebraic eigenclasses of $\mathrm{Frob}_{\mathfrak{p}}^*$ in $H^2_{\text{ét}}(X_{\mathfrak{p}}, \mathbb{Q}_\ell)$ come in pairs, corresponding to pairs of complex-conjugate eigenvalues which are not multiples of roots of unity by q (but algebraic integers of absolute value q by the Weil conjectures). In particular this would imply

$$\rho(X_{\mathfrak{p}}) \equiv b_2(X_{\mathfrak{p}}) \mod 2.$$

Thus, if we want to prove that some surface X has Picard number $\rho(X)$ of parity other than that of $b_2(X)$, one has to be more inventive. In the next section we sketch what has been done (and what might be done) for the prototype case of K3 surfaces with $\rho = 1$.

7 K3 Surfaces of Picard Number One

This section describes how to attack the computation of the Picard number of a K3 surface over some number field. We explain in some detail the prototype case of $\rho = 1$. The first one to exhibit an explicit K3 surface X with $\rho(X) = 1$ was van Luijk.

---

[1] Charles (to appear in Invent. Math.) and Madapusi Pera (arXiv: 1301.6326) have announced independent proofs of the Tate Conjecture for K3 surfaces outside characteristic 2 (and 3 in Charles' case).

7.1 van Luijk's Approach

In [29] van Luijk exhibited a K3 surface as quartic X over $\mathbb{Q}$ with two different primes $\mathfrak{p}, \mathfrak{p}'(= 2, 3)$ of good reduction where the point counting method from the previous section gave the upper bound

$$\rho(X) \leq \rho(X_\mathfrak{p}), \rho(X_{\mathfrak{p}'}) \leq 2. \tag{9}$$

Assuming that $\rho(X) = 2$, the embeddings of lattices

$$\mathrm{NS}(X_{\mathfrak{p}'}) \longleftrightarrow \mathrm{NS}(X) \hookrightarrow \mathrm{NS}(X_\mathfrak{p}) \tag{10}$$

would be of finite index. In particular, this would imply that the discriminants of all three lattices (i.e. the determinants of the Gram matrices for a basis) would be the same up to some square factors. This property, however, can lead to a contradiction by working out explicit basis for both $\mathrm{NS}(X_\mathfrak{p}), \mathrm{NS}(X_{\mathfrak{p}'})$ and verifying that the intersection forms are not compatible.

7.2 Kloosterman's Improvement

Subsequently Kloosterman noticed that it is possible to circumvent the determination of generators of $\mathrm{NS}(X_\mathfrak{p})$ and $\mathrm{NS}(X_{\mathfrak{p}'})$. In a similar situation in [23] he instead appealed to the Artin–Tate conjecture [59] which while equivalent to the Tate conjecture by [31], additionally predicts the square class of the discriminant of NS:

$$\mathrm{disc}\,\mathrm{NS}(X_\mathfrak{p}) \stackrel{?}{=} q \frac{P_2(T)}{(1 - qT)^{\rho(X_\mathfrak{p})}} \Big|_{T=1/q} \in \mathbb{Q}^*/(\mathbb{Q}^*)^2. \tag{11}$$

Note that this procedure does not actually require the validity of the Tate conjecture. For if the Tate conjecture were to be wrong for $X_\mathfrak{p}$ in the above setting, that is $\rho(X_\mathfrak{p}) < 2$, then automatically $\rho(X) = \rho(X_\mathfrak{p}) = 1$ by (9) which is exactly the original claim. On the other hand, if the Tate conjecture is valid for $X_\mathfrak{p}$ and $X_{\mathfrak{p}'}$, then we read off the square classes of $\mathrm{NS}(X_\mathfrak{p})$ and $\mathrm{NS}(X_{\mathfrak{p}'})$ from (11). If they do not agree, then we derive the desired contradiction to the assumption $\rho(X) = 2$.

7.3 Elsenhans–Jahnel's Work

The method pioneered by van Luijk takes a substantial amount of computation time and memory since point counting over fairly large finite fields is required for two suitable primes of good reduction (with no guarantee that 2 and 3 would work). With a view towards double sextics (which often have bad reduction at 2), Elsenhans and Jahnel modified van Luijk's method in such a way that point counting is only required at one suitable prime. Based on work by Raynaud [41] they show in [15] that the embedding (7) is primitive (i.e. the cokernel is torsion-free) in a wide range of cases including smooth surfaces over $\mathbb{Q}$ with good reduction at $\mathfrak{p} \neq 2\mathbb{Z}$. (In consequence the conjectural finite index embeddings in (10) would in fact be isometries of lattices.) In practice, this means the following:

Proposition 1 (Elsenhans–Jahnel). *In the above setup, assume* $\mathfrak{p} \neq 2\mathbb{Z}$. *If some divisor class does not lift from* $X_{\mathfrak{p}}$ *to* X, *then* $\rho(X) < \rho(X_{\mathfrak{p}})$.

In order to exhibit a K3 surface over some number field with $\rho = 1$, it thus suffices to find a single prime $\mathfrak{p}$ of good reduction such that

1. $\rho(X_{\mathfrak{p}}) \leq 2$ by inspection of the characteristic polynomial of $\mathrm{Frob}_{\mathfrak{p}}^*$ on $H^2_{\text{ét}}(X_{\mathfrak{p}}, \mathbb{Q}_{\ell})$
   and
2. Some divisor class on $X_{\mathfrak{p}}$ does not lift to X.

7.4 Outlook

Currently van Luijk and the author are working on an arithmetic deformation technique that would allow to construct explicit K3 surfaces with $\rho = 1$ (and other prescribed Picard numbers) without any point counting at all. The overall idea is to combine the above techniques with extra information which can be extracted from automorphisms (see Sect. 6). While we will actually mainly aim at Picard numbers of quintics and beyond, this method can be used, for instance, to prove that the following double sextic has $\rho(X) = 1$ over $\mathbb{C}$:

$$X : \quad w^2 = x^5 + xy^5 + 101y^4 + 1.$$

7.5 Feasibility

Having exhibited explicit K3 surfaces with $\rho = 1$, we shall now come to the problem whether the above techniques may be applied to all K3 surfaces over number fields. That is, we ask for an algorithm to compute the Picard number of a K3 surface which always terminates theoretically. This issue was taken up in a recent preprint by Charles [8].

In detail, it is shown using the endomorphism algebra E of the Hodge structure underlying the transcendental lattice (cf. (13)) that one cannot in general expect that there are primes $\mathfrak{p}$ such that $\rho(X_{\mathfrak{p}}) \leq \rho(X) + 1$. However, Charles proves that there are always infinitely many primes $\mathfrak{p}$ such that

$$\rho(X_{\mathfrak{p}}) = \rho(X) \quad \text{or} \quad \rho(X_{\mathfrak{p}}) = \rho(X) + \dim_{\mathbb{Q}} E.$$

Additionally, in the latter case, there are different discriminants turning up on the reductions. Once one knows E, these facts facilitate an algorithm which theoretically returns the Picard number of X. However, the determination of E (by similar methods) seems to require the validity of the Hodge conjecture for the self-product $X \times X$.

8 Hasse Principle for K3 Surfaces

Coming back to rational points on K3 surfaces, there are many more subtle problems to investigate. Here we comment on recent developments concerning the Hasse principle which in fact relate to the Picard number one problem as well.

Given a variety X over a number field K, one may wonder whether for the existence of a global point on X it suffices to have local points over K_v for every place v of K. This is called the *Hasse principle*, phrased in terms of the adèles $\mathbb{A}_K$:

$$X(\mathbb{A}_K) \neq \emptyset \overset{?}{\Longrightarrow} X(K) \neq \emptyset. \tag{12}$$

The classical case for the Hasse principle to hold consists in conics in $\mathbb{P}^2$, but already for cubics in $\mathbb{P}^2$ it may fail by an example due to Selmer [48]. Often this failure can be explained by the Brauer group

$$\mathrm{Br}(X) = H_{\text{ét}}^2(X, \mathbb{G}_m).$$

In fact, via local invariants any subset $S \subseteq \mathrm{Br}(X)$ gives rise to an intermediate set

$$X(K) \subseteq X(\mathbb{A}_K)^S \subseteq X(\mathbb{A}_K)$$

which may be empty even if $X(\mathbb{A}_K)$ is not (see [56, Sect. 5.2]). This is exactly the situation of a Brauer–Manin obstruction to the Hasse principle for X as pioneered by Manin in [30]. In practice one specifies two subgroups

$$\mathrm{Br}_0(X) \subseteq \mathrm{Br}_1(X) \subseteq \mathrm{Br}(X)$$

as follows:

$$\text{constant } \mathrm{Br}_0(X) = \mathrm{im}(\mathrm{Br}(k) \to \mathrm{Br}(X))$$
$$\text{algebraic } \mathrm{Br}_1(X) = \ker(\mathrm{Br}(X) \to \mathrm{Br}(X \otimes \bar{K}))$$

Class field theory shows for any $S \subseteq \mathrm{Br}_0(X)$ that $X(\mathbb{A}_K)^S = X(\mathbb{A}_K)$. Algebraic Brauer–Manin obstructions to the Hasse principle (and to the related concept of weak approximation, i.e. density of $X(K)$ in the product of all $X(K_v)$) have been studied extensively in the last 40 years (see the references in [18]). Meanwhile it is the *transcendental* elements in $\mathrm{Br}(X) \setminus \mathrm{Br}_1(X)$ that have resisted concrete realizations; in fact, most constructions in the last decade have only impacted weak approximation while relying in an essential way on elliptic fibrations (so that $\rho \geq 2$). This was rectified recently by a remarkable construction due to Hassett and Vàrilly-Alvarado [18]: they exhibit K3 surfaces with Picard number one (as double sextics over number fields) with explicit quaternion algebras in $\mathrm{Br}(K(X))$ giving rise to a transcendental Brauer–Manin obstruction to the Hasse principle. Their work builds on results of van Geemen on Brauer groups of K3 surfaces [17] and extends previous results which only applied to weak approximation [19].

9 Rational Curves on K3 Surfaces

To close our considerations about rational points of K3 surfaces, we briefly comment on the related topic of rational curves. The fundamental problem is:

Question 1. Does any K3 surface contain infinitely many rational curves?

We have already touched upon an answer for K3 surfaces with $\rho \geq 5$: these admit an elliptic fibration (Corollary 1) with infinitely many rational multisections (see the discussion in Sect. 4). The problem has seen amazing progress recently, starting from [6] and greatly extended in [26]. The main idea is to first reduce to K3 surfaces X over $\bar{\mathbb{Q}}$ and then use reduction mod p. Then the odd parity of $\rho(X)$ (and the Tate conjecture) implies the existence of additional curves on X_p—including infinitely many rational ones as one can show. Then one lifts back based on arguments going back to Bogomolov and Mumford.

Theorem 3 (Bogomolov–Hassett–Tschinkel [6], Li-Liedtke [26]). *Any K3 surface over $\bar{\mathbb{Q}}$ with odd ρ or $\rho \geq 5$ contains infinitely many rational curves (over $\bar{\mathbb{Q}}$).*

We remark that the theorem as it stands does not imply potential density because there is no control over the fields of definition of the rational curves.

10 Isogeny Notion for K3 Surfaces

Having said that Picard numbers are hard to compute, there is a big advantage when working with complex K3 surfaces. This files under a notion of isogeny introduced by Inose:

Proposition 2. *Let X, X' be complex K3 surfaces admitting a dominant rational map $X \dashrightarrow X'$. Then $\rho(X) = \rho(X')$.*

The proof is an easy exercise using Hodge structures. Essentially one only has to consider the blow-up $\tilde{X}$ of X along the locus of indeterminacy of the rational map:

$$
\begin{array}{ccc}
 & \tilde{X} & \\
\downarrow & & \searrow \\
X & \dashrightarrow & X'
\end{array}
$$

Then we can pull-back the transcendental Hodge structures from X and X' to $\tilde{X}$. Since the geometric genus is always 1, all these Hodge structures are determined as the smallest $\mathbb{Q}$-sub-Hodge structure of $H^2(\tilde{X}, \mathbb{Q})$ whose complexification contains $H^{2,0}(\tilde{X})$. In particular, they are isomorphic as $\mathbb{Q}$-Hodge structures. In consequence the Picard numbers of X and X' coincide. □

Remark 1. It was pointed out by Shioda that the analogue of Proposition 2 in positive characteristic fails in general, as there are unirational supersingular K3 surfaces (e.g. Kummer surfaces). There seems to be a proof, though, for K3 surfaces of finite height (such that the Tate conjecture holds true by [38]).

As an application, we can consider symplectic automorphisms of K3 surfaces, i.e. those leaving the regular 2-form invariant. Over $\mathbb{C}$, Nikulin proved in [35] that the fixed locus always consists of a certain finite number of isolated fixed points which only depends on the order of the automorphism (at most 8). On the quotient surface, the fixed points yield rational double point singularities whose resolution thus is again a K3 surface. By Proposition 2 the Picard numbers are the same.

Example 3. Let X be a K3 surface admitting a jacobian elliptic fibration with a torsion section. Then translation by the section defines an automorphism of the underlying surface. The quotient surface X' is naturally endowed with an elliptic fibration such that the quotient map of the K3 surfaces corresponds to an isogeny of the generic fibers as elliptic curves over the function field of $\mathbb{P}^1$. Here, of course, there is a dual isogeny $X' \dashrightarrow X$. (This does not hold in general for symplectic involutions of K3 surfaces.)

11 Singular K3 Surfaces

The arithmetic of K3 surfaces is conceivably best understood for big Picard number. In the workshop this could be witnessed in several talk, for instance by Bertin, Clingher, Elkies, Kumar and Whitcher. To begin with, we shall concentrate on the case of maximal Picard number over $\mathbb{C}$ in view of Lefschetz' bound in (5):

Definition 3. A complex K3 surface X is called *singular* if $\rho(X) = 20$.

Note that singular K3 surfaces are smooth by definition; the phrase "singular" is used in the sense of "exceptional". In fact, there is an analogy with elliptic curves with complex multiplication (CM) which will become clear in Sect. 12 (see also Remark 2). In the sequel, the Fermat quartic will serve as our guiding example:

Example 4 (Fermat quartic). The Fermat quartic surface

$$S = \{x_0^4 + x_1^4 + x_2^4 + x_3^4 = 0\} \subset \mathbb{P}^3$$

has $\rho(S) = 20$; thus it defines a singular K3 surface. There are several ways to see this. Intrinsically one could appeal to the general theory of Fermat varieties.

These come with a big automorphism group whose eigenspaces in cohomology can be described in combinatorial terms in such a way that one can read off which are algebraic [2]. Alternatively one could hand-pick the 48 lines on S, such as

$$x_0 + \sqrt[4]{-1}x_1 = x_2 + \sqrt[4]{-1}x_3 = 0,$$

and verify that their Gram matrix attains the maximum rank of 20.

We can construct a number of further singular K3 surfaces by applying symplectic automorphisms to S and quotienting as in Sect. 10. Presently, the alternating group A_4 acts symplectically by coordinate permutations, and we can also combine scalings by fourth roots of unity for symplectic automorphisms.

11.1 Torelli Theorem for Singular K3 Surfaces

The Torelli theorem states that K3 surfaces are essentially determined by the Hodge structure underlying the transcendental lattice

$$T(X) = \mathrm{NS}(X)^\perp \subset H^2(X, \mathbb{Z}). \tag{13}$$

More precisely, given two K3 surfaces X, X', any effective Hodge isometry of

$$H^2(X, \mathbb{Z}) \cong H^2(X', \mathbb{Z})$$

is induced from a unique isomorphism $X \cong X'$ (cf. [4, VIII.11]). For singular K3 surfaces this can be made very explicit as follows. In this situation $T(X)$ is a positive definite even lattice of rank 2 which comes with an orientation induced from the regular 2-form. We can thus identify $T(X)$ with a quadratic form

$$Q(X) = \begin{pmatrix} 2a & b \\ b & 2c \end{pmatrix} \tag{14}$$

with integer entries $a, c > 0$ and discriminant $d = b^2 - 4ac > 0$. The Torelli theorem can now be formulated as follows:

Theorem 4. *Two singular K3 surfaces X, X' are isomorphic if and only if there is an isometry $T(X) \cong T(X')$. Equivalently the quadratic forms $Q(X), Q(X')$ are conjugate under $SL(2, \mathbb{Z})$.*

Example 5 (Fermat quartic cont'd). The Fermat quartic S has transcendental lattice represented by the quadratic form

$$Q(S) = \begin{pmatrix} 8 & 0 \\ 0 & 8 \end{pmatrix}.$$

Proving this is a non-trivial task, and the first proper proof seems to go back to Mizukami [32]. (The proof in [40] relied on a claim by Demjanenko which was only later justified by Cassels in [7].) Generally this question can be reduced to the problem whether the lines generate NS(S) fully or only up to finite index (which can be solved using specialisation again, cf. [47]).

11.2 Surjectivity of the Period Map

In the moduli context it remains to discuss the surjectivity of the period map. With the formulation of the Torelli theorem at hand, this amounts to the question whether all positive definite quadratic forms as in (14) are attained by singular K3 surfaces. Here it is crucial to note that Kummer surfaces will not be sufficient since the quotient has the quadratic forms of the abelian surface multiplied by 2:

$$T(\text{Km}(A)) = T(A)[2], \quad \text{i.e.} \quad Q(\text{Km}(A)) = 2Q(A). \tag{15}$$

Here transcendental lattice and quadratic form of the abelian surface A are defined in complete analogy with (13), (14), and the above relation is valid for complex abelian surfaces of any Picard number. In essence, we find that Kummer surfaces have 2-divisible transcendental lattices; this prevents most quadratic forms as in (14) to be associated to singular K3 surfaces which are Kummer.

Despite this failure of Kummer surfaces to lead directly to the surjectivity of the period map for singular K3 surfaces, they still prove extremely useful. The first complete argument goes back to Shioda and Inose [53] who proceed in the following two steps:

1. Refer to work of Shioda and Mitani [54] where the corresponding Torelli theorem including the surjectivity of the period map is proven for singular abelian surfaces.
2. Prove that any singular K3 surface admits a rational map of degree 2 to a Kummer surface such that the transcendental lattices differ by the same factor of 2 as in (15).

Following Morrison [33], the construction forming the second step is nowadays often called a *Shioda–Inose structure*. We will review it in detail in the next section. Meanwhile we conclude this section with a brief discussion of the first step.

11.3 Singular Abelian Surfaces

In analogy with singular K3 surfaces, a complex abelian surface A is called singular if its Picard number attains the maximum $\rho(A) = 4$. In [54] Shioda and Mitani worked out the Torelli theorem including the surjectivity of the period map. We summarize their result in the following theorem:

Theorem 5 (Shioda–Mitani). *Isomorphism classes of singular abelian surfaces are in bijective correspondence with conjugacy classes of binary even positive definite quadratic forms.*

The heart of the proof is an explicit construction of singular abelian surfaces with given transcendental lattice. In terms of the quadratic form Q as in (14), one exhibits two elliptic curves as complex tori

$$E = \mathbb{C}/(\mathbb{Z} + \tau\mathbb{Z}), \quad \tau = \frac{-b + \sqrt{d}}{2a}, \qquad E' = \mathbb{C}/(\mathbb{Z} + \tau'\mathbb{Z}), \quad \tau' = \frac{b + \sqrt{d}}{2}. \quad (16)$$

Then Shioda and Mitani compute that the product $E \times E'$ has transcendental lattice exactly represented by Q.

Remark 2. The terminology "singular abelian/K3 surface" is indeed appropriate in the following sense: the elliptic curves E, E' in (16) have complex multiplication (CM) (in fact, they are isogenous); classically their j-invariants are called "singular".

12 Shioda–Inose Structures

Given a quadratic form Q as in (14), 11.3 exhibits elliptic curves E, E' such that $T(\mathrm{Km}(E \times E'))$ is represented by $2Q$. In order to recover the original quadratic form on a singular K3 surface, Shioda and Inose developed a geometric construction that applies generally to Kummer surfaces of product type. An instrumental ingredient consists in the so-called double Kummer pencil formed by 24 rational curves on $\mathrm{Km}(E \times E')$. In terms of the elliptic fibrations over $\mathbb{P}^1$ induced by the projections onto either factor, the curves constitute the four singular fibers (type I_0^*, 5 components each) and the four 2-torsion sections. We sketch the curves in the following figure where the fibrations could be thought of both in the horizontal or vertical direction (Fig. 1).

The key property for the considerations of [53] is that Kummer surfaces of product type admit several distinct elliptic fibrations. In the generic situation these were later classified by Oguiso [39]; Shioda and Inose exploit this feature by exhibiting a specific fibration on $\mathrm{Km}(E \times E')$ by singling out a divisor D of Kodaira type II^* in the double Kummer pencil. Recall from Sect. 4 that the divisor D (printed in blue in Fig. 2) induces indeed an elliptic fibration

$$\mathrm{Km}(E \times E') \to \mathbb{P}^1.$$

Moreover the rational curve C in green meets D transversally at exactly one point, so C will serve as a section of the new fibration.

There are six rational curves in the double Kummer pencil (printed in red in Fig. 2) which do not meet the divisor D of Kodaira type II^*. Hence they are components of other fibers. In fact, since both perpendicular 3-chains meet the section C,

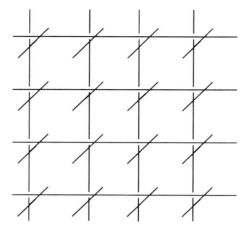

Fig. 1: Double Kummer pencil

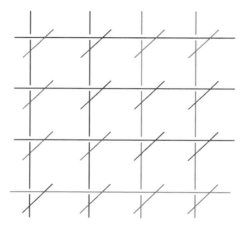

Fig. 2: Divisor of Kodaira type II^* in the double Kummer pencil

they are located in different fibers. A detailed analysis shows that the fiber types can only be I_0^*, I_1^* or IV^*. In fact, unless $E \cong E'$, both singular fibers have type I_0^*.

Shioda and Inose proceed by a quadratic base change that ramifies exactly at the above two singular fibers. Pull-back from $\text{Km}(E \times E')$ gives another jacobian elliptic surface X with the II^* fiber duplicated while the two fibers in the branch locus are generically replaced by smooth fibers. The Euler number $e(X) = 24$ reveals that X is again K3; in fact, since X dominates $\text{Km}(E \times E')$ by construction, we have $\rho(X) = \rho(\text{Km}(E \times E'))$ by Proposition 2. One concludes by verifying by an explicit calculation that

$$T(X) \cong T(\text{Km}(E \times E'))[1/2].$$

Together with (15), this gives

$$T(X) \cong T(A),$$

and the surjectivity of the period map for singular K3 surfaces follows from
Theorem 5.   □

To conclude this section, let us summarize the above construction: a K3 surface
X with a rational map of degree 2 to a Kummer surface which induces multiplication
by 2 on the transcendental lattices:

$$T(A) \cong T(X)$$

In generality, Morrison coined the terminology *Shioda–Inose structure* for such
a diagram. In [33] he proved that any K3 surface with $\rho \geq 19$ admits a Shioda–
Inose structure, and he worked out explicit criteria for K3 surfaces with $\rho = 17, 18$.
Shioda–Inose structures are a versatile tool for the study of K3 surfaces; they turned
up during the workshop especially in the context of classification and moduli problems (see also [46] and the references therein).

Before continuing our investigation of singular K3 surfaces, eventually aiming
for zeta functions and modularity, we take a little detour towards Mordell–Weil
ranks of elliptic K3 surfaces.

13 Mordell–Weil Ranks of Elliptic K3 Surfaces

By the Shioda–Tate formula (4), a jacobian elliptic K3 surface can have Mordell–
Weil rank at most 18 over $\mathbb{C}$. Recall from Sect. 4 how elliptic fibrations on K3 surfaces are governed by lattice theory. In order to see which MW-ranks are attained,
one can therefore appeal to the moduli theory of lattice polarised K3 surfaces, combined with the theory of Mordell–Weil lattices [50]. The solution was first given by
Cox [9]:

Theorem 6 (Cox). *Any integer between 0 and 18 occurs as Mordell–Weil rank of a
complex elliptic K3 surface.*

In the spirit of Sect. 5, it is still a delicate problem to exhibit an elliptic K3 surface with given MW-rank. Kuwata gave an almost complete answer in [25] based
on the Shioda–Inose structure related to $E \times E'$. Here we will briefly review his
construction.

Recall from the previous section that the Kummer surface $\mathrm{Km}(E \times E')$ is covered
by another K3 surface X which is equipped with an induced jacobian elliptic

fibration with two fibers of Kodaira type II^*. Much like what we did to the original elliptic fibration on $\mathrm{Km}(E \times E')$ one can try to apply base changes to X which remain K3. Here's the first example:

Example 6. A quadratic base change ramified at the two special fibers gives another K3 surface $X^{(2)}$ (by Euler number considerations). The induced elliptic fibration has two fibers of type IV^* instead of II^* (see the small table below). It is a non-trivial fact that $X^{(2)}$ indeed returns the Kummer surface $\mathrm{Km}(E \times E')$ (which thus sandwiches X, see [51]). The elliptic fibration can be understood in terms of (3) as projection onto $\mathbb{P}^1_y$ (i.e. a quadratic base change of a cubic pencil—the pencil is the corresponding quadratic twist of X).

In general, a fiber of type II^* (outside characteristics $2, 3$) pulls back as follows under a cyclic base change of degree n:

$n \mod 6$	0	1	2	3	4	5
Fiber type	I_0	II^*	IV^*	I_0^*	IV	II

Hence a simple Euler number computation allows us to determine all base changes of X which remain K3. This requires that the base change is cyclic of degree $n \leq 6$ and exactly ramifies at the two special fibers of type II^* (as in Example 6). By construction, the resulting K3 surface $X^{(n)}$ automatically comes with a rational map of degree n to X; hence by Proposition 2

$$\rho(X^{(n)}) = \rho(X) = \rho(\mathrm{Km}(E \times E')). \tag{17}$$

In consequence, the MW-rank of $X^{(n)}$ can be computed depending only on E and E'. Essentially this relies on E and E' being isogenous or not:

$$\rho(\mathrm{Km}(E \times E')) = 18 + \mathrm{rank}(\mathrm{Hom}(E, E')) = \begin{cases} 18, & \text{if } E \nsim E'; \\ 19, & \text{if } E \sim E' \text{ without CM}; \\ 20, & \text{if } E \sim E' \text{ with CM}. \end{cases}$$

In order to apply the Shioda–Tate formula (4), one further needs that the fibrations have additional reducible fibers if and only if E and E' are isomorphic. With the exception of the cases $j(E), j(E') \in \{0, 1728\}$, the results are summarised in Table 1.

The only MW-rank missing from Table 1 is 15. This gap was subsequently filled by Kloosterman. In [23], he exhibited an explicit elliptic K3 surface with MW-rank 15 much along the lines of Sect. 7. In brief he worked out a three-dimensional family of elliptic K3 surfaces with generic MW-rank 15 (again using base change from an appropriate elliptic K3 surface). Then the specialisation technique from Sects. 7.1, 7.2 enabled him to single out a general member of the family.

Remark 3. Extending (17), Shioda proved in [52] that $T(X^{(n)}) = T(X)[n]$.

One can enrich the arithmetic flavour of the MW-rank problem by specifying the ground field in consideration. Naturally one might wonder which MW-ranks are attained by elliptic K3 surfaces over $\mathbb{Q}$ (i.e. as elliptic curves over $\mathbb{Q}(t)$). The answer is due to Elkies (see [10]):

Table 1: Fiber configuration and Mordell–Weil rank of $X^{(n)}$

n	$E \cong E'$ Config	MW-rank	$E \ncong E'$ Config	$E \sim E'$ MW-rank	$E \nsim E'$ MW-rank
1	$2\,II^*, I_2, 2\,I_1$	$\begin{cases}1\\0\end{cases}$	$2\,II^*, 4\,I_1$	$\begin{cases}2\\1\end{cases}$	0
2	$2\,IV^*, 2\,I_2, 4\,I_1$	$\begin{cases}4\\3\end{cases}$	$2\,IV^*, 8\,I_1$	$\begin{cases}6\\5\end{cases}$	4
3	$2\,I_0^*, 3\,I_2, 6\,I_1$	$\begin{cases}7\\6\end{cases}$	$2\,I_0^*, 12\,I_1$	$\begin{cases}10\\9\end{cases}$	8
4	$2\,IV, 4\,I_2, 8\,I_1$	$\begin{cases}10\\9\end{cases}$	$2\,IV, 16\,I_1$	$\begin{cases}14\\13\end{cases}$	12
5	$2\,II, 5\,I_2, 10\,I_1$	$\begin{cases}13\\12\end{cases}$	$2\,II, 20\,I_1$	$\begin{cases}18\\17\end{cases}$	16
6	$6\,I_2, 12\,I_1$	$\begin{cases}12\\11\end{cases}$	$24\,I_1$	$\begin{cases}18\\17\end{cases}$	16

Theorem 7 (Elkies). *The maximal MW-rank of an elliptic K3 surface over $\mathbb{Q}$ is 17.*

On the existence side of Theorem 7, Elkies exhibits an explicit K3 surface of MW-rank 17 over $\mathbb{Q}$. The construction uses families of K3 surfaces and ingenious specialisation and lifting arguments (somewhat comparable to the concepts in Sect. 15). As an application, Elkies is able to specialise to elliptic curves over $\mathbb{Q}$ of even higher rank, pushing the previous record ranks as far as 28.

In order to explain the non-constructive part of Elkies' proof, we have to discuss fields of definition of singular K3 surfaces first.

14 Fields of Definition of Singular K3 Surfaces

Since singular K3 surfaces lie isolated in the moduli space, they are always defined over some number field. In this section, we will be more specific about the number field and also comment on known obstructions.

So far, we have not given explicit equations for singular K3 surfaces except for the Fermat quartic (Example 4). To remedy this, consider a singular K3 surface X with elliptic fibration coming from the Shioda–Inose structure. By Tate's algorithm [60], the two fibers of type II^* determine the Weierstrass equation of X almost completely. The remaining coefficients depend on Weber functions in terms of the j-invariants of the elliptic curves, as determined by Inose [21]:

$$X : \quad y^2 = x^3 - 3At^4 x + t^5(t^2 - 2Bt + 1) \tag{18}$$

where $A^3 = j(E)\,j(E')/12^6$, $B^2 = (1 - j(E)/12^3)\,(1 - j(E')/12^3)$. An easy twist reveals that the above fibration in fact admits a model over $\mathbb{Q}(j(E), j(E'))$ (see [43,

Proposition 4.1]). This field is related to class field theory as follows. Let d denote the discriminant of X, i.e. $d = -\det Q(X)$ where $Q(X)$ is the quadratic form representing $T(X)$. If $K = \mathbb{Q}(\sqrt{d})$, then class group theory attaches to d a ring class field $H(d)$ as maximal abelian extension of K with prescribed ramification. The Galois group of the extension $H(d)/K$ is canonically isomorphic to the class group $Cl(d)$, consisting of primitive quadratic forms of discriminant d as in (14) with Gauss composition. CM theory of elliptic curves states that

$$H(d) = K(j(E')) = K(j(E), j(E')).$$

In summary we find

Proposition 3. *Any singular K3 surface of discriminant d admits a model over the ring class field $H(d)$.*

As satisfying as the above canonical field of definition may be, in practice it is often far from optimal. As an illustration we return to the Fermat quartic once again.

Example 7. In Example 5 we have given the transcendental lattice of the Fermat quartic S. In the realm of Shioda–Inose structures, S thus arises from the elliptic curves with periods $\tau = \sqrt{-1}, \tau' = 4\sqrt{-1}$. Note the discrepancy that the latter CM curve is only defined over $\mathbb{Q}(\sqrt{2})$ (and so is Inose's model (18)) while the Fermat quartic has the obvious model over $\mathbb{Q}$. In fact, considering the Fermat quartic as a Kummer surface instead, it is associated to the elliptic curves with periods $\tau = \sqrt{-1}, \tau' = 2\sqrt{-1}$ which are indeed both defined over $\mathbb{Q}$.

Example 7 leads to the problem of working out obstructions for singular K3 surfaces to descend from the ring class field $H(d)$ to smaller fields, in particular to $\mathbb{Q}$. Essentially there are two obstructions known, one coming from the transcendental lattice (see [43]), the other imposed by the Néron–Severi lattice. Here we shall only consider the latter obstruction which was discovered independently by Elkies and the author (cf. [45]). In short, it states that the ring class field $H(d)$ is essentially preserved through the Galois action on the Néron–Severi lattice:

Theorem 8 (Elkies, Schütt). *Let X be a singular K3 surface of discriminant d. If $NS(X)$ has generators defined over some number field L, then*

$$L(\sqrt{d}) \supset H(d).$$

The theorem paves the way towards finiteness classifications. For instance, one readily deduces that there are only finitely many singular K3 surfaces over $\mathbb{Q}$ (up to $\bar{\mathbb{Q}}$-isomorphism) since the Néron–Severi lattice cannot admit Galois actions by arbitrarily large groups. To see this, note that the rank of the lattice is trivially constant and the hyperplane class is always preserved by Galois; this leaves a faithful Galois action on the negative-definite complement.

14.1 Mordell–Weil Ranks Over $\mathbb{Q}$

Returning to the problem of Mordell–Weil ranks over $\mathbb{Q}$, we note that rank 18 would imply that all of NS would be defined over $\mathbb{Q}$. By Theorem 8, the discriminant d could only have class number one, i.e.

$$d = -3, -4, -7, -8, -11, -12, -16, -19, -27, -28, -43, -67, -163.$$

Elkies continues to argue with the Mordell–Weil lattice M of the given elliptic fibration. By [50], this is an even positive-definite lattice of rank 18 and discriminant d without roots (i.e. $x^2 > 2$ for all $x \in M$). Note that with discriminant $|d| \le 163$ such a lattice would break the known density records for sphere packings, but this can only be regarded as evidence against Mordell–Weil rank 18 over $\mathbb{Q}$.

In order to rule out the existence of such a M, one appeals to general results of lattice theory developed by Nikulin [36]. Geared towards elliptic K3 surfaces, they imply that M admits a primitive embedding into some Niemeier lattice (i.e. one of the 24 unimodular positive-definite even lattices of rank 24, cf. [34]). Its orthogonal complement L (positive-definite of rank 6) is called the partner lattice of M. Often the partner lattice L can be determined a priori from NS or from the transcendental lattice $T(X)$. In fact, all jacobian elliptic fibrations on a given complex K3 surface X can be classified in terms of the primitive embeddings of L into Niemeier lattices. Full classifications have been established, for instance, by Nishiyama [37] for the singular K3 surfaces with discriminant -3 and -4. Here there are 6 resp. 13 jacobian fibrations, but the Mordell–Weil rank does never exceed 1.

Returning to the problem of Mordell–Weil rank 18, ruling this out for a single K3 surface amounts to proving that the partner lattice L does not admit a primitive embedding into any Niemeier lattice without any perpendicular roots. For the 13 discriminants of class number one, this is exactly what Elkies succeeds in doing, thus ruling out Mordell–Weil rank 18 over $\mathbb{Q}$. In fact, even Mordell–Weil rank 17 is only barely attained over $\mathbb{Q}$ by an elliptic K3 surface of discriminant $12 \cdot 79$ (see [10]).

15 Modularity of Singular K3 Surfaces

The modularity of elliptic curves over $\mathbb{Q}$ has been one of the biggest achievements in mathematics in the last 20 years (implying in particular Fermat's Last Theorem). Starting from an elliptic curve E over $\mathbb{Q}$, one considers the reductions E_p for all good primes. Counting points on E_p over $\mathbb{F}_p$, one defines the quantity

$$a_p = 1 + p - \#E_p(\mathbb{F}_p). \tag{19}$$

The Taniyama–Shimura–Weil conjecture states that the collection of integers $\{a_p\}$ comprise the Fourier coefficients of a single modular form (the eigenvalues of a

Hecke eigenform of weight 2). This is now a theorem, thanks to the work of Wiles [63] and others. Historically the first case to be settled comprised elliptic curves with CM. Generally over any number field, their zeta function was shown by Deuring to depend on certain Hecke characters ψ. For elliptic curves over $\mathbb{Q}$ with CM, this has an incarnation in terms of modular forms with CM.

For singular K3 surfaces, the Shioda–Inose structure allows one to express the zeta function in terms of the Hecke character ψ^2, but a priori only over some extension of $H(d)$ required for exhibiting the construction [53, Theorem 6]. Especially in view of the possible descent from $H(d)$ to some smaller fields as explored in Sect. 14, it is thus an independent task to prove modularity for singular K3 surfaces over $\mathbb{Q}$ (long predicted by standard conjectures for two-dimensional Galois representations). Here (19) is rephrased by virtue of Lefschetz' fixed point formula. For a singular K3 surface X over $\mathbb{Q}$ and its good reductions X_p, we define

$$\#X_p(\mathbb{F}_p) = 1 + b_p + hp + p^2 \qquad (20)$$

where the integer h encodes the action of Frob_p^* on $\mathrm{NS}(X_p \otimes \bar{\mathbb{F}}_p)$. In this context, the integers b_p ought to belong to a Hecke eigenform of weight 3. This was solved by Livné in a more general framework of orthogonal Galois representations [28]:

Theorem 9 (Livné). *Every singular K3 surface over $\mathbb{Q}$ is modular. Its zeta function is expressed in terms of a Hecke eigenform of weight 3 with CM by $\mathbb{Q}(d)$ where d denotes the discriminant of the K3 surface.*

Example 8. The zeta function of the Fermat quartic (the model S over $\mathbb{Q}$ from Example 4) can be expressed in terms of the eta product $\eta(4\tau)^6$ (the weight 3 eigenform of level 16 from [44, Table 1]). With Legendre symbols $\chi_{\bullet}$ accounting for the Galois action on the lines defined over $\mathbb{Q}(\sqrt{-1}, \sqrt{2})$, one finds

$$\zeta(S, s) = \zeta(s)\zeta(s-1)^5\zeta(\chi_{-1}, s-1)^3\zeta(\chi_2, s-1)^6\zeta(\chi_{-2}, s-1)^6 L(\eta(4\tau)^6, s)\zeta(s-2).$$

Spelling this out for $\#S_p(\mathbb{F}_p)$ as in (20), we find $h = 5 + 3\chi_{-1}(p) + 6(\chi_2(p) + \chi_{-2}(p))$ and Fourier coefficients b_p determined by the infinite product

$$\eta(4\tau)^6 = q \prod_{n\geq 1}(1 - q^{4n})^6 = \sum_{n\geq 1} b_n q^n.$$

We emphasize how the overall picture differs from the case of elliptic curves over $\mathbb{Q}$. There Shimura had shown in the 1950s how to associate an elliptic curve over $\mathbb{Q}$ as a factor of the Jacobian $J_0(N)$ to a Hecke eigenform of weight 2 and level N with eigenvalues in $\mathbb{Z}$, but the modularity remained open for more than 30 years. For singular K3 surfaces, quite on the contrary, modularity was proved first, but only a few years ago it became clear that they also sufficed in the opposite direction:

Theorem 10 (Elkies-Schütt [12]). *Every known Hecke eigenform of weight 3 with eigenvalues in $\mathbb{Z}$ is associated to a singular K3 surface over $\mathbb{Q}$.*

In view of Theorem 9 it is crucial to note that the CM property is by no means special, but rather forced in odd weight by real Hecke eigenvalues due to a result of Ribet [42]. Then the key step towards Theorem 10 consists in a classification of CM form with eigenvalues in $\mathbb{Z}$ which narrows the problem down to an essentially finite problem [44]. In practice it therefore suffices to exhibit a singular K3 surface over $\mathbb{Q}$ for any imaginary quadratic field whose class group has exponent 1 or 2. There are 65 such fields known, with at most one further field possible by [62]. Thus the restriction of Theorem 10 which, for instance, becomes vacuous if one is willing to assume the extended Riemann hypothesis for odd real Dirichlet characters.

We conclude this paper with a few words towards to construction of these singular K3 surfaces over $\mathbb{Q}$ in [12]. After exhausting the examples occurring in the literature, we started considering one-dimensional families of K3 surfaces with $\rho \geq 19$. These have dense specialisations with $\rho = 20$ over $\bar{\mathbb{Q}}$, parametrised by some modular or Shimura curve, but the determination of the CM points is a non-trivial task. For instance, one can deform the Fermat quartic to the so-called Dwork pencil

$$S_\lambda = \{x_0^4 + x_1^4 + x_2^4 + x_3^4 = \lambda x_0 x_1 x_2 x_3\} \subset \mathbb{P}^3. \tag{21}$$

This retains a big part of the automorphism group which partly explains why $\rho(S_\lambda) \geq 19$. We will briefly come back to this family below; for a detailed account, the reader is referred to the upcoming paper [13].

A common theme among the constructions is the use of jacobian elliptic fibration (again!), preferably with large contribution from the singular fibers (such that ideally the MW-rank would be zero generically). This has both advantages on the constructive side of exhibiting the families and for the decisive step of provably determining CM points (see below). Note that by Theorem 8, a singular K3 surface of big class number over $\mathbb{Q}$ has to admit a fairly big Galois action on NS, so the families necessarily become pretty complicated (despite the low MW-rank).

Finally we comment on the possible approaches to determine the CM points of a given family such as the Dwork pencil (21). One possibility to proceed would be to determine the moduli structure of the parametrising curve and compute the CM points explicitly. For instance, the parameter of the Dwork pencil can be interpreted as fourfold cover of the modular curve $X^*(2)$ parametrising pairs of 2-isogenous elliptic curves (where the CM-points can be calculated easily). Geometrically this can be achieved through a Shioda–Inose structure over $\mathbb{Q}$. However, with the families getting more and more complicated, the determination of the moduli curve soon becomes infeasible (let alone the calculation of CM points), and in fact these one-dimensional families of K3 surfaces have proved a versatile tool to work out explicit models of Shimura curves (see [11]). Here is how we proceeded instead in [12].

Experimentally one can count points over finite fields throughout the family and apply Lefschetz' fixed point formula (20). Fixing a target modular form, one can sieve those parameters modulo several primes p which would fit the Fourier coefficients b_p of the modular form. Once a collection of residue parameters is computed, one can try to lift them to a single parameter of small height in $\mathbb{Q}$. This procedure is surprisingly effective (though not for the Dwork pencil in the above form, because it admits an order 4 symmetry in the parameter λ). It remains to prove the CM points that one might have computed experimentally.

Proving the CM points amounts to exhibiting an extra divisor on the special member of the family. Here one takes advantage of the structure as jacobian elliptic surface, together with an extra bit of information extracted from Theorem 9. Namely the precise modular form that we are aiming at predicts the square class of the discriminant of the K3 surface X that we want to prove to have $\rho = 20$. Presently this usually requires the presence of an additional section. Thanks to the height pairing from the theory of Mordell–Weil lattices [50], information about the conjectural discriminant can be translated into details which fiber components have to be met by the section. If the height of the section is small enough, this gives enough information about the shape of the section to solve for it either directly or with p-adic methods (further simplified by the fact that we have already a candidate for the parameter of the family).

In the end, it turns out that the two obstructions—the families being rich enough and the height of the section being small enough to allow for explicit calculations—balance out just right to compute singular K3 surfaces for all but a handful of the known Hecke eigenforms (which require special treatment).

We end the paper by remarking that the analogous problem in higher dimension, realising all Hecke eigenforms of fixed weight with eigenvalues in $\mathbb{Z}$ by a single class of varieties such as Calabi–Yau manifolds, seems wide open even in dimension 3 where some modularity results are available (see Yui's introductory lectures).

Acknowledgements It is a great pleasure to thank the other organisers of the Fields workshop and particularly all the participants for creating such a stimulating atmosphere. Special thanks to the Fields Institute for the great hospitality and to the referee for his comments. These lecture notes were written down while the author enjoyed support from the ERC under StG 279723 (SURFARI) which is gratefully acknowledged.

References

1. T.G. Abbott, K. Kedlaya, D. Roe, Bounding Picard numbers of surfaces using p-adic cohomology, in *Arithmetic, Geometry and Coding Theory* (AGCT 2005). Séminaires et Congrès, vol. 21 (Societé Mathématique de France, Paris, 2009), pp. 125–159
2. N. Aoki, T. Shioda, Generators of the Néron–Severi group of a Fermat surface, in *Arithmetic and Geometry*, vol. I. Progress in Mathematics, vol. 35 (Birkhäuser, Boston, 1983), pp. 1–12
3. M. Artin, P. Swinnerton-Dyer, The Shafarevich–Tate conjecture for pencils of elliptic curves on K3 surfaces. Invent. Math. **20**, 249–266 (1973)
4. W. Barth, K. Hulek, C. Peters, A. van de Ven, *Compact Complex Surfaces*, 2nd edn. Erg. der Math. und ihrer Grenzgebiete, 3. Folge, Band 4 (Springer, Berlin, 2004)
5. F.A. Bogomolov, Y. Tschinkel, Density of rational points on elliptic K3 surfaces. Asian J. Math. **4**(2), 351–368 (2000)
6. F.A. Bogomolov, B. Hassett, Y. Tschinkel, Constructing rational curves on K3 surfaces. Duke Math. J. **157**, 535–550 (2011)

7. J.W.S. Cassels, A Diophantine equation over a function field. J. Aust. Math. Soc. Ser. A **25**(4), 489–496 (1978)
8. F. Charles, On the Picard number of K3 surfaces over number fields. Algebra Number Theor. Preprint (2011), arXiv: 1111.4117
9. D.A. Cox, Mordell Weil groups of elliptic curves over $C(t)$ with $p_g = 0$ or 1. Duke Math. J. **49**(3), 677–689 (1982)
10. N.D. Elkies, Three lectures on elliptic surfaces and curves of high rank. 840 Preprint (2007), arXiv: 0709.2908 841 211 844
11. N.D. Elkies, *Shimura Curve Computations Via K3 Surfaces of NS-Rank at Least 19*. ANTS VIII, 2008. Lecture Notes in Computer Science, vol. 5011 (Springer, Berlin, 2008), pp. 196–211
12. N.D. Elkies, M. Schütt, Modular forms and K3 surfaces. Preprint (2008), arXiv: 0809.0830
13. N.D. Elkies, M. Schütt, K3 families of high Picard rank, in preparation
14. J. Ellenberg, K3 surfaces over number fields with geometric Picard number one, in *Arithmetic of Higher-dimensional Algebraic Varieties*, Palo Alto, CA, 2002. Progress in Mathematics, vol. 226 (Birkhäuser, Boston, 2004), pp. 135–140
15. A.-S. Elsenhans, J. Jahnel, The Picard group of a K3 surface and its reduction modulo p. Algebra Number Theor. **5**, 1027–1040 (2011)
16. G. Faltings, Endlichkeitssätze für abelsche Varietäten über Zahlkörpern. Invent. Math. **73**(3), 349–366 (1983)
17. B. van Geemen, Some remarks on Brauer groups on K3 surfaces. Adv. Math. **197**, 222–247 (2005)
18. B. Hassett, A. Vàrilly-Alvarado, Failure of the Hasse principle on general K3 surfaces. J. Inst. Math. Jussieu. Preprint (2011), arXiv: 1110.1738
19. B. Hassett, A. Vàrilly-Alvarado, P. Varilly, Transcendental obstructions to weak approximation on general K3 surfaces. Adv. Math. **228**, 1377–1404 (2011)
20. K. Hulek, M. Schütt, Arithmetic of singular Enriques surfaces. Algebra Number Theor. **6**(2), 195–230. Preprint (2012), arXiv: 1002.1598
21. H. Inose, Defining equations of singular $K3$ surfaces and a notion of isogeny, in *Proceedings of the International Symposium on Algebraic Geometry*, Kyoto University, Kyoto, 1977 (Kinokuniya Book Store, Tokyo, 1978), pp. 495–502
22. I. Karzhemanov, One construction of a K3 surface with the dense set of rational points. 865 Preprint (2011), arXiv: 1102.1873
23. R. Kloosterman, Elliptic K3 surfaces with geometric Mordell–Weil rank 15. Can. Math. Bull. **50**(2), 215–226 (2007)
24. K. Kodaira, On compact analytic surfaces I–III. Ann. Math. **71**, 111–152 (1960); Ann. Math. **77**, 563–626 (1963); Ann. Math. **78**, 1–40 (1963)
25. M. Kuwata, Elliptic $K3$ surfaces with given Mordell–Weil rank. Comment. Math. Univ. St. Pauli **49**(1), 91–100 (2000)
26. J. Li, C. Liedtke, Rational curves on K3 surfaces. Invent. Math. **188**(3), 713–727 (2011)
27. M. Lieblich, D. Maulik, A. Snowden, Finiteness of K3 surfaces and the Tate conjecture. 874 Preprint (2011), arXiv: 1107.1221
28. R. Livné, Motivic orthogonal two-dimensional representations of $\mathrm{Gal}(\bar{\mathbb{Q}}/\mathbb{Q})$. Isr. J. Math. **92**, 149–156 (1995)
29. R. van Luijk, K3 surfaces with Picard number one and infinitely many rational points. Algebra Number Theor. **1**(1), 1–15 (2007)
30. Y.I. Manin, Le groupe de Brauer-Grothendieck en géométrie diophantienne, in *Actes du Congrès International des Mathématiciens*, Nice, 1970. Tome 1 (Gauthier-Villars, Paris, 1971), pp. 401–411
31. J. Milne, On a conjecture of Artin and Tate. Ann. Math. **102**, 517–533 (1975)
32. M. Mizukami, Birational mappings from quartic surfaces to Kummer surfaces (in Japanese), Master's Thesis, University of Tokyo, 1975
33. D.R. Morrison, On K3 surfaces with large Picard number. Invent. Math. **75**(1), 105–121 (1984)

34. H.-V. Niemeier, Definite quadratische Formen der Dimension 24 und Diskriminante 1. J. Number Theor. **5**, 142–178 (1973)
35. V.V. Nikulin, Finite groups of automorphisms of Kählerian K3 surfaces. Trudy Moskov. Mat. Obshch. **38**, 75–137 (1979)
36. V.V. Nikulin, Integral symmetric bilinear forms and some of their applications. Math. USSR Izv. **14**(1), 103–167 (1980)
37. K.-I. Nishiyama, The Jacobian fibrations on some K3 surfaces and their Mordell–Weil groups. Jpn. J. Math. **22**, 293–347 (1996)
38. N. Nygaard, A. Ogus, Tate's conjecture for K3 surfaces of finite height. Ann. Math. **122**, 461–507 (1985)
39. K. Oguiso, On Jacobian fibrations on the Kummer surfaces of the product of nonisogenous elliptic curves. J. Math. Soc. Jpn. **41**(4), 651–680 (1989)
40. I.I. Piatetski-Shapiro, I.R. Shafarevich, Torelli's theorem for algebraic surfaces of type K3. Izv. Akad. Nauk SSSR Ser. Mat. **35**, 530–572 (1971)
41. M. Raynaud, "p-torsion" du schéma de Picard. Astérisque **64**, 87–148 (1979)
42. K. Ribet, Galois representations attached to eigenforms with Nebentypus, in *Modular Functions of One Variable V*, ed. by J.-P. Serre, D.B. Zagier, Bonn, 1976. Lecture Notes in Mathematics, vol. 601 (Springer, Berlin, 1977), pp. 17–52
43. M. Schütt, Fields of definition of singular K3 surfaces. Comm. Number Theor. Phys. **1**(2), 307–321 (2007)
44. M. Schütt, CM newforms with rational coefficients. Ramanujan J. **19**, 187–205 (2009)
45. M. Schütt, K3 surfaces of Picard rank 20 over $\mathbb{Q}$. Algebra Number Theor. **4**(3), 335–356 (2010)
46. M. Schütt, Sandwich theorems for Shioda–Inose structures. Izvestiya Mat. **77**, 211–222 (2013)
47. M. Schütt, T. Shioda, R. van Luijk, Lines on the Fermat quintic surface. J. Number Theor. **130**, 1939–1963 (2010)
48. E. Selmer, The Diophantine equation $ax^3 + by^3 + cz^3 = 0$. Acta Math. **85**, 203–362 (1951)
49. T. Shioda, On the Picard number of a complex projective variety. Ann. Sci. École Norm. Sup.(4) **14**(3), 303–321 (1981)
50. T. Shioda, On the Mordell–Weil lattices. Comment. Math. Univ. St. Pauli **39**, 211–240 (1990)
51. T. Shioda, Kummer sandwich theorem of certain elliptic K3 surfaces. Proc. Jpn. Acad. Ser. A **82**, 137–140 (2006)
52. T. Shioda, K3 surfaces and sphere packings. J. Math. Soc. Jpn. **60**, 1083–1105 (2008)
53. T. Shioda, H. Inose, in *On Singular K3 Surfaces*, ed. by W.L. Baily Jr., T. Shioda. Complex Analysis and Algebraic Geometry (Iwanami Shoten, Tokyo, 1977), pp. 119–136
54. T. Shioda, N. Mitani, Singular abelian surfaces and binary quadratic forms, in *Classification of Algebraic Varieties and Compact Complex Manifolds*. Lecture Notes in Mathematics, vol. 412 (Springer, Berlin, 1974), pp. 259–287
55. J.H. Silverman, Rational points on K3 surfaces: a new canonical height. Invent. Math. **105**, 347–373 (1991)
56. A. Skorobogatov, in *Torsors and Rational Points*. Cambridge Tracts in Mathematics, vol. 144 (Cambridge University Press, Cambridge, 2001)
57. H. Sterk, Finiteness results for algebraic $K3$ surfaces. Math. Z. **189**(4), 507–513 (1985)
58. J. Tate, Algebraic cycles and poles of zeta functions, in *Arithmetical Algebraic Geometry*. Proceedings of Conference of Purdue University, 1963 (Harper & Row, New York, 1965), pp. 93–110
59. J. Tate, in *On the Conjectures of Birch and Swinnerton-Dyer and a Geometric Analog*, ed. by A. Grothendieck, N.H. Kuiper. Dix exposés sur la cohomologie des schemas (North-Holland, Amsterdam, 1968), pp. 189–214
60. J. Tate, Algorithm for determining the type of a singular fibre in an elliptic pencil, in *Modular Functions of One Variable IV*, Antwerpen, 1972. SLN, vol. 476 (Springer, Berlin, 1975), pp. 33–52

61. T. Terasoma, Complete intersections with middle Picard number 1 defined over $\mathbb{Q}$. Math. Z. **189**(2), 289–296 (1985)
62. P.J. Weinberger, Exponents of the class groups of complex quadratic fields. Acta Arith. **22**, 117–124 (1973)
63. A. Wiles, Modular elliptic curves and Fermat's Last Theorem. Ann. Math. (2) **141**(3), 443–551 (1995)

Modularity of Calabi–Yau Varieties: 2011 and Beyond

Noriko Yui

Abstract This paper presents the current status on modularity of Calabi–Yau varieties since the last update in 2003. We will focus on Calabi–Yau varieties of dimension at most three. Here modularity refers to at least two different types: arithmetic modularity and geometric modularity. These will include: (1) the modularity (automorphy) of Galois representations of Calabi–Yau varieties (or motives) defined over $\mathbb{Q}$ or number fields, (2) the modularity of solutions of Picard–Fuchs differential equations of families of Calabi–Yau varieties, and mirror maps (mirror moonshine), (3) the modularity of generating functions of invariants counting certain quantities on Calabi–Yau varieties, and (4) the modularity of moduli for families of Calabi–Yau varieties. The topic (4) is commonly known as geometric modularity.

Discussions in this paper are centered around arithmetic modularity, namely on (1), and (2), with a brief excursion to (3).

Key words: Elliptic curves, K3 surfaces, Calabi–Yau threefolds, CM type Calabi–Yau varieties, Galois representations, Modular (cusp) forms, Automorphic inductions, Geometry and arithmetic of moduli spaces, Hilbert and Siegel modular forms, Families of Calabi–Yau varieties, Mirror symmetry, Mirror maps, Picard–Fuchs differential equations

Mathematics Subject Classifications (2010): 11G42, 11F80, 11G40, 14J15, 14J32, 14J33

N. Yui (✉)
Department of Mathematics and Statistics, Queen's University,
Kingston, ON, Canada K7L 3N6
e-mail: yui@mast.queensu.ca

R. Laza et al. (eds.), *Arithmetic and Geometry of K3 Surfaces and Calabi–Yau Threefolds*,   101
Fields Institute Communications 67, DOI 10.1007/978-1-4614-6403-7_4,
© Springer Science+Business Media New York 2013

1 Introduction

These are notes of my introductory lectures at the Fields workshop on "Arithmetic and Geometry of K3 surfaces and Calabi–Yau threefolds", at the Fields Institute from August 16–25, 2011. My goal is twofold:

- Present an update on the recent developments on the various kinds of modularity associated to Calabi–Yau varieties. Here "recent" developments means various developments since my article [65] published in 2003.
- Formulate conjectures and identify future problems on modularity and related topics.

My hope is to motivate young (as well as mature) researchers to work in this fascinating area at the interface of arithmetic, geometry and physics around Calabi–Yau varieties.

1.1 Brief History Since 2003

The results and discoveries in the last 10 years on Calabi–Yau varieties, which will be touched upon in my lectures, are listed below.

- The modularity of the two-dimensional Galois representations associated to Calabi–Yau varieties defined over $\mathbb{Q}$.
- The modularity of highly reducible Galois representations associated to Calabi–Yau threefolds over $\mathbb{Q}$.
- The automorphy of higher dimensional Galois representations arising from CM type Calabi–Yau varieties (automorphic induction).
- Appearance of various types of modular forms in Mirror Symmetry as generating functions counting some mathematical/physical quantities.
- Modularity of families of Calabi–Yau varieties (solutions of Picard–Fuchs differential equations, monodromy groups, mirror maps).
- Moduli spaces of Calabi–Yau families, and higher dimensional modular forms (e.g., Siegel modular forms).

1.2 Plan of Lectures

Obviously, due to time constraints, I will not be able to cover all these topics in my two introductory lectures. Thus, for my lectures, I plan to focus on recent results on the first three (somewhat intertwined) items listed above, and with possibly very brief interludes to the rest if time permits.

This article includes my two lectures delivered at the workshop, as well as some subjects/topics which I was not able to cover in my lectures. However, I must emphasize that this note will not touch upon geometric modularity.

1.3 Disclaimer

Here I will make the disclaimer that the topics listed above are by no means exhaustive; it may be the case that I forget to mention some results, or not give proper attributions. I apologize for these oversights.

1.4 Calabi–Yau Varieties: Definition

Definition 1. Let X be a smooth projective variety of dimension d defined over $\mathbb{C}$. We say that X is a *Calabi–Yau* variety if

- $H^i(X, \mathcal{O}_X) = 0$ for every i, $0 < i < d$.
- The canonical bundle K_X is trivial.

We introduce the *Hodge numbers* of X:

$$h^{i,j}(X) := \dim_{\mathbb{C}} H^j(X, \Omega_X^i), \ 0 \le i, j \le d.$$

Then we may characterize a Calabi–Yau variety of dimension d in terms of its Hodge numbers.

A smooth projective variety X of dimension d over $\mathbb{C}$ is called a *Calabi–Yau* variety if

- $h^{i,0}(X) = 0$ for every i, $0 < i < d$.
- $K_X \simeq \mathcal{O}_X$, so that the geometric genus of X,

$$p_g(X) := h^{0,d}(X) = \dim_{\mathbb{C}} H^0(X, K_X) = \dim_{\mathbb{C}} H^0(X, \mathcal{O}_X) = 1.$$

Numerical characters of Calabi–Yau varieties of dimension d

- Betti numbers: For i, $0 \le i \le 2d$, the i-th Betti number of X is defined by

$$B_i(X) := \dim_{\mathbb{C}} H^i(X, \mathbb{C}).$$

There is Poincaré duality for $H^i(X, \mathbb{C})$; that is,

$$H^i(X, \mathbb{C}) \times H^{2d-i}(X, \mathbb{C}) \to \mathbb{C} \quad \text{for every } i, \ 0 \le i \le d$$

is a perfect pairing. This implies that

$$B_i(X) = B_{2d-i}(X), \text{for } i, \ 0 \le i \le d.$$

- Hodge numbers: They are defined in Definition 1 above. There is the symmetry of Hodge numbers: For $0 \leq i, j \leq d$,

$$h^{i,j}(X) = h^{j,i}(X) \quad \text{by complex conjugation,}$$

and

$$h^{i,j}(X) = h^{d-i,d-j}(X) \quad \text{by Serre duality.}$$

- There is a relation among Betti numbers and Hodge numbers, as a consequence of the Hodge decomposition:

$$B_k(X) = \sum_{i+j=k} h^{i,j}(X).$$

- The Euler characteristic of X is defined by

$$E(X) := \sum_{k=0}^{2d} (-1)^k B_k(X).$$

Example 1. (a) Let $d = 1$. The first condition is vacuous. The second condition says that $p_g(X) = 1$. So dimension 1 Calabi–Yau varieties are elliptic curves. The Hodge diamond of elliptic curves is rather simple.

$$
\begin{array}{ccc}
 & 1 & \\
1 & & 1 \\
 & 1 &
\end{array}
\qquad
\begin{array}{l}
B_0(X) = 1 \\
B_1(X) = 2 \\
B_2(X) = 1
\end{array}
$$

The Euler characteristic of X is given by

$$E(X) = B_0(X) - B_1(X) + B_2(X) = 0.$$

(b) Let $d = 2$. The first condition is $h^{1,0}(X) = 0$ and the second condition says that $h^{0,2}(X) = p_g(X) = 1$. So dimension 2 Calabi–Yau varieties are K3 surfaces. The Hodge diamond of K3 surfaces is of the form:

$$
\begin{array}{ccccc}
 & & 1 & & \\
 & 0 & & 0 & \\
1 & & 20 & & 1 \\
 & 0 & & 0 & \\
 & & 1 & &
\end{array}
\qquad
\begin{array}{l}
B_0(X) = 1 \\
B_1(X) = 0 \\
B_2(X) = 20 \\
B_3(X) = 0 \\
B_4(X) = 1
\end{array}
$$

The Euler characteristic of X is given by

$$E(X) = \sum_{k=0}^{4} (-1)^k B_k(X) = 1 + 22 + 1 = 24.$$

(c) Let $d = 3$. The first condition says that $h^{1,0}(X) = h^{2,0}(X) = 0$ and the second condition implies that $h^{0,3}(X) = p_g(X) = 1$. The Hodge diamond of X is given as follows:

$$
\begin{array}{ccccc}
 & & 1 & & \\
 & 0 & & 0 & \\
 0 & & h^{1,1}(X) & & 0 \\
 1 & h^{2,1}(X) & & h^{1,2}(X) & 1 \\
 0 & & h^{2,2}(X) & & 0 \\
 & 0 & & 0 & \\
 & & 1 & &
\end{array}
\qquad
\begin{array}{l}
B_0(X) = 1 \\
B_1(X) = 0 \\
B_2(X) = h^{1,1}(X) \\
B_3(X) = 2(1 + h^{2,1}(X)) \\
B_4(X) = h^{2,2}(X) = h^{1,1}(X) \\
B_5(X) = 0 \\
B_6(X) = 1
\end{array}
$$

The Euler characteristic is given by

$$
E(X) = \sum_{k=0}^{6} (-1)^k B_k(X) = 2(h^{1,1}(X) - h^{2,1}(X)).
$$

It is not known if there exist absolute constants that bound $h^{1,1}(X)$, $h^{2,1}(X)$, and hence $|E(X)|$. The currently known bound for $|E(X)|$ is 960.

(d) Here are some typical examples of families of Calabi–Yau varieties defined by hypersurfaces.

d	CY variety of dim d	CY varieties of dim d
1	$X_0^3 + X_1^3 + X_2^3 + 3\lambda X_0 X_1 X_2$	$y^2 = f_3(x)$
2	$X_0^4 + X_1^4 + X_2^4 + X_3^4 + 4\lambda X_0 X_1 X_2 X_3$	$z^2 = f_6(x, y)$
3	$X_0^5 + X_1^5 + X_2^5 + X_3^5 + X_4^5 + 5\lambda X_0 X_1 X_2 X_3 X_4$	$w^2 = f_8(x, y, z)$

The equations in the second column are generic polynomials in projective coordinates; while those in the third column are in affine coordinates and the f_i are smooth polynomials of degree i in affine coordinates.

We have a vast source of examples of Calabi–Yau threefolds via toric construction and other methods. The upper record for the absolute value of the Euler characteristic of all these Calabi–Yau threefolds is 960, though there is neither reason nor explanation for this phenomenon.

2 The Modularity of Galois Representations of Calabi–Yau Varieties (or Motives) Over $\mathbb{Q}$

We now consider smooth projective varieties defined over $\mathbb{Q}$ (say, by hypersurfaces, or complete intersections). We say that $X/\mathbb{Q}$ is a Calabi–Yau variety of dimension d over $\mathbb{Q}$, if $X \otimes_{\mathbb{Q}} \mathbb{C}$ is a Calabi–Yau variety of dimension d. A Calabi–Yau variety X over $\mathbb{Q}$ has a model defined over $\mathbb{Z}[\frac{1}{m}]$ (with some $m \in \mathbb{N}$), and this allows us to consider its reduction modulo primes. Pick a prime p such that $(p, m) = 1$, and define the reduction of X modulo p, denoted by $X \bmod p$. We say that p is a *good* prime if $X \bmod p$ is smooth over $\bar{\mathbb{F}}_p$, otherwise p is *bad*. For a good prime p, let Fr_p denote the Frobenius morphism on X induced from the p-th power map $x \mapsto x^p$.

Let ℓ be a prime different from p. Then, for each i, $0 \leq i \leq 2d$, Fr_p induces an endomorphism Fr_p^* on the i-th ℓ-adic étale cohomology group $H_{et}^i(\bar{X}_p, \mathbb{Q}_\ell)$, where $\bar{X}_p := X \otimes_{\mathbb{F}_p} \bar{\mathbb{F}}_p$. Grothendieck's specialization theorem gives an isomorphism $H_{et}^i(\bar{X}_p, \mathbb{Q}_\ell) \cong H_{et}^i(\bar{X}, \mathbb{Q}_\ell)$, where $\bar{X} := X \otimes_{\mathbb{Q}} \bar{\mathbb{Q}}$. Then the comparison theorem gives $H_{et}^i(\bar{X}, \mathbb{Q}_\ell) \otimes_{\mathbb{Q}_\ell} \mathbb{C} \cong H^i(X \otimes_{\mathbb{Q}} \mathbb{C}, \mathbb{C})$ so that $\dim_{\mathbb{Q}_\ell} H_{et}^i(\bar{X}, \mathbb{Q}_\ell) = B_i(X)$. There is Poincaré duality for $H_{et}^i(\bar{X}, \mathbb{Q}_\ell)$, that is,

$$H_{et}^i(\bar{X}, \mathbb{Q}_\ell) \times H_{et}^{2d-i}(\bar{X}, \mathbb{Q}_\ell) \to \mathbb{Q}_\ell \quad \text{for every } i,\ 0 \leq i \leq 2d$$

is a perfect pairing. Let

$$P_p^i(T) := \det(1 - \mathrm{Fr}_p^* \, T \mid H_{et}^i(\bar{X}, \mathbb{Q}_\ell))$$

be the reciprocal characteristic polynomial of Fr_p^*. (Here T is an indeterminate.) Then the Weil Conjecture (Theorem) asserts that

- $P_p^i(T) \in 1 + T\mathbb{Z}[T]$. Moreover, $P_p^i(T)$ does not depend on the choice of ℓ.
- $P_p^i(T)$ has degree $B_i(X)$.
- $P_p^{2d-i}(T) = \pm P_p^i(p^{d-\frac{i}{2}}T)$ for every i, $0 \leq i \leq d$.
- If we write

$$P_p^i(T) = \prod_{j=1}^{B_i}(1 - \alpha_{ij} T) \in \bar{\mathbb{Q}}[T]$$

  then α_{ij} are algebraic integers with $|\alpha_{ij}| = p^{i/2}$ for every i, $0 \leq i \leq 2d$.

Now we will bring in the absolute Galois group $G_{\mathbb{Q}} := \mathrm{Gal}(\bar{\mathbb{Q}}/\mathbb{Q})$. There is a continuous system of ℓ-adic Galois representations

$$\rho_{X,\ell}^i : G_{\mathbb{Q}} \to GL(H_{et}^i(\bar{X}, \mathbb{Q}_\ell))$$

sending the (geometric) Frobenius Fr_p^{-1} to $\rho^i(\mathrm{Fr}_p^{-1})$. The (geometric) Frobenius $\rho^i(\mathrm{Fr}_p^{-1})$ has the same action as the Frobenius morphism Fr_p^* on the étale cohomology $H_{et}^i(\bar{X}, \mathbb{Q}_\ell)$. We define its L-series $L(\rho_{X,\ell}^i, s) := L(H_{et}^i(\bar{X}, \mathbb{Q}_\ell), s)$ for each i, $0 \leq i \leq 2d$, where s is a complex variable.

We will now define the L-series of X.

Definition 2. The i-th (cohomological) L-series of X is defined by the Euler product

$$L_i(X, s) := L(H_{et}^i(\bar{X}, \mathbb{Q}_\ell), s)$$

$$:= (*) \prod_{p:p \neq \ell} P_p^i(p^{-s})^{-1} \times (\text{factor corresponding to } \ell = p)$$

where the product is taken over all good primes different from ℓ and $(*)$ corresponds to factors of bad primes. For $\ell = p$, we may choose another good prime $\ell \neq p$, or we can use some p-adic cohomology (e.g., crystalline cohomology) to define the factor.

For $i = d$, we write simply $L(X, s)$ for $L_d(X, s)$ if there is no danger of ambiguity.

Remark 1. We may define (for a good prime p) the zeta-function $\zeta(X_p, T)$, of a Calabi–Yau variety X_p defined over $\mathbb{F}_p$ by counting the number of rational points on all extensions of $\mathbb{F}_p$:

$$\zeta(X_p, T) := \exp\left(\sum_{n=1}^{\infty} \frac{\#X_p(\mathbb{F}_{p^n})}{n} T^n\right) \in \mathbb{Q}(T).$$

Then by Weil's conjecture, it has the form:

$$\zeta(X_p, T) = \frac{P_1(T)P_3(T)\cdots P_{2d-1}(T)}{P_0(T)P_2(T)\cdots P_{2d}(T)},$$

where we put (to ease the notation) $P_i(T) = P_p^i(T)$ for $i = 1, \ldots, 2d$.

Let X be a Calabi–Yau variety of dimension d defined over $\mathbb{Q}$. For a good prime p, Fr_p acts on $X \mod p$, and it will induce a morphism Fr_p^* on $H_{et}^i(\bar{X}_p, \mathbb{Q}_\ell) \simeq H_{et}^i(\bar{X}, \mathbb{Q}_\ell)$. Define the trace $t_i(p)$ by

$$t_i(p) := \mathrm{Trace}(\mathrm{Fr}_p^* \mid H_{et}^i(\bar{X}, \mathbb{Q}_\ell)), \quad \text{for } i,\, 0 \le i \le 2d.$$

Then $t_i(p) \in \mathbb{Z}$ for every i, $0 \le i \le 2d$. The Lefschetz fixed point formula gives a relation between the number of $\mathbb{F}_p$-rational points on X and traces:

$$\#X(\mathbb{F}_p) = \sum_{i=0}^{2d}(-1)^i t_i(p).$$

Example 2. (a) Let $d = 1$ and let E be an elliptic curve defined over $\mathbb{Q}$. For a good prime p,

$$\#E(\mathbb{F}_p) = \sum_{i=0}^{2}(-1)^i t_i(p) = t_0(p) - t_1(p) + t_2(p) = 1 + p - t_1(p),$$

where

$$|t_1(p)| \le 2p^{1/2}.$$

Then we have

$$P_p^0(T) = 1 - T,\ P_p^2(T) = 1 - pT,\ P_p^1(T) = 1 - t_1(p)T + pT^2.$$

The L-series of E is then given by

$$L(E, s) = (*) \prod_{p:\text{good}} P_p^1(p^{-s})^{-1}$$

$$= (*) \prod_{p:\text{good}} \frac{1}{1 - t_1(p)p^{-s} + p^{1-2s}}.$$

Expanding out, we may write

$$L(E, s) = \sum_{n=1}^{\infty} \frac{a(n)}{n^s} \quad \text{with } a_1 = 1 \text{ and } a(n) \in \mathbb{Z}.$$

So

$$a(p) = t_1(p) \quad \text{for every good prime } p.$$

(b) Let $d = 2$, and let X be a K3 surface defined over $\mathbb{Q}$. Let $NS(X)$ denote the Néron–Severi group of X generated by algebraic cycles. It is a free finitely generated abelian group, and

$$NS(X) = H^2(X, \mathbb{Z}) \cap H^{1,1}(X)$$

so that the rank of $NS(X)$ (called the Picard number of X), denoted by $\rho(X)$, is bounded above by 20. Let $T(X)$ be the orthogonal complement of $NS(X)$ in $H^2(X, \mathbb{Z})$ with respect to the intersection pairing. We call $T(X)$ the group of transcendental cycles of X. We have the decomposition

$$H^2(X, \mathbb{Z}) \otimes \mathbb{Q}_\ell = (NS(X) \otimes \mathbb{Q}_\ell) \oplus (T(X) \otimes \mathbb{Q}_\ell).$$

and this will enable us decompose the L-series of X as follows:

$$L(X, s) = L(H^2_{et}(\bar{X}, \mathbb{Q}_\ell), s) = L(NS(X) \otimes \mathbb{Q}_\ell, s) \times L(T(X) \otimes \mathbb{Q}_\ell, s).$$

We know that for a good prime p,

$$P_p^0(T) = 1 - T, P_p^4(T) = 1 - p^2 T, P_p^1(T) = P_p^3(T) = 1$$

and if

$$P_p^2(T) = \prod_{j=1}^{22} (1 - \alpha_j T) \in \bar{\mathbb{Q}}[T]$$

then

$$|\alpha_j| = p.$$

The L-series of $NS(X)$ is more or less understood by Tate's conjecture. The validity of the Tate conjecture for K3 surfaces in characteristic zero has been established (see Tate [57]). In fact, if we know that all the algebraic cycles generating $NS(X)$ are defined over some finite extension $\mathbb{K}$ of $\mathbb{Q}$ of degree r, then $\rho^2(\text{Fr}_{p^r})$ acts on $NS(X) \otimes \mathbb{Q}_\ell$ by multiplication by p^r so that the L-series may be expressed as

$$L(NS(X_\mathbb{K}) \otimes \mathbb{Q}_\ell, s) = \zeta_\mathbb{K}(s - 1)^{\rho(X)}$$

where $\zeta_\mathbb{K}(s)$ denotes the Dedekind zeta-function of $\mathbb{K}$.

Therefore, the remaining task is to determine the L-series $L(T(X) \otimes \mathbb{Q}_\ell, s)$ arising from the transcendental cycles $T(X)$, and we call this the motivic L-series of X.

(c) Let $d = 3$, and let X be a Calabi–Yau threefold defined over $\mathbb{Q}$. For a good prime p,

$$\#X(\mathbb{F}_p) = \sum_{i=0}^{6} (-1)^i t_i(p) = t_0(p) - t_1(p) + t_2(p) - t_3(p) + t_4(p) - t_5(p) + t_6(p)$$

$$= 1 + p^3 + (1 + p)t_2(p) - t_3(p).$$

Therefore,

$$t_3(p) = 1 + p^3 + (1 + p)t_2(p) - \#X(\mathbb{F}_p).$$

We have

$$t_0(p) = 1, \ t_6(p) = p^3, \ t_1(p) = t_5(p) = 0,$$

$$|t_2(p)| \le p h^{1,1}(X), \ t_4(p) = p t_2(p),$$

and

$$|t_3(p)| \le B_3 \, p^{3/2}.$$

Hence

$$P_p^0(T) = 1 - T, \ P_p^6(T) = 1 - p^3 T, \ P_p^1(T) = P_p^5(T) = 1,$$

$$P_p^4(T) = P_p^2(pT) \quad \text{and} \quad P_p^3(T) \in \mathbb{Z}[T] \quad \text{with degree } B_3(X).$$

Then the L-series of X is given by

$$L(X, s) = L(H_{et}^3(\bar{X}, \mathbb{Q}_\ell), s) = (*) \prod_{p \ good} P_p^3(p^{-s})^{-1}$$

where $(*)$ is the factor corresponding to bad primes.

We will now make a definition of what it means for a Calabi–Yau variety X defined over $\mathbb{Q}$ to be modular (or automorphic). This is a concrete realization of the conjectures known as the Langlands Philosophy.

In the appendix, we will briefly recall the definitions of various types of modular forms.

Now we will recall the Fontaine–Mazur conjectures, or rather some variant concentrated on Calabi–Yau varieties over $\mathbb{Q}$.

Definition 3. Let X be a Calabi–Yau variety of dimension $d \le 3$ defined over $\mathbb{Q}$. Let $L(X, s)$ be its L-series. We say that X is modular if there is a set of modular forms (or automorphic forms) such that $L(X, s)$ coincides with the L-series associated to modular forms (automorphic forms), up to a finite number of Euler factors.

Remark 2. In fact, when the Galois representation arising from X is reducible, then we will consider its irreducible factors and match their L-series with the L-series of modular forms. This is the so-called motivic modularity.

Now we define Calabi–Yau varieties X of *CM-type*. This involves a polarized rational Hodge structure h on the primitive cohomology $H^d_{\mathrm{prim}}(X, \mathbb{Q})$, where $d = \dim(X)$.

Let $\mathbb{S} := R_{\mathbb{C}/\mathbb{R}}\mathbb{G}_m$ be the real algebraic group obtained from $\mathbb{G}_m$ by restriction of scalars from $\mathbb{C}$ to $\mathbb{R}$. For $d = 1$, $H^1_{\mathrm{prim}}(X, \mathbb{Q}) = H^1(X, \mathbb{Q})$, and polarized rational Hodge structures on X are simple. For $d = 2$, The Hodge group of a polarized rational Hodge structure $h : \mathbb{S} \to GL(H^d_{\mathrm{prim}}(X, \mathbb{Q}) \otimes \mathbb{R})$ is the smallest algebraic group of $GL(H^2_{\mathrm{prim}}(X, \mathbb{Q}) \otimes \mathbb{R}$ defined over $\mathbb{Q}$ such that the real points $\mathrm{Hdg}(\mathbb{R})$ contain $h(U^1)$ where $U^1 := \{ z \in \mathbb{C}^* \mid z\bar{z} = 1 \}$. For details about Hodge groups, see Deligne [10], and Zarhin [66]. For $d = 3$, the Hodge structure on $H^3_{\mathrm{prim}}(X, \mathbb{Q}) = H^3(X, \mathbb{Q})$ is not simple for all Calabi–Yau threefolds X. (For instance, for the quintic threefold in the Dwork pencil, the Hodge structure would split into four-dimensional Hodge substructures, using the action of the $(\mathbb{Z}/5\mathbb{Z})^3$.)

Definition 4. A Calabi–Yau variety X of dimension $d \leq 3$ is said to be of CM *type* if the Hodge group $\mathrm{Hdg}(X)$ associated to a rational Hodge structure of weight k of $H^d_{\mathrm{prim}}(X, \mathbb{Q})$ is commutative. That is, the Hodge group $\mathrm{Hdg}(X)_{\mathbb{C}}$ is isomorphic to a copy of $\mathbb{G}_m \simeq \mathbb{C}^*$.

Hodge groups are very hard to compute in practice. Here are algebraic characterizations of CM type Calabi–Yau varieties of dimension d.

Proposition 1. • $d = 1$. *An elliptic curve E over $\mathbb{Q}$ is of CM type if and only if* $\mathrm{End}(E) \otimes \mathbb{Q}$ *is an imaginary quadratic field over $\mathbb{Q}$.*

• $d = 2$. *A K3 surface X over $\mathbb{Q}$ is of CM type if* $\mathrm{End}_{\mathrm{Hdg}}(T(X)) \otimes \mathbb{Q}$ *is a CM field over $\mathbb{Q}$ of degree equal to* rank $T(X)$.

• $d = 3$. *Let X be a Calabi–Yau threefold over $\mathbb{Q}$. A Calabi–Yau threefold X over $\mathbb{Q}$ is of CM type if and only if* $\mathrm{End}_{\mathrm{Hdg}}(X) \otimes \mathbb{Q}$ *is a CM field over $\mathbb{Q}$ of degree* $2(1 + h^{2,1}(X))$, *if and only if the Weil and Griffiths intermediate Jacobians of X are of CM type.*

For $d = 1$, this is a classical result. For $d = 2$, a best reference might be Zarhin [66], and for $d = 3$, see Borcea [3].

Later we will construct Calabi–Yau varieties of dimension 2 and 3 which are of CM type.

3 Results on Modularity of Galois Representations

3.1 Two-Dimensional Galois Representations Arising from Calabi–Yau Varieties Over $\mathbb{Q}$

We will focus on two-dimensional Galois representations arising from Calabi–Yau varieties over $\mathbb{Q}$.

First, for dimension 1 Calabi–Yau varieties over $\mathbb{Q}$, we have the celebrated theorem of Wiles et al.

Theorem 1. *(d* = 1*) Every elliptic curve E defined over $\mathbb{Q}$ is modular. More concretely, let E be an elliptic curve over $\mathbb{Q}$ with conductor N. Then there exists a Hecke eigen newform f of weight* 2 = 1 + d *on the congruence subgroup $\Gamma_0(N)$ such that*

$$L(E, s) = L(f, s).$$

That is, if we write $f(q) = \sum_{m=1}^{\infty} a_f(m)q^m$ with $q = e^{2\pi i z}$ and normalized by $a_f(1) = 1$, then $a(n) = a_f(n)$ for every n.

For dimension 2 Calabi–Yau varieties, namely, K3 surfaces, over $\mathbb{Q}$, there are naturally associated Galois representations. In particular, for a special class of K3 surfaces, the associated Galois representations are two-dimensional. Let X be a K3 surface defined over $\mathbb{Q}$ with Picard number $\rho(X_{\bar{\mathbb{Q}}}) = 20$. Such K3 surfaces are called *singular* K3 surfaces. Then the group (or lattice) $T(X)$ of transcendental cycles on X is of rank 2, and it gives rise to a two-dimensional Galois representations. Livné [40] has established the motivic modularity of X, that is, the modularity of $T(X)$.

Theorem 2. *(d* = 2*) Let X be a singular K3 surface defined over $\mathbb{Q}$. Then T(X) is modular, that is, there is a modular form f of weight* 3 = 1 + d *on some $\Gamma_1(N)$ or $\Gamma_0(N)$ with a character ε such that*

$$L(T(X) \otimes \mathbb{Q}_\ell, s) = L(f, s).$$

Remark 3. A representation theoretic formulation of the above theorem is given as follows. Let π be the compatible family of two-dimensional ℓ-adic Galois representations associated to $T(X)$ and let $L(\pi, s)$ be its L-series. Then there exists a unique, up to isomorphism, modular form of weight 3, level=conductor of π, and Dirichlet character $\varepsilon(p) = \left(\frac{-d}{p}\right)$ such that

$$L(\pi, s) = L(f, s).$$

Here $d = |\text{disc } NS(X)|$.

For dimension 3 Calabi–Yau varieties over $\mathbb{Q}$, we will focus on rigid Calabi–Yau threefolds. A Calabi–Yau threefold X is said to be *rigid* if $h^{2,1}(X) = 0$ so that $B_3(X) = 2$. This gives rise to a two-dimensional Galois representation.

Theorem 3. *(d* = 3*) Every rigid Calabi–Yau threefold X over $\mathbb{Q}$ is modular. That is, there exists a cusp form f of weight* 4 = 1 + d *on some $\Gamma_0(N)$ such that*

$$L(X, s) = L(f, s).$$

This theorem has been established by Gouvêa and Yui [23], and independently by Dieulefait [11]. Their proof relies heavily on the recent results on the modularity of Serre's conjectures about two-dimensional residual Galois representations by Khare–Wintenberger [29] and Kisin [30].

A list of rigid Calabi–Yau threefolds over $\mathbb{Q}$ can be found in the monograph of Meyer [44].

3.2 Modularity of Higher Dimensional Galois Representations Arising from K3 Surfaces Over $\mathbb{Q}$

We will first consider K3 surfaces X with $T(X)$ of rank ≥ 3.

The first result is for K3 surfaces with transcendental rank 3.

Theorem 4. *Let X be a K3 surface over $\mathbb{Q}$ with Picard number 19. Then X has a Shioda–Inose structure, that is, X has an involution ι such that X/ι is birational to a Kummer surface Y over $\mathbb{C}$.*

Suppose that the Kummer surface is given by the product $E \times E$ of a non-CM elliptic curve E over $\mathbb{Q}$. Then the Shioda–Inose structure induces an isomorphism of integral Hodge structures on the transcendental lattices, so, X and $Km(E \times E)$ have the same $\mathbb{Q}$-Hodge structure. In this case, the two-dimensional Galois representation ρ_E associated to E induces the three-dimensional Galois representation $Sym^2\rho_E$ on $T(X)$ over some number field. Consequently, $T(X)$ is **potentially modular** *in the sense that the L-series of $T(X)$ is determined over some number field K by the symmetric square of a modular form g of weight 2 associated to E, and over K,*

$$L(T(X) \otimes \mathbb{Q}_\ell, s) = L(Sym^2(g), s).$$

Remark 4. In the above theorem, we are not able to obtain the modularity results over $\mathbb{Q}$. Since the K3 surface S is defined over $\mathbb{Q}$, the representation on $T(X)$ is also defined over $\mathbb{Q}$, but the isomorphism to $Sym^2\rho_E$ may not be. Thus, we only have the potential modularity of X.

Can we say anything about the modularity of K3 surfaces with arbitrary large transcendental rank? The Galois representations associated to these K3 surfaces have large dimensions. The only result along this line is due to Livné–Schütt–Yui [41].

Consider a K3 surface X with non-symplectic automorphism. Let ω_X denote a holomorphic 2-form on X, fixed once and for all. Then $H^{2,0}(X) \simeq \mathbb{C}\omega_X$. Let $\sigma \in \mathrm{Aut}(X)$. Then σ induces a map

$$\sigma^* : H^{2,0}(X) \to H^{2,0}(X) \quad \omega_X \to \alpha\omega_X, \ \alpha \in \mathbb{C}^*.$$

We say that σ^* is non-symplectic if $\alpha \neq 1$. Let

$$H_X := \mathrm{Ker}(\mathrm{Aut}(X) \to O(NS(X))).$$

Then H_X is a finite cyclic group, and in fact, can be identified with the group of roots of unity μ_k for some $k \in \mathbb{N}$. Assume that $\det(T(X)) = \pm 1$ (that is, assume that $T(X)$ is unimodular.) Then we have the following possibilities for the values of k.

- $k \leq 66$ (Nikulin [47]).
- k is a divisor of $66, 44, 42, 36, 28, 12$ (Kondo [31]).
- If $\mathrm{rank}(T(X)) = \phi(k)$ (where ϕ is the Euler function), then $k = 66, 44, 42, 36, 28, 12$. Furthermore, there is a unique K3 surface X with given k (Kondo [31]). These results were first announced by Vorontsov [62], and proofs were given later by Nikulin [47] and Kondo [31].

We tabulate these six K3 surfaces.

k	$NS(X)$	$T(X)$	rank($T(X)$)
12	$U \oplus (-E_8)^2$	U^2	4
28	$U \oplus (-E_8)$	$U^2 \oplus (-E_8)$	12
36	$U \oplus (-E_8)$	$U^2 \oplus (-E_8)$	12
42	$U \oplus (-E_8)$	$U^2 \oplus (-E_8)$	12
44	U	$U^2 \oplus (-E_8)^2$	20
66	U	$U^2 \oplus (-E_8)^2$	20

Here $(-E_8)$ denotes the negative definite even unimodular lattice of rank 8.

Now we ought to realize these K3 surfaces over $\mathbb{Q}$. Here are explicit equations thanks to Kondo [31].

k	X	σ
12	$y^2 = x^3 + t^5(t^2 + 1)$	$(x,y,t) \mapsto (\zeta_{12}^2 x, \zeta_{12}^3 y, -t)$
28	$y^2 = x^3 + x + t^7$	$(x,y,t) \mapsto (-x, \zeta_{28}^7 y, \zeta_{28}^2 t)$
36	$y^2 = x^3 - t^5(t^6 - 1)$	$(x,y,t) \mapsto (\zeta_{36}^2 x, \zeta_{36}^3{}^3 y, \zeta_{36}^{30} t)$
42	$y^2 = x^3 + t^5(t^7 - 1)$	$(x,y,t) \mapsto (\zeta_{42}^2 x, \zeta_{42}^3 y, \zeta_{42}^{18} t)$
44	$y^2 = x^3 + x + t^{11}$	$(x,y,t) \mapsto (-x, \zeta_{44}^{11} y, \zeta_{44}^2 t)$
66	$y^2 = x^3 + t(t^{11} - 1)$	$(x,y,t) \mapsto (\zeta_{66}^2 x, \zeta_{66}^3 y, \zeta_{66}^6 t)$

Here ζ_k denotes a primitive k-th root of unity.

Now the main result of Livné–Schütt–Yui [41] is to establish the modularity of these K3 surfaces.

Theorem 5. *Let X be a K3 surface in the above table. Then for each k, the ℓ-adic Galois representation associated to $T(X)$ is irreducible over $\mathbb{Q}$ of dimension $\phi(k)$. Furthermore, this $G_{\mathbb{Q}}$-Galois representation is induced from a one-dimensional Galois representation of $\mathbb{Q}(\zeta_k)$.*

All these K3 surfaces are of CM type, and are modular (automorphic).

Proof.   • CM type is established by realizing them as Fermat quotients. Since Fermat surfaces are known to be of CM type, the result follows.

• Modularity (or automorphy) of the Galois representation is established by using automorphic induction. The restriction of the Galois representation to the cyclotomic field $\mathbb{Q}(\mu_k)$ is given by a one-dimensional Jacobi sum Grössencharakter. To get down to $\mathbb{Q}$, we take the Gal($\mathbb{Q}(\zeta_k)/\mathbb{Q}$)-orbit of the one-dimensional representation. This Galois group has order $\phi(k)$, and we obtain the irreducible Galois representation over $\mathbb{Q}$ of dimension $\phi(k)$.

• Those K3 surfaces corresponding to $k = 44$ and 66 are singular, so their modularity has already been established by Theorem 2.

Remark 5. When $T(X)$ is not unimodular, there are ten values of k such that rank $T(X) = \phi(k)$:

$$19, 17, 13, 11, 7, 25, 5, 27, 9, 3.$$

All these K3 surfaces are again dominated by Fermat surfaces, and hence they are all of CM type. We have also established their modularity (automorphy).

In the article of Goto–Livné–Yui [22], more examples of K3 surfaces of CM type are constructed. First we recall a classification result of Nikulin [48].

Let X be a K3 surface over $\mathbb{Q}$. Let $H^{2,0}(X) = \mathbb{C}\omega_X$ where we fix a nowhere vanishing holomorphic 2-form ω_X. Let σ be an involution on X such that $\sigma(\omega_X) = -\omega_X$. Let $= \operatorname{Pic}(X)^\sigma$ be the fixed part of $\operatorname{Pic}(X)$ by σ. Put $r = \operatorname{rank} \operatorname{Pic}(X)^\sigma$. Let $T(X)_0$ be the orthogonal complement of $\operatorname{Pic}(X)^\sigma$ in $H^2(X, \mathbb{Z})$. Then σ acts by -1 on $T(X)_0$. Consider the quotient groups $(\operatorname{Pic}(X)^\sigma)^*/\operatorname{Pic}(X)^\sigma)$ and $(T(X)_0^*/T(X)_0)$, where L^* denotes the dual lattice of a lattice L. Since $H^2(X, \mathbb{Z})$ is unimodular, the quotient abelian groups are canonically isomorphic

$$(\operatorname{Pic}(X)^\sigma)^*/\operatorname{Pic}(X)^\sigma \simeq (T(X)_0^*/T(X)_0).$$

Since σ acts as 1 on the first quotient; and as -1 on the second quotient, this forces these quotient groups to be isomorphic to $(\mathbb{Z}/2\mathbb{Z})^a$ for some positive integer $a \in \mathbb{Z}$.

The intersection pairing on $\operatorname{Pic}(X)$ induces a quadratic form q on the discriminant group with values in $\mathbb{Q}$ modulo $2\mathbb{Z}$; we put $\delta = 0$ if q has values only in $\mathbb{Z}$, and $\delta = 1$ otherwise.

Thus, we have a triplet of integers (r, a, δ) associated to a K3 surface X with the involution σ. A theorem of Nikulin [48] asserts that a pair (X, σ) is classified, up to deformation, by a triplet (r, a, δ).

Theorem 6. *There are* 75 *possible triplets* (r, a, δ) *that classify pairs* (X, σ) *of K3 surfaces* X *with involution* σ, *up to deformation.*

Now we will realize some of these 75 families of K3 surfaces with involution. We look for K3 surfaces defined by hypersurfaces. For this we use the famous 95 families of hypersurfaces in weighted projective 3-spaces determined by M. Reid [50] or Yonemura [64]. Let $[x_0 : x_1 : x_2 : x_3]$ denote weighted projective coordinates in a weighted projective 3-space with weight (w_0, w_1, w_2, w_3).

Theorem 7 (Goto–Livné–Yui [22]). *Among the* 95 *families of K3 surfaces, all but* 9 *families of K3 surfaces have an involution* σ, *satisfying the following conditions:*

(1) removing several monomials (if necessary) from Yonemura's hypersurface, a new defining hypersurface consists of exactly four monomials, i.e., it is of Delsarte type,

(2) the new defining hypersurface is quasi-smooth, and the singularity configuration should remain the same as the original defining equation of Yonemura,

(3) the new defining hypersurface contains only one monomial in x_0 of the form x_0^n, $x_0^n x_j$, $x_i^n + x_i x_j^m$ or $x_i^n x_k + x_i x_j^m$ for some j and k $(k \neq j)$ distinct from i.

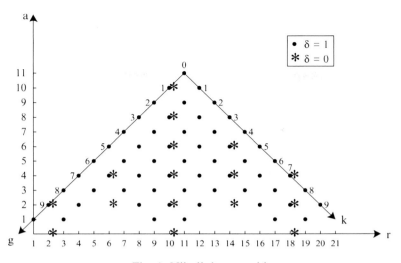

Fig. 1: Nikulin's pyramid

For 45 *(resp.* 41*) K3 surfaces, a defining equation of four monomials is of the form*

$$x_0^2 = f(x_1, x_2, x_3) \subset \mathbb{P}^3(w_0, w_1, w_2, w_3)$$

(resp.

$$F(x_0, x_1, x_2, x_3) = 0 \subset \mathbb{P}^3(w_0, w_1, w_2, w_3))$$

where f (resp. F) is a homogeneous polynomial in x_1, x_2, x_3 (resp. x_0, x_1, x_2, x_3) over $\mathbb{Q}$ of degree $\sum_{i=0}^{3} w_i$.

(4) For all $86 = 45 + 41$ K3 surfaces, there is an algorithm to compute the invariants r and a.

Remark 6. There are nine families for which the Theorem is not valid. The three families (#15, #53 and #54) do not have the required involution. Another different six families (#85, #94, #95) and (#90, #93, #91) cannot be realized as quasi-smooth hypersurfaces in four monomials. (We employ the numbering from Yonemura [64].)

Proposition 2 (Nikulin [48] and Voisin [61]). *Let (S, σ) be a K3 surface with involution σ. Let S^σ be the fixed part of S. Then for $(r, a, \delta) \neq (10, 10, 0), (10, 8, 0)$, let $S^\sigma = C_g \cup L_1 \cup \cdots \cup L_k$ where C_g is a smooth genus g curve and $L_1, \ldots, L_k$ are rational curves. Then*

$$r = 11 + k - g, \ a = 11 - g - k.$$

If $(r, a, \delta) = (10, 10, 0)$, then $S^\sigma = C_1 \cup C_2$ where C_i $(i = 1, 2)$ are elliptic curves, and if $(r, a, \delta) = (10, 8, 0)$, then $S^\sigma = \emptyset$.

Proposition 3 (Goto–Livné–Yui [22]). *Let (S, σ) be one of the 45 pairs in Theorem 7, which is given as the minimal resolution of a hypersurface $S_0 : x_0^2 = f(x_1, x_2, x_3) \subset \mathbb{P}^3(Q)$ where $Q = (w_0, w_1, w_2, w_3)$ and f is a homogeneous polynomial of degree $\sum_{i=0}^3 w_i$. Let $r(Q)$ denote the number of exceptional divisors in the resolution $S \to S_0$. Then r can be computed as follows:*

(i) *Suppose w_0 is odd. Then $r = r(Q) - w_i + 2$ if there is an odd weight $w_i \neq w_0$ such that $\gcd(w_0, w_i) = w_i \geq 2$, and $r = r(Q) + 1$ otherwise.*

(ii) *Suppose w_0 is even. Then $r = r(Q) + 1 - \sum_{i=1}^3 (d_i - 1)(\frac{2d_i}{w_i} - 1)$ where $d_i = \gcd(w_0, w_i)$.*

Remark 7. For non-Borcea type K3 surfaces defined by equations of the form $x_0^2 x_i = f(x_1, x_2, x_3)$ for some $i \in \{1, 2, 3\}$, the invariants r and a (or equivalently, g and k) are also computed.

Proposition 4 (Goto–Livné–Yui [22]). *At least the 39 triplets (r, a, δ) of integers are realized by the 86 families of K3 surfaces.*

Theorem 8 (Goto–Livné–Yui [22]). *For the 86 families of K3 surfaces, there are subfamilies of K3 surfaces which are of CM type, that is, there is a CM point in the moduli space of K3 surfaces. Indeed, at each CM point, the K3 surface is realized as a quotient of a Fermat (or Delsarte) surface. Consequently, it is modular (automorphic) by automorphic induction.*

Remark 8. Nikulin's Theorem 6 gives in total 75 triplets of integers (r, a, δ) that classify the isomorphism classes of pairs (S, σ) of K3 surfaces S with non-symplectic involution σ. With our choice of K3 surfaces of CM type, we can realize at least 39 triplets. This is based on our calculations of the invariants r and a. Since we have not yet computed the invariant δ for the 39 cases, this number may increase somewhat.

Remark 9. It is known (Borcea [4]) that over $\mathbb{C}$ the moduli spaces of Nikulin's K3 surfaces are arithmetic quotients of type IV (Shimura varieties). Recently, Ma [43] has shown the rationality of the moduli spaces for the 67 triplets. Our result above gives one explicit CM point in these moduli spaces. CM type implies that these families must be isolated points in the moduli space. We are not able to show the denseness of CM points, however.

It is notoriously difficult to compute Hodge groups, and the above theorem implies the commutativity of the Hodge group.

Corollary 1. *The Hodge groups of these 86 K3 surfaces are all commutative, i.e., copies of $\mathbb{G}_m$'s over $\mathbb{C}$.*

Proposition 5 (Goto–Livné–Yui [22]). *We classify the 86 hypersurfaces into four types:*

- *Forty-five families have the form: $x_0^2 = f(x_1, x_2, x_3)$ and $\sigma(x_0) = -x_0$.*

- *Thirty-three families cannot be put in the form (1) but defined by a hypersurface of the form $F(x_0, x_1, x_2, x_3) = 0$ with 4 monomials, and σ can be described explicitly.*
- *Four families, after changing term the x_0^3 to $x_0^2 x_1$ and then removing several terms, can be put into the form $F(x_0, x_1, x_2, x_3) = 0$ of four monimials equipped with an explicit involution σ.*
- *The remaining 4 families can be put in the form $F(x_0, x_1, x_2, x_3) = 0$ of four monomials equipped with a different kind of involution.*

Example 3. (1) Here is one of the 45 cases, #78 in Yonemura's list. The weight is $(11, 6, 4, 1)$, and Yonemura's hypersurface is

$$x_0^2 = x_1^2 x_2 + x_1^3 x_3^4 + x_1 x_2^4 + x_2^5 x_3^2 + x_3^{22}.$$

We can remove $x_1^3 x_3^4$ and $x_2^5 x_3^2$ so the new hypersurface is $x_0^2 = x_1^2 x_2 + x_1 x_2^4 + x_3^{22}$. The singularity is of type $A_1 + A_3 + A_5$.

(2) Here is one of the second cases, #19 in Yonemura's list. The weight is $(3, 2, 2, 1)$ and Yonemura's hypersurface is

$$F(x_0, x_1, x_2, x_3) = x_0^2 x_1 + x_0^2 x_2 + x_0^2 x_3^2 + x_1^4 + x_2^4 + x_3^4.$$

We can remove $x_0^2 x_1$ or $x_0^2 x_2$, $x_0^2 x_3^2$, so the new hypersurface is $x_0^2 x_1 + x_1^4 + x_2^4 + x_3^4$. The singularity is of type $4A_1 + A_2$.

(3) Here is one of the third cases, #18 in Yonemura's list. This hypersurface $x_0^3 + x_1^3 + x_0 x_2^3 + x_1 x_2^3 + x_2^4 x_3 + x_3^9$ acquires an involution if we replace x_0^3 by $x_0^2 x_1$ and remove the terms $x_0 x_2^3$ and $x_1 x_2^3$.

(4) The last case is #52 in Yonemura's list.

(5) #95 in Yonemura's list. This hypersurface has an involution, but cannot be realized by a quasi-smooth hypersurface with four monomials.

3.3 The Modularity of Higher Dimensional Galois Representations Arising from Calabi–Yau Threefolds Over $\mathbb{Q}$

There are several new examples of modular non-rigid Calabi–Yau threefolds X over $\mathbb{Q}$ with $B_3 \geq 4$. These examples were constructed after the article Yui [65], some of which have already been discussed in the article of E. Lee [36].

There are several approaches (with non-empty intersection) to produce these new modular examples:

1. Those non-rigid Calabi–Yau threefolds X over $\mathbb{Q}$ such that the semi-simplification of $H^3_{et}(\bar{X}, \mathbb{Q}_\ell)$ is highly reducible and splits into smaller dimensional irreducible Galois representations. For instance, most known cases are when the third cohomology group splits into two-dimensional or four-dimensional pieces. (Examples of E. Lee, Hulek and Verrill, Schütt, Cynk and Meyer, and a recent

example of Bini and van Geemen, and Schütt, and more.) The article of E. Lee [36] gives reviews on the modularity of non-rigid Calabi–Yau threefolds over $\mathbb{Q}$, up to 2008.

2. Those Calabi–Yau threefolds X over $\mathbb{Q}$ such that the ℓ-adic Galois representations arising from $H^3(\bar{X}, \mathbb{Q}_\ell)$ are irreducible and have small dimensions (e.g., 4, 6, or 8.) (Examples of Livné–Yui, Dieulefeit–Pacetti–Schütt, and more.)

3. Those Calabi–Yau threefolds over $\mathbb{Q}$ which are of CM type, thus the Galois representations are induced by one-dimensional representations. (Examples of Rohde, Garbagnati–van Geemen, Goto–Livné–Yui, and more.)

4. Given a Calabi–Yau threefold Y over $\mathbb{Q}$, if we can construct an algebraic correspondence defined over $\mathbb{Q}$ to some modular Calabi–Yau threefold X over $\mathbb{Q}$, the Tate conjecture asserts that their L-series should coincide. This will establish the modularity of Y.

(1) Calabi–Yau threefolds in the category (1), i.e, with highly reducible Galois representations

The general strategy is to consider Calabi–Yau threefolds which contain elliptic ruled surfaces. This is formulated by Hulek and Verrill [28].

Proposition 6 (Hulek and Verrill [28]). *Let X be a Calabi–Yau threefold over $\mathbb{Q}$. Suppose that X contains birational ruled elliptic surfaces S_j, $j = 1, \ldots, b$ over $\mathbb{Q}$ and whose cohomology classes span $H^{2,1}(X) \oplus H^{1,2}(X)$ (so $b = h^{2,1}(X)$.) Let ρ be the two-dimensional Galois representation given by the kernel U from the exact sequence*

$$0 \to U \to H^3_{et}(\bar{X}, \mathbb{Q}_\ell) \to \oplus H^3_{et}(\bar{S}_j, \mathbb{Q}_\ell) \to 0.$$

Then X is modular, that is,

$$L(X, s) = L(f_4, s) \prod_{j=1}^{b} L(g_2^j, s - 1)$$

where f_4 is a weight 4 modular form associated to ρ and g_2^j are the weight 2 modular forms associated to the base elliptic curves E_j of the birational ruled surfaces S_j.

The requirement that the third cohomology group splits as in the proposition is rather restrictive. Several examples of Calabi–Yau threefolds satisfying this condition are given by Hulek and Verrill [28], E. Lee [34], and Schütt [54]. They are constructed as resolutions of fiber products of semi-stable rational elliptic surfaces with section. Another series of examples along this line due to Hulek and Verrill [27] are toric Calabi–Yau threefolds associated to the root lattice A_4.

Cynk and Meyer [9] have established the modularity of 17 nonrigid double octic Calabi–Yau threefolds over $\mathbb{Q}$ with $B_3 = 6$. The Galois representations decompose into two- and four-dimensional sub-representations such that the L-series of each such sub-representation is of the form $L(g_4, s)$, $L(g_2, s - 1)$ or $L(g_2 \times g_3, s)$, where g_k is a weight k cusp form.

Now we will consider another construction due to E. Lee [34, 35] of Calabi–Yau threefolds associated to the Horrocks–Mumford vector bundle of rank 2. It is well known that Horrocks–Mumford quintics are determinantal quintics. The Schoen quintic $Q : \sum_{i=0}^{4} x_i^5 - 5 \prod_{i=0}^{4} x_i = 0$ is the early example of this type.

Lee has constructed more Calabi–Yau threefolds. Let $y \in \mathbb{P}^4$ be a generic point, and define the matrices

$$M_y(x) = \begin{pmatrix} x_0y_0 & x_3y_2 & x_1y_4 & x_4y_1 & x_2y_3 \\ x_3y_3 & x_1y_0 & x_4y_2 & x_2y_4 & x_0y_1 \\ x_1y_1 & x_4y_3 & x_2y_0 & x_0y_2 & x_3y_4 \\ x_4y_4 & x_2y_1 & x_0y_3 & x_3y_0 & x_1y_2 \\ x_2y_2 & x_0y_4 & x_3y_1 & x_1y_3 & x_4y_0 \end{pmatrix}$$

and

$$L_y(z) = \begin{pmatrix} z_0y_0 & z_2y_4 & z_4y_3 & z_1y_2 & z_3y_1 \\ z_4y_1 & z_1y_0 & z_3y_4 & z_0y_3 & z_2y_2 \\ z_3y_2 & z_0y_1 & z_2y_0 & z_4y_4 & z_1y_3 \\ z_2y_3 & z_4y_2 & z_1y_1 & z_3y_0 & z_0y_4 \\ z_1y_4 & z_3y_3 & z_0y_2 & z_2y_1 & z_4y_0 \end{pmatrix}$$

Note that $M_y(x)z = L_y(z)x$. Then

$$X_y := \{ \det M_y(x) = 0 \} \subset \mathbb{P}^4(x)$$

and

$$X'_y := \{ \det L_y(z) = 0 \} \subset \mathbb{P}^4(z)$$

are Horrocks–Mumford quintics. Define a threefold $\tilde{X}_y$ in $\mathbb{P}^4(x) \times \mathbb{P}^4(z)$ as a common partial resolution of $M_y(x)z = 0$. We ought to know the singularities of $\tilde{X}_y$. Let $\mathbb{P}^2_+ := \{ y : y_1 - y_4 = y_2 - y_3 = 0 \}$. Given $y \in \mathbb{P}^2_+$, the point $(1 : 0 : 0 : 0 : 0)$ is a singular point over $\mathbb{C}$ of X_y if and only if one of the coordinates of y is zero. For $y = (0 : 1 : -1 : -1 : 1), (0 : 2 : 3 \pm \sqrt{5} : 3 \pm \sqrt{5} : 2), (2 : -1 : 0 : 0 : -1), (2 : \pm\sqrt{5} - 1 : 0 : 0 : \pm\sqrt{5} - 1) \in \mathbb{P}^2_+$, X_y contains the Heisenberg orbits of $(1 : 0 : 0 : 0 : 0)$ and $(1 : 1 : 1 : 1 : 1)$ as nodes over $\mathbb{C}$.

Proposition 7 (Lee [34–36]).

(a) *Let $y = (0 : 1 : -1 : -1 : 1)$, and write $X_{(0:1:-1:-1:1)} = X$ for short. Then the Calabi–Yau threefold $\hat{X}$ obtained by crepant resolution of singularities has $B_3 = 6$. Furthermore,*

$$H^3_{et}(\bar{\hat{X}}, \mathbb{Q}_\ell) = V \oplus H^2(\bar{S}, \mathbb{Q}_\ell)(-1)$$

where V is two-dimensional and associated to the modular form of weight 4 and level 55, and $H^2(\bar{S}, \mathbb{Q}_\ell)$ is four-dimensional and is isomorphic to

$$Ind^{G_\mathbb{Q}}_{G_{\mathbb{Q}(i)}} H^1(\bar{E}, \mathbb{Q}_\ell)(-1)$$

where E is an elliptic curve over $\mathbb{Q}(i)$ coming from E over $\mathbb{Q}$.

(b) Let $y = (2 : -1 : 0 : 0 : -1)$, and let $X_{(-2:-1:0:0:-1)} = X'$ for short. Then the Calabi–Yau threefold $\hat{X}'$ obtained by crepant resolution of singularities has $B_3 = 4$. Furthermore

$$H_{et}^3(\bar{\hat{X}}', \mathbb{Q}_\ell) = V \oplus H^1(\bar{E}_2, \mathbb{Q}_\ell)(-1)$$

where V is the two-dimensional Galois representation associated to the modular form of weight 4 of level 55, and $H^1(\bar{E}_2, \mathbb{Q}_\ell)$ is associated with the modular form of weight 2 and level 550. Furthermore, the L-series of $\hat{X}'$ is given, up to Euler factors at the primes of bad reduction, by

$$L(\hat{X}', s) = L(f, s)L(g, s-1)$$

where f is the unique normalized cusp form of weight 4 and level 55 and g is the normalized cusp form of weight 2 and level 550.

(c) Now consider a smooth (big) resolution Z of the $(\mathbb{Z}/2\mathbb{Z})$-quotient of the Schoen quintic $Q : x_0^5 + x_1^5 + x_2^5 + x_3^5 + x_4^5 - 5x_0x_1x_2x_3x_4 = 0$. (There is the $(\mathbb{Z}/2\mathbb{Z})$-action on Q induced by the involution on $\mathbb{P}^4$ defined by $\iota[x_0 : x_1 : x_2 : x_3 : x_4] = [x_0 : x_4 : x_3 : x_2 : x_1]$.) Then Z is a Calabi–Yau threefold defined over $\mathbb{Q}$ with $B_3 = 4$. The Calabi–Yau threefold Z is modular, and up to Euler factors at primes of bad reduction at $p = 2$ and 5, $L(Z, s)$ is given by

$$L(Z, s) = L(f, s)L(g, s-1)$$

where f is the unique normalized cusp form of weight 4 and level 25, and g is a weight 2 cusp form of level 50.

A recent example due to Bini and van Geemen [2], and Schütt [55] is the Calabi–Yau threefold called Maschke's double octic, which arises as the double covering of $\mathbb{P}^3$ branched along Maschke's surface S. The Maschke octic surface S is defined by the homogeneous equation

$$S = \sum_{i=0}^{3} x_i^8 + 14 \sum_{i<j} x_i^4 x_j^4 + 168 x_0^2 x_1^2 x_2^2 x_3^2 = 0 \subset \mathbb{P}^3.$$

Now let X be the double cover of $\mathbb{P}^3$ along S. This is a smooth Calabi–Yau threefold defined over $\mathbb{Q}$. Let Y be the desingularization of the quotient of X by a suitable Hisenberg group. Then Y is also a smooth Calabi–Yau threefold defined over $\mathbb{Q}$. The results of Bini and van Geemen, and Schütt, are summarized in the following

Proposition 8. (a) The Maschke double octic Calabi–Yau threefolds X is modular over $\mathbb{Q}$. The third cohomology group $H_{et}^3(\bar{X}, \mathbb{Q}_\ell)$ has $B_3(X) = 300$. The Galois representation of $H_{et}^3(\bar{X}, \mathbb{Q}_\ell)$ decomposes completely over $\mathbb{Q}(i)$ into two-dimensional Galois representations which descend to $\mathbb{Q}$, and the latter correspond to modular forms of weight 4, or modular forms of weight 2.

(b) *Let Y be the desingularization of the quotient of X by a suitable Heisenberg group. Then Y is modular over $\mathbb{Q}$. The third cohomology group $H^3_{et}(\bar{Y}, \mathbb{Q}_\ell)$ has $B_3(Y) = 30$. The Galois representation of $H^3_{et}(\bar{Y}, \mathbb{Q}_\ell)$ decomposes completely over $\mathbb{Q}$ into two-dimensional Galois representations, and the latter correspond to modular forms of weight 4 and modular forms of weight 2.*

(c) *The Maschke surface S has Picard number $\rho(S) = 202$. The second cohomology group $H^2(\bar{S}, \mathbb{Q}_\ell)$ has $B_2(S) = 302$. The Galois representation of the transcendental part has dimension 100, which splits into 2 or three-dimensional Galois sub-representations over $\mathbb{Q}$, and the latter correspond to modular forms of weight 3, or modular forms of weight 2.*

(2) Calabi–Yau threefolds in the category (2), i.e, with irreducible Galois representations

Consani–Scholten [8] constructed a Calabi–Yau threefold over $\mathbb{Q}$ as follows. Consider the Chebyshev polynomial

$$P(y, z) = (y^5 + z^5) - 5yz(y^2 + z^2) + 5yz(y + z) + 5(y^2 + z^2) - 5(y + z)$$

and define an affine variety X in $\mathbf{A}^4$ by

$$X : P(x_1, x_2) = P(x_3, x_4)$$

and let $\bar{X} \subset \mathbb{P}^4$ be its projective closure. Then $\bar{X}$ has 120 ordinary double points. Let $\tilde{X}$ be its small resolution. Then $\tilde{X}$ is a Calabi–Yau threefold with $h^{1,1}(\tilde{X}) = 141$ and $h^{2,1}(\tilde{X}) = 1$. $\tilde{X}$ is defined over $\mathbb{Q}$ and the primes $2, 3, 5$ are bad primes. $H^3(\bar{\tilde{X}}, \mathbb{Q}_\ell)$ gives rise to a four-dimensional ℓ-adic Galois representation, ρ, which is irreducible over $\mathbb{Q}$. Let $F = \mathbb{Q}(\sqrt{5})$, and let $\lambda \in F$ be a prime above ℓ. Then the restriction $\rho|_{\mathrm{Gal}(\bar{\mathbb{Q}}/F)}$ is reducible as a representation to $GL(4, F_\lambda)$: There is a Galois representation $\sigma : \mathrm{Gal}(\bar{\mathbb{Q}}/F) \to GL(2, F_\lambda)$ such that $\rho = \mathrm{Ind}^{\mathbb{Q}}_F \sigma$. Consani–Scholten conjectured the modularity of $\tilde{X}$.

Theorem 9 (Dieulefait–Pacetti–Schütt [12]). *The Consani–Scholten Calabi–Yau threefold $\tilde{X}$ over $\mathbb{Q}$ is Hilbert modular. That is, the L-series associated to σ coincides with the L-series of a Hilbert modular newform $\mathfrak{f}$ on F of weight $(2, 4)$ and conductor $\mathfrak{c}_{\mathfrak{f}} = (30)$.*

The first example of Siegel modular varieties, as moduli spaces, of Calabi–Yau threefolds was given by van Geemen and Nygaard [60]. Recently, a series of articles by Freitag and Salvati Manni [19] on Siegel modular threefolds which admit Calabi–Yau models have appeared. The starting point is the van Geemen–Nygaard rigid Calabi–Yau threefold defined by a complete intersection Y of degree $(2, 2, 2, 2)$ in $\mathbb{P}^7$ by the equations:

$$
\begin{aligned}
Y_0^2 &= X_0^2 + X_1^2 + X_2^2 + X_3^2 \\
Y_1^2 &= X_0^2 - X_1^2 + X_2^2 - X_3^2 \\
Y_2^2 &= X_0^2 + X_1^2 - X_2^2 - X_3^2 \\
Y_3^2 &= X_0^2 - X_1^2 - X_2^2 + X_3^2
\end{aligned}
$$

A smooth small resolution of Y, denoted by X, is a Calabi–Yau threefold with $h^{1,1}(X) = 32$ and $h^{2,1}(X) = 0$, so X is rigid and hence is modular, indeed, the L-series of the Galois representation associated to $H^3(X, \mathbb{Q}_\ell)$ is determined by the unique weight 4 modular form on $\Gamma_0(8)$.

Several examples of non-rigid Calabi–Yau threefolds Z are obtained by quotienting X by finite group actions. Many of the resulting varieties Z are Calabi–Yau threefolds with small third Betti number.

Remark 10. For instance, Freitag–Salvati Manni has constructed such Calabi–Yau threefolds Z. The Galois representations associated to $H^3(Z, \mathbb{Q}_\ell)$ of their Calabi–Yau threefolds Z should be studied in detail. They should decompose into direct sum of those coming from the rigid Calabi–Yau threefold X, and those coming from elliptic surfaces or surfaces of higher genus arising from fixed points of finite groups in question.

In particular, this would imply that proper Siegel modular forms will not arise from these examples. So far as I know, we do not have examples of Calabi–Yau threefolds over $\mathbb{Q}$ with $B_3(X) = 4$ whose $L(X, s)$ comes from a Siegel modular form on $Sp(4, \mathbb{Z})$ or its subgroups of finite index.

(3) Calabi–Yau threefolds of CM type

We next consider Calabi–Yau threefolds which we will show to be of CM type. Then the Galois sub-representations associated to the third cohomology groups are induced by one-dimensional ones. Then, by applying the automorphic induction process, we will establish the automorphy of Calabi–Yau threefolds. These Calabi–Yau threefolds are realized as quotients of products of K3 surfaces and elliptic curves by some automorphisms. Rohde [51], Garbagnati–van Geemen [21], and Goto–Livné–Yui [22] produce examples of CM type Calabi–Yau threefolds with this approach.

Let S be a K3 surface with an involution σ acting on $H^{0,2}(S)$ by -1 discussed in Sect. 3.2. Let E be an elliptic curve with the standard involution ι. Consider the quotient of the product $E \times S / \iota \times \sigma$. This is a singular Calabi–Yau threefold having only cyclic quotient singularities. Resolving singularities we obtain a smooth crepant resolution X. Since the invariants of X are determined by a triplet of integers (r, a, δ) associated to S, we will write X as $X(r, a, \delta)$. The Hodge numbers and the Euler characteristic of X depend only on r and a:

$$h^{1,1}(X) = 5 + 3r - 2a, \; h^{2,1}(X) = 65 - 3r - 2a$$

and

$$e(X) = 2(h^{1,1}(X) - h^{2,1}(X)) = 6(r - 10).$$

Theorem 10 (Goto–Livné–Yui [22]). *Let (S, σ) be one of the K3 surfaces defined over $\mathbb{Q}$ in Theorem 7. Let E be an elliptic curve over $\mathbb{Q}$ with the standard involution ι. Let $X = X(r, a, \delta)$ be a smooth Calabi–Yau threefold. Then the following assertions hold:*

(1) X is of CM type if and only if E is of CM type.

(2) If X is of CM type, then the Jacobian variety $J(C_g)$ of C_g in S^σ is also of CM type, provided that the K3 surface component is of the form $x_0^2 = f(x_1, x_2, x_3)$ with involution $\sigma(x_0) = -x_0$.

(3) X is modular (automorphic).

Sketch of Proof:

1. The Hodge structure h_X of type $(3,0)$ of X is given by the tensor product $h_S \otimes h_E$ of the Hodge structures h_S of type $(2,0)$ and h_E of type $(1,0)$. Then h_X is of CM type if and only if both h_S and h_E are of CM type. Since S is already of CM type, we only need to require that E is of CM type.

2. When the K3 surface is defined by a hypersurface of the form $x_0^2 = f(x_1, x_2, x_3)$ and the involution σ takes x_0 to $-x_0$, then the curve C_g in the fixed locus S^σ is obtained by putting $x_0 = 0$, and hence it is also of Delsarte type. Hence the Jacobian variety $J(C_g)$ of C_g is also of CM type.

   When the hypersurface defining the K3 surface is not of the above form and the involution is more complicated, we ought to check each case whether C_g is of Delsarte type or not.

3. S is modular by Theorem 8, and E is modular by Wiles et al. Hence X is modular (automorphic).

Now we will discuss mirror Calabi–Yau threefolds of $X = X(r, a, \delta)$.

Proposition 9 (Borcea [4] and Voisin [61]). *Given a Calabi–Yau threefold $X = X(r, a, \delta)$, there is a mirror Calabi–Yau threefold $X^\vee$ such that $X^\vee$ is realized as a crepant resolution of a quotient of $E \times S / \iota \times \sigma$ and $X^\vee$ is characterized by the invariants $(20 - r, a, \delta)$. The Hodge numbers of $X^\vee$ are*

$$h^{1,1}(X^\vee) = 5 + 3(20 - r) - 2a = 65 - 3r - 2a = h^{2,1}(X),$$

$$h^{2,1}(X^\vee) = 65 - 3(20 - r) - 2a = 5 + 3r - 2a = h^{1,1}(X)$$

and the Euler characteristic is

$$e(X^\vee) = -12(r - 10) = -e(X).$$

In terms of g and k, $r = 11 - g + k$, $a = 11 - g - k$, and

$$h^{1,1}(X) = 1 + r + 4(k + 1), \quad h^{2,1}(X) = 1 + (20 - r) + 4g$$

and the Euler characteristic is $e(X) = 12(1 + k - g)$.

Remark 11. Mirror symmetry of Calabi–Yau threefolds of K3 $\times$ E do come from mirror symmetry of K3 surfaces. Mirror symmetry for K3 surfaces is the correspondence $r \leftrightarrow (20 - r)$. In fact, one can see this in Fig. 1: Nikulin's pyramid. Given a K3 surface S, we try to look for a mirror K3 surface $S^\vee$ satisfying this correspondence. This correspondence is established at a special CM point in the moduli space.

We know that the 95 families of K3 surfaces of Reid and Yonemura are not closed under mirror symmetry. Only 54 families of K3 surfaces with involution have mirror partners within the 95 families.

A recent article of Artebani–Boissière–Sarti [1] also considers this type of mirror symmetry for K3 surfaces.

Proposition 10. *For a mirror $X^\vee$, the assertion in Theorem 10 is true if the mirror K3 surface is also of Delsarte type.*

Proposition 11. *Let (S, σ) be one of the 86 families of K3 surfaces with involution σ. Let $X = X(r, a, \delta)$ and $X^\vee = X(20 - r, a, \delta)$ be mirror pairs of Calabi–Yau threefolds defined over $\mathbb{Q}$, where we suppose that the K3 surface component is of Delsarte type. Then X and $X^\vee$ have the same properties:*

- *X is of CM type if and only if $X^\vee$ is of CM type, and*
- *X is modular if and only if $X^\vee$ is modular.*

Rohde [51], Garbagnati and van Geemen [21] and Garbagnati [20] constructed Calabi–Yau threefolds which are quotients of the products of K3 surfaces and elliptic curves by non-symplectic automorphisms of higher order (than 2), that is, order 3, or order 4. These Calabi–Yau threefolds are parametrized by Shimura varieties.

Rohde [51] constructed families of Calabi–Yau threefolds as the desingularization of the quotient $S \times E$ by an automorphism of order 3 where E is the unique elliptic curve with an automorphism α_E of order 3, and S is a K3 surface with an automorphism α_S of order 3 which fixes k rational curves and $k + 3$ isolated points for some integer k, $0 \leq k \leq 6$.

Let ξ be a primitive cube root of unity. Choose the specific elliptic curve: $E = \mathbb{C}/\mathbb{Z} + \xi\mathbb{Z}$, which has a Weierstrass model $y^2 = x^3 - 1$ and $\alpha_E : (x, y) \mapsto (\xi x, y)$. Also choose some specific K3 surface S. Let

$$S = S_f : Y^2 = X^3 + f(t)^2, \ f = gh^2, \ \deg(f) = 6$$

where t is the coordinate on $\mathbb{P}^1$, and f has four distinct zeros. S_f has an automorphism of order 3: $\alpha_f : (X, Y, t) \mapsto (\xi X, Y, t)$. Now define a Calabi–Yau threefold X_f as the desingularization of $S_f \times E$ by the automorphism $\alpha := \alpha_f \times \alpha_E$. Note that S_f is birationally isomorphic to $(C_f \times E)/(\beta_f \times \alpha_E)$ where $C_f : v^3 = f(t)$ and $\beta_f : C_f \to C_f, (t, v) \mapsto (t, \xi v)$. Then X_f is birationally isomorphic to

$$(C_f \times E \times E)/H \quad \text{where} \quad H = <\beta_f \times \alpha_E \times 1_E, 1_{C_f} \times \alpha_E \times \alpha_E^{-1}> .$$

Rohde [51] worked out the case $\deg(g) = \deg(h) = 2$. Garbagnati–van Geemen [21] considered the other cases, i.e., $\deg(g) = 4, \deg(h) = 1$ and $\deg(g) = 6, \deg(h) = 0$. The Hodge numbers of the Calabi–Yau threefold X_f are computed and the results are tabulated as follows:

deg(g)	deg(h)	$g(C_f)$	$h^{2,1}(X_f)$	$h^{1,1}(X_f)$	k
6	0	4	3	51	3
4	1	3	2	62	4
2	2	2	1	73	5
0	3	1	0	84	6

For $k > 2$, the Calabi–Yau threefold is constructed by considering the curve $C_\ell : v^6 = \ell(t)$, deg(ℓ) = 12 such that $\ell(t)$ has five double zeros. It has the order 3 automorphism $\beta_\ell : (t, v) \mapsto (t, \xi v)$. The quotient $(C_\ell \times E)/(\beta_\ell \times \alpha_E)$ is a K3 surface S_ℓ which has the elliptic fibration with Weierstrass equation $Y^2 = X^3 + \ell(t)$. Then the desingularization of $(S_\ell \times E)/(\alpha_\ell \times \alpha_E)$ is a Calabi–Yau threefold X_ℓ with $h^{2,1}(X_\ell) = 4$ and $h^{1,1}(X_\ell) = 40$.

Similarly, for $k = 1$, one can find a Calabi–Yau threefold X with $h^{2,1}(X) = 5$ and $h^{1,1}(X) = 29$.

For details of these two cases, see Garbagnati–van Geemen [21] or Rohde [51].

We summarize the above discussion in the following form.

Proposition 12. *The Calabi–Yau threefold X_f (resp. X_ℓ) constructed above is of CM type if and only if the Jacobian variety $J(C_f)$ (resp. $J(C_\ell)$) is of CM type. In this case, X_f (resp. X_ℓ) is modular (automorphic).*

We will also mention results of Garbagnati [20], which generalize the method of Garbagnati–van Geemen [21] to automorphisms of order 4. In order to construct Calabi–Yau threefolds with a non-symplectic automorphism of order 4, start with the hyperelliptic curves

$$C_{f_g} : z^2 = t f_g(t^2), \ \deg(f_g) = g, \quad f_g \text{ without multiple roots.}$$

C_{f_g} has the automorphism $\alpha_C : (t, z) \mapsto (-r, iz)$. We consider the cases $g = 2$ or 3. Let $E_i : v^2 = u(u^2 + 1)$ be the elliptic curve and let $\alpha_E : (u, v) \mapsto (-u, iv)$ be the automorphism of E_i. Now take the quotient of the product $E_i \times C_{f_g}/\alpha_E \times \alpha_C$. Then the singularities are of A-D-E type, and the desingularization defines a K3 surface, S_{f_g}, with the automorphism $\alpha_S : (x, y, s) \mapsto (-x, iy, s)$. In fact, there is the elliptic fibration $\mathcal{E} : y^2 = x^3 + xs f_g(s)^2$ and the map $\pi : E_i \times C_{f_g} \to \mathcal{E}$ defined by $((u, v); (z, t)) \mapsto (x := uz^2, y := vz^3, s := t^2)$ is the quotient map $E_i \times C_{f_g} \to (E_i \times C_{f_g})/\alpha_E \times \alpha_C$. The K3 surface S_{f_g} thus obtained has large Picard number. Indeed, rank($T_{S_{f_g}}$) ≤ 4 if $g = 2$ and ≤ 6 if $g = 3$. Also, S_{f_g} admits the order 4 non-symplectic automorphism α_S, and the fixed loci of α_S and of α_S^2 contain no curves of genus > 0. The K3 surface (S_{f_g}, α_S^2) with involution α_S^2 indeed corresponds to the triplet $(18, 4, 1)$ for $g = 2$ and $(16, 6, 1)$ for $g = 3$.

Once we have a family of K3 surfaces with non-symplectic automorphism α_S of order 4, we can construct a family of Calabi–Yau threefolds as the quotient of the product of S_{f_g} with the elliptic curve E_i.

Proposition 13 (Garbagnati [20]). *There is a desingularization Y_{f_g} of $(E_i \times S_{f_g})/(\alpha_E^3 \times \alpha_S)$ which is a smooth Calabi–Yau threefold with*

$$h^{1,1}(Y_{f_g}) = \begin{cases} 73 & \text{if } g = 2 \\ 56 & \text{if } g = 3 \end{cases} \quad \text{and} \quad h^{2,1}(Y_{f_g}) = \begin{cases} 1 & \text{if } g = 2 \\ 2 & \text{if } g = 3 \end{cases}.$$

Proposition 14. *The Calabi–Yau threefold Y_{f_g} is of CM type if and only if the Jacobian variety $J(C_{f_g})$ is of CM type. When it is of CM type, Y_{f_g} is automorphic.*

4 The Modularity of Mirror Maps of Calabi–Yau Varieties, and Mirror Moonshine

- *Modularity of solutions of Picard–Fuchs differential equations*

  Let $n \in \mathbb{N}$ and let $M_n := U_2 \oplus (-E_8)^2 \oplus < -2n >$ be a lattice of rank 19. Here U_2 is the usual hyperbolic lattice of rank 2 and $-E_8$ is the unique negative-definite unimodular lattice of rank 8, and $< -2n >$ denotes the rank 1 lattice $\mathbb{Z}v$ with its bilinear form determined by $< v, v > = -2n$. We consider a one-parameter family of M_n polarized K3 surfaces X_t over $\mathbb{Q}$ with generic Picard number $\rho(X_t) = 19$. Here by a M_n-polarized K3 surfaces, we mean K3 surfaces X_t such that $T(X_t)$ is primitively embedded into $U_2 \oplus \mathbb{Z}u$ where u is a vector of height $2n$, $n \in \mathbb{N}$. The Picard–Fuchs differential equation of X_t is of order 3. It is shown by Doran [15] that for such a family of K3 surfaces X_t, there is a family of elliptic curves E_t such that the order 3 Picard–Fuchs differential equation of X_t is the symmetric square of the order 2 differential equation associated to the family of elliptic curves E_t. The existence of such a relation stems from the so-called Shioda–Inose structures of X_t (or by Dolgachev's result [14] which asserts that the coarse moduli space of M_n-polarized K3 surfaces is isomorphic to the moduli space of elliptic curves with level n structure). Long [42] gave an algorithm how to determine a family of elliptic curves E_t, up to projective equivalence.

  Yang and Yui [63] studied differential equations satisfied by modular forms of two variables associated to $\Gamma_1 \times \Gamma_2$ where $\Gamma_i (i = 1, 2)$ are genus zero subgroups of $SL(2, \mathbb{R})$ commensurable with $SL(2, \mathbb{Z})$. A motivation is to realize these differential equations satisfied by modular forms of two variables as Picard–Fuchs differential equations of K3 families with large Picard numbers, e.g., 19, 18, 17 or 16, thereby establishing the modularity of solutions of Picard–Fuchs differential equations. This goal was achieved for some of the families of K3 surfaces studied by Lian and Yau in [37, 39].

- *Monodromy of Picard–Fuchs differential equations of certain families of Calabi–Yau threefolds*

  Classically, it is known that the monodromy groups of Picard–Fuchs differential equations for families of elliptic curves and K3 surfaces are congruence subgroups of $SL(2, \mathbb{R})$. This modularity property of the monodromy groups ought to be extended to families of Calabi–Yau threefolds. For this, we will study the monodromy groups of Picard–Fuchs differential equations associated with one-parameter families of Calabi–Yau threefolds. In Chen–Yang–Yui [7] they considered 14 Picard–Fuchs differential equations of order 4 of hypergeometric type. They are of the form

$$\theta^4 - Cz(\theta + A)(\theta + 1 - A)(\theta + B)(\theta + 10B)$$

where $A, B, C \in \mathbb{Q}$.

Theorem 11. *In these 14 hypergeometric cases, the matrix representations of the monodromy groups relative to the Frobenius basis can be expressed in terms of the geometric invariants of the underlying Calabi–Yau threefolds. Here the geometric invariants are the degree d, the second Chern numbers, $c_2 \cdot H$ and the Euler number, c_3.*

Furthermore, under suitable change of basis, the monodromy groups are contained in certain congruence subgroups of $Sp(4, \mathbb{Z})$ of finite index (in $Sp(4, \mathbb{Z})$) and whose levels are related only to the geometric invariants.

However, finiteness of the index of the monodromy groups themselves in $Sp(4, \mathbb{Z})$ is not established.

Using the same idea for the hypergeometric cases, the monodromy groups of the differential equations of Calabi–Yau type that have at least one conifold singularity (not of the hypergeometric type) are computed. Our calculations verify numerically that if the differential equations come from geometry, then the monodromy groups are also contained in some congruence subgroups of $Sp(4, \mathbb{Z})$.

We should mention that van Enckevort and van Straten [59] numerically determined the monodromy for 178 Calabi–Yau equations of order 4 with a different method from ours, and speculated that these equations do come from geometry.

• Modularity of mirror maps and mirror moonshine

For a family of elliptic curves $y^2 = x(x - 1)(x - \lambda)$, the periods $\int_1^\infty \frac{dx}{\sqrt{x(x-1)(x-\lambda)}}$ satisfy the Picard–Fuchs differential equation

$$(1 - \lambda)\theta^2 f - \lambda\theta f - \frac{\lambda}{4}f = 0 \quad \left(\theta = \lambda\frac{d}{d\lambda}\right).$$

The monodromy group for this Picard–Fuchs differential equation is $\Gamma(2) \subset SL(2, \mathbb{R})$ of finite index. The periods can be expressed in terms of the hypergeometric function

$$_2F_1(\frac{1}{2}, \frac{1}{2}; 1; \lambda).$$

Now suppose that $y_0(\lambda) = 1 + \sum_{n\geq 1} a_n\lambda^n$ is the unique holomorphic solution at $\lambda = 0$ and $y_1(\lambda) = \lambda y_0(\lambda) + g(\lambda)$ be the solution with logarithmic singularity. Set $z = y_1(\lambda)/y_0(\lambda)$. Then λ, as a function of z, becomes a modular function for the modular group $\Gamma(2)$. This is called a *mirror map* of the elliptic curve family. That a mirror map is a Hauptmodul for a genus zero subgroup $\Gamma(2) \subset SL(2, \mathbb{R})$ is referred to as *mirror moonshine*.

For one-parameter families of K3 surfaces of generic Picard number 19, the Picard–Fuchs differential equations are of order 3. Since such families of K3 surfaces are equipped with Shioda–Inose structure (cf. Morrison [46]), the Picard–Fuchs differential equations are symmetric squares of differential equations of order 2. Hence the monodromy groups are realized as subgroups of $SL(2, \mathbb{R})$.

Classically, explicit period and mirror maps for the quartic K3 surface have been described by several articles, e.g., Hartmann [25] (and also see Lian and Yau [38]. Consider the deformation of the Fermat quartic: $F_t := \{x_0^4 + x_1^4 + x_2^4 + x_3^4 - 4tx_0x_1x_2x_3 = 0\} \subset \mathbb{P}^3$. Taking the quotient of F_t by some finite group and then resolving singularities, we obtain the Dwork pencil of K3 surfaces, denoted by X_t. Then X_t is a K3 surface with generic Picard number 19. The Picard–Fuchs differential equation is of order 3 and there is a unique holomorphic solution (at $t = 0$) of the form $w_0(t) = 1 + \sum_{n \geq 1} c_n t^n$, and another solution of logarithmic type: $w_1(z) = \log(t)w_0(t) + \sum_{n \geq 1} d_n t^n$. Now introduce the new variable z by $z := \frac{1}{2\pi i} \frac{w_1(t)}{w_0(t)}$, and put $q = e^{2\pi i z}$. The inverse $t = t(q)$ is the mirror map and is given by

$$t(q) = q - 104q^2 + 6444q^3 - 311744q^4 + 13018830q^5 + \cdots$$

This is the reciprocal of the Hauptmodul for $\Gamma_0(2)_+ \subset SL(2, \mathbb{R})$.

There are several more examples of one-parameter families of K3 surfaces with generic Picard number 19. Doran [15] has established the modularity of the mirror map for M_n-polarized K3 surfaces. However, for each n, the explicit description of mirror maps as modular functions are still to be worked out.

The situation will get more much complicated when we consider two-parameter families of K3 surfaces. Hashimoto and Terasoma [26] have studied the period and mirror maps of the two-parameter (in fact, projective one-parameter) family $\{\mathcal{X}_t\}$ $t = (t_0, t_1) \in \mathbb{P}^1$ of quartic family of K3 surfaces defined by

$$\mathcal{X}_t : x_1 + \cdots + x_5 = t_0(x_1^4 + \cdots + x_5^4) + t_1(x_1^2 + \cdots + x_5^2)^2 = 0$$

in $\mathbb{P}^4$ with homogeneous coordinates $(x_1 : \cdots : x_5)$. This family admits a symplectic group action by the symmetric group S_5. The Picard number of a generic fiber is equal to 19, and the Gram matrix of the transcendental lattice T is given by $\begin{pmatrix} 4 & 1 & 0 \\ 1 & 4 & 0 \\ 0 & 0 & -20 \end{pmatrix}$. The image of the period map of this family is a 1-dimensional subdomain Ω_T of the 19-dimensional period domain (the bounded symmetric domain of type IV). Let Ω_T° be a connected component of Ω_T. Since $O(T)$ has no cusp, there is a modular embedding $i : \Omega_T^\circ \to \mathbf{H}_2$ to the Siegel upper half-plane $\mathbf{H}_2$ of genus 2. This modular embedding is constructed using the Kuga–Satake construction. The inverse of the period map, that is, a mirror map, is constructed using automorphic forms of one variable on Ω_T°. In fact, automorphic forms are constructed as the pull-backs of the fourth power of theta constants of genus 2. This gives yet another example of a generalized mirror moonshine.

For a one-parameter family of Calabi–Yau threefolds, the mirror map is defined using specific solutions of the Picard–Fuchs differential equation of the family. At a point of maximal unipotent monodromy (e.g., $z = 0$), there is a unique holomorphic power series solution $\omega_0(z)$ with $\omega_0(0) = 1$, and a logarithmic power series solution $\omega_1(z) = \log(z)\omega_0(z) + g(z)$ where $g(z)$ is holomorphic near $z = 0$ with $g(0) = 0$. Now put $t := \frac{\omega_1(z)}{\omega_0(z)}$. We call the map defined by $q := e^{2\pi i t} = z e^{g(z)/\omega_0(z)}$ the *mirror map* of the Calabi–Yau family. (See, for instance, Lian and Yau [38]).

For some one-parameter families of Calabi–Yau threefolds, e.g., of hypergeometric type, the integrality of the mirror maps has been established by Krattenthaler and Rivoal [32, 33].

The modularity of mirror maps of Calabi–Yau families is getting harder to deal with in general. Doran [15] has considered certain one-parameter families of Calabi–Yau threefolds with $h^{2,1} = 1$. The Picard–Fuchs differential equations of these Calabi–Yau threefolds are of order 4. Under some some special constraints imposed by special geometry, and some conditions about a point $z = 0$ of maximal unipotent monodromy, there is a set of fundamental solutions (to the Picard–Fuchs differential equation) of the form $\{u, u \cdot t, u \cdot F', (tF' - 2F)\}$ where $u = u(z)$ is the fundamental solution locally holomorphic at $z = 0$, $t = t(z)$ is the mirror map, $F(z)$ is the prepotential and F' is the derivative of F with respect to z. When there are no instanton corrections, the Picard–Fuchs differential equation becomes the symmetric cube of some second-order differential equation. In this case, Doran has shown that the mirror map becomes automorphic. However, exhibiting automorphic forms explicitly remains an open problem. On the other hand, if there are instanton corrections, a necessary and sufficient condition is presented for a mirror map to be automorphic.

We should mention here a converse approach to the modularity question of solutions of Picard–Fuchs differential equations, along the line of investigation by Yang and Yui [63]. The starting point is modular forms and the differential equations satisfied by them. It may happen that these differential equations coincide with Picard–Fuchs differential equations of some families of Calabi–Yau varieties. Consequently, the modularity of solutions of Picard–Fuchs differential equations and mirror maps can be established.

5 The Modularity of Generating Functions of Counting Some Quantities on Calabi–Yau Varieties

Under this subtitle, topics included are enumerative geometry, Gromov–Witten invariants, and various invariants counting some mathematical/physical quantities, etc.

- Mirror symmetry for elliptic curves and quasimodular forms.

We consider the generating function, $F_g(q)$, counting simply ramified covers of genus $g \geq 1$ over a fixed elliptic curve with $2g - 2$ marked points.

Theorem 12. *For each $g \geq 2$, $F_g(q)$ (with $q = e^{2\pi i \tau}$, $\tau \in \mathfrak{H}$), is a quasimodular form of weight $6g - 6$ on $\Gamma = SL(2, \mathbb{Z})$. Consequently, $F_g(q)$ is a polynomial in $\mathbb{Q}[E_2, E_4, E_6]$ of weight $6g - 6$.*

This result is stated as the Fermion Theorem in Dijkgraaf [13], which is concerned with the A-model side of mirror symmetry for elliptic curves. A mathematically rigorous proof was given in the article of Roth–Yui [53]. The B-model (bosonic) counting constitutes the mirror side of the calculation. The bosonic counting will involve calculation with Feynman integrals of trivalent graphs. A mathematical rigorous treatment of the B-model counting is currently under way.

Further generalizations:

(a) The generating function of m-simple covers for any integer $m \geq 2$ of genus $g \geq 1$ over a fixed elliptic curve with $2g - 2$ marked points has been shown again to be quasimodular forms by Ochiai [49].
(b) The quasimodularity of the Gromov–Witten invariants for the three elliptic orbifolds with simple elliptic singularities $\tilde{E}_N$ ($N = 6, 7, 8$) has been established by Milanov and Ruan [45]. These elliptic orbifolds are realized as quotients of hypersurfaces of degree 3, 4 and 6 in weighted projective 2-spaces with weights $(1, 1, 1)$, $(1, 2, 2)$ and $(1, 2, 3)$, respectively.
(c) The recent article of Rose [52] has proved the quasimodularity of the generating function for the number of hyperelliptic curves (up to translation) on a polarized abelian surface.

6 Future Prospects

Here we collect some topics which we are not able to cover in this paper as well some problems for further investigation.

6.1 The Potential Modularity

The potential modularity of families of hypersurfaces. For the Dwork families of one-parameter hypersurfaces, the potential modularity has been established by R. Taylor and his collaborators. Extend the potential modularity to more general Calabi–Yau hypersurfaces, Calabi–Yau complete intersections, etc.

6.2 The Modularity of Moduli of Families of Calabi–Yau Varieties

Moduli spaces of lattice polarized K3 surfaces with large Picard number.

6.3 Congruences, Formal Groups

Congruences for Calabi–Yau families, formal groups.

6.4 The Griffiths Intermediate Jacobians of Calabi–Yau Threefolds

- Explicit description of the Griffiths intermediate Jacobians and their modularity. Let X be a Calabi–Yau threefold defined over $\mathbb{Q}$. Let

$$J^2(X) \simeq H^3(X, \mathbb{C})/F^2 H^3(X, \mathbb{C}) + H^3(X, \mathbb{Z}) \simeq H^3(X, \mathbb{C})^*/H_3(X, \mathbb{Z})$$

be the Griffiths intermediate Jacobian of X. There is the Abel–Jacobi map

$$CH^2(X)_{hom,\mathbb{Q}} \to J^2(X)_{\mathbb{Q}}.$$

A part of the Beilinson–Bloch conjecture asserts that this map is injective modulo torsion.

Now suppose that X is rigid. Then $J^2(X)$ is a complex torus of dimension 1 so that there is an elliptic curve E such that $J^2(X) \simeq E(\mathbb{C})$. We know that X is modular by [23].

Question: *Is it true that the Griffiths intermediate Jacobian $J^2(X)$ of a rigid Calabi–Yau threefold X over $\mathbb{Q}$ is defined over $\mathbb{Q}$ and hence modular?*

- Special values of L-series of Calabi–Yau threefolds over $\mathbb{Q}$.
Assuming a positive answer to the above question, we can consider a possible relation between the Birch and Swinnerton–Dyer conjecture for rational points on $J^2(X)_{\mathbb{Q}}$:

$$\operatorname{rank}_{\mathbb{Z}} J^2(X)_{\mathbb{Q}}(\mathbb{Q}) = \operatorname{ord}_{s=1} L(J^2(X)_{\mathbb{Q}}, s)$$

and the Beilinson–Bloch conjecture on the Chow group $CH^2(X)_{\mathbb{Q}}$ of X:

$$\operatorname{rank}_{\mathbb{Z}} CH^2(X)_{hom,\mathbb{Q}} = \operatorname{ord}_{s=2} L(X_{\mathbb{Q}}, s).$$

If the Abel–Jacobi map $CH^2(X)_{hom,\mathbb{Q}} \to H^2(X)_{\mathbb{Q}}$ is injective modulo torsion, then

$$\operatorname{ord}_{s=2} L(X_{\mathbb{Q}}, s) \leq \operatorname{ord}_{s=1} L(J^2(X)_{\mathbb{Q}}, s).$$

6.5 Geometric Realization Problem (the Converse Problem)

We know that every singular K3 surface X over $\mathbb{Q}$ is motivically modular in the sense that the transcendental cycles $T(X)$ corresponds to a newform of weight 3. For singular K3 surfaces over $\mathbb{Q}$, the converse problem asks:

Which newform of weight 3 with integral Fourier coefficients would correspond to a singular K3 surface defined over $\mathbb{Q}$?

This has been answered by Elkies and Schütt [18] (see also the article by Schütt [56] in this volume). Their result is the following theorem.

Theorem 13. *Every Hecke eigenform of weight 3 with eigenvalues in $\mathbb{Z}$ is associated to a singular K3 surface defined over $\mathbb{Q}$.*

Now we know that every rigid Calabi–Yau threefold over $\mathbb{Q}$ is modular (see Gouvêa–Yui [23]). The converse problem that has been raised, independently, by Mazur and van Straten is the so-called *geometric realization problem*, and is stated as follows:

Which newforms of weight 4 on some $\Gamma_0(N)$ with integral Fourier coefficients would arise from rigid Calabi–Yau threefolds over $\mathbb{Q}$? Do all such forms arise from rigid Calabi–Yau threefolds over $\mathbb{Q}$?

For Calabi–Yau threefolds, a very weak version of the above problem has been addressed in Gouvêa–Kiming–Yui [24].

Question: *Given a rigid Calabi–Yau threefold X over $\mathbb{Q}$ and a newform f of weight 4, for any non-square rational number d, there is a twist f_d by the quadratic character corresponding to the quadratic extension $\mathbb{Q}(\sqrt{d})/\mathbb{Q}$. Does f_d arise from a rigid Calabi–Yau threefold X_d over $\mathbb{Q}$?*

A result of Gouvêa–Kiming–Yui [24] in this volume is that the answer is positive if a Calabi–Yau threefold has an anti-symplectic involution. Let X be a rigid Calabi–Yau threefold over $\mathbb{Q}$. For a square-free $d \in \mathbb{Q}^{\times}$, let $K := \mathbb{Q}(\sqrt{d})$ and let σ be the non-trivial automorphism of K. We say that a rigid Calabi–Yau threefold X_d defined over $\mathbb{Q}$ is a *twist of X by d* if there is an involution ι of X which acts by -1 on $H^3_{et}(\bar{X}, \mathbb{Q}_\ell)$, and an isomorphism $\theta : (X_d)_K \cong X_K$ defined over K such that $\theta^\sigma \circ \theta^{-1} = \iota$.

Proposition 15. *Let X be a rigid Calabi–Yau threefold over $\mathbb{Q}$ and let f be the newform of weight 4 attached to X. Then, if X_d is twist by d of X, the newform attached to X_d is f_d, the twist of d by the Dirichlet character χ corresponding to K.*

Various types of modular forms have appeared in the physics literature. We wish to understand "conceptually" why modular forms play such pivotal roles in physics. Here we list some of the modular appearances in the physics literature.

6.6 Modular Forms and Gromov–Witten Invariants

Modular (automorphic) forms, quasi-modular forms, and Gromov–Witten invariants and generalized invariants.

6.7 Automorphic Black Hole Entropy

This has something to do with Conformal Field Theory. Mathematically, mock modular forms, Jacobi forms etc. will come into the picture. This is beyond the scope of this article.

6.8 M_{24}-Moonshine

Recently, some close relations between the elliptic genus of K3 surfaces and the Mathieu group M_{24} along the line of moonshine have been observed in the physics literature, e.g., [16, 17]. It has been observed that multiplicities of the non-BPS representations are given by the sum of dimensions of irreducible representations of M_{24} and furthermore, they coincide with Fourier coefficients of a certain mock theta function.

Appendix

In this appendix, we will recall modular forms of various kinds, e.g., classical modular forms, quasimodular forms, Hilbert modular forms, Siegel modular forms, and most generally, automorphic forms, which are relevant to our discussions. For details, the reader is referred to [6].

Definition 5. Dimension 1: Let $\mathfrak{H} := \{z \in \mathbb{C} \mid \mathrm{Im}(z) > 0\}$ be the complex upper-half plane. For a given integer $N > 0$, let

$$\Gamma_0(N) =: \left\{ \begin{pmatrix} a & b \\ c & d \end{pmatrix} \in SL(2, \mathbb{Z}) \mid c \equiv 0 \mod N \right\}$$

be a congruence subgroup of $SL(2, \mathbb{Z})$ (of finite index). A modular form of *weight k* and *level N* is a holomorphic function $f : \mathfrak{H} \to \mathbb{C}$ with the following properties:

(M1) For $\begin{pmatrix} a & b \\ c & d \end{pmatrix} \in \Gamma_0(N)$, $f\left(\frac{az+b}{cz+d}\right) = (cz + d)^k f(z)$;

(M2) f is holomorphic at the cusps.

Since $\begin{pmatrix} 1 & 1 \\ 0 & 1 \end{pmatrix} \in \Gamma_0(N)$, (a1) implies that $f(z + 1) = f(z)$, so f has the Fourier expansion $f(z) = f(q) = \sum_{n \geq 0} c(n)q^n$ with $q = e^{2\pi i z}$. f is a *cusp* form if it vanishes at all cusps.

If χ is a mod N Dirichlet character, we can define a modular form with character χ by replacing (a1) by $f\left(\frac{az+b}{cz+d}\right) = \chi(d)(cz + d)^k f(z)$.

The space $S_k(\Gamma_0(N))$ of all cusp forms of weight k and level N is a finite dimensional vector space, and similarly, so also is the space $M_k(\Gamma_0(N), \chi)$ of all modular forms of weight k and level N.

On $S_k(\Gamma_0(N), \chi)$ there are Hecke operators T_p for every prime p not dividing N. A cusp form f is a (normalized) Hecke *eigenform* if it is an eigenvector for all T_p, that is, $T_p(f) = c(p)f$. For such a normalized eigenform f, define the L-series $L(f, s)$ by

$$L(f, s) = \sum_{n \geq 1} c(n)n^{-s} = \prod_p \frac{1}{1 - c(p)p^{-s} + \chi(p)p^{k-1-2s}}$$

where $\chi(p) = 0$ if $p|N$.

Dimension 2: Let $F = \mathbb{Q}(\sqrt{d})$ be a totally quadratic field over $\mathbb{Q}$ where $d > 0$ and square-free, and let $\mathcal{O}_F$ be its ring of integers. The $SL(2, \mathcal{O}_F)$ can be embedded into $SL(2, \mathbb{R}) \times SL(2, \mathbb{R})$ via the two real embeddings of F to $\mathbb{R}$, and it acts on $\mathfrak{H} \times \mathfrak{H}$ via fractional linear transformations:

$$\begin{pmatrix} a & b \\ c & d \end{pmatrix} z = \left(\frac{az_1 + b}{cz_1 + d}, \frac{az_2 + b}{cz_2 + d}\right)$$

for $z = (z_1, z_2) \in \mathfrak{H} \times \mathfrak{H}$. The group

$$\Gamma(\mathcal{O}_F \oplus \mathfrak{a}) = \left\{\begin{pmatrix} a & b \\ c & d \end{pmatrix} \in SL(2, F), a, d \in \mathcal{O}_F, b\mathfrak{a}^{-1}, c \in \mathfrak{a}\right\}$$

is called *the Hilbert modular group* corresponding to a fractional ideal $\mathfrak{a}$ of F. If $\mathfrak{a} = \mathcal{O}_F$, put $\Gamma_F = \Gamma(\mathcal{O}_F \oplus \mathcal{O}_F) = SL(2, \mathcal{O}_F)$. Let $\Gamma \subset SL_2(F)$ be a subgroup commensurable with Γ_F, and let $(k_1, k_2) \in \mathbb{Z} \times \mathbb{Z}$.

A meromorphic function $f : \mathfrak{H} \times \mathfrak{H} \to \mathbb{C}$ is called a meromorphic *Hilbert modular form* of *weight* (k_1, k_2) for Γ if

$$f(\gamma z) = (cz_1 + d)^{k_1}(cz_2 + d)^{k_2} f(z)$$

for all $\gamma = \begin{pmatrix} a & b \\ c & d \end{pmatrix} \in \Gamma$ and $z = (z_1, z_2) \in \mathfrak{H}^2$. If f is holomorphic, then f is a holomorphic Hilbert modular form, and a holomorphic Hilbert modular form is *symmetric* if $f(z_1, z_2) = f(z_2, z_1)$. Further, f is a cusp form if it vanishes at cusps of Γ. A cusp form of weight $(2, 2)$ is identified with a holomorphic 2-form on the Hilbert modular surface $\mathfrak{H}^2/\Gamma$. The space of holomorphic Hilbert modular forms of weight

(k_1, k_2) for Γ is denoted by $M_k(\Gamma)$ and the cusp forms by $S_k(\Gamma)$. $M_k(\Gamma)$ is a finite dimensional vector space over $\mathbb{C}$.

A holomorphic Hilbert modular form has a Fourier expansion at ∞ of the following form. Let $M \subset F$ be a rank 2 lattice and $V \subset \mathcal{O}_F^*$ be a finite index subgroup acting on M in a suitable way. Then

$$f(z) = a_0 + \sum_{v \in M^\vee, v \geq 0} a_v e^{2\pi i \operatorname{tr}(vz)}$$

where v runs over the dual lattice $M^\vee$, and $\operatorname{tr}(v z) := v z_1 + v' z_2$.

The L-series of a Hilbert modular cusp form is defined by

$$L(f, s) = \sum_{\mathfrak{a} \subset \mathcal{O}_F} a(\mathfrak{a}) N(\mathfrak{a})^{-s}$$

where $\mathfrak{a}$ runs over principal ideals. For details, the reader should consult Bruinier [5].

Dimension 3: Let $g, N \in \mathbb{N}$. Define the Siegel upper-half plane by

$$\mathfrak{H}_g := \left\{ z \in M_{g \times g}(\mathbb{C}) \mid z^t = z, \operatorname{Im}(z > 0 \right\},$$

and the symplectic group $Sp(2g, \mathbb{Z})$ as the automorphism group of the symplectic lattice $\mathbb{Z}^{2g}$. That is,

$$Sp(2g, \mathbb{Z}) = \{ \gamma \in GL(2g, \mathbb{Z}) \mid \gamma^t J \gamma = J \}$$

where $J_g := \begin{pmatrix} 0 & I_g \\ -I_g & 0 \end{pmatrix}$. The group $Sp(2g, \mathbb{Z})$ acts on $\mathfrak{H}_g$ by

$$z \mapsto \gamma(z) = (Az + B)(Cz + D)^{-1} \quad \text{for} \quad \gamma = \begin{pmatrix} A & B \\ C & D \end{pmatrix} \in Sp(2g, \mathbb{Z}).$$

Let $\Gamma_g(N) := \left\{ \gamma \in Sp(2g, \mathbb{Z}) \mid \gamma \equiv I_{2g} \pmod{N} \right\}$ be a subgroup of $Sp(2g, \mathbb{Z})$. Then $\Gamma_g(N)$ acts freely when $N \geq 3$. Here I_{2g} is the identity matrix of order $2g$. If $N = 1$, $\Gamma_g(1) = Sp(2g, \mathbb{Z})$. The quotient space $\mathfrak{H}_g / \Gamma_g(N)$ (with $N \geq 3$) is a complex manifold of dimension $g(g + 1)/2$ (associated to a graded algebra of modular forms).

A holomorphic function $f : \mathfrak{H}^g \to \mathbb{C}$ is a *Siegel modular form of genus g, weight $k \in \mathbb{N}$* if

$$f(\gamma(z)) = \det(Cz + D)^k f(z)$$

for all $\gamma = \begin{pmatrix} A & B \\ C & D \end{pmatrix} \in Sp(2g, \mathbb{Z})$ an d all $z \in \mathfrak{H}_g$. The space of all holomorphic Siegel modular forms is a finitely generated graded algebra. The simplest examples of Siegel modular forms are given by theta constants.

A holomorphic Siegel modular form f has a Fourier expansion of the form

$$f(z) = \sum A(n) e^{2\pi \operatorname{tr}(nz)}$$

where the sum runs over all positive semi-definite integral matrices $n \in GL(g, \mathbb{Q})$. If the Fourier expansion is supported only on positive definite integral $g \times g$ matrices n, then f is called a cusp form.

There are at least two different L-series of a holomorphic Siegel modular form f: one is the spinor L-series, and the other is the standard L-series.

However, it is not clear how to associate these L-series to the Fourier expansion. For details, the reader should consult van der Geer [58].

Example 4. (a) For $N = 1$, the total space $M := \oplus_k M_k(SL(2, \mathbb{Z}))$, of all modular forms is generated by the Eisenstein series E_4 and E_6. The Eisenstein series E_2 is not modular but it may be called *quasimodular*. The space $\tilde{M}$ of all quasimodular forms for $SL(2, \mathbb{Z})$ is generated by the Eisenstein series E_2, E_4 and E_6, that is, $\tilde{M} = \mathbb{C}[E_2, E_4, E_6]$.

For $N > 1$, the space of modular forms of weight k for any congruence subgroup of level N is finite dimensional, and its basis can be determined.

Some properties of quasimodular forms for finite index subgroups of $SL(2, \mathbb{Z})$, dimension, basis, etc.

(b) There are Eisenstein series for $\Gamma_F = SL(2, \mathcal{O}_F)$ and k even given by

$$G_{k,B} = N(\mathfrak{b})^k \sum_{(c,d)\in\mathcal{O}_F^*\backslash\mathfrak{b}^2} N(cz + d)^{-k}$$

for B an ideal class of F.

If we put $g_k := \frac{1}{\zeta_F(k)}G_{k,\mathcal{O}_F}$, then g_2, g_6 and g_{10} generate the graded algebra $M_{2*}^{symm}(\Gamma_F)$ over $\mathbb{C}$, that is,

$$M_{2*}^{symm}(\Gamma_F) \cong \mathbb{C}[g_2, g_6, g_{10}].$$

(c) (Igusa) For $g = 2$, the graded algebra $\mathfrak{M}$ of classical Siegel modular forms of genus 2 is generated by the Eisenstein series E_4 and E_6, the Igusa cusp forms $C_{10}, C_{12},$ and C_{35} (where the subindex denote weights), and

$$\mathfrak{M} \cong \mathbb{C}[E_4, E_6, C_{10}, C12, C_{35}]/(C_{35}^2 = P(E_4, E_6, C_{10}, C_{12}))$$

where P is an explicit polynomial.

Acknowledgements I would like to thank Matthias Schütt for carefully reading the preliminary version of this paper and suggesting numerous improvements. I am indebted to Ron Livné for answering my questions about Galois representations.

I would also like to thank a number of colleagues for their comments and suggestions. This includes Jeng-Daw Yu, Ling Long, Ken-Ichiro Kimura, and Bert van Geemen.

We are grateful to V. Nikulin for allowing us to use the template of Nikulin's pryamid in Fig. 1.

Last but not least, my sincere thanks is to the referee for reading through the earlier versions of this article and for very helpful constructive criticism and suggestions for the improvement of the article. I would also like to thank Arther Greenspoon of Mathematical Reviews for copy-editing the article.

The article was completed while the author held visiting professorship at various institutions in Japan: Tsuda College, Kavli Institute for Physics and Mathematics of the Universe, and Nagoya University. I thank the hospitality of these instutions.

The author was supported in part by NSERC Discovery Grant.

References

1. M. Artebani, S. Boissière, A. Sarti, The Berglund–Hübsch–Chiodo–Ruan mirror symmetry for K3 surfaces [arXiv:1108.2780]
2. C. Bini, van B. Geemen, Geometry and arithmetic of Maschke's Calabi–Yau threefold. Comm. Number Theor. Phys. **5**(4), 779–826 (2011)
3. C. Borcea, Calabi–Yau threefolds and complex multiplication, in *Essays on Mirror Manifolds* (International Press, Boston, 1992), pp. 489–502
4. C. Borcea, K3 surfaces with involution and mirror pairs of Calabi–Yau manifolds, in *Mirror Symmetry*. AMS/IP Studies in Advanced Mathematics, vol. 1 (American Mathematical Society, Providence, 1997), pp. 717–743; **33**, 227–250 (1983)
5. J. Bruinier, Hilbert modular forms and their applications, in *The 1-2-3 of Modular Forms*, Universitext (Springer, Berlin, 2008), pp. 105–179
6. J. Bruiner, G. van der Geer, G. Harder, D. Zagier, *The 1-2-3 of Modular Forms*, Universitext (Springer, Berlin, 2008)
7. Y.-H. Chen, Y. Yang, N. Yui, Monodromy of Picard–Fuchs differential equations for Calabi–Yau threefolds. J. Reine Angew Math. **616**, 167–203 (2008)
8. C. Consani, J. Scholten, Arithmetic on a quintic threefold. Int. J. Math. **12**(8), 943–972 (2001)
9. S. Cynk, C. Meyer, Modularity of some nonrigid double octic Calabi–Yau threefolds. Rocky Mt. J. Math. **38**, 1937–1958 (2008)
10. P. Deligne, J.S. Milne, A. Ogus, K.-Y. Shih, in *Hodge Cycles, Motives and Shimura Varieties*. Lecture Notes in Mathematics, vol. 900 (Springer, Berlin, 1982)
11. L. Dieulefait, On the modularity of rigid Calabi–Yau threefolds. Zap. Nauchn. Sem. S-Peterburg. Otdel. Mat. Inst. Steklov. (POMI) **377**, 44–49 (2010) [Issledovaniya po teorii Chisel. 10]
12. L. Dieulefait, A. Pacetti, M. Schütt, *Modularity of the Consani–Scholten quintic*. Documenta Math. **17**, 953–987 (2012) [arXiv:1005.4523]
13. R. Dijkgraaf, Mirror symmetry and elliptic curves, in *The Moduli Space of Curves*, Texel Island, 1994. Progress in Mathematics, Birkhäuser Boston, vol. 129 (1995), pp. 149–163
14. I. Dolgachev, Mirror symmetry for lattice polarized K3 surfaces. Algebraic geometry, 4. J. Math. Sci **81**(3), 2599–2630 (1996)
15. C. Doran, Picard–Fuchs uniformization: modularity of the mirror map and mirror-moonshine, in *The Arithmetic and Geometry of Algebraic Cycles*, Banff, AB, 1998. CRM Proceedings of the Lecture Notes, vol. 24 (American Mathematical Society, Providence, 2000), pp. 257–281
16. T. Eguchi, K. Hikami, Superconformal algebras and mock theta functions. J. Phys. A. **42**(30), 304010 (2009)
17. T. Eguchi, K. Hikami, Superconformal algebras and mock theta functions 2. Rademacher expansion for K3 surface. Comm. Number Theor. Phys. **3**, 531–554 (2009)
18. N. Elkies, M. Schütt, Modular forms and K3 surfaces, Adv. Math. **240**, 106–131 (2013) [arXiv:0809.0830]
19. E. Freitag, R. Salvati Manni, Some Siegel threefolds with a Calabi–Yau model. Ann. Sc. Norm Super. Pisa Cl. Sci. **9**, 833–850 (2010); On Siegel threefolds with a projective Calabi–Yau model. Comm. Number Theor. Phys. **5**, 713–750 (2011)
20. A. Garbagnati, New examples of Calabi–Yau 3-folds without maximal unipotent monodromy. Manuscripta Math. **140**(3–4), 273–294 (2013)
21. A. Garbagnati, B. van Geemen, Examples of Calabi–Yau threefolds parametrized bu Shimura varieties. Rend. Sem. Mat. Univ. Pol. Torino **68**, 271–287 (2010)

22. Y. Goto, R. Livné, N. Yui, The modularity of certain K3 fibered Calabi–Yau threefolds over $\mathbb{Q}$ with involution, [arXiv:1212.4308]

23. F. Gouvêa, N. Yui, Rigid Calabi–Yau threefolds over $\mathbb{Q}$ are modular. Expos. Math. **29**, 142–149 (2011)

24. F. Gouvêa, I. Kiming, N. Yui, Quadratic twists of rigid Calabi–Yau threefolds over $\mathbb{Q}$. In this volume [arXiv: 1111.5275]

25. H. Hartmann, Period-and mirror maps for the quartic K3 (2011) [arXiv:1101.4601]

26. K. Hashimoto, T. Terasoma, Period maps of a certain K3 family with an S_5-action. J. Reine Angew Math. **652**, 1–65 (2011)

27. K. Hulek, H. Verrill, On modularity of rigid and non-rigid Calabi–Yau varieties associated to the root lattice A_4. Nagoya J. Math. **179**, 103–146 (2005)

28. K. Hulek, H. Verrill, On the modularity of Calabi–Yau threefolds containing elliptic ruled surfaces, in *Mirror Symmetry V*. AMS/IP Studies in Advanced Mathematics, vol. 38 (American Mathematical Society, Providence, 2006), pp. 19–34

29. C. Khare, J.-P. Wintenberger, Serre's modularity conjecture, I and II. Invent. Math. **178**, 485–504, 505–586 (2009)

30. M. Kisin, Modularity of 2-adic Barsotti–Tate representations. Invent. Math. **178**, 587–634 (2009)

31. S. Kondo, Automorphisms of algebraic K3 surfaces which act trivially on Picard groups. J. Math. Soc. Jpn. **44**, 75–98 (1992)

32. C. Krattenthaler, T. Rivoal, On the integrality of the Taylor coefficients of mirror maps, II. Comm. Number Theor. Phys. **3**(3), 555–591 (2009)

33. C. Krattenthaler, T. Rivoal, On the integrality of the Taylor coefficients of mirror maps, I. Duke J. Math. **151**(2), 175–218 (2010)

34. E. Lee, A modular non-rigid Calabi–Yau threefold, in *Mirror Symmetry V*. AMS/IP Studies in Advanced Mathematics, vol. 38 (American Mathematical Society, Providence, 2006), pp. 89–122

35. E. Lee, A modular quintic Calabi–Yau threefold of level 55. Can. J. Math. **63**, 616–633 (2011)

36. E. Lee, Update on modular non-rigid Calabi–Yau threefolds, in *Modular Forms and String Duality*. Fields Institute Communications, vol. 54, (American Mathematical Society, Providence, 2008), pp. 65–81

37. B.-H. Lian, S.-T. Yau, Mirror maps, modular relations and hypergeometric series, I, in *XIth International Congress of Mathematical Physics, Paris 1994* (International Press, Boston, 1995), pp. 163–184

38. B.-H. Lian, S-T. Yau, Arithmetic properties of mirror maps and quantum coupling. Comm. Math. Phys. **176**(1), 163–191 (1996)

39. B.-H. Lian, S.-T. Yau, Mirror maps, modular relations and hypergeometric series, II. Nucl. Phys. B Proc. Suppl. **46**, 248–262 (1996)

40. R. Livné, Motivic orthogonal two-dimensional representations of $Gal(\bar{\mathbb{Q}}/\mathbb{Q})$. Isr. J. Math. **92**, 149–156 (1995)

41. R. Livné, M. Schuett, N. Yui, The modularity of K3 surfaces with non-symplectic group action. Math. Ann. **348**, 333–355 (2010)

42. L. Long, On Shioda–Inose structures of one-parameter families of K3 surfaces. J. Number Theor. **109**, 299–318 (2004)

43. S. Ma, Rationality of the moduli spaces of 2-elementary K3 surfaces, J. Alg. Geom. (to appear) [arXiv:11105.5110]

44. C. Meyer, in *Modular Calabi–Yau Threefolds*. Fields Institute Monograph, vol. 22 (American Mathematical Society, Providence, 2005)

45. T. Milanov, Y. Ruan, Gromov–Witten theory of elliptic orbifold $\mathbb{P}^1$ and quasi-modular forms [arXiv:1106.2321]

46. D. Morrison, On K3 surfaces with large Picard number. Invent. Math. **75**, 105–121 (1984)

47. V. Nikulin, Integral symmetric bilinear forms and some of their geometric applications. Math. USSR-Izv. **14**, 103–167 (1980)

48. V. Nikulin, Discrete reflection groups in Lobachesvsky spaces and algebraic surfaces, in *Proceedings of the ICM*, Berkeley, CA, 1986, pp. 654–671

49. H. Ochiai, Counting functions for branched covers of elliptic curves and quasimodular forms. RIMS Kokyuroko **1218**, 153–167 (2001)
50. M. Reid, *Cananical 3-folds*, Proceedings of Algebraic Geometry, Anger (Sijthoff and Nordhoff, Alphen aan den Rijn, Netherlands, 1979), pp. 273–310
51. C. Rohde, in *Cyclic Coverings, Calabi–Yau Manifolds and Complex Multiplication*. Lecture Notes in Mathematics vol. 1975 (Springer, Berlin, 2009)
52. S. Rose, Counting hyperelliptic curves on an abelian surface with quasi-modular forms [arXiv:1202.2094]
53. M. Roth, N. Yui, Mirror symmetry for elliptic curves: the A-model (Fermionic) counting. Clay Math. Proc. **12**, 245–283 (2010)
54. M. Schütt, On the modularity of three Calabi–Yau threefolds with bad reduction at 11. Can. Math. Bull. **49**, 296–312 (2006)
55. M. Schütt, Modularity of Maschke's octic and Calabi–Yau threefold. Comm. Number Theor. Phys. **5**(4), 827–848 (2011)
56. M. Schütt, *Two Lectures on the Arithmetic of K3 Surfaces*, in this volume
57. J. Tate, Conjectures on algebraic cycles in ℓ-adic cohomology, in *Motives, Proceedings of Symposia in Pure Mathematics*, vol. 55, Part I (American Mathematical Society, Providence, 1994), pp. 71–83
58. G. van der Geer, Siegel modular forms and their applications, in *The 1-2-3 of Modular Forms*, Universitext (Springer, Berlin, 2008), pp. 181–245
59. C. van Enckevort, D. van Straten, Monodromy calculations of fourth order equations of Calabi–Yau type, in *Mirror Symmetry V*. Proceedings of BIRS Workshop on Calabi–Yau Varieties and Mirror Symmetry, 6–11 December, 2003. AMS/IP Studies in Advanced Mathematics, vol. 38 (2006), pp. 530–550
60. B. van Geemen, N. Nygaard, On the geometry and arithmetic of some Siegel modular threefolds. J. Number Theor. **53**, 45–87 (1995)
61. C. Voisin, Mirrors et involutions sur les surfaces K3. Astérisque **218**, 273–323 (1993)
62. S.P. Vorontsov, Automorphisms of even lattices arising in connection with automorphisms of algebraic K3 surfaces (Russian). Vestnik Moskov. Uni. Ser. I Mat. Mekh. (2), 19–21 (1983)
63. Y. Yang, N. Yui, Differential equations satisfied with modular forms and K3 surfaces. Ill. J. Math. **51**(2), 667–696 (2007)
64. H. Yonemura, Hypersurface simple K3 singularities. Tôhoku J. Math. **42**, 351–380 (1990)
65. N. Yui, Update on the modularity of Calabi–Yau varieties, in *Calabi–Yau Varieties and Mirror Symmetry*. Fields Institute Communications, vol. 38 (American Mathematical Society, Providence, 2003), pp. 307–362
66. Yu.G. Zarhin, Hodge groups of K3 surfaces J. Reine Angew. Math. **341**, 193–220 (1983)

Part II
Research Articles: Arithmetic and Geometry of K3, Enriques and Other Surfaces

Explicit Algebraic Coverings of a Pointed Torus

Ane S.I. Anema and Jaap Top

Abstract This note contains an application of the algebraic study by Schütt and Shioda of the elliptic modular surface attached to the commutator subgroup of the modular group. This is used here to provide algebraic descriptions of certain coverings of a *j*-invariant 0 elliptic curve, unramified except over precisely one point.

Key words: Covering, Elliptic surface, Torsion section, Potential stable reduction

Mathematics Subject Classifications (2010): Primary 14H30; Secondary 11G05, 14J27, 57M12

1 Introduction

This note was inspired by two recent papers of Jeroen Sijsling [8, 9]. Sijsling finds explicit equations for certain Shimura curves with the special property that they admit a morphism to a curve of genus one; moreover, this morphism is unramified except over precisely one point.

From a topological point of view, the description of such coverings is relatively simple (compare the Introduction of [3]): the fundamental group of a genus one curve minus a point is the free group F_2 on two generators. So finite coverings correspond to finite index subgroups H of F_2 and such a covering ramifies over the distinguished point, precisely when the intersection $N = \cap_{g \in F_2} gHg^{-1}$ of all conjugates of H in F_2 satisfies that F_2/N is nonabelian.

However, it is far from trivial to find algebraic equations for such a topological covering. In this note we use a particular elliptic surface described by Schütt and Shioda [6] in order to obtain some examples. Namely, they take the elliptic curve

A.S.I. Anema • J. Top (✉)
JBI-RuG, Nijenborgh 9, 9747 AG Groningen, The Netherlands
e-mail: a.s.i.anema@rug.nl; j.top@rug.nl

R. Laza et al. (eds.), *Arithmetic and Geometry of K3 Surfaces and Calabi–Yau Threefolds*, 143
Fields Institute Communications 67, DOI 10.1007/978-1-4614-6403-7_5,
© Springer Science+Business Media New York 2013

$B/\mathbb{C}$ with j-invariant 0 and an elliptic surface $\mathcal{E} \to B$ which has precisely one singular fibre. Observing that for any integer $n \neq 0$ the n-torsion subscheme $\mathcal{E}[n] \to B$ of $\mathcal{E} \to B$ is étale except possibly over the point corresponding to the singular fibre, we find coverings of B which are unramified away from this point. In what follows we calculate the ramification indices of such coverings, the degree and hence the genus, and we briefly describe some subcovers.

Most of the results described here were obtained as part of the master's thesis project [1] of the first author in 2011, supervised by the second author. We thank Jeroen Sijsling and Lenny Taelman for their interest in this project.

2 The Coverings

Throughout this note we denote by $B/\mathbb{C}$ the elliptic curve corresponding to the affine equation

$$B/\mathbb{C} \ : \ 4a^3 + 27b^2 = 1.$$

The function field $\mathbb{C}(B)$ of $B/\mathbb{C}$ is the quadratic extension $\mathbb{C}(a,b)$ of the rational function field $\mathbb{C}(a)$ given by $4a^3 + 27b^2 = 1$. The unique point of B where the functions a, b have a pole is denoted by O. As is well known, a has a pole of order 2 and b has a pole of order 3 at O.

The elliptic curve $E/\mathbb{C}(B)$ is defined by the equation

$$E/\mathbb{C}(B) \ : \ y^2 = x^3 + ax + b.$$

Then $E/\mathbb{C}(B)$ is the generic fibre of a unique elliptic surface

$$\mathcal{E} \longrightarrow B$$

defined over B. Up to some scaling factors, this is the elliptic surface studied in [6]. By construction the discriminant of the polynomial $x^3+ax+b \in \mathbb{C}(B)[x]$ is a nonzero constant. Hence $\mathcal{E} \to B$ has smooth fibers over all points in $B(\mathbb{C})$ except the point O. It is easy to check that the fiber over O is of type I_6^* in Kodaira's terminology.

Let ℓ be a prime number and denote by $\mathbb{C}(B)(E[\ell])$ the finite extension of $\mathbb{C}(B)$ obtained by adjoining all coordinates of all points on E of order ℓ. Then $\mathbb{C}(B)(E[\ell])$ is the function field of a unique smooth projective curve $C_\ell/\mathbb{C}$. Moreover, the inclusion $\mathbb{C}(B) \subset \mathbb{C}(B)(E[\ell])$ corresponds to a morphism

$$\pi_\ell \ : \ C_\ell \longrightarrow B.$$

Some properties of these coverings are described in the following result.

Theorem 1. *1. The morphism $\pi_\ell \ : \ C_\ell \to B$ is Galois and it is unramified away from O.*

   2. The morphism π_2 is unramified of degree 3.

   3. The morphism π_3 has degree 8 with Galois group the quaternion group $\{\pm 1, \pm i, \pm j, \pm k\}$, and its ramification index at every point over O equals 2.

4. *For $\ell > 3$, the Galois group of π_ℓ equals $SL_2(\mathbb{F}_\ell)$ and the ramification index at every point over O equals 2ℓ.*

The first claim here is part of the standard results on elliptic curves, see, e.g. [10, VII Proposition 4.1] and [7]. The other assertions will be proven in the next section. Note that the Galois group of π_ℓ is determined by its action on the points of order ℓ on the elliptic curve E, hence this Galois group is a subgroup of $GL_2(\mathbb{Z}/\ell\mathbb{Z})$. Moreover, since $\mathbb{C}(B)$ contains the ℓ-th roots of unity, the Galois invariance of the Weil e_ℓ-pairing implies that this Galois group is in fact a subgroup of $SL_2(\mathbb{Z}/\ell\mathbb{Z})$. Theorem 1 above implies in particular which subgroup we have depending on the prime ℓ: for $\ell = 2$ it is the 3-Sylow subgroup of $SL_2(\mathbb{Z}/2\mathbb{Z})$; for $\ell = 3$ it is the 2-Sylow subgroup of $SL_2(\mathbb{Z}/3\mathbb{Z})$, and for all primes $\ell > 3$ it is the full group $SL_2(\mathbb{Z}/\ell\mathbb{Z})$.

In the last section we briefly discuss some intermediate coverings of the ones described above. These correspond to the intersection of our Galois group with certain subgroups of $SL_2(\mathbb{Z}/\ell\mathbb{Z})$.

3 The Proofs

This section contains proofs of the assertions (2), (3) and (4) of Theorem 1.

3.1 2-Torsion

See also [7, Sect. 5.3a]. The discriminant of the polynomial $x^3 + ax + b \in \mathbb{C}(B)[x]$ is a square. Moreover, this cubic polynomial is irreducible: indeed, if $f \in \mathbb{C}(B)$ were a zero, then f would be regular at all points $\neq O$ of B and f would have a pole of order 1 at O. Since B has positive genus, such an f does not exist.

It follows that the splitting field over $\mathbb{C}(B)$ of the polynomial $x^3 + ax + b$ has degree 3. In particular, the Galois group of this splitting field over $\mathbb{C}(B)$ is cyclic of order 3.

This proves our claims concerning the curves $C_2 \to B$: this map is unramified since it is abelian. An affine curve birational over $\mathbb{C}$ to C_2 is the curve with coordinate ring

$$\mathbb{C}[a, b, x]/(4a^3 + 27b^2 - 1, x^3 + ax + b)$$
$$\cong$$
$$\mathbb{C}[a, x]/(4a^3 + 27x^6 + 54ax^4 + 27a^2x^2 - 1).$$

The map $(a, x) \mapsto (a, -x^3 - ax)$ from the curve with equation $4a^3 + 27x^6 + 54ax^4 + 27a^2x^2 - 1 = 0$ to B corresponds to the covering $C_2 \to B$.

3.2 3-Torsion

See also [7, 5.3b] for a discussion of point of order 3. In particular, since the discriminant of the elliptic curve E is a cube in $\mathbb{C}(B)$, the Galois group of $\pi_3 : C_3 \to B$ is contained in the (unique) 2-Sylow subgroup of $SL_2(\mathbb{Z}/3\mathbb{Z})$, which is isomorphic to the quaternion group of order 8.

Moreover, we claim that $C_3 \to B$ is ramified: indeed, recall that the elliptic curve E has additive reduction at the valuation corresponding to $O \in B$. Over $\mathbb{C}(C_3)$, the curve E cannot have additive reduction at any valuation. This is well-known; the argument runs as follows. Suppose K is the completion at a place v where E has additive reduction, and $E(K)$ contains all points of order $n \geq 3$. Since reduction modulo v:

$$E_0(K)[n] \longrightarrow \overline{E}_{ns}(\mathbb{C}) \cong (\mathbb{C}, +)$$

is injective on torsion points, additive reduction implies that $E_0(K)$ has trivial n-torsion. Hence $E(K)/E_0(K)$ contains a subgroup $\mathbb{Z}/n\mathbb{Z} \times \mathbb{Z}/n\mathbb{Z}$, contradicting the possible structures of this component group.

Since in the extension $\mathbb{C}(C_3) \supset \mathbb{C}(B)$ the reduction of E at (points over) O changes from additive to semi-stable, $C_3 \to B$ is ramified over O. It is unramified over all other points of B, hence we conclude that the Galois group of $C_3 \to B$ is not abelian.

Observing that all proper subgroups of the 2-Sylow subgroup of $SL_2(\mathbb{Z}/3\mathbb{Z})$ are abelian it follows that the Galois group of $C_3 \to B$ equals this 2-Sylow subgroup.

The x-coordinates of all points of order 3 on E generate the splitting field of

$$3x^4 + 6ax^2 + 12bx - a^2$$

over $\mathbb{C}(B)$. This splitting field is, by construction, unramified over all points $\neq O$ of B. It is unramified over O as well: indeed, by its construction, the splitting field has as Galois group over $\mathbb{C}(B)$ the image of $\mathrm{Gal}(\mathbb{C}(B)(E[3])/\mathbb{C}(B))$ under the canonical map

$$SL_2(\mathbb{Z}/3\mathbb{Z}) \longrightarrow PSL_2(\mathbb{Z}/3\mathbb{Z}).$$

Under this map, the image of the 2-Sylow subgroup of $SL_2(\mathbb{Z}/3\mathbb{Z})$ is $\cong \mathbb{Z}/2\mathbb{Z} \times \mathbb{Z}/2\mathbb{Z}$. This implies that the corresponding extension is unramified. Using

$$C_3 \longrightarrow C_3/(\pm 1) \longrightarrow B$$

in which the first map is cyclic of degree 2 and the second map is Galois and unramified, with Galois group $\mathbb{Z}/2\mathbb{Z} \times \mathbb{Z}/2\mathbb{Z}$, it follows that every ramified point of π_3 has ramification index 2.

This proves the assertions concerning π_3. An alternative, more computational proof of the assertion concerning the ramification index is presented in [1, Proposition 4.4].

3.3 ℓ-Torsion with ℓ ≥ 5

Now fix a prime $\ell \geq 5$ and consider the extension $\mathbb{C}(C_\ell) = \mathbb{C}(B)(E[\ell]) \supset \mathbb{C}(B)$. To prove the claims presented in Theorem 1 for this case, we recall a result of Igusa [2]; see also [5, Theorem 1].
Define

$$E' \ : \ y^2 = x^3 - \frac{27t}{t-1728}x - \frac{54t}{t-1728}.$$

This is an elliptic curve over $\mathbb{C}(t)$ with $j(E') = t$. The result of Igusa referred to above is

Proposition 1. *(Igusa)* *The extension $\mathbb{C}(t)(E'[\ell])/\mathbb{C}(t)$ is Galois with group* $SL_2(\mathbb{Z}/\ell\mathbb{Z})$.

To relate the curve $E'/\mathbb{C}(t)$ to the curve $E/\mathbb{C}(a,b)$, note that $j(E) = 6912a^3$. Hence identifying t with $j(E)$, which corresponds to the map $B \to \mathbb{P}^1$ given by $(a,b) \mapsto 6912a^3$, the curves E and E' are both defined over $\mathbb{C}(B)$ and moreover they have the same j-invariant. In fact,

$$-\frac{27 \cdot 6912a^3}{6912a^3 - 1728} = -\frac{27 \cdot 4a^3}{4a^3 - 1} = \left(\frac{2a}{b}\right)^2 a$$

and

$$-\frac{54 \cdot 6912a^3}{6912a^3 - 1728} = -\frac{54 \cdot 4a^3}{4a^3 - 1} = \left(\frac{2a}{b}\right)^3 b$$

This implies that over the quadratic extension $\mathbb{C}(a,b,c)$ of $\mathbb{C}(B)$ defined by $c^2 = 2a/b$, the curves E and E' are isomorphic. Therefore

$$\mathbb{C}(a,b,c,E[\ell]) = \mathbb{C}(a,b,c,E'[\ell]).$$

Consider the following diagram of field extensions:

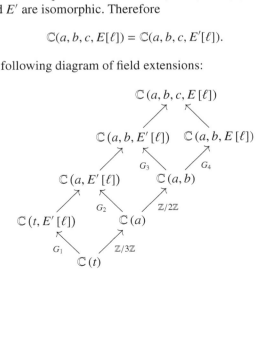

Here the arrows denote Galois extensions, and the corresponding Galois group is written below most of them. The groups G_i may be regarded as subgroups of $SL_2(\mathbb{Z}/\ell\mathbb{Z})$. We will study these groups.

By Proposition 1 we have $G_1 = SL_2(\mathbb{Z}/\ell\mathbb{Z})$. The extension $\mathbb{C}(a, E'[\ell])$ of $\mathbb{C}(t, E'[\ell])$ cannot have degree one, since this would imply that $G_1 = SL_2(\mathbb{Z}/\ell\mathbb{Z})$ would have a normal subgroup G_2 of index 3. This is not the case when $\ell > 3$ is prime, as is a well-known and classical fact; a recently published (standard) proof can be obtained from [4]. Hence the given extension has degree 3, implying that $G_2 = SL_2(\mathbb{Z}/\ell\mathbb{Z})$. A similar reasoning ($G_2$ does not have a subgroup of index 2 for $\ell \neq 2$) shows that $G_3 = SL_2(\mathbb{Z}/\ell\mathbb{Z})$.

Now if G_4 would not be the full group $SL_2(\mathbb{Z}/\ell\mathbb{Z})$, then

$$[\mathbb{C}(a, b, c, E[\ell]) : \mathbb{C}(a, b, E'[\ell])] < [\mathbb{C}(a, b, c, E[\ell]) : \mathbb{C}(a, b, E[\ell])] \leq 2,$$

implying that $\mathbb{C}(a, b, c, E[\ell]) = \mathbb{C}(a, b, E'[\ell])$. The conclusion is that G_4 would be a subgroup of $SL_2(\mathbb{Z}/\ell\mathbb{Z})$ of index 2, which is not true since $\ell \geq 5$. This shows that indeed $\mathbb{C}(a, b, E[\ell])/\mathbb{C}(a, b)$ has Galois group $G_4 = SL_2(\mathbb{Z}/\ell\mathbb{Z})$, which is one of the assertions in Theorem 1.

It remains to prove that for $\ell \geq 5$ the map $\pi_\ell : C_\ell \to B$ is ramified at every point over $O \in B$ with ramification index 2ℓ.

A uniformizer at O is $\pi := 2a/b \in \mathbb{C}(B)$. We consider the curve E over the completion $\mathbb{C}((\pi))$. Note that

$$a\pi^2 = -27 + b^{-2} \in -27 + \pi^6\mathbb{C}[[\pi]]$$

and

$$b\pi^3 = -54 + 2b^{-2} \in -54 + \pi^6\mathbb{C}[[\pi]].$$

Hence using $c^2 = \pi = 2a/b$ (which defines the quadratic covering of B used above, which ramifies over O since c is a uniformizer at the point over O), one finds that over $\mathbb{C}((c))$ the curve E is given as

$$y^2 = x^3 + c^{-4}(-27 + b^{-2})x + c^{-6}(-54 + 2b^{-2}).$$

This is equivalent to

$$y^2 = x^3 + (-27 + b^{-2})x - 54 + 2b^{-2}.$$

Modulo (c) the reduction of the above model is

$$y^2 = x^3 - 27x - 54 = (x + 3)^2(x - 6).$$

So we have an elliptic curve with multiplicative reduction over a local field with an algebraically closed residue field. By the theory of the Tate curve as, e.g., explained in [11, Chap. V Theorems 3.1 and 5.3], $q \in \mathbb{C}((c))$ exists such that for every finite extension L of $\mathbb{C}((c))$ we have a $\mathrm{Gal}(L/\mathbb{C}((c)))$-equivariant isomorphism

$$L^*/q^{\mathbb{Z}} \xrightarrow{\cong} E(L).$$

Moreover, the valuation of q equals the valuation of the discriminant of a minimal model of E. Such a model is given above, and the valuation of the discriminant is 12.

Since $\mathbb{C}((c))(E[\ell]) = \mathbb{C}((c))(\sqrt[\ell]{q})$ and $q \in c^{12}\mathbb{C}[[c]]^*$ and $\ell \geq 5$, one concludes that $\mathbb{C}((c))(E[\ell]) = \mathbb{C}((\sqrt[\ell]{c})) \supset \mathbb{C}((c))$ which is a totally ramified extension of degree ℓ. This yields the extensions

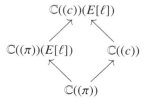

Now observe that any finite extension of $\mathbb{C}((\pi))$ over which the curve E has multiplicative reduction, must contain the element c. In particular, the fact that E has all its points of order ℓ rational over $\mathbb{C}((\pi))(E[\ell])$ implies that E has multiplicative reduction over this field, hence

$$c \in \mathbb{C}((\pi))(E[\ell]).$$

So this shows that

$$\mathbb{C}((\pi)) \subset \mathbb{C}((c)) \subset \mathbb{C}((c))(E[\ell]) = \mathbb{C}((\pi))(E[\ell]),$$

in which the first extension is ramified of degree 2 and the second one is ramified of degree ℓ. As a consequence, the extension $\mathbb{C}((\pi))(E[\ell])/\mathbb{C}((\pi))$ is ramified of degree 2ℓ.

This finishes the proof of Theorem 1. As a remark, one may prove part of the above result by combining a topological and a group theoretic argument. Namely, the topology of the torus implies that the inertia group at any ramified point over O is generated by a commutator $ABA^{-1}B^{-1}$ with $A, B \in \mathrm{SL}_2(\mathbb{Z}/\ell\mathbb{Z})$. Using some group theory one shows that such a commutator cannot have order ℓ, hence since its order is divisible by ℓ, it necessarily equals 2ℓ.

Using that $\#\mathrm{SL}_2(\mathbb{Z}/\ell\mathbb{Z}) = \ell^3 - \ell$ it is now straightforward to compute the genus of the curves C_ℓ. One finds

$$g(C_2) = 1,$$

$$g(C_3) = 3,$$

and

$$g(C_\ell) = 1 + (\ell^2 - 1)(2\ell - 1)/4$$

for any prime number $\ell \geq 5$.

4 Intermediate Coverings

Rather than adjoining the coordinates of all points of order ℓ on E to the field $\mathbb{C}(B)$, one can adjoin only the x-coordinates of these points, or only the coordinates of one point, or only the x-coordinate of one point. This results in an intermediate field between $\mathbb{C}(B)$ and $\mathbb{C}(C_\ell)$, hence with a subgroup of the corresponding Galois group. We briefly discuss this here.

4.1 All x-Coordinates

Let $\mathbb{C}(D_\ell)$ be the subfield of $\mathbb{C}(C_\ell)$ obtained by adjoining all x-coordinates of the points of order ℓ on E, to the field $\mathbb{C}(B)$. This extension corresponds to coverings of curves

$$C_\ell \longrightarrow D_\ell \longrightarrow B$$

and to the (normal) subgroup

$$\{\pm 1\} \cap \mathrm{Gal}(\mathbb{C}(C_\ell)/\mathbb{C}(B)),$$

Hence it is Galois over $\mathbb{C}(B)$, with Galois group equal to the image of $\mathrm{Gal}(\mathbb{C}(C_\ell)/\mathbb{C}(B))$ under the canonical map

$$\mathrm{SL}_2(\mathbb{Z}/\ell\mathbb{Z}) \longrightarrow \mathrm{PSL}_2(\mathbb{Z}/\ell\mathbb{Z}).$$

For $\ell = 2$ this canonical map is a bijection and $\mathbb{C}(D_\ell) = \mathbb{C}(C_\ell)$ (as is obvious since the y-coordinates of points of order 2 on E are zero).

For $\ell = 3$ the image is, as was already discussed earlier, the group $\mathbb{Z}/2\mathbb{Z} \times \mathbb{Z}/2\mathbb{Z}$. A consequence of this is that the polynomial

$$3x^4 + 6ax^2 + 12bx - a^2$$

is irreducible over $\mathbb{C}(B)$: indeed, this polynomial does not have a zero in $\mathbb{C}(B)$, since a root would be a function with a pole of order one at O and no other poles; such a function does not exist.

If the bi-quadratic polynomial would factor as a product of two quadratics f_1 and f_2, then in particular its discriminant (which is a nonzero constant) would equal the product of the discriminants of f_1 and f_2 times the square of the resultant of f_1 and f_2. Hence f_1 and f_2 would have the same splitting field, contradicting the fact that the bi-quadratic polynomial has a Galois group of order 4.

Finally, for $\ell \geq 5$ one has that $D_\ell \to B$ is Galois with group $\mathrm{PSL}_2(\mathbb{Z}/\ell\mathbb{Z})$. The image in this group of a cyclic subgroup in $\mathrm{SL}_2(\mathbb{Z}/\ell\mathbb{Z})$ of order 2ℓ, has order ℓ. As a result, $D_\ell \to B$ ramifies over $O \in B$ with ramification index ℓ.

For the genera of the curves D_ℓ, these observations imply

$$g(D_2) = 1 = g(D_3)$$

and

$$g(D_\ell) = 1 + (\ell^2 - 1)(\ell - 1)/4$$

for all primes $\ell \geq 5$.

4.2 One Point

Fix a point $P \in E$ of exact order ℓ. This yields a tower of extensions

$$\mathbb{C}(B) \subset \mathbb{C}(B)(P) \subset \mathbb{C}(C_\ell)$$

in which the last extension is Galois with group $\left(\begin{smallmatrix} 1 & * \\ 0 & 1 \end{smallmatrix}\right)$ of order ℓ (provided $\ell \geq 5$, otherwise one has $\mathbb{C}(B)(P) = \mathbb{C}(C_\ell)$.) In particular, for $\ell \geq 5$ it follows that the ramification index over O in any point of the first extension is one of 2 or 2ℓ. Moreover, in this case the inertia group at a point of C_ℓ over $O \in B$ is a cyclic subgroup of order 2ℓ in $SL_2(\mathbb{Z}/\ell\mathbb{Z})$. The group SL_2 permutes these ramified points in C_ℓ, and this yields the conjugation action of SL_2 on the set of cyclic subgroups of order 2ℓ. There are precisely $\ell + 1$ such subgroups (corresponding to the one-dimensional subspaces of $\mathbb{F}_\ell^2$). The number of ramified points in $C_\ell \to B$ over O equals $(\ell^3 - \ell)/(2\ell)$. Hence each of the subgroups of order 2ℓ appears $(\ell - 1)/2$ times as inertia group of a point in $C_\ell \to B$.

It follows that in $\mathbb{C}(B) \subset \mathbb{C}(B)(P)$ we have $(\ell - 1)/2$ points with ramification degree 2 (corresponding to the points in C_ℓ which have $\pm\left(\begin{smallmatrix} 1 & * \\ 0 & 1 \end{smallmatrix}\right)$ as inertia group), and $(\ell - 1)/2$ points with ramification index 2ℓ (corresponding to the remaining ramification points in $C_\ell \to B$).

As a consequence, the genus of the corresponding covering of B equals $1 + \ell$ $(\ell - 1)/2$. Note that for $\ell = 2$ and for $\ell = 3$ this genus is respectively 1 and 3, as was shown earlier.

4.3 One x-Coordinate

In a similar manner one may treat the extension

$$\mathbb{C}(B) \subset \mathbb{C}(B)(x(P))$$

obtained by adjoining the x-coordinate of one point of exact order ℓ.

For $\ell \leq 3$ this yields an unramified extension, hence a curve of genus one (these curves were already described above).

For $\ell \geq 5$ a prime number, the extension degree equals $(\ell^2 - 1)/2$ and over O one finds $(\ell - 1)/2$ unramified points and $(\ell - 1)/2$ points with ramification index ℓ. Consequently, the corresponding genus equals $1 + (\ell - 1)^2/4$.

References

1. A.S.I. Anema, Branched covering spaces of an elliptic curve that branch only above a single point, Master's thesis, Groningen, (2011) http://irs.ub.rug.nl/dbi/4e707a67dac82. Accessed 20 July 2012
2. J.-I. Igusa, Fibre systems of Jacobian varieties: (III. Fibre systems of elliptic curves). Am. J. Math. **81**, 453–476 (1959)
3. H.W. Lenstra, Galois theory for schemes. Lecture Notes (2008) http://websites.math.leidenuniv.nl/algebra/GSchemes.pdf. Accessed on 20 July (2012)
4. J. Maciel, $SL_n(F)$ equals its own derived group. Int. J. Algebra **2**, 585–594 (2008)
5. D.E. Rohrlich, Modular curves, Hecke correspondences, and L-functions, in *Modular Forms and Fermat's Last Theorem*, ed. by G.Cornell, J.H. Silverman, G. Stevens (Springer, New York, 1997), pp. 41–100
6. M. Schütt, T. Shioda, An interesting elliptic surface over an elliptic curve. Proc. Jpn. Acad. Sci. **83**, 40–45 (2007)
7. J.-P. Serre, Propriétés galoisiennes des points d'ordre fini des courbes elliptiques. Invent. Math. **15**, 259–331 (1972)
8. J. Sijsling, Arithmetic $(1; e)$-curves and Belyĭ maps. Math. Comp. **81**, 1823–1855 (2012)
9. J. Sijsling, Canonical models of arithmetic $(1; e)$-curves. Math. Zeitschrift **271**, 38 (2012)
10. J.H. Silverman, *The Arithmetic of Elliptic Curves*. Graduate Texts in Mathematics, vol. 106 (Springer, New York, 1986)
11. J.H. Silverman, *Advanced Topics in the Arithmetic of Elliptic Curves*. Graduate Texts in Mathematics, vol. 151 (Springer, New York, 1994)

Elliptic Fibrations on the Modular Surface Associated to $\Gamma_1(8)$

M.J. Bertin and O. Lecacheux

Abstract We give all the elliptic fibrations of the K3 surface associated to the modular group $\Gamma_1(8)$.

Key words: Modular Surfaces, Niemeier lattices, Elliptic fibrations of $K3$ surfaces

Mathematics Subject Classifications (2010): Primary 11F23, 11G05, 14J28; Secondary 14J27

1 Introduction

Stienstra and Beukers [27] considered the elliptic pencil

$$xyz + \tau(x + y)(x + z)(y + z) = 0$$

and the associated K3 surface $\mathcal{B}$ for $\tau = t^2$, double cover of the modular surface for the modular group $\Gamma_0(6)$. With the help of its L-series, they remarked that this surface should carry an elliptic pencil exhibiting it as the elliptic modular surface for $\Gamma_1(8)$ and deplored it was not visible in the previous model of $\mathcal{B}$.

Later on, studying the link between the logarithmic Mahler measure of some K3 surfaces and their L-series, Bertin considered in [3] K3 surfaces of the family previously studied by Peters and Stienstra [19]

$$X + \frac{1}{X} + Y + \frac{1}{Y} + Z + \frac{1}{Z} = k. \qquad (Y_k)$$

M.J. Bertin (✉) · O. Lecacheux
Université Pierre et Marie Curie (Paris 6), Institut de Mathématiques,
4 Place Jussieu, 75005 Paris, France
e-mail: bertin@math.jussieu.fr; lecacheu@math.jussieu.fr

R. Laza et al. (eds.), *Arithmetic and Geometry of K3 Surfaces and Calabi–Yau Threefolds*,   153
Fields Institute Communications 67, DOI 10.1007/978-1-4614-6403-7_6,
© Springer Science+Business Media New York 2013

For $k = 2$, Bertin proved that the corresponding K3 surface Y_2 is singular (i.e. its Picard rank is 20) with transcendental lattice $\left(\begin{smallmatrix} 2 & 0 \\ 0 & 4 \end{smallmatrix}\right)$.

Bertin noticed that Y_2 was nothing else than $\mathcal{B}$, corresponding to the elliptic fibration $X + Y + Z = s$ and $1/\tau = (s - 1)^2$. Its singular fibers are of Dynkin type A_{11}, A_5, $2A_1$ and Kodaira type I_{12}, I_6, $2I_2$, $2I_1$. Its Mordell–Weil group is the torsion group $\mathbb{Z}/6\mathbb{Z}$.

Using an unpublished result of Lecacheux (see also Sect. 7 of this paper), Bertin showed also that Y_2 carries the structure of the modular elliptic surface for $\Gamma_1(8)$. In that case, it corresponds to the elliptic fibration of Y_2 with parameter $Z = s$. Its singular fibers are of Dynkin type $2A_7$, A_3, A_1 and Kodaira type $2I_8$, I_4, I_2, $2I_1$. Its Mordell–Weil group is the torsion group $\mathbb{Z}/8\mathbb{Z}$.

Interested in K3 surfaces with Picard rank 20 over $\mathbb{Q}$, Elkies proved in [8] that their transcendental lattices are primitive of class number one. In particular, he gave in [9] a list of 11 negative integers D for which there is a unique K3 surface X over $\mathbb{Q}$ with Néron–Severi group of rank 20 and discriminant $-D$ consisting entirely of classes of divisors defined over $\mathbb{Q}$. For $D = -8$, he gave an explicit model of an elliptic fibration with $E_8 (= II^*)$ fibers at $t = 0$ and $t = \infty$ and an $A_1 (= I_2)$ fiber at $t = -1$

$$y^2 = x^3 - 675x + 27(27t - 196 + \frac{27}{t}).$$

For this fibration, the Mordell–Weil group has rank 1 and no torsion.

Independently, Schütt proved in [22] the existence of K3 surfaces of Picard rank 20 over $\mathbb{Q}$ and gave for the discriminant $D = -8$ an elliptic fibration with singular fibers A_3, E_7, E_8 (I_4, III^*, II^*) and Mordell–Weil group equal to (0). For such a model, you can refer to [21].

Recall also that Shimada and Zhang gave in [24] a list, without equations but with their Mordell–Weil group, of extremal elliptic K3 surfaces. In particular, there are 14 extremal elliptic K3 surfaces with transcendental lattice

$$\left(\begin{smallmatrix} 2 & 0 \\ 0 & 4 \end{smallmatrix}\right).$$

We mention Beukers and Montanus who worked out the semi-stable, extremal, elliptic fibrations of K3 surfaces [4].

As announced in the abstract, the aim of the paper is to determine all the elliptic fibrations with section on the modular surface associated to $\Gamma_1(8)$ and give for each fibration a Weierstrass model. Thus we recover all the extremal fibrations given by Shimada and Zhang and also fibrations of Bertin, Elkies, Schütt and Stienstra–Beukers mentioned above.

The paper is divided in two parts. In the first sections we use Nishiyama's method, as explained in [18, 23], to determine all the elliptic fibrations of K3 surfaces with a given transcendental lattice. The method is based on lattice theoretical ideas. We prove the following theorem

Theorem. *There are* 30 *elliptic fibrations with section, all distinct up to isomorphism, on the elliptic surface*

$$X + \frac{1}{X} + Y + \frac{1}{Y} + Z + \frac{1}{Z} = 2.$$

They are listed in Table 3 with the rank and torsion of their Mordell–Weil group. The list consists of 14 fibrations of rank 0, 13 fibrations of rank 1 and 3 fibrations of rank 2.

In the second part, i.e. Sects. 7–10, we first explain that Y_2 is the modular surface associated to the modular group $\Gamma_1(8)$. From one of its fibrations we deduce that it is the unique K3 surface X over $\mathbb{Q}$ with Néron–Severi group of rank 20 and discriminant -8, all of its classes of divisors being defined over $\mathbb{Q}$.

Then, for each fibration, we determine explicitly a Weierstrass model, with generators of the Mordell–Weil group.

We first use the 8-torsion sections of the modular fibration to construct the 16 first fibrations. Their parameters belong to a special group generated by 10 functions on the surface. This construction is similar to the one developed for $\Gamma_1(7)$ by Harrache and Lecacheux in [11]. The next fibrations are obtained by classical methods of gluing and breaking singular fibers. The last ones are constructed by adding a vertex to the graph of the modular fibration.

The construction of some of the fibrations can be done also for the other K3 surfaces Y_k of the family. Thus we hope to find for them fibrations of rank 0 and perhaps obtain more easily the discriminant of the transcendental lattice for singular K3 members.

2 Definitions

An integral symmetric bilinear form or a **lattice** of rank r is a free $\mathbb{Z}$-module S of rank r together with a symmetric bilinear form b. If S is a non-degenerate lattice, we write the signature of S, $\text{sign}(S) = (t_+, t_-)$. An indefinite lattice of signature $(1, t_-)$ or $(t_+, 1)$ is called an **hyperbolic** lattice. A lattice S is called **even** if $x^2 := b(x, x)$ is even for all x from S. For any integer n we denote by $\langle n \rangle$ the lattice $\mathbb{Z}e$ where $e^2 = n$. For every integer m we denote by $S[m]$ the lattice obtained from a lattice S by multiplying the values of its bilinear form by m. If $e = (e_1, \ldots, e_r)$ is a $\mathbb{Z}$-basis of a lattice S, then the matrix $G(e) = (b(e_i, e_j))$ is called the **Gram matrix** of S with respect to e.

A homomorphism of lattices $f : S \rightarrow S'$ is a homomorphism of the abelian groups such that $b'(f(x), f(y)) = b(x, y)$ for all $x, y \in S$. An injective (resp. bijective) homomorphism of lattices is called an **embedding** (resp. an isometry). The group of isometries of a lattice S into itself is denoted by $O(S)$ and called the orthogonal group of S. Two embeddings $i : S \rightarrow S'$ and $i' : S \rightarrow S'$ are called isomorphic if there exists an isometry $\sigma \in O(S')$ such that $i' = \sigma \circ i$. An embedding $i : S \rightarrow S'$ is called **primitive** if $S'/i(S)$ is a free group. A sublattice is a subgroup equipped with the induced bilinear form. A sublattice S' of a lattice S is called primitive if the identity map $S' \rightarrow S$ is a primitive embedding. The **primitive closure** of S inside S' is defined by $\overline{S} = \{x \in S'/mx \in S$ for some positive integer $m\}$. A lattice M is an **overlattice** of S if S is a sublattice of M such that the index $[M : S]$ is finite.

By $S_1 \oplus S_2$ we denote the orthogonal sum of two lattices defined in the standard way. We write S^n for the orthogonal sum of n copies of a lattice S. The **orthogonal complement of a sublattice** S of a lattice S' is denoted $(S)_{S'}^{\perp}$ and defined by $(S)_{S'}^{\perp} = \{x \in S'/b(x, y) = 0 \text{ for all } y \in S\}$.

3 Discriminant Forms

Let L be a non-degenerate lattice. The **dual lattice** L^* of L is defined by

$$L^* := \operatorname{Hom}(L, \mathbb{Z}) = \{x \in L \otimes \mathbb{Q}/\ b(x, y) \in \mathbb{Z} \text{ for all } y \in L\}.$$

and the **discriminant group** G_L by

$$G_L := L^*/L.$$

This group is finite if and only if L is non-degenerate. In the latter case, its order is equal to the absolute value of the lattice determinant $|\det(G(e))|$ for any basis e of L. A lattice L is **unimodular** if G_L is trivial.

Let G_L be the discriminant group of a non-degenerate lattice L. The bilinear form on L extends naturally to a $\mathbb{Q}$-valued symmetric bilinear form on L^* and induces a symmetric bilinear form

$$b_L : G_L \times G_L \to \mathbb{Q}/\mathbb{Z}.$$

If L is even, then b_L is the symmetric bilinear form associated to the quadratic form defined by

$$\begin{aligned} q_L : G_L &\to\ \mathbb{Q}/2\mathbb{Z} \\ q_L(x + L) &\mapsto x^2 + 2\mathbb{Z}. \end{aligned}$$

The latter means that $q_L(na) = n^2 q_L(a)$ for all $n \in \mathbb{Z}$, $a \in G_L$ and $b_L(a, a') = \frac{1}{2}(q_L(a + a') - q_L(a) - q_L(a'))$, for all $a, a' \in G_L$, where $\frac{1}{2} : \mathbb{Q}/2\mathbb{Z} \to \mathbb{Q}/\mathbb{Z}$ is the natural isomorphism. The pair (G_L, b_L) (resp. (G_L, q_L)) is called the **discriminant bilinear** (resp. **quadratic**) **form** of L.

4 Root Lattices

In this section we recall only what is needed for the understanding of the paper. For proofs and details one can refer to Bourbaki [5] or Martinet [14].

Let L be a negative-definite even lattice. We call $e \in L$ a **root** if $q_L(e) = -2$. Put $\Delta(L) := \{e \in L/q_L(e) = -2\}$. Then the sublattice of L spanned by $\Delta(L)$ is called the **root type** of L and is denoted by L_{root}. If $e \in \Delta(L)$, we call **reflection associated with** e the following isometry

$$R_e(x) = x + b(x, e)e.$$

The subgroup of $O(L)$ generated by R_e ($e \in \Delta(L)$) is called the **Weyl group** of L and is denoted by $W(L)$.

The **lattices** $A_n = \langle a_1, a_2, \ldots, a_n \rangle$ ($n \geq 1$), $D_l = \langle d_1, d_2, \ldots, d_l \rangle$ ($l \geq 4$), $E_p = \langle e_1, e_2, \ldots, e_p \rangle$ ($p = 6, 7, 8$) defined by the following **Dynkin diagrams** are called the **root lattices**. All the vertices a_j, d_k, e_l are roots and two vertices a_j and a'_j are joined by a line if and only if $b(a_j, a'_j) = 1$. We use Bourbaki's definitions [5] (Fig. 1).

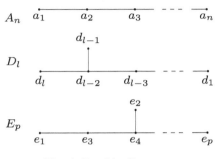

Fig. 1: Dynkin diagrams

Denote ϵ_i the vectors of the canonical basis of $\mathbb{R}^n$ with the negative usual scalar product.

4.1 A_n^*/A_n

We can represent A_n by the set of points in $\mathbb{R}^{n+1}$ with integer coordinates whose sum is zero. Set $a_i = \epsilon_i - \epsilon_{i+1}$ and define

$$\alpha_n = \epsilon_1 - \frac{1}{n+1} \sum_{j=1}^{n+1} \epsilon_j = \frac{1}{n+1} \sum_{j=1}^{n} (n - j + 1) a_j.$$

One can show that

$$A_n^* = \langle A_n, \alpha_n \rangle, \quad A_n^*/A_n \simeq \mathbb{Z}/(n+1)\mathbb{Z} \quad \text{and} \quad q_{A_n}(\alpha_n) = \left(-\frac{n}{n+1} \right).$$

4.2 D_l^*/D_l

We can represent D_l as the set of points of $\mathbb{R}^l$ with integer coordinates of even sum and define

$$\delta_l = \tfrac{1}{2}(\sum_{i=1}^{l} \epsilon_i) = \tfrac{1}{2}\left(\sum_{i=1}^{l-2} id_i + \tfrac{1}{2}(l-2)d_{l-1} + \tfrac{1}{2}ld_l\right)$$
$$\overline{\delta}_l = \epsilon_1 = \sum_{i=1}^{l-2} d_i + \tfrac{1}{2}(d_{l-1} + d_l)$$
$$\tilde{\delta}_l = \delta_l - \epsilon_l = \tfrac{1}{2}\left(\sum_{i=1}^{l-2} id_i + \tfrac{1}{2}ld_{l-1} + \tfrac{1}{2}(l-2)d_l\right).$$

One can show that

$$D_l^* = \langle \epsilon_1, \ldots, \epsilon_l, \delta_l \rangle.$$

Then, for l odd

$$D_l^*/D_l \simeq \mathbb{Z}/4\mathbb{Z} = \langle \delta_l \rangle, \quad \overline{\delta}_l \equiv 2\delta_l \text{ and } \tilde{\delta}_l \equiv 3\delta_l \text{ mod. } D_l$$

and for l even

$$D_l^*/D_l \simeq \mathbb{Z}/2\mathbb{Z} \times \mathbb{Z}/2\mathbb{Z},$$

the three elements of order 2 being the images of δ_l, $\tilde{\delta}_l$ and $\overline{\delta}_l$. Moreover,

$$q_{D_l}(\delta_l) = \left(-\frac{l}{4}\right), \quad q_{D_l}(\overline{\delta}_l) = (-1), \quad b_{D_l}(\delta_l, \overline{\delta}_l) = -\frac{1}{2}.$$

4.3 E_6^*/E_6

We can represent E_6 as a lattice in $\mathbb{R}^8$ generated by the six vectors
$e_1 = \tfrac{1}{2}(\epsilon_1 + \epsilon_8) - \tfrac{1}{2}(\sum_{i=2}^{7} \epsilon_i), \quad e_2 = \epsilon_1 + \epsilon_2, \quad e_i = \epsilon_{i-1} - \epsilon_{i-2}, \quad 3 \le i \le 6.$
If $\eta_6 = -\tfrac{1}{3}(2e_1 + 3e_2 + 4e_3 + 6e_4 + 5e_5 + 4e_6)$, then

$$E_6^* = \langle E_6, \eta_6 \rangle, \quad E_6^*/E_6 \simeq \mathbb{Z}/3\mathbb{Z}, \quad q_{E_6}(\eta_6) = \left(-\frac{4}{3}\right).$$

4.4 E_7^*/E_7

We can represent E_7 as a lattice in $\mathbb{R}^8$ generated by the six previous vectors e_i and
$e_7 = \epsilon_6 - \epsilon_5$. If $\eta_7 = -\tfrac{1}{2}(2e_1 + 3e_2 + 4e_3 + 6e_4 + 5e_5 + 4e_6 + 3e_7)$, then

$$E_7^* = \langle E_7, \eta_7 \rangle, \quad E_7^*/E_7 \simeq \mathbb{Z}/2\mathbb{Z}, \quad q_{E_7}(\eta_7) = \left(-\frac{3}{2}\right).$$

4.5 E_8^*/E_8

We can represent E_8 as the subset of points with coordinates ξ_i satisfying

$$2\xi_i \in \mathbb{Z}, \quad \xi_i - \xi_j \in \mathbb{Z}, \quad \sum_{i=1}^{\infty} \xi_i \in 2\mathbb{Z}.$$

Then $E_8^*/E_8 = (0)$.

5 Elliptic Fibrations

Before giving a complete classification of the elliptic fibrations on the K3 surface Y_2, we recall briefly some useful facts concerning K3 surfaces. For more details see [1, 29].

5.1 K3 Surfaces and Elliptic Fibrations

A K3 surface X is a smooth projective complex surface with

$$K_X = \mathcal{O}_X \quad \text{and} \quad H^1(X, \mathcal{O}_X) = 0.$$

If X is a K3 **surface**, then $H^2(X, \mathbb{Z})$ is torsion free. With the cup product, $H^2(X, \mathbb{Z})$ has the structure of an even lattice. By the Hodge index theorem it has signature $(3, 19)$ and by Poincaré duality it is unimodular. Moreover, as a lattice, $H^2(X, \mathbb{Z}) = U^3 \oplus E_8(-1)^2$.

The **Néron–Severi group** $NS(X)$ (i.e. the group of line bundles modulo algebraic equivalence), with the intersection pairing, if $\rho(X)$ is the Picard number of X, is a lattice of signature $(1, \rho(X) - 1)$. The natural embedding $NS(X) \hookrightarrow H^2(X, \mathbb{Z})$ is a primitive embedding of lattices. If C is a smooth projective curve over an algebraically closed field K, an **elliptic surface** Σ over C is a smooth surface with a surjective morphism

$$f : \Sigma \to C$$

such that almost all fibers are smooth curves of genus 1 and no fiber contains exceptional curves of the first kind. The morphism f defines an elliptic fibration on Σ. We suppose also that every elliptic fibration has a section and so a Weierstrass form. Thus we can consider the generic fiber as an elliptic curve E on $K(C)$ choosing a section as the zero section $\bar{O}$. In the case of K3 surfaces, $C = \mathbb{P}^1$.

The singular fibers were classified by Néron [15, Chap. 3] and Kodaira [11]. They are union of irreducible components with multiplicities; each component is a smooth rational curve with self-intersection -2. The singular fibers are classified in the following Kodaira types:

- Two infinite series $I_n (n > 1)$ and $I_n^* (n \geq 0)$.
- Five types $III, IV, II^*, III^*, IV^*$.

The dual graph of these components (a vertex for each component, an edge for each intersection point of two components) is an **extended Dynkin diagram** of type $\tilde{A}_n, \tilde{D}_l, \tilde{E}_p$. Deleting the zero component (i.e. the component meeting the zero section) gives the Dynkin diagram graph A_n, D_l, E_p. We draw the most useful diagrams, with the multiplicity of the components, the zero component being represented by a circle (Fig. 2).

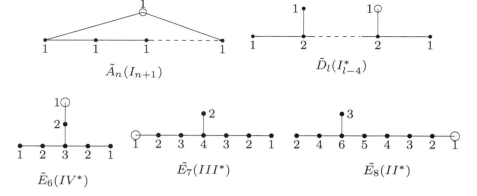

Fig. 2: Extended Dynkin diagrams

The **trivial lattice** $T(X)$ is the subgroup of the Néron–Severi group generated by the zero section and the fibers components. More precisely, the trivial lattice is the orthogonal sum

$$T(X) = < \bar{O}, F > \oplus_{v \in S} T_v$$

where $\bar{O}$ denotes the zero section, F the general fiber, S the points of C corresponding to the reducible singular fibers and T_v the lattice generated by the fiber components except the zero component. From this formula we can compute the determinant of $T(X)$. From Shioda's results on height pairing [25] we can define a positive-definite lattice structure on the Mordell–Weil lattice $MWL(X) := E(K(C))/E(K(C))_{tor}$ and get the following proposition.

Proposition 1. *Let X be a K3 surface or more generally any elliptic surface with section. We have the relation*

$$|disc(NS(X))| = disc(T(X))disc(MWL(X))/|E(K)_{tor}|^2.$$

Moreover since X is a K3 surface, the zero section has self-intersection $\bar{O}^2 = -\chi(X) = -2$. Hence the zero section $\bar{O}$ and the general fiber F generate an even unimodular lattice, called the hyperbolic plane U. The trivial lattice $T(X)$ of an elliptic surface is not always primitive in $NS(X)$. Its primitive closure $\overline{T(X)}$ is obtained by adding the torsion sections. The Néron–Severi lattice $NS(X)$ always contains an even sublattice of corank two, the **frame** $W(X)$

$$W(X) = \langle \bar{O}, F \rangle^{\perp} \subset NS(X).$$

Lemma 1. *For any elliptic surface X with section, the frame $W(X)$ is a negative-definite even lattice of rank $\rho(X) - 2$.*

Remark 1. In this paper X is really considered as a K3 surface with section, that is we look at its various elliptic fibrations f, $f : X \to \mathbb{P}^1$ with section σ. So $T(X)$ and $W(X)$ are in fact $T(f, \sigma)$ and $W(f, \sigma)$ defined for each (f, σ).

The Néron–Severi lattice of a K3 surface is an even lattice and one can read off the Mordell–Weil lattice, the torsion in the Mordell–Weil group MW and the type of singular fibers from $W(X)$ by

$$MWL(X) = W(X)/\overline{W(X)}_{\text{root}} \quad (MW)_{\text{tors}} = \overline{W(X)}_{\text{root}}/W(X)_{\text{root}}$$

$$T(X) = U \oplus W(X)_{\text{root}}.$$

We can also calculate the heights of points from the Weierstrass equation [10] and test if points generate the Mordell–Weil group, since $disc(NS(X))$ is independent of the fibration.

5.2 Nikulin and Niemeier's Results

Lemma 2. [17, Proposition 1.4.1] *Let L be an even lattice. Then, for an even over-lattice M of L, we have a subgroup M/L of $G_L = L^*/L$ such that q_L is trivial on M/L. This determines a bijective correspondence between even overlattices of L and subgroups G of G_L such that $q_L|_G = 0$.*

Lemma 3. [17, Proposition 1.6.1] *Let L be an even unimodular lattice and T a primitive sublattice. Then we have*

$$G_T \simeq G_{T^\perp} \simeq L/(T \oplus T^\perp), \quad q_{T^\perp} = -q_T.$$

In particular, $\det T = \det T^\perp = [L : T \oplus T^\perp]$.

Theorem 1. [17, Corollary 1.6.2] *Let L and M be non-degenerate even integral lattices such that*

$$G_L \simeq G_M, \quad q_L = -q_M.$$

Then there exists an unimodular overlattice N of $L \oplus M$ such that

1. *the embeddings of L and M in N are primitive*
2. *$L_N^\perp = M$ and $M_N^\perp = L$.*

Theorem 2. [17, Theorem 1.12.4] *Let there be given two pairs of nonnegative integers, $(t_{(+)}, t_{(-)})$ and $(l_{(+)}, l_{(-)})$. The following properties are equivalent:*

a) *every even lattice of signature $(t_{(+)}, t_{(-)})$ admits a primitive embedding into some even unimodular lattice of signature $(l_{(+)}, l_{(-)})$;*

b) *$l_{(+)} - l_{(-)} \equiv 0 \ (mod \ 8)$, $t_{(+)} \leq l_{(+)}$, $t_{(-)} \leq l_{(-)}$ and $t_{(+)} + t_{(-)} \leq \frac{1}{2}(l_{(+)} + l_{(-)})$.*

Theorem 3. [16] *A negative-definite even unimodular lattice L of rank 24 is determined by its root lattice L_{root} up to isometries. There are 24 possibilities for L and L/L_{root} listed in Table 1.*

Table 1: Niemeier lattices

L_{root}	L/L_{root}	L_{root}	L/L_{root}
E_8^3	(0)	$D_5^{\oplus 2} \oplus A_7^{\oplus 2}$	$\mathbb{Z}/4\mathbb{Z} \oplus \mathbb{Z}/8\mathbb{Z}$
$E_8 \oplus D_{16}$	$\mathbb{Z}/2\mathbb{Z}$	$A_8^{\oplus 3}$	$\mathbb{Z}/3\mathbb{Z} \oplus \mathbb{Z}/9\mathbb{Z}$
$E_7^{\oplus 2} \oplus D_{10}$	$(\mathbb{Z}/2\mathbb{Z})^2$	A_{24}	$\mathbb{Z}/5\mathbb{Z}$
$E_7 \oplus A_{17}$	$\mathbb{Z}/6\mathbb{Z}$	$A_{12}^{\oplus 2}$	$\mathbb{Z}/13\mathbb{Z}$
D_{24}	$\mathbb{Z}/2\mathbb{Z}$	$D_4^{\oplus 6}$	$(\mathbb{Z}/2\mathbb{Z})^6$
$D_{12}^{\oplus 2}$	$(\mathbb{Z}/2\mathbb{Z})^2$	$D_4 \oplus A_5^{\oplus 4}$	$\mathbb{Z}/2\mathbb{Z} \oplus (\mathbb{Z}/6\mathbb{Z})^2$
$D_8^{\oplus 3}$	$(\mathbb{Z}/2\mathbb{Z})^3$	$A_6^{\oplus 4}$	$(\mathbb{Z}/7\mathbb{Z})^2$
$D_9 \oplus A_{15}$	$\mathbb{Z}/8\mathbb{Z}$	$A_4^{\oplus 6}$	$(\mathbb{Z}/5\mathbb{Z})^3$
$E_6^{\oplus 4}$	$(\mathbb{Z}/3\mathbb{Z})^2$	$A_3^{\oplus 8}$	$(\mathbb{Z}/4\mathbb{Z})^4$
$E_6 \oplus D_7 \oplus A_{11}$	$\mathbb{Z}/12\mathbb{Z}$	$A_2^{\oplus 12}$	$(\mathbb{Z}/3\mathbb{Z})^6$
$D_6^{\oplus 4}$	$(\mathbb{Z}/2\mathbb{Z})^4$	$A_1^{\oplus 24}$	$(\mathbb{Z}/2\mathbb{Z})^{12}$
$D_6 \oplus A_9^{\oplus 2}$	$\mathbb{Z}/2\mathbb{Z} \oplus \mathbb{Z}/10\mathbb{Z}$	0	Λ_{24}

The lattices L defined in Table 1 are called **Niemeier lattices**.

5.3 Nishiyama's Method

Recall that a K3 surface may admit more than one elliptic fibration, but up to isomorphism, there is only a finite number of elliptic fibrations [26]. To establish a complete classification of the elliptic fibrations on the K3 surface Y_2, we use Nishiyama's method based on lattice theoretic ideas [18]. The technique builds on a converse of Nikulin's results.

Given an elliptic K3 surface X, Nishiyama aims at embedding the frames of all elliptic fibrations into a negative-definite lattice, more precisely into a Niemeier lattice of rank 24. For this purpose, he first determines an even negative-definite lattice M such that

$$q_M = -q_{NS(X)}, \quad \mathrm{rank}(M) + \rho(X) = 26.$$

By Theorem 1, $M \oplus W(X)$ has a Niemeier lattice as an overlattice for each frame $W(X)$ of an elliptic fibration on X. Thus one is bound to determine the (inequivalent) primitive embeddings of M into Niemeier lattices L. To achieve this, it is essential to consider the root lattices involved. In each case, the orthogonal complement of M into L gives the corresponding frame $W(X)$.

5.3.1 The Transcendental Lattice and Argument from Nishiyama Paper

Denote by $\mathbb{T}(X)$ the transcendental lattice of X, i.e. the orthogonal complement of $NS(X)$ in $H^2(X, \mathbb{Z})$ with respect to the cup-product,

$$\mathbb{T}(X) = NS(X)^{\perp} \subset H^2(X, \mathbb{Z}).$$

In general, $\mathbb{T}(X)$ is an even lattice of rank $t = 22 - \rho(X)$ and signature $(2, 20 - \rho(X))$. Let $t' = t - 2$. By Nikulin's Theorem 2, $\mathbb{T}(X)[-1]$ admits a primitive embedding into the following indefinite unimodular lattice:

$$\mathbb{T}(X)[-1] \hookrightarrow U^{t'} \oplus E_8.$$

Then define M as the orthogonal complement of $\mathbb{T}(X)[-1]$ in $U^{t'} \oplus E_8$. By construction, M is a negative-definite lattice of rank $2t' + 8 - t = t + 4 = 26 - \rho(X)$.

By Lemma 3 the discriminant form satisfies

$$q_M = -q_{\mathbb{T}(X)[-1]} = q_{\mathbb{T}(X)} = -q_{NS(X)}.$$

Hence M takes exactly the shape required for Nishiyama's technique.

5.3.2 Torsion Group

First we classify all the primitive embeddings of M into L_{root}. Let N be the orthogonal complement of M into L_{root} and W the orthogonal complement of M into L. If M satisfies $M_{\text{root}} = M$, we can apply Nishiyama's results [18]. In particular,

- M primitively embedded in $L_{\text{root}} \iff M$ primitively embedded in L,
- N/N_{root} is torsion-free.

Remark 2. Notice that the rank r of the Mordell–Weil group is equal to $\text{rk}(W) - \text{rk}(W_{\text{root}})$ and its torsion part is $\overline{W}_{\text{root}}/W_{\text{root}}$.

We need also the following lemma.

Lemma 4. [18, Lemma 6.6] *We have the following facts:*

1. *If* $\det N = \det M$, *then the Mordell–Weil group is torsion-free.*
2. *If* $r = 0$, *then the Mordell–Weil group is isomorphic to* W/N.
3. *In general, there are the following inclusions of groups:*

$$\overline{W}_{root}/W_{root} \subset W/N \subset L/L_{root}.$$

6 Elliptic Fibrations of Y_2

We follow Nishiyama's method. Since

$$\mathbb{T}(Y_2) = \begin{pmatrix} 2 & 0 \\ 0 & 4 \end{pmatrix},$$

we get, by Nishiyama's computation [18], $M = D_5 \oplus A_1$. Thus we have to determine all the primitive embeddings of M into the root lattices and their orthogonal complements.

6.1 The Primitive Embeddings of $D_5 \oplus A_1$ into Root Lattices

Proposition 2. *There are primitive embeddings of $D_5 \oplus A_1$ only into the following L_{root}:*

$$E_8^{\oplus 3}, \ E_8 \oplus D_{16}, \ E_7^{\oplus 2} \oplus D_{10}, \ E_7 \oplus A_{17}, \ D_8^{\oplus 3}, \ D_9 \oplus A_{15},$$

$$E_6^{\oplus 4}, \ A_{11} \oplus E_6 \oplus D_7, \ D_6^{\oplus 4}, \ D_6 \oplus A_9^{\oplus 2}, \ D_5^{\oplus 2} \oplus A_7^{\oplus 2}.$$

Proof. The assertion comes from Nishiyama's results [18]. The lattice A_1 can be primitively embedded in all A_n, D_l and E_p, $n \geq 1$, $l \geq 2$, $p = 6, 7, 8$. The lattice D_5 can be primitively embedded only in D_l, $l \geq 5$ and E_p, $p = 6, 7, 8$. The lattice $D_5 \oplus A_1$ can be primitively embedded only in D_l, $l \geq 7$, E_7 and E_8. The proposition follows from Theorem 3 and the previous facts. $\quad\square$

Proposition 3. *Up to the action of the Weyl group, the unique primitive embeddings are given in the following list*

- $A_1 = \langle a_n \rangle \subset A_n$
- $A_1 = \langle d_l \rangle \subset D_l, \ l \geq 4$
- $A_1 = \langle e_1 \rangle \subset E_p, \ p = 6, 7, 8$
- $D_5 = \langle d_{l-1}, d_l, d_{l-2}, d_{l-3}, d_{l-4} \rangle \subset D_l, \ l \geq 6$
- $D_5 = \langle e_2, e_5, e_4, e_3, e_1 \rangle \subset E_n, \ n \geq 6$
- $D_5 \oplus A_1 = \langle d_{l-1}, d_l, d_{l-2}, d_{l-3}, d_{l-4} \rangle \oplus \langle d_{l-6} \rangle \subset D_l, \ l \geq 7$
- $D_5 \oplus A_1 = \langle e_2, e_5, e_4, e_3, e_1 \rangle \oplus \langle e_7 \rangle \subset E_n, \ n \geq 7.$

Proof. The first five assertions follow from Nishiyama's computations [18]. Just be careful of the difference of notations between Nishiyama and us. The two last assertions follow from the lemmas below.

Lemma 5. *Up to isomorphism by an element of the Weyl group, there is a unique primitive embedding of $D_5 \oplus A_1$ into E_8 given by*

$$\langle e_2, e_5, e_4, e_3, e_1 \rangle \oplus \langle e_7 \rangle.$$

Proof. Up to isomorphism in E_8 there is a unique primitive embedding of D_5 into E_8 given by $\langle e_2, e_5, e_4, e_3, e_1 \rangle$ [18]. Moreover

$$(D_5)_{E_8}^{\perp} = \langle e_7, e_8, 3e_2 + 2e_1 + 4e_3 + 6e_4 + 5e_5 + 4e_6 + 3e_7 + 2e_8 \rangle = A_3$$

Thus we get three primitive embeddings of $D_5 \oplus A_1$ into E_8

(1) $\langle e_2, e_5, e_4, e_3, e_1 \rangle \oplus \langle e_7 \rangle$
(2) $\langle e_2, e_5, e_4, e_3, e_1 \rangle \oplus \langle e_8 \rangle$
(3) $\langle e_2, e_5, e_4, e_3, e_1 \rangle \oplus \langle x = 3e_2 + 2e_1 + 4e_3 + 6e_4 + 5e_5 + 4e_6 + 3e_7 + 2e_8 \rangle$.

The primitive embedding (2) is isomorphic to the primitive embedding (3) by the reflection $R = R_{3e_2+2e_1+4e_3+6e_4+5e_5+4e_6+3e_7+e_8}$, since $R(x) = e_8$, $R(e_i) = e_i$, $1 \leq i \leq 6$.

The primitive embedding (1) is isomorphic to the primitive embedding (2) by the isomorphism $R = R_{e_7} \circ R_{e_7+e_8}$, since $R(e_8) = e_7$ and $R(e_i) = e_i$, $1 \leq i \leq 5$. □

Lemma 6. *Up to isomorphism by an element of the Weyl group, there is a unique primitive embedding of $D_5 \oplus A_1$ into E_7 given by*

$$\langle e_2, e_5, e_4, e_3, e_1 \rangle \oplus \langle e_7 \rangle.$$

Proof. By Nishiyama [18], up to isomorphism, the unique primitive embedding of D_5 into E_7 is given by $\langle e_2, e_5, e_4, e_3, e_1 \rangle$. And its orthogonal into E_7 is $\langle e_7 \rangle \oplus \langle -4 \rangle$. □

Lemma 7. *Up to isomorphism by an element of the Weyl group, there is a unique primitive embedding of $D_5 \oplus A_1$ into D_l, $l \geq 7$, given by*

$$\langle d_{l-1}, d_l, d_{l-2}, d_{l-3}, d_{l-4} \rangle \oplus \langle d_{l-6} \rangle.$$

Proof. The unique primitive embedding, up to isomorphism, of D_5 into D_l, $l \geq 7$ is $\langle d_{l-1}, d_l, d_{l-2}, d_{l-3}, d_{l-4} \rangle$, its orthogonal in D_l being $\langle x, d_{l-6}, d_{l-5}, \ldots, d_1 \rangle$ with $x = d_{l-1} + d_l + 2(d_{l-2} + d_{l-3} + d_{l-4} + d_{l-5}) + d_{l-6}$.

It is sufficient to prove that the primitive embeddings

- $\langle d_{l-1}, d_l, d_{l-2}, d_{l-3}, d_{l-4} \rangle \oplus \langle x \rangle$
- $\langle d_{l-1}, d_l, d_{l-2}, d_{l-3}, d_{l-4} \rangle \oplus \langle d_{l-6} \rangle$

are isomorphic.

Let

$$R = R_x \circ R_{d_l+d_{l-2}+d_{l-3}+d_{l-4}+d_{l-5}+d_{l-6}} \circ R_{d_{l-1}+d_{l-2}+d_{l-3}+d_{l-4}+d_{l-5}+d_{l-6}}.$$

Now $R(x) = d_{l-6}$, $R(d_{l-1}) = d_l$, $R(d_l) = d_{l-1}$, $R(d_{l-i}) = d_{l-i}$, $2 \leq i \leq 5$. So R gives the isomorphism. □

□

Proposition 4. *We get the following results about the orthogonal complements of the previous embeddings. For notations we refer to Sect. 4.*

1. $(A_1)^\perp_{A_n} = L^2_{n-2} = \begin{pmatrix} -2 \times 3 & 2 & 0 & \dots & 0 \\ 2 & & & & \\ 0 & & & & \\ \vdots & & A_{n-2} & & \\ 0 & & & & \end{pmatrix}$

   with $\det L^2_{n-2} = 2(n+1)$

2. $(A_1)^\perp_{D_4} = A_1^{\oplus 3}$ and $(A_1)^\perp_{D_n} = A_1 \oplus D_{n-2},\ n \geq 5$

3. $(A_1)^\perp_{A_7} = (\langle a_7 \rangle)^\perp_{A_7} = \langle a_7 + 2a_6, a_5, a_4, a_3, a_2, a_1 \rangle$

   $\alpha_7 \in (A_1)^\perp_{A_7^*}$ but $k\alpha_7 \notin ((A_1)^\perp_{A_7}) root = A_5$ for all k

4. $(A_1)^\perp_{A_9} = (\langle a_9 \rangle)^\perp_{A_9} = \langle a_9 + 2a_8, a_7, a_6, a_5, a_4, a_3, a_2, a_1 \rangle$

   $\alpha_9 \in (A_1)^\perp_{A_9^*}$ but $k\alpha_9 \notin ((A_1)^\perp_{A_9}) root = A_7$ for all k

5. $(A_1)^\perp_{A_{11}} = \langle a_{11} + 2a_{10}, a_9, a_8, a_7, a_6, a_5, a_4, a_3, a_2, a_1 \rangle$

   $\alpha_{11} \in (A_1)^\perp_{A_{11}^*}$ but $k\alpha_{11} \notin ((A_1)^\perp_{A_{11}}) root = A_9$ for all k

6. $(A_1)^\perp_{D_6} = \langle d_5 \rangle \oplus \langle d_5 + d_6 + 2d_4 + d_3, d_3, d_2, d_1 \rangle = A_1 \oplus D_4$

   $\bar{\delta}_6$ and $\tilde{\delta}_6 \in (A_1)^\perp_{D_6^*},\quad \delta_6 \notin (A_1)^\perp_{D_6^*}$

7. $(A_1)^\perp_{D_7} = \langle d_7 \rangle^\perp_{D_7}$

   $= \langle d_6 \rangle \oplus \langle d_6 + d_7 + 2d_5 + d_4, d_4, d_3, d_2, d_1 \rangle = A_1 \oplus D_5$

   $3\delta_7 \in (A_1)^\perp_{D_7^*}$

8. $(A_1)^\perp_{D_{10}} = \langle d_{10} \rangle^\perp_{D_{10}}$

   $= \langle d_9 \rangle \oplus \langle d_9 + d_{10} + 2d_8 + d_7, d_7, d_6, d_5, d_4, d_3, d_2, d_1 \rangle$

   $= A_1 \oplus D_8$

   $2\bar{\delta}_{10} \in A_1 \oplus D_8$

9. $(A_1)^\perp_{E_6} = \langle e_1 + e_2 + 2e_3 + 2e_4 + e_5, e_6, e_5, e_4, e_2 \rangle = A_5$

   $3\eta_6 \in A_5$

10. $(A_1)^\perp_{E_7} = \langle e_1 + e_2 + 2e_3 + 2e_4 + e_5, e_7, e_6, e_5, e_4, e_2 \rangle = D_6$

    $2\eta_7 \in (A_1)^\perp_{E_7}$

11. $(A_1)^\perp_{E_8} = E_7$

12. $(D_5)^\perp_{D_l} = D_{l-5}$

    $D_1 = (-4)\quad D_2 = A_1^{\oplus 2}\quad D_3 = \begin{pmatrix} -2 & 0 & 1 \\ 0 & -2 & 1 \\ 1 & 1 & -2 \end{pmatrix} \simeq A_3$

13. $(D_5)^{\perp}_{D_6} = \langle d_5 + d_6 + 2d_4 + 2d_3 + 2d_2 + 2d_1 \rangle = \langle (-4) \rangle$

   $\delta_6 \quad and \quad \tilde{\delta}_6 \notin (D_5)^{\perp}_{D_6^*}, \quad \overline{\delta}_6 \in (D_5)^{\perp}_{D_6^*}$

14. $(D_5)^{\perp}_{E_6} = \langle e_2, e_5, e_4, e_3, e_1 \rangle^{\perp}_{E_6} = \langle 3e_2 + 2e_1 + 4e_3 + 6e_4 + 5e_5 + 4e_6 \rangle = \langle (-12) \rangle$

   $3\eta_6 = (-12)$

15. $(D_5)^{\perp}_{E_7} = \langle 2e_1 + 2e_2 + 3e_3 + 4e_4 + 3e_5 + 2e_6 + e_7 \rangle$

   $\qquad \oplus \langle e_2 + e_3 + 2e_4 + 2e_5 + 2e_6 + 2e_7 \rangle$

   $\qquad = A_1 \oplus (-4)$

   $\eta_7 \in (D_5)^{\perp}_{E_7}, \quad 2\eta_7 \notin A_1$

16. $(D_5)^{\perp}_{E_8} = A_3$

17. $(D_5 \oplus A_1)^{\perp}_{D_7} = A_1 = (d_6 + d_7 + 2d_5 + 2d_4 + 2d_3 + 2d_2 + d_1)$

18. $(D_5 \oplus A_1)^{\perp}_{D_8} = \langle d_7 + d_8 + 2d_6 + 2d_5 + 2d_4 + 2d_3 + d_2 \rangle \oplus$

   $\qquad \langle d_7 + d_8 + 2d_6 + 2d_5 + 2d_4 + 2d_3 + 2d_2 + 2d_1 \rangle$

   $\qquad = A_1 \oplus (-4)$

   $4\delta_8 \notin (D_5 \oplus A_1)^{\perp}_{D_8}, \quad \overline{\delta}_8 \notin A_1$

19. $(D_5 \oplus A_1)^{\perp}_{D_9} = A_1 \oplus A_1 \oplus A_1$

   $\qquad = (d_1) \oplus (d_8 + d_9 + 2d_7 + 2d_6 + 2d_5 + 2d_4 + d_3) \oplus$

   $\qquad \quad (d_8 + d_9 + 2d_7 + 2d_6 + 2d_5 + 2d_4 + 2d_3 + 2d_2 + d_1)$

   $(D_5 \oplus A_1)^{\perp}_{D_{10}} = A_1 \oplus A_3 \qquad (D_5 \oplus A_1)^{\perp}_{D_{12}} = A_1 \oplus D_5$

   $(D_5 \oplus A_1)^{\perp}_{D_{16}} = A_1 \oplus D_9 \qquad (D_5 \oplus A_1)^{\perp}_{D_{24}} = A_1 \oplus D_{17}$

20. $(D_5 \oplus A_1)^{\perp}_{E_7} = \langle 3e_2 + 2e_1 + 4e_3 + 6e_4 + 5e_5 + 4e_6 + 2e_7 \rangle = \langle (-4) \rangle$

   $2\eta_7 \notin (D_5 \oplus A_1)^{\perp}_{E_7}$

21. $(D_5 \oplus A_1)^{\perp}_{E_8} = A_1 \oplus (-4)$

   $\qquad = (3e_2 + 2e_1 + 4e_3 + 6e_4 + 5e_5 + 4e_6 + 3e_7 + 2e_8)$

   $\qquad \oplus (3e_2 + 2e_1 + 4e_3 + 6e_4 + 5e_5 + 4e_6 + 2e_7)$

Proof. The orthogonal complements are given in Nishiyama [18] and the rest of the proof follows immediately from the various expressions of η_7, η_6, δ_l, $\overline{\delta}_l$, $\tilde{\delta}$ and α_m given in Sect. 4. $\qquad\qquad\qquad\qquad\qquad\qquad\qquad\qquad\qquad\qquad\qquad \Box$

Once the different types of fibrations are known, we get the rank of the Mordell–Weil group by 2.

To determine the torsion part we need to know appropriate generators of L/L_{root}.

6.2 Generators of L/L_{root}

By Lemma 2, a set of generators can be described in terms of elements of L^*_{root}/L_{root}. We list in the Table 2 the generators fitting to the corresponding W. We restrict to relevant L_{root} according to Proposition 2.

For convenience, the generators are given modulo L_{root}.

Table 2: A set of generators of L/L_{root}

L_{root}	L/L_{root}
E_8^3	$\langle(0)\rangle$
$E_8 D_{16}$	$\langle \delta_{16}\rangle \simeq \mathbb{Z}/2\mathbb{Z}$
$E_7^2 D_{10}$	$\langle \eta_7^{(1)} + \delta_{10}, \eta_7^{(2)} + \bar{\delta}_{10}\rangle \simeq (\mathbb{Z}/2\mathbb{Z})^2$
$E_7 A_{17}$	$\langle \eta_7 + 3\alpha_{17}\rangle \simeq \mathbb{Z}/6\mathbb{Z}$
D_8^3	$\langle \delta_8^{(1)} + \bar{\delta}_8^{(2)} + \bar{\delta}_8^{(3)}, \bar{\delta}_8^{(1)} + \delta_8^{(2)} + \bar{\delta}_8^{(3)}, \bar{\delta}_8^{(1)} + \bar{\delta}_8^{(2)} + \delta_8^{(3)}\rangle \simeq (\mathbb{Z}/2\mathbb{Z})^3$
$D_9 A_{15}$	$\langle \delta_9 + 2\alpha_{15}\rangle \simeq \mathbb{Z}/8\mathbb{Z}$
E_6^4	$\langle \eta_6^{(1)} + \eta_6^{(2)} + \eta_6^{(3)}, 2\eta_6^{(1)} + \eta_6^{(3)} + \eta_6^{(4)}\rangle \simeq (\mathbb{Z}/3\mathbb{Z})^2$
$A_{11} E_6 D_7$	$\langle \alpha_{11} + \eta_6 + \delta_7\rangle \simeq \mathbb{Z}/12\mathbb{Z}$
D_6^4	$\langle \bar{\delta}_6^{(1)} + \bar{\delta}_6^{(4)}, \bar{\delta}_6^{(2)} + \bar{\delta}_6^{(3)}, \delta_6^{(1)} + \bar{\delta}_6^{(3)} + \bar{\delta}_6^{(4)}, \bar{\delta}_6^{(1)} + \bar{\delta}_6^{(2)}\rangle \simeq (\mathbb{Z}/2\mathbb{Z})^4$
$D_6 A_9^2$	$\langle \delta_6 + 5\alpha_9^{(2)}, \delta_6 + \alpha_9^{(1)} + 2\alpha_9^{(2)}\rangle \simeq \mathbb{Z}/2\mathbb{Z} \times \mathbb{Z}/10\mathbb{Z}$
$D_5^2 A_7^2$	$\langle \delta_5^{(1)} + \delta_5^{(2)} + 2\alpha_7^{(1)}, \delta_5^{(1)} + 2\delta_5^{(2)} + \alpha_7^{(1)} + \alpha_7^{(2)}\rangle \simeq \mathbb{Z}/4\mathbb{Z} \times \mathbb{Z}/8\mathbb{Z}$

Theorem 4. *There are* 30 *elliptic fibrations with section, unique up to isomorphism, on the elliptic surface* Y_2. *They are listed with the rank and torsion of their Mordell–Weil groups on Table 3.*

Proof. If the rank is 0, we apply Lemma 4(1) and (2) to determine the torsion part of the Mordell–Weil group. Thus we recover the 14 fibrations of rank 0 exhibited by Shimada and Zhang [24].

For the other 16 fibrations we apply Proposition 4 and Lemma 4(3). Recall that $\det W = 8$ and the torsion group is $\overline{W}_{root}/W_{root}$.

6.3 $L_{root} = E_8 D_{16}$

$$L/L_{root} = \langle \delta_{16} + L_{root}\rangle \simeq \mathbb{Z}/2\mathbb{Z}$$

Table 3: The elliptic fibrations of Y_2

L_{root}	L/L_{root}			Fibers	R	Tor.
E_8^3	(0)					
		$A_1 \subset E_8$	$D_5 \subset E_8$	$E_7 A_3 E_8$	0	(0)
		$A_1 \oplus D_5 \subset E_8$		$A_1 E_8 E_8$	1	(0)
$E_8 D_{16}$	$\mathbb{Z}/2\mathbb{Z}$					
		$A_1 \subset E_8$	$D_5 \subset D_{16}$	$E_7 D_{11}$	0	(0)
		$A_1 \oplus D_5 \subset E_8$		$A_1 D_{16}$	1	$\mathbb{Z}/2\mathbb{Z}$
		$D_5 \subset E_8$	$A_1 \subset D_{16}$	$A_3 A_1 D_{14}$	0	$\mathbb{Z}/2\mathbb{Z}$
		$A_1 \oplus D_5 \subset D_{16}$		$E_8 A_1 D_9$	0	(0)
$E_7^2 D_{10}$	$(\mathbb{Z}/2\mathbb{Z})^2$					
		$A_1 \subset E_7$	$D_5 \subset D_{10}$	$E_7 D_6 D_5$	0	$\mathbb{Z}/2\mathbb{Z}$
		$A_1 \subset E_7$	$D_5 \subset E_7$	$D_6 A_1 D_{10}$	1	(0)
		$A_1 \oplus D_5 \subset E_7$		$E_7 D_{10}$	1	$\mathbb{Z}/2\mathbb{Z}$
		$A_1 \oplus D_5 \subset D_{10}$		$E_7 E_7 A_1 A_3$	0	$\mathbb{Z}/2\mathbb{Z}$
		$D_5 \subset E_7$	$A_1 \subset D_{10}$	$A_1 A_1 D_8 E_7$	1	$\mathbb{Z}/2\mathbb{Z}$
$E_7 A_{17}$	$\mathbb{Z}/6\mathbb{Z}$					
		$A_1 \oplus D_5 \subset E_7$		A_{17}	1	$\mathbb{Z}/3\mathbb{Z}$
		$D_5 \subset E_7$	$A_1 \subset A_{17}$	$A_1 A_{15}$	2	(0)
D_{24}	$\mathbb{Z}/2\mathbb{Z}$					
		$A_1 \oplus D_5 \subset D_{24}$		$A_1 D_{17}$	0	(0)
D_{12}^2	$(\mathbb{Z}/2\mathbb{Z})^2$					
		$A_1 \subset D_{12}$	$D_5 \subset D_{12}$	$A_1 D_{10} D_7$	0	$\mathbb{Z}/2\mathbb{Z}$
		$A_1 \oplus D_5 \subset D_{12}$		$A_1 D_5 D_{12}$	0	$\mathbb{Z}/2\mathbb{Z}$
D_8^3	$(\mathbb{Z}/2\mathbb{Z})^3$					
		$A_1 \subset D_8$	$D_5 \subset D_8$	$A_1 D_6 A_3 D_8$	0	$(\mathbb{Z}/2)^2$
		$A_1 \oplus D_5 \subset D_8$		$A_1 D_8 D_8$	1	$\mathbb{Z}/2\mathbb{Z}$
$D_9 A_{15}$	$\mathbb{Z}/8\mathbb{Z}$					
		$A_1 \oplus D_5 \subset D_9$		$A_1 A_1 A_1 A_{15}$	0	$\mathbb{Z}/4\mathbb{Z}$
		$D_5 \subset D_9$	$A_1 \subset A_{15}$	$D_4 A_{13}$	1	(0)
E_6^4	$(\mathbb{Z}/3\mathbb{Z})^2$					
		$A_1 \subset E_6$	$D_5 \subset E_6$	$A_5 E_6 E_6$	1	$\mathbb{Z}/3\mathbb{Z}$
$A_{11} E_6 D_7$	$\mathbb{Z}/12\mathbb{Z}$					
		$A_1 \subset E_6$	$D_5 \subset D_7$	$A_5 A_1 A_1 A_{11}$	0	$\mathbb{Z}/6\mathbb{Z}$
		$A_1 \subset A_{11}$	$D_5 \subset D_7$	$A_9 A_1 A_1 E_6$	1	(0)
		$A_1 \oplus D_5 \subset D_7$		$A_{11} E_6 A_1$	0	$\mathbb{Z}/3\mathbb{Z}$
		$A_1 \subset A_{11}$	$D_5 \subset E_6$	$A_9 D_7$	2	(0)
		$D_5 \subset E_6$	$A_1 \subset D_7$	$A_{11} A_1 D_5$	1	$\mathbb{Z}/4\mathbb{Z}$
D_6^4	$(\mathbb{Z}/2\mathbb{Z})^4$					
		$A_1 \subset D_6$	$D_5 \subset D_6$	$A_1 D_4 D_6 D_6$	1	$(\mathbb{Z}/2)^2$
$D_6 A_9^2$	$\mathbb{Z}/2 \times \mathbb{Z}/10$					
		$D_5 \subset D_6$	$A_1 \subset A_9$	$A_7 A_9$	2	(0)
$D_5^2 A_7^2$	$\mathbb{Z}/4 \times \mathbb{Z}/8$					
		$D_5 \subset D_5$	$A_1 \subset D_5$	$A_1 A_3 A_7 A_7$	0	$\mathbb{Z}/8\mathbb{Z}$
		$D_5 \subset D_5$	$A_1 \subset A_7$	$D_5 A_5 A_7$	1	(0)

Fibration A_1D_{16} is obtained from the primitive embedding $D_5 \oplus A_1 \subset E_8$. Since by Proposition 4(21) $(D_5 \oplus A_1)^\perp_{E_8} = A_1 \oplus (-4)$, $\det N = 8 \times 4$, so $W/N \simeq \mathbb{Z}/2\mathbb{Z} \simeq L/L_{\text{root}} = \langle \delta_{16} + L_{\text{root}} \rangle$. Since $2\delta_{16} \in D_{16} = W_{\text{root}}$, thus $\delta_{16} \in \overline{W}_{\text{root}}$ and $\overline{W}_{\text{root}}/W_{\text{root}} = \mathbb{Z}/2\mathbb{Z}$.

6.4 $L_{root} = E_7^2 D_{10}$

$$L/L_{\text{root}} = \langle \eta_7^{(1)} + \delta_{10}, \eta_7^{(2)} + \bar{\delta}_{10} \text{ mod. } L_{\text{root}} \rangle \simeq (\mathbb{Z}/2\mathbb{Z})^2$$

Fibration $A_1D_6D_{10}$ is obtained from the primitive embeddings $A_1 \subset E_7^{(1)}$ and $D_5 \subset E_7^{(2)}$. Since by Proposition 4(10) and (15) $(A_1)^\perp_{E_7^{(1)}} = D_6$ and $(D_5)^\perp_{E_7^{(2)}} = A_1 \oplus (-4)$, we get $\det N = 8 \times 4^2$ and $W/N \simeq (\mathbb{Z}/2\mathbb{Z})^2 \simeq L/L_{\text{root}}$. By Proposition 4(15), $\eta_7(1) \in (D_5)^\perp_{E_7^{(2)}*} = A_1 \oplus (-4)$, but $2\eta_7 \notin A_1$ and by Proposition 4(10) $2\eta_7^{(1)} \in (A_1)^\perp_{E_7^{(1)}} = D_6$. So $\overline{W}_{\text{root}}/W_{\text{root}} = \langle \eta_7^{(1)} + \delta_{10} + W_{\text{root}} \rangle \simeq \mathbb{Z}/2\mathbb{Z}$.

Fibration E_7D_{10} is obtained from $D_5 \oplus A_1 \subset E_7^{(1)}$. Since by Proposition 4(20) $(D_5 \oplus A_1)^\perp_{E_7} = (-4)$, $\det N = 8 \times 4$ so $W/N \simeq \mathbb{Z}/2\mathbb{Z}$. Again by Proposition 4(20), $2\eta_7^{(1)} \notin (D_5 \oplus A_1)^\perp_{E_7^{(1)}}$ and we get $W/N = \langle \eta_7^{(2)} + \bar{\delta}_{10} + N \rangle$. Since $2\eta_7^{(2)} \in E_7$ and $2\bar{\delta}_{10} \in D_{10}$, it follows $\overline{W}_{\text{root}}/W_{\text{root}} \simeq \mathbb{Z}/2\mathbb{Z}$.

Fibration $2A_1D_8E_7$ is obtained from the primitive embeddings $A_1 \subset D_{10}$ and $D_5 \subset E_7^{(1)}$. By Proposition 4(8) and (15), we get $(A_1)^\perp_{D_{10}} = A_1 \oplus D_8$ and $(D_5)^\perp_{E_7^{(1)}} = A_1 \oplus (-4)$, so $\det N = 8 \times 4^2$ and $W/N \simeq (\mathbb{Z}/2\mathbb{Z})^2 \simeq L/L_{\text{root}}$. By Proposition 4(15), $2\eta_7^{(1)} \notin A_1$ and by Proposition 4(8) $2\bar{\delta}_{10} \in A_1 \oplus D_8$ so $\overline{W}_{\text{root}}/W_{\text{root}} \simeq \mathbb{Z}/2\mathbb{Z}$.

6.5 $L_{root} = E_7 A_{17}$

$$L/L_{\text{root}} = \langle \eta_7 + 3\alpha_{17} + L_{\text{root}} \rangle \simeq \mathbb{Z}/6\mathbb{Z}$$

Fibration A_{17} is obtained from the primitive embedding $D_5 \oplus A_1 \subset E_7$. By Proposition 4 20., $(D_5 \oplus A_1)^\perp_{E_7} = (-4)$, so $\det N = 8 \times 9$ and $W/N \simeq \mathbb{Z}/3\mathbb{Z} = \langle 6\alpha_{17} + N \rangle$. Moreover, since $18\alpha_{17} \in A_{17}$, $6\alpha_{17} \in \overline{W}_{\text{root}}$ so $\overline{W}_{\text{root}}/W_{\text{root}} \simeq \mathbb{Z}/3\mathbb{Z}$.

Fibration A_1A_{15} is obtained from the primitive embeddings $D_5 \subset E_7$ and $A_1 \subset A_{17}$. By Proposition 4(15) and (1), $(D_5)^\perp_{E_7} = A_1 \oplus (-4)$ and $(A_1)^\perp_{A_{17}} = L_{15}^2$ with $\det L_{15}^2 = 2 \times 18$, so $\det N = 8 \times 6^2$ and $W/N \simeq \mathbb{Z}/6\mathbb{Z} \simeq L/L_{\text{root}}$. But, by Lemma 2, $W_{\text{root}} = A_1A_{15}$ has no overlattice. Hence $\overline{W}_{\text{root}}/W_{\text{root}} \simeq (0)$.

6.6 $L_{root} = D_8^3$

$$L/L_{root} = \langle \delta_8^{(1)} + \overline{\delta}_8^{(2)} + \overline{\delta}_8^{(3)}, \overline{\delta}_8^{(1)} + \delta_8^{(2)} + \overline{\delta}_8^{(3)}, \overline{\delta}_8^{(1)} + \overline{\delta}_8^{(2)} + \delta_8^{(3)} \rangle \bmod. L_{root}$$
$$\simeq (\mathbb{Z}/2\mathbb{Z})^3$$

Fibration $A_1D_8D_8$ comes from the primitive embedding $D_5 \oplus A_1 \subset D_8^{(1)}$. By Proposition 4(18), $(D_5 \oplus A_1)_{D_8}^{\perp} = A_1 \oplus (-4)$ so $\det N = 8 \times 4^2$ and $W/N \simeq (\mathbb{Z}/2\mathbb{Z})^2$. By Proposition 4(18), $4\delta_8 \notin (D_5 \oplus A_1)_{D_8}^{\perp}$ so $W/N = \langle \overline{\delta}_8^{(1)} + \delta_8^{(2)} + \overline{\delta}_8^{(3)}, \overline{\delta}_8^{(1)} + \overline{\delta}_8^{(2)} + \delta_8^{(3)} \rangle$. Again by Proposition 4(18) $\overline{\delta}_8 \notin A_1$, so only $2(\overline{\delta}_8^{(2)} + \overline{\delta}_8^{(2)} + \overline{\delta}_8^{(3)} + \delta_8^{(3)}) \in W_{root}$ and $\overline{W}_{root}/W_{root} \simeq \mathbb{Z}/2\mathbb{Z}$.

6.7 $L_{root} = D_9A_{15}$

$$L/L_{root} = \langle \delta_9 + 2\alpha_{15} + L_{root} \rangle \simeq \mathbb{Z}/8\mathbb{Z}$$

Fibration D_4A_{13} comes from the primitive embeddings $D_5 \subset D_9$ and $A_1 \subset A_{15}$. By Proposition 4(12) and (1), $(D_5)_{D_9}^{\perp} = D_4$, $(A_1)_{A_{15}}^{\perp} = L_{13}^2$ with $\det L_{13}^2 = 2 \times 16$ so $\det N = 8 \times 16$ and $W/N \simeq \mathbb{Z}/4\mathbb{Z}$. But by Lemma 2, $W_{root} = D_4A_{13}$ has no overlattice since $q_{A_{13}}(\alpha_{13}) = \left(-\frac{1}{14}\right)$ and $q_{D_4}(\delta_4) \in \mathbb{Z}$. Hence $\overline{W}_{root}/W_{root} \simeq (0)$.

6.8 $L_{root} = E_6^4$

$$L/L_{root} = \langle \eta_6^{(1)} + \eta_6^{(2)} + \eta_6^{(3)}, 2\eta_6^{(1)} + \eta_6^{(3)} + \eta_6^{(4)} \rangle \bmod. L_{root} \simeq (\mathbb{Z}/3\mathbb{Z})^2$$

Fibration $A_5E_6E_6$ comes from the primitive embeddings $A_1 \subset E_6^{(1)}$ and $D_5 \subset E_6^{(2)}$. By Proposition 4(9) and (14), $(A_1)_{E_6}^{\perp} = A_5$ and $(D_5)_{E_6}^{\perp} = (-12)$ so $\det N = 8 \times 9^2$ and $W/N = (\mathbb{Z}/3\mathbb{Z})^2$. By Proposition 4(9), $3\eta_6^{(1)} \in A_5$ so $2\eta_6^{(1)} + \eta_6^{(3)} + \eta_6^{(4)} \in \overline{W}_{root}$ but $\eta_6^{(1)} + \eta_6^{(2)} + \eta_6^{(3)} \notin \overline{W}_{root}$ by Proposition 4(14). Hence $\overline{W}_{root}/W_{root} \simeq \mathbb{Z}/3\mathbb{Z}$.

6.9 $L_{root} = A_{11}E_6D_7$

$$L/L_{root} = \langle \alpha_{11} + \eta_6 + \delta_7 + L_{root} \rangle \simeq \mathbb{Z}/12\mathbb{Z}$$

Fibration $A_9A_1A_1E_6$ follows from the primitive embeddings $A_1 \subset A_{11}$ and $D_5 \subset D_7$. By Proposition 4(5) and (12), $(A_1)_{A_{11}}^{\perp} = L_9^2$, $\det L_9^2 = 2 \times 12$, $(D_5)_{D_7}^{\perp} = D_2 \simeq A_1^{\oplus 2}$ so $\det N = 8 \times 6^2$ and $W/N \simeq \mathbb{Z}/6\mathbb{Z} = \langle 2\alpha_{11} + 2\eta_6 + 2\delta_7 + N \rangle$. Since $k(2\alpha_{11}) \notin A_9$ by Proposition 4(5), we get $\overline{W}_{root}/W_{root} = (0)$.

Fibration A_9D_7. By Lemma 2, it follows that $W_{root} = A_9D_7$ has no overlattice since $q_{A_9}(\alpha_9) = \left(-\frac{1}{10}\right)$ and $q_{D_7} = \left(-\frac{7}{4}\right)$. Hence $\overline{W}_{root}/W_{root} = (0)$.

Fibration $A_{11}A_1D_5$ comes from the primitive embeddings $D_5 \subset E_6$ and $A_1 \subset D_7$. By Proposition 4(14) and (7), $(D_5)^{\perp}_{E_6} = (-12)$ and $(A_1)^{\perp}_{D_7} = A_1 \oplus D_5$, so $\det N = 8 \times 12^2$ and $W/N \simeq \mathbb{Z}/12\mathbb{Z} \simeq L/L_{root}$. Since $W_{root} \cap E_6 = \emptyset$, we get also $\overline{W}_{root} \cap E_6 = \emptyset$. Now $3\alpha_{11} + 3\delta_7 \in \overline{W}_{root}$ since $3\delta_7 \equiv \tilde{\delta}_7$ and $4(3\alpha_{11} + 3\delta_7) \in W_{root}$. Hence $\overline{W}_{root}/W_{root} \simeq \mathbb{Z}/4\mathbb{Z}$.

6.10 $L_{root} = D_6^4$

$$L/L_{root} = \langle \overline{\delta}_6^{(1)} + \overline{\delta}_6^{(4)}, \overline{\delta}_6^{(2)} + \overline{\delta}_6^{(3)}, \delta_6^{(1)} + \overline{\delta}_6^{(3)} + \delta_6^{(4)}, \overline{\delta}_6^{(1)} + \overline{\delta}_6^{(2)} \rangle \text{ mod. } L_{root}$$
$$\simeq (\mathbb{Z}/2\mathbb{Z})^4$$

Fibration $A_1D_4D_6D_6$ comes from the primitive embeddings $A_1 \subset D_6^{(1)}$ and $D_5 \subset D_6^{(2)}$. By Proposition 4(13) and 6, $(A_1)^{\perp}_{D_6} = A_1 \oplus D_4$, $(D_5)^{\perp}_{D_6} = (-4)$, so $\det N = 8 \times 8^2$ and $W/N \simeq (\mathbb{Z}/2\mathbb{Z})^3$. After enumeration of all the elements of L/L_{root}, since by Proposition 4(13) and 6. $\overline{\delta}_6^{(2)} \in (D_5)^{\perp}_{D_6^*}$ and only $\overline{\delta}_6^{(1)}$ or $\tilde{\delta}_6^{(1)} \in (A_1)^{\perp}_{D_6^*}$, we get

$$W/N =$$
$$\{\overline{\delta}_6^{(1)} + \overline{\delta}_6^{(4)}, \overline{\delta}_6^{(2)} + \overline{\delta}_6^{(3)}, \overline{\delta}_6^{(1)} + \overline{\delta}_6^{(2)}, \overline{\delta}_6^{(2)}$$
$$+\overline{\delta}_6^{(4)}, \overline{\delta}_6^{(1)} + \overline{\delta}_6^{(3)}, \overline{\delta}_6^{(3)} + \overline{\delta}_6^{(4)}, \overline{\delta}_6^{(1)}, \overline{\delta}_6^{(2)} + \overline{\delta}_6^{(3)} + \overline{\delta}_6^{(4)}\}$$
$$\simeq (\mathbb{Z}/2\mathbb{Z})^3$$

As $2\overline{\delta}_6^{(2)} \notin W_{root}$ and $2\overline{\delta}_6^{(1)} \in A_1 \oplus D_4$, it follows $\overline{W}_{root}/W_{root} = \{\overline{\delta}_6^{(1)} + \overline{\delta}_6^{(4)}, \overline{\delta}_6^{(1)} + \overline{\delta}_6^{(3)}, \overline{\delta}_6^{(3)} + \overline{\delta}_6^{(4)}, 0\} \simeq \mathbb{Z}/2\mathbb{Z} \times \mathbb{Z}/2\mathbb{Z}$.

6.11 $L_{root} = D_6A_9^2$

$$L/L_{root} = \langle \delta_6 + 5\alpha_9^{(2)}, \delta_6 + \alpha_9^{(1)} + 2\alpha_9^{(2)} \rangle \text{ mod.} L_{root} \simeq \mathbb{Z}/2\mathbb{Z} \oplus \mathbb{Z}/10\mathbb{Z}$$

Fibration A_7A_9 follows from the following primitive embeddings $D_5 \subset D_6$ and $A_1 \subset A_9^{(1)}$. By Proposition 4(13) and (4), $(D_5)^{\perp}_{D_6} = (-4)$, $(A_1)^{\perp}_{A_9} = L_7^2$, $\det L_7^2 = 2 \times 10$ so $\det N = 8 \times 10^2$ and $[W : N] = 10$. Enumerating the elements of L/L_{root} and since $\delta_6 \notin (D_5)^{\perp}_{D_6^*}$ by Proposition 4(13), we get

$$W/N = \{\alpha_9^{(1)} + 7\alpha_9^{(2)}, 2\alpha_9^{(1)} + 4\alpha_9^{(2)}, 3\alpha_9^{(1)} + \alpha_9^{(2)}, 4\alpha_9^{(1)} + 8\alpha_9^{(2)}, 5\alpha_9^{(1)} + 5\alpha_9^{(2)},$$
$$6\alpha_9^{(1)} + 2\alpha_9^{(2)}, 7\alpha_9^{(1)} + 9\alpha_9^{(2)}, 8\alpha_9^{(1)} + 6\alpha_9^{(2)}, 9\alpha_9^{(1)} + 3\alpha_9^{(2)}, 0\}$$
$$\simeq \mathbb{Z}/10\mathbb{Z}$$

Since $k\alpha_9^{(1)} \notin A_7$ by Proposition 4(4), it follows $\overline{W}_{root}/W_{root} = (0)$.

6.12 $L_{root} = D_5^2 A_7^2$

$$L/L_{\text{root}} = \langle 2\alpha_7^{(1)} + \delta_5^{(1)} + \delta_5^{(2)}, \alpha_7^{(1)} + \alpha_7^{(2)} + \delta_5^{(1)} + 2\delta_5^{(2)} \rangle \text{mod. } L_{\text{root}}$$
$$\simeq \mathbb{Z}/4\mathbb{Z} \times \mathbb{Z}/8\mathbb{Z}$$

Fibration $D_5 A_5 A_7$ follows from the primitive embeddings $D_5 \subset D_5^{(2)}$ and $A_1 \subset A_7^{(1)}$. By Proposition 4(3), $(A_1)_{A_7}^\perp = L_5^2$, $\det L_5^2 = 2 \times 8$ so $\det N = 8 \times 8^2$ and $[W : N] = 8$. Now enumerating the elements of L/L_{root} and since $\delta_5^{(2)}$ does not occur in W, we get $W/N = \langle 5\alpha_7^{(1)} + \alpha_7^{(2)} + 3\delta_5^{(1)} + N \rangle$. Since by Proposition 4(3) $k\alpha_7^{(1)} \notin A_5$, it follows $\overline{W}_{\text{root}}/W_{\text{root}} = (0)$. □

7 Equations of the Fibrations

In the next sections we give Weierstrass equations of all the elliptic fibrations. We use the following proposition

Proposition 5. [20, (pp. 559–560)] [23, Proposition 12.10] *Let X be a K3 surface and D an effective divisor on X that has the same type as a singular fiber of an elliptic fibration. Then X admits a unique elliptic fibration with D as a singular fiber. Moreover, any irreducible curve C on X with $D.C = 1$ induces a section of the elliptic fibration.*

First we show that one of the fibrations is the modular elliptic surface with base curve the modular curve $X_1(8)$ corresponding to modular group $\Gamma_1(8)$. As we see in the Table 3, it corresponds to the fibration $A_1, A_3, 2A_7$. The Mordell–Weil group is a torsion group of order 8. We draw a graph with the singular fibers $I_2, I_4, 2I_8$ and the 8-torsion sections. Most divisors used in the previous proposition can be drawn on the graph.

From this modular fibration we can easily write a Weierstrass equation of two other fibrations. From the singular fibers of these two fibrations we obtain the divisors of a set of functions on Y_2. These functions generate a group whose horizontal divisors correspond to the 8-torsion sections. These divisors lead to more fibrations.

If X is a K3 surface and

$$\pi : X \to C$$

an elliptic fibration, then the curve C is of genus 0 and we define an **elliptic parameter** as a generator of the function field of C. The parameter is not unique but defined up to linear fractional transformations.

From the previous proposition we can obtain equations from the linear system of D. Moreover if we have two effective divisors D_1 and D_2 for the same fibration we can choose an elliptic parameter with divisor $D_1 - D_2$. We give all the details for the fibrations of respective parameter t and ψ.

For each elliptic fibration we will give a Weierstrass model numbered from 1 to 30, generally in the two variables y and x. Parameters are denoted with small latin or greek letters. In most cases we give the change of variables that converts the defining equation into a Weierstrass form. Otherwise we use standard algorithms to obtain a Weierstrass form (see for example [6]). From a Weierstrass equation we get the singular fibers, using [28] for example, thus the corresponding fibration in Table 3; so we know the rank and the torsion of the Mordell–Weil group. If the rank is > 0 we give points and heights of points, which, using the formula of Proposition 1, generate the Mordell–Weil lattice. Heights are computed with Weierstrass equations as explained in [10]. Alternatively we can compute heights as in [11, 22].

7.1 Equation of the Modular Surface Associated to the Modular Group $\Gamma_1(8)$

We start with the elliptic surface

$$X + \frac{1}{X} + Y + \frac{1}{Y} = k.$$

From Beauville's classification [2], we know that it is the modular elliptic surface corresponding to the modular group $\Gamma_1(4) \cap \Gamma_0(8)$. Using the birational transformation

$$X = \frac{-U(U-1)}{V}, \qquad Y = \frac{V}{U-1} \qquad \text{with inverse}$$
$$U = -XY \quad V = -Y(XY+1)$$

we obtain the Weierstrass equation

$$V^2 - kUV = U(U-1)^2.$$

The point $Q = (U = 1, V = 0)$ is a 4–torsion point. If we want A with $2A = Q$ to be a rational point, then k must have the form $-s - 1/s + 2$. It follows

$$V^2 + (s + \frac{1}{s} - 2)UV = U(U-1)^2. \tag{1}$$

and

$$X + \frac{1}{X} + Y + \frac{1}{Y} + s + \frac{1}{s} = 2.$$

The point $A = (U = s, V = -1 + s)$ is of order 8. We obtain easily its multiples

	A	$2A$	$3A$	$4A$	$5A$	$6A$	$7A$
(X, Y)	$(-s, 1)$	$(\infty, 0)$	$\left(1, \frac{-1}{s}\right)$	$(0, 0)$	$\left(\frac{-1}{s}, 1\right)$	$(0, \infty)$	$(1, -s)$

Thus we get an equation for the modular surface Y_2 associated to the modular group $\Gamma_1(8)$

$$Y_2 : X + \frac{1}{X} + Y + \frac{1}{Y} + Z + \frac{1}{Z} - 2 = 0$$

and the elliptic fibration

$$(X, Y, Z) \mapsto Z = s.$$

Its singular fibers are

$$I_8 \ (s = 0), \ I_8 \ (s = \infty), \ I_4 \ (s = 1), \ I_2 \ (s = -1), \ I_1 \ (s = 3 \pm 2\sqrt{2}).$$

From now on, an expression such as $I_n \ (s = s_0)$ means a singular fiber of type I_n at $s = s_0$.

7.2 Construction of the Graph from the Modular Fibration

At $s = s_0$, we have a singular fiber of type I_{n_0}. We denote $\Theta_{s_0,j}$ with $s_0 \in \{0, \infty, 1, -1\}$, $j \in \{0, \ldots, n_0 - 1\}$ the components of a singular fiber I_{n_0} such that $\Theta_{i,j} . \Theta_{k,j} = 0$ if $i \neq k$ and

$$\Theta_{i,j} . \Theta_{i,k} = \begin{cases} 1 \text{ if } |k - j| = 1 \text{ or } |k - j| = n_0 - 1 \\ -2 \text{ if } k = j \\ 0 \text{ otherwise} \end{cases}$$

the dot meaning the intersection product. By definition, the component $\Theta_{k,0}$ intersects the zero section (0). The n_0-gon obtained can be oriented in two ways for $n_0 > 2$. For each s_0 we want to know which component is cut off by the section (A), i.e. the index $j(A, s_0)$ such that $A.\Theta_{s_0,j(A,s_0)} = 1$. For this, we compute the local height for the prime $s - s_0$ with a Weierstrass equation [10]. Since this height is also equal to $\frac{j(A,s_0)(n_{s_0} - j(A,s_0))}{n_{s_0}}$ we can give an orientation to the n_0-gon by choosing $0 \leq j(A, s_0) \leq \frac{n_{s_0}}{2}$. Hence we get the following results: $j(A, 0) = 3$, $j(A, s_0) = 1$ for $s_0 \neq 0$.

For the other torsion-sections (iA) we use the algebraic structure of the Néron model and get $(iA).\Theta_{0,j} = 1$ if $j = 3i \mod 8$, $(iA).\Theta_{0,j} = 0$ if $j \neq 3i$. For $s_0 \in \{\infty, 1, -1\}$ we have $(iA).\Theta_{s_0,j} = 1$ if $i = j \mod n_0$.

Remark 3. We can also compute $j(A, s_0)$ explicitly from the Néron model ([15] Theorem 1 and Proposition 5, p. 96).

Now we can draw the following graph. The vertices are the sections (iA) and the components $\Theta_{s_0,j}$ with $s_0 \in \{0, \infty, 1, -1\}$, $j \in \{0, 1.., n_0\}$. Two vertices B and C are linked by an edge if $B.C = 1$. For simplicity the two vertices $\Theta_{-1,0}$, $\Theta_{-1,1}$ and the edge between them are not represented. The edges joining $\Theta_{-1,0}$ and (jA), j even are suggested by a small segment from (jA), and also edges from $\Theta_{-1,1}$ to (iA), i odd.

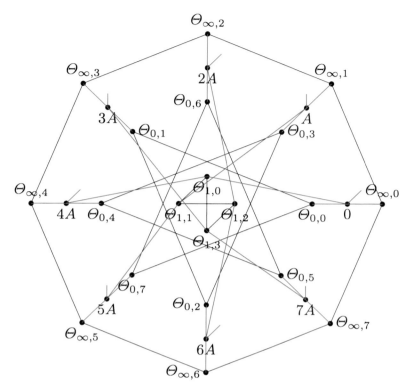

Fig. 3: Graph of singular fibers at $s = 0, \infty, 1, -1$ and torsion-sections

7.3 Two Fibrations

For the two fibrations to be considered we use the following factorizations of the equation of the surface:

$$(X + Y)(XY + 1)Z + XY(Z - 1)^2 = 0$$

$$(X + ZY)(XZ + Y) + (X + Y)(Y - 1)(X - 1)Z = 0.$$

7.3.1 Fibration of Parameter k

The parameter of the first one is $k = X + Y$. Eliminating for example X, we obtain an equation of bidegree 2 in Y and Z; easily we have the equation

$$y^2 - x\left(k^2 - 2k + 2\right)y = x(x - 1)\left(x - k^2\right) \tag{2}$$

with the birational transformation

$$Z = \frac{y}{k(x-1)}, Y = -\frac{yk}{-y+x^2-x}.$$

The singular fibers of this fibration are

$$I_1^* \ (k=0), \ I_{12} \ (k=\infty), \ I_2 \ (k=2), \ I_1 \ (k=4), \ I_1 \ (k=\pm 2i).$$

The rank of the Mordell–Weil group is one. The point $(x=1, y=0)$ is a non-torsion point of height $\frac{4}{3}$; the point (k,k) is of order 4, and its double is $(0,0)$.

7.3.2 Fibration of Parameter v

This fibration is obtained from the parameter $v = \frac{X+ZY}{Y-1}$. Eliminating Y and using the birational transformation

$$Z = -\frac{v(x-v^2(v+1))}{y}, X + v = \frac{Zx}{v(Z-(v+1))}$$

we get the equation

$$y^2 + (v+1)^2 yx - v^2(1+2v)y = (x-v)(x-v^2)(x-v^2-v^3). \tag{3}$$

The singular fibers of this fibration are

$$I_8 \ (v=0), \qquad I_{10} \ (v=\infty), \qquad I_1 \ (v=v_0).$$

where v_0 ranges over roots of the polynomial $t^6 - 5t^4 + 39t^2 + 2$. The Mordell–Weil group is of rank two; the two points $(0, v^3), (v, 0)$ are generators of the Mordell–Weil group (the determinant of the heights matrix is $\frac{1}{10}$). The Mordell–Weil torsion-group is 0.

7.4 Divisors

In this section we study the divisors of some functions. Using the elliptic fibration $(X, Y, Z) \mapsto Z = s$ we can compute the horizontal divisor of the following functions. We denote $(f)_h$ the horizontal divisor of f; then we have

$$
\begin{aligned}
(X)_h &= -(0) - (2A) + (4A) + (6A) & (X+s)_h &= -(0) - (2A) + 2(A) \\
(Y)_h &= -(0) - (6A) + (4A) + (2A) & (Y+s)_h &= -(0) - (6A) + 2(7A) \\
(X-1)_h &= -(0) - (2A) + (3A) + (7A) & (X+\tfrac{1}{s})_h &= -(0) - (2A) + 2(5A) \\
(Y-1)_h &= -(0) - (6A) + (A) + (5A) & (Y+\tfrac{1}{s})_h &= -(0) - (6A) + 2(3A) \\
(X+Y)_h &= -(2A) - (6A) + 2(4A) \\
(X+sY)_h &= -(0) - (2A) - (6A) + (A) + (3A) + (4A).
\end{aligned}
$$

Proposition 6. *The horizontal divisors of the 7 functions X, Y, $X - 1$, $Y - 1$, $Y + s$, $X + Y$, $X + sY$ generate the group of principal divisors with support in the 8-torsion sections.*

Proof. Let Λ the group of divisors of degree 0 with support in $\{(iA), 1 \le i \le 8\}$. If we write $f_1, f_2, ..$ for the seven functions of the proposition, then the determinant of the matrix $m_{i,j}$ with $m_{i,j} = ord_{(iA)}f_j$ with $1 \le j \le 7, 2 \le i \le 8$ is equal to 8. This shows the subgroup of Λ generated by divisors of functions f_i is of index 8 in Λ. Moreover the divisor $\sum_i k_i (iA)$ is principal iff $\sum_i ik_i \equiv 0 \bmod 8$. So, the subgroup of principal divisors of Λ is generated by $\{i(A) - (iA) - (i-1)(0)$ with $2 \le i \le 7\}$ and $8((A) - (0))$. This last subgroup is also of index 8 in Λ and the two subgroup are equal.   □

From this proposition we deduce the corollary used in [3].

Corollary 1. *The 8-order automorphism σ_8, of the surface Y_2, leaving invariant the fibration of parameter Z and defined by $M \mapsto M - A$ on the generic fiber, is given by*

$$\sigma_8 : (X, Y, Z) \mapsto \left(-\frac{Y + XZ}{X + YZ}, \frac{(Y + XZ)(1 + YZ)}{(X + YZ)(Y + Z)}, Z \right).$$

Proof. The image of X by σ_8 is the unique function of horizontal divisor $(7A) + (5A) - (A) - (3A)$ and equal to 1 at $(4A)$. Using the Proposition 6 we have $X^{\sigma_8} = \frac{Z(Y-1)(X-1)(X+Y)}{(X+YZ)^2} = -\frac{Y+XZ}{X+YZ}$. A similar argument gives the result for Y. We can notice that

$$\sigma_8^2 : (X, Y, Z) \mapsto (\frac{1}{Y}, X, Z).$$

Of course we can also compute the translation directly from a Weierstrass equation.   □

We use the following notations: $Div(f)$ for the divisor of the function f on the surface, $(f)_0$ for the divisor of the zeros of f and $(f)_\infty$ for the divisor of the poles. We get

$Div(Z) = \sum_{i=0}^{7} \Theta_{0,i} - \sum_{i=0}^{7} \Theta_{\infty,i}$
$(Z - 1)_0 = \sum_{i=0}^{3} \Theta_{1,i}$ $(Z + 1)_0 = \sum_{i=0}^{1} \Theta_{-1,i}.$

Since X, Y, Z play the same role, the elliptic fibrations $(X, Y, Z) \mapsto X$ and also $(X, Y, Z) \mapsto Y$ have the same property for the singular fibers: two singular fibers of type I_8 for $X = 0, \infty$ and $Y = 0, \infty$, one singular fiber of type I_4 for $X = 1, Y = 1$. Then we can represent on the graph the divisor of X, drawing two disjoint 8-gons going throught $(0), (2A)$ and $(4A), (6A)$ and a disjoint 4-gon throught $(3A), (7A)$. We have

$$Div(X) = -(0) - \Theta_{\infty,0} - \Theta_{\infty,1} - \Theta_{\infty,2} - (2A) - \Theta_{0,6} - \Theta_{0,7} - \Theta_{0,0}$$
$$+(4A) + \Theta_{\infty,4} + \Theta_{\infty,5} + \Theta_{\infty,6} + (6A) + \Theta_{0,2} + \Theta_{0,3} + \Theta_{0,4}.$$
$$(X - 1)_0 = (3A) + \Theta_{1,3} + (7A) + \Theta_{-1,1}.$$

A similar calculation for Y gives

$$Div(Y) = -(0) - \Theta_{\infty,0} - \Theta_{\infty,7} - \Theta_{\infty,6} - (6A) - \Theta_{0,2} - \Theta_{0,1} - \Theta_{0,0}$$
$$+(4A) + \Theta_{\infty,4} + \Theta_{\infty,3} + \Theta_{\infty,2} + (2A) + \Theta_{0,6} + \Theta_{0,5} + \Theta_{0,4}.$$
$$(Y-1)_0 = (A) + \Theta_{1,1} + (5A) + \Theta_{-1,1}.$$

The fibration $(X, Y, Z) \mapsto k = X + Y$ has singular fibers of type I_1^*, I_{12} at $k = 0$ and $k = \infty$, so we can write the divisor of $X + Y$. By permutation we have also the divisors of $Y + Z$ and $X + Z$

$$Div(X + Y) = -(2A) - \Theta_{\infty,2} - \Theta_{\infty,1} - \Theta_{\infty,0} - \Theta_{\infty,7} - \Theta_{\infty,6} - (6A)$$
$$-\Theta_{0,6} - \Theta_{0,7} - \Theta_{0,0} - \Theta_{0,1} - \Theta_{0,2} + \Theta_{\infty,4} + \Theta_{0,4} + 2(4A)$$
$$+2\Theta_{1,0} + \Theta_{1,1} + \Theta_{1,2}.$$

$$Div(X + Z) = -(0) - \Theta_{\infty,0} - \Theta_{\infty,7} - \Theta_{\infty,6} - \Theta_{\infty,5} - \Theta_{\infty,4} - \Theta_{\infty,3} - \Theta_{\infty,2}$$
$$-(2A) - \Theta_{0,6} - \Theta_{0,7} - \Theta_{0,0} + \Theta_{1,1} + \Theta_{-1,1} + 2(A)$$
$$+2\Theta_{0,3} + \Theta_{0,2} + \Theta_{0,4}.$$

$$Div(Y + Z) = -(0) - \Theta_{\infty,0} - \Theta_{\infty,1} - \Theta_{\infty,2} - \Theta_{\infty,3} - \Theta_{\infty,4} - \Theta_{\infty,5} - \Theta_{\infty,6}$$
$$-(6A) - \Theta_{0,0} - \Theta_{0,1} - \Theta_{0,2} + \Theta_{1,3} + \Theta_{-1,1} + 2(7A)$$
$$+2\Theta_{0,5} + \Theta_{0,4} + \Theta_{0,6}.$$

At last the fibration $(X, Y, Z) \mapsto v = \frac{(X+ZY)}{(Y-1)}$ has two singular fibers of type I_8, I_{12} at $v = 0$ and $v = \infty$; thus it follows

$$Div(\frac{X + ZY}{Y - 1}) = (3A) + \Theta_{1,0} + \Theta_{1,3} + 4A + \Theta_{0,4} + \Theta_{0,3} + \Theta_{0,2} + \Theta_{0,1}$$
$$-(2A) - \Theta_{\infty,2} - \Theta_{\infty,1} - \Theta_{\infty,0} - \Theta_{\infty,7} - \Theta_{\infty,6} - \Theta_{\infty,5}$$
$$-(5A) - \Theta_{0,7} - \Theta_{0,6}.$$

Remark 4. We can show that the following twenty elements form a basis of the Néron–Severi group: the eight torsion sections (nA) $0 \le n \le 7$, $\Theta_{\infty,i}$, $1 \le i \le 7$, $\Theta_{1,j}$ $1 \le j \le 3$, $\Theta_{-1,1}$, and the fibre. Just compute the Gram matrix using the graph (Fig. 3). Its determinant is equal to 8. So, we can recover the divisors of the previous functions by decomposition of others $\Theta_{i,j}$ in this basis.

8 Fibrations from the Modular Fibration

We give a first set of elliptic fibrations with elliptic parameters belonging to the multiplicative group of functions coming from Proposition 6 plus Z and $Z \pm 1$. The first ones come from some easy linear combination of divisors of functions. The others, like t, come from the following remark. We can draw, on Fig. 3, two disjoint subgraphs corresponding to singular fibers of the same fibration. We give all the details only in the case of parameter t.

8.0.1 Fibration of Parameter a

This fibration is obtained with the parameter $a = \frac{Z-1}{X+Y}$. Using the factorisation

$$(Z + X)(X + Y)(X - 1) = X(YZ + X)(X + Y + Z - 1)$$

it is easier, to have a Weierstrass equation, to use $a' = (a+1)^{-1} = \frac{X+Y}{X+Y+Z-1}$ and to eliminate Z. We do the birational transformation

$$X = \frac{x(x-a')}{y} \quad Y = -\frac{y}{x-a'} \quad \text{with inverse}$$
$$x = -XY \quad y = Y(XY + a')$$

Returning to parameter a we get

$$y^2 - \frac{(x-1)y}{(1+a)a} = x(x - \frac{1}{1+a})^2. \tag{4}$$

The singular fibers of this fibration are

$$I_8 \ (a = 0), \qquad I_1^* \ (a = \infty), \qquad I_6 \ (a = -1), \qquad I_1 \ (a = a_0),$$

where a_0 ranges over roots of the polynomial $16X^3 + 11X^2 - 2X + 1$.
The point $(x = \frac{1}{1+a}, y = 0)$ is of height $\frac{1}{24}$. The torsion-group of the Mordell–Weil group is 0.

8.0.2 Fibration of Parameter d

This fibration is obtained with parameter $d = XY$ which also is equal to $-x$ in the previous Weierstrass equation. Eliminating X and making the birational transformation
$$y = -(d+1)Y(d^2 - x), x = -ZYd(d+1)$$
we get

$$y^2 - 2d\,y\,x = x(x - d^2)(x - d(d+1)^2). \tag{5}$$

The singular fibers are

$$I_2^* \ (d = 0), \qquad I_2^* \ (d = \infty), \qquad I_2 \ (d = 1), \qquad I_0^* \ (d = -1).$$

The three points of abscisses $0, d + d^2, d^3 + d^2$ are two-torsion points. The point $(d^2, 0)$ is of height 1.

8.0.3 Fibration of Parameter p

This fibration is obtained with $p = \frac{(XY+1)Z}{X} = \frac{Vs}{U}$ which is also equal to x/d^2 with notation of the previous fibration. We start from the equation in U, V and eliminating V and making the birational transformation

$$s = \frac{xp(p+1)}{y+xp}, U = \frac{x}{p(p+1)}$$

we obtain

$$y^2 = x(x-p)(x-p(p+1)^2). \tag{6}$$

The singular fibers are

$$I_2^* \ (p = 0), \qquad I_4^* \ (p = \infty), \qquad I_2 \ (p = -2), \qquad I_4 \ (p = -1).$$

The Mordell–Weil group is isomorphic to $(\mathbb{Z}/2\mathbb{Z})^2$.

8.0.4 Fibration of Parameter w

Using the factorisation of the equation of Y_2

$$(Z+X)(X+Y)(X-1) = X(YZ+X)(X+Y+Z-1)$$

we put $w = X+Y+Z-1 = \frac{(X+Y)(X+Z)(X-1)}{X(YZ+X)}$. Eliminating Z in the equation of Y_2 and doing the birational transformation

$$x = -(1-Y+wY)(1-X+wX), y = -(w-1)Xx$$

we obtain the equation

$$y^2 + w^2(x+1)y = x(x+1)(x+w^2). \tag{7}$$

The singular fibers are

$$I_6 \ (w = 0), \qquad I_{12} \ (w = \infty), \qquad I_2 \ (w = 1), \qquad I_2 \ (w = -1), \qquad I_1 \ (w = \pm 2i\sqrt{2}).$$

The Mordell–Weil group is isomorphic to $\mathbb{Z}/6\mathbb{Z}$, generated by $(-w^2, 0)$.

8.0.5 Fibration of Parameter b

We put $b = \frac{XY}{Z}$. Eliminating Z in the equation of Y_2 and using the birational transformation

$$x = -\frac{b(Y+b)(X+b)}{X}, y = -\frac{Yb(x-(b+1)^2)}{X}$$

we obtain the equation

$$y^2 + 2b(b+1)xy + b^2(b+1)^2 y = x(x+b^2)(x-(b+1)^2) \tag{8}$$

or with $z = y + b^2 x$

$$z^2 + 2bzx + b^2(b+1)^2 z = x^3.$$

The singular fibers are

$$IV^* \ (b=0), \qquad IV^* \ (b=\infty), \qquad I_6 \ (b=-1), \qquad I_1 \ (b=b_0),$$

with b_0 ranges over the roots of the polynomial $27b^2 + 46b + 27$. The Mordell–Weil group is of rank 1, the point $(x = -b^2, y = 0)$ is of height $\frac{4}{3}$ and the torsion-group is of order 3.

8.0.6 Fibration of Parameter *r*

Let $r = \frac{(X+Z)(Y+Z)}{ZX}$, r is also equal to $\frac{-x}{b^2}$ with x from (8). Eliminating Y in the equation of Y_2 and doing the birational transformation.

$$x = -\frac{(X(r-1)-Z)\,r}{Z}, y = -\frac{X\left(x - r^3\right)x\,(r-1)}{-r^2 + x}$$

we obtain the equation

$$y^2 + 2(r-1)xy = x(x-1)(x-r^3). \tag{9}$$

The singular fibers are

$$I_2^* \ (r=0), \qquad I_6^* \ (r=\infty), \qquad I_2 \ (r=1), \qquad I_1 \ (r=\pm 2i).$$

The Mordell–Weil group is of rank 1, the point $(1, 0)$ is of height 1. The torsion group is of order 2, generated by $(0, 0)$.

8.0.7 Fibration of Parameter *e*

Let $e = \frac{YX}{(Y+Z)Z}$, e is also equal to $-\frac{x}{r^2}$, where x is from (9). Eliminating Y from the equation of Y_2 and doing the birational transformation

$$y = \frac{-\left(2e^2 + e + x\right)(e\,(x - 2e - 1)\,X - x(e+1))}{(2e+1)(e+1)}, x = \frac{-e\,(2e+1)(-Ze + X)}{X + Z}$$

we obtain the equation

$$y^2 = x(x^2 - e^2(e-1)x + e^3(2e+1)). \tag{10}$$

The singular fibers are

$$III^* \ (e = 0), \ I_4^* \ (e = \infty), \ I_2 \ (e = -1), \ I_2 \ (e = -\tfrac{1}{2}), \ I_1 \ (e = 4).$$

The Mordell–Weil group is of rank 1, the point $(e^3, e^3 + e^4)$ is of height 1. The torsion group is of order 2, generated by $(0, 0)$.

8.0.8 Fibration of Parameter f

Let $f = -\frac{Y(X+Z)^2(Z+Y)}{Z^3 X}$, f is also equal to x where x is from (9). We start from (9) and use the transformation

$$y = \frac{V'}{x(x-1)}, r = -\frac{U'}{x(x-1)}$$

we obtain the equation

$$V'^2 - 2fV'U' - 2f^2(f-1)V' = U'^3 + f^4(f-1)^3. \tag{11}$$

The singular fibers are

$$III^* \ (f = 0), \qquad II^* \ (f = \infty), \qquad I_4 \ (f = 1), \qquad I_1 \ (f = \tfrac{32}{27}).$$

The Mordell–Weil group is (0).

8.0.9 Fibration of Parameter g

Let $g = \frac{XY}{Z^2}$. Eliminating Y in the equation of Y_2 and using the birational transformation

$$y = -\frac{\left(g^2 - 1\right)\left(-gXZ - gZ^2 - X^2 + gXZ^2\right)g}{Z(X+Z)^2}, x = -\frac{g(g+1)(Zg+X)}{X+Z}$$

we obtain the equation

$$y^2 = x^3 + 4g^2x^2 + g^3(g+1)^2x. \tag{12}$$

The singular fibers are

$$III^* \ (g = 0), \qquad III^* \ (g = \infty), \qquad I_4 \ (g = -1), \qquad I_2 \ (g = 1).$$

The Mordell–Weil group is of order 2.

8.0.10 Fibration of Parameter h

Let $h = \frac{(Y+Z)YX^2}{Z^3(X+Z)}$, we can see that $h = \frac{x}{(g+1)}$, with x from (12). We start from (12) and if $y = (g+1)z$ we obtain a quartic equation in z and g with rational points $g = -1, z = \pm 2h$. Using standard transformation we obtain

$$y'^2 + (h - \frac{1}{h} - 8)x'y' - \frac{96}{h}y' = \left(x' - \frac{1}{4}(h^2 + \frac{1}{h^2}) + 4h + \frac{8}{h} + \frac{1}{2}\right)\left(x'^2 - \frac{256}{h}\right)$$

or also

$$y^2 = x^3 - \frac{25}{3}x + h + \frac{1}{h} - \frac{196}{27}. \tag{13}$$

The singular fibers are

$$II^* \ (h = 0), \qquad II^* \ (h = \infty), \qquad I_2 \ (h = -1), \qquad I_1 \ (h = h_0),$$

where h_0 ranges over the roots of the polynomial $27h^2 - 446h + 27$. The Mordell–Weil group is of rank 1 without torsion.

The point $(\frac{1}{16}(h^2 + \frac{1}{h^2}) + h + \frac{1}{h} + \frac{29}{24}, \frac{1}{64}\frac{(h-1)(h+1)(h^4+24\,h^3+126\,h^2+24\,h+1)}{h^3})$ is of height 4. We recover Elkies' result [9] cited in the introduction.

8.0.11 Fibration of Parameter t

On the graph (Fig. 3), we can see two singular fibers of type I_4^* of a new fibration (Fig. 4). They correspond to two divisors D_1 and D_2 with

$$D_1 = \Theta_{\infty,4} + (3A) + 2\Theta_{\infty,3} + 2\Theta_{\infty,2} + 2\Theta_{\infty,1} + 2\Theta_{\infty,0} + 2(0) + \Theta_{0,0} + \Theta_{1,0}$$
$$D_2 = (7A) + \Theta_{0,6} + 2\Theta_{0,5} + 2\Theta_{0,4} + 2\Theta_{0,3} + 2\Theta_{0,2} + 2(6A) + \Theta_{\infty,6} + \Theta_{1,2}.$$

We look for a parameter of the new fibration as a function t with divisor $D_1 - D_2$.

Let E_s and T_s be the generic fiber and the trivial lattice of the fibration of parameter s. For this fibration we write $D_i = \delta_i + \Delta_i$ with $i = 1,2$ and δ_i an horizontal divisor and Δ_i a vertical divisor. More precisely we have $\delta_1 = (3A) + 2(0)$ and $\delta_2 = (7A) + 2(6A)$, and the classes of δ_i and D_i are equal mod T_s. If $K = \mathbb{C}(s)$ recall the isomorphism: $E_s(K) \sim NS(Y_2)/T_s$. So the class of $\delta_1 - \delta_2$ is 0 in $NS(Y_2)/T_s$; thus there is a function $t_0 = \frac{X^2(Y+Z)}{(X-1)(X+Y)}$ with divisor $\delta_1 - \delta_2 \mod T_s$. We can choose for $t = t_0 Z^a(Z-1)^b$ where a and b are integer calculated using divisors of previous section. We find $a = 0, b = 1$ and we can take $t = \frac{X^2(Y+Z)(Z-1)}{(X-1)(X+Y)}$. We have also

$$t = -X\frac{Y\left(Z^2 - Z + YZ + 1\right)Z}{(Z-1)(YZ+1)} - \frac{Z^2(Y+1)}{(Z-1)(YZ+1)}.$$

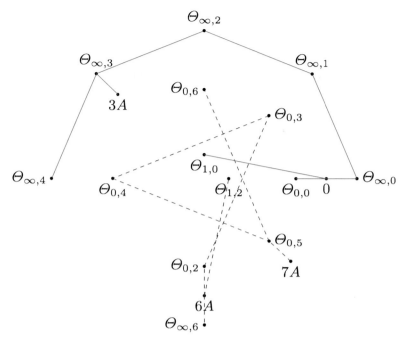

Fig. 4: Two singular fiber I_4^*

Eliminating X in the equation of Y_2 and then doing the birational transformation

$$Z = \frac{W}{WT + 1}, \quad Y = \frac{-(WT^2 + T + 1)}{WT + 1}$$

of inverse

$$T = \frac{Y + 1}{Z - 1}, \quad W = \frac{-Z(Z - 1)}{YZ + 1}$$

we obtain an equation of degree 2 in T. After some classical transformation we get a quartic with a rational point corresponding to $(T = -1 + t, W = 0)$ and then using a standard transformation we obtain

$$y^2 = x^3 + t(t^2 + 1 + 4t)x^2 + t^4 x. \tag{14}$$

The singular fibers are

$$I_4^* \ (t = 0), \qquad I_4^* \ (t = \infty), \qquad I_2 \ (t = -1), \qquad I_1 \ (t = t_0),$$

where t_0 ranges over roots of the polynomial $Z^2 + 6Z + 1$. The Mordell–Weil group is of rank one and the point $(-t^3, 2t^4)$ is of height 1. The torsion group is of order 2.

8.0.12 Fibration of Parameter l

Let $l = \frac{Z(YZ+X)}{X(1+YZ)}$. Eliminating Y in the equation of Y_2 and using the variable $W = \frac{Z+X}{X-1}$, we have an equation of bidegree 2 in W and Z; easily we obtain

$$y^2 - yx - 2l^3y = (x + l^3)(x + l^2)(x - l + l^3). \tag{15}$$

The singular fibers are

$$I_{10} \ (l = 0), \qquad I_3^* \ (l = \infty), \qquad I_1 \ (l = l_0),$$

where l_0 ranges over roots of the polynomial $16\,x^5 - 32\,x^4 - 24\,x^3 - 23\,x^2 + 12\,x - 2$. The Mordell–Weil group is of rank 2, without torsion; the two points $(-l^3, 0)$ and $(-l^2, 0)$ are independent and the determinant of the matrix of heights is equal to $\frac{1}{5}$.

9 A Second Set of Fibrations: Gluing and Breaking

9.1 Classical Examples

In the following we give fibrations obtained using Elkies' 2-neighbor method, given in [7] page 11 and explained in [13] Appendix A. If we have two fibrations with fiber F and F' satisfying $F \cdot F' = 2$ the authors explain how can be obtained a parameter from a Weierstrass equation of one fibration. Decomposing F' into vertical and horizontal component, $F' = F'_h + F'_v$ they use F'_h to construct a function on the generic fiber.

9.1.1 Fibration of Parameter o

Starting with a fibration with two singular fibers of type II^* and the (0) section we obtain a fibration with a singular fiber of type I_{12}^*. Starting from (13) we take x as new parameter. For simplicity, let $o = x + \frac{5}{3}$, we get

$$\tilde{y}^2 = \tilde{x}^3 + (o^3 - 5o^2 + 2)\tilde{x}^2 + \tilde{x}. \tag{16}$$

The singular fibers are

$$I_2 \ (o = 0), \ I_{12}^* \ (o = \infty), \ I_1 \ (o = 1), \ I_1 \ (o = 5), \ I_1 \ (o = o_0),$$

where o_0 range overs roots of $x^2 - 4x - 4$. The Mordell–Weil group is of rank 1, the torsion group is of order 2.

The point

$$(\frac{1}{16}(o-4)^2(o-2)^2 , \frac{1}{64}(o-4)(o-2)\left(o^4-4o^3-20o^2+96o-80\right))$$

is of height 4.

9.1.2 Fibration of Parameter q

We start with a fibration with singular fibers of type II^* and III^*, join them with the zero-section and obtain a singular fiber of type I_{10}^* of a new fibration. We transform (11) to obtain

$$y'^2 = x'^3 + (\frac{5}{3} - \frac{2}{f})x' + f + \frac{5}{3f} - \frac{70}{27}.$$

We take x' as the parameter of the new fibration; more precisely, for simplicity, let $q = x' - \frac{1}{3}$. We obtain

$$y^2 = x^3 + (q^3 + q^2 + 2q - 2)x^2 + (1-2q)x. \tag{17}$$

The singular fibers are

$$I_4 \ (q=0), \qquad I_{10}^* \ (q=\infty), \qquad I_2 \ (q=\tfrac{1}{2}), \qquad I_1 \ (q=q_0),$$

where q_0 ranges over roots of $X^2 + 2X + 5$. The Mordell–Weil group is of order 2.

9.2 Fibration with a Singular Fiber of Type I_n, n Large

We start with a fibration with two fibers I_n^* and I_m^* and a two-torsion section. Gluing them, we can construct a fibration with a singular fiber of type I_{n+m+8}. The parameter will be $\frac{y}{x}$ in a good model of the first fibration. We can also start from a fibration with two singular fibers of type I_n^* and I_2 and a two-torsion section, join them with the zero-section and obtain a new fibration with a singular fiber of type I_{n+6}.

9.2.1 Fibration of Parameter m

We start from the fibration with parameter t. With the two-torsion section and the two singular fibers of type I_4^*, we can form a singular fiber of type I_{16} of a new fibration (Fig. 5).
    From (14) we get

$$y'^2 = x'^3 + (t + \frac{1}{t} + 4)x'^2 + x'.$$

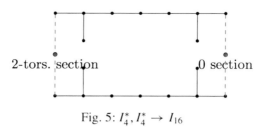

Fig. 5: $I_4^*, I_4^* \to I_{16}$

Let $m = \frac{y'}{x'}$, we obtain

$$y^2 + (m-2)(m+2)yx = x(x-1)^2. \tag{18}$$

The singular fibers are

$I_2\ (m=0),\qquad I_{16}\ (m=\infty),\qquad I_2\ (m=\pm 2),\qquad I_1\ (m=\pm 2\sqrt{2}).$

The Mordell–Weil group is cyclic of order 4 generated by $(x=1, y=0)$.

9.2.2 Fibration of Parameter n

With a similar method, we can start from a fibration with two singular fibers of type I_2^* and I_6^*, a two-torsion section and join them to have a fiber of type I_{16} (Fig. 6).

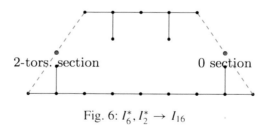

Fig. 6: $I_6^*, I_2^* \to I_{16}$

From (9) we obtain

$$y'^2 = x'^3 - (r - 1 + \frac{2}{r})x'^2 + \frac{x'}{r}.$$

Let $n = \frac{y'}{x'}$. The Weierstrass equation is

$$y^2 + (n^2 - 1)yx - y = x^3 - 2x^2. \tag{19}$$

The singular fibers are

$I_2\ (n=0),\qquad I_{16}\ (n=\infty),\qquad I_1\ (n=n_0),$

where n_0 ranges over roots of the polynomial $2x^6 - 9x^4 - 17x^2 + 125$. The Mordell–Weil group is of rank 2. The determinant of the matrix of heights of the two points $(1 \pm n, 1)$ is $\frac{1}{4}$.

9.2.3 Fibration of Parameter j

Instead of the two-torsion section we can use the section of infinite order $(-t, -2t)$ in the fibration of parameter t (Fig. 7).

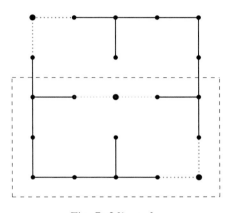

Fig. 7: $2I_4^* \rightarrow I_{12}$

In (14), let $U' = x + t$, $V' = y + 2t$ and take the parameter $\frac{V'}{U'}$. For simplification let $j = \frac{V'}{U'} - 2 = \frac{y-2x}{x+t}$; the new fibration obtained has a 3-torsion point which can be put in $(0, 0)$. So it follows

$$y^2 - (j^2 + 4j)xy + j^2y = x^3. \tag{20}$$

The singular fibers are

$$IV^* \; (j = 0), \qquad I_{12} \; (j = \infty), \qquad I_2 \; (j = -1), \qquad I_1 \; (j = j_0),$$

where j_0 ranges over roots of the polynomial $(x^2 + 10x + 27)$. The Mordell–Weil group is isomorphic to $\mathbb{Z}/3\mathbb{Z}$.

9.2.4 Fibration of Parameter c

We start from the fibration of parameter o with (16). For $o = 0$ we have a singular fiber of type I_2, the singular point of the bad reduction is $(x = 1, y = 0)$ so we put $x = 1 + u$ and obtain the equation

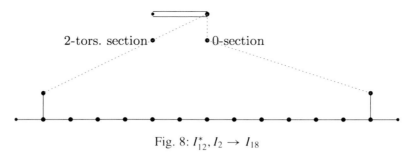

Fig. 8: $I_{12}^*, I_2 \to I_{18}$

$$y^2 = u^3 + (-1 - 5o^2 + o^3)u^2 + 2o^2(o - 5)u - o^2(o - 5).$$

The two-torsion section cut the singular fiber I_2 on the zero component, as shown on Fig. 8.

Let $c = \frac{y}{ox}$ with y, x from (16), we have easily the Weierstrass equation of the fibration. There is a 3-torsion point and after some translation we can suppose the three-torsion point is $(0, 0)$; we get

$$y^2 + (c^2 + 5)yx + y = x^3. \tag{21}$$

The singular fibers are

$$I_{18} \ (c = \infty) \qquad I_1 \ (c = c_0),$$

where c_0 ranges over roots of the polynomial $(x^2 + 2)(x^2 + x + 7)(x^2 - x + 7)$. The rank of the Mordell–Weil group is one. The height of $(\frac{-1}{4}(c^4 + c^2 + 1), \frac{1}{8}(c^2 - c + 1)^3)$ is equal to 4.

9.3 Fibrations with Singular Fibers of Type I_n^*

In this paragraph we obtain new fibrations by gluing two fibers of type I_p^* and I_q^* to obtain, with the zero section, a singular fiber of type I_{p+q+4}^* or I_{p+4}^*.

9.3.1 Fibration of Parameter u

We start from the fibration with parameter t. With the two singular fibers of type I_4^*, and the 0-section, we can form a singular fiber of type I_8^* of a new fibration (Fig. 9). From (14), it follows

$$\frac{y^2}{x^2 t^2} = \frac{x}{t^2} + t + \frac{t^2}{x} + \frac{1}{t} + 4.$$

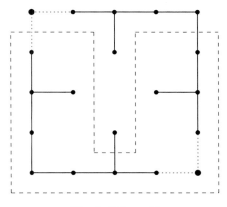

Fig. 9: $2I_4^* \rightarrow I_8^*$

Taking $u = \frac{x}{t^2} + t$ as the new parameter, we obtain

$$y'^2 = x'^3 + u(u^2 + 4u + 2)x'^2 + u^2 x'. \tag{22}$$

The singular fibers are

$$I_1^* \ (u = 0), \qquad I_8^* \ (u = \infty), \qquad I_2 \ (u = -2), \qquad I_1 \ (u = -4).$$

The Mordell–Weil group is of order 2.

9.3.2 Fibration of Parameter i

We start from the fibration of parameter u and from (22). With the two singular fibers of type I_8^* and I_1^*, and the 0 section, we can form a singular fiber of type I_{13}^* of a new fibration. We seek for a parameter of the form $\frac{x}{u^2} + ku$, with k chosen to have a quartic equation. We see that $k = 1$ is a good choice so the new parameter is $i = \frac{x}{u^2} + u$ and a Weierstrass equation is

$$y^2 = x^3 + \left(i^3 + 4i^2 + 2i\right)x^2 + \left(-2i^2 - 8i - 2\right)x + i + 4. \tag{23}$$

The singular fibers are

$$I_{13}^* \ (i = \infty), I_2 \ (i = -\tfrac{5}{2}), I_1 \ (i = i_0),$$

where i_0 ranges over roots of the polynomial $4x^3 + 11x^2 - 8x + 16$. The Mordell–Weil group is 0.

9.4 Breaking

In this paragraph we give a fibration obtained by breaking a singular fiber I_{18} and using the three-torsion points as for the fibration of parameter t.

9.4.1 Fibration of Parameter ψ

We start with the fibration (21) of parameter c and 3-torsion sections. We represent the graph of the singular fiber I_{18}, the zero section and two 3-torsion sections (Fig. 10). On this graph we can draw two singular fibers III^* and I_6^*. The function x of (21) has the horizontal divisor $-2(0) + P + (-P)$ if P denotes the 3-torsion point and we can take it as the parameter ψ of the new fibration. We get the equation

$$y^2 = x^3 - 5\,x^2\psi^2 - \psi\,x^2 - \psi^5 x. \tag{24}$$

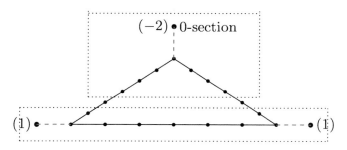

Fig. 10: A singular fiber $I_{18} \rightarrow III^*, I_6^*$

The singular fibers are

$$III^*\ (\psi = 0),\ I_6^*\ (\psi = \infty),\ I_1\ (\psi = -\tfrac{1}{4}),\ I_1\ (\psi = \psi_0),$$

with ψ_0 range overs roots of the polynomial $x^2 + 6x + 1$. The Mordell–Weil group is of rank 1. The height of the point

$$(1/4\left(\psi^2 + 3\psi + 1\right)^2,\ -1/8\left(\psi^2 + 3\psi + 1\right)\left(\psi^4 + 6\psi^3 + \psi^2 - 4\psi - 1\right))$$

is 4. The torsion-group is of order 2.

10 Last Set

From the first set of fibrations we see that not all the components of singular fibers defined on $\mathbb{Q}$ appear on the graph of the Fig. 3. We have to introduce some of them to construct easily the last fibrations. For example, we start with the fibration of

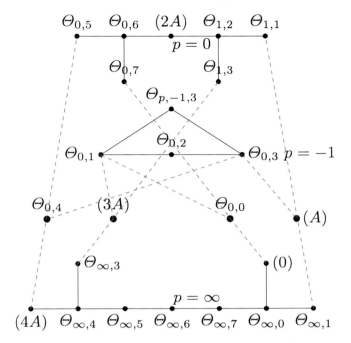

Fig. 11: Fibration of parameter p: three singular fibers

parameter p (6). Using the Fig. 3 we see only three components of the singular fiber I_4 for $p = -1$ i.e. $\Theta_{0,1}, \Theta_{0,2}, \Theta_{0,3}$. The fourth is the rational curve named $\Theta_{p,-1,3}$ parametrized by

$$X = 4\frac{(w+1)\,w}{1+3\,w^2},\ Y = 1/4\frac{1+3\,w^2}{(-1+w)\,w},\ Z = -2\frac{(-1+w)\,(w+1)}{1+3\,w^2}.$$

The four (rational) curves $\Theta_{0,0}, \Theta_{0,4}, (A), (3A)$ are sections of the fibration of parameter p which cut the singular fibers following the previous figure.

10.1 From Fibration of Parameter p

10.1.1 Fibration of Parameter δ

On the Fig. 11 we can see the divisor Δ,

$$\Delta = 6\Theta_{\infty,0} + 5\Theta_{\infty,7} + 4\Theta_{\infty,6} + 3\Theta_{\infty,5} + 2\Theta_{\infty,4} + \Theta_{\infty,3} + 3\Theta_{\infty,1} + 4(0) + 2\Theta_{0,0}.$$

The divisor Δ corresponds to a fiber of type II^*. Using (6) and the previous remark, we can calculate the divisors of $p, p + 1$ and $U p(p + 1)$. The poles of $\delta := U p(p + 1)$ give the divisor Δ. Note δ is equal to the x of (6). From the zeros of δ we get a fiber of type I_5^*

$$2A + 2\Theta_{1,2} + \Theta_{1,3} + \Theta_{1,1} + 2 \sum_{3}^{6} \Theta_{0,i} + \Theta_{0,2} + \Theta_{p,-1,3}.$$

After an easy calculation we get a Weierstrass equation of the fibration

$$y^2 = x^3 + \delta\,(1 + 4\,\delta)\,x^2 + 2\,\delta^4 x + \delta^7. \tag{25}$$

The singular fibers are

$$I_5^*\ (\delta = 0), \qquad II^*\ (\delta = \infty), \qquad I_2\ (\delta = -2),\ I_1 \qquad (\delta = -\tfrac{4}{27}).$$

The Mordell–Weil group is equal to 0.

10.1.2 Fibration of Parameter π

On the previous figure (Fig. 11) we can see the singular fiber

$$\Theta_{\infty,1} + \Theta_{\infty,7} + 2\Theta_{\infty,0} + 2(0) + 2\Theta_{0,0} + 2\Theta_{0,7} + 2\Theta_{0,6} + (2A) + 2\Theta_{1,2} + \Theta_{1,1} + \Theta_{1,3}.$$

Using the previous calculation for δ we see that it corresponds to a fibration of parameter $\pi = \frac{U(p+1)}{p}$. The zeros of π correspond to a fiber of type I_3^*. After an easy calculation we have a Weierstrass equation of the fibration

$$y^2 = x^3 + \pi(\pi^2 - 2\pi - 2)x^2 + \pi^2(2\pi + 1)x. \tag{26}$$

The singular fibers are

$$I_3^*\ (\pi = 0), \qquad I_6^*\ (\pi = \infty), \qquad I_2\ (\pi = -\tfrac{1}{2}), \qquad I_1\ (\pi = 4).$$

The Mordell–Weil group is isomorphic to $\mathbb{Z}/2\mathbb{Z}$.

10.1.3 Fibration of Parameter μ

From the fibration of parameter p we can also join the I_4^* and I_2^* fibers. Let $\mu = \frac{y}{p(x-p(p+1)^2)}$, with y, x from (6). After an easy calculation we obtain a Weierstrass equation of the fibration of parameter μ

$$y^2 + \mu^2(x - 1)y = x(x - \mu^2)^2. \tag{27}$$

The singular fibers are

$$IV^* \ (\mu = 0), \qquad I_{10} \ (\mu = \infty), \qquad I_2 \ (\mu = \pm 1), \qquad I_1 \ (\mu = \mu_0),$$

where μ_0 range overs roots of the polynomial $2x^2 - 27$.
The rank of the Mordell–Weil group is 1, the torsion group is 0. The height of point $(\mu^2, 0)$ is equal to $\frac{1}{15}$.

Remark 5. This fibration can also be obtained with the method of the first set and the parameter $\frac{X^2(Y-1)(Z-1)(YZ+X)}{(X-1)(X+Z)(X+Y)}$ or with parameter $\frac{(Us-1)s}{s-1}$.

10.1.4 Fibration of Parameter α

Let $\alpha = \frac{y}{p(x-p)}$. After an easy calculation we have a Weierstrass equation of the fibration of parameter α

$$y^2 + (\alpha^2 + 2)yx - \alpha^2 y = x^2(x - 1). \tag{28}$$

The singular fibers are

$$I_0^* \ (\alpha = 0), \qquad I_{14} \ (\alpha = \infty), \qquad I_1 \ (\alpha = \alpha_0),$$

where α_0 range overs roots of the polynomial $2x^4 + 13x^2 + 64$.
The Mordell–Weil group is of rank one, the torsion group is 0. The height of $(0, 0)$ is $\frac{1}{7}$.

10.2 From Fibration of Parameter δ

We redraw the graph of the components of the singular fibers and sections of the fibration of parameter δ, and look for subgraphs.

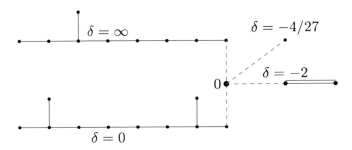

10.2.1 Fibration of Parameter β

We can see the subgraph corresponding to a singular fiber of type I_2^*.

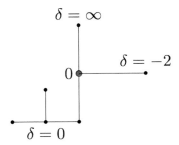

To get a parameter β corresponding to this fibration we do the transformation $x = u + \delta^3$ in (25) and obtain a new equation

$$y^2 = u^3 + \delta\,(\delta + 1)\,(3\,\delta + 1)\,u^2 + \delta^4\,(\delta + 2)\,(3\,\delta + 2)\,u + \delta^7\,(\delta + 2)^2.$$

The point $(0,0)$ is singular   mod δ and   mod $\delta + 2$. By calculation we see that $\beta = \frac{u}{\delta^2(\delta+2)}$ fits. We have a Weierstrass equation

$$y^2 = x^3 + 2\beta^2(\beta - 1)x^2 + \beta^3(\beta - 1)^2 x. \tag{29}$$

The Mordell–Weil group is of order 2. The singular fibers are

$$III^* \ (\beta = 0), \qquad I_2^* \ (\beta = \infty), \qquad I_1^* \ (\beta = 1).$$

10.2.2 Fibration of Parameter ϕ

We can see the subgraph corresponding to a singular fiber of type I_7^*.

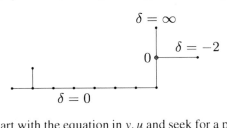

As previously, we start with the equation in y, u and seek for a parameter of the form $\phi' = \frac{u}{\delta^2(\delta+2)} + \frac{a'}{\delta}$. We choose a' to get an equation $y^2 = P(u)$ with P of degree ≤ 4; we find $a = \frac{1}{2}$. Let $\phi = \phi' + 1$, a Weierstrass equation is then

$$y^2 = x^3 + 2\,\phi^2\,(4\,\phi - 7)\,x^2 - 4\,\phi^3\left(-3\,\phi + 8\,\phi^2 - 4\right)x + 8\,(3 + 4\,\phi)\,\phi^6. \tag{30}$$

The singular fibers are

$$III^* \ (\phi = 0), \qquad I_7^* \ (\phi = \infty), \qquad I_1 \ (\phi = \phi_0),$$

where ϕ_0 range overs roots of the polynomial $8x^2 - 13x + 16$. The Mordell–Weil group is 0.

The next table gives the correspondence between parameters and elliptic fibrations.

Parameter	Singular fibers	Type of reductible fibers	Rank	Torsion
$1 - s$	$2I_8, I_4, I_2, 2I_1$	A_1, A_3, A_7, A_7	0	8
$2 - k$	$I_1^*, I_{12}, I_2, 3I_1$	A_{11}, A_1, D_5	1	4
$3 - v$	$I_8, I_{10}, 6I_1$	A_7, A_9	2	0
$4 - a$	$I_8, I_1^*, I_6, 3I_1$	D_5, A_5, A_7	1	0
$5 - d$	$2I_2^*, I_2, I_0^*$	$A_1, D_4, 2D_6$	1	2×2
$6 - p$	I_2^*, I_4^*, I_2, I_4	A_1, D_6, A_3, D_8	0	2×2
$7 - w$	$I_6, I_{12}, 2I_2, 2I_1$	$A_5, A_1, A_1 A_{11}$	0	6
$8 - b$	$2IV^*, I_6, 2I_1$	A_5, E_6, E_6	1	3
$9 - r$	$I_6^*, I_2^*, I_2, 2I_1$	D_6, A_1, D_{10}	1	0
$10 - e$	$III^*, I_4^*, 2I_2, I_1$	A_1, A_1, D_8, E_7	1	2
$11 - f$	III^*, II^*, I_4, I_1	E_7, A_3, E_8	0	0
$12 - g$	$2III^*, I_4, I_2$	E_7, E_7, A_1, A_3	0	2
$13 - h$	$2II^*, I_2, 2I_1$	A_1, E_8, E_8	1	0
$14 - t$	$2I_4^*, I_2, 2I_1$	A_1, D_8, D_8	1	2
$15 - l$	$I_{10}, I_3^*, 5I_1$	A_9, D_7	2	0
$16 - o$	$I_{12}^*, I_2, 4I_1$	A_1, D_{16}	1	2
$17 - q$	$I_{10}^*, I_4, I_2, 2I_1$	A_3, A_1, D_{14}	0	2
$18 - m$	$I_{16}, 3I_2, 2I_1$	A_1, A_1, A_1, A_{15}	0	4
$19 - n$	$I_{16}, I_2, 6I_1$	A_1, A_{15}	2	0
$20 - j$	$IV^*, I_{12}, I_2, 2I_1$	A_{11}, E_6, A_1	0	3
$21 - c$	$I_{18}, 6I_1$	A_{17}	1	3
$22 - u$	I_8^*, I_1^*, I_2, I_1	A_1, D_5, D_{12}	0	2
$23 - i$	$I_{13}^*, I_2, 3I_1$	$A_1 D_{17}$	0	0
$24 - \psi$	$III^*, I_6^*, 3I_1$	$E_7 D_{10}$	1	2
$25 - \delta$	I_5^*, II^*, I_2, I_1	$E_8, A_1 D_9$	0	0
$26 - \pi$	I_3^*, I_6^*, I_2, I_1	A_1, D_{10}, D_7	0	2
$27 - \mu$	$IV^*, I_{10}, 2I_2, 2I_1$	A_9, A_1, A_1, E_6	1	0
$28 - \alpha$	$I_0^*, I_{14}, 4I_1$	D_4, A_{13}	1	0
$29 - \beta$	III^*, I_2^*, I_1^*	E_7, D_6, D_5	0	2
$30 - \phi$	$III^*, I_7^*, 2I_1$	E_7, D_{11}	0	0

Acknowledgements We are grateful to Matthias Schütt for his suggestion to attack the problem and his many helpful comments. Our thanks go also to the organizers for the invitation to the workshop "Arithmetic and Geometry of K3 surfaces and Calabi–Yau threefolds" at the Fields Institute in August 2011. And special thanks to Noriko Yui for the stimulating atmosphere and the great hospitality.

References

1. W. Barth, K. Hulek, C. Peters, A. Van de Ven, *Compact Complex Surfaces*, 2nd edn. (Springer, Berlin, 2004)
2. A. Beauville, Les familles stables de courbes elliptiques sur $\mathbb{P}^1$ admettant quatre fibres singulières C. R. Math. Acad. Sci. (Paris) Sér. I Math. **294**, 657–660 (1982)
3. M.-J. Bertin, Mahler's measure and L-series of K3 hypersurfaces, in *Mirror Symmetry V*, *Proceedings of the BIRS Workshop on Calabi-Yau Varieties and Mirror Symmetry*, ed. by S.-T. Yau. Studies in Advanced Mathematics (American Mathematical Society, International Press (AMS/IP))
4. F. Beukers, H. Montanus, Explicit calculation of elliptic K3-surfaces and their Belyi-maps, in *LMS Lecture Notes Series, vol. 352* (Cambridge University Press, Cambridge, 2008), pp. 33–51
5. N. Bourbaki, *Groupes et algèbres de Lie, Chaps. 4–6* (Masson, Paris, 1981)
6. J.W.S. Cassels, Lectures on elliptic curves, in *London Mathematical Society Student Texts*, vol. 24 (Cambridge University Press, Cambridge, 1991)
7. N.D. Elkies, Three lectures on elliptic surfaces and curves of high rank. Arxiv preprint arXiv:0709.2908v1 [math. NT], 18 Sep. 2007
8. N.D. Elkies, The maximal Mordell-Weil rank of an elliptic K3 surface over $\mathbb{Q}(t)$, in *Talk at the Conference on Birational Automorphisms of Compact Complex Manifold and Dynamical Systems at Nagoya University*, 28 Aug 2007
9. N.D. Elkies, Mordell-Weil generators for singular Shioda-Inose surfaces over $\mathbb{Q}$, http://www.math.harvard.edu/~elkies/K3_20SI.html
10. T. Harrache, Elkies and Kodaira O. Lecacheux, Etude des fibrations elliptiques d'une surface K3. J. Théor. Nombres Bordeaux **23**(1), 183–207 (2011)
11. K. Kodaira, On compact analytic surfaces I-III. Ann. Math. **71**, 111–152 (1960); **77**, 563–626 (1963); **78**, 1–40 (1963)
12. A. Kumar, Elliptic fibration on a generic Jacobian Kummer surface, Arxiv preprint arXiv:1105.1715v2 [math. AG], 5 Aug 2012
13. M. Kuwata, The canonical height and elliptic surfaces. J. Number Theor. **36**(2), 201–211 (1990)
14. J. Martinet, *Les réseaux parfaits des espaces euclidiens* (Masson, Paris, 1996)
15. A. Néron, Modèles minimaux des variétés abéliennes sur les corps locaux et globaux. Inst. Hautes Études Sci. Publ. Math. **21**, 5–128 (1964)
16. H.-V. Niemeier, Definite quadratische Formen der Dimension 24 und Diskriminante 1. J. Number Theor. **5**, 142–178 (1973)
17. V. Nikulin, Integral symmetric bilinear forms and some of their applications. Math. USSR Izv. Math. **14**, 103–167 (1980)
18. K.-I. Nishiyama, The Jacobian fibrations on some K3 surfaces and their Mordell-Weil groups. Jpn. J. Math. **22**, 293–347 (1996)
19. C. Peters, J. Stienstra, A pencil of K3-surfaces related to Apéry's recurrence for $\zeta(3)$ and Fermi surfaces for potential zero, in *Arithmetic of Complex Manifolds*, Erlangen, 1988, ed. by W.-P. Barth, H. Lange. Lecture Notes in Mathematics, vol. 1399 (Springer, Berlin, 1989), pp. 110–127
20. I.-I. Piatetski-Shapiro, I.-R. Shafarevich, Torelli's theorem for algebraic surfaces of type K3. Math. USSR Izv. **35**, 530–572 (1971)
21. M. Schütt, Elliptic fibrations of some extremal K3 surfaces. Rocky Mt. J. Math. **37**(2), 609–652 (2007)
22. M. Schütt, K3 surfaces with Picard rank 20 over $\mathbb{Q}$. Algebra Number Theor. **4**, 335–356 (2010)
23. M. Schütt, T. Shioda, Elliptic surfaces, in *Algebraic Geometry in East Asia - Seoul 2008*. Advanced Studies in Pure Mathematics, vol. 60 (Mathematical Society of Japan, Tokyo, 2010), pp. 51–160
24. I. Shimada, D.Q. Zhang, Classification of extremal elliptic K3 surfaces and fundamental groups of open K3 surfaces. Nagoya Math. J. **161**, 23–54 (2001)

25. T. Shioda, On the Mordell-Weil lattices. Comment. Math. Univ. St. Paul. **39**, 211–240 (1990)
26. H. Sterk, Finiteness results for algebraic K3 surfaces. Math. Z. **189**, 507–513 (1985)
27. J. Stienstra, F. Beukers, On the Picard-Fuchs equation and the formal Brauer group of certain elliptic K3 surfaces. Math. Ann. **271**(2), 269–304 (1985)
28. J. Tate, Algorithm for determining the type of a singular fibre in an elliptic pencil, in *Modular Functions of One Variable IV*, Antwerpen, 1972. Lecture Notes in Mathematics, vol. 476 (Springer, Berlin, 1975), pp. 33–52
29. N. Yui, in *Arithmetic of Certain Calabi-Yau Varieties and Mirror Symmetry*, ed. by B. Conrad and K. Rubin. It is a co-publication of the American Mathematical Society and IAS/Park City Mathematics Institute. vol. 9 (2001), pp. 509–569

Universal Kummer Families Over Shimura Curves

Amnon Besser and Ron Livné

Abstract We give a number of examples of an isomorphism between two types of moduli problems. The first classifies elliptic surfaces over the projective line with five specified singular fibers, of which four are fixed and one gives the parameter; the second classifies $K3$ surfaces with a specified isogeny to an abelian surface with quaternionic multiplication.

Key words: K3 surfaces, Elliptic surfaces, Shimura curves, Discriminant forms

Mathematics Subject Classifications (2010): Primary 14J28, 14J27; Secondary 14J15, 14K02, 14K10, 11G10, 11G15, 11F32

1 Introduction

In this article we will write down explicit families of complex K3 surfaces associated with universal families of abelian varieties of quaternionic multiplication (QM) type. One can view our results as a two-dimensional analog of the identifications of various families of elliptic curves as universal families over modular curves. For example the Legendre family

$$y^2 = x(x - 1)(x - \lambda)$$

A. Besser (✉)
Department of Mathematics, Ben Gurion University of the Negev,
Be'er Sheva 84105 02138, Israel
e-mail: bessera@math.bgu.ac.il

R. Livné
Institute of Mathematics, Hebrew University of Jerusalem Givat-Ram,
Jerusalem 91904, Israel
e-mail: rlivne@math.huji.ac.il

R. Laza et al. (eds.), *Arithmetic and Geometry of K3 Surfaces and Calabi–Yau Threefolds*,   201
Fields Institute Communications 67, DOI 10.1007/978-1-4614-6403-7_7,
© Springer Science+Business Media New York 2013

can be identified as a universal family of elliptic curves with certain level structure, intermediate between level 2 and level 4, over a modular curve isomorphic to $Y(2) = \mathbb{P}^1 \setminus \{0, 1, \infty\}$ with coordinate λ.

The interest in such families, both in the 1- and the higher-dimensional cases comes from several sources. One is counting points modulo primes: such examples are QM-modular families, and thus the number of rational points of the total space modulo primes can be expressed in terms of Fourier coefficients of modular forms [12]. In addition, if we fix a prime p, the number of points modulo p on each fiber can be explicitly described as well: see [19, Chap. IV.4], [11] for the Legendre family, where the results are of interest in cryptography. These results were vastly generalized by Dwork: see e.g. [22]. A completely different place where such families occur is in irrationality proofs of numbers such as $\zeta(2)$ and $\zeta(3)$ (see [6, 7]). Physicists have also studied such families of Calabi–Yau manifolds in connection with mirror symmetry and related phenomena [16].

As is well known, the moduli space of abelian varieties with given polarization, level, and endomorphism structure is a quasi-projective varieties, and if the level structure is fine enough there is a universal family over it. Over $\mathbb{C}$ such families can be described via transcendental means (for all this see e.g. [35] or [13]). However, a projective construction is hard, partly because the theta functions give an embedding into a projective space of a very high-dimension [26–28]. In this respect the 1-dimensional case is misleading: for example, an algorithm of Tate constructs the universal family over $Y_1(N)$ (see [23]) in Weierstrass form, but in the higher dimensional cases no examples seem to be known.

The situation improves if we restrict to the (principally polarized) two-dimensional case *and* content ourselves with the universal family of Kummer surfaces $A/\langle \pm 1 \rangle$ of the relevant abelian varieties. For one thing, a universal family exists as soon as full 2-level structure is added. In addition, the theta functions embed the Kummer surfaces as quartics in $\mathbb{P}^3$. This poses a problem for Shimura curves associated to non-split rational quaternion algebras, since they are compact and hence there is no easy way to construct their associated theta functions. One approach to the problem is through a nice form of the equation for the universal family over the resulting Siegel modular threefold $S(2)$ ([39], see also [4]). If endomorphisms are added to the moduli problem one gets Hilbert modular surfaces or Shimura curves mapping into $S(2)$ and the universal Kummer family can be pulled back to them. In addition, a variant of this method (using a construction of Humbert) was used in [20] to construct genus 2 curves whose jacobians have for endomorphisms a maximal order in the rational quaternion algebras of (reduced) discriminants 6 and 10. However even the Kummer surfaces get more and more complicated to write down, rendering the equations less useful.

Our approach here extends another approach found in [4], where it was proved that the Kummer variety of a QM abelian surface (with multiplication by a maximal order in a rational quaternion algebra) admits a special elliptic fibration when the quaternion algebra has discriminant 6 or 15. More precisely, Besser takes specific elliptic surfaces over $\mathbb{P}^1$ with four singular fibers, and makes a quadratic twist of them at one of the singular fibers and a moving non-singular fiber. Such a description appears particularly useful in connection with certain "modular" threefolds: compare e.g. [15, 16] where examples related to the split quaternion algebra $\text{Mat}_{2\times2}(\mathbb{Q})$ are discussed. See also the recent [38].

In this work we give results of three related kinds. Firstly, we extend Besser's result by showing that in eleven cases (given below in Table 1), each Kummer surface of a specific QM type is isogenous to a K3 surface with an explicit elliptic fibration. Of these eleven cases, two are already in [4], and the isogeny is in fact an isomorphism. In the other nine cases the isogeny is not an isomorphism, but in eight of them (all except case 4 in Table 1) we determine the isogeny up to an isomorphism. Secondly, our elliptically fibered K3 surfaces arise, as in Besser's case, by making a quadratic twist of a fixed elliptic fibration over $\mathbb{P}^1$ with four singular fibers (and a section). In [21] Herfurtner classified all such elliptic families. The twist corresponds to a double cover which is ramified at two points: one point above which there is a singular fiber and another point above which the fiber is regular. This identifies the base space of the resulting K3 families with the base space of the fixed elliptic fibration, or sometimes with a simple quotient of it. The moduli interpretation of Shimura curves implies that this last base space (always a projective line) is therefore related via some correspondence to a Shimura curve. An issue left open in [4], was to determine this correspondence explicitly. In fact we prove in all 11 cases that the $\mathbb{P}^1$ base of our family is canonically an explicit Shimura curve. Thirdly, in all but case 4 in Table 1 we actually prove an isomorphism of families even with some additional level structure.

Our method consists of three steps: we first rigidify the moduli problems by adding level structure on the universal QM side, and going to a certain monodromy cover in the elliptic fibration side. Next we use Morisson's results on Nikulin involutions and Shioda–Inose structures [25] and Nikulin's techniques on discriminant forms to show that the resulting rigidifications are isomorphic. Lastly, we take appropriate quotients to prove that the original parameter spaces are isomorphic. Ultimately our results are based on the Piatecki–Shapiro–Shafarevich Torelli theorem for K3 surfaces (see e.g. [1]), and on Nikulin's detailed study (see [30]) of discriminant forms.

The merit of our approach is twofold. Firstly, the equations we give are extremely simple. This renders the geometry transparent and helps keep calculations easy. Even more interestingly, our method establishes the isomorphism of two moduli problems which a-priori are unrelated: abelian surfaces with a specific QM type, and specific elliptic fibrations. This isomorphism comes with an explicit functorial correspondence between the universal families, even with some auxiliary level structure. Such examples are rare. For instance, there exists a unique genus 3 curve with 168 automorphism. This curve classifies generalized elliptic curves with 7-level structure, but it also classifies certain higher dimensional abelian varieties by virtue of its also being a compact Shimura curve (see [35, 3.18–19]). But in this case a relationship between the corresponding classified objects, elliptic curves with a level 7 structure and these higher dimensional abelian varieties is not even known to exist.

In future work we will discuss the Picard–Fuchs equations associated with our families. These are prominent in the applications to Number Theory, cryptography, and Physics. In particular, we will compute the Picard–Fuchs equations both for the family of K3 surfaces and for the family of Kummer surfaces over the relevant Shimura curve. These computations are particularly easy and transparent for our families because of the simple form in which they are given. They show that the parameter spaces are isomorphic not merely as genus 0 Riemann surfaces with four marked points, but also as parameter spaces for appropriate variations of Hodge structures. This provides independent verification for the computations in this work.

All in all, our work highlights the power of the Torelli theorem for K3 surfaces combined with Nikulin's technique of discriminant forms, but it also shows its limitations. It enables us to prove the existence of isomorphisms of moduli spaces and universal families without actually exhibiting them. In fact, our moderate efforts to exhibit the isomorphisms of corresponding classified surfaces explicitly were unsuccessful. (In contrast, there are a few cases in Herfurtner's list which give rise, by the same quadratic twist construction, to the split quaternion algebra, and here matters can be made explicit.) Also, we only prove that our models are isomorphic over $\mathbb{C}$ and not over $\mathbb{Q}$ as one would have preferred. Here one could hope to pass to an isomorphism over $\mathbb{Q}$, for example by studying the reduction modulo a few primes (for a computation of this type see eg [37, Sect. 5]), but we have not done this. For a different approach see [18].

We thank Ron Donagi for a helpful suggestion concerning isogenies of $K3$ surfaces and to the referee for a careful reading of the manuscript and for making many valuable corrections and suggestions. In particular, the references to [32] are due to the referee. This research was supported by a joint grant from the Israel Science Foundation.

2 K3 Surfaces with Picard Number 19 and Twists of Elliptic Surfaces

In this section we explain the method for obtaining our examples: families of twists of elliptic surfaces. Relevant facts about rational quaternion algebras used in this section and in the rest of this work can be found in.

2.1 Lattices

We recall a few basic notions on lattices. A *lattice* is a finite rank $\mathbb{Z}$-module with an integral valued non-degenerate symmetric bilinear form. An isomorphism of lattices is just a bijective isometry. The *discriminant* disc(L) of a lattice L is the absolute value of the determinant of a representing matrix for the form with respect to any $\mathbb{Z}$ basis. A lattice is *even* if the associated quadratic form takes only even value. If L is a lattice and $0 \neq n \in \mathbb{Z}$, we denote by $L[n]$ the lattice with the same underlying module and with form multiplied by n. For a positive n we also write L^n for the orthogonal direct sum of L with itself n times.

The lattices L and T are *isogenous* if there exists a linear isomorphism of $L \otimes \mathbb{Q}$ with $T \otimes \mathbb{Q}$ which is compatible with the respective bilinear forms up to a scalar.

Certain special lattices will occur. The lattice E_i for $i = 6, 7, 8$ are the even positive definite lattices associated with Dynkin diagrams $\tilde{E}_i$ (see [1, Sect. I.2]). The hyperbolic plane U has generators e, f with $e^2 = f^2 = 0$ and $e \times f = 1$. Letting $s = f - e$ we see that U is also generated by e and s with $e \times s = 1$ and $s^2 = -2$.

Finally, we recall that if X is a compact complex surface, then the second cohomology group $H^2(X) := H^2(X, \mathbb{Z})$ modulo torsion, together with the bilinear form furnished by the cup product, is a lattice. If X is algebraic, the Néron–Severi lattice of $H^2(X)$ is a sublattice denoted $\mathrm{NS}(X)$. Its orthogonal complement is the transcendental lattice $T(X)$.

2.2 Elliptic Surfaces

An elliptic surface, always considered over $\mathbb{P}^1$, is a smooth and connected compact complex algebraic surface E, together with a surjective morphism $\pi : E \to \mathbb{P}^1$, such that the generic fiber is a curve of genus 1. We will always assume that the fibration is relatively minimal and has a given section, denoted 0.

For all but a finite number of points $s \in \mathbb{P}^1$, the fiber $E_s = \pi^{-1}(s)$ is an elliptic curve. The singular locus $\Sigma = \Sigma(E)$ of the fibration is the (finite) subset of $\mathbb{P}^1$ over which the fibers are singular (namely π is not everywhere smooth). Kodaira [24] classified all possible types of singular fibers (see also [1, Chap. V.7]). All but Kodaira's types I_1 and II consist of a configuration of smooth rational curves, each one with self intersection -2. For such a type t we denote by L_t the lattice spanned by all such projective lines which do not intersect the 0-section with the intersection pairing. Types which do not consist of smooth rational curves are irreducible and their associated lattices are trivial by definition.

Following [4, p. 284] we will make the following

Definition 1. The type of an elliptic fibration is the collection of Kodaira types of singular fibers, counted with multiplicity, except that we do not distinguish between the following pairs of Kodaira types: I_1 and II, I_2 and III, I_3 and IV.

The reason for this definition will be seen in Remark 4 below. We note that unlike loc. cit. we do not neglect I_1 fibers for clarity. Since we will need sometimes to refer to the type as a collection of Kodaira types we will call this collection the *strong type* of the fibration.

Definition 2. We denote by H_T the moduli space of elliptic fibrations of type T (taken up to isomorphism as fibrations over $\mathbb{P}^1$).

We will see later how to construct this moduli space in the cases we consider.

Elliptic surfaces are completely determined by two invariants. The functional invariant is the meromorphic function $J(s)$ giving the J-invariant of each fiber E_s. The homological invariant is the local system of the first homology groups on

$\mathbb{P}^1 - \Sigma(E)$. It can be interpreted as an $SL_2(\mathbb{Z})$-valued representation of the fundamental group $\pi_1(\mathbb{P}^1 - \Sigma)$. It is well known, and immediate from the list of local monodromies (ibid), that the associated *projective* representation is already determined by the functional invariant.

A theorem of Shioda [36, Theorem 1.1] describes the Néron–Severi lattice of an elliptic surface. We will use the following version of Shioda's result.

Proposition 1. *For an elliptic fibration $\pi : E \to C$ over a curve C (as always, with a section) the natural map $L_T \oplus U \to \mathrm{NS}(E)$ is an injective isometry with cokernel canonically isomorphic to the Mordell–Weil group of E over the function field of C.*

Remark 1. We only consider cases where the Néron–Severi lattice has rank 19. Then the lattices $(L_T \oplus H)$ and $\mathrm{NS}(E)$ become isomorphic over $\mathbb{Q}$, a result we exploit in this section. In most of our cases these lattices are actually isomorphic, see Proposition 36, and see Sect. 8.4 for the other cases.

2.3 Quadratic Twists

Given two distinct points a and b in $\mathbb{P}^1$, the quadratic twist $E_{a,b}$ at these points can be described in two ways. Algebraically, write E in Weierstrass equation $y^2 = f(x)$. If a and b are finite points, then $E_{a,b}$ has the equation

$$\frac{t-a}{t-b}y^2 = f(x)(= 4x^3 - g_2(t)x - g_3(t)) . \tag{1}$$

Analytically, $E_{a,b}$ can be described as follows. Take the double cover $S \to \mathbb{P}^1$ ramified at a and b and let E' be the pullback surface. Now quotient E' by the transformation which identifies the two fibers above each fiber of E with sign -1. It follows that $(E_{a,b})_{a,b} = E$.

Remark 2. For any $0 \ne \alpha \in \mathbb{C}$ the quadratic twist $E_{a,b}$ is also given by the equation $\alpha\frac{t-a}{t-b}y^2 = f(x)$. The resulting surfaces is isomorphic to $E_{a,b}$ of course, but the isomorphism is only canonical up to ± 1 because it involves a choice of a square root of α.

A quadratic twist has no effect on the functional invariant. On the homological invariant it reverses the sign of the local monodromy around the two points a and b where the twist is done. If the fiber above one of these points is of type I_n, the fiber of the twisted surface is therefore of type I_n^*. In particular, twisting at a non-singular fiber gives a fiber of type I_0^*.

Definition 3. Let $E \to \mathbb{P}^1$ be an elliptic surface with a singular locus $\Sigma = \Sigma(E)$. Fix $s \in \Sigma$. For $\lambda \in \mathbb{P}^1 - \Sigma$ let $E_{s,\lambda}$ be the twisted family at s and at λ. These surfaces vary in a family $\mathscr{TW}_s(E)$ over the λ-*line* $\mathbb{P}^1(\lambda) - \Sigma$.

Remark 3. The twisted surfaces also make sense for $\lambda \in \Sigma$, but over our chosen base the variation of the part of the second cohomology coming from the singular fibers is nice.

Lemma 1. *In the notation of definition 3 suppose that $E_{s,\lambda}$ is a K3 surface with at least 4 singular fibers. Then the rank r_T of the abelian group L_T is ≤ 17. If equality holds, then the Mordell–Weil rank r_λ of $E_{s,\lambda}$ is 0 for general λ.*

Proof. Observe first that L_T is independent of $\lambda \notin \Sigma$, since the fiber types are constant outside of Σ. Let ρ_λ be the rank of $NS(E_{s,\lambda})$. By Shioda's formula 1.2.1, $r_\lambda + r_T + 2 = \rho_\lambda$, which is at most 20 for a K3 surface. Hence, if the assertion of the Lemma is false we must have, in either case, that ρ_λ is 20 for a general λ. In this case the isomorphism type of the K3 surface is in a countable set, hence $E_{s,\lambda}$ is constant as λ varies. Localizing in λ we can trivialize the local system $NS(E_{s,\lambda})$, and in particular assume that the fiber divisor of elliptic pencil is constant. Then the subset of $t \in \mathbb{P}^1$ above which the fibers of $E_{s,\lambda}$ are singular is constant in moduli. However this set consists of one free-moving point and at least three fixed ones. Such a set cannot be constant in moduli (it has moving cross ratios). This contradiction proves the Lemma.

2.4 The Basic Construction

In [21] Herfurtner classified rational elliptic fibrations with four singular fibers, giving explicit equations for each fibration. Applying the construction $\mathcal{T}\mathcal{W}$ described above to any fibration on his list yields an explicit one-parameter family of surfaces. The cases when the general member of this family is a K3 surface with Picard number 19 can be singled out by the following

Proposition 2. *Let E be an rational elliptic surface with four singular fibers, one of which is at $s \in \mathbb{P}^1$. Then, for a generic $\lambda \in \mathbb{P}^1$, the surface $E_{s,\lambda}$ is a K3 surface whose lattice L_T has rank ≥ 17 if and only if E has an unstarred singular fiber at s (namely of type I_n, $n \geq 1$, II, III, or IV) and the other singular fibers are semistable (namely each is of type I_n for some n).*

Proof. A quadratic twist exchanges starred and unstarred fibers, adding 6 to the Euler number of an unstarred fiber and subtracting 6 from the Euler number of a starred one. Hence the twists of the rational fibration E (which has Euler number 12) are K3 fibrations (which have Euler number 24), if and only if the fiber of E at s is starless.

For $1 \leq i \leq 4$ let us denote by e_i (respectively n_i) the Euler number (respectively the number of components not meeting the 0 section) of the ith singular fiber of E, where $i = 4$ corresponds to the fiber at s. Let m denote the number of fibers of type I_n (any n) among the first three singular fibers. It remains to show that $m = 3$ if and only if $\text{rank} L_T \geq 17$. By Lemma 1 this is equivalent to the assertion that $m = 3$ if and only if $\text{rank} L_T = 17$.

Notice now that the Euler number of a singular fiber is 2 more than the number of components of the fiber not meeting the 0-section, unless the fiber is of type I_n, in which case it is only 1 more. Since the fiber of $E_{s,\lambda}$ at s is starred, it has

$(e_4 + 6) - 2$ components not meeting the 0-section. It follows that rank$L_T = n_1 + n_2 + n_3 + (e_4 + 4) + 4$, where the I_0^* fiber at λ contributes the last term 4. Equivalently, rank$L_T = [(e_1 - 2) + (e_2 - 2) + (e_3 - 2) + m] + e_4 + 8 = 14 + m$, which implies the assertion and concludes the proof of the Proposition.   □

2.5 The Néron–Severi and the Transcendental Lattices

A K3 surface X with Picard number 19 is isogenous to the Kummer surface associated with an abelian surface A (see [25]). The transcendental lattices of X and A are then isogenous. The Néron–Severi lattice NS(A) is then isogenous to both (in fact NS(A) $\simeq T(A)[-1]$). In particular NS(A) has rank 3. By [29, Chap. 4], the rational endomorphism algebra of A is an indefinite rational quaternion algebra $B_A = B_X$ which we wish to determine (our considerations in this subsection are a precursor to much more precise considerations in the remainder of this work).

Let X be a K3 surface with an elliptic fibration whose Mordell–Weil rank is 0. Our methods apply to this case in general, but we will restrict attention to the cases which interest us here, where the elliptic fibration has five singular fibers of types I_a, I_b, I_c, I_0^*, and J, where J is one of I_d^*, II^*, III^*, or IV^*. The ranks of the forms L_{t_i}'s are then $a - 1$, $b - 1$, $c - 1$, 4, and $r(J) = d + 4$, 8, 7, or 6 respectively. Set $\rho = \rho(J) = 4, 1, 2,$ or 3 respectively (ρ is discL_J). We will prove the following

Lemma 2. *In the situation above let B_X be the quaternion algebra associated to X. Then a prime p divides the discriminant of B_X if and only if*

$$(-\rho, abc)_p (a, b)_p (a, c)_p (b, c)_p = -1,$$

where $(x, y)_p$ is the Hilbert symbol at p.

Proof. Since we are assuming that the Mordell–Weil rank of the fibration is 0, we have by Proposition 1 a rational isometry

$$\text{NS}(X) \otimes \mathbb{Q} \cong \left(\bigoplus_i L_{t_i} \oplus U \right) \otimes \mathbb{Q},$$

with U the hyperbolic plane, and we further have a rational isometry $H^2(X, \mathbb{Z}) \otimes \mathbb{Q} \cong (\text{NS}(X) \oplus T) \otimes \mathbb{Q}$, with $T = T(X)$. Since $H^2(X, \mathbb{Z}) \simeq E_8[-1]^2 \oplus U^3$ (see e.g. [1, Proposition VIII.3.3.ii], with L as defined in Chap. VIII.1 there), we have an isomorphism of rational quadratic forms:

$$(T \oplus \bigoplus_i L_{t_i} \oplus U) \otimes \mathbb{Q} \simeq (E_8[-1]^2 \oplus U^3) \otimes \mathbb{Q}.$$

We may add $T[-1]$ on both sides and, recalling that T has rank 3 and using Lemma 24, we find by Witt's cancellation Proposition 37, multiplying the entire expression by -1,

$$\left(\bigoplus_i L_{t_i}[-1] \oplus U\right) \otimes \mathbb{Q} \simeq (T \oplus E_8^2) \otimes \mathbb{Q}.$$

By the Theorem 8 the lattice $L_{II^*}[-1] = E_8$ has discriminant 1 and local ϵ invariants (see Definition 35 for the definition of these) $\epsilon_p(E_8) = 1$ for any (finite) prime p. It follows from Lemma 23 that the discriminant and epsilon invariants of the right hand side are the same as those for T. By Theorem 8 and Lemma 23 we compute $\mathrm{disc}(T) = -abc\rho$ and

$$\epsilon_p(T) = (-1, \rho abc)_p (\rho, abc)_p (a, b)_p (a, c)_p (b, c)_p (-1, a)_p (-1, b)_p (-1, c)_p$$
$$= (-1, \rho abc)_p (-\rho, abc)_p (a, b)_p (a, c)_p (b, c)_p.$$

The lattice T is isometric to $\mathrm{NS}(A)[-2]$, and the assumptions imply by [3] (see Lemma 4) that $\mathrm{NS}(A)$, hence T, are multiples of the quadratic form Nm described in Sect. A.1.3 associated with a rational quaternion algebra $B(\alpha, \beta)$. According to Proposition 38, the multiplying factor for T should be $\mathrm{disc}(T)$ and so the same Proposition implies the formula

$$(\alpha, \beta)_p = \epsilon_p(T)(-\mathrm{disc}(T), -1)_p$$
$$= (-1, \rho abc)_p (-\rho, abc)_p (a, b)_p (a, c)_p (b, c)_p (-1, abc)_p$$
$$= (-\rho, abc)_p (a, b)_p (a, c)_p (b, c)_p,$$

proving the lemma. □

We can now go through all cases in the list of elliptic fibrations E found by Herfurtner [21]. We first restrict attention to those fibrations satisfying the conditions of Proposition 2, so that the twists $E_{s,\lambda}$ generically have Picard number ≥ 19. We use Lemma 2 to determine the associated quaternion algebra. The cases that yield the split quaternions are $(3*I_1, I_9)$, $(2*I_1, I_2, I_8)$, (I_1, I_2, I_3, I_6), $(2*I_1, 2*I_5)$, $(2*I_2, 2*I_4)$, and $(4*I_3)$. In fact, for all these examples $\rho = 1$, and Lemma 2 shows that no primes are ramified (We do not know why these cases are characterized by the fact that all the singular fibers, before the twist, are of type I_n.) The remaining cases are listed in Table 1.

Let us consider for example the first entry in Table 1, the bad fibers (after the twist) are $(2*I_1, I_8, IV^*, I_0^*)$. Here Lemma 2 gives ramification at $p = 2$ or 3, so B_X has discriminant 6. The other cases in Table 1 are done similarly.

The analysis of the families of fibrations arising from the fibrations listed in this table are the main focus of our work. We have listed some further information in this table. For each of our examples we give

- A reference number; [4] we do not know the exact isogeny between the abelian surface and the elliptic fibration because the techniques developed in Sect. 7 do not apply (we can however determine with reasonable certainty the relation between the moduli spaces using Picard–Fuchs equations techniques [5]).

Table 1: The examples

#					
1	I_1 I_1 I_8 II γ $\bar\gamma$ ∞ 0 2 2 2 6				$V_6(2)/S_3 \times \langle w_3\rangle$
2	I_1 I_2 I_7 II $-9/4$ $-8/9$ ∞ 0 2 2 2 6				$V_{6,7}/\langle w_2,w_3,w_7\rangle$
3	I_1 I_4 I_5 II -10 0 ∞ $1/8$ 2 2 2 6				$V_{15}/\langle w_3,w_5\rangle$
4	I_2 I_3 I_5 II $-5/9$ 0 ∞ 3 2 2 2 6				$V_{10,3}/\langle w_2,w_3,w_5\rangle$
5	I_1 I_1 I_7 III δ $\bar\delta$ ∞ 0 2 2 2 4				$V_{14}/\langle w_2,w_7\rangle$
6	I_1 I_2 I_6 III 4 1 ∞ 0 2 2 2 4				$V_6(\pi_2)/\langle w_2,w_3\rangle$
7	I_1 I_3 I_5 III $-25/3$ 0 ∞ $1/5$ 2 2 2 4				$V_{6,5}/\langle w_2,w_3,w_5\rangle$
8	I_2 I_3 I_4 III $-1/3$ 0 ∞ 1 2 2 2 4				$V_6(\pi_2)/\langle w_2,w_3\rangle$
9	I_1 I_1 I_6 IV 1 -1 ∞ 0 2 2 2 3				$V_6/\langle w_2,w_3\rangle$ $\mathbb{P}^1_\lambda \simeq V_6/\langle w_6\rangle$
10	I_1 I_2 I_5 IV $-27/4$ $-1/2$ ∞ 0 2 2 2 3				$V_{10}/\langle w_2,w_5\rangle$
11	I_3 I_3 I_2 IV ∞ 0 -1 1 2 2 2 3				$V_6/\langle w_2,w_3\rangle$ $\mathbb{P}^1_\lambda \simeq V_6/\langle w_6\rangle$

- The (strong) types of the four singular fibers, their locations, and their ramification indices (see [1, V.7] for types of singular fibers and just after Definition 1 for the terminology of *strong type*).
- H_T (see Definition 2) as a Shimura curve. When $\mathbb{P}^1_\lambda \not\simeq H_T$ both are given as Shimura curves.

In the table $\gamma = \dfrac{-(1+\sqrt{-2})^4}{3}$ and $\delta = \dfrac{(1+\sqrt{-7})^7}{512}$. Conjugates for such elements are over $\mathbb{Q}$. In what follows we will refer to the nth item in Table 1 as Table 1(n).

Our notation for Shimura curves is as follows:

Definition 4. Let $\mathcal{M} = \mathcal{M}_{D,N}$ be an Eichler order of conductor N in an indefinite rational quaternion algebra B of discriminant D. Both B and $\mathcal{M}$ are unique up to inner automorphisms, so their choice does not matter; moreover N is necessarily prime to D. For n prime to N we will denote the moduli space parameterizing abelian surfaces with $\mathcal{M}_{D,N}$-action, full level-n structure, and a certain compatible polarization (of type $(1, N)$) by $V_{D,N}(n)$. When N or n are 1 they will be omitted from the notation, so that for example $V_D = V_{D,1} = V_{D,1}(1)$. Analytically we have

$$V_{D,N}(n) = B^\times \backslash (\mathcal{H}^\pm \times (B_f^\times / K(n))).$$

Here $K(n) = K_{D,N}(n)$ denotes the principal congruence subgroup of level n in $\mathcal{M} \otimes \hat{\mathbb{Z}}$, and $B_f = (B \otimes \mathbb{A}_f)$ is the finite adèles of B. For a prime p dividing D there is a unique two-sided ideal $\pi_p \mathcal{M} \subset \mathcal{M}$ of norm p. It is principal, so that the level n has the form $n' \prod_{p|D} \pi_p^{k_p} \mathcal{M}$, where each k_p is 0 or 1. For every $d|DN$ there exists a modular involution w_d acting on $V_{D,N}$. Lastly, $\mathcal{H}^\pm = \mathbb{P}^1(\mathbb{C}) \setminus \mathbb{P}^1(\mathbb{R})$ is the union of the upper half plane $\mathcal{H}$ and the lower half plane.

3 The Moduli Map for Discriminants 6 and 15

As was explained, a general elliptic surface appearing in one of the families in Table 1, is isogenous to an abelian variety whose rational endomorphism algebra is a rational quaternion algebra of known discriminant. This gives rise to a correspondence between the base space of the family of elliptic surfaces and the Shimura curve classifying such abelian surfaces. In [4] the first named author proved that for the families of Table 1(1) and (3) the general elliptic surface is in fact isomorphic to the Kummer surface of an abelian surface whose endomorphism ring is a maximal order in the corresponding quaternion algebra. In this section we will describe the resulting correspondence completely for these cases. In fact we will show that a specific covering $\tilde{H}_T^0$ of the moduli space H_T of elliptic fibrations of a certain type is isomorphic to the corresponding Shimura curve with full 2-level structure. We will also identify the forgetful map $\tilde{H}_T^0 \to H_T$ in terms of Shimura curves. This is partly a warm-up for a the more complicated analysis required for some of the other cases.

For $D = 6$ or $D = 15$ let A be an abelian surface together with a ring isomorphism $\mathcal{M}_D \xrightarrow{\sim} \text{End}(A)$. By [3, Theorem 3.10], the rational Néron–Severi lattice, NS(A)$\otimes\mathbb{Q}$, of A is identified with the traceless elements of B. Moreover, NS(A) is constant as A varies over the Shimura curve parameterizing this data by loc. cit. Now let ϕ be a full level-two structure on A. This data then determines an isometry $f : N \xrightarrow{\sim} \text{NS}(X)$ between a certain lattice N, determined by the discriminant, and the Néron–Severi lattice NS(X) of the Kummer surface $X = \text{Km}(A)$. There furthermore exists a fixed class $d \in N$ such that $f(d)$ is an ample class on X. It is shown in loc. cit., relying of ideas of [33], that d determines in N a class e such that $f(e)$ is the class of the fiber of an elliptic fibration on X, with prescribed singular fibers as in Table 1(1) and (3). Since these fibrations turn out to be twists of a fixed fibration we obtain a well defined value corresponding to the place of the twist. The correspondence can be described graphically as follows:

$$(A + \text{level } \phi) \longleftrightarrow (X + f : N \to \text{NS}(X)) \qquad (2)$$

$$(A) \qquad\qquad (X + \text{fibration}) \longleftrightarrow \lambda \in H_T$$

where H_T is usually the λ-line but could also be a quotient of it (see Proposition 3 for the precise relation). In fact the horizontal maps turn out to be bijective, while the vertical maps are not. The data $(A + \text{level } \phi)$ is classified by the Shimura curve $V_D[2]$. It follows (but not stated explicitly in loc. cit.) that there exists a rational map $V_D[2] \to H_T$. To do this we need the notion of markings.

3.1 Marked Elliptic Fibrations and Moduli Spaces

Definition 5. Let t be a type of a singular fiber. The associated Dynkin graph G_t has a vertex for each component which does not meet the zero section and two vertices are connected if and only if the corresponding components intersect. If T is a type of singular fibers the associated graph G_T is the disjoint union of the graphs for all $t \in T$. If E is an elliptic fibration the graph G_E of E is defined in the same way.

Remark 4. An isomorphism of elliptic fibrations clearly induces an isomorphism of the associated graphs. Notice that the graph depends only on the type and not on the strong type (this is in fact the reason for the definition of type).

Definition 6. A marking of type T for an elliptic fibration $E \to \mathbb{P}^1$ is an isomorphism $\alpha : G_T \to G_E$.

Let us now discuss the types of moduli spaces that we will consider in this work more carefully. We assume we are given an elliptic fibration with four singular fibers that appears in Table 1. Thus, it has three singular fibers of type I_n, and one fiber, at $s_0 \in \mathbb{P}^1$, which is of type II, III or IV. Let Σ be the singular locus. There is a corresponding type T where there is an additional I_0^* fiber and the non I_n fiber is starred.

The first moduli space is the λ-line $\mathbb{P}^1(\lambda) - \Sigma$ which carries the family of twists $E_{s_0,\lambda}$.

Lemma 3. *The λ-line is the coarse moduli space of elliptic fibrations of type T up to isomorphisms of fibrations which on the base fix the location of the I_0 fiber.*

Proof. The maps between the set of isomorphism classes of elliptic fibrations as above and the λ-line and back are obvious, sending an elliptic fibration to $\lambda =$ the location of the I_0-fiber and λ to the elliptic fibration $E_{s_0,\lambda}$. That these maps are inverse to each other is obvious in one direction. In the opposite direction, let E' be an elliptic fibration of type T. Then the quadratic twist at the two starred fibers gives an elliptic fibration with four singular fibers of the types we started from. The tables in [21] shows that this fibration is moreover unique in each of our cases, and hence must be isomorphic to E. Twisting back shows that E' is isomorphic to the appropriate $E_{s_0,\lambda}$. $\square$

The second moduli space is the moduli space H_T of elliptic fibrations of type T up to isomorphisms. Having the λ-line it is trivial to construct it.

Proposition 3. *For a type T arising from Table 1 let D_T be the group of all the automorphisms of the λ-line preserving the singular locus S which send each $s \in S$ to a point with the same (strong) type of fiber. Then $\mathbb{P}^1_\lambda / D_T \simeq H_T$.*

Proof. The natural map from the λ line to H_T, which sends λ to the isomorphism type of $E_{s,\lambda}$, clearly factors through the action of D_T, and induces a morphism $\phi : \mathbb{P}^1_\lambda / D_T \to H_T$. It is injective, since we have accounted precisely for the isomorphisms between the fibrations. $\qquad\square$

Note that in most of the cases in Table 1 the above group D_T is trivial.

The next moduli space is the space Λ_T of pairs (λ, α) where λ belongs to the λ-line and α is a marking for the elliptic fibration $E_{s_0,\lambda}$. Here it is important, because of Remark 2, that $E_{s_0,\lambda}$ means the twist given by (1), rather than any quadratic twist of E ramified at s_0 and λ. Obviously, Λ_T is an unramified covering of the λ-line which is Galois with group $\mathrm{Aut}G_T$, and may well fail to be connected.

The next moduli space is $\tilde{H}_T$. It is the coarse moduli space of marked elliptic fibrations of type T. Note that an elliptic fibration always has the action of -1, which changes the marking. It gives a well define automorphism of G_T. The moduli space $\tilde{H}_T$ is easily seen to be the quotient of Λ_T by the action of D_T and by the action of -1. It is thus sometimes useful to also consider the space Λ'_T obtained by dividing out Λ_T only by the action of -1, which is still an unramified covering of the λ-line. The following obvious consequence is sometimes useful.

Proposition 4. *The covering $\tilde{H}_T / H_T$ can only be ramified over fixed points of D_T*

Finally, as the moduli spaces above may well fail to be connected, we consider connected components of them. A connected component $\tilde{H}^0_T$ of $\tilde{H}_T$ corresponds to the subgroup M_T of $\mathrm{Aut}G_T$ generated by the local monodromies around the singular fibers. The forgetful map from $\tilde{H}^0_T$ to H_T is a Galois covering, whose Galois group is the *monodromy group* Mon_T, equal to the quotient of M_T by the action of ± 1. Indeed, on the level of the corresponding stacks we clearly have a Galois covering with Galois group M_T. However when considering the coarse moduli spaces, we must take into account the fact that the action of $d \in M_T$ on the marking arises from an isomorphic fibration if and only if d acts as ± 1.

We will also denote by the same notation the natural compactifications of all of the moduli spaces considered above, coming from the obvious compactifications of the λ-line. All results about ramification need to be modified then to mean that ramification can further occur over points corresponding to singular fibers.

3.2 Types of Marked K3 Surfaces

We now identify a marked elliptic fibration of type T with another type of object.

We begin by recalling some facts about K3 surfaces and the Torelli theorem.

Definition 7. We follow [1, VIII.1]. Let X be a K3 surface. The set $\{x \in H^{1,1}(X, \mathbb{R})|$ $(x, x) > 0\}$ consists of two convex cones ($H^{1,1}(X, \mathbb{R})$ is $H^{1,1}(X) \cap H^2(X, \mathbb{R})$). One of these cones contains the Kähler classes on X. We call this cone the positive cone and denote it by $\mathscr{C}_X$. The Kähler classes form a convex sub-cone of $\mathscr{C}_X$, called the Kähler cone. The classes in $H^2(X, \mathbb{Z}) \cap H^{1,1}(X, \mathbb{R})$ of the effective divisors also span a convex sub-cone of $\mathscr{C}_X$. An isometry $H^2(X, \mathbb{Z}) \to H^2(X', \mathbb{Z})$ is called an *effective Hodge isometry* if

1. It preserves the cup product;
2. Its $\mathbb{C}$-linear extension preserves the Hodge decomposition; and
3. It preserves the positive cone and the classes of effective divisors.

The following is the global Torelli Theorem for Kähler K3 surfaces

Theorem 1 ([1, VIII, Theorem 11.1, Proposition 11.3]). *An effective Hodge isometry between K3 surfaces X and X' is induced by a unique isomorphism.*

Let S be an even lattice. The positive cone of S is decomposed into chambers by the planes orthogonal to the -2-classes in S. Let d be one fixed chamber.

Definition 8. An (S, d) marked K3 surface consists of a K3 surface X together with an isometric embedding $\tilde{\alpha} : S \to \mathrm{NS}(X)$ (called marking) with finite cokernel which is good in the sense that $\tilde{\alpha}(d)$ is the ample cone of X. An isomorphism of (S, d) marked K3 surfaces is an isomorphism commuting with the marking. The marking is called strong if its cokernel is 0.

We will show that marked elliptic surfaces of a certain type T are the same as K3 surfaces marked by a certain pair (S, d) which we now define.

Definition 9. For a graph G let S_G be the lattice with one generator for each vertex of G with square -2 and with intersections according to the edges. For a type T let $S_T = S_{G_T} \oplus U$. Recall that U is the hyperbolic plane, generated by classes f with square 0 and s with square -2, with $f \times s = 1$, to be thought of respectively as the fiber and the zero section of an elliptic fibration.

In particular, for every $t \in T$ the lattice S_T contains the elements $F_t = f - \sum_{g \in G_t} n_g g$, where n_g is the multiplicity of g in the singular fiber as determined by Kodaira (see [1, Sect. V.7, Table 3]). We now describe a certain chamber in S_T.

Definition 10. Let d_T be any chamber in S_T containing the element w constructed as follows: For every $t \in T$ let x_t be a linear combination with positive integer coefficients of $g \in G_t$ which has a positive intersection with every $g \in G_t$. Let $x = \sum_{t \in T} x_t$. Let $y = x + n \times s$ with n sufficiently large so that $y \times F_t > 0$ for every $t \in T$. Let $z = y + mf$ with m sufficiently large so that $z \times s > 0$. Finally, let w be the sum of all conjugates of z over all automorphisms of G_T.

Proposition 5. *Let (E, α) be a marked elliptic fibration of type T with Mordell–Weil rank 0. Let $\tilde{\alpha} : S_T \to \mathrm{NS}(E)$ be the map sending f to the class of a fiber of E, s to the class of the zero section, and for every $g \in G_T$ the class of the component of singular fiber $\alpha(g)$. Then $(E, \tilde{\alpha})$ is a marked (S_T, d_T) surface.*

Proof. By Shioda's work (Proposition 1) it suffices to prove that $\tilde{\alpha}(z)$, where z is the element constructed as part of Definition 10, is ample. By the Nakai–Moishezon criterion one has to show that it has positive intersection with every effective divisor. By construction, $\tilde{\alpha}(z)$ has a positive intersection with the fiber, zero section and all the irreducible components of the singular fibers of E (for each singular fiber there is one component which does not appear in the graph, which is the one intersecting the zero section, but the intersection with it is also positive because the intersection with the fiber is positive). Any other effective divisor is a multi-section, and since we have taken the fiber with a positive multiplicity the intersection here is positive as well.   □

The converse theorem is also true, essentially part of [4].

Proposition 6. *Let T be a type of elliptic fibrations. Let S be a finite over-lattice of S_T occurring as a sublattice of the Néron–Severi lattice of an elliptic fibration of type T with Mordell–Weil rank 0 on a K3 surface. Let d_T be the chamber described above. Let $(X, \tilde{\alpha})$ be a strongly marked (S, d_T) surface. Then $\tilde{\alpha}(f)$ is the fiber of an elliptic fibration on X, with Mordell–Weil rank 0. Furthermore, for any $g \in G_T$, $\tilde{\alpha}(g)$ is the class of a unique component of a singular fiber and sending g to this component gives a marking of type T on the elliptic fibration. This construction is reverse to the previous construction.*

Proof. This is a refinement of [4, Theorem 1]. The lattice S, class f and the chamber d_T have the following basic property: If $x \in S$ satisfies $x^2 = -2$ and $x \times d_T > 0$ then $x \times f \geq 0$. Indeed, since S is contained in the Néron–Severi lattice of a K3 surface, it follows from [33] (see also [1, Proposition VIII.3.7.2]) that either x or $-x$ is effective. But the assumption $x \times d_T > 0$ implies that it is x which is effective and it is then clear it has a non-negative intersection with the class f of an elliptic curve. We now consider Proposition 2.4 in [4]. The proposition gives conditions on $e = \tilde{\alpha}(f)$ to be the class of a fiber of an elliptic fibration on X. All conditions are easily satisfied except for condition 4. However, that condition only enters in the proof there of Lemma 2.5 that e has non-negative intersection with any effective divisor, and one sees that one ends up using only that e has non-negative intersection with any effective divisor of square -2, but that is exactly the condition we established. Thus, $\tilde{\alpha}(f)$ is indeed the class of the fiber of an elliptic fibration. Finally, from the data of the class f and the class z the classes of the irreducible components of the singular fibers can also be constructed. Namely, the orthogonal complement to the hyperbolic plane consisting of the fiber and the section splits into the direct sum of negative definite lattices which are related to root systems and the irreducible classes are simply the elements of the root system determined by the functional defined by intersection with z. It follows that the x_i are mapped to the irreducible components of the singular fibers. Note that there can be no other components of singular fibers, coming from the over-lattice S, because that would increase the Picard number. Thus, when we perform the construction of z as above we indeed recover z.   □

3.3 Abelian Surfaces with Quaternionic Multiplication

Let $\mathcal{M}$ be an order in a quaternion algebra B over $\mathbb{Q}$.

Definition 11. An abelian surface with multiplication by $\mathcal{M}$ is an abelian surface A together with an injection $j : \mathcal{M} \hookrightarrow \operatorname{End}(A)$

We will essentially only be considering the "general" case, where j is an isomorphism.

In this work, we will prefer to study such abelian surfaces, not by their endomorphism algebra but rather by their Néron–Severi lattices. These two can, in some cases, be determined one from the other, as we recall now.

Definition 12. Given the order $\mathcal{M}$, a lattice $N_{\mathcal{M}}$ is defined as follows:

$$N_{\mathcal{M}} := \{b \in B, \ \operatorname{tr}(b) = 0, \ \operatorname{tr}(bx) \in \mathbb{Z} \text{ for all } x \in \mathcal{M}\}$$

The form is obtained by making the following embedding of $N_{\mathcal{M}}$ into $\wedge^2 \mathcal{M}^*$ into an isometry: $x \in N_{\mathcal{M}}$ maps to the symplectic form given by $x(l_1, l_2) = \operatorname{tr}(xl_1 l_2')$, where $'$ is the canonical involution of B.

We note that the embedding is clearly primitive.

Lemma 4. *The form on $N_{\mathcal{M}}$ is a constant multiple of the form $a \times b = \operatorname{tr}(ab')$. If $\mathcal{M}$ is an Eichler order in a rational quaternion algebra B of discriminant D and level N, then this constant is DN.*

Proof. The first statement is [3, Theorem 3.10]. That the constant for an Eichler order is DN is [3, Proposition 2.18]. □

Proposition 7. *Suppose that every left ideal of $\mathcal{M}$ whose left order is $\mathcal{M}$ is principal. Then, a multiplication $j : \mathcal{M} \xrightarrow{\sim} \operatorname{End}(A)$ on an abelian surface A determines in a unique way an isometry of lattices $N_{\mathcal{M}} \xrightarrow{\sim} \operatorname{NS}(A)$.*

Proof. This is an easy extension of the proof of Proposition 3.2 in [4]. We just sketch the main points: By our assumptions on $\mathcal{M}$, the group $H_1(A, \mathbb{Z})$ is isomorphic to $\mathcal{M}$ as an $\mathcal{M}$-module. Thus, we have an isometry $\wedge^2 \mathcal{M}^* \xrightarrow{\sim} H^2(A, \mathbb{Z})$ and as a consequence a primitive embedding $N_{\mathcal{M}} \hookrightarrow H^2(A, \mathbb{Z})$. One shows that the image of this embedding is of type $(1, 1)$ hence contained in the Néron–Severi lattice. One also knows that the rank of the Néron–Severi lattice is 3 in this case, and since the embedding is primitive we obtain an equality. Finally, one shows easily that the embedding of $N_{\mathcal{M}}$ (thought not of the entire $\wedge^2 \mathcal{M}^*$) is independent of the identification of $\mathcal{M}$ with $H_1(A, \mathbb{Z})$. □

Definition 13. Let N be a lattice of rank 3 and let n_0 be an element of N with $n_0 \times n_0 > 0$. An N-marked (or more precisely an (N, n_0)-marked) abelian surface is an abelian surface A together with an embedding $N \hookrightarrow \operatorname{NS}(A)$ such that every divisor of A representing the image of n_0 is very ample.

Thus, by fixing $n_0 \in N_{\mathscr{M}}$, a multiplication on an abelian surface A determines in a unique way, under our condition on $\mathscr{M}$ from Proposition 7, an (N, n_0) marking on A.

What about the reverse determination? Rationally, the Néron–Severi lattice always determines the endomorphism algebra. More precisely, if (N, n_0) is as above, and B is the Clifford algebra of the orthogonal complement to n_0 in $N \otimes \mathbb{Q}$, then the marking induces a canonical isomorphism $B \to \text{End}(A) \otimes \mathbb{Q}$ on any marked (N, n_0) abelian surface. This makes the following definition very natural.

Definition 14. The pair (N, n_0) is said to determine the endomorphism ring if for some subring $\mathscr{M}_N \subset B$ we have that for any (N, n_0)-marked abelian surface A the isomorphism $B \xrightarrow{\sim} \text{End}(A) \otimes \mathbb{Q}$ induces an isomorphism $\mathscr{M}_N \xrightarrow{\sim} \text{End}(A)$.

We have the following basic result.

Proposition 8. *If the rank of the lattice N is ≤ 3 and $n_0^2 = 2$, then the pair (N, n_0) determines the endomorphism ring. In other words, if A is a principally polarized abelian surface with Picard number ≤ 3, then $\text{End}(A)$ is determined by $\text{NS}(A)$.*

Proof. This is essentially contained in [4, Proposition 4.3]. We recall the proof since we will need the details later. Let A be an abelian surface, marked by $\alpha : N \to \text{NS}(A)$. Let $L := H_1(A, \mathbb{Z})$. We may interpret $\alpha(n_0)$ as a symplectic form on L and the assumption on n_0 implies that this form is perfect. Let $x \in \text{NS}(A)$. We may similarly interpret $\alpha(x)$ as a symplectic form. There is then an endomorphism $T_x : L \to L$ with the property that $\alpha(x)(l_1, l_2) = \alpha(n_0)(T_x l_1, l_2)$. The content of loc. cit. is that the T_x are endomorphisms of A and that they generate $\text{End}(A)$. Furthermore, the inclusion of the subring generated by the T_x with x in the orthogonal complement to n_0 factor via the Clifford algebra of this complement. Since T_{n_0} is clearly the identity, it is easy to obtain the ring $\mathscr{M}_N$ in such a way that sending x to T_x extends to the required isomorphism. $\qquad\square$

Let $\mathscr{M}$ be a maximal order. By a theorem of Drinfel'd (see e.g. [10, III Proposition 1.5]) and Lemma 4, $N_{\mathscr{M}}$ contains an element n_0 with square 2, which we fix once and for all.

Corollary 1. *The pair $(N_{\mathscr{M}}, n_0)$ determines the endomorphism ring, and $\mathscr{M}_N = \mathscr{M}$.*

Remark 5. We will later (Proposition 20) show that the same is true if $\mathscr{M}$ is an Eichler order of prime level.

We now want to introduce a level 2 structure. This is defined as follows:

Definition 15. Let A be an abelian surface with multiplication by $\mathscr{M}$. A level 2 structure on A is a map $\phi : \mathscr{M}/2 \to A$ which is a map of $\mathscr{M}$-modules.

Our main object of investigation is a certain moduli space associated with these objects.

Definition 16. The moduli curve $V_{\mathscr{M}}[2]$ parameterizes abelian surfaces with multiplication by $\mathscr{M}$ and a level 2 structure.

Remark 6. Suppose that $\mathcal{M}$ is maximal. In this case we let the group $\Gamma_{\mathcal{M}}[2]$ be the group of units of norm 1 in $\mathcal{M}$ which are also congruent to 1 modulo 2. The quotient $\mathcal{H}/\Gamma$ is an algebraic curve which has a model $V_{\mathcal{M}}[2]$ over $\mathbb{Q}$.

For completeness we give the analogous definition if we are using the Néron–Severi lattice.

Definition 17. If (N, n_0) determines the endomorphism ring, a level 2 structure on a marked (N, n_0) abelian surface A is a map of $\mathcal{M}_N$ modules $\mathcal{M}_N/2 \to A$. Suppressing n_0 from the notation, the moduli space of all of these will be denoted $V_N[2]$.

3.4 The Associated Kummer Surface

Suppose that (N, n_0) determines the endomorphism ring $\mathcal{M}_N$ and that $\mathcal{M}_N$ has a unique ideal whose order is $\mathcal{M}$ as in Proposition 7. Let $W := \mathcal{M}_N/2$. As we have seen in Definition 12, there is a canonical embedding of N into the space of integral quadratic symplectic forms on $\mathcal{M}_N$. By reduction each $x \in N$ gives rise to a symplectic form on W (i.e., a bilinear form with $(w, w) = 0$). According to [1, VIII, Lemma 4.3], this is the same thing as a polynomial function g_x of degree 2 on W defined up to a polynomial function of degree ≤ 1. Let (A, α, ϕ) be an (N, n_0)-marked abelian surface with a level two structure $\phi : W = \mathcal{M}_N/2 \to A$. Let $X = \mathrm{Km}(A)$ be the corresponding Kummer surface. We now show how the additional data allows us to mark $\mathrm{NS}(X)$.

Definition 18. The Kummer lattice K associated to W is generated by orthogonal vectors e_v with square -2, for $v \in W$, together with $(\sum_{f(v)=0} e_v)/2$ for every affine linear functional $f : W \to \mathbb{Z}/2$.

Definition 19. We construct an over-lattice S_N of $N[2] \oplus K$ by adding all the vectors of the form

$$\frac{1}{2}\left(\left(\sum_{g_x(w)=0} e_w\right) \oplus x\right) \tag{3}$$

for all x, where g_x is the polynomial function defined before.

Note that each such vector is well defined modulo $N[2] \oplus K$. We distinguish in S_N the class e_0 and the class corresponding to n_0.

Definition 20. Let d_N be the chamber in S_N containing the element $m \times n_0 - \sum_{w \in W} e_w$ where $m \geq 20$.

Proposition 9. *The marking and level structure on A induce in a canonical way an (S_N, d_N) marking on X, sending $x \in N$ to the image of $\alpha(x)$ in $\mathrm{NS}(X)$ and sending a class e_w to the blowup of the singular point at $\phi(w)$.*

Proof. The fact that the map above $K \oplus N[2] \to NS(X)$ extends to S_N is a consequence of [1, Proposition VIII.5.8]. It remains to check that it sends d_N to the ample cone in NS(X). This is a general instance of the following more general fact.   □

Lemma 5. *Let X be a surface, and let $\iota : X \to X$ be an involution with m fixed points. Let $q : X \to X/\iota$ be the quotient map, and let $\pi : Y \to X/\iota$ be the resolution of the m double points, with $E_1, \ldots, E_m$ the corresponding exceptional curves. If D is very ample on X, then $D_n'' = n\pi^*\iota_* - \sum_i E_i$ is very ample on Y for $n \geq m + 3$.*

Proof. The divisor $D' = \iota_* D$ is very ample on X/ι. Let $f : X/\iota \to \mathbb{P}^r$ be the immersion to $\mathbb{P}^r$ it defines. Then the linear system of hypersurfaces on $\mathbb{P}^r$ of degree $n \geq m + 3$ through the m ordinary double points of $f(X)$ defines the blowup of $\mathbb{P}^r$ at these points, because the hypersurfaces separate points and tangent vectors as well as 2-jets at the singular points of $f(X)$. This linear system restricts to D_n'' on Y, hence D_n'' is very ample on Y.   □

Proposition 10. *Suppose, in addition to the assumptions of this subsection, that $n_0 \times n_0 = 2$. Then, the construction of Proposition 9 is a bijection between isomorphism classes of marked (N, n_0) abelian surface with a 2 level structure and S_N-marked K3 surfaces.*

We first prove the following lemma, which shows that the effectivity of a class and the irreducibility of a -2-class are lattice theoretic.

Lemma 6. *Let (S, d) be a pair consisting of a lattice with a chamber d, as in Definition 8. Consider all (S, d)-marked K3 surfaces (X, v). Then*

1. *For a -2-class z, $v(z)$ is effective for all (resp. one) (X, v) if and only if its product with one (equivalently all) the elements of (the interior of) d is > 0. Call these classes effective -2-classes.*
2. *For any class z, $v(z)$ is effective for all (resp. one) (X, v) if and only if z is in the cone spanned by the effective -2-classes and the classes in the closure of d. Call these classes effective classes.*
3. *For a -2-class z, $v(z)$ is represented by an irreducible curve for all (resp. one) (X, v), if and only if it is effective and cannot be decomposed into the sum of effective classes.*

Proof. The first claim is immediate. The second is in [1, VIII Proposition 3.7]. The third is immediate from the previous two.   □

Proof of Proposition 10. The construction above shows that there exists a K3 surface marked by (S_N, d_N) such that the marking sends the classes $e_v \in K \subset S_N$, for $v \in W$, to the classes of irreducible -2-curves. By Lemma 6 it now follows that for any (S_N, d_N)-marked K3 surface X the images of the classes e_v, for $v \in W$, in NS(X) are represented by irreducible -2-curves. Let X be such a surface. Let $\tilde{A}$ be the double cover of ramified along these -2-curves. It exists since $\sum_{w \in W} w$ is divisible by 2 in K; it is unambiguously defined since X is simply connected. Let A be the blow down of the resulting -1-curves on $\tilde{A}$ to points $\phi(v)$. This construction is the reverse to the Kummer one, and choosing $0 = \phi(0)$, the surface A becomes an abelian surface [1, Proposition VIII.6.1] with 0 being the neutral element for the group law; X is the Kummer surface of A. In particular, $A[2]$ is a vector space over $\mathbb{Z}/2$. By construction, the map ϕ sends 0 to 0 and sends three-dimensional affine subspaces of W to those of $A[2]$. This implies that ϕ is linear. The Néron–Severi lattice of A is isomorphic to the orthogonal complement of the lattice generated by the 16 exceptional divisors on X, with the form divided by 2. Since our marking gives an isomorphism of $N[2]$ with this complement it induces the required marking $N \to \mathrm{NS}(A)$. It remains to show that the map ϕ is a map of $\mathscr{M}_N$-modules. For this we return to the proof of Proposition 8. Let $x \in N$. Define an operator $t_x : W \to W$ as follows: The condition that the vector (3) belongs to S_N determines g_x on W as a polynomial function of degree 2 modulo those of degree 1, hence determines a symplectic form Q_x on W. The form Q_{n_0} is perfect. Thus, there is a unique $t_x : W \to W$ such that $Q_x(w_1, w_2) = Q_{n_0}(t_x w_1, w_2)$. By the definition of the endomorphism T_x in the proof of Proposition 8 it is clear that ϕ interwinds t_x with T_x modulo 2, and the result is now clear.   □

For future use we will need a slight strengthening of this Proposition, showing that the conclusion sometimes holds even if we have no principal polarization.

Proposition 11. *The conclusion of Proposition 10 continues to hold if the assumption $n_0^2 = 2$ is replaced by the weaker assumption that n_0 induces a bilinear form on $\mathscr{M}_n$ which is perfect modulo 2.*

Proof. The argument is almost entirely the same. The only problem is that with n_0 which is not principal, not every $x \in N$ gives an endomorphism T_x. However, the proof that the T_x generate the entire endomorphism ring is completely local. Thus, those x's for which T_x is defined generate a subring of the Endomorphism ring of index prime to 2. As before, the action of these T_x on the order 2 points is completely determined, hence so is the action of the entire endomorphism ring.   □

3.5 The Basic Isomorphism

As a corollary, we get the following

Theorem 2. *Let $\mathscr{M}$ be a maximal order in a quaternion algebra. Suppose that the lattice $S_{N_{\mathscr{M}}}$, where $N_{\mathscr{M}}$ is defined in Definition 12, is isometric to the lattice S_T associated with a type T of an elliptic fibration as in Definition 9. Then, the covering $\tilde{H}_T$ is isomorphic to the Shimura curve $V_{\mathscr{M}}[2]$.*

Proof. We have essentially constructed the isomorphism between the moduli problems. The only difficulty is that we assumed throughout that we are in the generic case. Thus we must work generically. Above a generic point of a component of $\tilde{H}_T$ lies a K3 surface X marked by (S_T, d_T). By assumption we have an isometry $S_T \cong S_N$. A-priori, this may not map d_T to d_N. However, as every two chambers can be moved one to the other by an element of the Weyl group, it follows that we may find an isometry which does map d_T to d_N. It follows that X is also marked by (S_N, d_N). By Proposition 10 X is the Kummer surface associated with an abelian surface A marked by the lattice N together with a level 2 structure. By Proposition 1 A has multiplication by $\mathcal{M}$ and thus corresponds to a point on some component of $\tilde{V}_{\mathcal{M}}[2]$. Since X is determined by A this point must be generic and we therefore obtain a birational map between these two components, hence between a component of $\tilde{H}_T$ and $V_{\mathcal{M}}[2]$. Since both are non-singular curves, this is an isomorphism. Since X is determined by A it also follows that $\tilde{H}_T$ has only one component, hence the theorem. $\square$

In this connection we have the following evident

Remark 7. The connected components of $\tilde{H}_T$ are in an bijection with the set $(\mathrm{Aut} G_T / \pm 1)/\mathrm{Mon}_T$.

For the two examples we consider in this section we will show that $\mathrm{Mon}_T = \mathrm{Aut} G_T / \pm 1$, hence that $\tilde{H}_T = \tilde{H}_T^0$ is connected.

3.6 Local Monodromies

To determine a connected component we now consider the monodromy representation associated with the covering. It is easiest to consider the monodromies for the covering Λ_T of the λ-line $\mathbb{P}^1(\lambda) - \Sigma$, whose fundamental group is generated by loops γ_s for each $s \in \Sigma$. To determine the monodromy group we would like to know the conjugacy class of the action of γ_s on the marking $\alpha|_{G_t}$ for each $t \in T$. It will act by composition with an automorphism of the graph G_t. In the following table we list the graphs with non-trivial automorphism groups and the group itself. We also list the action of the automorphism -1 of the fibration on this graph.

t	$I_n, n \geq 3$	I_0^*	$I_n^*, n > 0$	IV^*
$\mathrm{Aut}(G_t)$	± 1	S_3	± 1	± 1
-1 acts as	-1	1	$(-1)^n$	-1

The fibers on which the monodromy acts are the fiber of type I_0^* arising from the twisting of a non-singular fiber, and the fibers at $u \in \Sigma$.

Proposition 12. *The local monodromy of the loop γ_s around $s \in \Sigma$ on the singular fibers is given as follows: On the I_0^* fiber, the image of γ_s in $G_{I_0^*} = S_3$ is the action of γ_s on the non-zero two-torsion points of the original fibration E. On the fiber at $u \in \Sigma$, if $u \neq \infty$ then γ_s acts as -1 if $s = u$ or if $s = \infty$ and as the identity in all other cases. If $u = \infty$ then all monodromies act as the identity.*

Proof. The vertices of $G_{I_0^*}$ correspond to the non-zero two-torsion points of E_s as follows: The I_0^* fiber is obtained by first taking the quotient of E_s by ± 1 and then blowing up the four singular points at the fixed points. The component corresponding to 0 intersects the section and the 3 others give 3 vertices of G_s (there is an additional vertex, corresponding to the quotient itself, which is fixed by all automorphisms). This proves the first statement.

For the other statements, we first consider the fiber of type t at a point $u \in \Sigma$ different from s_0. Suppose the Weierstrass equation of this fiber is $y^2 = f(x, u)$. The quadratic twist (1) at s_0 and λ changes the equation of the fiber at u to

$$y^2(u - \lambda)(u - s_0).$$

This equation is of course isomorphic to the original equation, upon choosing a square root of $(u - \lambda)/(u - s_0)$. Since this choice forces a choice of square roots for neighboring u's, we obtain an isomorphism of $\pi^{-1}(D)$ and $\pi_\lambda^{-1}(D)$, where D is an open disc around u and $\pi : E \to \mathbb{P}^1$ and $\pi_\lambda : E_{s_0,\lambda} \to \mathbb{P}^1$ are the respective projections. To be precise, E and $E_{s_0,\lambda}$ are obtained from the Weierstrass equations by blowing up the singular points (of the surface) on the singular fibers, but the blowup depends only on a neighborhood of the fiber, so this isomorphism carries over. In particular, we have a canonical isomorphism of the singular fibers at u. As λ moves in the loop γ_s (which, we recall, does not intersect the singular locus) and we make a continuous choice of this root we end up with either the same root or its negative, and correspondingly the monodromy is therefore either the identity or the action of -1. It remains to check when we get the identity and when -1, that is to say, when the loop $(u - \lambda)/(u - s_0)$ goes an even or odd number or times around 0. But when λ moves in a small loop around $s \neq \infty$ we find $(u - \lambda)/(u - s_0)$ moving in a small loop around $(u - s)/(u - s_0)$, so this goes once around 0 if and only if $u = s$. When $s = \infty$ the loop can be taken to be a circle of sufficiently large radius and $(u - \lambda)/(u - s_0)$ does go around 0 and the monodromy is the action of -1 again.

We next notice that the restriction $u \neq s_0$ is not really required. Indeed, fix a general $b \in \mathbb{P}^1$. The twist at s_0 and λ is isomorphic, upon choosing a square root to $(u - \lambda)/(u - b)$ to the twist at s_0 and b. Thus, the same analysis applies.

There remains the case $u = \infty$. But the previous analysis shows that the only loop which may have a non-trivial monodromy on the fiber there is the one around ∞. But since the product of the monodromies has to be 1 we see that this monodromy is trivial as well.   □

We now consider the monodromies for Λ_T'. The local monodromies are the same as for Λ_T, except that they should be taken modulo the action of -1. This allows for the following more symmetric description.

Corollary 2. *The local monodromies for the loop γ_s is given as follows (modulo the action of -1). On the fiber of type I_0^* it is the same as in Proposition 12. Otherwise, on the fiber at u the action is that of -1 if $s = u$ and is the identity otherwise.*

Proof. We only need to observe that the action of γ_∞ is the identity on the fiber at ∞ and is -1 on all other fibers, so composing with the action of -1 it is -1 on the fiber at ∞ and the identity on the other fibers, i.e., the same behavior as the other loops.   □

The monodromy on the two torsion points is precisely the reduction modulo 2 of the homological invariant. There are known tables for the local monodromy of the homological invariant [1, V.10, Table 6]. From this we obtain the following table of monodromy. The unique conjugacy invariant for a permutation on three elements is the order of the permutation, which is either 1 (identity), 2 (transposition) or 3 (a cycle of length 3). So it is sufficient to indicate this number.

I_n, I_n^*	I_n, I_n^*	II	III	IV
$2 \| n$	$2 \nmid n$	IV^*	III^*	II^*
1	2	3	2	3

For the computations to follow we will also need to observe the following.

Remark 8. The order of the local monodromy at a point of H_T is the ramification index of the covering $\tilde{H}_T^0 / H_T$ above it.

3.7 Case Number 1 on the List

To handle Table 1(1) we will need the following

Proposition 13. *1. The curve $V_6(2)$ is hyperelliptic of genus 3, with w_3 the hyperelliptic involution. In the quotient $P = V_6(2)/w_3 \simeq \mathbb{P}^1$ the 8 Weierstrass points can be viewed as the vertices of a cube.*
    2. The automorphisms of $V_6(2)$ are all "modular", coming from the modular involutions w_2, w_3 and from changing the level 2 structure. We have $\mathrm{Aut}V_6(2) \simeq S_2 \times S_4$ with the S_2 factor generated by w_3.

Proof. (sketch) The genus is 3 by the standard genus formula for Shimura curves (with level structure). One can alternatively use the Riemann-Hurwitz formula for the covering $V_6(2) \to V_6$.

Next observe that $\mathrm{Aut}V_6(2)$ contains the modular automorphisms. To see they give $S_2 \times S_4$ as the factor at 3 times the factor at 2, let $\mathscr{B}_2$ be "the" division quaternion algebra over the field $\mathbb{Q}_2$ of 2-adic numbers (see eg [Chap. 2] for the following facts). Let $\mathbb{Q}_4$ denote the quadratic unramified extension of $\mathbb{Q}_2$ and let $\mathbb{Z}_4$ be its ring of integers. Then $\mathscr{B}_2 = \mathbb{Q}_4 \oplus \pi_2 \mathbb{Q}_4$, where $\pi_2^2 = -2$, and moreover π_2 normalizes $\mathbb{Q}_4$ and induces on it the nontrivial automorphism via conjugation. The unique maximal order $\mathcal{M}_2$ of $\mathscr{B}_2$ (its order of integers) is $\mathbb{Z}_4 \oplus \pi_2 \mathbb{Z}_4$. admits a filtration by the two-sided ideals $\wp^i$, where $\wp = \pi_2 \mathcal{M}_2$, with $\mathcal{M}_2 = \mathcal{M}_D \otimes \mathbb{Z}_2$). We have $\wp^i / \wp^{i+1} \simeq \mathbb{F}_4$, the field with four elements, and one also routinely checks that

$$(\mathcal{M}_D/2\mathcal{M}_D)^\times \simeq (\mathcal{M}_2/2\mathcal{M}_2)^\times \simeq \mathcal{M}_2^\times/(1 + \wp^2) \simeq A_4,$$

the alternating group on 4 letters. The filtration induced on A_4 from $\wp \subset \mathcal{M}_2$ is $V_4 \subset A_4$, with $V_4 \simeq (\mathbb{Z}/2\mathbb{Z})^2$ the Klein four-group. Together with w_2 one gets an action of $\mathscr{B}^\times/((\mathbb{Q}_2^\times)(1 + 2\mathcal{M}_2)) \simeq S_4$, the symmetric group of 4 letters, on $V_D(2)$. The modular automorphisms at 3 are clearly $\langle w_3 \rangle$.

An easy analysis shows that the quotient P by w_3 is a $\mathbb{P}^1$ (equivalently, has eight fixed points). It is well-known that the resulting S_4 action on P is the symmetries of a cube for the standard metric on $\mathbb{P}^1$.

To see $V_6(2)$ has no other automorphisms observe that the arithmetic congruence subgroup $\Gamma_6(2)$ uniformizing $V_6(2)$, taken modulo ± 1, is torsion-free. Let $N(\Gamma_6(2))$ denote the normalizer of $\Gamma_6(2)$ in the automorphism group $\mathrm{Aut}\mathscr{H}$ of the upper half plane $\mathscr{H}$. It then follows that $\mathrm{Aut}V_6(2) \simeq N(\Gamma_6(2))/\Gamma_6(2)$ is as asserted. Alternatively one can apply the Riemann–Hurwitz formula to the covering $V_6(2) \to \mathrm{Aut}V_6(2)\backslash V_6(2)$ to deduce that a curve of genus 3 cannot have an automorphism group which strictly contains $S_2 \times S_4$.   □

Returning to the twists of the fibration of type $T = (I_1, I_1, I_8, II)$, the algebra has discriminant 6, and the local monodromies are given in the following table, which we justify in the comments that follow.

	I_8	IV^*	I_0^*
γ_{I_1}	1	1	(12)
$\gamma_{I_1'}$	1	1	(23)
γ_{I_8}	-1	1	1
γ_{II}	1	-1	(123)
	-1	-1	-1	1

Everything affirmed in the table is clear except for the precise types of the permutations in the last column. The cycle length of the permutations are as indicated and their product is 1 so we may assume the situation is as described.

From the table we see that M_T is $\mathrm{Aut}G_T \simeq S_2 \times S_2 \times S_3$, and hence that $\mathrm{Mon}_T \simeq S_2 \times S_3$. Then $\tilde{H}_T^0 \simeq V_6(2)$ by Theorem 2 and the discussion thereafter. We have the following

Lemma 7. *1. All subgroups H of $\mathrm{Aut}V_6(2)$ isomorphic to $S_2 \times S_3$ are conjugate. In particular the S_2 factor of H is $\langle w_3 \rangle$.*
2. $H_T \simeq V_6(2)/S_2 \times S_3$.

Proof. The first part is an easy exercise, and the second part follows from it.   □

By Elkies [17, Sect. 3.1] we know that the Shimura curve $V_6/\langle w_2, w_3 \rangle$ has genus 0 and three elliptic points of ramification indices 2, 4, and 6 respectively relative to the Fuchsian uniformization by the upper-half plane. We let t be the coordinate on this curve sending these points respectively to 1, ∞ and 0. We will prove the following

Lemma 8. *The natural map $H_T = \mathbb{P}_\lambda^1 \to V_6/\langle w_2, w_3 \rangle$ is given (in the coordinates above) by $t = \frac{-256}{27}\lambda/(\lambda - 1)^4$.*

Proof. Choose points P_0, P_1, and P_∞ of $Z = V_6(2)$ above $t = 0$, $t = 1$, and $t = \infty$ respectively, where t is our standard coordinate for $X = V_6/\langle w_2, w_3 \rangle$. Set $Z = V_6(2)$ and $Y = V_6(2)/(\langle w_3 \rangle \times S_3)$. We will determine the ramification indices of Z/Y over the points of Y by computing the ramification pattern of Y/X. Of course this can be done using the uniformization by arithmetic groups, but for variety we will choose another method.

Since the quotient map $\mathcal{H} \to V_6(2)$ is unramified, the ramification of $Z \to X$ has orders 6, 2, and 4 respectively at these points. Let S_P denote the stabilizer of P in $G = \mathrm{Aut}Z \simeq S_2 \times S_4$. Then the points of Y above P are in bijection with the double classes $H \backslash G / S_P$, where $H = \langle w_3 \rangle \times S_3$. Let $y \in Y$, above $P \in X$, correspond to a double class HgS_P. The ramification of y above P is of order $[C_P : (C_P \cap g^{-1}Hg)]$. By Proposition 13 Z is a hyperelliptic curve of genus 3, with w_3 the hyperelliptic involution. Hence the fixed points of w_3 are the Weierstrass points. These form one orbit under S_4, so that they are above 0, and we may assume $S_{P_0} = S_2 \times A_3$. (In fact, since S_4 permutes the Weierstrass points their images in $\mathbb{P}^1 = Z/w_3$ can be viewed as the vertices of a cube.) The stabilizers S_{P_1} and S_{P_∞} do not contain w_3, and may be viewed, up to an (outer) automorphism, as cyclic subgroups of S_4. this forces S_∞ to be conjugate to $\langle (1234) \rangle$. For S_{P_1} there are a-priori two possibilities up to conjugacy—$C_2' = \langle (12) \rangle$ or $C_2'' = \langle (12)(34) \rangle$. Computing the double classes shows that we get two points P_0', P_0'' of Y, of respective ramification indices 3 and 1 over $t = 0$. Likewise we get one point P_∞' of Y over $t = \infty$ of ramification index 4. If $S_{P_1} = C_2''$, we would have over $t = 1$ two points P_1' and P_1' of ramification indices 2. This would contradict the Riemann–Hurwitz formula

$$2 = 2 - 2g(Y) \neq 4(2 - 2g(X)) - 2 - 3 - 2 = 1.$$

Hence $S_{P_1} = C_2'$, and there are in Y over $t = 1$ two points, P_1'' and P_1''', of ramification index 1, and one point, P_1', of ramification index 2.

Now let η be the coordinate on Y such that $\eta(P_0') = 0$, $\eta(P_0'') = \infty$, and $\eta(P_\infty') = 1$. (Notice that this coordinate is defined over $\mathbb{Q}$, because it is defined in a Galois-invariant way.) Set $\beta = \eta(P_1')$, $\gamma = \eta(P_1'')$, and $\overline{\gamma} = \eta(P_1''')$. Then $t = c\eta/(\eta - 1)^4$, and the double root at β gives $(\beta - 1)^4 - c\beta = 4(\beta - 1)^3 - c = 0$. Solving gives $\beta = -1/3$ and $c = -256/27$, so γ and $\overline{\gamma}$ are the roots of $(3\eta + 7)^2 + 32 = 0$, hence agree with their namesakes in Table 1(1). Notice also that the covering Z/Y has ramification indices 6, 2, 2, and 2 at $\eta = \infty$, 0, γ, and $\overline{\gamma}$ respectively. This proves that the covering $\tilde{H}_T^0/H_T$ is isomorphic to Z/Y as claimed.   □

3.8 Case Number 3 on the List

Here the bad fibers are I_1, I_4, I_5, and II, so that after twisting we have I_1, I_4, I_5, IV^*, and I_0^*. The algebra has discriminant 15. The local monodromies are

	I_4	I_5	IV^*	I_0^*
γ_{I_1}	1	1	1	(12)
γ_{I_4}	−1	1	1	1
γ_{I_5}	1	−1	1	(23)
γ_{II}	1	1	−1	(123)
	−1	−1	−1	1

It follows that the monodromy group is the quotient of $S_2 \times S_2 \times S_2 \times S_3$ by ± 1, and it embeds into $\mathrm{Aut} V_{15}(2)$ by Theorem 2. As in Proposition 13, this curve is uniformized as $\Gamma_{15}(2) \backslash \mathscr{H}$, where $\Gamma_{15}(2)$ is torsion free except for ± 1. Here however 2 is prime to 15, hence $\mathrm{Aut} V_{15}(2) \simeq \langle w_3 \rangle \times \langle w_5 \rangle \times \mathrm{GL}_2(\mathbb{F}_2)$. This means that the monodromy is everything. Since the $\mathbb{P}_\lambda^1$ has no fiber-preserving automorphisms, we get the following

Lemma 9. *The curve* $\tilde{H}_T = \tilde{H}_T^0$ *is connected and* $\mathbb{P}_\lambda^1 \simeq H_T \simeq V_{15} / \langle w_3, w_5 \rangle$.

In [17, Sect. 5.2] Elkies gives $X = V_{15} / \langle w_3, w_5 \rangle$ as $\mathbb{P}^1$ with 4 ramification points (of the cover $\mathscr{H} \to X$) P_2, P_2', P_2'', and P_6 of ramification indices 2, 2, 2, and 6 respectively. He also gives a coordinate t on X for which $t(P_2) = 0$, $t(P_2') = 81$, $t(P_2'') = 1$, and $t(P_6) = \infty$. We now have the following simpler analog of Lemma 8, whose proof is left to the reader.

Lemma 10. *The natural isomorphism* $V_{15} / \langle w_3, w_5 \rangle \to \mathbb{P}_\lambda^1$ *is given by* $\lambda = \frac{t-81}{8t}$.

Remark 9. For both cases Table 1(1), (3) on the list we have checked that the Picard–Fuchs equations of the family of Kummer surfaces over the Shimura curve, and of the family of twists over H_T, agree. This will appear elsewhere [5]

4 Isogenies Between Abelian Surfaces and Discriminant Forms

In what we have done so far we have seen how one can start from an elliptic surface and identify that it is a Kummer surface associated with an abelian surface with Quaternionic multiplication. In more general cases the abelian surface will only be isogenous to one with QM by a maximal order in a certain rational quaternion algebra. We would like to develop tools to allow us to recognize such a situation. Note that in our case we know which quaternion algebra it should be because of the computations in Sect. 2. We address more generally the question of determining an isogeny between two abelian surfaces in terms of their Néron–Severi lattices and even more precisely in terms of their discriminant forms.

4.1 The Theory of Discriminant Forms

Recall from [30] that a lattice is a free abelian group S together with a $\mathbb{Z}$-valued symmetric bilinear form. The lattice is called even if the associated quadratic form takes only even values. For a lattice S the lattice $S[n]$ will denote the lattice with the same underlying group but with the form multiplied by n. For an even lattice S the discriminant group A_S is the (finite) quotient group S^*/S, where S^* is the $\mathbb{Z}$-dual of S and $S \hookrightarrow S^*$ using the form. The quadratic form induces on A_S a quadratic form q_S with values in $\mathbb{Q}/2\mathbb{Z}$ called the discriminant form of S. We will often shorthand the pair (A_S, q_S) to simply q_S. We will always assume that the lattice is even, unless it is explicitly stated otherwise.

Nikulin [30, Propositions 1.8.1 and 1.8.2] gives a complete classification of discriminant forms. There is an obvious $\mathbb{Z}_p$ version of a lattice and a notion of a discriminant form associated with such a lattice. Nikulin's classification goes hand in hand with a classification of these $\mathbb{Z}_p$-lattices as well.

Theorem 3 (Nikulin). *A $\mathbb{Z}_p$-lattice is a direct sum of lattices of the following form:*

- $K_\theta^{(p)}(p^k)$ *for any prime p and any $k \geq 0$. This lattice is generated by a single element with square θp^k, where $\theta \in (\mathbb{Z}_p^\times)^2/\mathbb{Z}_p^\times$.*
- $U^{(2)}(2^k)$ *and $V^{(2)}(2^k)$ when $p = 2$ and $k \geq 0$—these are two dimensional lattices corresponding to the matrices $\begin{pmatrix} 0 & 2^k \\ 2^k & 0 \end{pmatrix}$ and $\begin{pmatrix} 2^{k+1} & 2^k \\ 2^k & 2^{k+1} \end{pmatrix}$ respectively.*

Any finite discriminant form is a direct sum of forms of the following types:

- *The discriminant form $q_\theta^{(p)}(p^k)$ of $K_\theta^{(p)}(p^k)$ for any prime p and $k > 0$.*
- *The discriminant forms $u_+^{(2)}(2^k)$ of $U^{(2)}(2^k)$ and $v_+^{(2)}(2^k)$ of $V^{(2)}(2^k)$ for $p = 2$ and $k > 0$.*

There is a rather long list of relations between these objects. We refer to [30, Proposition 1.8.2]

Definition 21. The signature of a discriminant form q is the signature of a lattice M with $q = q_M$ taken modulo 8.

Theorem 4 ([30, 1.11.2]). *The signature of q is well defined. It can be computed as follows:*

1. $\mathrm{sign}(q_\theta^{(p)}(p^k)) \equiv k^2(1 - p) + 4k\eta \pmod 8$ *if $p \neq 2$ where η is defined by $\left(\frac{\theta}{p}\right) = (-1)^\eta$ and $(\frac{\cdot}{\cdot})$ is the Legendre symbol.*
2. *For $p = 2$ we have*

$$\mathrm{sign}(q_\theta^{(2)}(2^k)) \equiv \theta + k\frac{\theta^2 - 1}{2}, \quad \mathrm{sign}(v_+^{(2)}(2^k)) \equiv 4k,$$

$$\mathrm{sign}(u_+^{(2)}(2^k)) \equiv 0 \pmod 8.$$

Under some fairly mild conditions, Nikulin shows that an even lattice is characterized by its signature and discriminant forms. For a discriminant form q, let $l(q)$ be the minimal number of generators for the underlying group.

Theorem 5 ([30, Theorems 1.13.2 and 1.14.2]). *An even lattice T is uniquely determined by its discriminant form q, rank r and signature if the following conditions are satisfied:*

1. *It is indefinite and $r \geq 3$.*
2. *For a prime $p \neq 2$ either (a) $r \geq 2 + l(q_p)$ or (b) $q_p \cong q_{\theta_1}^{(p)}(p^k) \oplus q_{\theta_2}^{(p)}(p^k) \oplus q_p'$.*
3. *Either $r \geq 2 + l(q_2)$ or $q_2 = q_2' \oplus q_2''$ with q_2'' isomorphic to (a) $u_+^{(2)}(2^k)$, $v_+^{(2)}(2^k)$, or (b) $q_{\theta_1}^{(2)}(2^k) \oplus q_{\theta_2}^{(2)}(2^{k+1})$.*

If conditions (1) and (2a) are satisfied, and in addition either $r \geq 1 + l(q_2)$ or (3a) is satisfied, with $k = 1$, then the map $\mathrm{Aut}(T) \to \mathrm{Aut}(q_T)$ is surjective.

For future reference we record here several results on discriminant forms associated with singular fibers and with abelian surfaces.

Proposition 14 ([4, Lemma 2.8]). *For each type t of a singular fiber, the discriminant forms q_t of the associated lattice S_{G_t} is given by Tables 2 and 3. For a singular fiber of type I_b the discriminant group has one generator of order b whose square is $b^{-1} - 1$.*

Table 2: Discriminant forms of singular fibers

t	II*	III*	IV*
q_t	$\{0\}$	$q_1^{(2)}(2)$	$q_{-1}^{(3)}(3)$

Table 3: Discriminant forms of I_b^* fibers

b (mod 8)	$2 \nmid b$	± 2	4	0
$q_{I_b^*}$	$q_{4-b}^{(2)}(4)$	$q_{2-b/2}^{(2)}(2) \oplus q_{2-b/2}^{(2)}(2)$	$u_+^{(2)}(2)$	$v_+^{(2)}(2)$

4.2 Rank 4 Lattices and Discriminant Forms

Let L and L_1 be oriented rank 4 lattices. An orientation on L means that a choice is given for a generator of $\wedge^4 L$. Our goal is to analyze isogenies between these lattices (i.e., orientation preserving embeddings with finite cokernels) using as much as possible only the $\wedge^2$ of the lattices, together with the intersection form. Moreover, we are really interested in the case where L and L_1 are the homology or cohomology of abelian surfaces. In this case, we are supplied with primitive sublattices inside these $\wedge^2$, corresponding to the Néron–Severi lattice, and we want to force a certain behavior on these lattices. For simplicity we only treat the case of an isogeny of prime degree. More complicated cases can then be deduced by composition.

We let $H = \wedge^2 L$. The choice of orientation provides H with an even quadratic form given by $x \wedge x = x \times x\,x$ orientation.

Let $L^* = \mathrm{Hom}(L, \mathbb{Z})$. The choice of orientation on L determines a choice of orientation on L^*, hence gives $H^* = \wedge^2 L^*$ an even quadratic form as well. Note that H^* is the $\mathbb{Z}$-dual of H based on the pairing $\langle x \wedge y, z \wedge w \rangle = x(z) \times y(w)$. Since H together with its quadratic form is a unimodular lattice, we obtain an isomorphism $H \to H^*$, which is easily seen to be an isometry. We can write this explicitly as follows: Let $\{e_1 \ldots, e_4\}$ be a basis of L and assume that the orientation is given by $e_1 \wedge \cdots \wedge e_4$. Let $\{e_1^*, \ldots, e_4^*\}$ be the dual basis. Suppose that i, j, k and l are 4 distinct indices and that they are oriented. Then the isometry $H \to H^*$ takes $e_i \wedge e_j$ to $e_k^* \wedge e_l^*$.

Let

$$f : L \hookrightarrow L_1 \qquad (4)$$

be an orientation preserving embedding of lattices such that $L_1/f(L) \cong \mathbb{Z}/p$. Let $H_1 = \wedge^2 L_1$ and $H_1^* = \wedge^2 L_1^*$. The isogeny f induces a map $\wedge^2 f : H \to H_1$. The map $\wedge^4 f$ takes the orientation on L to p-times the orientation on L_1. This implies that the map $\wedge^2 f$ multiplies the quadratic form by p. It thus provides an isometric embedding $\wedge^2 f : H[p] \hookrightarrow H_1$ with finite cokernel.

We now recall more of Nikulin's theory. An over-lattice S' of a lattice S is an isometric inclusion $S \subset S'$ of finite index.

Proposition 15 (Nikulin [30, Proposition 1.4.1]). *Such an over-lattice of S is determined by an isotropic subgroup of q_S, $W_{S':S} \subset q_S$, which is defined as*

$$W_{S':S} = S'/S \subset S^*/S = q_S .$$

Further, $W_{S':S}$ determines $q_{S'}$ as

$$q_{S'} = W_{S':S}^{\perp} / W_{S':S} \qquad (5)$$

where $\perp$ denotes the orthogonal complement with respect to the bilinear form induced by q_S.

Several consequences of this observation will be needed later on. The first is an easy extension of an argument of Nikulin.

Lemma 11. *If S is a primitive sublattice inside a lattice H, with orthogonal complement $S^{\perp}$, then the group $W_{H:S \oplus S^{\perp}}$ is the graph of an isometry between a subgroup A in $q_{S^{\perp}}$ and a subgroup B in $q_{S[-1]}$. We have the equality*

$$|A|^2 = |B|^2 = \frac{|A_S| \times |A_{S^{\perp}}|}{|A_H|} .$$

Proof. Except for the estimate on the size of A, this is simply Proposition 1.5.1 of [30]. The size estimate follows since $|A| = |W_{H:S \oplus S^{\perp}}|$, while by the nondegeneracy of the discriminant form we have $|W_{H:S \oplus S^{\perp}}| \times |W_{H:S \oplus S^{\perp}}^{\perp}| = |A_{S \oplus S^{\perp}}| = |A_S| \times |A_{S^{\perp}}|$ and so (5) gives

$$|A_H| = \frac{|W_{H:S \oplus S^{\perp}}^{\perp}|}{|W_{H:S \oplus S^{\perp}}|} = \frac{|A_S| \times |A_{S^{\perp}}|/|A|}{|A|} ,$$

giving the result. $\qquad\qquad\qquad\qquad\qquad\qquad\qquad\qquad\qquad\qquad\qquad\qquad\quad\square$

Corollary 3 ([30, Corollary 1.6.2]). *If, in the notation of the above Lemma, the lattice H is unimodular, then the group $W_{H:S \oplus S^{\perp}}$ is the graph of an isometry*

$$q_{S^{\perp}} \xrightarrow{\sim} q_{S[-1]}.$$

Proof. As $|A_H| = 1$, and as $|A|$, $|B| \leq |A_S|$, $|A_{S^\perp}|$, we easily obtain that all four numbers are the same, and the required isometry. □

Proposition 16. *For the lattice* S_N *of Definition 19 we have the equality of discriminant forms* $q_{S_N} = q_{N[2]}$.

Proof. This can be checked starting from the definition by a direct computation, but there is an easier way. Let N be the Néron–Severi lattice of an abelian surface A with an associated Kummer surface X with Néron–Severi lattice S_N, and let $T(A)$ and $T(X)$ be the transcendental lattices of A and X respectively. We will see later in Corollary 9 that we have $T(A) \cong N[-1]$ and by [1, Chap. VIII] one has that $T(X) \cong T(A)[2] \cong N[-2]$. It now follows from Corollary 3 that

$$q_{S_N} \cong q_{T(X)[-1]} \cong q_{N[2]}.$$

□

We now return to applications to the situation at hand. The over-lattice H_1 of $H[p]$ determines a lagrangian subgroup $W_{H_1:H[p]} \subset q_{H[p]}$ inside the discriminant form of $H[p]$, where, by definition, $W_{H_1:H[p]}$ is the image of H_1, acting as functionals on $H[p]$ inside $H[p]^*/H[p]$. Since H is unimodular, $H[p]^*/H[p]$ may be identified with H/pH, and thus also with H^*/pH^*, i.e., with $\wedge^2(L/pL)$ or with $\wedge^2(L^*/pL^*)$.

An alternative way of viewing the above identifications is the following: After tensoring everything with $\mathbb{Q}$ we may think of L_1 as sitting in the sequence $\frac{1}{p}L \supset L_1 \supset L$. Correspondingly, we have

$$\wedge^2 \frac{1}{p}L \supset \frac{1}{p} \wedge^2 L \supset \wedge^2 L_1 \supset \wedge^2 L$$

where the second from the right inclusion arises because L_1/L is one dimensional. The form identifies $\frac{1}{p} \wedge^2 L$ with $H[p]^*$, hence $H[p]^*/H[p]$ with $\frac{1}{p} \wedge^2 L/ \wedge^2 L$, and finally this group with $\wedge^2(L/pL)$ via the map

$$\frac{1}{p} \wedge^2 L/ \wedge^2 L \xrightarrow{p} \wedge^2 L/p \wedge^2 L \to \wedge^2(L/pL).$$

The subgroup $W_{H_1:H[p]}$ is just the image of $\wedge^2 L_1$ under this map.

One question we would like to address now is: Given a lagrangian subspace $W \subset q_{H[p]}$, does it in fact come from an embedding of lattices f as in (4). We first recall a general fact about lagrangian subspaces in this situation.

Proposition 17 ([14]). *Let* V *be a four-dimensional space over a field* F *and let* $Q = \wedge^2 V$ *be equipped with a quadratic form (defined up to a non-zero constant) coming from the wedge product. Let* $W \subset Q$ *be a totally isotropic subspace. Then*

1. The subspace W *is of one of two forms:*

    a. $W = \wedge^2 V'$ *where* V' *is a three-dimensional subspace of* V.
    b. $W = v \wedge V$ *where* $v \in V$ *is a non-zero vector.*

2. *If W_1 and W_2 are two such subspaces then* $\dim W_1 \cap W_2$ *is odd if W_1 and W_2 are of the same type and even otherwise.*

3. *Let V^* be the dual space to V and let $Q^* := \wedge V^*$, identified with the dual of Q in the obvious way. Let $W^\perp$ be the annihilator of W in Q^*. Let W^* be the image of W under the isometry $Q \to Q^*$ described (in the integral case) before. Then $W^\perp$ and W^* are of the complementary type to W.*

Definition 22. We will call a subspace $W \subset Q$ of type 1 (respectively type 2) if it is of the form described in (1a) (respectively (1b)) of Proposition 17.

We can now fairly easily decide what the type of $W_{H_1:H[p]}$ above is.

Proposition 18. *The subspace $W_{H_1:H[p]} \subset \wedge^2(L/pL)$ is of type 2 in the situation described above. Conversely, if W is of type 2, then there exist a map $f : L \to L_1$ with $W = W_{H_1:H[p]}$.*

Proof. Suppose we have a basis $\{e_1, \ldots, e_4\}$ for L such that L_1 is given by adding e_1/p. Then $\wedge^2 L_1$ is obtained from $\wedge^2 L$ by adding $\frac{1}{p}e_1 \wedge e_i$ for $i > 1$ and the image in $\wedge^2(L/pL)$ is $e_1 \wedge (L/pL)$, which is of type 2. Conversely, if W is of type 2 corresponding to a vector $v \in L/pL$, then clearly L_1 generated over L by $\frac{1}{p}v$ gives the correct W. □

Note that to know the type of W it suffices to know the type of just one space W_0, for then the type of W is determined, via Proposition 17, from $\dim W \cap W_0$. Also, if we want to know the existence of a subspace of type 2 it suffices to find two subspaces with even intersection.

A useful easy corollary of the preceding proposition is the following.

Corollary 4. *Suppose that $W \subset \wedge^2(L/pL)$ is of type 1. Then there exists a map $g : L_1 \to L$ such that the induced map*

$$H \xrightarrow{\sim} H^* \xrightarrow{\wedge^2 g^*} H_1^* \xrightarrow{\sim} H_1 \tag{6}$$

gives an isometric embedding $H[p] \hookrightarrow H_1$ corresponding to W.

Proof. We consider $W^* \subset \wedge^2(L^*/pL^*)$ which is of type 2 by Proposition 17. By Proposition 18 W^* corresponds to a map $g^* : L^* \to L_1^*$ and we recover g from this. □

Note that if we view in the proposition and the corollary the subspaces W as contained in $\wedge^2(L^*/pL^*)$ then the conditions about the type should be reversed.

4.3 Applications to Isogenies of Abelian Varieties

If $L = H_1(A)$, $L_1 = H_1(A_1)$, where A and A_1 are abelian surface, then an isogeny $f : A \to A_1$ determines an orientation preserving map $f_* : L \to L_1$. Conversely, if A is given, an orientation preserving $L \to L_1$ is obtained as f_* for a uniquely determined abelian surface A_1 and an isogeny $f : A \to A_1$. Note that if the orientation on L_1 is not specified, then we can always choose an orientation to make the map orientation preserving. There is a dual picture where $L = H^1(A)$, $L_1 = H^1(A_1)$ and the map is f^* for an isogeny $f : A_1 \to A$.

The following is an easy consequence of the discussion above and of Part 3 of Proposition 17

Corollary 5. *Let A be an abelian surface, let $H = H^2(A, \mathbb{Z})$ and let $W \subset q_{H[p]}$ be a lagrangian subspace. Then the embedding $H \subset H_1$ controlled by W is realized as either f^* for an isogeny $A_1 \to A$ or as f_* for an isogeny $f : A \to A_1$, both isogenies of degree p.*

We now want to have an analysis of isogenies of abelian surfaces based on their Néron–Severi lattices. Suppose A is an abelian surface with Néron–Severi lattice N. We want to know if there is an isogeny of degree p connecting A with an abelian surface with Néron–Severi lattice N', given the relation between N and N', or even a relation between their discriminant forms.

Suppose that such an isogeny, say $f : A \to A'$ exists. Then the map f_* is an isometry $N[p] \hookrightarrow N'$ and the over-lattice N' of $N[p]$ is described in terms of $W = W_{N':N[p]} \subset q_{N[p]}$ as before. Suppose conversely that W is given. We would like to know if it indeed comes from an isogeny as above, and if possible say something about this isogeny, or the number of possible isogenies.

Consider the following situation. Let $i : S \hookrightarrow R$ be a primitive sublattice. Let $S^\perp$ be the orthogonal complement of S in R. Then we have an over-lattice situation $S \oplus S^\perp \subset R$ which is controlled by the subgroup $W_{R:S\oplus S^\perp} \subset A_S \oplus A_{S^\perp}$.

Suppose now that $R \subset R'$ is an over-lattice, controlled by the subgroup $W_{R':R} \subset A_R$. The lattice $S' = \mathbb{Q}S \cap R'$ is an over-lattice of S. We would like to describe the subgroup $W_{S':S} \subset A_S$ determined by it.

There is no direct functoriality for discriminant forms. However, we can make the following definition.

Definition 23. The image of A_S in A_R is the subgroup $i_* A_S$ of A_R generated by all functionals on R which vanish on $S^\perp$.

The following lemma is easy.

Lemma 12. *Under the isomorphism* (5) *the subgroup $i_* A_S$ is identified with*

$$(A_S \cap W^\perp_{R:S\oplus S^\perp})/W_{R:S\oplus S^\perp} \, ,$$

where A_S sits as the first summand in $A_{S\oplus S^\perp} = A_S \oplus A_{S^\perp}$.

Lemma 13. *We have $W_{S':S} \subset A_S \cap W^\perp_{R:S\oplus S^\perp}$ and the identification of the last Lemma induces an isomorphism $W_{S':S} \cong i_*(A_S) \cap W_{R':R}$.*

Proof. The intersection corresponds in the previous description of the image to those functionals that actually come from an element of R'. The following two maps are clearly inverse to each other and provide the required isomorphism: Start first with an element of S'. Then, seen as an element of R' it induces a functional on R which vanishes on $S^{\perp}$ hence gives an element of the image. Elements of S clearly give 0 this way. The map on the other direction starts with an element of R' that kills $S^{\perp}$. Then it is clearly in $\mathbb{Q}S$ and therefore in S'. $\square$

The considerations above lead to the following procedure for analyzing an isogeny of abelian surfaces in terms of discriminant forms.

Proposition 19. *Let A be an abelian surface and let H be its second cohomology (respectively homology) $N \subset H$ is the Néron–Severi lattice (if H is the homology, then we identify H with the cohomology as above). Let $T = N^{\perp}$ be the lattice of transcendental cycles. Let N' be an over-lattice of $N[p]$ corresponding to an isotropic subgroup $W \subset q_{N[p]}$. Then, for the map $N \to N'$ to be f^* (resp. f_*) for an isogeny of degree p, $f : A' \to A$ (resp. $f : A \to A'$), the following conditions are sufficient and necessary:*

- $W \subset W^{\perp}_{H[p]:N[p] \oplus T[p]}$ *and $W \cap W_{H[p]:N[p] \oplus T[p]} = 0$ (this implies that W can be found inside $i_* q_{N[p]}$).*
- *There exists a lagrangian subgroup $W_{H':H[p]}$ such that $W = i_*(q_{N[p]}) \cap W_{H[p]:H}$.*
- *The subgroup $W_{H':H[p]}$ is of type 2 (resp. 1).*

Proof. Indeed, f^* is $\wedge^2$ of the map f^* on H^1, so corresponds to a subgroup of type 2 by Proposition 18. An f_* corresponds to a subgroup of type 1 by Corollary 4. $\square$

Corollary 6. *Suppose that we have an isogeny of degree p, $A \to A'$, or $A' \to A$, of abelian surfaces. Suppose that A_1 is an abelian surface and that there is an isometry $q_{NS(A_1)} \cong q_{NS(A)}$. Then there exists an abelian surface A'_1 with $q_{NS(A'_1)} \cong q_{NS(A')}$ such that there exists an isogeny of degree p either $A_1 \to A'_1$ or $A'_1 \to A_1$.*

Proof. The first two conditions of Proposition 19 are immediately satisfied by the conditions (using the proposition again). The third condition only determines the direction of the isogeny. $\square$

4.4 Abelian Varieties with Multiplication by Eichler Orders

Our first application of the above considerations is as follows: Let $\mathcal{M}_{D,n}$ be an Eichler order of discriminant D and level n. In [4, Proposition 3.2] the Néron–Severi lattice of a general abelian surface with multiplication by $\mathcal{M}_{D,n}$ is determined. In fact, there is a lattice $N_{D,n}$ (this lattice is described in loc. cit.) such that an injection $\mathcal{M}_{D,n} \hookrightarrow \operatorname{End}(A)$ determines an injection $N_{D,n} \hookrightarrow \operatorname{NS}(A)$. It was further proved in [4, Corollary 4.4] that if $n = 1$, i.e., when the order is maximal, then conversely an isometry $N_{D,n} \cong \operatorname{NS}(A)$ determines an isomorphism $\mathcal{M}_{D,n} \hookrightarrow \operatorname{End}(A)$. We now show the same in the case that n is prime.

Proposition 20. *Let $N = N_{D,p}$ be the Néron–Severi lattice of an abelian surface with multiplication by the Eichler order $\mathcal{M} = \mathcal{M}_{D,p}$ of discriminant D and prime level p. Suppose that A is an abelian surface whose Néron–Severi lattice is isometric to N. Then A has multiplication by $\mathcal{M}$.*

Proof. We know that there is an isogeny of degree p between abelian surfaces with multiplication by $\mathcal{M}$ and those by $\mathcal{M}_{D,1}$. By Corollary 6 we see that there exist an abelian surface A' whose Néron–Severi lattice has the same discriminant form as $N_{D,1}$, and such that there is either an isogeny $A \to A'$ or an isogeny $A' \to A$. Nikulin's theory implies that $N_{D,1}$ is determined by its discriminant form and signature and therefore $\mathrm{NS}(A') \cong N_{D,1}$, hence by what was said before A' has multiplication by the maximal order $\mathcal{M}' = \mathcal{M}_{D,1}$.

Suppose we have an isogeny $A \to A'$. We find $H_1(A) \subset H_1(A') \cong \mathcal{M}'$, with a quotient of order p. The endomorphism ring of A, which we now denote by $\mathcal{M}$ and must show is Eichler, is the subring of $\mathcal{M}' \otimes \mathbb{Q}$ stabilizing $H_1(A)$. This will defer from $\mathcal{M}'$ at most at p, so we may now tensor everything with $\mathbb{Z}_p$, where $\mathcal{M}' = M_2(\mathbb{Z}_p)$. Since $H_1(A)$ will contain an invertible matrix we may assume (by multiplying $H_1(A)$ from the left) that $\mathcal{M} \subset \mathcal{M}'$. It remains to compute the reduction of $\mathcal{M}$ inside $M_2(\mathbb{F}_p)$. Using the form $\mathrm{tr}(C \times B)$ on $M_2(\mathbb{F}_p)$ we see that the reduction of $H_1(A)$ modulo p will be the subspace annihilated by a single non-zero matrix C. This matrix has either rank 1 or 2. If it has rank 1, then the reduction of $\mathcal{M}$ fixes a single vector and $\mathcal{M}$ is an Eichler order as required. If C has rank 2 we may assume that it is the identity matrix. In this case $H_1(A)$ consists of matrices whose trace is divisible by p and the reduction of $\mathcal{M}$ consists of scalar matrices. To rule out this option we compute the p-part of the discriminant form of the Néron–Severi lattice in this case and we show that it not the same as for an Eichler order. In fact, we claim that in this case we have $\mathrm{NS}(A) = \mathrm{NS}(A')[p]$ and the discriminant is divisible by p^3, while for an Eichler order it should only be divisible by p. Indeed, an element of both $\mathrm{NS}(A)$ and $\mathrm{NS}(A')$ is described by a matrix $b \in B_0$ giving rise to a symplectic form $\mathrm{tr}(bxy')$ and it belongs to $\mathrm{NS}(A)$ (respectively $\mathrm{NS}(A')$) if and only if it takes integral values on $H_1(A)$ (respectively $H_1(A')$) and we must show that these two conditions are in fact equivalent. This is easily seen from the fact that products of two traceless matrices in $M_2(\mathbb{F}_p)$ span all of $M_2(\mathbb{F}_p)$.   □

4.5 Further Analysis

To end this section we return to an arbitrary abelian surface and examine the situation a bit more closely.

Lemma 14. *Let S be a lattice with a discriminant form q_S having an underlying group A_S. Then $A_S = pA_{S[p]}$, i.e., is embedded in $A_{S[p]}$ as the subgroup of elements divisible by p. The relation between the quadratic forms is given by the formula $q_S(px) = p \times q_{S[p]}(x)$ for $x \in A_{S[p]}$ and $px \in A_S$.*

Proof. The map $i : A_S \to A_{S[p]}$ is the right vertical map in the following commutative diagram with exact rows

$$
\begin{array}{ccccccccc}
0 & \longrightarrow & S & \longrightarrow & S^* & \longrightarrow & A_S & \longrightarrow & 0 \\
 & & {\scriptstyle id}\downarrow & & {\scriptstyle p}\downarrow & & \downarrow & & \\
0 & \longrightarrow & S[p] & \longrightarrow & S[p]^* & \longrightarrow & A_{S[p]} & \longrightarrow & 0
\end{array}
$$

An easy diagram chase shows that this map is injective and its image is $pA_{S[p]}$. Finally, if $x \in S[p]^*$, then $px \pmod{S[p]} = i(x \pmod{S})$, hence we view $p(x \pmod{S[p]}) \in A_S$ as represented by x and $q_S(x \pmod{S[p]}) = x \times_{S^*} x = px \times_{S[p]^*} x$. $\qquad\qquad\qquad\qquad\qquad\qquad\qquad\qquad\qquad\qquad\qquad\qquad\qquad\qquad\qquad\qquad\square$

Corollary 7. *Let A be an abelian surface with cohomology or homology lattice H as before and with Néron–Severi and transcendental lattices N and T respectively. The inclusion $N \oplus T \subset H$ determines an isomorphism $q_T \cong q_N[-1]$. Suppose that this is induced from an isometry $T \cong N[-1]$. In this case we also have $q_{T[p]} \cong q_{N[p]}[-1]$ and identifying the underlying groups under this isometry we have*

$$
W_{H[p]:N[p]\oplus T[p]} = \{(px, px) : \ x \in q_{N[p]}\}
$$

and

$$
W^{\perp}_{H[p]:N[p]\oplus T[p]} = \{(x, y) : \ p(x - y) = 0\}
$$

hence

$$
A_{H[p]} = \frac{\{(x, y) : \ p(x - y) = 0\}}{\{(px, px) : \ x \in q_{N[p]}\}}
$$

5 Isogenies Related to Abelian Surfaces with Quaternionic Multiplication

In the previous section we have seen that isogenies between abelian surfaces can be to a large extent determined solely from the discriminant form of the Néron–Severi lattice. In fact, Proposition 19 shows that the only delicate part is to determine the direction of the isogeny. This requires understanding of the relation between the Néron–Severi lattice and the cohomology or homology. Such information is available if the abelian surface in question has quaternionic multiplication. In this section we will develop some tools for finding isogenies. In the following section we will demonstrate these techniques by describing an isogeny or order 2 from an abelian surface A with multiplication by an order in a quaternion algebra ramified at 2 when the order is maximal at 2.

5.1 A Special Subgroup

In the notation of the previous section the isogeny is determined by a lagrangian subspace

$$W \subset q_{H[p]} \cong W^{\perp}_{H[p]:N[p] \oplus T[p]} / W_{H[p]:N[p] \oplus T[p]} \, ,$$

where $W_{H[p]:N[p] \oplus T[p]}$ and $W^{\perp}_{H[p]:N[p] \oplus T[p]}$ are described in Corollary 7. Whether the isogeny is to or from A depends on the type of W. To determine this type it suffices, by Proposition 17, to know the type of a single lagrangian subspace W_0. The type of any other subspace W is determined based on the dimension of the intersection $W \cap W_0$. A natural W_0 to consider is the space, again in the notation of the previous section,

$$W_0 := \frac{\{(x,x) : \ x \in A_{N[p]}\}}{\{(px,px) : \ x \in A_{N[p]}\}}$$

5.2 The Integral Cohomology of a QM Abelian Surface

We now determine the type of W_0 in the following situation: A has multiplication by an order $\mathcal{M}$ in a quaternion algebra B ramified at p and $\mathcal{M}$ is maximal at p. The first homology of A is an ideal of $\mathcal{M}$ and we further assume that it is principal and can therefore be identified with $\mathcal{M}$. This assumption is true if $\mathcal{M}$ is either maximal or Eichler.

We first recall the description of the Néron–Severi lattice and the second cohomology of A as found in [3]. The cohomology group $H^2(A, \mathbb{Q})$ can be identified with the collection of rational valued antisymmetric quadratic forms on $B \cong H_1(A, \mathbb{Q})$ and the integral homology with the lattice of forms taking integral values on $\mathcal{M}$.

Let $B_0 := \{b \in B : \ \mathrm{tr}(b) = 0\}$. In [3, Theorem 3.10] it is shown that there are two embeddings of B_0 into $H^2(A, \mathbb{Q})$ with orthogonal images given by sending $v \in B_0$ to the forms on B defined via $(x,y) \mapsto \mathrm{tr}(bx'y)$ and $(x,y) \mapsto \mathrm{tr}(bxy')$ respectively. Moreover, the rational Néron–Severi lattice is simply the image of the second embedding. Thus we can identify $H^2(A, \mathbb{Q})$ with $B_0 \oplus B_0$. Let $\mathscr{D}^{-1}$ be the inverse different of $\mathcal{M}$, defined as the lattice of all $x \in B$ such that $\mathrm{tr}(x\mathcal{M}) \subset \mathbb{Z}$. The integral Néron–Severi lattice is the image of $\mathscr{D}^{-1}$ under the second embedding and the transcendental lattice is the image under the first embedding. The full integral cohomology is determined as follows.

Proposition 21. *Let $\mathcal{M}'$ be the lattice of all $b \in B$ such that $\mathrm{tr}(b(xy - yx)) \in \mathbb{Z}$ for all $x, y \in \mathcal{M}$. Then, under the identification above*

$$H^2(A, \mathbb{Z}) = \{(b, b + \delta) : \ b \in \mathcal{M}', \ \delta \in \mathscr{D}^{-1}\} \, .$$

Proof. We have

$$H^2(A, \mathbb{Z}) = \{(b_1, b_2) \in B_0 \oplus B_0 : \ \mathrm{tr}(b_1 x'y) + \mathrm{tr}(b_2 xy') \in \mathbb{Z}, \ x, y \in \mathcal{M}\} \, .$$

We first test the condition with $y = 1$ and get for all $x \in \mathcal{M}$

$$\mathrm{tr}(b_1 x' + b_2 x) = \mathrm{tr}(b_1' x + b_2 x) = \mathrm{tr}((b_2 - b_1)x) \in \mathbb{Z} \, .$$

It follows that $\delta := b_2 - b_1$ belongs to the inverse different $\mathscr{D}^{-1}$. We now substitute $b_2 = b_1 + \delta$ into the original condition and get

$$\mathrm{tr}(b_1(x'y + xy') + \delta xy') \in \mathbb{Z} \, .$$

Since the trace of the second summand is already in $\mathbb{Z}$ we see that the condition is $\mathrm{tr}(b_1(x'y + xy')) \in \mathbb{Z}$. This simplifies a bit since $\mathrm{tr}(b_1 xy') = \mathrm{tr}(-b_1 yx')$ so we see that the condition is equivalent to the one in the statement of the proposition. □

Corollary 8. *The $\mathbb{Z}$-dual of* NS(A) *can be identified via the cup product with* $\{(0, x), \ x \in \mathcal{M}'\}$.

Proof. If $x \in \mathscr{D}^{-1}$ and $y \in \mathcal{M}'$ then $(x, 0) \cup (y, 0) = (x, 0) \cup (y, y) \in \mathbb{Z}$ since $(y, y) \in H^2(A, \mathbb{Z})$. Conversely, since $\mathscr{D}^{-1} \hookrightarrow H^2(A, \mathbb{Z})$ is primitive, any $\mathbb{Z}$-valued functional on $\mathscr{D}^{-1}$ can be extended to $H^2(A, \mathbb{Z})$, and this, being unimodular, is given by cup product with another element. But $(x, 0) \cup (y, z) = (x, 0) \cup (y, 0)$. □

5.3 The Type of the Special Subgroup

It follows from the considerations above that $A_N \oplus A_T$ is identified with the space

$$\frac{\{(x, y), \ x, y \in \mathcal{M}'\}}{\{(x, y), \ x, y \in \mathscr{D}^{-1}\}} \, .$$

Lemma 15. *The space $W_{H:N+T} \subset A_N \oplus A_T$ is the image of the diagonal* $\{(x, x), \ x \in \mathcal{M}'\}$.

Proof. Indeed, elements of H are congruent modulo $\mathscr{D}^{-1}$ to diagonal elements. □

Corollary 9. *The identification $q_N \cong -q_T$ is induced by the isometry $N \cong T[-1]$ induced by the identity map on $\mathcal{M}'$. In particular, the assumptions of Corollary 7 are satisfied.*

Proposition 22. *The subspace W_0 is of type 1.*

Proof. We need to describe an element in W_0 as a quadratic form modulo p on $\mathcal{M}$. It is easy to see that this is the functional obtained from an appropriate diagonal element (b, b), giving the quadratic form $(x, y) \mapsto \mathrm{tr}(b(xy' + x'y))$ which clearly vanishes when $x = 1$ regardless of b. Thus W_0 annihilates $1 \wedge \mathcal{M}/p$. The result follows from Proposition 17. □

5.4 Discriminant Forms Associated with QM Abelian Surfaces

Here we record some results of [4] about the discriminant forms associated to the Néron–Severi lattices of QM abelian surfaces

Proposition 23. *Let A be an abelian surface whose endomorphism ring is isomorphic to the Eichler order $\mathcal{M}_{D,N}$. The $\mathbb{Z}_p$ lattice $\mathrm{NS}(A) \otimes \mathbb{Z}_p$ is isomorphic to:*

1. *When $2 \neq p | D$—$K^{(p)}_{-2vDN/p}(p) \oplus K^{(p)}_{-2DN/p}(1) \oplus K^{(p)}_{2vDN/p}(1)$;*
2. *When $2 \neq p \nmid D$—$K^{(p)}_{-2DN/p^n}(p^n) \oplus K^{(p)}_1(1) \oplus K^{(p)}_{-1}(1)$;*
3. *When $p = 2 | D$—$K^{(2)}_{-5DN/2}(4) \oplus V^{(2)}(1)$;*
4. *When $p = 2 \nmid D$—$K^{(2)}_{-DN/2^n}(2^{n+1}) \oplus U^{(2)}(1)$,*

where $n = \mathrm{ord}_p(N)$.

Corollary 10. *The discriminant form of the lattice $\mathrm{NS}(A)[2]$ is*

$$\bigoplus_{2 \neq p | D} q^{(p)}_{-v_p DN/p}(p) \oplus \bigoplus_{2 \neq p | N} q^{(p)}_{-DN/p^{n_p}}(n_p) \oplus \begin{cases} q^{(2)}_{-5DN/2}(8) \oplus v^{(2)}_+(2) & if\ 2|D; \\ q^{(2)}_{-DN/2^{n_2}}(n_2 + 2) \oplus u^{(2)}_+(2) & if\ 2 \nmid D. \end{cases}$$

6 A Special Isogeny

We now specialize the theory of the previous section even further, to analyze in great detail a particular isogeny of degree 2. Throughout this section, $B = B_D$ is a quaternion algebra of discriminant D, where we assume $2|D$. We consider an Eichler order $\mathcal{M} = \mathcal{M}_{D,p}$ in B of prime level p, where $p = 1$ means the order is maximal. We will denote by π_2 an element of $\mathcal{M}$ of norm 2 which normalizes $\mathcal{M}$—such an element exists by a theorem of Eichler (see e.g. [Example III.5.5]). In addition, by Proposition 20 there is an associated lattice $N = N_{D,p}$ such that for an abelian surface A an isomorphism $\mathcal{M} \xrightarrow{\sim} \mathrm{End}(A)$ determines and is determined by an isometry $N \xrightarrow{\sim} \mathrm{NS}(A)$.

6.1 The Isogeny

By Corollary 10 the 2-primary part of $q_{N[2]}$ is $q^{(2)}_\theta(8) \oplus v^{(2)}_+(2)$ for some θ. The underlying group is $\mathbb{Z}/8\mathbb{Z} \oplus (\mathbb{Z}/2\mathbb{Z})^2$.

Definition 24. Let $N' \supset N[2]$ be the over-lattice determined by the lagrangian subspace $\Upsilon \subset q_{N[2]}$ generated by $4 \oplus (0 \oplus 0)$.

We have $W^\perp_{H[2]:N[2]\oplus T[p]} = \{(x,y) : x,y \in A_{N[2]}, \ 2(x-y) = 0\}$. The intersection with $A_{N[2]} \oplus 0$ consists exactly of $(\Upsilon \oplus v^{(2)}_+(2)) \oplus 0$ and Lemma 12 tells us that this last group is $i_* A_{N[2]}$. In particular, Υ is mapped isomorphically into this image.

Let $\varepsilon \in \mathcal{M}/2\mathcal{M}$ be the reduction modulo 2 of π. Then $\mathcal{M}/2\mathcal{M} \simeq \mathbb{F}_4[\varepsilon]$, and we have $\varepsilon^2 = 0$ and $\alpha\varepsilon = \varepsilon\bar{\alpha}$ for $\alpha \in \mathbb{F}_4$. The ideal $\mathbb{F}_4\varepsilon$ is canonically the unique two sided nilpotent ideal. The stabilizer in $\mathbb{F}_4[\varepsilon]$ of any one-dimensional subspace of $\mathbb{F}_4\varepsilon$ is $\mathbb{F}_2 + \varepsilon\mathbb{F}_4$ Let $\mathcal{M}' \subset \mathcal{M}$ be the order which is the pre-image of $\mathbb{F}_2 + \varepsilon\mathbb{F}_4$.

Let A be an abelian surface with $\mathcal{M} \hookrightarrow \mathrm{End}(A)$. The algebra $\mathcal{M}/2\mathcal{M} = \mathbb{F}_4[\varepsilon]$ acts on the two torsion subgroup $A[2]$ and the kernel of ε there is well defined. We consider an isogeny of degree 2, $f : A \to A'$, whose kernel is contained in $\mathrm{Ker}\varepsilon$.

Lemma 16. *We have $\mathcal{M}' \xrightarrow{\sim} \mathrm{End}(A')$.*

Proof. The $\mathcal{M}'$ action on A clearly preserves $\mathrm{Ker}\varepsilon$ and acts trivially on it. Hence $\mathcal{M}'$ acts on A'. Since the only orders of B containing $\mathcal{M}'$ are $\mathcal{M}'$ and $\mathcal{M}$, we must show that $\mathcal{M}$ does not act on A'. This is clear, since any non-zero element of $\mathrm{Ker}\varepsilon$ generates $\mathrm{Ker}\varepsilon$ as an $\mathcal{M}$-module. $\square$

Proposition 24. *Suppose A is an abelian surface with $\mathcal{M} \xrightarrow{\sim} \mathrm{End}(A)$. Let $f : A \to A'$ be an isogeny of order 2. Then $\mathrm{Ker}(f) \subset \mathrm{Ker}(\varepsilon)$ if and only if we have an isometry $N' \xrightarrow{\sim} \mathrm{NS}(A')$ and f_* induces the isometry $N[2] \hookrightarrow N'$.*

Proof. By Proposition 19 and by Corollary 7, an isogeny f such that $N' \xrightarrow{\sim} \mathrm{NS}(A')$ is determined by a lagrangian subspace

$$W \subset A_{H[2]} = \frac{\{(x,y) : x,y \in A_{N[2]}, 2(x-y) = 0\}}{\{(2x, 2x) : x \in A_{N[2]}\}}$$

satisfying the following two properties:

1. Its intersection with $i_* q_{N[2]}$ should equal Υ.
2. It is of type 1.

In $q_{H[2]}$ we found the subspace

$$W_0 := \frac{\{(x,x) : x \in A_{N[2]}\}}{\{(2x, 2x) : x \in A_{N[2]}\}},$$

which is of type 1 by Proposition 22. It follows from Proposition 17 that the second condition can be replaced by the condition that W has an intersection of odd dimension with W_0.

We consider now in more detail the discriminant form on $A_{H[2]}$. Clearly, it depends only on the 2-primary part in $A_{N[2]}$. Hence we may replace $A_{N[2]}$ by its 2-primary part and we have

$$A_{H[2]} = \frac{\{(x_1, x_2), x_i \in \mathbb{Z}/8, 2(x_1 - x_2) = 0\}}{\{(2x, 2x), x \in \mathbb{Z}/8\}} \oplus (\mathbb{Z}/2)^4$$

The last four factors give two copies of $v_+^{(2)}(2)$. The quotient group is generated by the element $e_1 = (1,1)$ and $e_2 = (4,0)$, both of order 2. The square of these two elements is 0 and together they generate a copy of $u_+^{(2)}(2)$ (indeed we have

$q_{H[2]} \cong (u_+^{(2)}(2))^3 \cong v_+^{(2)}(2) \oplus v_+^{(2)}(2) \oplus u_+^{(2)}(2))$. The three-dimensional subspace W_0 is generated by the graph of the identity morphism $v_+^{(2)}(2) \to v_+^{(2)}(2)$ together with e_1. By Lemma 12 the image $i_* q_{[N[2]]}$ of $q_{N[2]}$ is generated by the first copy of $v_+^{(2)}(2)$ together with e_2.

The lagrangian subspace W of $v_+^{(2)}(2)^2 \oplus u_+^{(2)}(2)$ should have intersection with $i_* q_{[N[2]]}$ equal to the subspace generated by e_2, and an intersection of odd dimension with the subspace W_0. Since W_0 does not contain e_2, the intersection must be one-dimensional. Since $e_1 \times e_2 = 1/2$ we see that W is the direct sum of e_2 with the intersection of W with $v_+^{(2)}(2) \oplus v_+^{(2)}(2)$, which is a lagrangian subspace with trivial intersection with the first $v_+^{(2)}(2)$ and with 1-dimensional intersection with the graph of the identity. This is exactly the graph of an isometry from the second $v_+^{(2)}(2)$ to the first which has a 1-dimensional fixed space. The isometries of $v_+^{(2)}(2)$ are all permutations of the non-zero elements hence there are three isometries of the required type. There is a non-canonical identification of $A[2]$ with $\mathbb{F}_4[\varepsilon]$ as $\mathbb{F}_4[\varepsilon]$-modules. Under the action of $\mathbb{F}_4[\varepsilon]^\times$ there are two orbits of non-zero elements, $\mathbb{F}_4 \varepsilon - 0$ and the other elements $\mathbb{F}_4[\varepsilon] - \mathbb{F}_4 \varepsilon$. Quotients by elements in the same orbit give isomorphic abelian surfaces. Thus, the three possibilities for W must correspond to the three non-zero elements in $\mathbb{F}_4 \varepsilon$.   □

6.2 A Converse Theorem

We have the following converse to Proposition 24:

Proposition 25. *Let A' be an abelian surface. The following conditions on A' are equivalent:*

1. *There exists an abelian surface A, an isometry $N \xrightarrow{\sim} \mathrm{NS}(A)$ (and as a consequence an isomorphism $\mathcal{M} \xrightarrow{\sim} \mathrm{End}(A)$) and an isogeny $f : A \to A'$ of degree 2 whose kernel is contained in the kernel of ε.*
2. *There exists an isometry $N' \xrightarrow{\sim} \mathrm{NS}(A')$.*

Proof. That (1) implies (2) follows from Proposition 24. It also follows from this proposition, together with Corollary 5 the over-lattice $N' \supset N[2]$ is induced by either an isogeny $f : A \to A'$, which then must be of the appropriate type, or by an isogeny $A' \to A$. We only need to consider this second case. In this case there is an isogeny $g : A \to A'$ with $\mathrm{Ker}(g) \cong (\mathbb{Z}/2)^3$ such that $g \circ f$ is multiplication by 2. We will show that there are exactly 3 isogenies from A to an abelian variety with Néron–Severi lattice isometric to N'. As in the proof of Proposition 13, the 2-adic completion of B is "the" division quaternion algebra over $\mathbb{Q}_2$, and it contains an element π_2 of norm 2 which normalizes its unique maximal order (compare Sect. 6). The compositions of π_2 with the quotient by the three non-zero vectors in the kernel of π_2 are 3 such isogenies. Thus, g must be of this type and this will complete the proof.

Let L and L' be the homology lattices of A and A', we have $L \subset L'$ with $L'/L \cong (\mathbb{Z}/2)^3$. This induces the isometric embedding $\wedge^2 L[8] \subset \wedge^2 L'$ where the quotient group is isomorphic to $(\mathbb{Z}/4)^3 \oplus (\mathbb{Z}/2)^3$. This shows that $\wedge^2 L$ can be divided by 2 in $\wedge^2 L'$. Thus we find instead an isometric embedding $\wedge^2 L[2] \subset \wedge^2 L'$ where the quotient is isomorphic to $(\mathbb{Z}/2)^3$ and is $\wedge^2$ of L'/L. This situation is completely analogous to the one in the proof of Proposition 24, except that now we are looking for a subspace of type 2. Proceeding as in that proof we see that the possible isogenies g are determined by isometries of $v_+^{(2)}(2)$ whose graph has an even intersection with the graph of the identity, i.e., which are either the identity or have no fixed points. There are exactly 3 such isometries corresponding to the possible cyclic shifts on the three non-zero vectors.   □

Remark 10. Note that if either of the equivalent conditions of Proposition 25 is satisfied, then there exists an isomorphism $\mathscr{M}' \xrightarrow{\sim} \mathrm{End}(A')$ by Lemma 16. We do not know if this condition is also equivalent to the above two. This indicates that it is in some sense better to work with the Néron–Severi lattice than with the endomorphism ring.

6.3 Level Structures

In applications, we will need to endow A with a level structure which corresponds to a full level 2 structure on A', i.e., an injection of $\mathscr{M}'$-modules $\mathscr{M}'/2\mathscr{M}' \to A'$. We will see that a level $2\pi_2$-structure on A is such a level structure.

Definition 25. A level $2\pi_2$-structure on A is an injection of $\mathscr{M}$-modules

$$\phi : \mathscr{M}/2\pi_2\mathscr{M} \to A.$$

Suppose we have an abelian surface A with an isomorphism $\mathscr{M} \xrightarrow{\sim} \mathrm{End}(A)$ and a $2\pi_2$-level structure ϕ. The kernel of π_2 (or of its reduction ε) on the 2-torsion points of A contains $\phi(2)$. Consider the abelian surface $A' = A/\langle \phi(2) \rangle$. By Proposition 24 and Lemma 16, A' has multiplication by $\mathscr{M}'$ and we have an isometry $N' \xrightarrow{\sim} \mathrm{NS}(A')$. Restricting ϕ to $\mathbb{Z}_2 + \pi_2\mathscr{M}/2\pi_2\mathscr{M} = \mathscr{M}'/2\mathscr{M}$ and then passing to corresponding quotients (by 2 and $\phi(2)$ respectively) gives the required level 2-structure on A', namely an injection of $\mathscr{M}'$-modules

$$\phi' : \mathscr{M}'/2\mathbb{Z}_2 + 2\pi_2\mathscr{M} = \mathscr{M}'/2\mathscr{M}' \to A/\langle \phi(2) \rangle = A'$$

Proposition 26. *The map $(A, \phi) \to (A', \phi')$ is a bijection on isomorphism classes between abelian surfaces A together with an isomorphism $\mathscr{M} \xrightarrow{\sim} \mathrm{End}(A)$ and a $2\pi_2$-level structure ϕ, and abelian surfaces A' with $\mathrm{NS}(A') \xrightarrow{\sim} N'$ (see Remark 10 regarding the $\mathscr{M}'$-action) and a full 2-level structure.*

Proof. Let (A', ϕ') be an abelian surface with $N' \cong \mathrm{NS}(A')$ and with a full level 2 structure. According to Proposition 25 there exists an abelian surface A with multiplication by $\mathcal{M}$ and an isogeny $f : A \to A'$ of degree 2, with $\mathrm{Ker}(f) = \langle x \rangle \subset \mathrm{Ker}(\varepsilon)$. We would like to reconstruct a $2\pi_2$-level structure $\phi : \mathcal{M}/2\pi_2\mathcal{M} \to A$ on A giving rise to ϕ'. To do this it suffices to determine $\phi(1)$. Since $f(\phi(1)) = \phi'(1)$ there are only two options for $\phi(1)$, say y and $\tilde{y} = y - x$. But since $2\phi'(1) = 0$ we see that $f(2y) = 0$, and as $2y \neq 0$, we must have $2y = x$, so $\tilde{y} = -y$. It follows that the two options are ϕ and $-\phi$ and these two level structures are isomorphic. It now follows immediately that A can be reconstructed as $A'/\phi'(2\rho)$, where $\rho \in \mathcal{M}/2\pi_2\mathcal{M}$ reduces to an element of $\mathbb{F}_4 - \mathbb{F}_2$ and this completes the proof. $\square$

7 Isogenies and the Morrison Correspondence

Aside from the two examples we already dealt with in Sect. 3, in all other examples in Table 1 it can be checked (as we will see) that the elliptic fibration we obtain is not itself a Kummer surface. However, the computations in Sect. 2 and computations in [5] showing that the Picard–Fuchs equation is a symmetric square suggests that these elliptic fibrations will have an isogeny to a Kummer surface associated with a QM abelian surface. In this section we first recall a construction, based on the work of Nikulin and Morrison, that produces such an isogeny. In fact, Morrison's work [25] shows that for K3 surfaces with Picard number 19, such as the elliptic fibrations we are considering, there is always an isogeny to a Kummer surface associated with some abelian surface. We can then use the methods developed in the previous sections to show that this abelian surface is isogenous to a QM abelian surface. Thus, we obtain the following diagram, a more complicated version of diagram (2).

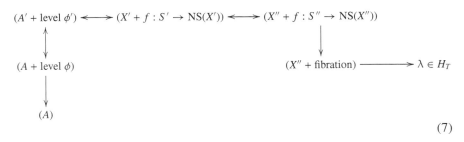

$$(7)$$

7.1 The Morrison Correspondence

Definition 26. Let X be a K3 surface. A Nikulin involution on X is an involution ι on X with the property that $\iota^*\omega = \omega$ for every $\omega \in H^{2,0}(X)$.

By [31, Sect. 5] every Nikulin involution on X has eight isolated fixed points.

Theorem 6 ([25, Theorems 5.7 and 6.3]). *Let X'' be a K3 surface. Suppose there exists an embedding $E_8^2[-1] \hookrightarrow \mathrm{NS}(X'')$. Let g be the involution of $\mathrm{NS}(X'')$ that fixes the orthogonal complement of $E_8^2[-1]$ and switches the two copies of $E_8[-1]$. Then there exists a Nikulin involution ι on X'' such that ι^* is conjugate to g by an element of the Weyl group of $\mathrm{NS}(X'')$. Let X' be the quotient X''/ι blown up once at the 8 quotient singularities. Then X' is a Kummer surface $X' = \mathrm{Km}(A')$ and the quotient maps $A' \to X' \leftarrow X''$ induce an isomorphism $T_{X''} \cong T_{A'}$.*

Definition 27. We call the association $X'' \mapsto A'$ the Morrison construction applied to X''.

Corollary 11. *If the Morrison construction on the surface X'' yields $\mathrm{Km}(A')$, then $q_{\mathrm{NS}(A')} \cong q_{\mathrm{NS}(X'')}$.*

Proof. Both are minus the discriminant form of the corresponding transcendental lattices. $\square$

7.2 Nikulin Markings

We now begin to describe the correspondence (7). We consider a general situation and make the necessary assumptions. We will later consider the examples we have and see how they fit the general pattern.

In Sect. 3 we constructed a bijection between isomorphism classes of marked elliptic fibrations and marked K3 surfaces. Suppose (E, α) is a marked elliptic surface of type T and let $(X'', \tilde{\alpha}'')$ be the corresponding marked $(S'', d'') = (S_T, d_T)$ surface.

We make the following assumptions:

Assumption 1. The 2-primary part of $q_{S''}$ is $q_\theta^{(2)}(2) \oplus v_+^{(2)}(2)$ with $\theta = \pm 1$.

Assumption 2. We have a decomposition

$$S'' = E_8[-1]^2 \oplus S_1 \tag{8}$$

with S_1 of rank 3.

We fix θ as in Assumptions 1.

Lemma 17. *The 2-primary part of $q_{S_1[2]}$ is $q_\theta^{(2)}(4) \oplus v_+^{(2)}(4)$.*

Proof. From the two assumptions it follows that the 2-primary part of q_{S_1} is $q_\theta^{(2)}(2) \oplus v_+^{(2)}(2)$. Since S_1 has rank 3, it easily follows from Theorem 3 that $S_1 \otimes \mathbb{Z}_2$ is determined from its discriminant form to be $K_\theta^{(2)}(2) \oplus V^{(2)}(2)$, hence $S_1[2] \otimes \mathbb{Z}_2 \cong K_\theta^{(2)}(4) \oplus V^{(2)}(4)$ and the result follows immediately from Theorem 3 again. $\square$

To a decomposition (8) there corresponds an involution which fixes the S_1 component and switches the two $E_8[-1]$ components. We now reprove part of Morrison's Theorem 6 in a slightly more precise way, following the proof of [31, Theorem 4.3]

Lemma 18. *There exists a decomposition* (8) *with the property that for any marked* (S'', d'') *K3 surface* $(X'', \tilde{\alpha}'')$ *the induced involution is the restriction of the involution* ι^* *for a unique Nikulin involution* ι *on* X''.

Proof. Start with any decomposition whose existence is guaranteed by Assumption 2 and let ι' be the associated involution. Since $E_8[-1]$ is negative definite the lattice S_1 has signature $(1, -2)$ so it has a positive vector, say x, which is fixed by ι'. Let β be an element of the Weyl group sending the chamber of x to d''. Then the conjugate involution $\iota'' = \beta \iota' \beta^{-1}$ fixes d'' and corresponds to a different decomposition. Under the marking, this involution will be an effective Hodge isometry in the sense of Definition 7. It follows from the global Torelli Theorem for Kähler K3 surfaces (Theorem 1) that it is indeed induced by a unique Nikulin involution. □

From now on we fix a decomposition as above. By the uniqueness of the induced Nikulin involution it is clear that an isomorphism of marked K3 surfaces commutes with the Nikulin involutions.

Definition 28. A Nikulin marked (S'', d'') K3 surface is a marked (S'', d'') K3 surface $(X'', \tilde{\alpha}'')$ together with a numbering of the fixed points of the induced Nikulin involution w, i.e., an injective map β from $\{1, 2, \ldots, 8\}$ to X'' into the set of fixed points of w. An isomorphism of Nikulin marked (S'', d'') K3 surface is an isomorphism of marked K3 surfaces that commutes with the β's.

7.3 The Néron–Severi Lattice of the Quotient Surface

Definition 29. The Nikulin lattice Nik is generated by orthogonal classes y_i of square -2 for $i = 1, \ldots, 8$ together with $\frac{1}{2} \sum y_i$.

The following Lemma is obvious and is given for notational purposes.

Lemma 19. *The discriminant form q_{Nik} is given as follows: Consider a group generated by two torsion elements x_i, for $i = 1, \ldots, 8$. The underlying group A_{Nik} for q_{Nik} is the subquotient*

$$A_{\mathrm{Nik}} \cong \frac{\{\sum a_i x_i, \quad \sum a_i = 0\}}{(\sum x_i)}$$

The form is determined by the conditions $x_i^2 = 1/2$ and $x_i \times x_j = 0$ for $i \neq j$.

Let $(X'', \tilde{\alpha}'', \beta)$ be a Nikulin marked K3 surface. Let X' be the resolution of singularities of X''/w, where w is the canonical involution resulting from Lemma 18. The quotient map induces a map on Néron–Severi lattices, after the $\pi_* \mathrm{NS}(X'') \to \mathrm{NS}(X')$. Let Nik' be the primitive closure of the sublattice of $\mathrm{NS}(X')$ generated by the exceptional curves. Then by [31, Theorem 4.15] (see [25, Sect. 5] for details) it follows that the map ($y_i \mapsto$ resolution of singularities of the point $\beta(i)$) extends to an isometry $\gamma : \mathrm{Nik} \cong \mathrm{Nik}'$.

Let $R' = E_8[-1] \oplus S_1[2]$ and let $\pi_* : S'' \to R'$ be the map, corresponding to the decomposition (8), sending the two copies of $E_8[-1]$ to the single copy of $E_8[-1]$ and the S_1 on itself.

Lemma 20. *Let $(X'', \tilde{\alpha}'')$ be a marked K3 surface. Then there is a commutative diagram, defining the bottom map*

$$
\begin{array}{ccc}
S'' & \xrightarrow{\tilde{\alpha}''} & \mathrm{NS}(X'') \\
\pi_* \downarrow & & \downarrow \pi_* \\
R' & \xrightarrow{\delta} & \pi_*(\mathrm{NS}(X''))
\end{array}
$$

the horizontal maps are isometric isomorphisms and $\pi_*(\mathrm{NS}(X''))$ is primitive in $\mathrm{NS}(X')$.

Proof. See [25], Theorem 5.7 and its proof.   □

For a marked K3 surface $(X'', \tilde{\alpha}'')$ the over-lattice $\mathrm{NS}(X') \supset \mathrm{Nik}' \oplus \pi_*(\mathrm{NS}(X''))$ determines a totally isotropic subgroup of $q_{\mathrm{Nik}' \oplus \pi_*(\mathrm{NS}(X''))}$, by Proposition 15. For a Nikulin marked K3 surface $(X'', \tilde{\alpha}'', \beta)$ there is also an isomorphism

$$
\mathrm{Nik} \oplus R' \xrightarrow{\gamma \oplus \delta} \mathrm{Nik}' \oplus \pi_*(\mathrm{NS}(X''))
$$

inducing an isomorphism on discriminant forms. Thus we obtain a totally isotropic subgroup $D' \in q_{\mathrm{Nik} \oplus R'}$.

Definition 30. We call D' the type of the Nikulin marked K3 surface $(X'', \tilde{\alpha}'', \beta)$.

Definition 31. For a type D', let S' be the over-lattice of $\mathrm{Nik} \oplus R'$ determined by D'.

Let $(X'', \tilde{\alpha}'', \beta)$ be a Nikulin marked K3 surface of type D'. Then we obtain an isometry $\eta : S' \cong \mathrm{NS}(X')$.

Lemma 21. *There exists a chamber $d' \subset S'$ such that for any Nikulin marked K3 surface of type D' $\eta(d')$ is in the ample cone of $\mathrm{NS}(X')$.*

Proof. By Lemma 5 with $m = 8$ the number of fixed points of a Nikulin involution, it suffices to take $11D'' - \sum_{i=1}^{8} E_i$, where D'' is the pull back to the blowup X' of X''/w of a class in d'', and E_i are the exceptional curves.   □

7.4 The Precise Correspondence

Fix a chamber d' as above. We obtain from the Nikulin marked K3 surface of type D', $(X'', \tilde{\alpha}'', \beta)$, an (S', d')-marked K3 surface (X', η).

Proposition 27. *The above construction establishes a bijection between isomorphism classes of Nikulin marked* (S'', d'') *K3 surface of type* D' *and* (S', d')*-marked K3 surface.*

Proof. Start from the (S', d')-marked K3 surface (X', η). Consider the eight classes $y_i \in$ Nik $\subset S'$. It follows from Lemma 6 and the same argument that was used for the proof of Proposition 10 that the marking sends them to the cohomology classes of -2-curves. Let Y be the double cover of X' ramified along these eight curves. This exists because the sum of the curves is divisible by 2 since $\sum y_i/2 \in$ Nik. The -2-curve corresponding to y_i becomes a -1 curve in Y and it can be blown down to a point which we can denote by $\beta(i)$. Doing this for $i = 1, \ldots, 8$ we obtain a surface X''. We want to obtain a marking of NS(X''). The part of NS(X'') invariant under the automorphism of the covering maps isomorphically on the $S_1[2]$ part of NS(X') this isomorphism is in fact an isometry of this part onto S_1. Consider now the part of NS(X') isometric to $E_8[-1]$. This is generated by -2-curves. In Y the pre-image of each of these curves consists of two -2-curves. The resulting 16 curves do not intersect because Nik is disjoint from $E_8[-1]$. Since the graph of E_8 is connected, a choice of a pre-image for one of these -2-curves determines a unique choice for the other curves in such a way that the resulting graph is connected again. Thus, there are two choices for markings of X''. But these two choices are interchanged by the Nikulin involution. Thus, the inverse map on isomorphism classes is well defined.   □

The following Proposition classifies the possible types

Proposition 28. *A type* D' *can occur if and only if it is the graph of an antiembedding of the two-torsion in* $q_{R'}$ *into* q_{Nik}. *It is uniquely determined by an ordered triple of pairwise disjoint subsets of size 2 in* $\{1, \ldots, 8\}$. *For any type that can occur we have the equality*

$$q_{S'} = q_{S_1[2]} \tag{9}$$

Proof. If the type D' occurs, then we have that NS$(X') \cong S'$. But in the Morrison construction, we have the equality $T(X') = T(X'')[2]$, and we always have

$$q_{T(X'')} = q_{\text{NS}(X'')[-1]} = q_{S''[-1]} = q_{S_1[-1]} \; .$$

This immediately implies that the prime to p parts of $q_{T(X'')[2]}$ and $q_{S_1[-2]}$ agree, and the fact that the 2-primary parts agree as well follows easily from the same argument used to prove Lemma 17, because both S_1 and $T(X'')$ have rank 3. This shows (9). The 2-primary part of $q_{S''}$ is $q_\theta^{(2)}(2) \oplus v_+^{(2)}(2)$ by Assumption 1. It follows from Lemma 17 that the 2-primary part of the discriminant of R' and of S' is $q_\theta^{(2)}(4) \oplus v_+^{(2)}(4)$, which is also equal to $q_{3\theta}^{(2)}(4)^3$, by the relation [30, Proposition 1.8.2, d]. Now, the type is to be an isotropic subgroup of $q_{R'} \oplus q_{\text{Nik}}$ in such a way that the over-lattice which it determines has discriminant form $q_{S'}$. By Lemma 11 the subgroup is the graph of an isometry between subgroups of q_{Nik} and $q_{R'}$, hence is two torsion since A_{Nik} is, and the square of its order, equal to the square of the two-part of its order, is given by the products of the size of the two parts of A_{Nik} and $A_{R'}$, both equal to 2^6, divided by the two part of $|A_{S'}|$, again of order 2^6. The subgroup therefore has order 8. Since A_{Nik} is 2-torsion, the subgroup in $A_{R'}$ is two-torsion as well, and since the two-torsion in $A_{R'}$ has order 8 it must be the entire two-torsion. The two torsion of $q_{R'} = q_{3\theta}^{(2)}(4)^3$ is generated by three orthogonal elements of square 1 (mod 2) (this is true for both possible values of θ). Thus, an isometry of this into q_{Nik} is equivalent to finding 3 such elements in q_{Nik}. Elements with square 1 correspond to subsets of $\{1, \ldots, 8\}$ of size 2 (or a subset of size 6, which is equivalent to the element corresponding to its complement of size 2) and these elements are orthogonal if and only if the subsets are disjoint. Finally, since 3 disjoint subsets are permuted by S_8 it is clear that all types occur.   □

We can finally realize the correspondence suggested by diagram (7). To state the result we introduce one further notion.

Definition 32. A Nikulin marked elliptic fibration of type (T, D') is an elliptic fibration of type T which is a K3-surface together with a Nikulin marking on the associated marked K3 surface which is of type D'. The notion of isomorphism is obvious.

Theorem 7. *Let T be a type of an elliptic fibration. Let $S'' = S_T$ be the associated lattice. Let $2|D$ be a discriminant and $p \nmid D$ a prime number, or $p = 1$. Let $\mathcal{M}$ be an Eichler order of discriminant D and level p, and let $N = N_{\mathcal{M}}$ be the associated lattice. Suppose that the rank of S'' is 19, that the prime to two parts of $q_{N[2]}$ and $q_{S''}$ are the same and that the two part of $q_{S''}$ is $q_\theta^{(2)}(2) \oplus v_+^{(2)}(2)$ with $\theta \equiv -Dp/2$ (mod 4). Then, the constructions above provide a bijection between isomorphism classes of abelian surfaces A with $\mathcal{M} \xrightarrow{\sim} \mathrm{End}(A)$ and a $2\pi_2$-level structure and isomorphism classes of Nikulin marked elliptic fibrations of type (T, D').*

Proof. The assumptions imply that S'' satisfies Assumption 1. Let N' be the over-lattice of $N[2]$ of Definition 24. From Corollary 10 It is easy to compute that the 2-primary part of $q_{N'}$ and $q_{S''}$ agree, and therefore that $q_{N'} = q_{S''}$. Since S'' is of rank 19 and is indefinite, it follows from Theorem 5 that S'' is determined by its signature and discriminant form, hence we must have $S'' \cong E_8[-1]^2 + N'$ and assumption 2 also holds with $S_1 = N'$. Combine the bijection of Proposition 26 sending A to A' with $\mathrm{NS}(A') \xrightarrow{\sim} N'$ and a 2-level structure, with the bijection of Proposition 10 sending A' to K3 surface X' with a marking by the lattice $S_{N'}$ of Definition 19. Now, let S' be the lattice of Definition 31. We have

$$q_{S'} \cong q_{S_1[2]} \qquad \text{by (9)}$$
$$\cong q_{N'[2]} \qquad \text{since } S_1 = N'$$
$$\cong q_{S_{N'}} \qquad \text{by Proposition 16}$$

Since S' and $S_{N'}$ have the same signature, it follows from Theorem 5 that they are isometric. Thus, we may compose with the bijection of Proposition 27, sending X' to a Nikulin Marked (S'',d'') K3 surface X'' of type D'. Finally, compose with the bijection of Proposition 6, giving X'' the structure of an elliptic fibration of type T and remembering the Nikulin marking.   □

As in Sect. 3.5, we can deduce from the theorem above an isomorphism of curves.

Definition 33. Fix a type D'. The curve $\tilde{\tilde{H}}_T$ is the coarse moduli space of Nikulin marked elliptic fibrations of type D', where the fibration is of type T. We denote a connected component of $\tilde{\tilde{H}}_T$ by $\tilde{\tilde{H}}_T^0$.

Remark 11. Let us analyze more closely the structure of the moduli space $\tilde{\tilde{H}}_T$. We have a Galois cover, with Galois group S_8, of the curve $\tilde{H}_T$, of all Nikulin marked elliptic fibrations. Each component has its own type and $\tilde{\tilde{H}}_T$ consists of the union of all components of the specified type D'. Thus, it is clearly a Galois cover of $\tilde{H}_T$.

To compare these moduli spaces with Shimura curves, we analyze the structure of these first. These results work in the following generality: we assume that we have an Eichler order $\mathcal{M} = \mathcal{M}_{D,N}$ as in Definition 4 where $2|D$.

Definition 34. The curve $V_\mathcal{M}(2\pi_2)$ is the moduli space of abelian surface with multiplication by $\mathcal{M}$ and a $2\pi_2$ level structure (see Definitions 4 and 25).

To express $V_\mathcal{M}(2\pi_2)$ in "classical" terms, let $\Gamma(2)$ denote the principal congruence subgroup of level 2 in the group $\mathcal{M}_1^\times$ of norm 1 units in $\mathcal{M}$. It acts as usual on the upper half plane $\mathcal{H}$, and we have the following:

Proposition 29. *The curve $V_\mathcal{M}(2\pi_2)$ consists of two connected components, each isomorphic analytically to $\Gamma(2)\backslash\mathcal{H}$.*

Proof. the analytic space $\mathcal{H}^\pm = \mathbb{P}^1(\mathbb{C}) - \mathbb{P}^1(\mathbb{R})$ consists of the upper and the lower half planes. The standard adelic description of Shimura curves gives

$$V_\mathcal{M}(2\pi_2)^{an} = B^\times\backslash(\mathcal{H}^\pm \times (B^{f,\times}/K(2\pi_2))),$$

where the level subgroup $K(2\pi_2) \subset B^{f,\times}$ corresponds to our moduli problem, namely it consists of the elements of $\hat{\mathcal{M}}^\times = \mathcal{M} \otimes \hat{\mathbb{Z}}$ congruent to the identity modulo $2\pi_2$. This level subgroup is a product $K(2\pi_2) = \prod_p K_p(2\pi_2)$ over the primes, where each $K_p(2\pi_2)$ is maximal except for $K_2(2\pi_2)$. The surjectivity of the norm for (indefinite rational) quaternion algebras and strong approximation give the set of connected components of $V_\mathcal{M}(2\pi_2)$ as the adelic set

$$\pi_0(V_\mathcal{M}(2\pi_2)) = \mathbb{Q}^\times\backslash\mathbb{A}^\times/\mathrm{Nm}(K(2\pi_2))\mathbb{R}^{>0} \simeq \mathbb{Z}_2^\times/\mathrm{Nm}(K_2(2\pi_2)).$$

Lemma 22. $\mathrm{Nm}(K_2(2\pi_2)) = 1 + 4\mathbb{Z}_2$ *and* $\mathrm{Nm}(K_2(2)) = \mathbb{Z}_2^\times$.

Proof. (of the lemma) The group $\mathrm{Nm}(K_2(2\pi_2))$ is clearly open in $\mathbb{Z}_2^\times$ and is contained in $1 + 4\mathbb{Z}_2$. On the other hand, the 2-adic completion $B_2 = B \otimes \mathbb{Q}_2$ is isomorphic to the 2-adic completion of the standard (rational) Hamilton quaternions. In particular, the standard quaternions $1, \hat{i}, \hat{j}$, and $\hat{i}\hat{j}$, satisfying the standard relations $\hat{i}^2 = \hat{j}^2 = 1$ and $\hat{i}\hat{j} = -\hat{j}\hat{i}$. The completion $\mathcal{M}_2 = \mathcal{M} \otimes \mathbb{Z}_2$ is the unique maximal order of B_2, and it contains the element $\hat{u} = (1 + \hat{i} + \hat{j} + \hat{i}\hat{j})/2$. Then $1 + 4\hat{u}$ is in K_2 and $\mathrm{Nm}\hat{u} = 21$. Therefore $\mathrm{Nm}K_2 = 1 + 4\mathbb{Z}_2$ as asserted. The second assertion of the lemma follows from the first. Alternatively, it also follows from the fact that $\mathcal{M}_2^\times$ contains the ring of integers of the quadratic unramified extension of $\mathbb{Q}_2$, whose norm surjects onto $\mathbb{Z}_2^\times$. □

We return to the proof of the Proposition. It follows from the Lemma that there are two connected components. The description of these components shows that they are represented by elements $\beta_\pm$ of B_2, of norms ± 1 respectively. The corresponding components of $V_\mathcal{M}(2\pi_2)$ are $\Gamma_{\beta_\pm} \backslash \mathcal{H}$, where $\Gamma_{\beta_\pm}$ consists of the norm 1 elements in $B^\times \cap \beta_\pm K(2\pi_2) \beta_\pm^{-1}$. However, $B_2^\times$ and hence our $\beta_\pm$'s normalizes $K(2\pi_2)$, so the two components are isomorphic. We must prove that each is a copy of $\Gamma(2) \backslash \mathcal{H}$. Taking $\beta_+ = 1$ (as we may), it will suffice to prove that $\Gamma(2) = \langle \Gamma_{\beta_+} \cdot \langle \pm 1 \rangle$. Indeed, $K(2\pi_2)$ has index 4 in $K(2)$ (the quotient is isomorphic to the additive group of the field with four elements); by the previous lemma, intersecting with the norm 1 elements gives a subgroup $K(2\pi_2)_1$ of index 2 in the respective intersection $K(2)_1$. Since $K(2)_1$ contains -1, the images in $\mathrm{PSL}_2(\mathbb{R})$ of $\Gamma(2)$ and of $\Gamma_{\beta_\pm}$ agree. □

Corollary 12. *Under the assumptions of Theorem 7 there is an isomorphism $\tilde{H}_T^0 \cong V_\mathcal{M}(2\pi_2)$.*

Proof. This follows from Theorem 7 in exactly the same way that Theorem 2 is deduced from Proposition 10. □

In order to identify the quotient moduli space $\mathcal{H}_T$ with a Shimura curve we will need the following

Proposition 30. *Let $N(\mathcal{M})$ denote the normalizer of $\mathcal{M}$ in $B_+^\times$, the elements of B with positive norm. Then the automorphism group $\mathrm{Aut} V_\mathcal{M}(2)$ of $V_\mathcal{M}(2)$ is identified with the quotient $N(\mathcal{M})/\mathbb{Q}^\times \Gamma(2)$. This group is isomorphic to*

$$S_4 \times \prod_{2 \neq p | DN} S_2,$$

where the S_2 factor at p is generated by the modular involution w_p, $A_4 \subset S_4$ which is isomorphic to $(\mathcal{M}/2\mathcal{M})^\times$ is the automorphism of the level two structure and $S_4 = A_4 \cup w_2 A_4$. In particular, the order of $\mathrm{Aut} V_\mathcal{M}(2)$ is 12×2^r, where r is the number of factors of DN.

Proof. The image of $\Gamma(2)$ in $\mathrm{PSL}_2(\mathbb{R})$ is torsion free and hence the automorphism group of the hyperbolic surface $V_\mathcal{M}(2)$ is finite. We get an extension Γ of $\Gamma(2)$ by this automorphism group in $\mathrm{PSL}_2(\mathbb{R})$. The arithmetic group Γ certainly contains the image of $N(\mathcal{M})$ in $\mathrm{PSL}_2(\mathbb{R})$. This shows that $\mathrm{Aut} V_\mathcal{M}(2)$ is at least as big as claimed. But in the other direction, a theorem of Borel [8] gives that this image of $N(\mathcal{M})$ in $\mathrm{PSL}_2(\mathbb{R})$ is maximal discrete. The asserted equality follows. □

The following easy consequence will be frequently used in the examples.

Corollary 13. *Suppose that G is a subgroup of automorphisms of $V_\mathcal{M}(2\pi_2)$ satisfying the following conditions*

- *It permutes the two components of $V_{\mathcal{M}}(2\pi_2)$.*
- *No element of G fixes a component of $V_{\mathcal{M}}(2\pi_2)$ pointwise.*

Then the quotient $V_{\mathcal{M}}(2\pi_2)/G$ is isomorphic to the quotient of $V_{\mathcal{M}}(2)$ by a subgroup of automorphisms which is a subgroup of index 2 in G. In particular, if the order of G is 24×2^r, where r is the number of factors of DN, this quotient is exactly $V^*_{\mathcal{M}}$.

To end this section we analyze the Galois cover structure resulting from the Nikulin marking

Proposition 31. *The covering $\tilde{\tilde{H}}^0_T / \tilde{H}^0_T$ is Galois with Galois group S^4_2.*

Proof. It is easy to see that the covering is Galois and that the Galois group is precisely the group of permutations of $1, \ldots 8$ which fixes pointwise the image of the type D' in q_N. This image is generated by the 4 sums $x_{2i-1} + x_{2i}$ (with the relation that their sum is 0). The Galois group is therefore generated by the four transpositions (12), (34), (56) and (78). $\square$

7.5 A Transcendental Description

It is possible to give the Morrison correspondence above a purely transcendental Hodge theoretic description. This is useful for proving several results which we were not able to prove algebraically.

Proposition 32. *Let S be a lattice with signature $(1, 18)$, d a chamber in S, T a lattice with signature $(2, 1)$. Then there exists a bijection between the set of isomorphism classes of marked (S, d) K3 surfaces (X, α) such that the orthogonal complement of $\alpha(S)$ in $H^2(X, \mathbb{Z})$ is isometric to T with the quotient $\mathcal{H}_T \times \mathcal{ISO}_{T,S}/\mathrm{Aut}(T)$, where $\mathcal{H}_T$ is the space of all pure Hodge structures of weight 2 on T, having the property that $T^{1,1}_{\mathbb{C}} \cap T = \{0\}$, while $\mathcal{ISO}_{T,S}$ is the set of all isomorphisms $q_T \cong q_{S[-1]}$.*

Proof. Given a marked K3 surface (X, α), we choose an isometry $\beta : T \cong \alpha(S)^\perp$. Since $H^2(X, \mathbb{Z})$ is unimodular, the embedding α induces an isometry $q_{\alpha(S)^\perp} \cong q_{S[-1]}$. Pulling back via the isometry induced by β on discriminant forms we obtain the element of $\mathcal{ISO}_{T,S}$ and pulling back the induced Hodge structure on $\alpha(S)^\perp$ we obtain the required element of $\mathcal{H}_T$, as verifying the conditions on the pulled back Hodge structure is easy. Modifying β by an isometry of T acts in the obvious way. Conversely, given a pair $(h, i) \in \mathcal{H}_T \times \mathcal{ISO}_{T,S}$ consider the over-lattice L of $S \oplus T$ determined by the graph of i and put on it the Hodge structure which on T is h and for each the entire $S \otimes \mathbb{C}$ is of type $(1, 1)$. L is unimodular of signature $(19, 3)$ and there is just one such lattice up to isomorphism. By the surjectivity of the period mapping [1, Chap. VIII] the lattice L, together with its distinguished cone d and its Hodge structure, correspond to a marked K3 surface, unique up to isomorphism. It is finally clear that applying an element of $\mathrm{Aut}(T)$ results in a surface connected by a signed Hodge isometry with the original surface, hence with an isomorphic surface by Torelli. $\square$

Proposition 33. *Suppose that $G \subset \mathrm{Aut}(S)$ is a finite group of isometries fixing d. There is a clear action of G on isomorphism classes of (S, d)-marked K3 surfaces. The identification of Proposition 32 translates this to the obvious action on $\mathcal{ISO}_{T,S}$. The kernel of this action is the kernel of the map $G \to \mathrm{Aut}(q_S)/\pm 1$.*

Proof. The first statement is clear. Consequently, the action factors via $\mathrm{Aut}(q_S)$. Still, an isometry of S would act trivially on marked K3 surfaces if the action on $\mathcal{ISO}_{T,S}$ would correspond to the action of an isometry of T fixing all Hodge structures. The only possible isometry of this type is clearly -1. $\qquad\square$

Suppose now that S is the lattice S'' from Sect. 7.2. Let T be a lattice with signature $(2, 1)$ and discriminant form $q_{S[-1]}$. By Assumption 1 and Theorem 5 the lattices T and $T[2]$ are determined by their signature and discriminant form. It follows that for an (S'', d'') marked K3 surface (X, α'') we automatically have $\alpha''(S'')^\perp \cong T$. Consequently, Proposition 32 gives a bijection between $\mathcal{H}_T \times \mathcal{ISO}_{T,S''}/\mathrm{Aut}(T)$ and isomorphism classes of all (S'', d'')-marked K3 surfaces. Similarly, replacing S by the lattice S' of Definition 31 and T by $T[2]$ we obtain a bijection between isomorphism classes of (S', d')-marked K3 surfaces and $\mathcal{H}_T \times \mathcal{ISO}_{T[2],S'}/\mathrm{Aut}(T)$.

Proposition 34. *The map which associates, as in Proposition 27, to an (S', d')-marked K3 surface the associated Nikulin marked (S'', d'') K3 surface of type D, and then forgets the Nikulin marking, is given, in the description of Proposition 32, as follows: The Hodge structure on $T[2]$ is mapped to the same Hodge structure on T. The isomorphism $q_{T[2]} \xrightarrow{\sim} q_{S'} \xrightarrow{\sim} q_{S_1[2]}$ is sent to the restriction on the elements divisible by 2, $q_T \cong q_{S_1}$, composed with the isomorphism $q_{S_1} \cong q_{S''}$. The map is a Galois covering with Galois group*

$$G' = \{g \in \mathrm{Aut}(q_{T[2]}) \mid g = 1 \text{ on } q_T\}$$

Proof. Here, by a Galois covering, we mean that the fiber is a simple G'-space for a sufficiently general point. The result is essentially obvious from the description, in Theorem 6, of the Morrison correspondence. The only non-trivial point is that it is indeed a covering. This translates to the fact that the map $\mathrm{Aut}(q_{T[2]}) \to \mathrm{Aut}(q_T)$ is surjective. This in turn follows since $\mathrm{Aut}(T) \to \mathrm{Aut}(q_T)$ is surjective by Theorem 5. $\qquad\square$

By essentially the same argument we get the following result

Proposition 35. *The map sending an isomorphism class of (S'', d'')-marked K3 surface to the corresponding isomorphism class of elliptic fibrations of a prescribed type is Galois, with Galois group*

$$G'' = \{g'' \in \mathrm{Aut}(q_{S[2]}) \mid g''|_{q_S} = \pm g \text{ for some } g \in G\}/\pm 1,$$

where G is the automorphism group of the graph corresponding to the type of fibration.

Corollary 14. *An element of G'' cannot fix a component of $\tilde{H}_T$ pointwise.*

Proof. Let (h, i) be an element of $\mathcal{H}_T \times \mathcal{ISO}_{T[2],S'}$ and suppose that $g \in G''$ fixes the image of (h, i) in $\tilde{H}_T$. This implies that there exists an isometry t of T which fixes h and such that $h \circ i \circ g = id$. Since g is non-trivial, so must be t; but such t cannot fix a sufficiently general Hodge structure on T, hence no sufficiently general point can be fixed by g. $\square$

Corollary 15. *The space $\tilde{H}_T$ has two components which are permuted by the monodromy group G''. Thus, H_T is isomorphic to the quotient of either component of $\tilde{H}_T^0$ by a group which is of index 2 in G'', acting effectively on $\tilde{H}_T^0$.*

Proof. The space H_T is connected so G'' must permute the components. The action is effective by Corollary 14. $\square$

8 Explicit Computations

We now use the theory developed in the preceding sections to treat in detail several examples from Table 1, namely, those corresponding to numbers 9, 5, 10, 6, 2, 7, 8 and 11 on this table. The result will give us the precise correspondence between the base of the family of twists and the corresponding Shimura curve.

As a first step, we consider our list of examples and consider the possibility of a torsion section.

Proposition 36. *Consider an elliptic fibration, arising in one of the families described in Table 1 and having Mordell–Weil rank 0. If the fibration comes from either number 6 or number 8 on the list, then its Mordell–Weil group is torsion of order 2. In all the other examples, the Mordell–Weil group has no torsion.*

Proof. It is well known that the torsion subgroup of the Mordell–Weil injects into the Kodaira–Néron model of each fiber. When a singular fiber has additive reduction its torsion subgroup is just the group of reduced components. This is $(\mathbb{Z}/2)^2$ for the I_0^* fiber which is always present. Hence we only need to check 2-torsion, which, being invariant under quadratic twists is visible in Herfurtner's model. Hence in (1) we need to look for rational solutions $x = x(t)$ for $f(x, t) = 0$. In fact one can cut on the work as follows. When one has in addition a fiber of type II^* (respectively IV^*), arising by twisting from a fiber of type IV (respectively II), where the group of components is 0 (respectively $\mathbb{Z}/3$), there is no torsion. In the case where a fiber of type III^* is present, the group of components is $\mathbb{Z}/2$ so a $\mathbb{Z}/2$ torsion (at most) is possible in the cases 5, 6, 7, 8 and in all the other cases there is no torsion. In the cases 6 and 8 a torsion section will be directly exhibited (see the relevant subsections). It suffices then to rule out a two torsion section in cases 5 and 7. Such a torsion section cannot exist (in the untwisted family) by [32, Main Theorem], No. 47 for example 5, No. 56 for example 7. $\square$

It turns out that none of the elliptic surfaces we are considering is Kummer itself (hence one must apply the Morrison construction to get a Kummer surface). We have the following easy method for testing this: Suppose that our elliptic surface X'' is a Kummer, associated with an abelian surface A'', and let T be the transcendental lattice of A'', The discriminant form of NS(X'') is the same as of $T(X'') = T[-2]$. We can compute the discriminant form of T by "multiplying that of NS(X'') by $-1/2$". In fact, it is easy to see that the discriminant form of $T[-2]$ determines that of T. This is clear on the prime to 2 part and on the 2-torsion part follows from Lemma 14. It is then obvious that the procedure for recovering q_T from $q_{T[-2]}$ takes $q_\theta^{(p)}(p^k)$ to $q_{-\theta/2}^{(p)}(p^k)$ for $p \neq 2$ and multiplying by -1 and removing a power of 2 on the 2-torsion part. Once this is done we can check, using Theorem 4, if the resulting discriminant form has signature as it should be, i.e., 1 modulo 8. If this is not the case X'' is not Kummer (in fact, as the referee pointed out, it suffices to note that an even lattice of odd rank has even discriminant).

8.1 Number 5 on the List

Here the base elliptic surface has singular fibers $(2I_1, I_7, III)$, so the twists have singular fibers of type $T = (2I_1, I_7, III^*, I_0^*)$. This gives a discriminant form $q_{S_T} = q_1^{(7)}(7) \oplus q_1^{(2)}(2) \oplus v_+^{(2)}(2)$. We first check if X'' is a Kummer. "Multiplying the discriminant form of S_T by $-1/2$" as above we find the discriminant form $q_{-1}^{(7)}(7)$ which has signature -2. Thus X'' is not a Kummer.

We therefore attempt to apply a Morrison transformation and to appeal to Theorem 7 and Corollary 12

From Table 1 we learn that X'' should be isogenous to an abelian surface with multiplication by a maximal order of a quaternion algebra of discriminant 14, so $D = 14$ while $p = 1$. By Corollary 10 the prime to 2 part of $q_{N[2]}$ is $q_1^{(7)}(7)$ while $-Dp/2 = -7 \equiv 1 \pmod{4}$.

From Corollary 12 it follows that there is an isomorphism between the Shimura curve $V_{14}(2\pi_2)$ and the moduli curve $\tilde{\tilde{H}}_T^0$.

We now describe this curve as a covering of H_T. By Proposition 31 the intermediate covering $\tilde{H}_T^0$ is such that the covering $\tilde{\tilde{H}}_T^0/\tilde{H}_T^0$ is Galois with group S_2^4. On the other hand, the analysis of the covering $\tilde{H}_T^0/H_T$ is done in the same way as in Sect. 3. The corresponding table is

	I_7	I_0^*
γ_{I_1}	1 (12)	
γ_{I_1}	1 (12)	
γ_{I_7}	-1 (23)	
γ_{III}	1 (23)	
-1	-1	1

To see this, notice that we know that on the last column the first four rows give transpositions. The global monodromy is surjective since there is no section of order 2. It is a simple combinatorial exercise to see that we may assume that they are of the indicated types. But in fact for our present purpose we only need that the monodromy on S_3 is surjective. Thus, the Galois group here is $S_3 \times S_2 / \langle \pm 1 \rangle \simeq S_3$.

To sum up, we see that $\tilde{H}_T^0 \cong V_{14}(2\pi_2)$ is a covering of H_T with Galois group which is an extension of S_3 by S_2^4, hence has order 24×2^2. By Corollaries 15 and 13 we immediately have $H_T \cong V_{14}^*(1)$. Here this moduli space is the λ-line itself since it has no automorphisms.

8.2 Case Number 9 on the List

This case starts out very much like the previous one. The type is $(2I_1, I_6, IV)$ so each fiber in our family of twists has type $(2I_1, I_6, II^*, I_0^*)$. Let X'' be one such fiber. We have $q_{S_T} = q_{-1}^{(3)}(3) \oplus q_1^{(2)}(2) \oplus v_+^{(2)}(2)$. Multiplying this discriminant form by $-1/2$ as before we find $q_2^{(3)}(3)$ of signature 2 (mod 8) so again X'' cannot be Kummer. Table 1 encourages us to consider an abelian surface A with multiplication by the maximal order in the quaternion algebra of discriminant 6. The prime to two part on both q_{S_T} and $q_{N[2]}$ are $q_{-1}^{(3)}(3)$ and $-Dp/2 = -3 \equiv 1$ (mod 4). Thus, as before we find an isomorphism $V_6(2\pi_2) \cong \tilde{H}_T^0$. To compute the covering $\tilde{H}_T^0/H_T$ we first compute the covering $\tilde{H}_T^0/H_T$. The relevant table reads

	I_6	I_0^*
γ_{I_1}	1	(12)
γ_{I_1}	1	(23)
γ_{I_6}	-1	1
γ_{IV}	1	(123)
-1	-1	1

More precisely, we know that the first two rows on the last column are transpositions, and the fourth row has order 3. Hence we may assume that they are of the indicated types. Thus, the Galois group here is S_3. As in the previous example we therefore get an isomorphism $H_T \cong V_6/\langle w_2, w_3 \rangle$. Note however that, unlike the previous case, H_T is not the λ-line but is rather its quotient by the involution $\lambda \to -\lambda$. In fact, the group D_T, defined in Proposition 3 as the group of automorphisms of the Λ-line which preserves the set of singular fibers and the types of fibers, consists in our case of the identity and the involution above as it replaces the two points ± 1 where the two I_1 fibers are, and fixes the points 0 and ∞ where the two other singular fibers are.

We would now like to discover the modular interpretation of the λ-line, and to compare our results with those of Elkies [17].

For this, consider first the covering $\Lambda_T' \cong \mathbb{P}_\lambda^1 \times_{H_T} \tilde{H}_T^0$ of the λ-line. The table of local monodromies is exactly the table above. Thus, we see that the covering is connected and is Galois of order 6 and that the ramifications are as follows: There are three points, each ramified to order 2 above the I_1 fibers at $\lambda = \pm 1$, two points of order 3 above the IV fiber at $\lambda = 0$ and 6 unramified points over the I_6 fiber at $\lambda = \infty$. An easy computation using the Hurwitz formula implies that Λ_T' has genus 0.

Now we consider the covering $\tilde{H}_T^0$ of H_T, which is obtained from Λ_T' by dividing by an involution above the involution $\lambda \to -\lambda$ of the λ-line. Note that this involution can only have two fixed points as the genus of Λ_T' is 0. Since H_T is obtained from the λ-line by dividing out by ± 1, it is natural to choose on it the coordinate $z = \lambda^2$. We now analyze the ramifications for the covering.

Above $z = \infty$ we see the quotient by the involution of the six points above $\lambda = \infty$. None of these six points can be a fixed point by symmetry. Thus, there are exactly three points above $z = \infty$ and each is ramified to order 2. None of the points above $\lambda = \pm 1$ can be fixed under the involution, as points above 1 are carried to points above -1 and vice versa. It follows that there are exactly three points above $z = 1$, each again ramified to order 2.

Above $\lambda = 0$ we have two points and the involution can either fix or can permute them. If the second case holds the involution has no fixed points whatsoever, which is impossible. Therefore, above $z = 0$ we have 2 points, each of order 3.

We know that H_T is isomorphic to $V_6/\langle w_2, w_3 \rangle$. This has a coordinate t described by Elkies in [17, 3.1] by the condition that the elliptic points are above $t = 0$, $t = 1$ and $t = \infty$ with orders 2, 4, and 6 respectively. Now, $\tilde{H}_T$ is covered by the modular curve $V_6(2)$, which is covered in turn in an unramified way by the upper half plane. The map from $V_6(2)$ to $\tilde{H}_T$ is a $(\mathbb{Z}/2)^3$-covering. Thus, it is clear that the three points $z = 0, 1, \infty$ must correspond to the three points $t = 0, 1, \infty$ and further that $z = 0$ must correspond to $t = \infty$. This leaves the two options $t = 1/z$ or $t = 1 - 1/z$, which cannot be distinguished by these considerations.

To proceed, we now follow Elkies in describing the covering $V_6 \xrightarrow{\pi} V_6/\langle w_2, w_3 \rangle$. According to Elkies, the modular involutions w_2, w_3, w_6 are represented by elements s_4, s_6, s_2 respectively, with s_i having order i. These fix the corresponding elliptic points. Thus, it is easy to see (loc. cit.), that $\pi^{-1}(0)$ (respectively $\pi^{-1}(1)$, $\pi^{-1}(\infty)$) consists of two points which are fixed by w_6 (respectively w_2, w_3). Thus, the covering V_6/w_2 (respectively V_6/w_3, V_6/w_6) is ramified at $t = 0, \infty$ (respectively $t = 0, 1$, $t = 1, \infty$). In any case, it is clear that the λ-line is one of these quotients, as one of them will be a ramified cover of $\mathbb{P}^1$ ramified at the same points.

To determine the precise nature of the λ-line, as well as the relation between t and z, we are now forced to use our knowledge of the Picard–Fuchs equation (see [5]). It follows from the above that the map from the upper half plane to H_T factors via the λ-line. Using the Picard–Fuchs equation we can determine the elliptic points for the λ-line. It turns out that $\lambda = \infty, 1, -1$ have above them elliptic points of order 2 while 0 has above it an elliptic point of order 3. Thus, $z = 1$ has above it an elliptic point of order 2 while $z = \infty$ a point of order 4. This implies that $t = 1 - 1/z$. the λ-line is ramified over $t = 1, \infty$ and consequently is isomorphic with V_6/w_6.

8.3 Number 10 on the List

In this example the base fibration has singular fibers (I_1, I_2, I_5, IV), leading to X'' of type $T = (I_1, I_2, I_5, II^*, I_0^*)$ and to discriminant form $q_{S_T} = q_1^{(5)}(5) \oplus q_{-1}^{(2)}(2) \oplus v_+^{(2)}(2)$. Multiplying by $-1/2$ again we get $q_2^{(5)}(5)$, which has signature 0, so X'' is not Kummer.

Table 1 predicts a quaternion algebra with discriminant 10 and an abelian surface A with multiplication by its maximal order has $q_{NS(A)[2]} = q_1^{(5)}(5) \oplus q_{-1}^{(2)}(8) \oplus v_+^{(2)}(2)$. Thus, in exactly the same way as before we find again an isomorphism $V_{10}(2\pi_2) \cong \tilde{H}_T^0$.

By Proposition 31 the Galois group of $\tilde{\tilde{H}}_T^0$ over $\tilde{H}_T^0$ is again S_2^4. We now compute the Galois group of $\tilde{H}_T^0$ over H_T. We have the following table

	I_5	I_0^*
γ_{I_1}	1	(12)
γ_{I_2}	1	1
γ_{I_5}	-1	(23)
γ_{IV}	1	(123)
	-1	-1	1

Here we know that on the last column we get transpositions, and we may assume that they are of the indicated types as in the first case. Thus, the Galois group here is again S_3. This implies as before that $V_{10}/\langle w_2, w_5 \rangle \cong H_T$.

8.4 Number 6 on the List

This example is slightly complicated by the presence of a 2-torsion section on our elliptic fibration. The base fibration has type (I_1, I_2, I_6, III) so X'' has type $(I_1, I_2, I_6, III^*, I_0^*)$ and the corresponding discriminant form is $q_{-1}^{(2)}(2) \oplus q_2^{(3)}(3) \oplus q_1^{(2)}(2) \oplus q_1^{(2)}(2) \oplus v_+^{(2)}(2)$ (here it will be important to keep track of the relation with the fibers, so we note that the first summand corresponds to I_2, the second and third to I_6, the fourth to III^* and the last to I_0^*). However, this is not the discriminant form of the fibration because this elliptic fibration has a torsion section. By the proof of Proposition 36 the torsion subgroup is at most of order 2. On the other hand we do find a torsion section of order 2: for this we solve the equation $4x^3 - g_2 x - g_3 = 0$, where

$$g_2(s) = 12s(-3 + 9s - 6s^2 + s^3)$$
$$g_3(s) = 4s^2(27 - 63s + 54s^2 - 18s^3 + 2s^4)$$

[21, p. 337]. (Notice Herfurtner says he is giving the G_i when in fact he is giving the g_i). We find a section with $x = 3s - s^2$ and $y = 0$. It is clear, as was already said, that this 2-torsion section is inherited by all the twists. We compute where it intersects the singular fibers as follows:

- The I_2 fiber is at 1. The equation there is $y^2 = 4(x - 2)(x + 1)^2$ and the section has $x = 2$, so it intersects at the identity component.
- The I_6 fiber is at ∞. To compute the equation there we write the equation as $(y/s^3)^2 = 4(x/s^2)^3 - (g_2/s^4)(x/s^2) - (g_3/s^6)$. So the equation at ∞ is $y^2 = 4x^3 - 12x - 8 = 4(x - 2)(x + 1)^2$ and the value of the section is -1 so this time it intersects a non-identity component.
- The other intersections are forced by the additive reduction to be not at the identity component.

Writing the three non-zero elements in $v_+^{(2)}(2)$ as e_1, e_2, e_3 (with $e_3 = e_1 + e_2$), we find that the torsion section gives an over-lattice corresponding to the isotropic subspace generated by $0 \oplus 0 \oplus 1 \oplus 1 \oplus e_1$, summands in the respective components. We want to compute the quotient of the orthogonal complement by this subspace. We may concentrate on the last three components, and we may fix representatives by insisting that the third component is 0. Thus, we get $q_{-1}^{(2)}(2) \oplus q_2^{(3)}(3) \oplus$ the subgroup of $q_1^{(2)}(2) \oplus v_+^{(2)}(2)$ orthogonal to $1 \oplus e_1$. Since $1^2 = 1/2$ while $e_1^2 = 0$ (modulo 1) and $e_i \times e_i = 1/2$ for other i we see that this group contains 0, $0 \oplus e_1$, $1 \oplus e_2$ and $1 \oplus e_3$. The last two are orthogonal and have square $-1/2$ each. Thus we find this group isomorphic to $q_{-1}^{(2)}(2)^2$ and overall we have that the discriminant form of the Néron–Severi lattice of our elliptic fibration is $q_{NS(X'')} = q_2^{(3)}(3) \oplus q_{-1}^{(2)}(2)^3 = q_{-1}^{(3)}(3) \oplus q_1^{(2)}(2) \oplus v_+^{(2)}(2)$, where in this last equality we have used the relation (d) from [30, Proposition 1.8.2]. This is the same as in Sect. 8.2 so by the same considerations we find an isomorphism $V_6(2\pi_2) \cong \tilde{H}_T^0$. Furthermore, the Galois group of $\tilde{H}_T^0$ over $\tilde{H}_T^0$ is again S_2^4.

For the computation of $\mathrm{Gal}(\tilde{H}_T^0/H_T)$ we use the table

	I_6	I_0^*
γ_{I_1}	1	(12)
γ_{I_2}	1	1
γ_{I_6}	-1	1
γ_{III}	1	(12)
-1	-1	1

by the same arguments as before, since here we know that the monodromy on the last column is S_2 because of the 2-torsion section. Thus, the Galois group here is $S_2 \times S_2/\langle \pm 1 \rangle \simeq S_2$. Therefore, H_T is $V_6(2\pi_2)/G$, where G has order 32. Thus G is the 2Sylow subgroup of the group $\mathrm{Gal}(V_6(\pi_2^3)/V_6^*(1))$ of order 96, and we can simplify this to $V_6(\pi_2)/\langle w_2, w_3 \rangle$.

8.5 Number 2 on the List

In this example the base fibration has singular fibers of type (I_1, I_2, I_7, II), leading to a family of twists with fiber X'' of type $(I_1, I_2, I_7, IV^*, I_0^*)$ and to discriminant form

$q_{NS(X'')} = q_1^{(7)}(7) \oplus q_{-1}^{(2)}(2) \oplus q_{-1}^{(3)}(3) \oplus v_+^{(2)}(2)$. Multiplying by $-1/2$ again we get $q_{-1}^{(7)}(7) \oplus q_{-1}^{(3)}(3)$, which has signature 0, so X'' cannot be Kummer again.

Looking at Table 1 we expect a relation with a Shimura curve associated to a quaternion algebra with discriminant 6. The presence of $q_{-1}^{(7)}(7)$ suggests some kind of an isogeny of order 7 (at least). Instead of analyzing what happens when we apply an isogeny of order 7 we can use the known information about the Néron–Severi lattice of an abelian surface A with multiplication by an Eichler order $\mathcal{M}_{6,7}$ of level 7. By [4, Proposition 3.3] we have $q_{NS(A)[2]} = q_1^{(7)}(7) \oplus q_{-1}^{(2)}(8) \oplus q_{-1}^{(3)}(3) \oplus v_+^{(2)}(2)$. Thus, Corollary 12 (since this applies to Eichler orders as well) implies an isomorphism $V_6(14\pi_2) \cong \tilde{H}_T^0$.

The computation of the Galois group $\mathrm{Gal}(\tilde{H}_T^0/H_T)$ is done using the following table.

	I_7	IV^*	I_0^*
γ_{I_1}	1	1	(12)
γ_{I_2}	1	1	1
γ_{I_7}	-1	1	(23)
γ_{II}	1	-1	(123)
-1	-1	-1	1

Since the square of γ_{II} is of order 3 on I_0^* and trivial on the other two fibers we can produce S_3 on that fiber and it is easy to see that the monodromy group is $\pm 1 \times \pm 1 \times S_3 / \pm 1 \cong S_3 \times S_2$. This gives a total monodromy group of order $6 \times 2 \times 16$. This is the full automorphism group: by Proposition 30 we obtain an isomorphism $H_T \cong V_{6,7} / \langle w_2, w_3, w_7 \rangle$.

8.6 Number 7 on the List

The base fibration has fibers (I_1, I_3, I_5, III), leading to $(I_1, I_3, I_5, III^*, I_0^*)$ and to discriminant form $q_{NS(X'')} = q_1^{(3)}(3) \oplus q_1^{(5)}(5) \oplus q_1^{(2)}(2) \oplus v_+^{(2)}(2)$. Multiplying by $-1/2$ again we get $q_1^{(3)}(3) \oplus q_2^{(5)}(5)$, which has signature -6, so it cannot be Kummer again.

The corresponding quaternion algebra has discriminant 6. Because of the presence of 5 we try an abelian surface with multiplication by an Eichler order of level 5. Such a surface A has, according to [4, Proposition 3.3] $q_{NS(A)[2]} = q_1^{(3)}(3) \oplus q_1^{(5)}(5) \oplus q_1^{(2)}(8) \oplus v_+^{(2)}(2)$. Thus, as in the previous case we find an isomorphism $V_6(10\pi_2) \cong \tilde{H}_T^0$

We next compute the Galois group $\mathrm{Gal}(\tilde{H}_T^0/H_T)$. Recalling that there are no torsion sections by Proposition 36 the monodromy table looks as follows:

	I_3	I_5	I_0^*
γ_{I_1}	1	1	(12)
γ_{I_3}	-1	1	(ab)
γ_{I_5}	1	-1	(cd)
γ_{III}	1	1	(ef)
-1	-1	-1	1

Now, the fact that there is no torsion shows that the projection on the S_3 part is onto. Since the cube of the product of two transpositions is always 1 we easily see that the monodromy group contains $1 \times -1 \times 1$ and $-1 \times 1 \times 1$. It follows that the monodromy group is full, equal to $(\pm 1 \times \pm 1 \times S_3)/\pm 1$. The rest of the analysis is the same as in the previous case, showing that $H_T \cong V_{6,5}/\langle w_2, w_3, w_5 \rangle$

8.7 Number 8 on the List

For this case we show an isogeny with the elliptic fibration given at number 6 on the list. In this example, as well as the next one, we need to take quotients with respect to torsion sections, for which we have used the computer algebra system MAGMA (although these are just standard formulas).

Consider the elliptic fibration which is no. 6 on the list. In Sect. 8.4 we showed it has a rational two torsion section provided by $x = 3s - s^2$. Taking the quotient by this two torsion and base changing by $s \mapsto 1 - 1/s$ we obtain exactly the fibration which was no. 8 on the list. The dual isogeny is obtained as follows: The equation for fibration no. 6 is $y^2 = 4x^3 - g_2 x - g_3$, where

$$g_2(s) = 3(s-1)(16s^3 - 3s - 1)$$
$$g_3(s) = (s-1)^2(64s^4 + 32s^3 + 6s^2 + 5s + 1)$$

[21, p. 338]. By setting $y = 0$ we find a 2-torsion section with $x = 1 + s - 2s^2$. Taking the quotient by this we get the dual isogeny.

This isogeny extends to an isogeny between the twisted surfaces and consequently the two λ-lines have the same moduli interpretation.

8.8 Number 11 on the List

This case will turn out to be isogenous to the elliptic surface at no. 9 on the list. This is given by the equation $y^2 = 4x^3 - g_2 x - g_3$, where

$$g_2(s) = 3s^2(9s^2 - 8)$$
$$g_3(s) = s^2(27s^4 - 36s^2 + 8)$$

[21, p. 338]. It has a rational subgroup of order 3 given by the equation $-2x = 3s^2$. Taking the quotient by this subgroup and base changing via $s \mapsto (s-1)/(s+1)$ we obtain exactly the fibration which is no. 11 on the list, given by the equation $y^2 = 4x^3 - g_2 x - g_3$, where

$$g_2(s) = 3(s-1)^2(9s^2 + 14s + 9)$$
$$g_3(s) = (s-1)^2(27s^4 + 36s^3 + 2s^2 + 36s + 27)$$

[21, p. 338]. Thus, the λ-lines are again the same.

Appendix

A.1 Rational Invariants of Quadratic Forms Associated with Singular Fibers

A.1.1 Quadratic Forms Over $\mathbb{Q}_p$

In this appendix p is a fixed finite prime and K is a completion of $\mathbb{Q}$, either at p or at ∞. Let q be a non-degenerate quadratic form K.

Definition 35. The ϵ invariant of a diagonal form $\sum_i a_i x_i^2$ is $\prod_{i<j}(a_i, a_j)_p$ where $(x, y) = (x, y)_p$ is the Hilbert symbol at p

$$(\bullet, \bullet) : (\mathbb{Q}_p^\times/(\mathbb{Q}_p^\times)^2 \times (\mathbb{Q}_p^\times/(\mathbb{Q}_p^\times)^2 \to \pm 1.$$

The ϵ invariant of a the quadratic form q is the ϵ invariant of any diagonal form of q.

Proposition 37. *1. The quadratic form q is characterized by its discriminant, $\mathrm{disc}(q)$, equal to the determinant of a representing matrix in $K^*/(K^*)^2$, and either its signature when $K \simeq \mathbb{R}$, or its $\epsilon = \epsilon_p$ invariant when $K \simeq \mathbb{Q}_p$.*
2. $q_1 \oplus q \cong q_2 \oplus q$ implies $q_1 \cong q_2$.

Proof. For the first part see [34, Chap. 1] or [9, Chap. 1]. The second is Witt's Cancellation Theorem.

In what follows we will make repeated use of the standard properties of the symbol (x, y): symmetry and bilinearity, $(x, -x) = 1$, and $(x, 1 - x) = 1$.

Lemma 23. *We have*

1. $\mathrm{disc}(q_1 \oplus q_2) = \mathrm{disc}(q_1) \times \mathrm{disc}(q_2)$,
2. $\epsilon(q_1 \oplus q_2) = \epsilon(q_1) \times \epsilon(q_2) \times (\mathrm{disc}(q_1), \mathrm{disc}(q_2))$,
3. $\mathrm{disc}(\lambda q) = \lambda^n \mathrm{disc}(q)$ and
4. $\epsilon(\lambda q) = \epsilon(q) \times (\lambda, \mathrm{disc}(q))^{n-1} \times (\lambda, -1)^{n(n-1)/2}$.

Proof. All assertions are immediate except perhaps for the last one, which one obtains as follows: If q is diagonalized as $\sum_{i=1}^n a_i x_i$, then we have

$$\epsilon(\lambda q) = \prod_{i<j}(\lambda a_i, \lambda a_j) = \epsilon(q) \times \prod_{i<j}(\lambda, a_i a_j) \times (\lambda, \lambda)^{n(n-1)/2}$$
$$= \epsilon(q) \times (\lambda, \prod_{i<j} a_i a_j) \times (\lambda, \lambda)^{n(n-1)/2}$$
$$= \epsilon(q) \times (\lambda, \mathrm{disc}(q))^{n-1} \times (\lambda, \lambda)^{n(n-1)/2}$$
$$= \epsilon(q) \times (\lambda, \mathrm{disc}(q))^{n-1} \times (\lambda, -1)^{n(n-1)/2}$$

$\square$

The hyperbolic plane U is the binary quadratic form $x^2 - y^2$, or equivalently xy. Clearly U is equivalent to $-U$. The following is well known.

Lemma 24. *For a quadratic form q of rank r we have $q \oplus (-q) \simeq U^r$.*

A.1.2 Quadratic Forms of Singular Fibers

For a singular elliptic fiber of type t, we denote by q_t the lattice freely spanned by the components not meeting the 0-section with the intersection form. We will prove the following

Theorem 8. *The forms $L_t[-1]$ are positive definite, and U is indefinite. Their rank, discriminant, and p-adic ϵ-invariant are given in the following table:*

t	I_n	I_n^*	II^*	III^*	IV^*	U
rank$L_t[-1]$	$n-1$	$n+3$	8	7	6	2
disc($L_t[-1]$)	n	1	1	2	3	-1
$\epsilon(L_t[-1])$	$(-1,n)_p$	1	1	1	1	1

Proof. Let $DI_n(\alpha_1, \ldots, \alpha_n)$ denote the $n \times n$ matrix with diagonal entries α_1 to α_n, with -1 on the diagonals just above and below the main diagonal, and with zeroes elsewhere. We write (α) for $DI_1(\alpha)$. Notice that adding the first row multiplied by $1/\alpha_1$ to the second row and then doing the same on columns gives the following relation:

$$DI_n(\alpha_1, \ldots, \alpha_n) \equiv (\alpha_1) \oplus DI_{n-1}(\alpha_2 - 1/\alpha_1, \alpha_3, \ldots, \alpha_n) \qquad (10)$$

For $i \geq 2$ the form $L_{I_n}[-1] = 2 \sum_{i=1}^{n-1} x_i^2 - 2 \sum_{1 \leq i \leq n-2} x_i x_{i+1}$ is associated with the matrix $DI_{n-1}(2, 2, \ldots, 2)$. Applying (10) successively, we see that it is equivalent to $(2/1) \oplus (3/2) \oplus \cdots \oplus (n/(n-1))$. By Lemma 23 this form has discriminant n. We compute the ϵ invariant inductively: for $n = 2$ the formula holds since $\epsilon = 1$ for a form of rank 1. Assuming the formula for n we get from Lemma 23 and the induction assumption

$$\epsilon_{L_{I_{n+1}}[-1]} = \epsilon_{-L_{I_n}}(n, (n+1)/n) = (-1, n)(n, -n)(n, -n-1) = (n, n+1) =$$
$$= (-n, n+1)(-1, n+1) = (-1, n+1)$$

as asserted.

Next we consider

$$L_{I_n^*}[-1] = L_{I_{n+3}}[-1] + 2x_{n+4}^2 - 2x_{n+4}x_{n+2}.$$

Applying the same procedure as for $L_{I_{n+3}}$ to diagonalize the first $n+1$ rows and columns we find that this form is equivalent to the form associated with the matrix

$$(2/1) \oplus (3/2) \oplus \cdots \oplus ((n+2)/(n+1)) \oplus \begin{pmatrix} (n+3)/(n+2) & -1 & -1 \\ -1 & 2 & 0 \\ -1 & 0 & 2 \end{pmatrix}$$

Diagonalizing further, the last 3×3 matrix is found to be equivalent to $\mathrm{diag}(1/(n+2), 2, 2)$. Hence $L_{I_n^*}[-1]$ is equivalent to $L_{I_{n+2}}[-1] \oplus \mathrm{diag}(1/(n+2), 2, 2)$. From this it easily follows that the discriminant of $L_{I_n^*}[-1]$ is 4 (hence 1). The ϵ invariant is calculated as before, using Lemma 23:

$$\begin{aligned} \epsilon_{L_{I_n^*}[-1]} &= (-1, n+2) \times \epsilon(\mathrm{diag}(n+2, 2, 2)) \times (n+2, n+2) \\ &= \epsilon(\mathrm{diag}(n+2, 2, 2)) = (2, 2) = (-1, 2) = 1 \;. \end{aligned}$$

The forms $L_t[-1]$ fibers of type $t = II^*$, $t = III^*$, and $t = IV^*$ correspond to the forms $L_n = L_{I_n}[-1] + 2x_n^2 - 2x_n x_{n-3}$ for $n = 8, 7$, and 6 respectively. Similar to the I_n^* computation this is equivalent to the form associated with

$$(2/1) \oplus (3/2) \oplus \cdots \oplus ((n-3)/(n-4)) \oplus \begin{pmatrix} \frac{n-2}{n-3} & -1 & 0 & -1 \\ -1 & 2 & -1 & 0 \\ 0 & -1 & 2 & 0 \\ -1 & 0 & 0 & 2 \end{pmatrix}.$$

Killing the off diagonal terms in the last two rows and columns we find this last 4×4 matrix to be equivalent to

$$(2) \oplus (2) \oplus \begin{pmatrix} \frac{n-2}{n-3} & -\frac{1}{2} & -1 \\ -1 & & \frac{3}{2} \end{pmatrix}.$$

Therefore

$$L_n \simeq (2/1) \oplus \cdots \oplus ((n-3)/(n-4)) \oplus (3/2) \oplus (2) \oplus (2) \oplus \begin{pmatrix} \frac{n-2}{n-3} & -\frac{1}{2} & -\frac{2}{3} \end{pmatrix}.$$

The first $n-1$ terms have discriminant $6(n-3)$ and ϵ invariant

$$\begin{aligned} &\epsilon(L_{I_{n-3}}[-1]) \times \epsilon(\mathrm{diag}(3/2, 2, 2) \times (n-3, 6) \\ &= (-1, n-3)(2, 2) \times (n-3, 6) = (n-3, -6). \end{aligned}$$

The last entry is $(9-n)/(6(n-3))$. Thus we find the total discriminant to be $9-n$ and the ϵ invariant to be

$$\begin{aligned} &(n-3, -6) \times (6(n-3), (9-n) \times 6(n-3)) \\ &= (n-3, -6) \times (-1, 6(n-3)) \times (6(n-3), 9-n) \\ &= (n-3, -6) \times (6(n-3), n-9) \end{aligned}$$

which we check separately for $n = 6, 7, 8$. For $n = 6$ we get

$$(3, -6) \times (18, -3) = (3, 2) \times (2, -3) = (2, -1) = 1$$

For $n = 7$ we get

$$(4, -6) \times (24, -2) = (6, -2) = (3, -2) \times (2, -2) = 1$$

and finally for $n = 8$ we get

$$(5, -6) \times (30, -1) = (5, -6) \times (5, -1) \times (6, -1) = (5, 6) \times (6, -1) = (6, -5) = 1$$

$\square$

A.1.3 Ternary Forms of Quaternion Algebras

Let B be a quaternion algebra over $\mathbb{Q}$ of discriminant D, let B_0 be the space of traceless elements of B, and let $\mathrm{Nm} : B \to \mathbb{Q}$ denote the (reduced) norm. For $d \in \mathbb{Q}^*$ we will compute the invariants of the ternary quadratic form $q(x) = d\mathrm{Nm}(x)$, $x \in B_0$. Write B in standard form $B(\alpha, \beta)$, with $\hat{\imath}^2 = \alpha$, $\hat{\jmath}^2 = \beta$, with α, β in $\mathbb{Q}^*$ and $\hat{\imath}\hat{\jmath} = -\hat{\jmath}\hat{\imath}$. Recall that a prime p divides the discriminant of $B(\alpha, \beta)$ if and only if $(\alpha, \beta)_p = -1$. Then in terms of the standard basis $\hat{\imath}$, $\hat{\jmath}$, and $\hat{\imath}\hat{\jmath}$ of B_0 the form q is

$$d\mathrm{Nm}(x_1, x_2, x_3) = -\alpha dx_1^2 - \beta dx_2^2 + \alpha\beta dx_3^2.$$

Hence $\mathrm{disc}(d\mathrm{Nm}) = d$ and

$$\epsilon_p(d\mathrm{Nm}) = (-d\alpha, -d\beta)_p \times ((-d\alpha)(-d\beta), d\alpha\beta)_p = (-d, -d)_p(\alpha, \beta)_p$$
$$= (-d, -1)_p(\alpha, \beta)_p.$$

We have proved the following.

Proposition 38. *If* Nm *is the quadratic form associated with the quaternion algebra* $B(\alpha, \beta)$, *then for any integer* d *we have*

$$\mathrm{disc}(d\mathrm{Nm}) = d, \quad (\alpha, \beta)_p = \epsilon_p(d\mathrm{Nm})(-d, -1)_p. \tag{11}$$

Notice that the left hand side of this last equation is independent of the scaling factor d. In fact it is not hard to see that the rank and the right hand side for all p are complete invariants of rational ternary forms up to scaling. We will not need this fact, but it underlies our application in the article of (38): by [2, Sect. 5.3, formula (1)], we have $d = DN$ for the Néron–Severi lattice $\mathrm{NS}(A)$ of a QM abelian surface A whose ring of endomorphisms is an Eichler order of conductor N in a quaternion algebra $B = B_A$ over $\mathbb{Q}$ of discriminant D. More generally, any isogenous abelian or K3 surface X has a Néron–Severi lattice $\mathrm{NS}(X)$ isomorphic (rationally) to B_0 up to scaling (for example, $\mathrm{NS}(\mathrm{Km}(A)) \simeq 2\mathrm{NS}(A)$). Furthermore, $T(X) \simeq -\mathrm{NS}(X)$. We can therefore recover the isomorphism type of B_X from any of these lattices. The determination of the degrees of the isogenies is much more delicate and requires the integral methods developed by Nikulin, as is done in several places in the article.

Acknowledgements A. Besser was partially supported by ISF grant. R. Livné was partially supported by an ISF grant.

References

1. W.P. Barth, K. Hulek, C.A.M. Peters, A. Van de Ven, Compact complex surfaces, in *Ergebnisse der Mathematik und ihrer Grenzgebiete. 3. Folge*. A Series of Modern Surveys in Mathematics, vol. 4, 2nd edn. [Results in Mathematics and Related Areas, 3rd Series. A Series of Modern Surveys in Mathematics] (Springer, Berlin, 2004)
2. A. Besser, Universal families over shimura curves. Ph.D. thesis, Tel Aviv University, 1993
3. A. Besser, CM cyles over Shimura curves. J. Algebr. Geom. **4**, 659–693 (1995)
4. A. Besser, Elliptic fibrations of $K3$ surfaces and QM Kummer surfaces. Math. Z. **228**(2), 283–308 (1998)
5. A. Besser, R. Livne, Picard-Fuchs equations of families of QM abelian surfaces. Preprint (2012), URL http://arxiv.org/abs/1202.2808
6. F. Beukers, A note on the irrationality of $\zeta(2)$ and $\zeta(3)$. Bull. Lond. Math. Soc. **11**(3), 268–272 (1979)
7. F. Beukers, C.A.M. Peters, A family of $K3$ surfaces and $\zeta(3)$. J. Reine Angew. Math. **351**, 42–54 (1984)
8. A. Borel, Commensurability classes and volumes of hyperbolic 3-manifolds. Ann. Sc. Norm. Sup. Pisa Cl. Sci. (4) **8**(1), 1–33 (1981)
9. A.I. Borevich, I.R. Shafarevich, *Number Theory* (Academic, New York, 1966)
10. J.F. Boutot, H. Carayol, Uniformisation p-adique des courbes de Shimura: les théorèmes de Čerednik et de Drinfel'd. Astérisque 196–197 (1991); **7**, 45–158 (1992); Courbes modulaires et courbes de Shimura, Orsay, 1987/1988
11. C.H. Clemens, A scrapbook of complex curve theory, in *Graduate Studies in Mathematics*, vol. 55, 2nd edn. (American Mathematical Society, Providence, 2003)
12. P. Deligne, Formes modulaires ét representations ℓ-adiques, in *Seminaire Bourbaki*, vol. 355. Lecture Notes in Mathematics, vol. 179 (Springer, Heidelberg, 1971), pp. 139–172
13. P. Deligne, Travaux de Shimura, in *Séminaire Bourbaki*, 23ème année (1970/1971), Exp. No. 389. Lecture Notes in Mathematics, vol. 244 (Springer, Berlin, 1971), pp. 123–165
14. J. Dieudonné, Sur les groupes classiques (Hermann, Paris, 1973). Troisième édition revue et corrigée, Publications de l'Institut de Mathématique de l'Université de Strasbourg, VI, Actualités Scientifiques et Industrielles, No. 1040
15. I.V. Dolgachev, Mirror symmetry for lattice polarized $K3$ surfaces. Algebraic geometry, 4. J. Math. Sci. **81**(3), 2599–2630 (1996);
16. C.F. Doran, Picard-Fuchs uniformization and modularity of the mirror map. Comm. Math. Phys. **212**(3), 625–647 (2000)
17. N.D. Elkies, Shimura curve computations, in *Algorithmic Number Theory*, Portland, OR (Springer, Berlin, 1998), pp. 1–47
18. N.D. Elkies, Shimura curve computations via $K3$ surfaces of Néron-Severi rank at least 19, in *Algorithmic Number Theory*. Lecture Notes in Computer Science, vol. 5011 (Springer, Berlin, 2008), pp. 196–211
19. R. Hartshorne, in *Algebraic Geometry*. Graduate Texts in Mathematics, No. 52 (Springer, New York, 1977)
20. K. Hashimoto, N. Murabayashi, Shimura curves as intersections of Humbert surfaces and defining equations of QM-curves of genus two. Tohoku Math. J. (2) **47**(2), 271–296 (1995)
21. S. Herfurtner, Elliptic surfaces with four singular fibers. Math. Ann. **291**, 319–342 (1991)
22. N. Katz, Travaux de Dwork, in *Séminaire Bourbaki*, 24ème année (1971/1972), Exp. No. 409. Lecture Notes in Mathematics, vol. 317 (Springer, Berlin, 1973), pp. 167–200
23. A.W. Knapp, Elliptic curves, in *Mathematical Notes*, vol. 40 (Princeton University Press, Princeton, 1992)

24. K. Kodaira, On compact analytic surfaces: II. Ann. Math. **77**, 563–626 (1962)
25. D.R. Morrison, On $K3$ surfaces with large Picard number. Invent. Math. **75**(1), 105–121 (1984)
26. D. Mumford, On the equations defining abelian varieties, I. Invent. Math. **1**, 287–354 (1966)
27. D. Mumford, On the equations defining abelian varieties, II. Invent. Math. **3**, 75–135 (1967)
28. D. Mumford, On the equations defining abelian varieties, III. Invent. Math. **3**, 215–244 (1967)
29. D. Mumford, *Abelian Varieties* (Oxford University Press, Oxford, 1970)
30. V. Nikulin, Integral symmetric bilinear forms and some of their applications. Izv. Akad. Nauk SSSR **43**, 75–137 (1979)
31. V.V. Nikulin, Finite groups of automorphisms of Kählerian $K3$ surfaces. Trudy Moskov. Mat. Obshch. **38**, 75–137 (1979)
32. K. Oguiso, T. Shioda, The Mordell-Weil lattice of a rational elliptic surface. Comment. Math. Univ. St. Paul. **40**(1), 83–99 (1991)
33. I.I. Piatetski-Shapiro, I.R. Shafarevic, A Torelli theorem for algebraic surfaces of type K3. Izv. Akad. Nauk SSSR **35**, 503–572 (1971)
34. J.P. Serre, *A Course in Arithmetic* (Springer, New York, 1973) [Translated from the French, Graduate Texts in Mathematics, No. 7]
35. G. Shimura, On analytic families of polarized abelian varieties and automorphic functions. Ann. Math. (2) **78**, 149–192 (1963)
36. T. Shioda, On elliptic modular surfaces. J. Math. Soc. Jpn. **4**, 20–59 (1972)
37. T. Shioda, M. Schütt, An interesting elliptic surface over an elliptic curve. Proc. Jpn. Acad. Ser. A Math. Sci. **83**(3), 40–45 (2007)
38. J.P. Smith, Picard-Fuchs differential equations for families of K3 surfaces. Ph.D. thesis, University of Warwick, 2006. URL http://arxiv.org/abs/0705.3658
39. G. van der Geer, On the geometry of a Siegel modular threefold. Math. Ann. **260**, 317–350 (1982)

Numerical Trivial Automorphisms of Enriques Surfaces in Arbitrary Characteristic

Igor V. Dolgachev

To the memory of Torsten Ekedahl

Abstract We extend to arbitrary characteristic some known results on automorphisms of complex Enriques surfaces that act identically on the cohomology or the cohomology modulo torsion.

Key words: Enriques surfaces, Automorphism groups, Positive characteristic

Mathematics Subject Classifications (2010): Primary 14J28; Secondary 14G17, 20F55

1 Introduction

Let S be algebraic surface over an algebraically closed field $\Bbbk$ of characteristic $p \geq 0$. An automorphism σ of S is called *numerically trivial* (resp., *cohomologically trivial*) if it acts trivially on $H^2_{\acute{e}t}(S, \mathbb{Q}_\ell)$ (resp. $H^2_{\acute{e}t}(S, \mathbb{Z}_\ell)$). In the case when S is an Enriques surface, the Chern class homomorphism $c_1 : \mathrm{Pic}(S) \to H^2_{\acute{e}t}(S, \mathbb{Z}_\ell)$ induces an isomorphism $\mathrm{NS}(S) \otimes \mathbb{Z}_\ell \cong H^2_{\acute{e}t}(S, \mathbb{Z}_\ell)$, where $\mathrm{NS}(S)$ is the Néron–Severi group of S isomorphic to the Picard group $\mathrm{Pic}(S)$. Moreover, it is known that the torsion subgroup of $\mathrm{NS}(S)$ is generated by the canonical class K_S. Thus, an automorphism σ is cohomologically (resp. numerically) trivial if and only if it acts identically on $\mathrm{Pic}(S)$ (resp. $\mathrm{Num}(S) = \mathrm{Pic}(S)/(K_S)$). Over the field of complex numbers, the classification of numerically trivial automorphisms can be found in [12, 13]. We have

I.V. Dolgachev (✉)
Department of Mathematics, University of Michigan, 525 E. University Avenue,
Ann Arbor, MI 49109, USA
e-mail: idolga@umich.edu

R. Laza et al. (eds.), *Arithmetic and Geometry of K3 Surfaces and Calabi–Yau Threefolds*,
Fields Institute Communications 67, DOI 10.1007/978-1-4614-6403-7_8,
© Springer Science+Business Media New York 2013

Theorem 1. *Assume* $\Bbbk = \mathbb{C}$. *The group* $\mathrm{Aut}(S)_{ct}$ *of cohomologically trivial automorphisms is cyclic of order* ≤ 2. *The group* $\mathrm{Aut}(S)_{nt}$ *of numerically trivial automorphisms is cyclic of order* 2 *or* 4.

The tools in the loc.cit. are transcendental and use the periods of the K3-covers of Enriques surfaces, so they do not extend to the case of positive characteristic.

Our main result is that Theorem 1 is true in any characteristic.

The author is grateful to S. Kondō, J. Keum and the referee for useful comments to the paper.

2 Generalities

Recall that an Enriques surface S is called *classical* if $K_S \neq 0$. The opposite may happen only if $\mathrm{char}(\Bbbk) = 2$. Enriques surfaces with this property are divided into two classes: μ_2-surfaces or α_2-surfaces. They are distinguished by the property of the action of the Frobenius on $H^2(S, \mathcal{O}_S) \cong \Bbbk$. In the first case, the action is non-trivial, and in the second case it is trivial. They also differ by the structure of their Picard schemes. In the first case it is isomorphic to the group scheme μ_2, in the second case it is isomorphic to the group scheme α_2. Obviously, if S is not classical, then $\mathrm{Aut}(S)_{nt} = \mathrm{Aut}(S)_{ct}$.

It is known that the quadratic lattice $\mathrm{Num}(S)$ of numerical equivalence divisor classes on S is isomorphic to $\mathrm{Pic}(S)/(K_S)$. It is a unimodular even quadratic lattice of rank 10 and signature $(1, 9)$. As such it must be isomorphic to the orthogonal sum $E_{10} = E_8 \oplus U$, where E_8 is the unique negative definite even unimodular lattice of rank 8 and U is a hyperbolic plane over $\mathbb{Z}$. One can realize E_{10} as a primitive sublattice of the standard unimodular odd hyperbolic lattice

$$\mathbb{Z}^{1,10} = \mathbb{Z}\mathbf{e}_0 + \mathbb{Z}\mathbf{e}_1 + \cdots + \mathbb{Z}\mathbf{e}_{10}, \tag{1}$$

where $\mathbf{e}_0^2 = 1, \mathbf{e}_i^2 = -1, i > 0, \mathbf{e}_i \cdot \mathbf{e}_j = 0, \ i \neq j$. The orthogonal complement of the vector

$$\mathbf{k}_{10} = -3\mathbf{e}_0 + \mathbf{e}_1 + \cdots + \mathbf{e}_{10}$$

is isomorphic to the lattice E_{10}.

Let

$$\mathbf{f}_j = -\mathbf{k}_{10} + \mathbf{e}_j, \ j = 1, \ldots, 10.$$

The 10 vectors $\mathbf{f}_j$ satisfy

$$\mathbf{f}_j^2 = 0, \quad \mathbf{f}_i \cdot \mathbf{f}_j = 1, \ i \neq j.$$

Under an isomorphism $E_{10} \to \mathrm{Num}(S)$, their images form an *isotropic sequence* $(f_1, \ldots, f_{10})$, a sequence of 10 isotropic vectors satisfying $f_i \cdot f_j = 1, i \neq j$. An isotropic sequence generates an index 3 sublattice of $\mathrm{Num}(S)$.

A smooth rational curve R on S (a (-2)-*curve*, for brevity) does not move in a linear system and $|R + K_S| = \varnothing$ if $K_S \neq 0$. Thus we can and will identify R with its class $[R]$ in $\mathrm{Num}(S)$. Any (-2)-curve defines a reflection isometry of $\mathrm{Num}(S)$

$$s_R : x \mapsto x + (x \cdot R)R.$$

Any numerical divisor class in $\mathrm{Num}(S)$ of non-negative norm represented by an effective divisor can be transformed by a sequence of reflections s_R into the numerical divisor class of a nef divisor. Any isotropic sequence can be transformed by a sequence of reflections into a *canonical* isotropic sequence, i.e. an isotropic sequence $(f_1, \ldots, f_{10})$ satisfying the following properties

- $f_{k_1}, \ldots, f_{k_c}$ are nef classes for some $1 = k_1 < k_2 < \ldots < k_c \leq 10$.
- $f_j = f_{k_i} + \mathcal{R}_j$, $k_i < j < k_{i+1}$, where $\mathcal{R}_j = R_{i,1} + \cdots + R_{i,s_j}$ is the sum of $s_j = j - k_i$ classes of (-2)-curves with intersection graph of type A_{s_j} such that $f_{k_i} \cdot \mathcal{R}_j = f_j \cdot R_{i,1} = 1$.

Any primitive isotropic numerical nef divisor class f in $\mathrm{Num}(S)$ is the class of nef effective divisors F and $F' \sim F + K_S$. The linear system $|2F| = |2F'|$ is base-point-free and defines a fibration $\phi : S \to \mathbb{P}^1$ whose generic fiber S_η is a regular curve of arithmetic genus one. If $p \neq 2$, S_η is a smooth elliptic curve over the residue field of the generic point η of the base. In this case, ϕ is called an elliptic fibration. The divisors F and F' are *half-fibers* of ϕ, i.e. $2F$ and $2F'$ are fibers of ϕ.

The following result by J.-P. Serre [15] about lifting to characteristic 0 shows that there is nothing new if $p \neq 2$.

Theorem 2. *Let $W(\Bbbk)$ be the ring of Witt vectors with algebraically closed residue field $\Bbbk$, and let X be a smooth projective variety over $\Bbbk$, and let G be a finite automorphism group of X. Assume*

- *$\#G$ is prime to $\mathrm{char}(\Bbbk)$;*
- *$H^2(X, \mathcal{O}_X) = 0$;*
- *$H^2(X, \Theta_X) = 0$, where Θ_X is the tangent sheaf of X.*

Then the pair (X, G) can be lifted to $W(\Bbbk)$, i.e. there exists a smooth projective scheme $\mathcal{X} \to \mathrm{Spec}\ W(\Bbbk)$ with special fiber isomorphic to X and an action of G on $\mathcal{X}$ over $W(\Bbbk)$ such that the induced action of G in X coincides with the action of G on X.

We apply this theorem to the case when $G = \mathrm{Aut}_{nt}(S)$, where S is an Enriques surface over a field $\Bbbk$ of characteristic $p \neq 2$. We will see later that the order of $G = \mathrm{Aut}_{nt}(S)$ is a power of 2, so it is prime to p. We have an isomorphism $H^2(S, \Theta_S) \cong H^0(S, \Omega_S^1(K_S))$. Let $\pi : X \to S$ be the K3-cover. Since the map $\pi^* : H^0(S, \Omega_S^1(K_S)) \to H^0(X, \Omega_X^1) \cong H^0(X, \Theta_X)$ is injective and $H^0(X, \Theta_X) = 0$, we obtain that all conditions in Serre's Theorem are satisfied. Thus, there is nothing new in this case. We can apply the results of [12, 13] to obtain the complete classification of numerically trivial automorphisms. However, we will give here another, purely geometric, proof of Theorem 1 that does not appeal to K3-covers nor does it uses Serre's lifting theorem.

3 Lefschetz Fixed-Point Formula

We will need a Lefschetz fixed-point formula comparing the trace of an automorphism σ of finite order acting on the l-adic cohomology $H^*_{\acute{e}t}(X, \mathbb{Q}_l)$ of a normal projective algebraic surface X with the structure of the subscheme X^σ of fixed points of σ.

The subscheme of fixed points X^σ is defined as the scheme-theoretical intersection of the diagonal with the graph of σ. Let $\mathcal{J}(\sigma)$ be the ideal sheaf of X^σ. If $x \in X^\sigma$, then the stalk $\mathcal{J}(\sigma)_x$ is the ideal in $\mathcal{O}_{X,x}$ generated by elements $a - \sigma^*(a), a \in \mathcal{O}_{X,x}$. Let $\mathrm{Tr}_i(\sigma)$ denote the trace of the linear action of σ on $H^i_{\acute{e}t}(X, \mathbb{Q}_l)$. The following formula was proved in [9], Proposition 3.2:

$$\sum (-1)^i \mathrm{Tr}_i(\sigma) = \chi(X, \mathcal{O}_{X^\sigma}) + \chi(X, \mathcal{J}(\sigma)/\mathcal{J}(\sigma)^2) - \chi(X, \Omega^1_X \otimes \mathcal{O}_{X^\sigma}). \quad (2)$$

If σ is *tame*, i.e. its order is prime to p, then X^σ is reduced and smooth [8], and the Riemann–Roch formula easily implies

$$\mathrm{Lef}(\sigma) := \sum (-1)^i \mathrm{Tr}_i(\sigma) = e(X^\sigma), \quad (3)$$

where $e(X^\sigma)$ is the Euler characteristic of X^σ in étale l-adic cohomology. This is the familiar Lefschetz fixed-point formula from topology.

The interesting case is when σ is *wild*, i.e. its order is divisible by p. We will be interested in application of this formula in the case when σ is of order 2 equal to the characteristic and X is an Enriques surface S.

Let $\pi : S \to Y = S/(\sigma)$ be the quotient map. Consider an $\mathcal{O}_Y$-linear map

$$T = 1 + \sigma : \pi_* \mathcal{O}_S \to \mathcal{O}_Y.$$

Its image is an ideal sheaf $\mathcal{I}_Z$ of a closed subscheme Z of Y and the inverse image of this ideal in $\mathcal{O}_S$ is equal to $\mathcal{J}(\sigma)$.

Theorem 3. *Let S be a classical Enriques surface and let σ be a wild automorphism of S of order 2. Then S^σ is non-empty and is connected.*

As was first observed by J.-P. Serre, the first assertion follows from the Woods Hole Lefschetz fixed-point formula for cohomology with coefficients in a coherent sheaf [7] (we use that $\sum (-1)^i \mathrm{Tr}(g|H^i(S, \mathcal{O}_S)) = 1$ and hence the right-hand side of the formula is not zero).

The proof of the second assertion is a modification of arguments from [5]. In the case of classical Enriques surfaces, the proof can be found in [10].[1]

Proposition 1. *Assume that S^σ consists of one point s_0. Then* $\mathrm{Lef}(\sigma) = 4$.

Proof. Let $\pi : S \to Y = S/\langle\sigma\rangle$ be the quotient morphism and $y = \pi(s_0)$. It follows from [2] that (10) implies that the formal completion of the local ring $\mathcal{O}_{Y,y}$ is a rational double point of type $D_4^{(1)}$ isomorphic to $\Bbbk[[x, y, z]]/(z^2 + xyz + x^2y + xy^2)$ (see [5], Remark 2.6). Moreover, identifying $\hat{\mathcal{O}}_{Y,y}$ with the ring of invariants of $\hat{\mathcal{O}}_{X,x_0} = \Bbbk[[u, v]]$, we have

$$x = u(u + y), \; y = v(v + x), z = xu + yv.$$

This implies that the ideal $\mathcal{J}(\sigma)_{s_0}$ generates the ideal (u^2, v^2) in $\Bbbk[[u, v]]$. Applying (2), we easily obtain

$$\mathrm{Lef}(\sigma) = \dim_\Bbbk \Bbbk[[u, v]]/(u^2, v^2) + \dim_\Bbbk (u^2, v^2)/(u^4, v^4, u^2v^2)$$

$$-2\dim_\Bbbk \Bbbk[[u, v]]/(u^2, v^2) = 4 + 8 - 8 = 4.$$

Since, for any $\sigma \in \mathrm{Aut}_{ct}(S)$, we have $\mathrm{Lef}(\sigma) = 12$, we obtain the following.

Corollary 1. *Let σ be a wild cohomologically trivial automorphism of order 2 of a classical Enriques surface S. Then S^σ is a connected curve.*

Remark 1. After we resolve minimally the singular point of the quotient surface, we obtain an Enriques surface Y' with a singular fiber of type $\tilde{D}_4$. A possible scenario is the following. The automorphism of the surface S is defined by a 2-torsion element of the Mordell–Weil group of the jacobian elliptic fibration. The only fixed point is a cusp of its unique singular fiber of additive type (see [10], Proposition 4.2).

Proposition 2. *Assume that S^σ is a connected curve with $(S^\sigma)_{\mathrm{red}}$ contained in a fiber F of a genus one fibration on S. Then $(S^\sigma)_{\mathrm{red}} = F_{\mathrm{red}}$ or $(S^\sigma)_{\mathrm{red}} = F_{\mathrm{red}} - R$, where R is an irreducible component of F.*

Proof. Since σ fixes each irreducible component of F, and has one fixed point on each component which is not contained in S^σ, the structure of fibers show that there is only one such component.

4 Cohomologically Trivial Automorphisms

Let $\phi : S \to \mathbb{P}^1$ be a genus one fibration defined by a pencil $|2F|$. Let D be an effective divisor on S. We denote by D_η its restriction to the generic fiber S_η. If D is of relative degree d over the base of the fibration, then D_η is an effective divisor

---

[1] The assertion is not true for non-classical Enriques surfaces. The analysis of this case reveals a missing case in [5]: $X^3\sigma$ may consist of an isolated fixed point and a connected curve.

of degree d on S_η. In particular, if D is irreducible, the divisor D_η is a point on S_η of degree d. Since ϕ has a double fiber, the minimal degree of a point on S_η is equal to 2.

Lemma 1. *S admits a genus one fibration* $\phi : S \to \mathbb{P}^1$ *such that* $\sigma \in \mathrm{Aut}_{\mathrm{ct}}(S)$ *leaves invariant all fibers of ϕ and at least 2 (3 if $K_S \neq 0$) points of degree two on* S_η.

Proof. By Theorem 3.4.1 from [3] (we will treat the exceptional case when S is extra E_8-special in characteristic 2 in the last section), one can find a canonical isotropic sequence $(f_1, \ldots, f_{10})$ with nef classes $f_1, f_{k_2}, \ldots, f_{k_c}$ where $c \geq 2$.

Assume $c \geq 3$. Then we have three genus one fibrations $|2F_1|$, $|2F_{k_2}|$, and $|2F_{k_3}|$ defined by f_1, f_{k_2}, f_{k_3}. The restriction of F_{k_2} and F_{k_3} to the general fiber S_η of the genus one fibration defined by the pencil $|2F_1|$ are two degree 2 points. If $K_S \neq 0$, then the half-fibers $F'_{k_2} \in |F_{k_2} + K_S|$ and $F'_{k_3} \in |F_{k_3} + K_S|$ define two more degree two points.

Assume $c = 2$. Let $f_1 = [F_1], f_2 = [F_{k_2}]$. By definition of a canonical isotropic sequence, we have the following graph of irreducible curves

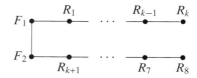

Assume $k \neq 0$. Let $\phi : S \to \mathbb{P}^1$ be a genus one fibration defined by the pencil $|2F_1|$. Then the curves F_2 and R_1 define two points of degree two on S_η. If S is classical, we have the third point defined by a curve $F'_2 \in |F_2 + K_S|$. Since σ is cohomologically trivial, it leaves the half-fibers F_1, F_2, and F'_2 invariant. It also leaves invariant the (-2)-curve R_1. If $k = 0$, we take for ϕ the fibration defined by the pencil $|2F_{k_2}|$ and get the same result.

The next theorem extends the first assertion of Theorem 1 from the Introduction to arbitrary characteristic.

Theorem 4. *The order of* $\mathrm{Aut}_{\mathrm{ct}}(S)$ *is equal to 1 or 2.*

Proof. By the previous lemma, $\mathrm{Aut}_{\mathrm{ct}}(S)$ leaves invariant a genus one fibration and 2 or 3 degree two points on its generic fiber. For any $\sigma \in \mathrm{Aut}_{\mathrm{ct}}(S)$, the automorphism σ^2 acts identically on the residue fields of these points. If $p \neq 2$ (resp. $p = 2$), we obtain that σ, acting on the geometric generic fiber $S_{\bar\eta}$, fixes 6 (resp. 4) points. The known structure of the automorphism group of an elliptic curve over an algebraically closed field of any characteristic (see [16], Appendix A) shows that this is possible only if σ is the identity.

So far, we have shown only that each non-trivial element in $\mathrm{Aut}_{\mathrm{ct}}(S)$ is of order 2. However, the previous argument also shows that any two elements in the group share a common orbit in $S_{\bar\eta}$ of cardinality 2. Again, the known structure of the automorphism group of an elliptic curve shows that this implies that the group is of order 2.

Lemma 2. *Let F be a singular fiber of a genus one fibration on an elliptic surface. Let σ be a non-trivial tame automorphism of order 2 that leaves invariant each irreducible component of F. Then*

$$e(F^\sigma) = e(F). \qquad (4)$$

Remark 2. Formula (4) agrees with the Lefschetz fixed-point formula whose proof in the case of a reducible curve I could not find.

Proof. The following pictures exhibit possible sets of fixed points. Here the star denotes an irreducible component in F^σ, the red line denotes the isolated fixed point that equal to the intersection of two components, the red dot denotes an isolated fixed point which is not the intersection point of two components.

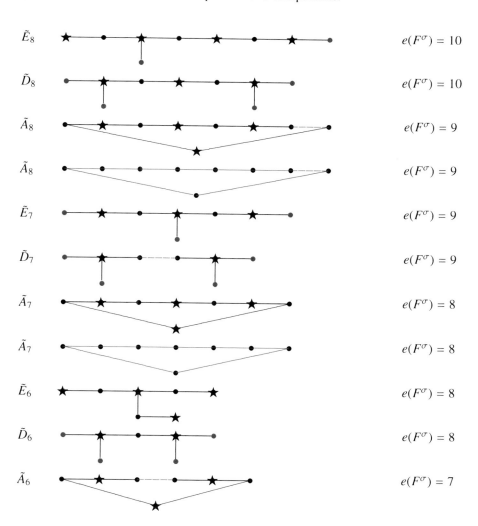

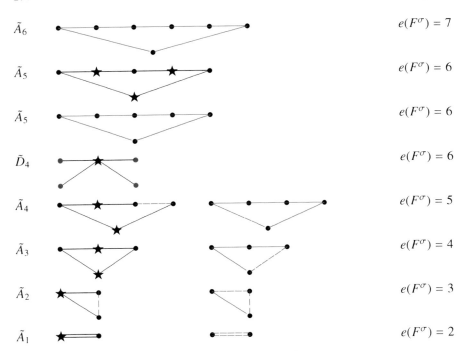

$\tilde{A}_6$ $e(F^\sigma) = 7$

$\tilde{A}_5$ $e(F^\sigma) = 6$

$\tilde{A}_5$ $e(F^\sigma) = 6$

$\tilde{D}_4$ $e(F^\sigma) = 6$

$\tilde{A}_4$ $e(F^\sigma) = 5$

$\tilde{A}_3$ $e(F^\sigma) = 4$

$\tilde{A}_2$ $e(F^\sigma) = 3$

$\tilde{A}_1$ $e(F^\sigma) = 2$

Also, if F is of type $\tilde{A}_2^*(IV)$ (resp. $\tilde{A}_2^*(III)$, resp. $\tilde{A}_1^*(II)$, resp. $\tilde{A}_0(I_1)$, resp. $\tilde{A}^{**}(II)$), we obtain that F^σ consists of four (resp. 3, resp. 2, resp. 1) isolated fixed points. Observe that the case $\tilde{D}_5$ is missing. It does not occur. The equality $e(F^\sigma) = e(F)$ is checked case by case.

Theorem 5. *Assume that S is a classical Enriques surface. A cohomologically trivial automorphism σ leaves invariant any genus one fibration and acts identically on its base.*

Proof. The first assertion is obvious. Suppose σ does not act identically on the base of a genus one fibration $\phi : S \rightarrow \mathbb{P}^1$. By assumption $K_S \neq 0$, hence a genus one fibration has two half-fibers. Since σ is cohomologically trivial, it fixes the two half-fibers F_1 and F_2 of ϕ. Assume $p = 2$. Since σ acts on the base with only one fixed point, we get a contradiction. Assume $p \neq 2$. Then σ has exactly two fixed points on the base. In particular, all non-multiple fibers must be irreducible, and the number of singular non-multiple fibers is even. By Lefschetz fixed-point formula, we get

$$e(S^\sigma) = e(F_1^\sigma) + e(F_2^\sigma) = 12.$$

Since $p \neq 2$, F_i is either smooth or of type $\tilde{A}_{n_i}, i = 1, 2$. Suppose that F_1 and F_2 are singular fibers. Since σ fixes any irreducible component of a fiber, Lemma 2 implies that $e(F_i^\sigma) = e(F_i) = n_i$. So, we obtain that $n_1 + n_2 = 12$. However, F_1, F_2 contribute $n_1 + n_2 - 1$ to the rank of the sublattice of Num(S) generated by components of fibers. The rank of this sublattice is at most 9. This gives us a contradiction. Next we assume that one of the half-fibers is smooth. Then a smooth fiber has four fixed points, hence the other half-fiber must be of type $\tilde{A}_7$. It is easy to see that a smooth relatively minimal model of the quotient $S/(\sigma)$ has singular fibers of type $\tilde{D}_4$ and $\tilde{A}_7$. Since the Euler characteristics of singular fibers add up to 12, this is impossible.

Remark 3. The assertion is probably true in the case when S is not classical. However, I could prove only that S admits at most one genus one fibration on which σ does not act identically on the base. In this case $(S^\sigma)_{\text{red}}$ is equal to the reduced half-fiber.

We also have the converse assertion.

Proposition 3. *Any numerically trivial automorphism σ that acts identically on the base of any genus one fibration is cohomologically trivial.*

This follows from Enriques's Reducibility Lemma [3], Corollary 3.2.2. It asserts that any effective divisor on S is linearly equivalent to a sum of irreducible curves of arithmetic genus one and smooth rational curves. Since each irreducible curve of arithmetic genus one is realized as either a fiber or a half-fiber of a genus one fibration, its class is fixed by σ. Since σ fixes also the class of a smooth rational curve, we obtain that it acts identically on the Picard group.

5 Numerically Trivial Automorphisms

Here we will be interested in the group $\text{Aut}_{\text{nt}}(S)/\text{Aut}_{\text{ct}}(S)$. Since $\text{Num}(S)$ coincides with $\text{Pic}(S)$ for a non-classical Enriques surface S, we may assume that $K_S \neq 0$.

Let $O(\text{NS}(S))$ be the group of automorphisms of the abelian group $\text{NS}(S)$ preserving the intersection product. It follows from the elementary theory of abelian groups that

$$O(\text{NS}(S)) \cong (\mathbb{Z}/2\mathbb{Z})^{10} \rtimes O(\text{Num}(S)).$$

Thus

$$\text{Aut}_{\text{nt}}(S)/\text{Aut}_{\text{ct}}(S) \cong (\mathbb{Z}/2\mathbb{Z})^a. \qquad (5)$$

The following theorem extends the second assertion of Theorem 1 to arbitrary characteristic.

Theorem 6.
$$\text{Aut}_{\text{nt}}(S)/\text{Aut}_{\text{ct}}(S) \cong (\mathbb{Z}/2\mathbb{Z})^a, \ a \leq 1.$$

Proof. Assume first that $p \neq 2$. Let $\sigma \in \text{Aut}_{\text{nt}}(S) \setminus \text{Aut}_{\text{ct}}(S)$. By Proposition 3, there exists a genus one fibration $\phi : S \to \mathbb{P}^1$ such that σ acts non-trivially on its base. Since $p \neq 2$, σ has two fixed points on the base. Let F_1 and F_2 be the fibers over these points. Obviously, σ must leave invariant any reducible fiber, hence all fibers

$F \neq F_1, F_2$ are irreducible. On the other hand, the Lefschetz fixed-point formula shows that one of the fixed fibers must be reducible. Let G be the cyclic group generated by (σ). Assume there is $\sigma' \in \mathrm{Aut}_{nt}(S) \setminus G$. Since $\mathrm{Aut}_{nt}(S)/\mathrm{Aut}_{ct}(S)$ is an elementary 2-group, the actions of σ' and σ on the base of the fibration commute. Thus σ' either switches F_1, F_2 or it leaves them invariant. Since one of the fibers is reducible, σ' must fix both fibers. We may assume that F_1 is reducible. By looking at all possible structure of the locus of fixed points containing in a fiber (see the proof of Lemma 2), we find that σ and σ' (or $\sigma \circ \sigma'$) fixes pointwisely the same set of irreducible components of F_1. Thus $\sigma \circ \sigma'$ (or σ') acts identically on F_1. Since the set of fixed points is smooth, we get a contradiction with the assumption that $\sigma' \neq \sigma$.

Next we deal with the case $p = 2$. Suppose the assertion is not true. Let σ_1, σ_2 be two representatives of non-trivial cosets in $\mathrm{Aut}_{nt}(S)/\mathrm{Aut}_{ct}(S)$. Let ϕ_i be a genus fibration such that σ_i does not act identically on its base. Since, we are in characteristic 2, σ_i has only one fixed point on the base. Let F_i be the unique fiber of ϕ_i fixed by σ_i. Replacing σ_2 with $\sigma_1 \circ \sigma_2$, if needed, we may assume that σ_2 does not act identically on the base of ϕ_1. It follows from Proposition 1 that $S^{\sigma_i} = (F_i)_{\mathrm{red}} - R_i$ for some (-2)-curves R_i. Then $\sigma_3 = \sigma_1 \circ \sigma_2$ acts identically on the bases of ϕ_1 and ϕ_2, and hence contains 2-sections of ϕ_1 and ϕ_2. It is easy to see that they coincide with R_1 and R_2. Now S^{σ_3} contains $(F_1)_{\mathrm{red}} + R_2 = (F_2)_{\mathrm{red}} + R_1$ and this cannot be contained in F_3. This contradiction proves the assertion.

6 Examples

In this section we assume that $p \neq 2$.

Example 1. Let us see that the case when $\mathrm{Aut}_{nt}(S) \neq \mathrm{Aut}_{ct}(S)$ is realized. Consider $X = \mathbb{P}^1 \times \mathbb{P}^1$ with two projections p_1, p_2 onto $\mathbb{P}^1$. Choose two smooth rational curves R and R' of bidegree $(1, 2)$ such that the restriction of p_1 to each of these curves is a finite map of degree two. Assume that R is tangent to R' at two points x_1 and x_2 with tangent directions corresponding to the fibers L_1, L_2 of p_1 passing through these points. Counting parameters, it is easy to see that this can be always achieved. Let x_1', x_2' be the points infinitely near x_1, x_2 corresponding to the tangent directions. Let L_3, L_4 be two fibers of p_1 different from L_1 and L_2. Let

$$R \cap L_3 = \{x_3, x_4\}, \ R \cap L_4 = \{x_5, x_6\}, \ R' \cap L_3 = \{x_3', x_4'\}, \ R' \cap L_4 = \{x_5', x_6'\}.$$

We assume that all the points are distinct. Let $b : X' \to X$ be the blow-up of the points $x_1, \ldots, x_6, x_1', \ldots, x_6'$. Let R_i, R_i' be the corresponding exceptional curves, $\bar{L}_i, \bar{R}, \bar{R}'$ be the proper transforms of L_i, R, R'. We have

$$D = \bar{R} + \bar{R}' + \sum_{i=1}^{4} \bar{L}_i + R_1 + R_2$$

$$\sim 2b^*(3f_1 + 2f_2) - 2\sum_{i=1}^{6}(R_i + R_i') - 4(R_1' + R_2'),$$

where f_i is the divisor class of a fiber of the projection $p_i : X \to \mathbb{P}^1$. Since the divisor class of D is divisible by 2 in the Picard group, we can construct a double cover $\pi : S' \to X'$ branched over D. We have

$$K_{X'} = b^*(-2f_1 - 2f_2) + \sum_{i=1}^{6}(R_i + R_i') + R_1' + R_2',$$

hence

$$K_{S'} = \pi^*(K_{X'} + \frac{1}{2}D) = (b \circ \pi)^*(f_1 - R_1' - R_2').$$

We have $\bar{L}_1^2 = \bar{L}_2^2 = R_1^2 = R_2^2 = -2$, hence $\pi^*(\bar{L}_i) = 2A_i, i = 1, 2$, and $\pi^*(R_i) = 2B_i, i = 1, 2$, where A_1, A_2, B_1, B_2 are (-1)-curves. Also $\bar{R}^2 = \bar{R}'^2 = \bar{L}_3^2 = \bar{L}_4^2 = -4$, hence $\pi^*(\bar{R}) = 2\tilde{R}, \pi^*(\bar{R}') = 2\tilde{R}', \pi^*(\bar{L}_3) = 2\tilde{L}_3, \pi^*(\bar{L}_4) = 2\tilde{L}_4$ where $\tilde{R}, \tilde{R}', \tilde{L}_3, \tilde{L}_4$ are (-2)-curves. The curves $\bar{R}_i = \pi^*(R_i)$, $\bar{R}_i' = \pi^*(R_i'), i = 3, 4, 5, 6$, are (-2)-curves. The preimages of the curves R_1' and R_2' are elliptic curves F_1', F_2'. Let $\alpha : S' \to S$ be the blowing down of the curves A_1, A_2, B_1, B_2. Then the preimage of the fibration $p_1 : X \to \mathbb{P}^1$ on S is an elliptic fibration with double fibers $2F_1, 2F_2$, where $F_i = \alpha(F_i')$. We have $K_S = 2F_1 - F_1 - F_2 = F_1 - F_2$. So, S is an Enriques surface with rational double cover $S \dashrightarrow \mathbb{P}^1 \times \mathbb{P}^1$. The elliptic fibration has two fibers of types $\tilde{D}_4$ over L_3, L_4 and two double fibers over L_1 and L_2.

The following diagram pictures a configuration of curves on S.

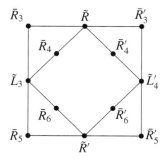

Let us see that the cover automorphism is numerically trivial but not cohomologically trivial (see other treatment of this example in [12]). Consider the pencil of curves of bidegree $(4, 4)$ on X generated by the curve $G = R + R' + L_3 + L_4$ and $2C$, where C is a unique curve of bidegree $(2, 2)$ passing through the points $x_4, x_4', x_6, x_6', x_1, x_1', x_2, x_2'$. These points are the double base points of the pencil. It is easy to see that this pencil defines an elliptic fibration on S with a double fiber of type $\tilde{A}_7$ formed by the curves $\bar{R}_3, \tilde{L}_3, \bar{R}_5, \bar{R}', \bar{R}'_5, \tilde{L}_4, \bar{R}'_3, \tilde{R}$ and the double fiber $2\bar{C}$, where $\bar{C}$ is the preimage of C on S. If $g = 0, f = 0$ are local equations of the curves G and C, the local equation of a general member of the pencil is $g + \mu f^2 = 0$, and the local equation of the double cover $S \dashrightarrow X$ is $g = z^2$. It clear that the pencil splits. By Proposition 3 the automorphism is not cohomologically trivial.

Note that the K3-cover of S has four singular fibers of type $\tilde{D}_4$. It is a Kummer surface of the product of two elliptic curves. This is the first example of a numerically trivial automorphism due to David Lieberman (see [13], Example 1). Over $\mathbb{C}$, a special case of this surface belongs to Kondō's list of complex Enriques surfaces with finite automorphism group [11]. It is a surface of type III. It admits five elliptic fibrations of types

$$\tilde{D}_8, \ \tilde{D}_4 + \tilde{D}_4, \ \tilde{D}_6 + \tilde{A}_1 + \tilde{A}_1, \ \tilde{A}_7 + \tilde{A}_1, \tilde{A}_3 + \tilde{A}_3 + \tilde{A}_1 + \tilde{A}_1.$$

Example 2. Let $X = \mathbb{P}^1 \times \mathbb{P}^1$ be as in the previous example. Let R' be a curve of bidegree $(3, 4)$ on X such that the degree of p_1 restricted to R' is equal to 4. It is a curve of arithmetic genus 6. Choose three fibers of L_1, L_2, L_3 of the first projection and points $x_i \in L_i$ on it no two of which lie on a fiber of the second projection. Let $x'_i \succ x_i$ be the point infinitely near x_i in the tangent directions defined by the fiber L_i. We require that R' has double points at $x_1, x_2, x'_2, x_3, x'_3$ and a simple point at x'_1 (in particular R' has a cusp at x_1 and has tacnodes at x_2, x_3). The dimension of the linear system of curves of bidegree $(3, 4)$ is equal to 19. We need five conditions to have a cusp at x_1 as above, and six conditions for each tacnode. So, we can always find R'.

Consider the double cover $\pi : Y \to X$ branched over $R' + L_1 + L_2 + L_3$. It has a double rational point of type E_8 over x_1 and simple elliptic singularities of degree 2 over x_2, x_3. Let $r : S' \to Y$ be a minimal resolution of singularities. The composition $f' = p_1 \circ r \circ \pi : S' \to \mathbb{P}^1$ is a non-minimal elliptic fibration on S'. It has a fiber F'_1 of type $\tilde{E}_8$ over L_1. The preimage of L_2 (resp. L_2) is the union of an elliptic curve F'_2 (resp. F'_3) and two disjoint (-1)-curves A_2, A'_2 (resp. A_3, A'_3), all taken with multiplicity 2. Let $S' \to S$ be the blow-down of the curves A_2, A'_2, A_3, A'_3. It is easy to check that S is an Enriques surface with a fiber F_1 of type $\tilde{E}_8$ and two half-fibers F_2, F_3, the images of F'_2, F'_3.

The following picture describes the incidence graph of irreducible components of F_1.

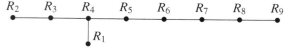

Under the composition of rational maps $\pi : S \dashrightarrow S' \to Y \to X$, the image of the component R_8 is equal to L_1, the image of the component R_9 is the intersection point $x_0 \neq x_1$ of the curves R' and L_1. Let σ be the deck transformation of the cover π (it extends to a biregular automorphism because S is a minimal surface).

Consider a curve C on X of bidegree $(1, 2)$ that passes through the points $x_1, x_2, x'_2, x_3, x'_3$. The dimension of the linear system of curves of bidegree $(1, 2)$

is equal to 5. We have five condition for C that we can satisfy. The proper transform of C on S is a (-2)-curve R_0 which intersects the components R_8 and R_2. We have the following graph which is contained in the incidence graph of (-2)-curves on S:

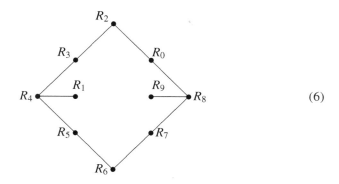

$$(6)$$

One computes the determinant of the intersection matrix $(R_i \cdot R_j))$ and obtains that it is equal to -4. This shows that the curves $R_0, \ldots, R_9$ generate a sublattice of index 2 of the lattice $\mathrm{Num}(S)$. The class of the half-fiber F_2 does not belong to this sublattice, but $2F_2$ belongs to it. This shows that the numerical classes $[F_2], [R_0], \ldots, [R_9]$ generate $\mathrm{Num}(S)$. We also have a section $s : \mathrm{Num}(S) \to \mathrm{Pic}(S)$ of the projection $\mathrm{Pic}(S) \to \mathrm{Num}(S) = \mathrm{Pic}(S)/(K_S)$ defined by sending $[R_i]$ to R_i and $[F_2]$ to F_2. Since the divisor classes R_i and F_2 are σ-invariant, we obtain that $\mathrm{Pic}(S) = K_S \oplus s(\mathrm{Num}(S))$, where the both summands are σ-invariant. This shows that σ acts identically on $\mathrm{Pic}(S)$, and, by definition, belongs to $\mathrm{Aut}_{\mathrm{ct}}(S)$.

Remark 4. In fact, we have proven the following fact. Let S be an Enriques surface such that the incidence graph of (-2)-curves on it contains the subgraph (6). Assume that S admits an involution σ that acts identically on the subgraph and leaves invariant the two half-fibers of the elliptic fibration defined by a subdiagram of type $\tilde{E}_8$. Then $\sigma \in \mathrm{Aut}_{\mathrm{ct}}(S)$. The first example of such a pair (S, σ) was constructed in [4]. The surface has additional (-2)-curves R_1' and R_9' forming the following graph.

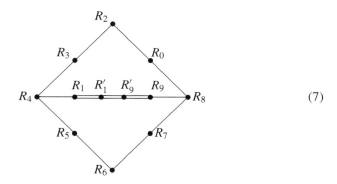

$$(7)$$

All smooth rational curves are accounted in this diagram. The surface has a finite automorphism group isomorphic to the dihedral group D_4 of order 4. It is a surface of type I in Kondō's list. The existence of an Enriques surface containing the diagram (7) was first shown by E. Horikawa [6]. Another construction of pairs (S, σ) as above was given in [13] (the paper has no reference to the paper [4] that had appeared in the previous issue of the same journal).

Observe now that in the diagram (6) the curves $R_0, \ldots, R_7$ form a nef isotropic effective divisor F_0 of type $\tilde{E}_7$. The curve R_9 does not intersect it. This implies that the genus one fibration defined by the pencil $|F_0|$ has a reducible fiber with one of its irreducible components equal to R_9. Since the sum of the Euler characteristics of fibers add up to 12, we obtain that the fibration has a reducible fiber or a half-fiber of type $\tilde{A}_1$. Let R_9' be its another irreducible component. Similarly, we consider the genus one fibration with fiber $R_0, R_2, R_3, R_5, \ldots, R_9$ of type $\tilde{E}_7$. It has another fiber (or a half-fiber) of type $\tilde{A}_1$ formed by R_1 and some other (-2)-curve R_1'.

So any surface S containing the configuration of curves from (6) must contain a configuration of curves described by the following diagram.

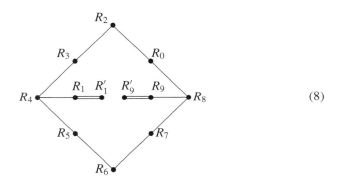

$$(8)$$

Note that our surfaces S depend on two parameters. A general surface from the family is different from the Horikawa surface. For a general S, the curve R_9' originates from a rational curve Q of bidegree $(1, 2)$ on X which passes through the points x_0 and x_2, x_2', x_3, x_3'. It intersects R_8 with multiplicity 1. The curve R_1' originates from a rational curve Q' of bidegree $(5, 6)$ of arithmetic genus 20 which has a 4-tuple point at x_1 and two double points infinitely near x_1. It also has four triple points at x_2, x_2', x_3, x_3'. It intersects R_4 with multiplicity 1. In the special case when one of the points x_2 or x_3 is contained in a curve $(0, 1)$ Q_0 of bidegree $(0, 1)$ containing x_0, the curve Q becomes reducible, its component Q_0 defines the curve R_9' which does not intersect R_8. Moreover, if there exists a curve Q_0' of bidegree $(2, 3)$ which has multiplicity 2 at x_2, multiplicity 1 at x_2', x_3, x_4, and has a cusp at x_1 intersecting R' at this point with multiplicity 7, then Q_0' will define a curve R_1' that does not intersect R_4. The two curves R_1' and R_9' will intersect at two points on the half-fibers of the elliptic fibration $|2F|$. This gives us the Horikawa surface.

Example 3. Let $\phi : X \to \mathbb{P}^1$ be a rational elliptic surface with reducible fiber F_1 of type IV and F_2 of type $I_0^* = \tilde{D}_4$ and one double fiber $2F$. The existence of such surface follows from the existence of a rational elliptic surface with a section with the same types of reducible fibers. Consider the double cover $X' \to X$ branched over F_1 and the union of the components of F_2 of multiplicity 1. It is easy to see that X' is birationally equivalent to an Enriques surface with a fiber of type $\tilde{E}_6$ over F_1 and a smooth elliptic curve over F_2. The locus of fixed points of the deck transformation σ consists of four components of the fiber of type $\tilde{E}_6$ and four isolated points on the smooth fiber. Thus the Lefschetz number is equal to 12 and σ is numerically trivial. Over $\mathbb{C}$, this is Example 1 from [12] which was overlooked in [13]. A special case of this example can be found in [11]. It is realized on a surface of type V in Kondō's list of Enriques surfaces with finite automorphism group.

7 Extra Special Enriques Surfaces

In this section we will give examples of cohomologically trivial automorphisms which appear only in characteristic 2.

An Enriques surface is called *extra special* if there exists a root basis B in $\mathrm{Num}(S)$ of cardinality ≤ 11 that consists of the classes of (-2)-curves such that the reflection subgroup G generated by B is of finite index in the orthogonal group of $\mathrm{Num}(S)$. Such a root basis was called *crystallographic* in [3]. By a theorem of E. Vinberg [17], this is possible if and only if the Coxeter diagram of the Coxeter group (G, B) has the property that each affine subdiagram is contained in an affine diagram, not necessary connected, of maximal possible rank (in our case equal to 8).

One can easily classify extra special Enriques surfaces. They are of the following three kinds.

An extra $\tilde{E}_8$-special surface with the crystallographic basis of (-2)-curves described by the following diagram:

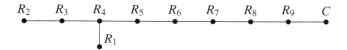

It has a genus one fibration with a half-fiber of type $\tilde{E}_8$ with irreducible components $R_1, \ldots, R_9$ and a smooth rational 2-section C.

An extra $\tilde{D}_8$-special surface with the crystallographic basis of (-2)-curves described by the following diagrams:

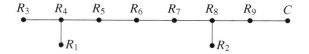

It has a genus one fibration with a half-fiber of type $\tilde{D}_8$ with irreducible components $R_1, \ldots, R_9$ and a smooth rational 2-section C.

An extra $\tilde{E}_7 + \tilde{A}_1$-special Enriques surface with the crystallographic basis of (-2)-
curves described by the following diagram:

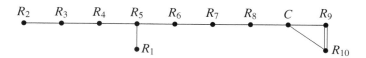

or

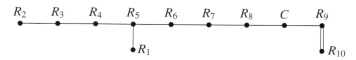

It has a genus one fibration with a half-fiber of type $\tilde{E}_7$ with irreducible components $R_1, \ldots, R_8$ and a fiber or a half-fiber of type $\tilde{A}_1$ with irreducible components
R_9, R_{10}. The curve C is a smooth rational 2-section.

It follows from the theory of reflection groups that the fundamental polyhedron
for the Coxeter group (G, B) in the nine-dimensional Lobachevsky space is of finite
volume. Its vertices at infinity correspond to maximal affine subdiagrams and also
to G-orbits of primitive isotropic vectors in $\mathrm{Num}(S)$. The root basis B is a maximal
crystallographic basis, so the set of the curves R_i, C is equal to the set of all (-2)-
curves on the surface and the set of nef primitive isotropic vectors in $\mathrm{Num}(S)$ is
equal to the set of affine subdiagrams of maximal rank. Thus the number of genus
one fibrations on S is finite and coincides with the set of affine subdiagrams of
rank 8.

It is not known whether an extra $\tilde{D}_8$-special Enriques surface exists. However,
examples of extra-special surfaces of types $\tilde{E}_8$, or $\tilde{E}_7 + \tilde{A}_1$ are given in [14]. They
are either classical Enriques surfaces or α_2-surfaces. The surfaces are constructed as
separable double covers of a rational surface, so they always admit an automorphism
σ of order 2.

Suppose that S is an extra $\tilde{E}_8$-special surface. Then we find that the surface has
only one genus one fibration. It is clear that σ acts identically on the diagram. This
allows one to define a σ-invariant splitting $\mathrm{Pic}(S) \cong \mathrm{Num}(S) \oplus K_S$. It implies that
σ is cohomologically trivial.

Assume that S is extra $\tilde{E}_7 \oplus \tilde{A}_1$-special surface. The surface has a unique genus
one fibration with a half-fiber of type $\tilde{A}_7$. It also has two fibrations in the first case
and one fibration in the second case with a fiber of type $\tilde{E}_8$. It implies that the
curves $R_1, \cdots, R_8, C$ are fixed under σ. It follows from Salomonsson's construction
that $\sigma(R_9) = R_{10}$ in the first case. In the second case, R_9 and R_{10} are σ-invariant on
any extra special $\tilde{E}_7 \oplus \tilde{A}_1$-surface.

References

1. M. Artin, On isolated rational singularities of surfaces. Am. J. Math. **88**, 129–136 (1966)
2. M. Artin, Wildly ramified Z/2 actions in dimension two. Proc. Am. Math. Soc. **52**, 60–64 (1975)
3. F. Cossec, I. Dolgachev, *Enriques Surfaces. I*. Progress in Mathematics, vol. 76 (Birkhäuser, Boston, 1989)
4. I. Dolgachev, On automorphisms of Enriques surfaces. Invent. Math. **76**, 163–177 (1984)
5. I. Dolgachev, J. Keum, Wild p-cyclic actions on K3-surfaces. J. Algebr. Geom. **10**, 101–131 (2001)
6. E. Horikawa, On the periods of Enriques surfaces. II. Math. Ann. **235**, 217–246 (1978)
7. L. Illusie, in *Formule de Lefschetz, par A. Grothendieck*. Cohomologie l-adique et fonctions L. Séminaire de Géometrie Algébrique du Bois-Marie 1965–1966 (SGA 5). Edité par Luc Illusie. Lecture Notes in Mathematics, vol. 589 (Springer, Berlin, 1977), pp. 73–137
8. B. Iversen, A fixed point formula for action of tori on algebraic varieties. Invent. Math. **16**, 229–236 (1972)
9. K. Kato, S. Saito, T. Saito, Artin characters for algebraic surfaces. Am. J. Math. **110**, 49–75 (1988)
10. J. Keum, Wild p-cyclic actions on smooth projective surfaces with $p_g = q = 0$. J. Algebra **244**, 45–58 (2001)
11. S. Kondō, Enriques surfaces with finite automorphism groups. Jpn. J. Math. (N.S.) **12**, 191–282 (1986)
12. S. Mukai, Numerically trivial involutions of Kummer type of an Enriques surface. Kyoto J. Math. **50**, 889–902 (2010)
13. S. Mukai, Y. Namikawa, Automorphisms of Enriques surfaces which act trivially on the cohomology groups. Invent. Math. **77**, 383–397 (1984)
14. P. Salomonsson, Equations for some very special Enriques surfaces in characteristic two [math.AG/0309210]
15. J.-P. Serre, *Le groupe de Cremona et ses sous-groupes finis*. Séminaire Bourbaki. Volume 2008/2009. Exposés 997–1011. Astérisque No. 332 (2010), Exp. No. 1000, pp. 75–100
16. J. Silverman, *The Arithmetic of Elliptic Curves*. Graduate Texts in Mathematics, vol. 106 (Springer, New York, 1992)
17. E. Vinberg, in *Some Arithmetical Discrete Groups in Lobachevsky Spaces*. Discrete Subgroups of Lie Groups and Applications to Moduli, International Colloquium, Bombay, 1973 (Oxford University Press, Bombay, 1975), pp. 323–348

Picard–Fuchs Equations of Special One-Parameter Families of Invertible Polynomials

Swantje Gährs

Abstract In this article we calculate the Picard–Fuchs equation of hypersurfaces defined by certain one-parameter families associated to invertible polynomials. For this we deduce the Picard–Fuchs equation from the GKZ system. As consequences of our work and facts from the literature, we show a relation between the Picard–Fuchs equation, the Poincaré series and the monodromy in the space of period integrals.

Key words: Picard–Fuchs equation, Invertible polynomials, Griffiths–Dwork method, GKZ systems, Poincaré series, Monodromy

Mathematics Subject Classifications (2010): Primary 14J33; Secondary 32G20, 32S40, 34M35, 14J32

1 Introduction

In this article we investigate the Picard–Fuchs equation of certain one-parameter families of Calabi–Yau varieties. Calabi–Yau varieties have been studied in much detail, especially in Mirror Symmetry. Much of the early interest in this field focused on Calabi–Yau varieties arising as hypersurfaces in toric varieties. This is mostly due to Batyrev [3], who showed that for hypersurfaces in toric varieties duality in the sense of Mirror Symmetry can be reduced to polar duality between polytopes of toric varieties. This was the starting point for many achievements in Mirror Symmetry of Calabi–Yau varieties. The work of Batyrev, however, does not cover the families that we consider in this article. In particular, Batyrev requires that the ambient space is Gorenstein. This implies that every weight divides the sum of

S. Gährs (✉)
Institut für algebraische Geometrie, Leibniz Universität Hannover,
Welfengarten 1, 30167 Hannover, Germany
e-mail: swantjegaehrs@gmx.de

R. Laza et al. (eds.), *Arithmetic and Geometry of K3 Surfaces and Calabi–Yau Threefolds*,   285
Fields Institute Communications 67, DOI 10.1007/978-1-4614-6403-7_9,
© Springer Science+Business Media New York 2013

all weights. The only overlap between Batyrev's and our work are the polynomials of Brieskorn–Pham type.

The hypersurfaces we investigate in this article are defined by invertible polynomials. These are weighted homogeneous polynomials $g(x_1, \ldots, x_n) \in \mathbb{C}[x_1, \ldots, x_n]$, which are a sum of exactly n monomials, such that the weights $q_1, \ldots, q_n$ of the variables $x_1, \ldots, x_n$ are unique up to a constant and the affine hypersurface defined by the polynomial has an isolated singularity at 0. The class of invertible polynomials includes all polynomials of Brieskorn–Pham type, but is much bigger. These polynomials were already studied by Berglund and Hübsch [5], who showed that a mirror manifold is related to a dual polynomial. For an invertible polynomial $g(x_1, \ldots, x_n) = \sum_{j=1}^{n} \prod_{i=1}^{n} x_i^{E_{ij}}$ the transpose or dual polynomial $g^t(x_1, \ldots, x_n)$ is defined by transposing the exponent matrix $E = (E_{ij})_{i,j}$ of the original polynomial, so $g^t(x_1, \ldots, x_n) = \sum_{j=1}^{n} \prod_{i=1}^{n} x_i^{E_{ji}}$. If the polynomial is of Brieskorn–Pham type then the polynomial is in the above sense self-dual. This work was made precise by Krawitz et al. (cf. [24, 25]), where an isomorphism is given between the FJRW-ring of the polynomial (cf. [17]) and a quotient of the Milnor ring of the dual polynomial. In addition Chiodo and Ruan [8] have made progress by stating the so-called Landau–Ginzburg/Calabi–Yau correspondence for invertible polynomials. Among other things this includes the statement that the Chen–Ruan orbifold cohomology of the mirror partners interchange. Recently, Borisov [6] developed a theory combining his work with Batyrev on toric varieties [4] in mirror symmetry and the work of Krawitz on invertible polynomials in mirror symmetry [24].

In this article we analyse the Picard–Fuchs equations of a special one-parameter family of hypersurfaces. The Picard–Fuchs equation is a differential equation that is satisfied by the periods of the family, i.e. the integrals of a form over a basis of cycles. These differential equations have been studied by many people and this led to several aspects of mirror symmetry. For example, Morrison [27] used the Picard–Fuchs equations of hypersurfaces to calculate the mirror map and Yukawa couplings for mirror manifolds. In [7] Chen, Yang and Yui study the monodromy for Picard–Fuchs equations of Calabi–Yau threefolds in terms of monodromy groups. These give two potential applications of the results of this article to further research.

We consider a special one dimensional deformation of an invertible polynomial and calculate the Picard–Fuchs equation for this family. More precisely we start with an invertible polynomial $g(x_1, \ldots, x_n)$, such that the weights $q_1, \ldots, q_n$ of g add up to the degree d of g. This is called the Calabi–Yau condition, because this condition implies that the canonical bundle of the hypersurface $\{g(x_1, \ldots, x_n) = 0\} \subset \mathbb{P}(q_1, \ldots, q_n)$ is trivial (cf. [13]). The special one-parameter family we are dealing with is given by

$$f(x_1, \ldots, x_n) := g(x_1, \ldots, x_n) + s \prod_{i=1}^{n} x_i,$$

where s is a parameter. In her paper [30] Noriko Yui already investigated some of these special one-parameter families, e.g. for the Fermat quintic, and mentioned the general case as an interesting problem. It is possible to calculate the Picard–Fuchs equation for this one-parameter family by using the Griffiths–Dwork method, which provides an algorithm to calculate the Picard–Fuchs equation (cf. [10]). Unfortunately, this method of calculation can be quite computationally expensive. But it is also possible to compute the GKZ system satisfied by the periods and in this way we can prove a general formula for the Picard–Fuchs equation. For the one-parameter family f defined above the Picard–Fuchs equation is given by

$$0 = \prod_{i=1}^{n} \widehat{q_i}^{\widehat{q_i}} s^{\widehat{d}} \prod_{i=1}^{n} \prod_{j=0}^{\widehat{q_i}-1} (\delta + \frac{j \cdot \widehat{d}}{\widehat{q_i}}) \prod_{\ell \in I} (\delta + \ell)^{-1} - (-\widehat{d})^{\widehat{d}} \prod_{j=0}^{\widehat{d}-1} (\delta - j) \prod_{\ell \in I} (\delta - \ell)^{-1},$$

where $\delta = s\frac{\partial}{\partial s}$, $\widehat{q_1}, \ldots, \widehat{q_n}$ are the weights of the dual polynomial g^t, $\widehat{d}$ is the degree of g^t, and $I = \{0, \ldots, \widehat{d} - 1\} \cap \bigcup_{i=1}^{n} \left\{0, \frac{\widehat{d}}{\widehat{q_i}}, \frac{2\widehat{d}}{\widehat{q_i}}, \ldots, \frac{(\widehat{q_i}-1)\widehat{d}}{\widehat{q_i}}\right\}$.

One interesting observation is that the Picard–Fuchs equation is determined only by the data given by the dual polynomial, namely the dual weights and the dual degree. As pointed out to us by Stienstra, this Picard–Fuchs equation was already obtained in a work by Corti and Golyshev [9] in the context of local systems and Landau–Ginzburg pencils. We will show the relation to our work and in addition we are able to show for certain values of the parameter a 1-1 correspondence between the roots of the Picard–Fuchs equation of f, the Poincaré series of the dual polynomial g^t and the monodromy in the solution space of the Picard–Fuchs equation.

One important class that will be studied in detail in this article is the case of the 14 exceptional unimodal hypersurface singularities that are part of Arnold's strange duality [1]. The duality between these singularities was known before mirror symmetry, but was shown to fit into the language of mirror symmetry (cf. [14]). We will not only calculate the Picard–Fuchs equation here, but also investigate the structure of the cohomology which is used in the calculations for the Picard–Fuchs equation.

2 Preliminaries on Invertible Polynomials

We start this section by defining invertible polynomials and proving some properties we need later.

Definition 1. Let

$$g(\underline{x}) = \sum_{j=1}^{m} c_j \prod_{i=1}^{n} x_i^{E_{ij}} \in \mathbb{C}[\underline{x}]$$

be a quasihomogeneous polynomial with weights $q_1, \ldots, q_n \in \mathbb{Z}$, where $E_{ij} \in \mathbb{N}$, $\underline{x} = (x_1, \ldots, x_n)$ and $c_j \neq 0$ for all j. Then $g(\underline{x})$ is an invertible polynomial if the following conditions hold:

(i) $n = m$,
(ii) $\det(E_{ij}) \neq 0$ and
(iii) $V(g) \subset \mathbb{P}(q_1, \ldots, q_n)$ is quasi-smooth.

From now on we assume that the coefficients c_j are all equal to 1. This can always be achieved by an easy coordinate transformation. Additionally we require the weights to be reduced, i.e. $\gcd(q_1, \ldots, q_n) = 1$. In this way the weights are unique.

Remark 1. We want to make some remarks for the article:

- Some authors call the polynomial $g(\underline{x})$ invertible if the first two conditions are satisfied, and a non-degenerate invertible polynomial if $g(\underline{x})$ satisfies all three conditions.
- The weights are also defined by the smallest numbers $q_1, \ldots, q_n \in \mathbb{N}$ and $d \in \mathbb{N}$ satisfying the equation

$$\begin{pmatrix} E_{11} & \cdots & E_{1n} \\ \vdots & & \vdots \\ E_{n1} & \cdots & E_{nn} \end{pmatrix} \begin{pmatrix} q_1 \\ \vdots \\ q_n \end{pmatrix} = \begin{pmatrix} d \\ \vdots \\ d \end{pmatrix}$$

or concisely $E \cdot \underline{q} = \underline{d}$. We call E the *exponent matrix*.

M. Kreuzer and H. Skarke showed that the polynomials which are invertible are a composition of only two types.

Theorem 1. (Kreuzer and Skarke [26]) *Every invertible polynomial is a sum of polynomials with distinct variables of the following two types*

$$\text{loop: } x_1^{k_1} x_2 + x_2^{k_2} x_3 + \cdots + x_{m-1}^{k_{m-1}} x_m + x_m^{k_m} x_1 \qquad \text{for } m \geq 2$$

$$\text{chain: } x_1^{k_1} x_2 + x_2^{k_2} x_3 + \cdots + x_{m-1}^{k_{m-1}} x_m + x_m^{k_m} \qquad \text{for } m \geq 1$$

Example 1. We want to list two very famous classes of examples here:

(i) A polynomial is of Brieskorn–Pham type if it is of the form $g(\underline{x}) = \sum_{i=1}^{n} x_i^{k_i}$ with $k_i \in \mathbb{N}$. In this case the polynomial is always invertible and the exponent matrix is a diagonal matrix with the exponents k_i on the diagonal. It follows that $q_i = \frac{\mathrm{lcm}(k_1, \ldots, k_n)}{k_i}$ and $d = \mathrm{lcm}(k_1, \ldots, k_n)$.

(ii) For the 14 exceptional unimodal singularities, invertible polynomials can be chosen. Table 1 lists their names, invertible polynomials, reduced weights and degrees in the first four columns. In the last column the dual singularity due to Arnold [1] is listed. In the next definition we will see how this duality fits into the context of invertible polynomials which also explains the rest of the table. The example of Arnold's strange duality will be studied in detail in Sect. 3.4.

In their paper [5] P. Berglund and T. Hübsch proposed a way to define dual pairs of invertible polynomials by transposing the exponent matrix.

Definition 2. If $g(\underline{x}) = \sum_{j=1}^{n} \prod_{i=1}^{n} x_i^{E_{ij}}$ is an invertible polynomial then the Berglund–Hübsch transpose is given by

$$g^t(\underline{x}) = \sum_{j=1}^{n} \prod_{i=1}^{n} x_i^{E_{ji}}.$$

Example 2. In their paper [16] Ebeling and Takahashi showed that one can choose invertible polynomials defining the 14 exceptional unimodal hypersurface singularities in such a way that the dual singularities are defined by the transposed polynomials. Here dual singularities are given if the Dolgachev number of the one singularity is equal to the Gabrielov number of the other singularity and vice versa.

Table 1: Arnold's strange duality

Name	$g(x, y, z)$	Weights	Deg	$g^t(x, y, z)$	Dual
E_{12}	$x^7 + y^3 + z^2$	(6,14,21)	42	$x^7 + y^3 + z^2$	E_{12}
E_{13}	$x^5 y + y^3 + z^2$	(4,10,15)	30	$x^5 + xy^3 + z^2$	Z_{11}
Z_{11}	$x^5 + xy^3 + z^2$	(6,8,15)	30	$x^5 y + y^3 + z^2$	E_{13}
E_{14}	$x^4 z + y^3 + z^2$	(3,8,12)	24	$x^4 + y^3 + xz^2$	Q_{10}
Q_{10}	$x^4 + y^3 + xz^2$	(6,8,9)	24	$x^4 z + y^3 + z^2$	E_{14}
Z_{12}	$x^4 y + xy^3 + z^2$	(4,6,11)	22	$x^4 y + xy^3 + z^2$	Z_{12}
W_{12}	$x^5 + y^2 z + z^2$	(4,5,10)	20	$x^5 + y^2 + yz^2$	W_{12}
Z_{13}	$x^3 z + xy^3 + z^2$	(3,5,9)	18	$x^3 y + y^3 + xz^2$	Q_{11}
Q_{11}	$x^3 y + y^3 + xz^2$	(4,6,7)	18	$x^3 z + xy^3 + z^2$	Z_{13}
W_{13}	$x^4 y + y^2 z + z^2$	(3,4,8)	16	$x^4 + xy^2 + yz^2$	S_{11}
S_{11}	$x^4 + y^2 z + xz^2$	(4,5,6)	16	$x^4 z + y^2 + yz^2$	W_{13}
Q_{12}	$x^3 z + y^3 + xz^2$	(3,5,6)	15	$x^3 z + y^3 + xz^2$	Q_{12}
S_{12}	$x^3 y + y^2 z + xz^2$	(3,4,5)	13	$x^3 z + xy^2 + yz^2$	S_{12}
U_{12}	$x^4 + y^2 z + yz^2$	(3,4,4)	12	$x^4 + y^2 z + yz^2$	U_{12}

Remark 2. Notice that taking the transpose does not change the type of the polynomial. The exponent matrix is a direct sum of matrices, where every summand belongs to a polynomial of chain or loop type. Therefore we can transpose every chain and loop separately:

$$g(\underline{x}) = x_1^{k_1} x_2 + \cdots + x_{m-1}^{k_{m-1}} x_m + x_m^{k_m} x_1$$
$$\Rightarrow g^t(\underline{x}) = x_m x_1^{k_1} + x_1 x_2^{k_2} \cdots + x_{m-1} x_m^{k_m}$$

and

$$g(\underline{x}) = x_1^{k_1} x_2 + \cdots + x_{m-1}^{k_{m-1}} x_m + x_m^{k_m}$$
$$\Rightarrow g^t(\underline{x}) = x_1^{k_1} + x_1 x_2^{k_2} \cdots + x_{m-1} x_m^{k_m}.$$

Definition 3. Let $g(\underline{x})$ be an invertible polynomial. We set $f(\underline{x})$ to be the one-parameter family associated to $g(\underline{x})$ via

$$f(\underline{x}) = g(\underline{x}) + s \prod_{i=1}^{n} x_i,$$

where s denotes the parameter.

This one-parameter family $f(\underline{x})$ will be one of the main objects of interest in this article. Because we still want this family to be quasihomogeneous, we require that the weights of $g(\underline{x})$ add up to the degree of $g(\underline{x})$. In [13] I. Dolgachev showed that this is the condition for a quasihomogeneous polynomial to define a Calabi–Yau hypersurface.

Proposition 1. ([13]) *Let $g(\underline{x})$ be a quasihomogeneous polynomial with weights $q_1, \ldots, q_n$. Then $g(\underline{x})$ defines a hypersurface in the weighted projective space that is Calabi–Yau if*

$$\sum_{i=1}^{n} q_i = d = \deg g(\underline{x}).$$

Lemma 1. *If the Calabi–Yau condition holds for the weights of an invertible polynomial then it also holds for the weights of the transposed polynomial.*

Notation 2. *For an invertible polynomial $g(\underline{x})$ we denote the reduced weights by $q_1, \ldots, q_n$ and $\deg g = d$. For the dual polynomial g^t the weights are $\widehat{q_1}, \ldots, \widehat{q_n}$ and $\deg g^t = \widehat{d}$. The diagonal entries of the exponent matrix E are $k_1, \ldots, k_n$. Notice that these are the same for g and g^t.*

Proof. The Calabi–Yau condition is equivalent to the condition

$$\det \begin{pmatrix} 1 & & \\ \vdots & E & \\ 1 & \cdots & 1 \end{pmatrix} = 0.$$

This is due to the fact that the weights are unique up to scaling and therefore the linear relations between the rows of the above matrix have to be given by multiplying with the vector $(-d, q_1, \ldots, q_n)^t$. Now it is obvious that if the above condition holds for E it also holds for E^t.

3 The Picard–Fuchs Equation for Invertible Polynomials and Consequences

In this section we focus on the Picard–Fuchs equation of the one-parameter family $f(\underline{x})$ and discuss some consequences. In the first subsection of this section we calculate the GKZ system and in the second subsection we see how this proves Theorem 6, which yields the Picard–Fuchs equation for the one-parameter family $f(\underline{x})$. In Sect. 3.2 we will also see how this relates to a paper by Corti and Golyshev [9], where the same differential equation appears. This is also the starting point for Sect. 3.3, where we concentrate on relations between the cohomology of the hypersurface defined by the one-parameter family $f(\underline{x})$ and the cohomology of the solution space of the Picard–Fuchs equation. In Sect. 3.4 we will discuss the results in an important class of examples given by Arnold's strange duality. This was also the starting point of the research done in this article. Finally, in the last section we cover the relation between the zero sets of the Picard–Fuchs equation of f for special choices of the parameter, the Poincaré series of the dual polynomial g^t and the monodromy in the solution space of the Picard–Fuchs equation. We will use the same notation as before, but we want to recall it again here and use it throughout this section without further notice.

Notation 3. *Let $g(\underline{x}) = g(x_1, x_2, \ldots, x_n) := \sum_{i=1}^{n} \prod_{i=1}^{n} x_i^{E_{ij}}$ be an invertible polynomial with reduced weights $q_1, q_2, \ldots, q_n$ and $\deg g = d$ for which the Calabi–Yau condition, $d = \sum_{i=1}^{n} q_i$, holds. The diagonal entries of the exponent matrix $E = (E_{ij})_{i,j}$ are defined as $k_1, \ldots, k_n$. We denote by $g^t(\underline{x})$ the transposed polynomial of g, the dual reduced weights belonging to g^t are denoted by $\widehat{q}_1, \widehat{q}_2, \ldots, \widehat{q}_n$ and the degree by $\deg g^t = \widehat{d}$.*
The invertible polynomial consists of loops and chains of arbitrary length. For a variable x_i we always take x_{i-1} and x_{i+1} to be the neighbouring variables in the loop or chain. The indices are without further notice taken modulo the length of the loop or chain.
We always denote by $f(x_1, \ldots, x_n)$ the one-parameter family with parameter s defined by $f(\underline{x}) = f(x_1, \ldots, x_n) := g(x_1, \ldots, x_n) + s \prod_{i=1}^{n} x_i$.

3.1 The GKZ System for Invertible Polynomials

This section is devoted to GKZ systems. We will give a short introduction to GKZ systems and do the calculations for invertible polynomials afterwards.

3.1.1 Introduction to GKZ Systems

In this first part we want to give a short introduction to GKZ systems as far as we need it. The theory on GKZ systems is much larger than the part we present here. Good references for an introduction as well as an overview on several aspects of GKZ systems are the article by Stienstra [29], which has a large part on solutions of GKZ systems, the book by Cox and Katz [10], which among other things embeds GKZ systems in a bigger context, and the article of Hosono [23], which focuses on the case of toric varieties. The theory of GKZ systems was originally established by a series of articles of Gelfand, Kapranov and Zelevinsky [19–22] as a generalisation of hypergeometric differential equations.

Notation 4. *Let $\mathscr{A} \subset \mathbb{Z}^n$ be a finite subset which generates $\mathbb{Z}^n$ as an abelian group and for which there exists a group homomorphism $h : \mathbb{Z}^n \to \mathbb{Z}$ such that $h(\mathcal{A}) = 1$,*

i.e. $\mathcal{A}$ lies in a hypersurface. Let $\gamma \in \mathbb{C}^n$ be an arbitrary vector.
Let $|\mathcal{A}| = N$, then $\mathbb{L} := \{(l_1, \ldots, l_N) \in \mathbb{Z}^N : l_1 a_1 + \cdots + l_N a_N = 0, a_i \in \mathcal{A}\}$ denotes the lattice of linear relations among $\mathcal{A}$. Because of $\mathcal{A}$ lying in a hypersurface, $\sum l_i = 0$ holds for $(l_1, \ldots, l_N) \in \mathbb{L}$.

Remark 3. We will calculate the GKZ system for the one-parameter family $f(\underline{x})$ later. Keep in mind that for these calculations $\mathcal{A}$ will be the set of all exponent vectors of our one-parameter family. The reasons for this will also become clear later.

Definition 4. The GKZ system (sometimes also called $\mathcal{A}$ system) for $\mathcal{A}$ and γ is a system of differential equations for functions Φ of N variables $v_1, \ldots, v_N$ given by

$$\prod_{l_i > 0} \left(\frac{\partial}{\partial v_i}\right)^{l_i} \Phi = \prod_{l_i < 0} \left(\frac{\partial}{\partial v_i}\right)^{-l_i} \Phi \text{ for every } l \in \mathbb{L} \text{ and} \tag{1}$$

$$\sum_{i=1}^{N} a_{ij} v_i \frac{\partial \Phi}{\partial v_i} = \gamma_j \Phi \text{ for all } j = 1, \ldots, k+1 \text{ and } (a_{i1}, \ldots, a_{ik+1}) \in \mathcal{A}. \tag{2}$$

The above definition gives a system of partial differential equations.

3.1.2 Calculation of the GKZ System for Invertible Polynomials

We will now start calculating the GKZ system for the one-parameter family $f(\underline{x}) = g(\underline{x}) + s \prod_i x_i$, where $g(\underline{x})$ is an invertible polynomial. The notation in this section is the same as before and can be found in 3 and 4. In addition we will define some extra notation:

Notation 5. *We define $\underline{e}_i = (e_{i1}, \ldots, e_{in})$ for $i = 1, \ldots, n$ to be the rows of the exponent matrix E. Then we can write $g(\underline{x})$ as $g(\underline{x}) = \sum_{i=1}^{n} \underline{x}^{\underline{e}_i}$, where $\underline{x}^{\underline{e}_i} = \prod_{j=1}^{n} x_j^{e_{ij}}$. Now we define a general $(n+1)$-parameter family $f_{\underline{v}}(\underline{x}) = f_{v_1, \ldots, v_n}(\underline{x}) = \sum_{i=1}^{n} v_i \underline{x}^{\underline{e}_i} + s \underline{x}^{(1, \ldots, 1)}$ with parameters $v_1, \ldots, v_n$ and s. So in the previously used notation we have $N = n + 1$ and we set $v_{n+1} := s$. In this way the notation is consistent with the previous sections, because we have that*

$$f_{1, \ldots, 1}(\underline{x}) = \sum_{i=1}^{n} \underline{x}^{\underline{e}_i} + s \underline{x}^{(1, \ldots, 1)} = g(\underline{x}) + s \prod_{i=1}^{n} x_i = f(\underline{x}).$$

For the set $\mathcal{A} = \{\underline{e}_1^t, \ldots, \underline{e}_n^t, (1, \ldots, 1)^t\} \subset \mathbb{Z}^n$ and $\gamma = (-1, \ldots, -1)^t$ we will now start calculating the GKZ system. The reason for the choice of γ will become clear when we look at the solutions of the GKZ system. For the first equation (1) we need to calculate the lattice of linear relations $\mathbb{L}$ among the vectors in $\mathcal{A}$. If we define A to be the matrix with columns $\underline{e}_1^t, \ldots, \underline{e}_n^t, (1, \ldots, 1)^t$, then A is an $n \times (n+1)$-matrix and $\mathbb{L}$ is one-dimensional. We know that

$$
A \cdot \begin{pmatrix} \widehat{q_1} \\ \vdots \\ \widehat{q_n} \\ -\widehat{d} \end{pmatrix} = E^t \cdot \begin{pmatrix} \widehat{q_1} \\ \vdots \\ \widehat{q_n} \end{pmatrix} - \begin{pmatrix} \widehat{d} \\ \vdots \\ \widehat{d} \end{pmatrix} = \begin{pmatrix} 0 \\ \vdots \\ 0 \end{pmatrix}
$$

and therefore $\mathbb{L} = \langle \widehat{q_1}, \ldots, \widehat{q_n}, -\widehat{d})^t \rangle$. Now we are able to write down equation (1) for this lattice $\mathbb{L}$:

$$
\left(\frac{\partial}{\partial s} \right)^{\widehat{d}} \Phi = \left(\frac{\partial}{\partial v_1} \right)^{\widehat{q_1}} \cdots \left(\frac{\partial}{\partial v_n} \right)^{\widehat{q_n}} \Phi. \tag{3}
$$

In the end we want to compare the GKZ system to the Picard–Fuchs equation from Theorem 6. To do this we will write the GKZ system with the differential operators $\delta = s \frac{\partial}{\partial s}$ and $\delta_i = v_i \frac{\partial}{\partial v_i}$ for $i = 1, \ldots, n$ by inserting $s^{-1} \delta = \frac{\partial}{\partial s}$ and $v_i^{-1} \delta_i = \frac{\partial}{\partial v_i}$.

$$
\left(s^{-1} \delta \right)^{\widehat{d}} \Phi = \left(v_1^{-1} \delta_1 \right)^{\widehat{q_1}} \cdots \left(v_n^{-1} \delta_n \right)^{\widehat{q_n}} \Phi.
$$

Now we move s^{-1} and v_i^{-1} to the front and the product rule gives us an easy way to interchange the differential operators δ, δ_i with the variables s, v_i:

$$
\begin{aligned}
\delta s^p &= s^p (\delta + p) \quad \text{for } p \in \mathbb{Z} \text{ and} \\
\delta_i v_i^p &= v_i^p (\delta_i + p) \quad \text{for } i = 1, \ldots, n \text{ and } p \in \mathbb{Z}.
\end{aligned} \tag{4}
$$

Using these equations we can move every s and every v_i to the front of the equation:

$$
(s^{-1} \delta)^{\widehat{d}} = s^{-1} \delta s^{-1} \delta \cdots s^{-1} \underbrace{\delta s^{-1}}_{s^{-1}(\delta - 1)} \delta
$$

$$
= \ldots
$$

$$
= s^{-\widehat{d}} (\delta - (\widehat{d} - 1)) \cdots (\delta - 1) \delta
$$

and in the same way we get

$$
(v_i^{-1} \delta_i)^{\widehat{q_i}} = v_i^{-1} \delta_i v_i^{-1} \delta_i \cdots v_i^{-1} \underbrace{\delta v_i^{-1}}_{v_i^{-1}(\delta_i - 1)} \delta_i
$$

$$
= \ldots
$$

$$
= v_i^{-\widehat{q_i}} (\delta_i - (\widehat{q_i} - 1)) \cdots (\delta_i - 1) \delta_i.
$$

Putting this together the first equation of the GKZ system is given by

$$
s^{-\widehat{d}} (\delta - (\widehat{d} - 1)) \cdots (\delta - 1) \delta \Phi = \prod_{i=1}^{n} v_i^{-\widehat{q_i}} (\delta_i - (\widehat{q_i} - 1)) \cdots (\delta_i - 1) \delta_i \Phi. \tag{5}
$$

We will work with this equation later on and calculate the second part (2) of the GKZ system next. The second system of equations of the GKZ system is given by putting $\gamma = (-1, \ldots, -1)^t$ in (2):

$$
A \cdot \begin{pmatrix} v_1 \frac{\partial}{\partial v_1} \\ \vdots \\ v_n \frac{\partial}{\partial v_n} \\ s \frac{\partial}{\partial s} \end{pmatrix} \Phi = A \cdot \begin{pmatrix} \delta_1 \\ \vdots \\ \delta_n \\ \delta \end{pmatrix} \Phi = E^t \begin{pmatrix} \delta_1 \\ \vdots \\ \delta_n \end{pmatrix} \Phi + \begin{pmatrix} 1 \\ \vdots \\ 1 \end{pmatrix} \delta\Phi = \begin{pmatrix} -1 \\ \vdots \\ -1 \end{pmatrix} \Phi
\tag{6}
$$

Before we do any further calculations, we focus on solutions of the GKZ system. There is a whole theory on solutions of GKZ systems which, for example, is explained in [29]. We however, do not need the full strength of this, because to compare the GKZ system to the Picard–Fuchs equation in Theorem 6, it is enough to know that the form $\omega = \frac{s\Omega_0}{f(\underline{x})}$, where $\Omega_0 = \sum_{i=1}^n (-1)^i q_i x_j dx_1 \wedge \cdots \wedge \widehat{dx_i} \wedge \cdots \wedge dx_n$, is a solution of the GKZ system shown in (5) and (6). This is the goal, but we will start with a slightly different solution in the next lemma.

Lemma 2. *The form* $\Phi = \frac{\Omega_0}{f_{\underline{v}}(\underline{x})}$ *is a solution for the above GKZ system.*

Proof. We will calculate the differentials for $\Phi = \frac{\Omega_0}{f_{\underline{v}}(\underline{x})}$ and see that (3) and (6) hold. For (3) we need the partial derivatives with respect to s and v_i for $i = 1, \ldots, n$. They are easy to calculate:

$$
\left(\frac{\partial}{\partial s}\right)^{\widehat{d}} \Phi = (-1)^{\widehat{d}} \widehat{d}! \left(\prod x_i\right)^{\widehat{d}} \frac{\Omega_0}{\left(f_{\underline{v}}(\underline{x})\right)^{\widehat{d}+1}}
$$

$$
\frac{\partial}{\partial v_i} \Phi = -\underline{x}^{\underline{e}_i} \frac{\Omega_0}{\left(f_{\underline{v}}(\underline{x})\right)^2}
$$

$$
\prod_{i=1}^n \left(\frac{\partial}{\partial v_i}\right)^{\widehat{q}_i} \Phi = (-1)^{\sum \widehat{q}_i} \left(\sum \widehat{q}_i\right)! \underline{x}^{\sum \widehat{q}_i \underline{e}_i} \frac{\Omega_0}{\left(f_{\underline{v}}(\underline{x})\right)^{1+\sum \widehat{q}_i}}.
$$

Because of the Calabi–Yau condition we have $\sum \widehat{q}_i = \widehat{d}$ and from the definition of the dual weights and degree we get $\sum \widehat{q}_i \underline{e}_i = E^t \cdot (\widehat{q}_1, \ldots, \widehat{q}_n)^t = (\widehat{d}, \ldots, \widehat{d})^t$. Therefore we have

$$
\left(\frac{\partial}{\partial s}\right)^{\widehat{d}} \Phi = (-1)^{\widehat{d}} \widehat{d}! \left(\prod x_i\right)^{\widehat{d}} \frac{\Omega_0}{\left(f_{\underline{v}}(\underline{x})\right)^{1+\widehat{d}}}
$$

$$
= (-1)^{\sum \widehat{q}_i} \left(\sum \widehat{q}_i\right)! \underline{x}^{\sum \widehat{q}_i \underline{e}_i} \frac{\Omega_0}{\left(f_{\underline{v}}(\underline{x})\right)^{1+\sum \widehat{q}_i}}
$$

$$
= \prod_{i=1}^n \left(\frac{\partial}{\partial v_i}\right)^{\widehat{q}_i} \Phi.
$$

This proves that Φ is a solution of (3). Now we check the second equation, where we need $\delta\Phi$ and $\delta_i\Phi$ for $i = 1, \ldots, n$, because the system of equations is given by

$$\begin{pmatrix} 1 \\ \vdots \\ 1 \end{pmatrix} \delta\Phi + E^t \begin{pmatrix} \delta_1 \\ \vdots \\ \delta_n \end{pmatrix} \Phi + \begin{pmatrix} 1 \\ \vdots \\ 1 \end{pmatrix} \Phi = \begin{pmatrix} 0 \\ \vdots \\ 0 \end{pmatrix} \Phi.$$

So for every $j = 1, \ldots, n$ we have the following equation:

$$\delta\Phi + \sum_{i=1}^n e_{ij}\delta_i\Phi + \Phi = -s\underline{x}^{(1,\ldots,1)} \frac{\Omega_0}{\left(f_{\underline{v}}(\underline{x})\right)^2} + \sum_{i=1}^n e_{ij}(-v_i\underline{x}^{\underline{e_i}}) \frac{\Omega_0}{\left(f_{\underline{v}}(\underline{x})\right)^2} + \frac{\Omega_0}{f_{\underline{v}}(\underline{x})}$$

$$= -\frac{\left(s\underline{x}^{(1,\ldots,1)} + \sum_{i=1}^n e_{ij}v_i\underline{x}^{\underline{e_i}}\right)\Omega_0}{\left(f_{\underline{v}}\right)^2(\underline{x})} + \frac{\Omega_0}{f_{\underline{v}}(\underline{x})}$$

$$= -\frac{x_j\frac{\partial}{\partial x_j}f_{\underline{v}}(\underline{x})\Omega_0}{\left(f_{\underline{v}}(\underline{x})\right)^2} + \frac{\Omega_0}{f_{\underline{v}}(\underline{x})}$$

$$= 0,$$

where the last expression is an exact form due to the Griffiths formula and is therefore zero.

As mentioned before Φ is not the solution we want to have. A solution that would fit our purposes would be $\omega_{\underline{v}} = \frac{s\Omega_0}{f_{\underline{v}}(\underline{x})}$, because $\omega_{1,\ldots,1} = \frac{s\Omega_0}{f(\underline{x})} = \omega$. So we insert $\Phi = s^{-1}\omega_{\underline{v}}$ in (5) and (6). So (5) leads to

$$s^{-\widehat{d}}(\delta - (\widehat{d} - 1)) \cdots (\delta - 1)\delta s^{-1}\omega_{\underline{v}}$$

$$= \prod_{i=1}^n v_i^{-\widehat{q_i}}(\delta_i - (\widehat{q_i} - 1)) \cdots (\delta_i - 1)\delta_i s^{-1}\omega_{\underline{v}}.$$

We can use (4) as before to move the variable s to the front and get the following equation.

$$s^{-\widehat{d}}(\delta - \widehat{d}) \cdots (\delta - 1)\omega_{\underline{v}} = \prod_{i=1}^n v_i^{-\widehat{q_i}}(\delta_i - (\widehat{q_i} - 1)) \cdots (\delta_i - 1)\delta_i\omega_{\underline{v}}. \qquad (7)$$

By putting $\Phi = s^{-1}\omega_{\underline{v}}$ in (6) and using (4) again we get

$$\begin{pmatrix} 1 \\ \vdots \\ 1 \end{pmatrix} \delta\omega_{\underline{v}} + E^t \begin{pmatrix} \delta_1 \\ \vdots \\ \delta_n \end{pmatrix} \omega_{\underline{v}} = \begin{pmatrix} 0 \\ \vdots \\ 0 \end{pmatrix} \omega_{\underline{v}}.$$

Solving this equation for $(\delta_1, \ldots, \delta_n)^t$ gives

$$\begin{pmatrix} \delta_1 \\ \vdots \\ \delta_n \end{pmatrix} = -(E^t)^{-1} \begin{pmatrix} 1 \\ \vdots \\ 1 \end{pmatrix} \delta = \begin{pmatrix} \frac{\widehat{q_1}}{d} \\ \vdots \\ \frac{\widehat{q_n}}{d} \end{pmatrix} \delta.$$

In other words we can write each of the differential operators $\delta_1, \ldots, \delta_n$ in terms of δ. For all $i = 1, \ldots, n$ we have

$$\delta_i = -\frac{\widehat{q_i}}{d}\delta.$$

We can use this equation to write (7) as an ordinary differential equation with differential operator δ:

$$s^{-\widehat{d}}(\delta - \widehat{d}) \cdots (\delta - 1)\omega_{\underline{v}}$$

$$= \prod_{i=1}^{n} v_i^{-\widehat{q_i}}(\delta_i - (\widehat{q_i} - 1)) \cdots (\delta_i - 1)\delta_i \omega_{\underline{v}}$$

$$= \prod_{i=1}^{n} \left(-\frac{\widehat{q_i}}{d}v_i^{-1}\right)^{\widehat{q_i}} \left(\delta + \frac{(\widehat{q_i} - 1)\widehat{d}}{\widehat{q_i}}\right) \cdots \left(\delta + \frac{\widehat{d}}{\widehat{q_i}}\right)\delta\omega_{\underline{v}}.$$

Now we set $v_i = 1$, which brings us back to our one-parameter family $f(\underline{x})$. Because the solutions of the differential equation before are given by $\omega_{\underline{v}}$, we get a differential equation for $\omega = \frac{s\Omega_0}{f(\underline{x})}$. So our final expression is given by

$$s^{-\widehat{d}}(\delta - \widehat{d}) \cdots (\delta - 1)\omega = \prod_{i=1}^{n} \left(-\frac{\widehat{q_i}}{\widehat{d}}\right)^{\widehat{q_i}} \left(\delta + \frac{(\widehat{q_i} - 1)\widehat{d}}{\widehat{q_i}}\right) \cdots \left(\delta + \frac{\widehat{d}}{\widehat{q_i}}\right)\delta\omega.$$

or to have the same appearance as in Theorem 6:

$$0 = s^{\widehat{d}} \prod_{i=1}^{n} (\widehat{q_i})^{\widehat{q_i}} \delta \left(\delta + \frac{\widehat{d}}{\widehat{q_i}}\right) \cdots \left(\delta + \frac{(\widehat{q_i} - 1)\widehat{d}}{\widehat{q_i}}\right)\omega$$

$$- (-\widehat{d})^{-\widehat{d}}(\delta - 1) \cdots (\delta - \widehat{d})\omega. \quad (8)$$

3.2 The Picard–Fuchs Equation

Now we want to state the Picard–Fuchs equation of $f(\underline{x})$, which should divide the above GKZ system. If we look at examples such as those in Sect. 3.4, we can also conjecture what the Picard–Fuchs equation looks like. We use the GKZ system that we calculated in the last section to confirm that this is true.

Theorem 6. *Let $g(x_1, \ldots, x_n)$ be an invertible polynomial with weighted degree* $\deg g = d$ *and reduced weights* $q_1, \ldots, q_n$ *for which the Calabi–Yau condition,* $d = \sum q_i$, *holds. Let* $g^t(x_1, \ldots, x_n)$ *be the transposed polynomial with reduced weights* $\widehat{q}_1, \ldots, \widehat{q}_n$ *and degree* $\deg g^t = \widehat{d}$. *Then the Picard–Fuchs equation for the one-parameter family* $f(x_1, \ldots, x_n) = g(x_1, \ldots, x_n) + s \prod x_i$ *is given by*

$$0 = \prod_{i=1}^{n} \widehat{q}_i^{\,\widehat{q}_i} s^{\widehat{d}} \prod_{i=1}^{n} \prod_{j=0}^{\widehat{q}_i - 1} (\delta + \frac{j \cdot \widehat{d}}{\widehat{q}_i}) \prod_{\ell \in I} (\delta + \ell)^{-1} - (-\widehat{d})^{\widehat{d}} \prod_{j=0}^{\widehat{d}-1} (\delta - j) \prod_{\ell \in I} (\delta - \ell)^{-1},$$

where $I = \{0, \ldots, \widehat{d} - 1\} \cap \bigcup_{i=1}^{n} \left\{ 0, \frac{\widehat{d}}{\widehat{q}_i}, \frac{2\widehat{d}}{\widehat{q}_i}, \ldots, \frac{(\widehat{q}_i - 1)\widehat{d}}{\widehat{q}_i} \right\}.$

Proof. From Lemma 2 we know that $\omega = \frac{s\Omega}{f(\underline{x})}$ is a solution of (8). It follows that all period integrals are solutions of (8) and therefore the Picard–Fuchs equation divides

$$0 = s^{\widehat{d}} \prod_{i=1}^{n} (\widehat{q}_i)^{\widehat{q}_i} \delta \left(\delta + \frac{\widehat{d}}{\widehat{q}_i} \right) \cdots \left(\delta + \frac{(\widehat{q}_i - 1)\widehat{d}}{\widehat{q}_i} \right) \omega$$
$$- (-\widehat{d})^{-\widehat{d}} (\delta - 1) \cdots (\delta - \widehat{d}) \omega.$$

Now one can look for common factors in the summand until the equation is irreducible. The other possibility is to show what the order of the Picard–Fuchs equation has to be. This was done in an extended version of this article [18]. There it is shown that the order of the Picard–Fuchs equation of $f(\underline{x})$ is given by

$$u = \widehat{d} - \left| \{0, 1, \ldots, \widehat{d} - 1\} \cap \bigcup_{i=1}^{n} \left\{ 0, \frac{\widehat{d}}{\widehat{q}_i}, \ldots, \frac{(\widehat{q}_i - 1)\widehat{d}}{\widehat{q}_i} \right\} \right|.$$

So now we know there are $\widehat{d} - u$ common factors in the summands of (8). If we multiply (8) by $s^{-\widehat{d}}$ and use the commutation relations (4) to pass it through the differential operators we get

$$0 = \prod_{i=1}^{n} (\widehat{q}_i)^{\widehat{q}_i} \delta \left(\delta + \frac{\widehat{d}}{\widehat{q}_i} \right) \cdots \left(\delta + \frac{(\widehat{q}_i - 1)\widehat{d}}{\widehat{q}_i} \right) \omega$$
$$- (-\widehat{d})^{-\widehat{d}} (\delta + (\widehat{d} - 1)) \cdots (\delta + 1) \delta s^{-\widehat{d}} \omega.$$

Now it is easy to see that every linear factor $\delta + j$ with $j \in \{0, 1, \ldots, \widehat{d} - 1\} \cap \bigcup_{i=1}^{n} \{0, \frac{\widehat{d}}{\widehat{q}_i}, \ldots, \frac{(\widehat{q}_i - 1)\widehat{d}}{\widehat{q}_i}\}$ is in both summands and can therefore be deleted. This leads us to the equation

$$0 = \prod_{i=1}^{n} \widehat{q}_i^{\,\widehat{q}_i} s^{\widehat{d}} \prod_{i=1}^{n} \prod_{j=0}^{\widehat{q}_i - 1} (\delta + \frac{j \cdot \widehat{d}}{\widehat{q}_i}) \prod_{\ell \in I} (\delta + \ell)^{-1} - (-\widehat{d})^{\widehat{d}} \prod_{j=0}^{\widehat{d}-1} (\delta - j) \prod_{\ell \in I} (\delta - \ell)^{-1}$$

where $I = \{0, \ldots, \widehat{d} - 1\} \cap \bigcup_{i=1}^{n} \left\{0, \frac{\widehat{d}}{\widehat{q}_i}, \frac{2\widehat{d}}{\widehat{q}_i}, \ldots, \frac{(\widehat{q}_i - 1)\widehat{d}}{\widehat{q}_i}\right\}$.

Finally, this equation is divisible by the Picard–Fuchs equation, because ω is still a solution of this differential equation, and it is irreducible, because there are no common factors or equivalently, due to Theorem 2.8 in [18]. Therefore this is the Picard–Fuchs equation of the family of hypersurfaces defined by $f(\underline{x})$.

We give another class of examples here, which are the simple elliptic singularities. There are only three examples and their Picard–Fuchs equation is known.

Example 3. In the following table we can see three polynomials that define the simple elliptic singularities, their weights, the degree and the Picard–Fuchs equation, which can easily be calculated with Theorem 6.

Table 2: Simple elliptic singularities and their Picard–Fuchs equations

Name	$g(x, y, z)$	Deg	Weights	Picard–Fuchs equation
$\tilde{E}_6$	$x^3 + y^3 + z^3$	3	$(1, 1, 1)$	$s^3\delta^2 + 3^3(\delta - 1)(\delta - 2)$
$\tilde{E}_7$	$x^4 + y^4 + z^2$	4	$(1, 1, 2)$	$s^4\delta^2 - 4^3(\delta - 1)(\delta - 3)$
$\tilde{E}_8$	$x^6 + y^3 + z^2$	6	$(1, 2, 3)$	$s^6\delta^2 - 2 \cdot 6^3(\delta - 1)(\delta - 5)$

A similar result, but approached by different methods, can be found in a paper by Corti and Golyshev [9]. In this paper the differential equation that they look at is the same as our Picard–Fuchs equation, but they start with a local system, which is given in the following way:

$$Y = \begin{cases} \prod_{i=1}^{n} y_i^{w_i} = \lambda \\ \sum_{i=1}^{n} y_i = 1 \end{cases} \subset (\mathbb{C}^*)^n \times \mathbb{C}^* \tag{9}$$

If we insert $y_i = -s^{-1}\underline{x}^{\underline{e}_i - (1,\ldots,1)}$ and $w_i = \widehat{q}_i$, then we get that Y consists of the following two equations:

$$\lambda = \prod_{i=1}^{n} y_i^{w_i} = \prod_{i=1}^{n} \left(-s^{-1}\underline{x}^{\underline{e}_i - (1,\ldots,1)}\right)^{\widehat{q}_i} = \left((-s)^{-\sum \widehat{q}_i}\right)\underline{x}^{\sum \widehat{q}_i\underline{e}_i - (\sum \widehat{q}_i, \ldots, \sum \widehat{q}_i)}$$

$$= (-s)^{-\widehat{d}}\underline{x}^{(\widehat{d},\ldots,\widehat{d}) - (\widehat{d},\ldots,\widehat{d})} = (-s)^{-\widehat{d}}$$

$$1 = \sum_{i=1}^{n} y_i = \sum_{i=1}^{n} \left(-s^{-1}\underline{x}^{\underline{e}_i - (1,\ldots,1)}\right) = -s^{-1}\underline{x}^{-(1,\ldots,1)} \sum_{i=1}^{n} \underline{x}^{\underline{e}_i}.$$

So, from the first equation we get $(-s)^{-\widehat{d}} = \lambda$ and the second equation can easily be rewritten as

$$0 = \sum_{i=1}^{n} \underline{x}^{\underline{e}_i} + s\underline{x}^{(1,\ldots,1)} = f(\underline{x}).$$

This shows the direct connection to our hypersurface $V(f)$. It is very easy to write the Picard–Fuchs equation with differential operator $\mathcal{D} = \lambda\frac{\partial}{\partial\lambda}$, because the relation between $\mathcal{D}$ and δ is just given by

$$\delta = s\frac{\partial}{\partial s} = -\widehat{d}(-s)^{-\widehat{d}}\frac{\partial}{\partial(-s)^{-\widehat{d}}} = -\widehat{d}\lambda\frac{\partial}{\partial\lambda} = -\widehat{d}\mathcal{D}.$$

So in terms of $\mathcal{D}$ the Picard–Fuchs equation is given by

$$0 = \prod_{i=1}^{n}\widehat{q}_i^{\widehat{q}_i}\, s^{\widehat{d}} \prod_{i=1}^{n}\prod_{j=0}^{\widehat{q}_i-1}(\delta + \frac{j\cdot\widehat{d}}{\widehat{q}_i})\prod_{\ell\in I}(\delta+\ell)^{-1} - (-\widehat{d})^{\widehat{d}}\prod_{j=0}^{\widehat{d}-1}(\delta - j)\prod_{\ell\in I}(\delta-\ell)^{-1}$$

$$= \prod_{i=1}^{n}\widehat{q}_i^{\widehat{q}_i} \prod_{i=1}^{n}\prod_{j=0}^{\widehat{q}_i-1}(\mathcal{D} - \frac{j}{\widehat{q}_i})\prod_{\ell\in I}(\mathcal{D}-\frac{\ell}{\widehat{d}})^{-1} - \widehat{d}^{\widehat{d}}\lambda\prod_{j=0}^{\widehat{d}-1}(\mathcal{D}+\frac{j}{\widehat{d}})\prod_{\ell\in I}(\mathcal{D}+\frac{\ell}{\widehat{d}})^{-1} \quad (10)$$

which agrees with formula (1) in [9].

In Theorem 1.1 of the article [9] it is stated that the solutions of the Picard–Fuchs equation come from the local system (9) and in Conjecture 1.4 and Proposition 1.5 the Hodge numbers for the solution space are given. This brings us to the next section where we will investigate this in detail.

3.3 Statements on the Cohomology of the Solution Space

We want to relate already known statements to the work we have done so far. First we continue the last section. We will relate the results done in [18], which is an extension of the article, to work of Corti and Golyshev [9]. In their paper there is a result that calculates the Hodge numbers of the solution space of the Picard–Fuchs equation. We will state their result in a form which is compatible with our setting.

Proposition 2. ([9] Conjecture 1.4 and Proposition 1.5) *Consider the sets*

$$A := \bigsqcup_{i=1}^{n}\{0, \frac{\widehat{d}}{\widehat{q}_i}, 2\frac{\widehat{d}}{\widehat{q}_i}, \ldots, (\widehat{q}_i - 1)\frac{\widehat{d}}{\widehat{q}_i}\}\ and$$
$$D := \{0, 1, 2, \ldots, \widehat{d} - 1\}.$$

Set $\{\alpha_1, \ldots, \alpha_u\} := A \setminus (A \cap D)$ *with* $\alpha_i \le \alpha_{i+1}$ *for all i and* $\{\beta_1, \ldots, \beta_u\} := D \setminus (A \cap D)$ *with* $\beta_i < \beta_{i+1}$ *for all i. Notice that* $\{\alpha_1, \ldots, \alpha_u\}$ *is not a set with distinct elements, because entries will appear multiple times. Now consider the differential equation (Sect. 3.2), which is with the above notation given by*

$$s^{\widehat{d}}\prod_{i=1}^{n}\widehat{q}_i^{\widehat{q}_i}\prod_{i=1}^{u}(\delta + \alpha_i) - (-\widehat{d})^{\widehat{d}}\prod_{i=1}^{u}(\delta - \beta_i) = 0.$$

Now define the following function:

$$p(k) := |\{j \mid \alpha_j < \beta_k\}| - (k-1) \quad for \quad k = 1, \ldots, u$$

and let $p_+ := \max\{p(k)\}$ and $p_- := \min\{p(k)\}$.
Then the local system of solutions of the ordinary differential equation above supports a real polarised variation of Hodge structure of weight $p_+ - p_-$ and Hodge numbers

$$h^{j-p_-,p_+-j} = |p^{-1}(j)|.$$

Remark 4. One can show with the calculations in [18] that the following numbers coincide:

- $p_+ = p(1) = n - 1$
- $p_- = p(u) = 1$
- $\sum_{j=1}^{n-1} h^{j-1,n-1-j} = \sum_{j=1}^{n-1} |p^{-1}(j)| = u$
- $p(k) + p(u - k + 1) = n$ for all $1 \leq k \leq u$ and therefore $h^{i-1,n-i-1} = h^{n-i-1,i-1}$.

In [18] we are able to make the relation between u and the above Hodge numbers even more precise. In order state the relation we define $H_{special} \subseteq H^{n-2}(V(f))$ to be the part of the cohomology of $V(f(\underline{x}))$ that is spanned by the forms $\omega, \delta\omega, \delta^2\omega, \ldots$. It is shown in [18] Theorem 2.8 that the dimension of $H_{special}$ is u. If B is a basis of $H_{special}$, denote by $u_i := |B \cap H^{n-i-2,i}(V(f))|$. Then u_i is the number of basis elements where the monomial in the numerator has degree $i \cdot d$. Now it is proven in [18] that these numbers coincide with the Hodge numbers above.

Proposition 3. *The Hodge numbers of the solution space of the Picard–Fuchs equation of f are in one-to-one correspondence with the basis elements of the part of the cohomology spanned by the derivatives of ω:*

$$u_{i-1} = h^{i-1,n-i-1} = h^{i-p_-,p_+-i}.$$

In [9] one can also find a more detailed description of the Hodge numbers that appear here. This mainly relies on the work of Danilov [11] on Deligne–Hodge numbers and Newton polyhedra.

Remark 5. The Hodge numbers $h^{i-1,n-i-1} = u_i$ that appear in [18] as well as in [9] are the Deligne–Hodge numbers of the cohomology with compact support of a hypersurface defined by a Laurent polynomial with Newton polyhedron Δ, where $\Delta = \left\langle \left(\frac{\widehat{q_1}}{d}, \ldots, \frac{\widehat{q_n}}{d}\right), (1, 0 \ldots, 0), \ldots, (0, \ldots, 0, 1) \right\rangle$. In particular from this viewpoint the u_i are Deligne–Hodge numbers of a toric variety with polytope Δ in the lattice $\mathbb{Z}\left(\frac{\widehat{q_1}}{d}, \ldots, \frac{\widehat{q_n}}{d}\right) + \mathbb{Z}^n$.

3.4 The Case of Arnold's Strange Duality

In this section we will show the results and some details for the 14 exceptional unimodal hypersurfaces singularities.

Table 3: Compactifications of the 14 exceptional unimodal hypersurface singularities

Name	$g(w, x, y, z)$	Deg	Weights	Dual
E_{12}	$w^{42} + x^7 + y^3 + z^2$	42	$(1,6,14,21)$	E_{12}
E_{13}	$w^{30} + x^5 y + y^3 + z^2$	30	$(1,4,10,15)$	Z_{11}
Z_{11}	$w^{30} + x^5 + xy^3 + z^2$	30	$(1,6,8,15)$	E_{13}
E_{14}	$w^{24} + x^4 z + y^3 + z^2$	24	$(1,3,8,12)$	Q_{10}
Q_{10}	$w^{24} + x^4 + y^3 + xz^2$	24	$(1,6,8,9)$	E_{14}
Z_{12}	$w^{22} + x^4 y + xy^3 + z^2$	22	$(1,4,6,11)$	Z_{12}
W_{12}	$w^{20} + x^5 + y^2 z + z^2$	20	$(1,4,5,10)$	W_{12}
Z_{13}	$w^{18} + x^3 z + xy^3 + z^2$	18	$(1,3,5,9)$	Q_{11}
Q_{11}	$w^{18} + x^3 y + y^3 + xz^2$	18	$(1,4,6,7)$	Z_{13}
W_{13}	$w^{16} + x^4 y + y^2 z + z^2$	16	$(1,3,4,8)$	S_{11}
S_{11}	$w^{16} + x^4 + y^2 z + xz^2$	16	$(1,4,5,6)$	W_{13}
Q_{12}	$w^{15} + x^3 z + y^3 + xz^2$	15	$(1,3,5,6)$	Q_{12}
S_{12}	$w^{13} + x^3 y + y^2 z + xz^2$	13	$(1,3,4,5)$	S_{12}
U_{12}	$w^{12} + x^4 + y^2 z + yz^2$	12	$(1,3,4,4)$	U_{12}

These 14 exceptional unimodal hypersurface singularities have been studied first by Arnold in [1] where he, among other things, discovered that there is a duality among this 14 exceptional unimodal hypersurfaces singularities which is now known as Arnold's strange duality. One can define Gabrielov and Dolgachev numbers for every one of these hypersurface singularities and he showed that for every one of the 14 exceptional unimodal hypersurface singularities there is another singularity in this list with interchanged Dolgachev and Gabrielov numbers. The consequences of this duality have been studied by a number of people. An overview on a lot of aspects of this duality can be found in a paper by Ebeling [15]. These examples were also the starting point for the analysis of the Picard–Fuchs equations in this article. We want to concentrate in this section on the duality between the invertible polynomials that arise here. In Table 3 we listed some important data of these singularities we need in this section. In particular we look at the compactification that comes from compactifying the Milnor fibres in the weighted projective space with one extra dimension which has weight one. The table consists of the polynomial defining the compactification of the hypersurface singularity, the degree, the weights and the dual singularity due to Arnold. As in Table 1 there is still the duality between the invertible polynomials. We will now list the sets $\{\alpha_1, \ldots, \alpha_u\}$ and $\{\beta_1, \ldots, \beta_u\}$, as defined in Proposition 2, and the resulting order of the Picard–Fuchs equation u. Notice that the sets Q_i and D are defined via the dual weights and that $\delta + \alpha_i$ and $\delta - \beta_i$ are the linear factors in the two summands of the Picard–Fuchs equation. Together with the dual weights and the dual degree they completely determine the Picard–Fuchs equation.

Table 4: The sets defining the linear factors of the Picard–Fuchs equation

Name	u	$\alpha_1, \ldots, \alpha_u$	$\beta_1, \ldots, \beta_u$
E_{12}	12	$0, 0, 0, 6, 12, 14,$ $18, 21, 24, 28, 30, 36$	$1, 5, 11, 13, 17, 19,$ $23, 25, 29, 31, 37, 41$
E_{13}	12	$0, 0, 0, \frac{15}{4}, \frac{15}{2}, 10,$ $\frac{45}{4}, 15, \frac{75}{4}, 20, \frac{45}{2}, \frac{105}{4}$	$1, 3, 7, 9, 11, 13,$ $17, 19, 21, 23, 27, 29$
Z_{11}	10	$0, 0, 0, 6, \frac{15}{2}, 12, 15, 18, \frac{45}{2}, 24$	$1, 5, 7, 11, 13, 17, 19, 23, 25, 29$
E_{14}	12	$0, 0, 0, \frac{8}{3}, \frac{16}{3}, 8,$ $\frac{32}{3}, 12, \frac{40}{3}, 16, \frac{56}{3}, \frac{64}{3}$	$1, 2, 5, 7, 10, 11,$ $13, 14, 17, 19, 22, 23$
Q_{10}	8	$0, 0, 0, 6, 8, 12, 16, 18$	$1, 5, 7, 11, 13, 17, 19, 23$
Z_{12}	10	$0, 0, 0, \frac{11}{3}, \frac{11}{2}, \frac{22}{3}, 11, \frac{44}{3}, \frac{33}{2}, \frac{55}{3}$	$1, 3, 5, 7, 9, 13, 15, 17, 19, 21$
W_{12}	8	$0, 0, 0, 4, 8, 10, 12, 16$	$1, 3, 7, 9, 11, 13, 17, 19$
Z_{13}	12	$0, 0, 0, \frac{18}{7}, \frac{9}{2}, \frac{36}{7},$ $\frac{54}{7}, 9, \frac{72}{7}, \frac{90}{7}, \frac{27}{2}, \frac{108}{7}$	$1, 2, 4, 5, 7, 8,$ $10, 11, 13, 14, 16, 17$
Q_{11}	9	$0, 0, 0, \frac{18}{5}, 6, \frac{36}{5}, \frac{54}{5}, 12, \frac{72}{5}$	$1, 3, 5, 7, 9, 11, 13, 15, 17$
W_{13}	12	$0, 0, 0, \frac{8}{3}, \frac{16}{5}, \frac{16}{3},$ $\frac{32}{5}, 8, \frac{48}{5}, \frac{32}{3}, \frac{64}{5}, \frac{40}{3}$	$1, 2, 3, 5, 6, 7,$ $9, 10, 11, 13, 14, 15$
S_{11}	8	$0, 0, 0, 4, \frac{16}{3}, 8, \frac{32}{3}, 12$	$1, 3, 5, 7, 9, 11, 13, 15$
Q_{12}	8	$0, 0, 0, \frac{5}{2}, 5, \frac{15}{2}, 10, \frac{25}{2}$	$1, 2, 4, 7, 8, 11, 13, 14$
S_{12}	12	$0, 0, 0, \frac{13}{5}, \frac{13}{4}, \frac{13}{3},$ $\frac{26}{5}, \frac{13}{2}, \frac{39}{5}, \frac{26}{3}, \frac{39}{4}, \frac{52}{5}$	$1, 2, 3, 4, 5, 6,$ $7, 8, 9, 10, 11, 12$
U_{12}	6	$0, 0, 0, 3, 6, 9$	$1, 2, 5, 7, 10, 11$

We want to mention that $\{\beta_1, \ldots, \beta_u\}$ contains all numbers $1 \le b \le \widehat{d}$ which are coprime to $\widehat{d}$. The set $\{\alpha_1, \ldots, \alpha_u\} \cap \mathbb{Z}$ on the other hand contains only elements which are not coprime to $\widehat{d}$. With Table 4 and Proposition 2 we are able to state how many basis elements for the cohomology spanned by the derivatives of ω that appear in the Picard–Fuchs equation we need in every degree. In [18] it is shown how this basis can be computed directly by using a diagrammatic version of the Griffiths–Dwork method. Here we will only state the result. In the examples we are looking at, we have always exactly 1 basis element in degree 1 and $2d$, so we only give the $u - 2$ basis elements in degree d. With this construction we are able to calculate the basis of the part of the middle cohomology that is used in the calculations for all examples, and this is listed in the following table. So the table includes the name of the singularity, the number $h^{1,1} = u - 2$ from Proposition 2 and a basis for the part of the Milnor ring in degree d, which gives also a basis of the part of the middle cohomology we are using in our calculations. We want to mention again that we are considering the Milnor ring of the one-parameter family.

Table 5: Basis elements for the middle cohomology

Name	$u - 2$	Basis elements in the Milnor ring in degree d
E_{12}	10	$wxyz, w^{36}x, w^{30}x^2, w^{24}x^3, w^{18}x^4, w^{12}x^5, w^6x^6, w^{28}y,$ $w^{14}y^2, w^{21}z$
E_{13}	10	$wxyz, w^{26}x, w^{22}x^2, w^{18}x^3, w^{20}y, w^{10}y^2, w^{15}z, w^{11}xz,$ w^7x^2z, w^3x^3z
Z_{11}	8	$wxyz, w^{24}x, w^{18}x^2, w^{12}x^3, w^6x^4, w^{22}y, w^{15}z, w^7yz$
E_{14}	10	$wxyz, w^{21}x, w^{18}x^2, w^{16}y, w^8y^2, w^{12}z, w^{13}xy, w^{10}x^2y,$ $w^5xy^2, w^2x^2y^2$
Q_{10}	6	$wxyz, w^{18}x, w^{12}x^2, w^6z^2, w^{16}y, w^8y^2$
Z_{12}	8	$wxyz, w^{18}x, w^{14}x^2, w^{16}y, w^{11}z, w^7xz, w^3x^2z, w^5yz$
W_{12}	6	$wxyz, w^{16}x, w^{12}x^2, w^8x^3, w^4x^4, w^{10}y^2$
Z_{13}	10	$wxyz, w^{15}x, w^{12}x^2, w^{13}y, w^9z, w^{10}xy, w^7x^2y, w^5xy^2,$ $w^2x^2y^2, w^4yz$
Q_{11}	7	$wxyz, w^{14}x, w^{10}x^2, w^{12}y, w^6y^2, w^7xz, w^4z^2$
W_{13}	10	$wxyz, w^{13}x, w^{10}x^2, w^7x^3, w^{12}y, w^9xy, w^6x^2y, w^3x^3y,$ w^5xz, w^2x^2z
S_{11}	6	$wxyz, w^{12}x, w^8x^2, w^4x^3, w^{10}z, w^5yz$
Q_{12}	6	$wxyz, w^{12}x, w^{10}y, w^5y^2, w^7xy, w^3xy^2$
S_{12}	10	$wxyz, w^{10}x, w^7x^2, w^9y, w^8z, w^6xy, w^3x^2y, w^5xz,$ w^2x^2z, w^4yz
U_{12}	4	$wxyz, w^9x, w^6x^2, w^3x^3$

Of course from the previous work we can immediately calculate the Picard–Fuchs equation by inserting in the α_i and β_j as linear factors as in Theorem 6. The output for all singularities we investigated in this section can therefore be calculated with Table 4. As stated in Proposition 2 the Picard–Fuchs equation is given by

$$ s^{\widehat{d}} \prod_{i=1}^n \widehat{q_i}^{\widehat{q_i}} \prod_{i=1}^u (\delta + \alpha_i) - (-\widehat{d})^{\widehat{d}} \prod_{i=1}^u (\delta - \beta_i) = 0. $$

This means for example that the Picard–Fuchs equation for the one-parameter family associated to Q_{10} is given by

$$ 0 = s^{24}\delta^3(\delta + 6)(\delta + 8)(\delta + 12)(\delta + 16)(\delta + 18) $$

$$ - 2^{24}3^9(\delta - 1)(\delta - 5)(\delta - 7)(\delta - 11)(\delta - 13)(\delta - 17)(\delta - 19)(\delta - 23) $$

We give another viewpoint on the Picard–Fuchs equation. From Theorem 6 we know that the Picard–Fuchs equation always consists of exactly two summands. They can be separated by setting $s^{\widehat{d}} = 0$ or $s^{\widehat{d}} = \infty$. We already know that if we view the Picard–Fuchs equation as a polynomial with variable δ then the zeroes of these polynomials after setting $s^{\widehat{d}} = 0$ are given by $\beta_1, \ldots, \beta_u$ and the zeroes for $s^{\widehat{d}} = \infty$ are given by $-\alpha_1, \ldots, -\alpha_u$. Now we want to focus on a polynomial that has related zeroes. Namely, we define χ_0 to be the polynomial with zeroes $\exp\left(2\pi i \frac{\beta_i}{d}\right)$ for $i = 0, \ldots, n$ and χ_∞ the polynomial with roots $\exp\left(2\pi i \frac{\alpha_i}{d}\right)$ for $i = 0, \ldots, n$ and notice that multiple roots in the Picard–Fuchs equation lead to multiple roots of χ_∞ and χ_0. Equivalently, we can first write the Picard–Fuchs equation for the variable $\lambda = (-s)^{-\widehat{d}}$ and then start with the zeroes of this equation for $\lambda = \infty$ and $\lambda = 0$.

Table 6: The functions χ_0 and χ_∞ for the 14 exceptional unimodal hypersurface singularities

Name	Deg	Weights	χ_0	χ_∞
E_{12}	42	(1,6,14,21)	$2 \cdot 3 \cdot 7 \cdot 42/1 \cdot 6 \cdot 14 \cdot 21$	$2 \cdot 3 \cdot 7$
E_{13}	30	(1,4,10,15)	$3 \cdot 30/6 \cdot 15$	$1 \cdot 3 \cdot 8$
Z_{11}	30	(1,6,8,15)	$5 \cdot 30/10 \cdot 15$	$1 \cdot 4 \cdot 5$
E_{14}	24	(1,3,8,12)	$2 \cdot 24/6 \cdot 8$	$1 \cdot 2 \cdot 9$
Q_{10}	24	(1,6,8,9)	$4 \cdot 24/8 \cdot 12$	$1 \cdot 3 \cdot 4$
Z_{12}	22	(1,4,6,11)	$1 \cdot 22/2 \cdot 11$	$1 \cdot 1 \cdot 4 \cdot 6/2$
W_{12}	20	(1,4,5,10)	$2 \cdot 20/4 \cdot 10$	$1 \cdot 2 \cdot 5$
Z_{13}	18	(1,3,5,9)	$18/6$	$1 \cdot 4 \cdot 7$
Q_{11}	18	(1,4,6,7)	$18/9$	$1 \cdot 3 \cdot 5$
W_{13}	16	(1,3,4,8)	$16/4$	$1 \cdot 5 \cdot 6$
S_{11}	16	(1,4,5,6)	$16/8$	$1 \cdot 3 \cdot 4$
Q_{12}	15	(1,3,5,6)	$1 \cdot 15/3 \cdot 5$	$1 \cdot 1 \cdot 6$
S_{12}	13	(1,3,4,5)	$13/1$	$3 \cdot 4 \cdot 5$
U_{12}	12	(1,3,4,4)	$1 \cdot 12/3 \cdot 4$	$1 \cdot 1 \cdot 4$

Notation 7. *We will shorten the notation for a rational function with only roots of unity as zeroes and poles. We will write $v_1 \cdots v_{m_1}/\eta_1 \cdots \eta_{m_2}$ for the rational function*

$$\chi(t) = \frac{(1 - t^{v_1}) \cdots (1 - t^{v_{m_1}})}{(1 - t^{\eta_1}) \cdots (1 - t^{\eta_{m_2}})}$$

With this notation we are able to write down the functions χ_0 and χ_∞ in the form shown in Table 6. The functions χ_0 and χ_∞ are in all cases somewhat different, but the interesting thing is that the quotient of the two functions is always the same.

Remark 6. The rational functions χ_0 and χ_∞ described in Table 6 always have the property that

$$\frac{\chi_0(t)}{\chi_\infty(t)} = \frac{(1 - t^{\widehat{d}})}{(1 - t^{\widehat{q}_1})(1 - t^{\widehat{q}_2})(1 - t^{\widehat{q}_3})(1 - t^{\widehat{q}_4})}$$

In the next section we will look at this phenomenon in more generality and we will also see that the roots of χ_0 and χ_∞ are the eigenvalues of the local monodromy around $(-1)^{\widehat{d}}\lambda^{-1} = s^{\widehat{d}} = 0$ and $(-1)^{\widehat{d}}\lambda^{-1} = s^{\widehat{d}} = \infty$ respectively.

3.5 Relations to the Poincaré Series and Monodromy

In this section we want to relate the numbers in the Picard–Fuchs equation of $f(\underline{x})$ to the Poincaré series of $g^t(\underline{x})$ and to the monodromy around 0 and ∞ in the solution space of the Picard–Fuchs equation. The last remark in the previous section already showed us the direction.

3.5.1 Poincaré Series

First we want to investigate the relation to the Poincaré series. Therefore we consider the Picard–Fuchs equation in the form of (10) which is a differential equation with parameter $\lambda = (-s)^{-\widehat{d}}$. If we view this differential equation as a polynomial with variable $\mathscr{D}$, then we can immediately read off the zeroes for $\lambda = 0$ and $\lambda = \infty$:

$$\lambda = 0: \quad \frac{\alpha_1}{\widehat{d}}, \dots, \frac{\alpha_u}{\widehat{d}}$$

$$\lambda = \infty: \quad -\frac{\beta_1}{\widehat{d}}, \dots, -\frac{\beta_u}{\widehat{d}}$$

Remark 7. Due to the symmetry of the α_j and β_j, the sets $\left\{\exp\left(2\pi i \frac{\alpha_j}{d}\right)\right\}$ and $\left\{\exp\left(2\pi i \frac{\beta_j}{d}\right)\right\}$ are closed under complex conjugation.

We will now relate these numbers α_j and β_j or $\exp\left(2\pi i \frac{\alpha_j}{d}\right)$ and $\exp\left(2\pi i \frac{\beta_j}{d}\right)$, respectively, to the Poincaré series of $g^t(\underline{x})$. Let us recall first how the Poincaré series is defined.

Definition 5. Let $A := \mathbb{C}[\underline{x}]/(g(\underline{x}))$ be the coordinate algebra of the hypersurface $\{g(\underline{x}) = 0\}$. Then A admits naturally a grading $A = \bigoplus_{m=0}^{\infty} A_m$, where A_m is generated by the monomials in A of weighted degree m. The Poincaré series for this hypersurface is given by

$$p_A(t) := p_g(t) := \sum_{m=0}^{\infty} \dim_{\mathbb{C}} A_m t^m$$

Remark 8. (cf. [2]) If $g(\underline{x})$ is a quasihomogeneous polynomial with weights $q_1, \ldots, q_n$ and weighted degree d, then the Poincaré series is given by

$$p_g(t) = \frac{(1 - t^d)}{(1 - t^{q_1}) \cdots (1 - t^{q_n})}$$

A rational function of this form is of course uniquely determined by the set of poles and zeroes. So we study these sets for the Poincaré series of $g^t(\underline{x})$, because as mentioned before that will be related to the Picard–Fuchs equation of $f(\underline{x})$. So we study the zeroes and poles of the function

$$p_{g^t}(t) = \frac{(1 - t^{\widehat{d}})}{(1 - t^{\widehat{q_1}}) \cdots (1 - t^{\widehat{q_n}})}.$$

The zeroes of $(1 - t^{\widehat{d}})$ are given by the set

$$Z_{\widehat{d}} := \left\{ \exp\left(2\pi i \frac{j}{\widehat{d}}\right) \mid 0 \le j \le \widehat{d} - 1 \right\}$$

and the zeroes of $(1 - t^{\widehat{q_1}}) \cdots (1 - t^{\widehat{q_n}})$ are given by the set

$$\bigcup_{k=1}^{n} Z_{\widehat{q_k}} := \bigcup_{k=1}^{n} \left\{ \exp\left(2\pi i \frac{j}{\widehat{q_k}}\right) \mid 0 \le j \le \widehat{q_k} - 1 \right\}.$$

So putting this together, the zeroes of the Poincaré series of $g^t(\underline{x})$ are given by

$$Z_{\widehat{d}} \setminus \left(Z_{\widehat{d}} \cap \bigcup_{k=1}^{n} Z_{\widehat{q_k}} \right) = \left\{ \exp\left(2\pi i \frac{\beta_j}{\widehat{d}}\right) \mid j = 0, \ldots, u \right\}$$

and the poles are given by the set

$$\bigsqcup_{k=1}^{n} Z_{\widehat{q_k}} \setminus \left(Z_{\widehat{d}} \cap \bigcup_{k=1}^{n} Z_{\widehat{q_k}} \right) = \left\{ \exp\left(2\pi i \frac{\alpha_j}{\widehat{d}}\right) \mid j = 0, \ldots, u \right\},$$

where the disjoint union indicates that poles occur in this set counted with multiplicity.

In the above we can see clearly the relation between the zeroes of the Picard–Fuchs equation of $f(\underline{x})$ for $\lambda = 0$ and $\lambda = \infty$ and the Poincaré series of $g^t(\underline{x})$. We summarize this in the following corollary.

Corollary 1. *The zeroes of the Poincaré series of $g^t(\underline{x})$ are in 1-1 correspondence with the zeroes of the Picard–Fuchs equation of $f(\underline{x})$ for $\lambda = \infty$ or $s = 0$ and the poles of the Poincaré series of $g^t(\underline{x})$ are in 1-1 correspondence with the zeroes of the Picard–Fuchs equation of $f(\underline{x})$ for $\lambda = 0$ or $s = \infty$.*

Equivalently the same holds for the Picard–Fuchs equation of $f'(\underline{x}) = g'(\underline{x}) + s \prod x_i$ and the Poincaré series of $g(\underline{x})$.

3.5.2 Monodromy

Now we want to explain why the roots of the Picard–Fuchs equation for $\lambda = (-s)^{-\hat{d}} = 0$ and $\lambda = (-s)^{-\hat{d}} = \infty$ are in 1-1 correspondence with the eigenvalues of the local monodromy around 0 and ∞ in the solution space of the Picard–Fuchs equation, i.e. the space of the period integrals. More precisely, the eigenvalues of the monodromy around 0 and ∞ are equal to the poles and zeroes of the Poincaré series respectively. First we recall monodromy in the context of Picard–Fuchs equations in as much generality as we need. References for the relation between monodromy and the Picard–Fuchs equation are [10, 27] and [12].

In this subsection we will always regard the Picard–Fuchs equation in $\mathcal{D} = \lambda \frac{\partial}{\partial \lambda}$, so we are working with the differential equation (10)

$$0 = \prod_{i=1}^{n} \widehat{q_i}^{\widehat{q_i}} \prod_{i=1}^{n} \prod_{j=0}^{\widehat{q_i}-1} (\mathcal{D} - \frac{j}{\widehat{q_i}}) \prod_{\ell \in I} (\mathcal{D} - \frac{\ell}{d})^{-1} \Phi - \widehat{d}^{\widehat{d}} \lambda \prod_{j=0}^{\widehat{d}-1} (\mathcal{D} + \frac{j}{d}) \prod_{\ell \in I} (\mathcal{D} + \frac{\ell}{d})^{-1} \Phi.$$

Due to [12] this Picard–Fuchs equation has only regular singular points. This can for example be seen by the fact that in the Picard–Fuchs equation, written as

$$\mathcal{D}^u \Phi + \sum_{i=0}^{u-1} h_i(\lambda) \mathcal{D}^i \Phi = 0, \tag{11}$$

all coefficients $h_i(\lambda)$ are holomorphic functions of λ. Now we can define the residue matrix for λ.

Definition 6. Let $\omega_1, \ldots, \omega_u$ be a basis of the solution space of the Picard–Fuchs equation and define the connection matrix $(\Gamma)_{ij}$ via $\mathcal{D}\omega_i = \sum_j \Gamma_{ij}\omega_j$. Then the residue matrix is given by $\mathrm{Res} = \mathrm{Res}_{\lambda=0}\left((\Gamma)_{ij}\right)$.

Remark 9. In the cases we consider $(\Gamma)_{ij}$ has no poles at $\lambda = 0$, so the residue matrix is just given by $\mathrm{Res} = \left((\Gamma)_{ij}\right)_{\lambda=0}$.

Theorem 8. ([12]) *The following relations between the residue matrix and the monodromy around $\lambda = 0$ in the solution space of the Picard–Fuchs equation hold.*

(i) η is an eigenvalue of Res $\Leftrightarrow$ $\exp(2\pi i\eta)$ is an eigenvalue of the monodromy.
(ii) $\exp(-2\pi i\mathrm{Res})$ is conjugate to the monodromy.
(iii) The monodromy is unipotent $\Leftrightarrow$ Res is nilpotent.

We cannot be sure that $\omega, \mathcal{D}\omega, \ldots, \mathcal{D}^{u-1}\omega$, with ω a solution of the Picard–Fuchs equation, is a basis for the solution space, but we can easily write down the connection matrix for this basis:

$$\Gamma = \begin{pmatrix} 0 & 1 & 0 & \cdots & & 0 \\ 0 & 0 & 1 & 0 \cdots & & 0 \\ \vdots & & \ddots & \ddots & & \vdots \\ 0 & \cdots & & 0 & 1 & 0 \\ 0 & 0 & \cdots & & 0 & 1 \\ -h_1(\lambda) & -h_2(\lambda) & -h_3(\lambda) \cdots & & & -h_{u-1}(\lambda) \end{pmatrix}$$

A theorem by Morrison gives a condition for the elements $\omega, \mathcal{D}\omega, \ldots, \mathcal{D}^{u-1}\omega$ to be a basis of the solution space. The condition depends on the eigenvalues of the matrix Γ.

Theorem 9. ([27]) *Let* $\mathcal{D}\omega(\lambda) = \Gamma\omega(\lambda)$ *be a system of ordinary differential equations with a regular singular point at* $\lambda = 0$. *If distinct eigenvalues of* $\Gamma_{\lambda=0}$ *do not differ by integers, then* $\omega_1, \ldots, \omega_u$ *with* $\omega = (\omega_1, \ldots, \omega_u)$ *is a basis for the solution space of the system of ordinary differential equations.*

So we calculate the eigenvalues of $\Gamma_{\lambda=0}$. For this purpose we only have to remember that (11) or equally (10) has the following solutions for $\lambda = 0$:

$$\bigsqcup_{i=1}^{n} \left\{ 0, \frac{1}{\widehat{q_i}}, \ldots, \frac{\widehat{q_i} - 1}{\widehat{q_i}} \right\} \setminus \left(\left\{ 0, \frac{1}{\widehat{d}}, \ldots, \frac{\widehat{d} - 1}{\widehat{d}} \right\} \cap \bigcup_{i=1}^{n} \left\{ 0, \frac{1}{\widehat{q_i}}, \ldots, \frac{\widehat{q_i} - 1}{\widehat{q_i}} \right\} \right).$$

This means that no distinct eigenvalues differ by an integer and therefore $\Gamma_{\lambda=0} = $ Res is a residue matrix by Theorem 9. In addition it follows from Theorem 8 that for every eigenvalue η of Γ we get an eigenvalue $\exp(2\pi i\eta)$ of the monodromy. So, together with Corollary 1, we get the following statement.

Corollary 2. *The poles of the Poincaré series of* $g^t(\underline{x})$ *are the eigenvalues of the monodromy around* $\lambda = (-s)^{-\widehat{d}} = 0$ *in the solution space of the Picard–Fuchs equation of* $f(\underline{x}) = g(\underline{x}) + s \prod_i x_i$ *and the zeroes of the Poincaré series of* $g^t(\underline{x})$ *are the eigenvalues of the monodromy around* $\lambda = (-s)^{-\widehat{d}} = \infty$.

The second part of this statement is proved analogously to the first part, with only the substitution of λ by λ^{-1}.

Remark 10. For the calculations in the last section this means that the eigenvalues of the monodromy around $\lambda = 0$ are given by the roots of χ_∞ and the eigenvalues of the monodromy around $\lambda = \infty$ are given by the roots of χ_0.

Remark 11. Notice that the monodromy around 0 and ∞ is not unipotent, but it is quasi-unipotent, i.e. a power of the monodromy is unipotent. This agrees with Theorem 2.3 in [12].

We want to mention that the points 0 and ∞ are not the only points with monodromy. At $\lambda = \prod \widehat{q_i}^{\widehat{q_i}} / \widehat{d}^{\widehat{d}}$ the Picard–Fuchs equation degenerates and therefore we can consider monodromy around this point in the solution space as well. But the monodromy around this point is just a combination of the monodromy around the other two points. This can be seen from the fact that the parameter space can be regarded as a projective line (cf. [28]).

Also, we want to mention that the critical points of λ in the solution space of the Picard–Fuchs equation apart from $\lambda = \infty$ are in 1-1 correspondence with the critical values of $f(\underline{x})$ in s. Namely $\lambda = (-s)^{-\widehat{d}} = 0$ and $\lambda = (-s)^{-\widehat{d}} = \frac{\Pi \widehat{q_i}^{\widehat{q_i}}}{\widehat{d}^{\widehat{d}}}$ are the critical values of $f(\underline{x})$ in s.

Acknowledgements This work was supported in part by DFG-RTG 1463. This article was part of my Ph.D. thesis and I would like to thank my supervisor Prof. Wolfgang Ebeling for his support. During my research I was supported by DFG RTG 1463. I would also like to thank Prof. Noriko Yui and Prof. Ragnar-Olaf Buchweitz, who were gave me input during my stay at the Fields Institute in Toronto. In addition I would like to thank the referee for helpful comments.

References

1. V.I. Arnold, Critical points of smooth functions, and their normal forms. Uspehi Mat. Nauk **30**(5(185)), 3–65 (1975)
2. V.I. Arnold, S.M. Guseĭn-Zade, A.N. Varchenko, in *Singularities of Differentiable Maps. Vol. I*. Monographs in Mathematics, vol. 82 (Birkhäuser, Boston, 1985)
3. V.V. Batyrev, Dual polyhedra and mirror symmetry for Calabi–Yau hypersurfaces in toric varieties. J. Algebr. Geom. **3**(3), 493–535 (1994)
4. V.V. Batyrev, L.A. Borisov, Dual cones and mirror symmetry for generalized Calabi–Yau manifolds, in *Mirror Symmetry, II*. AMS/IP Studies in Advanced Mathematics (American Mathematical Society, Providence, 1997), pp. 71–86
5. P. Berglund, T. Hübsch, A generalized construction of mirror manifolds, in *Essays on Mirror Manifolds* (International Press, Hong Kong, 1992), pp. 388–407
6. L.A. Borisov, Berglund–Hübsch mirror symmetry via vertex algebras. Preprint (2010), arXiv:1007.2633v3
7. Y.-H. Chen, Y. Yang, N. Yui, Monodromy of Picard–Fuchs differential equations for Calabi–Yau threefolds. J. Reine Angew. Math. **616**, 167–203 (2008)
8. A. Chiodo, Y. Ruan, LG/CY correspondence: the state space isomorphism. Adv. Math. **7**(2), 57–218 (2011)
9. A. Corti, V. Golyshev, Hypergeometric equations and weighted projective spaces. Sci. China Math. **54**(8), 1577–1590 (2011)
10. D.A. Cox, S. Katz, in *Mirror Symmetry and Algebraic Geometry*. Mathematical Surveys and Monographs, vol. 68 (American Mathematical Society, Providence, 1999)
11. V.I. Danilov, A.G. Khovanskiĭ, Newton polyhedra and an algorithm for calculating Hodge–Deligne numbers. Izv. Akad. Nauk SSSR Ser. Mat. **50**(5), 925–945 (1986)
12. P. Deligne, in *Equations différentielles à points singuliers réguliers*. Lecture Notes in Mathematics, vol. 163 (Springer, Berlin, 1970)
13. I. Dolgachev, Weighted projective varieties, in *Group Actions and Vector Fields*, Vancouver, BC, 1981. Lecture Notes in Mathematics, vol. 956 (Springer, Berlin, 1982), pp. 34–71
14. I.V. Dolgachev, Mirror symmetry for lattice polarized $K3$ surfaces. J. Math. Sci. **81**(3), 2599–2630 (1996)
15. W. Ebeling, Strange duality, mirror symmetry, and the Leech lattice, in *Singularity Theory*, Liverpool, 1996. London Mathematical Society Lecture Note Series, vol. 263 (Cambridge University Press, Cambridge, 1999), pp. 55–77

16. W. Ebeling, A. Takahashi, Strange duality of weighted homogeneous polynomials. Compos. Math., **147**, 1413–1433 (2011)
17. H. Fan, T. Jarvis, Y. Ruan, The Witten equation, mirror symmetry and quantum singularity theory. Preprint (2009), arXiv:0712.4021v3
18. S. Gährs, Picard–Fuchs equations of special one-parameter families of invertible polynomials. Preprint (2011), arXiv:1109.3462
19. I.M. Gel'fand, A.V. Zelevinskiĭ, M.M. Kapranov, Hypergeometric functions and toric varieties. Funktsional. Anal. i Prilozhen. **23**(2), 12–26 (1989)
20. I.M. Gel'fand, A.V. Zelevinskiĭ, M.M. Kapranov, Generalized Euler integrals and A-hypergeometric functions. Adv. Math. **84**(2), 255–271 (1990)
21. I.M. Gel'fand, A.V. Zelevinskiĭ, M.M. Kapranov, Hypergeometric functions, toric varieties and Newton polyhedra, in *Special Functions*, Okayama, 1990. ICM-90 Satellite Conference Proceedings (Springer, Tokyo, 1991), pp. 104–121
22. I.M. Gel'fand, A.V. Zelevinskiĭ, M.M. Kapranov, Correction to "Hypergeometric functions and toric varieties". Funktsional. Anal. i Prilozhen. **27**(4), 91 (1993)
23. S. Hosono, GKZ systems, Gröbner fans, and moduli spaces of Calabi–Yau hypersurfaces, in *Topological Field Theory, Primitive Forms and Related Topics*, Kyoto, 1996. Progress in Mathematics, vol. 160 (Birkhäuser, Boston, 1998), pp. 239–265
24. M. Krawitz, FJRW rings and Landau–Ginzburg Mirror Symmetry. Preprint (2009), arXiv:0906.0796
25. M. Krawitz, N. Priddis, P. Acosta, N. Bergin, H. Rathnakumara, FJRW-rings and mirror symmetry. Comm. Math. Phys. **296**, 145–174 (2010)
26. M. Kreuzer, H. Skarke, On the classification of quasihomogeneous functions. Comm. Math. Phys. **150**(1), 137–147 (1992)
27. D.R. Morrison, Picard–Fuchs equations and mirror maps for hypersurfaces, in *Essays on Mirror Manifolds* (International Press, Hong Kong, 1992), pp. 241–264
28. D.R. Morrison, Geometric aspects of mirror symmetry, in *Mathematics Unlimited—2001 and Beyond* (Springer, Berlin, 2001), pp. 899–918
29. J. Stienstra, GKZ hypergeometric structures, in *Arithmetic and Geometry Around Hypergeometric Functions*. Progress in Mathematics, vol. 260 (Birkhäuser, Basel, 2007), pp. 313–371
30. N. Yui, Arithmetic of certain Calabi–Yau varieties and mirror symmetry, in *Arithmetic Algebraic Geometry*, Park City, UT, 1999. IAS/Park City Mathematics Series, vol. 9 (American Mathematical Society, Providence, 2001), pp. 507–569

A Structure Theorem for Fibrations on Delsarte Surfaces

Bas Heijne and Remke Kloosterman

Abstract In this paper we study a special class of fibrations on Delsarte surfaces. We call these fibrations Delsarte fibrations. We show that after a specific cyclic base change, the fibration is the pullback of a fibration with three singular fibers and that this second-base change is completely ramified at two points where the fiber is singular. As a corollary we show that every Delsarte fibration of genus 1 with nonconstant j-invariant occurs as the base change of an elliptic surface from Fastenberg's list of rational elliptic surfaces with $\gamma < 1$.

Key words: Delsarte surfaces, Elliptic surfaces

Mathematics Subject Classifications (2010): Primary 14J27; Secondary 14J25

1 Introduction

A Delsarte surface S is a surface in $\mathbf{P}^3$ defined by the vanishing of a polynomial F consisting of four monomials. Let A be the exponent matrix of F; then a Delsarte surface is the quotient of a Fermat surface if and only if $\det(A) \neq 0$. Shioda used this observation in [7] to present an algorithm to determine the Lefschetz number of any smooth surface that is birationally equivalent with S.

B. Heijne
Institut für algebraische Geometrie, Leibniz Universität Hannover, Welfengarten 1,
30167 Hannover, Germany
e-mail: heijne@math.uni-hannover.de

R. Kloosterman (✉)
Institut für Mathematik, Humboldt-Universität zu Berlin, Unter den Linden 6,
10099 Berlin, Germany
e-mail: klooster@math.hu-berlin.de

R. Laza et al. (eds.), *Arithmetic and Geometry of K3 Surfaces and Calabi–Yau Threefolds*,   311
Fields Institute Communications 67, DOI 10.1007/978-1-4614-6403-7_10,
© Springer Science+Business Media New York 2013

Fix now two disjoint lines ℓ_1, ℓ_2 in $\mathbf{P}^3$. The projection with center ℓ_1 onto ℓ_2 yields a rational map $S \dashrightarrow \mathbf{P}^1$. Resolving the indeterminacies of this map yields a fibration $\tilde{S} \to \mathbf{P}^1$. If the genus of the general fiber is one and this morphism has a section, then Shioda's algorithm together with the Shioda–Tate formula allows one to determine the Mordell–Weil rank of the group of sections. Shioda applied this to the surface

$$y^2 + x^3 + 1 + t^n$$

and showed in [8] that the maximal Mordell–Weil rank (by varying n) is 68. In [1] this result is reproven by completely different methods.

If both lines ℓ_i are intersections of two coordinate hyperplanes, then we call the obtained fibration a *Delsarte fibration*. We will introduce the notion of a *Delsarte base change*. Roughly said, this is a base change $\mathbf{P}^1 \to \mathbf{P}^1$ completely ramified over 0 and ∞. In particular, the pullback of a Delsarte fibration under a Delsarte base change is again a Delsarte fibration. The first author determined in his Ph.D. thesis [4] the maximal Mordell–Weil rank under Delsarte base changes of any Delsarte fibration such that the general fiber has genus one. In this way he showed that Shioda's example has the highest possible rank among Delsarte fibration of genus one.

In [2, 3] Fastenberg calculated the maximal Mordell–Weil rank under base changes $t \mapsto t^n$ for a special class of elliptic surfaces, i.e., elliptic curves over $\mathbf{C}(t)$ with nonconstant j-invariant such that a certain invariant γ is smaller than 1. It turned out that all the ranks that occur for Delsarte surfaces with nonconstant j-invariant also occur in Fastenberg's list. In [4, Chap. 6] it is shown that for every Delsarte fibration of genus one, there exist integers m, n such that the Delsarte base change of degree m of the Delsarte fibration is isomorphic to a base change of the form $t \mapsto t^n$ of one of the surfaces in Fastenberg's list.

In this paper we present a more conceptual proof for this phenomenon: first we study the configuration of singular fibers of a Delsarte fibration. We show that for any Delsarte fibration, each two singular fibers over points $t \neq 0, \infty$ are isomorphic. Then we show that after a base change of the form $t \mapsto t^n$, the Delsarte fibration is a base change of the form $t \mapsto t^m$ of a fibration with at most one singular fiber away from $0, \infty$.

If there is no singular fiber away from $0, \infty$, then the fibration becomes split after a base change of $t \mapsto t^m$. If there is at least one singular fiber away from $0, \infty$, then we show that there are three possibilities, namely, the function field extension $K(S)/K(\mathbf{P}^1) = K(x, y, t)/K(t)$ is given by $m_1 + m_2 + (1 + t)m_3$, where the m_i are monomials in x and y, or this extension is given by $y^a = x^b + x^c + tx^d$, where b, c, d are mutually distinct, or the singular fiber away from 0 and ∞ has only nodes as singularities and is therefore semistable. See Proposition 3.

In the case of a genus one fibration we can use this classification to check almost immediately that any Delsarte fibration of genus one admits a base change of the form $t \mapsto t^n$ such that the pulled back fibration is the pullback of a fibration with $\gamma < 1$ or has a constant j-invariant. See Corollary 2. This procedure is carried out in Sect. 2.

The techniques used in the papers by Fastenberg use the fact that the fibration is not isotrivial and it seems very hard to extend these techniques to isotrivial fibrations. In Sect. 3 we consider an example of a class of isotrivial Delsarte fibrations. Shioda's algorithm yields the Lefschetz number of any Delsarte surface with $\det(A) \neq 0$. Hence it is interesting to see how it works in the case where Fastenberg's method breaks down. Let p be an odd prime number, and a a positive integer. We consider the family of surfaces

$$S : y^2 = x^p + t^{2ap} + s^{2ap}$$

in $\mathbf{P}(2a, ap, 1, 1)$. Then S is birational to a Delsarte surface. After blowing up $(1 : 1 : 0 : 0)$ we obtain a smooth surface $\tilde{S}$ together with a morphism $\tilde{S} \to \mathbf{P}^1$. The general fiber of this morphism is a hyperelliptic curve of genus $(p - 1)/2$. We show that if $p > 7$, then $\rho(\tilde{S}) = 2 + 6(p - 1)$; in particular the Picard number is independent of a. Two of the generators of the Néron–Severi group of $\tilde{S}$ can be easily explained: the first one is the pullback of the hyperplane class on S and the second class is the exceptional divisor of the morphism $\tilde{S} \to S$. In Example 3 we give also equations for some further curves on $\tilde{S}$ which generate a subgroup of finite index of $\mathrm{NS}(\tilde{S})$.

If we take $p = 3$, then we recover Shioda's original example. However, in Shioda's example it turns out that $\rho(\tilde{S})$ depends on $\gcd(a, 60)$. Similarly for $p = 5$ and for $p = 7$, one observes that the Picard number of $\tilde{S}$ depends on a, in contrast with the situation for $p > 7$.

2 Delsarte Surfaces

In this section we work over an algebraically closed field K of characteristic zero.

Definition 1. A surface $S \subset \mathbf{P}^3$ is called a *Delsarte surface* if S is the zero set of a polynomial of the form

$$F := \sum_{i=0}^{3} c_i \prod_{j=0}^{3} X_i^{a_{i,j}},$$

with $c_i \in K^*$ and $a_{i,j} \in \mathbf{Z}_{\geq 0}$. The 4×4 matrix $A := (a_{i,j})$ is called the *exponent matrix* of S.

A *Delsarte fibration of genus g* on a Delsarte surface S consists of the choice of two disjoint lines ℓ_1, ℓ_2 such that both the ℓ_i are the intersection of two coordinate hyperplanes and the generic fiber of the projection $S \dashrightarrow \ell_2$ with center ℓ_1 is an irreducible curve of geometric genus g.

A *Delsarte birational map* is a birational map $\varphi : \mathbf{P}^3 \dashrightarrow \mathbf{P}^3$ such that $\varphi(X_0 : \cdots : X_3) = (\prod_j X_j^{b_{0j}} : \cdots : \prod_j X_j^{b_{3j}})$, i.e., φ is a birational monomial map.

Remark 1. Since K is algebraically closed, we can multiply each of the four coordinates X_i by a nonzero constant such that all four constants in F coincide; hence, without loss of generality we may assume that $c_i = 1$.

After permuting the coordinates, if necessary, we may assume that ℓ_1 equals $V(X_2, X_3)$ and ℓ_2 equals $V(X_0, X_1)$. Then the projection map $S \dashrightarrow \ell_2$ is just the map $[X_0 : X_1 : X_2 : X_3] \to [X_2 : X_3]$. Let $f(x, y, t) := F(x, y, t, 1) \in K(t)[x, y]$. Then the function field extension $K(S)/K(\ell_2)$ is isomorphic to the function field extension $\mathrm{Quot}(K[x, y, t]/f)$ over $K(t)$.

We call a Delsarte fibration with $\ell_1 = V(X_2, X_3)$ and $\ell_2 = V(X_0, X_1)$ the *standard fibration* on S.

Definition 2. Let n be a nonzero integer. A *Delsarte base change of degree* $|n|$ of a Delsarte fibration $\varphi : S \dashrightarrow \mathbf{P}^1$ is a Delsarte surface S_n, together with a Delsarte fibration $\varphi_n : S_n \dashrightarrow \mathbf{P}^1$ and a Delsarte rational map $S_n \dashrightarrow S$ of degree $|n|$, such that there exists a commutative diagram

$$
\begin{array}{ccc}
S_n & \dashrightarrow & S \\
\downarrow & & \downarrow \\
\mathbf{P}^1 & \longrightarrow & \mathbf{P}^1
\end{array}
$$

and $K(S_n)/K(\mathbf{P}^1)$ is isomorphic to the field extension $\mathrm{Quot}(K[x, y, s]/(f(x, y, s^n))$ over $K(s)$.

Remark 2. Note that n is allowed to be negative. If n is negative, then a base change of degree $-n$ is the composition of the automorphism $t \mapsto 1/t$ of $\mathbf{P}^1$ with the usual degree $-n$ base change $t \mapsto t^{-n}$. In many cases we compose a base change with a Delsarte birational map which respects the standard fibration. In affine coordinates such a map is given by $(x, y, s) \mapsto (xs^a, ys^b, s^n)$ for some integers a, b.

Lemma 1. *Let S be a Delsarte surface with exponent matrix A. Suppose there is a nonzero vector $\mathbf{v} = (a, b, 0, 0)^T$ in $\mathbf{Z}^4$ such that $A\mathbf{v} \in \mathrm{span}(1, 1, 1, 1)^T$. Then the generic fiber of the standard fibration $\varphi : S \to \mathbf{P}^1$ is a rational curve.*

Proof. After interchanging x and y, if necessary, we may assume that a is nonzero. Consider now $f_0 := f(x^a, x^b y, t)$. The exponents of x in the four monomials of f_0 are precisely the entries of $A\mathbf{v}$. Since $A\mathbf{v} = e(1, 1, 1, 1)^T$ for some integer e, we have that $f_0 = x^e g(y, t)$. This implies that the generic fiber of φ is dominated by a finite union of rational curves. Since the generic fiber is irreducible, it follows that the generic fiber of φ is a rational curve.

Lemma 2. *Let S be a Delsarte surface with exponent matrix A. Suppose there is a nonzero vector $\mathbf{v} = (a, b, c, 0)^T$ in $\mathbf{Z}^4$ such that $c \neq 0$ and $A\mathbf{v} \in \mathrm{span}(1, 1, 1, 1)^T$. Then there is a curve C and a Delsarte base change of degree $|c|$ such that the pullback of the standard fibration on S is birational to a product $C \times \mathbf{P}^1 \to \mathbf{P}^1$.*

Proof. Consider now $f_0 := f(xt^a, yt^b, t^c)$. The exponents of t in the four monomials of f_0 are precisely the entries of $A\mathbf{v}$. Since $A\mathbf{v} = e(1, 1, 1, 1)^T$ for some integer e, we have that $f_0 = t^e g(x, y)$. Let S' be the projective closure of $g = 0$ in $\mathbf{P}^3$. Then S' is a cone over the plane curve $g = 0$, in particular S' is birational to $C \times \mathbf{P}^1$ and the standard fibration on S' is birational to the projection $C \times \mathbf{P}^1 \to \mathbf{P}^1$. Here C is the curve given by the equation g. Now S' is birational to the surface S_c, the projective closure of $f(x, y, t^c) = 0$. Hence $S_c \to \mathbf{P}^1$ is birational to $C \times \mathbf{P}^1 \to \mathbf{P}^1$.

Lemma 3. *Let A be the exponent matrix of a Delsarte surface S. There exists a nonzero vector $\mathbf{v} = (a, b, c, 0)^T$ in $\mathbf{Z}^4$ such that $A\mathbf{v} \in \mathrm{span}(1, 1, 1, 1)^T$ if and only if $\det(A) = 0$.*

Proof. Since each row sum of A equals d, the degree of the i-th monomial in F, it follows that $A(1, 1, 1, 1)^T = d(1, 1, 1, 1)^T$. Suppose first that $\det(A) \neq 0$. Then from $A(1, 1, 1, 1)^T = d(1, 1, 1, 1)^T$ it follows that $A^{-1}(1, 1, 1, 1)^T \in \mathrm{span}(1, 1, 1, 1)^T$, which does not contain a nonzero vector with vanishing fourth coordinate.

Suppose now that $\det(A) = 0$. Denote with A_i the i-th column of A. From the fact that each row sum of A equals d, we get that $A_1 + A_2 + A_3 + A_4 = d(1, 1, 1, 1)^T$. Since $\det(A) = 0$, there exists a nonzero vector (a_1, a_2, a_3, a_4) such that $\sum a_i A_i = 0$. From this we obtain

$$(a_4 - a_1)A_1 + (a_4 - a_2)A_2 + (a_4 - a_3)A_3 = a_4(A_1 + A_2 + A_3 + A_4) = a_4 d(1, 1, 1, 1)^T.$$

That is, $\mathbf{v} = (a_4 - a_1, a_4 - a_2, a_4 - a_3, 0)^T$ is a vector such that $A\mathbf{v} \in \mathrm{span}(1, 1, 1, 1)^T$. We need to show that $\mathbf{v}$ is nonzero. Suppose the contrary, then also $A\mathbf{v} = a_4 d$ $(1, 1, 1, 1)^T$ is zero and therefore $a_4 = 0$. Substituting this in $\mathbf{v}$ yields that $\mathbf{v} = (-a_1, -a_2, -a_3, 0) = (0, 0, 0, 0)$ holds, which contradicts our assumption that (a_1, a_2, a_3, a_4) is nonzero.

Remark 3. We want to continue to investigate the singular fibers of a Delsarte fibration, in particular the singular fibers over points $t = t_0$ with $t_0 \neq 0, \infty$. If $\det(A) = 0$, then either the generic fiber has geometric genus 0 or after a Delsarte base change the fibration is split, i.e., the fibration is birational to a product. In the latter case all the fibers away from 0 and ∞ are smooth. Hence from now on we restrict to the case where $\det(A) \neq 0$.

Lemma 4. *Let S be a Delsarte surface with $\det(A) \neq 0$, such that the generic fiber has positive geometric genus. Let $\varphi : S \to \mathbf{P}^1$ be the standard Delsarte fibration. Then there exists a Delsarte base change of φ that is birational to the standard fibration on a Delsarte surface S' with affine equation of the form $m_1 + m_2 + m_3 + t^n m_4$, where each m_i is a monomial in x and y.*

Proof. Let $\mathbf{e}_0 = (1, 1, 1, 1)^T$ and $\mathbf{e}_i$ be the i-th standard basis vector of $\mathbf{Q}^4$. Let V_i be the vector space spanned by $\mathbf{e}_0$ and $\mathbf{e}_i$. Since $A^{-1}\mathbf{e}_0 = \frac{1}{d}\mathbf{e}_0$, it follows that $A^{-1}V_i$ is not contained in $\mathrm{span}\{\mathbf{e}_1, \mathbf{e}_2, \mathbf{e}_3\}$. In particular, $\dim A^{-1}V_i \cap \mathrm{span}\{\mathbf{e}_1, \mathbf{e}_2, \mathbf{e}_3\} = 1$.

Let ℓ_i be the line $A^{-1}V_i \cap \mathrm{span}\{\mathbf{e}_1, \mathbf{e}_2, \mathbf{e}_3\}$ and let $\mathbf{v}_i$ be a vector spanning ℓ_i. We can scale $\mathbf{v}_i$ such that $A\mathbf{v}_i = \mathbf{e}_i + t_i \mathbf{e}_0$ for some $t_i \in K$. Since $\mathbf{e}_0, \mathbf{e}_1, \mathbf{e}_2, \mathbf{e}_3$ are linearly independent, it follows that $\{\mathbf{e}_i + t_i \mathbf{e}_0\}_{i=1}^3$ are linearly independent and therefore $\mathbf{v}_1, \mathbf{v}_2, \mathbf{v}_3$ are linearly independent. Hence $\mathrm{span}\{\mathbf{v}_1, \mathbf{v}_2, \mathbf{v}_3\}$ is three-dimensional and there is at least one $\mathbf{v}_i = (a_i, b_i, c_i, 0)$ with $c_i \neq 0$. Then the rational map defined by $(x, y, t) \mapsto (xt^{a_i}, yt^{b_i}, t^{c_i})$ is a composition of a Delsarte base change and a Delsarte rational map.

Now three of the four entries of $A\mathbf{v}_i$ coincide, say they equal e. The exponent of t in the four monomials of $f_0 := f(xt^{a_i}, yt^{b_i}, t^{c_i})$ are the entries of $A\mathbf{v}_i$. In particular, in precisely three of the four monomials the exponents of t equal the same constant e. Therefore $g := f_0/t^e$ consists of four monomials of which precisely one contains a t. If the exponent of t in this monomial is negative, then we replace t by $1/t$ in g. Then $g = 0$ is an affine polynomial equation for the surface S'.

Recall that we investigate the singular fibers of a Delsarte fibration, in particular the singular fibers over points $t = t_0$ with $t_0 \neq 0, \infty$. If we have a Delsarte fibration and take a Delsarte base change, then the type of singular fiber over $t = 0, \infty$ may change, since the base change map is ramified over these points. Over points with $t \neq 0, \infty$, the base change map is unramified, and therefore the type of singular fibers remains the same. Hence to describe the possible types of singular fibers over points with $t \neq 0, \infty$, it suffices by Lemma 4 to study Delsarte surfaces such that only one monomial contains a t, i.e., we may restrict ourselves to Delsarte surfaces with affine equation $m_1 + m_2 + m_3 + t^n m_4$. If $n = 0$, then the fibration is split and there are no singular fibers. If $n \neq 0$, then the possible types of singular fibers are already determined at $n = 1$, i.e., it suffices to consider Delsarte surfaces with affine equation $m_1 + m_2 + m_3 + tm_4$.

Definition 3. We call the standard fibration on a Delsarte surface a *minimal Delsarte fibration* if the following conditions hold:

1. The affine equation for the standard fibration is of the form $m_1 + m_2 + m_3 + tm_4$, where the m_i are monomials in x and y.
2. The exponent matrix A of the corresponding surface $S \subset \mathbf{P}^3$ satisfies $\det(A) \neq 0$.

Remark 4. In the function field $K(S) = \text{Quot}(K[x, y, t]/f)$ we have the relation $t = (-m_1 - m_2 - m_3)/m_4$. In particular $K(S) = K(x, y)$ and therefore S is a rational surface.

Consider now the defining polynomial for S, i.e., $M_1 + M_2 + M_3 + X_2 M_4$, where the M_i are monomials in X_0, X_1, X_3, the degrees of M_1, M_2, and M_3 are the same, say d, and the degree of M_4 equals $d - 1$.

The Delsarte fibration is induced by the map $(X_0 : X_1 : X_2 : X_3) \mapsto (X_2 : X_3)$. If S contains the line $\ell_1 : X_2 = X_3 = 0$, then this rational map can be extended to a morphism on all of S; otherwise, we blow up the intersection of this line with S and obtain a morphism $\tilde{S} \to \mathbf{P}^1$, such that each fiber is a plane curve of degree d.

There is a different way to obtain this family of plane curves. Define N_i' as follows:

$$N_i' := M_i(X_0, X_1, X_2, X_2) \text{ for } i = 1, 2, 3 \text{ and } N_4' = X_2 M_4(X_0, X_1, X_2, X_2)$$

Now the four N_i' have a nontrivial greatest common divisor if and only if $X_3 \mid M_i$ for $i = 1, 2, 3$. The later condition is equivalent to the condition that the line ℓ_1 is contained in S. Moreover, if the greatest common divisor is nontrivial, then it equals X_2. Now set $N_i = N_i'$ if $\ell_1 \not\subset S$ and set $N_i = N_i'/X_2$ if $\ell_1 \subset S$. Then $\lambda(N_1 + N_2 + N_3) + \mu N_4$ is a pencil of plane curves of degree d or $d - 1$, and the generic member of this pencil is precisely the generic fiber of the standard fibration on S.

We can consider the generic member of this family as a projective curve C over $K(t)$ with defining polynomial $G := N_1 + N_2 + N_3 + tN_4 \in K(t)[X_0, X_1, X_2]$. Let A' be the exponent matrix of C (considered as a curve in $\mathbf{P}^2_{K(t)}$). Set

$$B := \begin{pmatrix} 1 & 0 & 0 \\ 0 & 1 & 0 \\ 0 & 0 & 1 \\ 0 & 0 & 1 \end{pmatrix} \text{ if } \ell_1 \not\subset S \text{ and } B := \begin{pmatrix} 1 & 0 & \frac{-1}{d} \\ 0 & 1 & \frac{-1}{d} \\ 0 & 0 & \frac{d-1}{d} \\ 0 & 0 & \frac{d-1}{d} \end{pmatrix} \text{ otherwise.}$$

Then $A' = AB$. Since A is invertible and B has rank 3, it follows that rank $A' = 3$. Moreover the first three rows of A' are linearly independent, since the upper 3×3 minor of A' equals the upper 3×3 minor of A times the upper 3×3 minor of B.

In particular, there is a vector $\mathbf{k}$, unique up to scalar multiplication, such that $\mathbf{k}A' = 0$. Since the upper three rows of A' are linearly independent, it follows that the fourth entry k_4 of $\mathbf{k}$ is nonzero. We can make the vector $\mathbf{k}$ unique, by requiring that $k_4 > 0$, $k_i \in \mathbf{Z}$ for $i = 1, \ldots, 4$ and $\gcd(k_1, k_2, k_3, k_4) = 1$. Moreover from rank $A' = 3$ and the fact that $(0, 0, 1, -1)B$ vanishes, it follows that $\mathbf{k} \in \mathrm{span}\{(0, 0, 1, -1)A'^{-1}\}$.

Since none of the rows of A' is zero, there are at least two nonzero entries in $\mathbf{k}$. Suppose that there are precisely two nonzero entries, say k_i and k_4. Then $-k_i$ times the i-th row of A' equals k_4 times the fourth row of A'. Each row sum of A' equals the degree of C, say d. From this it follows that $k_i d = -k_4 d$ and hence that $k_i = -1, k_4 = 1$. In particular, the i-th row and the fourth row coincide. After permuting m_1, m_2, m_3, if necessary, we may assume that the affine equation for the standard fibration is of the form $m_1 + m_2 + (1 + t)m_3$.

Hence if the four monomials m_1, m_2, m_3, m_4 (in x, y) are distinct, then at least three of the four entries of $\mathbf{k}$ are nonzero.

Let A'_i be the i-th row of A'. Recall that each row sum of A'_i equals d. Since $\sum k_i A_i$ equals zero, it follows that $0 = \sum_i k_i \sum_j A'_{i,j} = \sum_i k_i d$ and hence $\sum k_i = 0$. Let p be a prime number dividing one of the k_i. Since $\gcd(k_1, \ldots, k_4) = 1$, there is a j such that $p \nmid k_j$. From $\sum k_i = 0$, it follows that there is a $j' \neq j$ such that $p \nmid k_{j'}$. Hence, each prime number p does not divide at least two of the entries of $\mathbf{k}$.

Proposition 1. *Let $S \to \mathbf{P}^1$ be a minimal Delsarte fibration. Let A', $\mathbf{k}$, and N_i be as above. Suppose that the fiber over $t = t_0$ is singular and $t_0 \neq 0, \infty$, then*

$$t_0^{k_4} - \prod_{i:k_i \neq 0} k_i^{k_i} = 0$$

or two of the N_i coincide.

Proof. Let $G \in K(t)[X_0, X_1, X_2]$ be as above. Then G defines a pencil of plane curves in $\mathbf{P}^2_K$. Assume that no two of the N_i coincide. We aim at determining the singular members of the pencil defined by G. Let B_t be the matrix obtained from A' by multiplying the fourth row by t. Let us consider the matrix B_{t_0} for some $t_0 \in K^*$. Since the upper 3×3 minor of B_{t_0} equals the upper 3×3 minor of A', and this minor is nonzero, it follows that rank $B_{t_0} = 3$. Hence the kernel of right multiplication by B_{t_0} is one-dimensional and is generated by $(k_1, k_2, k_3, \frac{k_4}{t_0})$.

Consider now the closure of the image of the rational map $M : K^3 \dashrightarrow K^4$ sending (x, y, z) to (N_1, N_2, N_3, N_4). Let z_1, z_2, z_3, z_4 be the coordinates on K^4. Then by the definition of the vector (k_1, k_2, k_3, k_4), one has that $\prod N_i^{k_i} = 1$ holds, i.e., on the image of M, one has

$$\prod z_i^{k_i} = 1$$

Since the greatest common divisor of the k_i equals one, this defines an irreducible hypersurface $\overline{V}$ in K^4. Moreover, from the fact that rank A' equals 3, it follows that M has finite fibers; hence, the image of M is three-dimensional and the closure of the image of M is precisely the closure of $\prod z_i^{k_i} = 1$.

We want now to determine the values t_0 for which the corresponding member of the pencil of plane curves is singular. Hence we want to find $(x_0 : y_0 : z_0) \in \mathbf{P}^2$ and $t_0 \in K^*$ such that for $(t, X_0, X_1, X_2) = (t_0, x_0, y_0, z_0)$, the vector $(G_{X_0}, G_{X_1}, G_{X_2})$ is zero. In particular, the vector $(X_0 G_{X_0}, X_1 G_{X_1}, X_2 G_{X_2})$ is zero. A direct calculation shows that the latter vector equals $(N_1, N_2, N_3, t N_4) A'$, which in turn equals $(N_1, N_2, N_3, N_4) B_t$. Hence if (x_0, y_0, z_0) is a singular point of a fiber over $t = t_0$, then $M(x_0, y_0, z_0)$ is contained in ker $B_{t_0} \cap \overline{V}$.

We consider first the case where $M(x_0, y_0, z_0)$ is nonzero and $t_0 \neq 0$. Then $\prod_{i : k_i \neq 0} z_i^{k_i} = 1$ and (z_1, z_2, z_3, z_4) is a multiple of $(k_1, k_2, k_3, k_4/t_0)$. In particular,

$$\frac{\prod_{i : k_i \neq 0} k_i^{k_i}}{t_0^{k_4}} = 1$$

holds, which finishes the case where $M(x_0, y_0, z_0)$ is nonzero.

To finish we show that if $t_0 \neq 0$ and ker $B_{t_0} \cap \overline{V}$ consists only of $(0, 0, 0, 0)$, then the fiber over t_0 is smooth. Since ker $B_{t_0} \cap \overline{V}$ consists only of $(0, 0, 0, 0)$, each singular point of the fiber satisfies $N_1 = N_2 = N_3 = N_4 = 0$. In particular at least two of the X_i are zero. Without loss of generality we may assume that the point $(0 : 0 : 1)$ is singular. Consider now $G(x, y, 1)$ and write this as $m_1 + m_2 + m_3 + t m_4$.

Since all the four N_i are distinct, we have that $m_1 + m_2 + m_3 + t m_4 = 0$ is an equisingular deformation of $m_1 + m_2 + m_3 + t_0 m_4$ for t in a small neighborhood of t_0. Hence we can resolve this singularity simultaneously for all t in a neighborhood of t_0. Therefore all fibers in a neighborhood of t_0 are smooth and, in particular, the fiber over t_0 is smooth.

Lemma 5. *Let $\varphi : S \to \mathbf{P}^1$ be a minimal Delsarte fibration. Then there is an automorphism $\sigma : S \to S$, mapping fibers of φ to fibers, such that its action on the base curve is $t \mapsto \zeta_{k_4} t$.*

Proof. Let d be the smallest integer such that $D := dA^{-1}$ has integral coefficients. Let $T = \{\sum X_i^d = 0\} \subset \mathbf{P}^3$ be the Fermat surface of degree d. Then there is a rational map $T \dashrightarrow S$ given by $(X_0 : X_1 : X_2 : X_3) \mapsto (\prod X_j^{d_{0j}} : \cdots : \prod X_j^{d_{3j}})$. On T there is a natural action of $(\mathbf{Z}/d\mathbf{Z})^3$, given by $(X_0 : X_1 : X_2 : X_3) \mapsto (\zeta_d^{a_1} X_0 : \zeta_d^{a_2} X_1 : \zeta_d^{a_3} X_2 : X_3)$. On the affine chart $X_3 \neq 0$ with coordinates x, y, t, this action is given by $(x, y, t) \mapsto (\zeta_d^{a_1} x, \zeta_d^{a_2} y, \zeta_d^{a_3} t)$.

The rational map $T \dashrightarrow S$ is given (in affine coordinates) by

$$(x, y, t) \mapsto \left(\frac{x^{d_{00}} y^{d_{01}} t^{d_{02}}}{x^{d_{30}} y^{d_{31}} t^{d_{32}}}, \frac{x^{d_{10}} y^{d_{11}} t^{d_{12}}}{x^{d_{30}} y^{d_{31}} t^{d_{32}}}, \frac{x^{d_{20}} y^{d_{21}} t^{d_{22}}}{x^{d_{30}} y^{d_{31}} t^{d_{32}}} \right).$$

The action of $(\mathbf{Z}/d\mathbf{Z})^3$ descents to S and respects the standard fibration. Let $t = X_2/X_3$ be a coordinate on the base of the standard fibration. Then $(a_1, a_2, a_3) \in (\mathbf{Z}/d\mathbf{Z})^3$ acts as $t \mapsto \zeta_d^e t$ with $e \equiv (d_{20} - d_{30})a_1 + (d_{21} - d_{31})a_2 + (d_{22} - d_{32})a_3 \mod d$. Since $\mathbf{k}$ as defined in Remark 4 is proportional to $(0, 0, 1, -1)A^{-1}$, it follows that $\mathbf{k}$ is proportional to $(d_{20} - d_{30}, d_{21} - d_{31}, d_{22} - d_{32}, d_{23} - d_{33})$, i.e., there is an $m \in \mathbf{Z}$ such that $mk_i = d_{2i} - d_{3i}$. In particular, setting $d' = d/m$ it follows that (a_0, a_1, a_2) acts as $t \mapsto \zeta_{d'}^e t$ with $e \equiv k_1 a_1 + k_2 a_2 + k_3 a_3 \mod d'$.

Let p be a prime number and suppose that p^m divides k_4. Since k_4 is a divisor of d and the greatest common divisor of the k_i equals one, it follows that p^m also divides d'. Since the greatest common divisor of the k_i equals one, it follows that at least one of the k_i is not divisible by p. Without loss of generality we may assume that k_1 is invertible modulo p. From this it follows that we can choose a_1 in such a way that $a_1 k_1 + a_2 k_2 + a_3 k_3 \equiv 1 \mod p^m$.

The corresponding automorphism σ'_{p^m} of S maps t to ζt where ζ is a primitive $p^t n$-th root of unity. Take now $\sigma_{p^m} := (\sigma'_{p^m})^n$. Then σ_{p^m} multiplies t with a primitive p^t-root of unity. Write now $k_4 = \prod p_i^{t_i}$. Then $\sigma := \prod_i \sigma_{p_i^{t_i}}$ multiplies t with a primitive k_4-th root of unity.

Proposition 2. *Let $S \to \mathbf{P}^1$ be a Delsarte fibration with $\det(A) \neq 0$; then there exists a Delsarte base change $S_n \to \mathbf{P}^1$ of $S \to \mathbf{P}^1$ which is isomorphic to the base change of a genus g fibration $S_0 \to \mathbf{P}^1$ with at most one singular fiber outside $0, \infty$.*

Proof. From Lemma 4 it follows that we may assume that the Delsarte fibration is a minimal Delsarte fibration, i.e., we have an affine equation for the generic fiber of the form $m_1 + m_2 + m_3 + tm_4$, where the m_i are monomials in x and y. On a minimal Delsarte fibration $\varphi : S \to \mathbf{P}^1$, there is an automorphism of order k_4 that acts on the t-coordinate as $t \mapsto \zeta_{k_4} t$. In particular, all the fixed points of this automorphism are in the fibers over 0 and ∞.

Consider next $\psi : S/\langle \sigma \rangle \to \mathbf{P}^1/\langle \sigma \rangle \cong \mathbf{P}^1$. Now the singular fibers of φ are possibly at $t = 0, \infty$ and at $t^{k_4} = \prod k_i^{k_i}$; hence, the singular fibers of ψ are possibly at $t = 0, \infty$ and $t = \prod k_i^{k_i}$.

Proposition 3. *Let $\varphi : S \to \mathbf{P}^1$ be a minimal Delsarte fibration with affine equation $m_1 + m_2 + m_3 + tm_4$ such that the general fiber has positive geometric genus. Then one of the following happens:*

- *m_4 equals one of m_1, m_2, m_3. In this case the fibration is isotrivial.*
- *S is Delsarte birational to a Delsarte surface with equation of the form $y^a = f(x, t)$.*
- *Every singular fiber over $t = t_0$ with $t_0 \neq 0, \infty$ is semistable.*

Proof. Assume that all four m_i are distinct. Let N_i be as in Remark 4. Let $t_0 \in K^*$ be such that the fiber over $t = t_0$ is singular. Let $P = (X_0 : X_1 : X_2) \in \mathbf{P}^2$ be a singular point of the fiber. From the proof of Proposition 1, it follows that at least one of the N_i is nonzero and that $(N_1 : N_2 : N_3 : N_4) = (k_1 : k_2 : k_3 : \frac{k_4}{t_0})$ holds. From Remark 4, it follows that at most one of the k_i is zero.

Suppose first that one of the k_i, say k_1, is zero. This implies that N_1 vanishes and that the other N_i are nonzero. Therefore, one of the coordinate of P has to be zero (in order to have $N_1 = 0$). If two of the coordinates of P are zero, then from $\det(A) \neq 0$, it follows that there is some $i \neq 1$ such that $N_i = 0$, which contradicts the fact that at most one N_i vanishes. Hence without loss of generality we may assume $P = (\alpha : 0 : 1)$ with $\alpha \neq 0$, $X_1 \mid N_1$ and $X_1 \nmid N_i$ for $i = 1, 2, 3$. In particular, we have an affine equation for the fibration of the form $m_1 + m_2 + m_3 + t m_4$, where y divides m_1 and m_2, m_3, m_4 are of the form x^{a_i}. Multiply the equation with a power of x such that m_1 is of the form $x^{ab} y^b$ and set $y_1 = y/x^a$. Then we obtain an equation of the form $y_1^b = f(x, t)$, where $f(x, t)$ is of the form $x^a + x^b + t x^c$. This yields the second case.

It remains to consider the case where all the N_i are nonzero. Let $P \in S$ be a point where the fiber over $t = t_0$ singular. Let f be an affine equation for S. We prove below that if we localize $K[x, y, z, t]/(f_x, f_y, f_z, t - t_0)$ at P, then this ring is isomorphic to $k[x]$. Hence the scheme defined by the Jacobian ideal of fiber at $t = t_0$ has length one at the point P. Equivalently, the Milnor number of the singularity of the fiber at $t = t_0$ at the point P equals one. In particular, the singularity of the fiber at P is an ordinary double point.

Consider now the rational map $\tau : \mathbf{P}^2 \setminus V(X_0 X_1 X_2) \to \mathbf{P}^3$ given by $(X_0 : X_1 : X_2) \mapsto (N_1 : N_2 : N_3 : N_4)$. The map τ is unramified at all points $Q \in \mathbf{P}^2$ such that $\tau(Q) \notin V(X_0 X_1 X_2 X_3)$.

Since we assumed that all the N_i are nonzero, it follows that also all the X_i are nonzero. Hence the length of $V(f_{X_0}, f_{X_1}, f_{X_2}, t - t_0)$ at P equals the length of $V(X_0 f_{X_0}, X_1 f_{X_1}, X_2 f_{X_2}, t - t_0)$ at P. From the proof of Proposition 1, it follows that $V(X_0 f_{X_0}, X_1 f_{X_1}, X_2 f_{X_2}, t - t_0)$ is the scheme-theoretic intersection of $\ker B_{t_0}$ and $V(\prod z_i^{k_i} - 1)$ and that this intersection is locally given by $V(k_4 Z_0 - t_0 k_1 Z_3, k_4 Z_1 - t_0 k_2 Z_3, k_4 Z_2 - t_0 k_3 Z_4, Z_2 - t_0 Z_4)$, whence the length of the scheme equals one, and therefore the local Milnor number equals one, and the singularity is an ordinary double point.

Theorem 1. *Suppose $S \to \mathbf{P}^1$ is a Delsarte fibration of genus 1 with nonconstant j-invariant. Then every singular fiber at $t \neq 0, \infty$ is of type I_ν.*

Proof. Without loss of generality we may assume that the fibration is a Delsarte minimal fibration. In particular we have an affine equation for this fibration of the form described in the previous proposition.

In the first case the fibration is isotrivial and therefore the j-invariant is constant; hence, we may exclude this case. If we are in the third case, then each singular fiber at $t = t_0$ is semistable and, in particular, is of type I_ν.

It remains to consider the second case. In this case we have an affine equation of the form $y^a = f(x, t)$. Suppose first that $a > 2$ holds. Then the generic fiber has an automorphism of order a with fixed points. This implies that the j-invariant of the generic fiber is either 0 or 1,728. In particular, the j-invariant is constant and that the fibration is isotrivial. Hence we may assume $a = 2$. In this case we have an affine equation $y^2 = f(x, t)$. Without loss of generality we may assume $x^2 \nmid f$. Since the generic fiber has genus 1, it follows that $\deg_x(f) \in \{3, 4\}$. Since S is a Delsarte surface, it follows that f contains three monomials.

Suppose first that $\deg_x(f) = 3$ and that at $t = t_0$, there is a singular fiber of type different from I_v. Then $f(x, t_0)$ has a triple root, i.e., $f(x, t_0) = (x - t_0)^3$. This implies that $f(x, t_0)$ consists of either one or four monomials in x. This contradicts the fact that $f(x, t)$ consists of three monomials and $t_0 \neq 0$.

If $\deg_x(f) = 4$, then we may assume (after permuting coordinates, if necessary) that $f = x^4 + x^a + t$ or $f = x^4 + tx^a + 1$. If the fiber type at $t = t_0$ is different from I_v, then $f(x, t_0)$ consists of three monomials and $y^2 = f(x, t_0)$ has at singularity different from a node. In particular, $f(x, t_0)$ has a zero or order at least 3 and therefore $f(x, t_0) = (x - a)^4$ or $f(x, t_0) = (x - a)(x - b)^3$. In the first case $f(x, t_0)$ contains five monomials, contradicting the fact that it has three monomials. In the second case note that the constant coefficient of $f(x, t_0)$ is nonzero and hence $ab \neq 0$. Now either the coefficient of x or of x^3 is zero. From this it follows that either $b = -3a$ or $a = -3b$ holds. Substituting this in $f(x, t_0)$ and the fact that $f(x, t_0)$ has at most three monomials yields $b = 0$, contradicting $ab \neq 0$.

Corollary 1. *Let $\varphi : S \to \mathbf{P}^1$ be an elliptic Delsarte surface; then there exists a cyclic base change of φ ramified only at 0 and ∞ that is isomorphic to a cyclic base change, ramified only at 0 and ∞, of an elliptic surface with at most one singular fiber away from 0 and ∞ and this fiber is of type I_v.*

Let $\pi : E \to \mathbf{P}^1$ be an elliptic surface (with section). Define $\gamma(\pi)$ to be

$$\gamma(\pi) := \sum_{t \neq 0, \infty} \left(f_t - \frac{e_t}{6} \right) - \frac{n_0}{6} - \frac{n_\infty}{6},$$

where f_t is the valuation at t of the conductor of the generic fiber of π, e_t the Euler number of $\pi^{-1}(t)$ and n_p is zero unless the fiber at p is of type I_n or I_n^* and in this cases $n_p = n$.

In [2, 3] Fastenberg studies rational elliptic surfaces with $\gamma < 1$. She determines the maximal Mordell–Weil rank of such elliptic surfaces under cyclic base changes of the form $t \mapsto t^n$.

We will now show that each Delsarte fibration of genus 1 with nonconstant j-invariant becomes after a Delsarte base change the base change of a rational elliptic surface with $\gamma < 1$. In particular, the maximal Mordell–Weil ranks for Delsarte fibrations of genus 1 under cyclic base change (as presented in [4, Sect. 3.4] and [5]) can also be obtained from [3].

Corollary 2. *Let $\pi : S \to \mathbf{P}^1$ be a minimal Delsarte fibration of genus 1 with nonconstant j-invariant. Then S is the base change of a rational elliptic surface with $\gamma < 1$.*

Proof. From Theorem 1 it follows that π is the base change of an elliptic fibration $\pi' : S' \to \mathbf{P}^1$ with at most one singular fiber away from 0 and ∞ and this fiber is of type I_ν. Since the j-invariant is nonconstant, it follows that π' has at least three singular fibers; hence, there is precisely one singular fiber away from 0 and ∞. Since this fiber is of type I_ν, it follows that $f_t = 1$ for this fiber. Hence

$$\gamma = 1 - \frac{e_t + n_0 + n_\infty}{6} < 1.$$

Remark 5. The converse statement to this results is also true: let $\pi : S \to \mathbf{P}^1$ be a rational elliptic surface with $\gamma < 1$, only one singular fiber away from 0 and ∞ and this fiber is of type I_ν. Then there exists a base change of the form $t \mapsto t^n$ such that the pullback of $\pi : S \to \mathbf{P}^1$ along this base change is birational to the standard fibration on a Delsarte surface. One can obtain this result by comparing the classification of elliptic Delsarte surfaces from [4, Chap. 3] with the tables in [2, 3].

Example 1. The second line of entry 14 of [3, Table 4] exhibits the existence of an elliptic surface with a IV-fiber at $t = 0$, an I_1-fiber at $t = \infty$ and one further singular fiber that is of type I_1^*. In [3] it is shown that the maximal rank under base changes of the form $t \mapsto t^n$ is 9. Such a fibration has a nonconstant j-invariant and $\gamma < 1$. Corollary 1 now implies that this fibration is not a Delsarte fibration.

If we twist the I_1^* fiber and one of the fibers at $t = 0$ or $t = \infty$, then we get the following fiber configurations $IV; I_1^*; I_1$ or $II^*; I_1; I_1$. Both these configuration occur also in [3, Table 4]. The maximal rank under base changes of the form $t \mapsto t^n$ equals 9 in both cases. Now $y^2 = x^3 + x^2 + t$ has singular fibers of type I_1 at $t = 0$ and $t = -4/27$ and of type II^* at $t = \infty$ and $y^2 = x^3 + tx + t^2$ has a IV-fiber at $t = 0$, a I_1 fiber at $t = -4/27$ and a I_1^* fiber at $t = \infty$. Hence both fibration occur as Delsarte fibrations.

Example 2. Consider the elliptic Delsarte surface that corresponds to

$$Y^2 = X^3 + X^2 + tX.$$

We can easily compute the discriminant and j-invariant of this fibration:

$$\Delta = -64t^3 - 16t^2 \text{ and } j = 256\frac{(3t-1)^3}{4t^3 - t^2}.$$

From this we can see that there are three singular fibers. Over $t = 0$ there is I_2-fiber, over $t = \infty$ there is a III-fiber, and over $t = -1/4$ there is a I_1-fiber. We then check that this corresponds to the second entry in the list of [3].

Remark 6. The approaches to determine the maximal Mordell–Weil ranks under cyclic base change in [3] and in [4] are quite different. The former relies on studying the local system coming from the elliptic fibration, whereas the latter purely relies on Shioda's algorithm to determine Lefschetz numbers of Delsarte surfaces. This explains why Fastenberg can deal with several base changes where the "minimal" fibration has four singular fibers (which cannot be covered by Shioda's algorithm because of Proposition 2) but cannot deal with fibrations with constant j-invariant. Instead Shioda's algorithm can handle some of them.

3 Isotrivial Fibrations

Using Proposition 3 one easily describes all possible isotrivial minimal Delsarte fibrations.

Proposition 4. *Suppose the standard fibration on S is isotrivial and that the genus of the generic fiber is positive. Then there is a Delsarte base change and a Delsarte birational map such that the pullback of the standard fibration is of the form $m_1 + m_2 + (1 + t^n)m_3$, $y^3 + x^3 + x^2 + t^n$, or $y^a + x^2 + x + t^n$.*

Proof. Suppose the affine equation for S is of the third type of Proposition 3. Then S admits a semistable fiber and in particular the fibration cannot be isotrivial. If the affine equation for S is of the first type of Proposition 3, then the generic fiber is (after an extension of the base field) isomorphic to $m_1 + m_2 + m_3$; in particular, each two smooth fibers of the standard fibration are isomorphic and therefore this fibration is isotrivial. In this case S is the pullback of $m_1 + m_2 + (1 + t^n)m_3$.

Hence we may restrict ourselves to the case where we have an affine equation of the form $y^a = x^b f(x, t)$ where f consists of three monomials, $f(0, t)$ is not zero, and the exponent of x in each of the three monomials in f is different. Moreover, after a Delsarte birational map we may assume that $b < a$.

The surface S is birational to a surface $y^a = x^b z^{c+\deg(f)} f(x/z, t)$ in $\mathbf{P}(1, w, 1)$, with $0 \le c < a$ and $w = (b + c + \deg(f))/a \in \mathbf{Z}$. The standard fibration on S is isotrivial if and only if the moduli of the zero set of $x^b z^{c+\deg(f)} f(x/z, t)$ in $\mathbf{P}^1_{(x:z)}$ are independent of t. We will now consider this problem.

We cover first the case where $d' := \deg_x(f) > 2$ holds.

After swapping the role of x and z, if necessary, we may assume that the coefficient of $xz^{d'-1}$ is zero. We claim that after a map of the form $y = t^{c_1} y, x = t^{c_2} x, z = z, t = t^{c_3}$, we may assume that $f = x^{d'} + x^{c'} z^{d'-c'} + t z^{d'}$. To see this, take an affine equation for S of the form $y^a = x^b(a_1 x^{d'} + a_2 x^{c'} + a_3)$, where $a_i \in \{1, t\}$ and two of the a_i equal 1. If $a_1 = t$, then we need to take an integer solution of $ac_1 = bc_2 + d'c_2 + c_3 = bc_2 + c'c_2$ and if $a_2 = t$, then we need to take an integer solution of $ac_1 = bc_2 + d'c_2 + c_3 = bc_2 + c'c_2$. In both cases we obtain an affine equation of the form $y^a = x^b(x^{d'} + x^{c'} + t^n)$. This fibration is isotrivial if and only if $y^a = x^b(x^{d'} + x^{c'} + t)$ is isotrivial, which proves the claim.

Hence from now on we assume that f is of the form $x^{d'} + x^{c'} z^{d'-c'} + t z^{d'}$ with $d' > 2$ and $c' > 1$.

Let s denote the number of distinct zeroes of $g(x, z) := x^b z^{c+\deg(f)} f(x/z, t)$ for a general t-value. We say that the fiber at $t = t_0$ is bad if $x^b z^{c+\deg(f)} f(x/z, t)$ has at most $s - 1$ distinct zeroes. The main result from [6] yields that if the fiber at $t = t_0$ is bad, then $g(x, z)$ has at most 3 distinct zeroes. We are first going to classify all g satisfying this condition. Then we will check case-by-case whether the moduli of the zeroes of g are independent of t.

Consider the fiber over $t = 0$. From $c' > 1$ it follows that $x = 0$ is a multiple zero of $f(x, 0)$. Hence that the fiber over $t = 0$ is bad. If c is positive, then the criterion from [6] implies that $g(x, 0)$ can have at most one further zero and hence $d' = c' + 1$. If $c = 0$, then g can have at most two further zeroes and therefore $d' - c' \in \{1, 2\}$.

Suppose first $d' = c' + 1$. Consider $f'(x,t) := \frac{\partial}{\partial x} f(x,t)$. Our assumption on f implies that $f'(x,t)$ is a polynomial only in x. The fiber at $t = t_0$ is bad if and only if $f'(x, t_0)$ and $f(x, t_0)$ have a common zero. From $c' = d' - 1$, it follows that $f'(x,t)$ has a unique zero different from 0, say x_0, and x_0 is a simple zero of $f'(x,t)$. Now $f(x_0, t)$ is a linear polynomial in t. Hence there is a unique nonzero t-value t_0 over which there is a bad fiber. Since x_0 is a simple zero of $f'(x, t_0)$, it follows that $(x - x_0)^2$ divides $f(x, t_0)$ and that there are $d' - 2$ further distinct zeroes, all different from 0. Using that g has at most 3 zeroes, it follows that if both b and c are nonzero, then $d' - 2 = 0$; if one of b, c is zero, then $d' - 2 \leq 1$; and if both b and c are zero, then $d' - 2 \leq 2$. Using that we assumed that d' is at least 3, we obtain the following possibilities for g: $x^b(x^3 + x^2z + tz^3)$, $z^c(x^3 + x^2z + tz^3)$, $x^3 + x^2z + tz^3$ and $x^4 + x^3z + tz^4$.

Suppose now $c = 0$ and $d' = c' + 2$. Then f' is of the form $\beta(x^2 + \alpha)x^{d'-3}$. In particular, there are two possible x-values for a bad point in a bad fiber. If they occur in the same fiber and $b = 0$, then $d' \in \{4, 5\}$, otherwise $d' \in \{3, 4\}$. Since $2 \leq c' = d' - 2$, we may exclude $d' = 3$ and we obtain that the two polynomials $x^4 + x^2 + t$ and $x^5 + x^3 + t$ are the only possibilities for f. We can exclude $x^5 + x^3 + t$, since it has bad fibers at $t^2 = \frac{-3125}{108}$ and a necessary condition to have $d' = 5$ is that there is at most one bad fiber with $t \neq 0, \infty$.

If $b > 0$, then $d' \leq 4$; in particular, we have only $x^b(x^4 + x^2 + t)$ to check.

Actually only in one of the above cases the moduli are independent of t, namely, $g = x^3 + x^2z + tz^3$.

Note that the j-invariants of the elliptic curves $y^2 = x^3 + x^2 + t$ and $y^2 = tz^3 + z + 1$ are not constant; hence, the moduli of the zeros of $x^b(x^3 + x^2z + tz^3)$ and of $z^c(x^3 + x^2z + tz^3)$ depend on t (if $b > 0$ resp. $c > 0$ holds). Since $x^3 + x^2z + tz^3$ has degree 3, the moduli of its zeroes are obviously constant.

The family of genus one curves $y^2 = x^4 + x^2 + t$ has a semistable fiber at $t = \frac{1}{4}$, and the family of genus one curves $y^2 = x^4 + x^3z + tz^4$ has a semistable fiber at $t = \frac{27}{256}$. Hence the moduli of the zeroes of $x^b(x^4 + x^2 + t)$ for $b \geq 0$ and of $x^4 + x^3z + tz^4$ depend on t.

Consider now the final case $d' = 2$. Then $f = x^2 + x + t$, and therefore automatically two of the three possibilities for g, namely, $z^c(x^2 + xz + tz^2)$ and $x^b(x^2 + xz + tz^2)$, have constant moduli since they define three points in $\mathbf{P}^1$. Now $y^{b+2} = z^b(x^2 + xz + tz^2)$ and $y^{b+2} = x^b(x^2 + xz + tz^2)$ are birationally equivalent up to a Delsarte base change, e.g., take $((x : y : z), t) \mapsto ((z : \frac{y}{t} : \frac{x}{t^{b+2}}), t^{b+2})$. Hence these two cases yield only one case up to isomorphism. We may assume that the affine equation equals $y^b + x^2 + x + t^n$. If $b = c = 0$ holds, then the generic fiber is a cyclic cover of $\mathbf{P}^1$ ramified at two points, and in particular has genus 0. Hence we can exclude this case. Finally, $x^b z^c(x^2 + xz + tz^2)$ does not have constant moduli since the j-invariant of $y^2 = x^3 + x^2 + tx$ is nonconstant.

Remark 7. In the case of $y^a + x^2 + x + t$, we may complete the square. This yields a surface that is isomorphic to $y^a + x^2 + 1 + t$; in particular, the fibration is birationally equivalent to a fibration of the first kind. However, they are not Delsarte birational.

In [4, Sect. 3.5.1] it is shown that $y^3 + x^3 + x^2 + t$ is birational to $y^2 + x^3 + t^3 + 1$; however, the given birational map is not a Delsarte birational map.

Hence both exceptional cases are fibrations that are birational to a fibration of the first type.

From the previous discussion, it follows that almost all minimal isotrivial Delsarte fibrations are of the form $m_1 + m_2 + (1 + t)m_3$.

We will calculate the Picard numbers for one class of such fibrations and consider the behavior of the Picard number under Delsarte base change, i.e., base changes of the form $t \mapsto t^a$.

Example 3. Let $p = 2g + 1$ be a prime number. Consider the isotrivial fibration $y^2 = x^p + t^{2ap} + s^{2ap}$ of genus g-curves over $\mathbf{P}^1_{(s:t)}$. This equation defines a quasi-smooth surface S of degree $2ap$ in $\mathbf{P}(2a, ap, 1, 1)$. The surface S has one singular point, namely, at $(1 : 1 : 0 : 0)$. A single blowup of this suffices to obtain a smooth surface $\tilde{S}$. The Lefschetz number of $\tilde{S}$ can be computed by using Shioda's algorithm, which we do below. The exceptional divisor of $\tilde{S} \to S$ is a smooth rational curve. In particular, using the Mayer–Vietoris sequence, one easily obtains that $h^2(\tilde{S}) = h^2(S) + 1$ and $\rho(\tilde{S}) = \rho(S) + 1$. From the fact that S is quasi-smooth, it follows that the mixed Hodge structure on $H^2(S)$ is a pure weight 2 Hodge structure. To determine the Hodge numbers of this Hodge structure, we use a method of Griffiths and Steenbrink. Note first that $\dim H^2(S)_{\mathrm{prim}} = h^2(S) - 1 = h^2(\tilde{S}) - 2$.

Let R be the Jacobian ring of S, i.e.,

$$R = \mathbf{C}[x, y, s, t]/\left(\frac{\partial f}{\partial x}, \frac{\partial f}{\partial y}, \frac{\partial f}{\partial s}, \frac{\partial f}{\partial t}\right) = \mathbf{C}[x, y, s, t]/(x^{p-1}, y, t^{2p-1}, s^{2p-1}).$$

This is a graded ring with weights $(2a, ap, 1, 1)$. Let $d = 2ap$ be the degree of S and $w = ap + 2a + 2$ the sum of the weights.

From Griffiths–Steenbrink [9], it follows that $H^{2-q,q}(S)_{\mathrm{prim}}$ is isomorphic with $R_{(q+1)d-w}$. The set

$$B_0 := \{x^i t^j s^k \mid 2ai + j + k = (q + 1)d - w, 0 \le i < p - 1, 0 \le j, k < 2ap - 1\}$$

is a basis for $R_{(q+1)d-w}$. From now on we consider the elements of B_0 as elements of $\mathbf{C}[x, y, s, t]$. Multiplying each element of B_0 with $yxts$ shows that the elements of B_0 are in a one-to-one correspondence with the elements of

$$B := \{yx^i t^j s^k \mid ap + 2ai + j + k = (q + 1)d, 0 < i \le p - 1, 0 < j, k \le 2ap - 1\}.$$

We associate to an element $yx^i t^j s^k \in B$ the vector $\mathbf{v} = \left(\frac{1}{2}, \frac{i}{p}, \frac{j}{2ap}, \frac{k}{2ap}\right) \in \mathbf{Q}^4$. From $yx^i t^j s^k \in B$, it follows that each entry of $\mathbf{v}$ is a rational number α such that $0 < \alpha < 1$ holds. In particular, $\mathbf{v}$ is determined by its image in $(\mathbf{Q}/\mathbf{Z})^4$, and none of the coordinates of $\mathbf{v}$ is zero in $\mathbf{Q}/\mathbf{Z}$. Summarizing we have constructed a one-to-one correspondence between a basis for $H^{2-q,q}_{\mathrm{prim}}(S)$ and the elements of

$$\left\{\left(\frac{1}{2}, \frac{i}{p}, \frac{j}{2ap}, \frac{k}{2ap}\right) \in (\mathbf{Q}/\mathbf{Z})^4 \,\middle|\, \begin{array}{l} i, j, k \in \mathbf{Z}; 0 < i < p; 0 < j, k < 2ap; \\ \frac{1}{2} + \frac{i}{p} + \frac{j}{2ap} + \frac{k}{2ap} = q + 1 \end{array}\right\}. \quad (1)$$

(Here we consider $\frac{1}{2} + \frac{i}{p} + \frac{j}{2ap} + \frac{k}{2ap}$ as an element of $\mathbf{Q}$.)

In [4, Sect. 2.1] a variant of Shioda's algorithm [7] is presented. This algorithm calculates the Lefschetz number of a resolution of singularities of $\tilde{T}$ a Delsarte surface T in $\mathbf{P}^3$. In our case we apply this algorithm to the surface $T \subset \mathbf{P}^4$ given by

$$-Y^2 Z^{2ap-2} + X^p Z^{2ap-p} + W^{2ap} + Z^{2ap}.$$

Since the Lefschetz number is a birational invariant, one has that the Lefschetz numbers of $\tilde{S}$ and $\tilde{T}$ coincide.

To determine the Lefschetz number, we follow [4]. Let

$$A = \begin{pmatrix} 0 & 2 & 0 & 2ap-2 \\ p & 0 & 0 & (2a-1)p \\ 0 & 0 & 2ap & 0 \\ 0 & 0 & 0 & 2ap \end{pmatrix}$$

be the exponent matrix of T. Define $\mathbf{v}_1, \mathbf{v}_2, \mathbf{v}_3$ as follows:

$$\mathbf{v}_1 := (1,0,0,-1)A^{-1}, \mathbf{v}_2 := (1,0,0,-1)A^{-1}, \text{ and } \mathbf{v}_3 := (0,0,1,-1)A^{-1}.$$

In our case this yields the vectors

$$\mathbf{v}_1 = \left(0, \frac{1}{p}, 0, \frac{-1}{p}\right), \mathbf{v}_2 = \left(\frac{1}{2}, 0, 0, \frac{-1}{2}\right), \text{ and } \mathbf{v}_3 = \left(0, 0, \frac{1}{2ap}, \frac{-1}{2ap}\right).$$

Consider now the set $L := i\mathbf{v}_1 + k\mathbf{v}_2 + j\mathbf{v}_3 \in \mathbf{Q}/\mathbf{Z}$. These are precisely the vectors of the form

$$\left\{ \left(\frac{k}{2}, \frac{i}{p}, \frac{j}{2ap}, \frac{-apk - 2ai - j}{2ap}\right) \in (\mathbf{Q}/\mathbf{Z})^4 \,\middle|\, i,j,k \in \mathbf{Z} \right\}.$$

For an element $\alpha \in \mathbf{Q}/\mathbf{Z}$ denote with $\{\alpha\}$ the fractional part, i.e., the unique element $\beta \in \mathbf{Q} \cap [0,1)$ such that $\alpha - \beta \equiv 0 \mod \mathbf{Z}$ and with $\mathrm{ord}_+(\alpha)$ the smallest integer $k > 0$ such that $k\alpha \in \mathbf{Z}$.

Let $L_0 \subset L$ be the set of vectors $\mathbf{v} \in L$ such that none of the entries of $\mathbf{v}$ equals 0 modulo $\mathbf{Z}$, i.e.,

$$\left\{ \left(\frac{1}{2}, \frac{i}{p}, \frac{j}{2ap}, \frac{-ap - 2ai - j}{2ap}\right) \in (\mathbf{Q}/\mathbf{Z})^4 \,\middle|\, \begin{array}{l} i,j \in \mathbf{Z}, 0 < i < p, 0 < j < 2ap, \\ j \not\equiv -ap - 2ai \mod 2ap \end{array} \right\}.$$

For an element $\mathbf{v} = (\alpha_1, \alpha_2, \alpha_3, \alpha_4) \in (\mathbf{Q}/\mathbf{Z})^4$ define $q(\mathbf{v}) := \{a_1\} + \{a_2\} + \{a_3\} + \{a_4\}$. Note that for $\mathbf{v} \in L_0$ we have $q(\mathbf{v}) \in \{1,2,3\}$. Moreover for $q = 0,1,2$, we have that the set $\{\mathbf{v} \in L_0 \mid q(\mathbf{v}) = q + 1\}$ corresponds to a basis of $H^{2-q,q}(S)_{\mathrm{prim}}$ by (1). In particular, $\#L_0$ is precisely $h^2(S)_{\mathrm{prim}}$.

Define $\Lambda \subset L_0$ as the set of elements $(\alpha_1, \alpha_2, \alpha_3, \alpha_4) \in L_0$ such that there is a $t \in \mathbf{Z}$ for which $\mathrm{ord}_+(\alpha_k t) = \mathrm{ord}_+(\alpha_k)$ holds for $k = 1,2,3,4$ and $\{t\alpha_1\} + \{t\alpha_2\} + \{t\alpha_3\} + \{t\alpha_4\} \neq 2$. The condition $\mathrm{ord}_+(\alpha_k t) = \mathrm{ord}_+(\alpha_k)$ for $k = 1,2,3,4$ is equivalent with t being invertible modulo $2a'p$, where $a' = a/\gcd(a, j)$. Then the Lefschetz number $\lambda = h^2(\tilde{T}) - \rho(\tilde{T})$ equals $\#\Lambda$.

Since $\lambda(\tilde{S}) = \lambda(\tilde{T})$ and $h^2(\tilde{S}) = 2 + \#L_0$, it follows that $\rho(S)$ equals

$$2 + \#\left\{(\alpha_1, \alpha_2, \alpha_3, \alpha_4) \in L_0 \,\middle|\, \begin{array}{l} \{t\alpha_1\} + \{t\alpha_2\} + \{t\alpha_3\} + \{t\alpha_4\} = 2 \text{ for } t \in \mathbf{Z} \\ \text{such that } \mathrm{ord}_+(t\alpha_k) = \mathrm{ord}_+(\alpha_k), k = 1, 2, 3, 4 \end{array}\right\}.$$

We now determine this set.

Consider now a vector $\mathbf{v}$ from L_0, i.e., a vector

$$\left(\frac{1}{2}, \frac{i}{p}, \frac{j}{2ap}, \frac{ap - 2ai - j}{2ap}\right)$$

with $i, j \in \mathbf{Z}$, $i \not\equiv 0 \bmod p$, $j \not\equiv 0 \bmod 2ap$, $ap - 2api - j \not\equiv 0 \bmod 2ap$.

Take $t \in \{1, \ldots, 2a'p - 1\}$ such that $\gcd(t, 2a'p) = 1$ and $t \equiv i^{-1} \bmod p$. Then $\mathbf{v} \in \Lambda$ if and only if $t\mathbf{v} \in \Lambda$. Hence, to determine whether a vector is in Λ, it suffices to assume $i \equiv 1 \bmod p$.

Suppose now that $p > 7$. In Lemma 6 we show that $\mathbf{v} \notin \Lambda$ if and only if the fractional part $\{\frac{j}{2ap}\}$ is in the set $\left\{\frac{p-1}{2p}, \frac{1}{2}, \frac{p+2}{2p}, \frac{2p-4}{2p}, \frac{2p-2}{2p}, \frac{2p-1}{2p}\right\}$. Each of the six values for j yields $(p - 1)$ elements in $L_0 \setminus \Lambda$; hence, $\rho(\tilde{S}) = 2 + 6(p - 1)$.

One can easily find several divisors on $\tilde{S}$. We will describe a subgroup of finite index in $\mathrm{NS}(\tilde{S})$. Since the maximal Picard number is attained for $a = 1$, we may assume that a equals 1. We remarked in the introduction that the pullback of the hyperplane class H and the exception divisor E yield two independent classes in $\mathrm{NS}(\tilde{S})$. We give now $6(p - 1)$ further independent classes: Let $\alpha_\pm$ be p-th roots of $\pm\sqrt{2}$. Consider now the curves C_1 given by $x - t^2 = y - s^p = 0$, and the curves $C_{2,3}$ given by $x = \alpha_\pm t, y = t^p \pm s^p$. Let D_i be the curve obtained by swapping s and t.

An easy argument using the intersection pairing shows that H, E, C_1, C_2, C_3 are independent in $\mathrm{NS}(\tilde{S})$. Let ζ be a primitive p-th root of unity. Let σ be the automorphism $t \mapsto \zeta t$, let $G = \langle\sigma\rangle$. Consider now the subgroup of $\mathrm{NS}(\tilde{S})$ generated by H, E, C_1, C_2, C_3 and its conjugates under the powers of σ. This subgroup has a natural structure $\mathbf{Z}^r \oplus Z[\zeta]^s$. Since H and E are fixed under G, it follows that $r \geq 2$. Since C_1, C_2 and C_3 are linearly independent and not mapped to an algebraically equivalent divisor under σ, it follows that $s \geq 3$. For each i we have $[\sum_{\sigma \in G} C_i^\sigma] = [H]$ in $\mathrm{NS}(\tilde{S})$; hence, the conjugates of H, E, C_1, C_2, C_3 generate a subgroup of $\mathrm{NS}(\tilde{S})$ or rank at most $2 + 3(p - 1)$. Hence $r = 2$ and $s = 3$.

Now the above constructed subgroup of $\mathrm{NS}(\tilde{S})$ is invariant under the automorphism τ mapping s to ζs. However this automorphism does map each D_i to a divisor that is non-algebraically equivalent to D_i. As above one can show that the H, E, C_i, D_i and all their conjugates generate a rank $2 + 6(p - 1)$ subgroup of $\mathrm{NS}(\tilde{S})$.

Lemma 6. *Suppose $p > 7$. Let*

$$\mathbf{v} = \left(\frac{1}{2}, \frac{1}{p}, \frac{j}{2ap}, \frac{-2a - ap - j}{2ap}\right) \in (\mathbf{Q}/\mathbf{Z})^4$$

such that $j \not\equiv 0 \bmod 2ap, 2a + j + ap \not\equiv 0 \bmod 2ap$. Then $\mathbf{v} \notin \Lambda$ if and only if

$$\frac{j}{2ap} \in \left\{ \frac{p-1}{2p}, \frac{1}{2}, \frac{p+2}{2p}, \frac{2p-4}{2p}, \frac{2p-2}{2p}, \frac{2p-1}{2p} \right\}.$$

Proof. Without loss of generality we may assume that $\gcd(a, j) = 1$.

We start by proving that if a prime $\ell \geq 5$ divides a, then $v \in \Lambda$. For this it suffices to give a t, invertible modulo $2ap$ such that

$$\left\{ \frac{t}{2} \right\} + \left\{ \frac{t}{p} \right\} + \left\{ \frac{tj}{2ap} \right\} + \left\{ \frac{(-2a - ap - j)t}{2ap} \right\} = 1.$$

Since the left hand side is an integer for any choice of t and each summand is smaller than one, it suffices to prove that

$$\left\{ \frac{t}{2} \right\} + \left\{ \frac{t}{p} \right\} + \left\{ \frac{tj}{2ap} \right\} \leq 1.$$

Consider the value

$$t = 1 + ck\frac{2ap}{\ell},$$

with $c \equiv j^{-1} \bmod \ell$ and $k \in \mathbf{Z}$ such that $k \not\equiv (c\frac{2ap}{\ell})^{-1} \bmod \ell$ and k in the interval

$$\left(-\frac{\ell j}{2ap}, -\frac{\ell j}{2ap} + \frac{\ell(p-2)}{2p} \right).$$

Note that we have to assume $p > 7$ or $\ell \geq 5$ to ensure the existence of such a k. Then $\left\{ \frac{t}{2} \right\} = \frac{1}{2}$ and

$$\left\{ \frac{t}{p} \right\} = \left\{ \frac{1}{p} + ck\frac{2a}{\ell} \right\} = \frac{1}{p}.$$

Moreover, we have that

$$\left\{ \frac{tj}{2ap} \right\} = \left\{ \frac{(1 + ck\frac{2ap}{\ell})j}{2ap} \right\} = \left\{ \frac{j}{2ap} + \frac{k}{\ell} \right\} \leq \frac{(p-2)}{2p}.$$

From this it follows that

$$\left\{ \frac{t}{2} \right\} + \left\{ \frac{t}{p} \right\} + \left\{ \frac{tj}{2ap} \right\} \leq 1$$

holds, which finishes this case.

Suppose now that the only primes dividing a are 2 or 3. If $p = 11, 13, 17$ and $a = 3$, then one can find by hand a t-value such that

$$\left\{ \frac{t}{2} \right\} + \left\{ \frac{t}{p} \right\} + \left\{ \frac{tj}{2ap} \right\} + \left\{ \frac{(-2a - ap - j)t}{2ap} \right\} = 1$$

holds. For all other combinations (a, j, p) with $a > 1$ we give a value for t in Table 1 such that the above formula holds.

Table 1: t-values for the case $a = 2^{v_2} 3^{v_3}$, $a \neq 1$

a	$\frac{j}{2ap} \in I$		t
$4 \mid a$	$(0, \frac{p-2}{2p})$		1
$p > 3$	$(\frac{1}{2}, \frac{p-1}{p})$		$1 + ap$
	$(0, \frac{p-4}{4p}) \cup (\frac{3}{4}, 1)$	$j \equiv 1 \bmod 4$	$1 + \frac{ap}{2}$
	$(\frac{1}{4}, \frac{3p-4}{4p})$	$j \equiv 1 \bmod 4$	$1 + \frac{3ap}{2}$
	$(0, \frac{p-4}{4p}) \cup (\frac{3}{4}, 1)$	$j \equiv 3 \bmod 4$	$1 + \frac{3ap}{2}$
	$(\frac{1}{4}, \frac{3p-4}{4p})$	$j \equiv 3 \bmod 4$	$1 + \frac{ap}{2}$
$2 \mid a, 4 \nmid a'$	$(0, \frac{p-2}{2p})$		1
$p > 7$	$(\frac{1}{2}, \frac{p-1}{p})$		$1 + ap$
	$(0, \frac{1}{8} - \frac{1}{p}) \cup (\frac{3}{8}, \frac{5}{8} - \frac{1}{p}) \cup (\frac{7}{8}, 1)$	$j \equiv 1 \bmod 4$	$2 + \frac{ap}{2}$
	$(\frac{3}{8}, \frac{5}{8} - \frac{1}{p})$	$j \equiv 3 \bmod 4$	$2 + \frac{3ap}{2}$
$9 \mid a$	$(0, \frac{p-2}{2p})$		1
$p > 5$	$(\frac{1}{3}, \frac{5}{6} - \frac{1}{p})$	$j \equiv 2 \bmod 3$	$1 + \frac{2ap}{3}$
	$(0, \frac{1}{3} - \frac{1}{p}) \cup (\frac{2}{3}, 1)$	$j \equiv 2 \bmod 3$	$1 + \frac{4ap}{3}$
	$(\frac{1}{3}, \frac{5}{6} - \frac{1}{p})$	$j \equiv 1 \bmod 3$	$1 + \frac{4ap}{3}$
	$(0, \frac{1}{3} - \frac{1}{p}) \cup (\frac{2}{3}, 1)$	$j \equiv 1 \bmod 3$	$1 + \frac{2ap}{3}$
$a = 3$	$(0, \frac{p-2}{2p})$		1
$p \equiv 1 \bmod 3$	$(\frac{1}{3}, \frac{5}{6} - \frac{1}{p})$	$j \equiv 1 \bmod 3$	$1 + 4p$
$p > 18$	$(\frac{8}{9}, \frac{19}{18} - \frac{1}{p})$	$j \equiv 1 \bmod 3$	$3 + 2p$
	$(\frac{7}{9}, \frac{17}{18} - \frac{1}{p})$	$j \equiv 1 \bmod 3$	$3 + 4p$
	$(\frac{2}{3}, 1)$	$j \equiv 2 \bmod 3$	$1 + 4p$
	$(\frac{4}{9}, \frac{11}{18} - \frac{1}{p})$	$j \equiv 2 \bmod 3$	$3 + 2p$
	$(\frac{5}{9}, \frac{13}{18} - \frac{1}{p})$	$j \equiv 2 \bmod 3$	$3 + 4p$
$a = 3$	$(0, \frac{p-2}{2p})$		1
$p \equiv 2 \bmod 3$	$(\frac{2}{3}, 1)$	$j \equiv 1 \bmod 3$	$1 + 2p$
$p > 18$	$(\frac{5}{9}, \frac{13}{18} - \frac{1}{p})$	$j \equiv 1 \bmod 3$	$3 + 2p$
	$(\frac{4}{9}, \frac{11}{18} - \frac{1}{p})$	$j \equiv 1 \bmod 3$	$3 + 4p$
	$(\frac{1}{3}, \frac{5}{6} - \frac{1}{p})$	$j \equiv 2 \bmod 3$	$1 + 2p$
	$(\frac{7}{9}, \frac{17}{18} - \frac{1}{p})$	$j \equiv 2 \bmod 3$	$3 + 2p$
	$(\frac{8}{9}, \frac{19}{18} - \frac{1}{p})$	$j \equiv 2 \bmod 3$	$3 + 4p$

The only case left to consider is the case $a = 1$. If $p \leq 30$, then one can easily find an appropriate t-value by hand. Hence we may assume that $p > 30$. If we take $t = 1$, then we see that $v \in \Lambda$ whenever

$$\frac{j}{2ap} = \frac{j}{2p} \in \left(0, \frac{1}{2} - \frac{1}{p}\right).$$

We will consider what happens if $\frac{j}{2p} > \frac{p-2}{2p}$.

Suppose $t < p$ is an odd integer and k is an integer such that $k \leq \frac{tj}{2p} < k + 1$. Then we have

$$\left\{\frac{t}{2}\right\} + \left\{\frac{t}{p}\right\} + \left\{\frac{tj}{2p}\right\} = \frac{1}{2} + \frac{t}{p} + \frac{tj}{2p} - k$$

The right hand side is at most 1 if

$$\frac{j}{2p} \leq \frac{1 + 2k}{2t} - \frac{1}{p}.$$

Hence if

$$\frac{j}{2p} \in \left(\frac{k}{t}, \frac{1 + 2k}{2t} - \frac{1}{p}\right),$$

then $\mathbf{v} \in \Lambda$.

If we take $k = t - 1$, then we get the interval

$$I_t := \left(1 - \frac{1}{t}, 1 - \frac{1}{2t} - \frac{1}{p}\right)$$

and if we take $k = (t + 1)/2$, then we get

$$I'_t := \left(\frac{1}{2} + \frac{1}{2t}, \frac{1}{2} + \frac{1}{t} - \frac{1}{p}\right).$$

Note that $I_3 = I'_3$.

We claim that if $p > 30$ and $5 \leq t \leq \frac{p-1}{2} - 3$, then $I'_t \cap I'_{t-2} \neq \emptyset$ and $I_t \cap I_{t-2} \neq \emptyset$. For this, it suffices to check that

$$\frac{1}{2} + \frac{1}{2(t-2)} < \frac{1}{2} + \frac{1}{t} - \frac{1}{p} \quad \text{and} \quad 1 - \frac{1}{2(t-2)} - \frac{1}{p} > 1 - \frac{1}{t}.$$

Both conditions are equivalent with

$$2t^2 - (p + 4)t + 4p < 0. \tag{2}$$

The smallest value to check is $t = 5$; then the above formula yields that $p > 30$, which is actually the case. For fixed p we have that the above bound is equivalent with $t \in (\frac{1}{4}p + 1 - \frac{1}{4}\sqrt{p^2 - 24p + 16}, \frac{1}{4}p + 1 + \frac{1}{4}\sqrt{p^2 - 24p + 16})$. The previous argument already shows that the left boundary of this interval is smaller than 5. Substituting $t = \frac{p-1}{2} - 3$ in (2) yields that for $p > 77/3$ the boundary point on the right is bigger than $\frac{p-1}{2} - 3$. In particular, if $p > 30$, t is odd, then $I'_t \cap I'_{t-2} \neq \emptyset$ and $I_t \cap I_{t-2} \neq \emptyset$. Take now the union of I'_t and I_t for all odd t with $3 < t < \frac{p-1}{2} - 5$. This yields an interval $I = (\alpha, \beta)$ such that for all $\frac{j}{2p} \in I$, we have that $\mathbf{v} \in \Lambda$. The maximal t-value is either $\frac{p-1}{2} - 3$ or $\frac{p-1}{2} - 4$ (depending on $p \bmod 4$). Hence we know only that the maximal t is at least $\frac{p-1}{2} - 4$. From this it follows that

$$I \supset \left(\frac{1}{2} + \frac{1}{p-9}, 1 - \frac{1}{p} - \frac{1}{p-9} \right).$$

Note that $p - 9 > \frac{2}{3}p$ and hence $\frac{1}{p} + \frac{1}{p-9} \leq \frac{5}{2p}$. Hence the only possibilities for $\frac{j}{2p} \notin I$ and $p - 2 < j < 2p$ are

$$\left\{ \frac{p-1}{2p}, \frac{p}{2p}, \frac{p+1}{2p}, \frac{p+2}{2p}, \frac{2p-4}{2p}, \frac{2p-3}{2p}, \frac{2p-2}{2p}, \frac{2p-1}{2p} \right\}.$$

If $\frac{j}{2p} \in \{ \frac{p+1}{2p}, \frac{2p-3}{2p} \}$, then we have that $\mathbf{v}$ is in Λ. This can be verified by taking $t = p - 2$. Hence we have shown that for all but six values for $\frac{j}{2p}$, the corresponding vector is in Λ.

It remains to show that for the remaining values of $\frac{j}{2p}$, we have that $\mathbf{v} \notin \Lambda$. If $\frac{j}{2p} \in \{ \frac{1}{2}, \frac{2p-2}{2p} \}$, then two coordinates α, β of $\mathbf{v}$ equal $\frac{1}{2}$. Hence for any admissible t we have

$$\left\{ \frac{t}{2} \right\} + \left\{ \frac{t}{p} \right\} + \left\{ \frac{tj}{2ap} \right\} + \left\{ \frac{(-2a - ap - j)t}{2ap} \right\} > \frac{1}{2} + \frac{1}{2} = 1.$$

Since the left hand side is an integer, it is at least 2.

In the other four cases we have two entries α, β such $\alpha = \beta + \frac{1}{2}$. Since t is odd, we have then that $|\{t\alpha\} - \{t\beta\}| = \frac{1}{2}$ and therefore

$$\left\{ \frac{t}{2} \right\} + \left\{ \frac{t}{p} \right\} + \left\{ \frac{tj}{2p} \right\} + \left\{ \frac{(-2 - p - j)t}{2p} \right\} > \left\{ \frac{1}{2} \right\} + \{t\alpha\} + \{t\beta\} > 1.$$

Summarizing we have that for all t that are invertible modulo $2p$ that

$$\left\{ \frac{t}{2} \right\} + \left\{ \frac{t}{p} \right\} + \left\{ \frac{tj}{2p} \right\} + \left\{ \frac{(-2 - p - j)t}{2p} \right\} \geq 2$$

holds. Using the symmetry of the coordinates it follows that for all t that are invertible modulo $2p$, we have

$$\left\{ \frac{t}{2} \right\} + \left\{ \frac{t}{p} \right\} + \left\{ \frac{tj}{2p} \right\} + \left\{ \frac{(-2 - p - j)t}{2p} \right\} \leq 2;$$

hence $\mathbf{v} \notin \Lambda$, which finishes the proof.

Remark 8. The description of the generators for $\mathrm{NS}(\tilde{S})$ we gave at the end of Example 3 does not require that p is prime, i.e., if we drop the assumption that p is prime, we get that $\rho(\tilde{S}) \geq 2 + 6(p - 1)$.

For the primes $p = 3, 5, 7$, one can calculate the maximal Picard numbers. It turns out that if $p = 7$ and $3|a$ holds, then $\rho(\tilde{S})$ equals $2 + 14(p - 1)$; if $p = 5$ and $6|a$ holds, then $\rho(\tilde{S})$ equals $2 + 18(p - 1)$; and if $p = 3$ and $60|a$ holds, then $\rho(\tilde{S})$ equals $2 + 30(p - 1)$.

Acknowledgements This paper is inspired by the results of [4, Chap. 6] and by some discussions which took place on the occasion of the Ph.D. defense of the first author at the University of Groningen. Part of the research was done while the first author held a position at the University of Groningen. His position was supported by a grant of the Netherlands Organization for Scientific Research (NWO). The research is partly supported by ERC Starting grant 279723 (SURFARI). The second author acknowledges the hospitality of the University of Groningen and the Leibniz Universität Hannover, where most of the work was done.

We thank the referee for providing many comments to improve the exposition.

References

1. J. Chahal, M. Meijer, J. Top, Sections on certain $j = 0$ elliptic surfaces. Comment. Math. Univ. St. Paul. **49**(1), 79–89 (2000)
2. L.A. Fastenberg, Computing Mordell–Weil ranks of cyclic covers of elliptic surfaces. Proc. Am. Math. Soc. **129**(7), 1877–1883 (electronic) (2001)
3. L.A. Fastenberg, Cyclic covers of rational elliptic surfaces. Rocky Mt. J. Math. **39**(6), 1895–1903 (2009)
4. B. Heijne, Elliptic Delsarte surfaces, Ph.D. thesis, Rijksuniversiteit Groningen, 2011
5. B. Heijne, The maximal rank of elliptic Delsarte surfaces. Math. Comp. **81**(278), 1111–1130 (2012)
6. R. Kloosterman, O. Tommasi, Locally trivial families of hyperelliptic curves: the geometry of the Weierstrass scheme. Indag. Math. (N.S.) **16**(2), 215–223 (2005)
7. T. Shioda, An explicit algorithm for computing the Picard number of certain algebraic surfaces. Am. J. Math. **108**(2), 415–432 (1986)
8. T. Shioda, Some remarks on elliptic curves over function fields. *Astérisque*, **209**(12), 99–114, 1992. Journées Arithmétiques, 1991 (Geneva)
9. J. Steenbrink, Intersection form for quasi-homogeneous singularities. Compos. Math. **34**(2), 211–223 (1977)

Fourier–Mukai Partners and Polarised K3 Surfaces

K. Hulek and D. Ploog

Abstract The purpose of this note is twofold. We first review the theory of Fourier–Mukai partners together with the relevant part of Nikulin's theory of lattice embeddings via discriminants. Then we consider Fourier–Mukai partners of K3 surfaces in the presence of polarisations, in which case we prove a counting formula for the number of partners.

Key words: K3 surfaces, Fourier–Mukai partners, Torelli theorem, Lattice embeddings

Mathematics Subject Classifications: Primary 14J28; Secondary 11E12, 18E30

The theory of Fourier-Mukai (FM) partners has played a crucial role in algebraic geometry and its connections to string theory in the last 25 years. Here we shall concentrate on a particularly interesting aspect of this, namely, the theory of FM partners of K3 surfaces. We shall survey some of the most significant results in this direction. Another aspect, and this has been discussed much less in the literature, is the question of Fourier–Mukai partners in the presence of polarisations. We shall also investigate this in some detail, and it is here that the paper contains some new results.

To begin with, we review in Sect. 1 the use of derived categories in algebraic geometry focusing on Fourier–Mukai partners. In Sects. 2 and 3 we then give a self-contained introduction to lattices and lattice embeddings with emphasis on indefinite, even lattices. This contains a careful presentation of Nikulin's theory as well as some enhancements which will then become important for our counting formula. From Sect. 4 onwards we will fully concentrate on K3 surfaces. After recalling the classical as well as Orlov's derived Torelli theorem for K3 surfaces, we describe

K. Hulek (✉) • D. Ploog

Institut für Algebraische Geometrie, Leibniz Universität Hannover, Welfengarten 1, 30167 Hannover, Germany

e-mail: hulek@math.uni-hannover.de;ploog@math.uni-hannover.de

R. Laza et al. (eds.), *Arithmetic and Geometry of K3 Surfaces and Calabi–Yau Threefolds*, Fields Institute Communications 67, DOI 10.1007/978-1-4614-6403-7_11, © Springer Science+Business Media New York 2013

the counting formula for the FM number of K3 surfaces given by Hosono et al. [22]. In Sect. 5 we discuss polarised K3 surfaces and their moduli. The relationship between polarised K3 surfaces and FM partners was discussed by Stellari in [42, 43]. Our main result in this direction is a counting formula given in Sect. 7 in the spirit of [22].

In a number of examples we will discuss the various phenomena which occur when considering Fourier–Mukai partners in the presence of polarisations.

Conventions: We work over the field $\mathbb{C}$.

We will denote bijections of sets as $A \overset{1:1}{=} B$. Also, all group actions will be left actions. In particular, we will denote the sets of orbits by $G \backslash A$ whenever G acts on A. However, factor groups are written G/H.

If we have group actions by G and G' on a set A which are compatible (i.e. they commute), then we consider this as a $G \times G'$-action (and not as a left–right bi-action). In particular, the total orbit set will be written as $G \times G' \backslash A$ (and not $G \backslash A / G'$).

1 Review Fourier–Mukai Partners of K3 Surfaces

For more than a century algebraic geometers have looked at the classification of varieties up to birational equivalence. This is a weaker notion than biregular isomorphism which, however, captures a number of crucial and interesting properties.

About two decades ago, a different weakening of biregularity has emerged in algebraic geometry: derived equivalence. Roughly speaking, its popularity stems from two reasons: on the one hand, the seemingly ever-increasing power of homological methods in all areas of mathematics and, on the other hand, the intriguing link, which derived categories provide to other mathematical disciplines such as symplectic geometry and representation theory as well as to theoretical physics.

1.1 History: Derived Categories in Algebraic Geometry

Derived categories of abelian categories were introduced in the 1967 thesis of Grothendieck's student Verdier [45]. The goal was to set up the necessary homological tools for defining duality in greatest generality—which meant getting the right adjoint of the push-forward functor f_*. This adjoint cannot exist in the abelian category of coherent sheaves; if it did, f_* would be exact. Verdier's insight was to embed the abelian category into a bigger category with desirable properties, the derived category of complexes of coherent sheaves. The reader is referred to [21] for an account of this theory.

In this review, we will assume that the reader is familiar with the basic theory of derived categories [16, 47]. An exposition of the theory of derived categories in algebraic geometry can be found in two textbooks, namely, by Huybrechts [23] and by Bartocci et al. [3]. We will denote by $D^b(X)$ the bounded derived category of coherent sheaves. This category is particularly well behaved if X is a smooth, projective variety. Later on we will consider K3 surfaces, but in this section, we review some general results.

We recall that two varieties X and Y are said to be *derived equivalent* (sometimes shortened to *D-equivalent*) if there is an exact equivalence of categories $D^b(X) \cong D^b(Y)$.

It should be mentioned right away that the use of the *derived* categories is crucial: a variety is uniquely determined by the abelian category of coherent sheaves, due to a theorem of Gabriel [15]. Thus, the analogous definition using abelian categories does not give rise to a new equivalence relation among varieties.

After their introduction, derived categories stayed in a niche, mainly considered as a homological bookkeeping tool. They were used to combine the classical derived functors into a single derived functor, or to put the Grothendieck spectral sequence into a more conceptual framework. The geometric use of derived categories started with the following groundbreaking result:

Theorem (Mukai [29]). *Let A be an abelian variety with dual abelian variety $\hat{A}$. Then A and $\hat{A}$ are derived equivalent.*

Since an abelian variety and its dual are in general not isomorphic (unless they are principally polarised) and otherwise never birationally equivalent, this indicates a new phenomenon. For the proof, Mukai employs the Poincaré bundle $\mathcal{P}$ on $A \times \hat{A}$ and investigates the functor $D^b(A) \to D^b(\hat{A})$ mapping $E \mapsto R\hat{\pi}_*(\mathcal{P} \otimes \pi^*E)$ where $\hat{\pi}$ and π denote the projections from $A \times \hat{A}$ to $\hat{A}$ and A, respectively.

Mukai's approach was not pursued for a while. Instead, derived categories were used in different ways for geometric purposes: Beilinson et al. [5] introduced perverse sheaves as certain objects in the derived category of constructible sheaves of a variety in order to study topological questions. The school around Rudakov introduced exceptional collections (of objects in the derived category), which under certain circumstances leads to an equivalence of $D^b(X)$ with the derived category of a finite-dimensional algebra [39]. It should be mentioned that around the same time, Happel introduced the use of triangulated categories in representation theory [19].

1.2 Derived Categories as Invariants of Varieties

Bondal and Orlov started considering $D^b(X)$ as an *invariant* of X with the following highly influential result:

Theorem (Bondal, Orlov, [7]). *Let X and Y be two smooth, projective varieties with $D^b(X) \cong D^b(Y)$. If X has ample canonical or anti-canonical bundle, then $X \cong Y$.*

In other words, at the extreme ends of the curvature spectrum, the derived category determines the variety. Note the contrast with Mukai's result, which provides examples of non-isomorphic, derived equivalent varieties with zero curvature (trivial canonical bundle). This begs the natural question: which (types of) varieties can possibly be derived equivalent? The philosophy hinted at by the theorems of Mukai, Bondal and Orlov is not misleading.

Proposition. *Let X and Y be two smooth, projective, derived equivalent varieties. Then the following hold true:*

1. *X and Y have the same dimension.*
2. *The singular cohomology groups $H^*(X, \mathbb{Q})$ and $H^*(Y, \mathbb{Q})$ are isomorphic as ungraded vector spaces; the same is true for Hochschild cohomology.*
3. *If the canonical bundle of X has finite order, then so does the canonical bundle of Y and the orders coincide; in particular, if one canonical bundle is trivial, then so is the other.*
4. *If the canonical (or anti-canonical) bundle of X is ample (or nef), the same is true for Y.*

The proposition is the result of the work of many people; see [23, Sects. 4–6]. Stating it here is ahistorical because some of the statements rely on the notion of Fourier–Mukai transform which we turn to in the next section. It should be said that our historical sketch is very much incomplete: for instance, developments like spaces of stability conditions [10] or singularity categories (Buchweitz, 1986, Maximal Cohen–Macaulay modules and Tate cohomology over Gorenstein rings, tspace.library.utoronto.ca/handle/1807/16682, unpublished), [36] are important but will not play a role here.

1.3 Fourier–Mukai Partners

Functors between geometric categories defined by a "kernel", i.e. a sheaf on a product (as in Mukai's case), were taken up again in the study of moduli spaces: if a moduli space M of sheaves of a certain type on Y happens to possess a (quasi-)universal family $\mathcal{E} \in \mathrm{Coh}(M \times Y)$, then this family gives rise to a functor $\mathrm{Coh}(M) \to \mathrm{Coh}(Y)$, mapping $A \mapsto p_{Y*}(\mathcal{E} \otimes p_M^* A)$, where p_M and p_Y are the projections from $M \times Y$ to M and Y, respectively. In particular, skyscraper sheaves of points $[E] \in M$ are sent to the corresponding sheaves E. This (generally non-exact!) functor does not possess good properties, and it was soon realised that it is much better to consider its derived analogue, which we define below. Sometimes, for example, the functors between derived categories can be used to show birationality of moduli spaces.

In the following definition, we denote the canonical projections of the product $X \times Y$ to its factors by p_X and p_Y, respectively.

Definition. Let X and Y be two smooth, projective varieties and let $K \in D^b(X \times Y)$. The *Fourier–Mukai functor* with *kernel K* is the composition

$$\mathsf{FM}_K \colon D^b(X) \xrightarrow{\ p_X^*\ } D^b(X \times Y) \xrightarrow{\ K \overset{\mathsf{L}}{\otimes}\ } D^b(X \times Y) \xrightarrow{\ \mathsf{R}p_{Y*}\ } D^b(Y)$$

of pullback, derived tensor product with K and derived push-forward. If FM_K is an equivalence, then it is called a *Fourier–Mukai transform*.

X and Y are said to be *Fourier–Mukai partners* if a Fourier–Mukai transform exists between their derived categories. The set of all Fourier–Mukai partners of X up to isomorphisms is denoted by $\mathsf{FM}(X)$.

Remarks. This important notion warrants a number of comments.

1. Fourier–Mukai functors should be viewed as classical correspondences, i.e. maps between cohomology or Chow groups on the level of derived categories. In particular, many formal properties of correspondences as in [14, Sect. 14] carry over verbatim: the composition of Fourier–Mukai functors is again such with the natural "convoluted" kernel; the (structure sheaf of the) diagonal gives the identity. In fact, a Fourier–Mukai transform induces correspondences on the Chow and cohomological levels, using the Chern character of the kernel.

2. Neither notation nor terminology is uniform. Some sources mean "Fourier–Mukai transform" to be an equivalence whose kernel is a sheaf, for example. Notationally, often used is $\Phi_K^{X \to Y}$ which is inspired by Mukai's original article [30]. This notation, however, has the drawback of being lengthy without giving additional information in the important case $X = Y$.

Fourier–Mukai transforms play a very important and prominent role in the theory due to the following basic and deep result:

Theorem (Orlov, [34]). *Given an equivalence* $\Phi\colon D^b(X) \xrightarrow{\sim} D^b(Y)$ *(as $\mathbb{C}$-linear, triangulated categories) for two smooth, projective varieties X and Y, then there exists an object $K \in D^b(X \times Y)$ with a functor isomorphism $\Phi \cong \mathsf{FM}_K$. The kernel K is unique up to isomorphism.*

By this result, the notions "derived equivalent" and "Fourier–Mukai partners" are synonymous.

The situation is very simple in dimension 1: two smooth, projective curves are derived equivalent if and only if they are isomorphic. The situation is a lot more interesting in dimension 2: apart from the abelian surfaces already covered by Mukai's result, K3 and certain elliptic surfaces can have non-isomorphic FM partners. For K3 surfaces, the statement is as follows (see Sect. 4 for details):

Theorem (Orlov, [34]). *For two projective K3 surfaces X and Y, the following conditions are equivalent:*

1. *X and Y are derived equivalent.*
2. *The transcendental lattices T_X and T_Y are Hodge isometric.*
3. *There exist an ample divisor H on X, integers $r \in \mathbb{N}$, $s \in \mathbb{Z}$ and a class $c \in H^2(X, \mathbb{Z})$ such that the moduli space of H-semistable sheaves on X of rank r, first Chern class c and second Chern class s is non-empty, fine and isomorphic to Y.*

In general, it is a conjecture that the number of FM partners is always finite. For surfaces, this has been proven by Bridgeland and Maciocia [8]. The next theorem implies finiteness for abelian varieties, using that an abelian variety has only a finite number of abelian subvarieties up to isogeny [17].

Theorem (Orlov, Polishchuk [35, 38]). *Two abelian varieties A and B are derived equivalent if and only if $A \times \hat{A}$ and $B \times \hat{B}$ are symplectically isomorphic, i.e. there is an isomorphism $f = \left(\begin{smallmatrix} \alpha & \beta \\ \gamma & \delta \end{smallmatrix}\right) \colon A \times \hat{A} \xrightarrow{\sim} B \times \hat{B}$ such that $f^{-1} = \left(\begin{smallmatrix} \hat{\delta} & -\hat{\beta} \\ -\hat{\gamma} & \hat{\alpha} \end{smallmatrix}\right)$.*

The natural question about the number of FM partners has been studied in greatest depth for K3 surfaces. The first result was shown by Oguiso [33]: a K3 surface with a single primitive ample divisor of degree $2d$ has exactly $2^{p(d)-1}$ such partners, where $p(d)$ is the number of prime divisors of d. In [22], a formula using lattice counting for general projective K3 surfaces was given. In Sect. 4, we will reprove this result and give a formula for polarised K3 surfaces. We want to mention that FM partners of K3 surfaces have been linked to the so-called Kähler moduli space; see Ma [26] and Hartmann [20].

1.4 Derived and Birational Equivalence

We started this review by motivating derived equivalence as a weakening of isomorphism, like birationality is. This naturally leads to the question whether there is an actual relationship between the two notions. At first glance, this is not the case: since birational abelian varieties are already isomorphic, Mukai's result provides examples of derived equivalent but not birationally equivalent varieties. And in the other direction, let Y be the blowing up of a smooth projective variety X of dimension at least two in a point. Then X and Y are obviously birationally equivalent but never derived equivalent by a result of Bondal and Orlov [6].

Nevertheless, some relation is expected. More precisely,

Conjecture (Bondal, Orlov [6]). If X and Y are smooth, projective, birationally equivalent varieties with trivial canonical bundles, then X and Y are derived equivalent.

Kawamata suggested a generalisation using the following notion: two smooth, projective varieties X and Y are called *K-equivalent* if there is a birational correspondence $X \xleftarrow{p} Z \xrightarrow{q} Y$ with $p^*\omega_X \cong q^*\omega_Y$. He conjectures that K-equivalent varieties are D-equivalent.

The conjecture is known in some cases, for example, the standard flop (Bondal, Orlov [6]), the Mukai flop (Kawamata [25], Namikawa [31]), Calabi–Yau threefolds (Bridgeland [9]) and Hilbert schemes of K3 surfaces (Ploog [37]).

2 Lattices

Since the theory of K3 surfaces is intricately linked to lattices, we provide a review of the lattice theory as needed in this note. By a lattice, we always mean a free abelian group L of finite rank equipped with a nondegenerate symmetric bilinear pairing $(\cdot, \cdot) \colon L \times L \to \mathbb{Z}$. The lattice L is called *even* if $(v, v) \in 2\mathbb{Z}$ for all $v \in L$. We shall assume all our lattices to be even.

Sometimes, we denote by L_K the K-vector space $L \otimes K$, where K is a field among $\mathbb{Q}, \mathbb{R}, \mathbb{C}$. The pairing extends to a symmetric bilinear form on L_K. The *signature* of L is defined to be that of $L_{\mathbb{R}}$.

The lattice L is called *unimodular* if the canonical homomorphism $d_L \colon L \to L^{\vee} = \mathrm{Hom}(L, \mathbb{Z})$ with $d_L(v) = (v, \cdot)$ is an isomorphism. Note that d_L is always injective, as we have assumed $(\cdot, \cdot)$ to be nondegenerate. This implies that for every element $f \in L^{\vee}$, there is a natural number $a \in \mathbb{N}$ such that af is in the image of d_L. Thus $L^{\vee}$ can be identified with the subset $\{w \in L \otimes \mathbb{Q} \mid (v, w) \in \mathbb{Z} \; \forall v \in L\}$ of $L \otimes \mathbb{Q}$ with its natural $\mathbb{Q}$-valued pairing.

We shall denote the *hyperbolic plane* by U. A *standard basis* of U is a basis e, f with $e^2 = f^2 = 0$ and $(e, f) = 1$. The lattice E_8 is the unique positive definite even unimodular lattice of rank 8, and we denote by $E_8(-1)$ its negative definite opposite. For an integer $n \neq 0$ we denote by $\langle n \rangle$ the rank one lattice where both generators square to n. Finally, given a lattice L, then aL denotes a direct sum of a copies of the lattice L.

Given any non-empty subset $S \subseteq L$, the *orthogonal complement* is $S^{\perp} := \{v \in L \mid (v, S) = 0\}$. A submodule $S \subseteq L$ is called *primitive* if the quotient group L/S is torsion free. Note the following obvious facts: $S^{\perp} \subseteq L$ is always a primitive submodule; we have $S \subseteq S^{\perp\perp}$; and S is primitive if and only if $S = S^{\perp\perp}$. In particular, $S^{\perp\perp}$ is the *primitive hull* of S.

A vector $v \in L$ is called *primitive* if the lattice $\mathbb{Z}v$ generated by it is primitive.

The *discriminant group* of a lattice L is the finite abelian group $D_L = L^{\vee}/L$. Since we have assumed L to be even, it carries a natural quadratic form q_L with values in $\mathbb{Q}/2\mathbb{Z}$. By customary abuse of notation, we will often speak of a quadratic form q (or q_L), suppressing the finite abelian group it lives on. Finally, for any lattice L, we denote by $l(L)$ the minimal number of generators of D_L.

2.1 Gram Matrices

We make the above definitions more explicit using the matrix description. After choosing a basis, a lattice on $\mathbb{Z}^r$ is given by a symmetric $r \times r$ matrix G (often called Gram matrix), the pairing being $(v, w) = v^t G w$ for $v, w \in \mathbb{Z}^r$. To be precise, the (i, j)-entry of G is $(e_i, e_j) \in \mathbb{Z}$ where $(e_1, \ldots, e_r)$ is the chosen basis.

Changing the matrix by symmetric column-and-row operations gives an isomorphic lattice; this corresponds to $G \mapsto SGS^t$ for some $S \in \mathrm{GL}(r, \mathbb{Z})$. Since our pairings are nondegenerate, G has full rank. The lattice is unimodular if the Gram matrix has determinant ± 1. It is even if and only if the diagonal entries of G are even.

The inclusion of the lattice into its dual is the map $G \colon \mathbb{Z}^r \hookrightarrow \mathbb{Z}^r$, $v \mapsto v^t G$. Considering a vector $\varphi \in \mathbb{Z}^r$ as an element of the dual lattice, there is a natural number a such that $a\varphi$ is in the image of G, i.e. $v^t G = a\varphi$ for some integral vector v. Then $(\varphi, \varphi) = (v, v)/a^2 \in \mathbb{Q}$.

The discriminant group is the finite abelian group with presentation matrix G, i.e. $D \cong \mathbb{Z}^r/\mathrm{im}(G)$. Elementary operations can be used to diagonalise it. The quadratic form on the discriminant group is computed as above, only now taking values in $\mathbb{Q}/2\mathbb{Z}$.

The *discriminant* of L is defined as the order of the discriminant group. It is the absolute value of the determinant of the Gram matrix: $\mathrm{disc}(L) := \#D_L = |\det(G_L)|$. Classically, discriminants (of quadratic forms) are defined with a factor of ± 1 or $\pm 1/4$; see Example 2.3.

2.2 Genera

Two lattices L and L' of rank r are said to be *in the same genus* if they fulfil one of the following equivalent conditions:

1. The localisations L_p and L'_p are isomorphic for all primes p, including $\mathbb{R}$.
2. The signatures of L and L' coincide and the discriminant forms are isomorphic: $q_L \cong q_{L'}$.
3. The matrices representing L and L' are *rationally equivalent without essential denominators*, i.e. there is a base change in $\mathrm{GL}(r, \mathbb{Q})$ of determinant ± 1, transforming L into L' and whose denominators are prime to $2 \cdot \mathrm{disc}(L)$.

For details on localisations, see [32]. The equivalence of (1) and (2) is a deep result of Nikulin [32, 1.9.4]. We elaborate on (2): a map $q\colon A \twoheadrightarrow \mathbb{Q}/2\mathbb{Z}$ is called a quadratic form on the finite abelian group A if $q(na) = n^2 q(a)$ for all $n \in \mathbb{Z}, a \in A$ and if there is a symmetric bilinear form $b\colon A \times A \to \mathbb{Q}/\mathbb{Z}$ such that $q(a_1 + a_2) = q(a_1) + q(a_2) + 2b(a_1, a_2)$ for all $a_1, a_2 \in A$. It is clear that discriminant forms of even lattices satisfy this definition. Two pairs (A, q) and (A', q') are defined to be isomorphic if there is a group isomorphism $\varphi\colon A \xrightarrow{\sim} A'$ with $q(a) = q'(\varphi(a))$ for all $a \in A$.

The history of the equivalence between (1) and (3) is complicated: using analytic methods, Siegel [41] proved that L and L' are in the same genus if and only if for every positive integer d there exists a rational base change $S_d \in \mathrm{GL}(r, \mathbb{Q})$ carrying L into L' and such that the denominators of S_d are prime to d (and he called this property rational equivalence without denominators). There are algebraic proofs of that statement, e.g. [24, Theorem 40] or [46, Theorem 50]. These references also contain (3) above, i.e. the existence of a single $S \in \mathrm{GL}(r, \mathbb{Q})$ whose denominators are prime to $2 \cdot \mathrm{disc}(L)$.

For binary forms, all of this is closely related to classical number theory. In particular, the genus can then also be treated using the ideal class group of quadratic number fields. See [12] or [48] for this. Furthermore, there is a strengthening of (3) peculiar to *field discriminants* (see [12, Sect. 3.B]):

4. Let $L = \left(\begin{smallmatrix} 2a & b \\ b & 2c \end{smallmatrix}\right)$ and $L' = \left(\begin{smallmatrix} 2a' & b' \\ b' & 2c' \end{smallmatrix}\right)$ be two binary even, indefinite lattices with $\gcd(a, b, c) = \gcd(a', b', c') = 1$ and of the same discriminant $D := b^2 - 4ac$ such that either $D \equiv 1 \mod 4$, D square-free, or $D = 4k$, $k \not\equiv 1 \mod 4$, k square-free. Then L and L' are in the same genus if and only if they are rationally equivalent, i.e. there is a base change $S \in \mathrm{GL}(2, \mathbb{Q})$ taking L to L'.

The genus of L is denoted by $\mathcal{G}(L)$ and it is a basic but non-trivial fact that $\mathcal{G}(L)$ is a finite set. We will also have to specify genera in other ways, using a quadratic form $q\colon D_q \to \mathbb{Q}/2\mathbb{Z}$ on a finite abelian group D_q, as follows:

$\mathcal{G}(t_+, t_-, q)$ lattices with signature (t_+, t_-) and discriminant form q,
$\mathcal{G}(\mathrm{sgn}(K), q)$ lattices with same signature as K and discriminant form q.

Example 2.3. We consider binary forms, that is, lattices of rank 2. Clearly, a symmetric bilinear form with Gram matrix $\left(\begin{smallmatrix} a & b \\ b & c \end{smallmatrix}\right)$ is even if and only if both diagonal terms are even.

Note that many classical sources use quadratic forms instead of lattices. We explain the link for binary forms $f(x, y) = ax^2 + bxy + cy^2$ (where $a, b, c \in \mathbb{Z}$). The associated bilinear form has Gram matrix $G = \frac{1}{2}\left(\begin{smallmatrix} 2a & b \\ b & 2c \end{smallmatrix}\right)$—in particular, it need not be integral. An example is $f(x, y) = xy$. In fact, the bilinear form, i.e. G, is integral if and only if b is even (incidentally, Gauß always made that assumption). Note that the quadratic form $2xy$ corresponds to our hyperbolic plane $\left(\begin{smallmatrix} 0 & 1 \\ 1 & 0 \end{smallmatrix}\right)$. The discriminant of f is classically defined to be $D := b^2 - 4ac$ which differs from our definition (i.e. $|\det(G)| = \#D$) by a factor of ± 4.

We proceed to give specific examples of lattices as Gram matrices. Both $A = \left(\begin{smallmatrix} 2 & 4 \\ 4 & 0 \end{smallmatrix}\right)$ and $B = \left(\begin{smallmatrix} 0 & 4 \\ 4 & 0 \end{smallmatrix}\right)$ are indefinite, i.e. of signature $(1, 1)$, and have discriminant 16, but the discriminant groups are not isomorphic: $D_A = \mathbb{Z}/2\mathbb{Z} \times \mathbb{Z}/8\mathbb{Z}$ and $D_B = \mathbb{Z}/4\mathbb{Z} \times \mathbb{Z}/4\mathbb{Z}$. Thus, A and B are not in the same genus.

Another illuminating example is given by the forms A and $C = \left(\begin{smallmatrix} -2 & 4 \\ 4 & 0 \end{smallmatrix}\right)$. We first notice that these forms are not isomorphic: the form A represents 2, but C does not, as can be seen by looking at the possible remainders of $-2x^2 + 8xy$ modulo 8. The two forms have the same signature and discriminant groups, but the discriminant forms are different. To see this we note that D_A is generated by the residue classes of $t_1 = e_1/2$ and $t_2 = (2e_1 + e_2)/8$, whereas D_C is generated by the residue classes of $s_1 = e_1/2$ and $s_2 = (-2e_1 + e_2)/8$. The quadratic forms q_A and q_C are determined by $q_A(\overline{t_1}) = 1/2$, $q_A(\overline{t_2}) = 3/8$ and $q_C(\overline{s_1}) = -1/2$, $q_C(\overline{s_2}) = -3/8$. The forms cannot be isomorphic, for the subgroup of D_A of elements of order 2 consists of $\{0, t_1, 4t_2, t_1 + 4t_2\}$ (this is the Klein four group) and the values of q_A on these elements in $\mathbb{Q}/2\mathbb{Z}$ are $0, 1/2, 4^2 \cdot 3/8 = 0, 42/4 = 1/2$. Likewise, the values of q_C on the elements of order 2 in D_C are 0 and $-1/2$. Hence (D_A, q_A) and (D_C, q_C) cannot be isomorphic.

Zagier's book also contains the connection of genera to number theory and their classification using ideal class groups [48, Sect. 8]. An example from this book [48, Sect. 12] gives an instance of lattices in the same genus which are not isomorphic: the forms $D = \left(\begin{smallmatrix} 2 & 1 \\ 1 & 12 \end{smallmatrix}\right)$ and $E = \left(\begin{smallmatrix} 4 & 1 \\ 1 & 6 \end{smallmatrix}\right)$ are positive definite of field discriminant -23. They are in the same genus (one is sent to the other by the fractional base change $-\frac{1}{2}\left(\begin{smallmatrix} 1 & 1 \\ -3 & 1 \end{smallmatrix}\right)$) but not equivalent: D represents 2 as the square of $(1, 0)$, whereas E does not represent 2 as $4x^2 + 2xy + 6y^2 = 3x^2 + (x + y)^2 + 5y^2 \geq 4$ if $x \neq 0$ or $y \neq 0$.

Unimodular, indefinite lattices are unique in their genus, as follows from their well-known classification. A generalisation is given by [32, Corollary 1.13.3]:

Lemma 2.4 (Nikulin's criterion) *An indefinite lattice L with* $rk(L) \geq 2 + l(L)$ *is unique within its genus. This holds in particular when L contains a hyperbolic plane.*

Recall that $l(L)$ denotes the minimal number of generators of the finite group D_L. Since always $rk(L) \geq l(L)$, Nikulin's criterion only fails to apply in two cases, namely, if $l(L) = rk(L)$ or $l(L) = rk(L) - 1$. As a corollary of Nikulin's criterion, $L \oplus U$ is unique within its lattice for any L.

For a lattice L, we denote its group of isometries by $O(L)$. An isometry of lattices $f: L \xrightarrow{\sim} L'$ gives rise to $f_{\mathbb{Q}}: L_{\mathbb{Q}} \xrightarrow{\sim} L'_{\mathbb{Q}}$ and hence to $D_f: D_L \xrightarrow{\sim} D_{L'}$. In particular, there is a natural homomorphism $O(L) \to O(D_L)$ which is used to define the *stable isometry group* as

$$\tilde{O}(L) := \ker(O(L) \to O(D_L)).$$

Finally, we state a well-known result of Eichler [13, Sect. 10]. It uses the notion of the *divisor* $div(v)$ of a vector $v \in L$, which is the positive generator of the ideal (v, L). Note that this is the largest positive integer a such that $v = av'$ for some element $v' \in L^{\vee}$.

Lemma 2.5 (Eichler's criterion) *Suppose that an even lattice L contains* $U \oplus U$ *as a direct summand. The* $O(L)$*-orbit of a primitive vector* $v \in L$ *is determined by the length* v^2 *and the element* $v/div(v) \in D(L)$ *of the discriminant group.*

3 Overlattices

In this section, we elaborate on Nikulin's theory of overlattices and primitive embeddings [32]; we also give some examples. Eventually, we generalise slightly to cover a setting needed for the Fourier–Mukai partner counting in the polarised case.

We fix a lattice M with discriminant form $q_M: D_M \to \mathbb{Q}/2\mathbb{Z}$.

By an *overlattice* of M, we mean a lattice embedding $i: M \hookrightarrow L$ with M and L of the same rank. Note that we have inclusions

$$M \xrightarrow[d_M]{i} L \xrightarrow{d_L} L^{\vee} \xrightarrow{i^{\vee}} M^{\vee}$$

with $d_L: L \hookrightarrow L^{\vee}$ and $d_M: M \hookrightarrow M^{\vee}$ the canonical maps. (For now, we will denote these canonical embeddings just by d and later not denote them at all.) From this, we get a chain of quotients:

$$L/iM \xrightarrow{d} L^{\vee}/diM \xrightarrow{i^{\vee}} M^{\vee}/i^{\vee}diM = D_M.$$

We call the image $H_i \subset D_M$ of L/iM the *classifying subgroup* of the overlattice. Note that D_M is equipped with a quadratic form, so we can also speak of the orthogonal complement $H_i^\perp$. We will consider $L^\vee/diM$ as a subgroup of D_M in the same way via $i^\vee$.

We say that two embeddings $i: M \hookrightarrow L$ and $i': M \hookrightarrow L'$ *define the same overlattice* if there is an isometry $f: L \xrightarrow{\sim} L'$ such that $fi = i'$:

$$
\begin{array}{ccc}
M & \xrightarrow{\ i\ } & L \\
\| & & \downarrow f \\
M & \xrightarrow{\ i'\ } & L'
\end{array}
$$

This means in particular that within each isomorphism class, we can restrict to looking at embeddings $i: M \hookrightarrow L$ into a *fixed* lattice L.

Lemma 3.1 [32, Proposition 1.4.1] *Let $i: M \hookrightarrow L$ be an overlattice. Then the subgroup H_i is isotropic in D_M, i.e. $q_M|_{H_i} = 0$. Furthermore, $H_i^\perp = L^\vee/diM$ and there is a natural identification $H_i^\perp/H_i \cong D_L$ with $q_M|_{H_i^\perp/H_i} = q_L$.*

We introduce the following sets of overlattices L of M and quotients L/M, respectively, where we consider L/M as an isotropic subgroup of the discriminant group D_M:

$$\mathcal{O}(M) := \{(L, i) \mid L \text{ lattice}, i: M \hookrightarrow L \text{ overlattice}\}$$
$$\mathcal{Q}(M) := \{H \subset D_M \text{ isotropic}\}.$$

We also use the notation $\mathcal{O}(M, L)$ to specify that the target lattice is isomorphic to L. With this notation we can write $\mathcal{O}(M)$ as a disjoint union

$$\mathcal{O}(M) = \coprod_L \mathcal{O}(M, L)$$

where L runs through all isomorphism classes of possible overlattices of M.

Example 3.2. The set $\mathcal{Q}(M)$ is obviously finite. On the other hand, an overlattice $i: M \hookrightarrow L$ can always be modified by an isometry $f \in O(M)$ to yield an overlattice $if: M \hookrightarrow L$. However, if $f \in \tilde{O}(M)$ is a stable isometry, then it can be extended to an isometry of L and hence i and if define the same overlattice. This shows that $\mathcal{O}(M)$ is also finite.

The following lemma is well known and implicit in [32].

Lemma 3.3 *There is a bijection between $\mathcal{O}(M)$ and $\mathcal{Q}(M)$.*

Proof. We use the maps

$$\mathsf{H}: \mathcal{O}(M) \to \mathcal{Q}(M), \quad (L, i) \mapsto H_i,$$
$$\mathsf{L}: \mathcal{Q}(M) \to \mathcal{O}(M), \quad H \mapsto (L_H, i_H)$$

where, for $H \in \mathcal{Q}(M)$, we define $L_H := \{\varphi \in M^\vee \mid [\varphi] \in H\} = \pi^{-1}(H)$ where $\pi \colon M^\vee \to D_M$ is the canonical projection. The canonical embedding $d \colon M \hookrightarrow M^\vee$ factors through L_H, giving an injective map $i_H \colon M \to L_H$. All of this can be summarised in a commutative diagram of short exact sequences

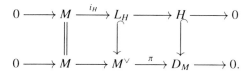

The abelian group L_H inherits a $\mathbb{Q}$-valued form from $M^\vee$. This form is actually $\mathbb{Z}$-valued because of $q_M|_H = 0$. Furthermore, the bilinear form on L_H is even since the quadratic form on D_M is $\mathbb{Q}/2\mathbb{Z}$-valued. Hence, L_H is a lattice and i_H is obviously a lattice embedding.

It is immediate that $\mathsf{HL} = \mathrm{id}_{\mathcal{Q}(M)}$. On the other hand, the overlattices $\mathsf{LH}(L, i)$ and (L, i) are identified by the embedding $L \to M^\vee$, $v \mapsto \langle v, i(\cdot)\rangle_M$ which has precisely $\mathsf{LH}(L, i)$ as image. $\qquad \square$

We want to refine this correspondence slightly. For this we fix a quadratic form (D, q) which occurs as the discriminant of some lattice (and forget L) and set

$$\mathcal{O}(M, q) := \{(L, i) \in \mathcal{O}(M) \mid [L] \in \mathcal{G}(\mathrm{sgn}(M), q)\},$$
$$\mathcal{Q}(M, q) := \{H \in \mathcal{Q}(M) \mid q_M|_{H^\perp/H} \cong q\}.$$

The condition $q_M|_{H^\perp/H} \cong q$ here includes $H^\perp/H \cong D$.

Lemma 3.4 *There is a bijection between $\mathcal{O}(M, q)$ and $\mathcal{Q}(M, q)$.*

Proof. We only have to check that the maps $\mathsf{H} \colon \mathcal{O}(M, q) \to \mathcal{Q}(M)$ and $\mathsf{L} \colon \mathcal{Q}(M, q) \to \mathcal{O}(M)$ have image in $\mathcal{Q}(M, q)$ and $\mathcal{O}(M, q)$, respectively. For H, this is part of Lemma 3.1. For L, we have $\mathrm{sgn}(L_H) = \mathrm{sgn}(M)$ and the discriminant form of L_H is $D_M|_{H^\perp/H} \cong q$, by assumption on H. $\qquad \square$

In the course of our discussions we have to distinguish carefully between different notions of equivalence of lattice embeddings. The following notion is due to Nikulin ([32, Proposition 1.4.2]):

Definition 3.5 *Two embeddings $i, i' \colon M \hookrightarrow L$ define* isomorphic overlattices, *denoted $i \simeq i'$, if there exists an isometry $f \in \mathrm{O}(L)$ with $f i(M) = i'(M)$—inducing an isometry $f|_M \in \mathrm{O}(M)$—or, equivalently, if there is a commutative diagram:*

$$
\begin{array}{ccc}
M & \xrightarrow{\ i\ } & L \\
{\scriptstyle f|_M}\downarrow & & \downarrow{\scriptstyle f} \\
M & \xrightarrow{\ i'\ } & L
\end{array}
.
$$

Note that this definition also makes sense if M and L do not necessarily have the same rank. Two embeddings of lattices $i, i' : M \hookrightarrow L$ of the same rank defining the *same* overlattice are in particular isomorphic.

Definition 3.6 *Two embeddings $i, i' : M \hookrightarrow L$ are* stably isomorphic, *denoted $i \approx i'$, if there exists a stable isometry $f \in \tilde{O}(L)$ with $fi(M) = i'(M)$, i.e. there is a commutative diagram*

$$
\begin{array}{ccc}
M & \xrightarrow{\ i\ } & L \\
{\scriptstyle f|_M} \downarrow & & \downarrow {\scriptstyle f \text{ stable}} \\
M & \xrightarrow{\ i'\ } & L
\end{array}\ .
$$

We note that embeddings of lattices of the same rank defining the same overlattice are not necessarily stably isomorphic.

We can put this into a broader context. For this we consider the set

$$
\mathcal{E}(M, L) := \{i : M \hookrightarrow L\}
$$

of embeddings of M into L where, for the time being, we do not assume M and L to have the same rank. The group $O(M) \times O(L)$ acts on this set by $(g, \tilde{g}) : i \mapsto \tilde{g} i g^{-1}$. Instead of the action of $O(M) \times O(L)$ on $\mathcal{E}(M, L)$, one can also consider the action of any subgroup, and we shall see specific examples later when we discuss Fourier–Mukai partners of K3 surfaces. If M and L have the same rank, then the connection with our previously considered equivalence relations is the following:

$$
\mathcal{O}(M, L) = (\{\mathrm{id}_M\} \times O(L)) \backslash \mathcal{E}(M, L).
$$

The set of all isomorphic overlattices of M isomorphic to L is given by $(O(M) \times O(L)) \backslash \mathcal{E}(M, L)$, whereas stably isomorphic embeddings are given by $(O(M) \times \tilde{O}(L)) \backslash \mathcal{E}(M, L)$.

We now return to our previous discussion of the connection between overlattices and isotropic subgroups.

Lemma 3.7 *Let $i, i' : M \hookrightarrow L$ be embeddings of lattices of the same rank. Then $i \simeq i'$ if and only if there exists an isometry $g \in O(M)$ such that $D_g(H_i) = H_{i'}$.*

Proof. Given $f \in O(L)$ with $fi(M) = i'(M)$, then $g := f|_M$ will have the correct property.

Given g, recall that the lattices are obtained from their classifying subgroups as $\pi^{-1}(H_i)$ and $\pi^{-1}(H_{i'})$. Then, $D_g(H_i) = H_{i'}$ implies that the map $g^\vee : M^\vee \to M^\vee$ induced from g sends L to itself, and $f = g^\vee|_L$ gives the desired isomorphism. $\square$

Note that an isometry $g \in O(M)$ with $D_g(H_i) = H_{i'}$ induces an isomorphism $H_i^\perp \xrightarrow{\sim} H_{i'}^\perp$ and hence an isomorphism of the quotients. Recall that there is a natural identification $H_i^\perp / H_i = D_L$.

Lemma 3.8 *Let $i, i' : M \hookrightarrow L$ be embeddings of lattices of the same rank. $i \approx i'$ if and only if there exists an isometry $g \in O(M)$ such that $D_g(H_i) = H_{i'}$ and the induced map $D_L = H_i^\perp / H_i \to H_{i'}^\perp / H_{i'} = D_L$ is the identity.*

Proof. Just assuming $D_g(H_i) = H_{i'}$, we get a commutative diagram

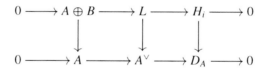

which, together with the proof of Lemma 3.7, shows the claim.   □

3.9 Overlattices from Primitive Embeddings

A natural source of overlattices is $M := T \oplus T^{\perp} \subset L$ for any sublattice $T \subset L$. If T is moreover a primitive sublattice of L, then the theory sketched above can be refined, as we explain next. We start with an elementary lemma.

Lemma 3.10 *Let $A, B \subset L$ be two sublattices such that $i: A \oplus B \hookrightarrow L$ is an overlattice, i.e. A and B are mutually orthogonal and $\mathrm{rk}(A \oplus B) = \mathrm{rk}(L)$. Then $p_A: H_i \hookrightarrow D_{A\oplus B} \twoheadrightarrow D_A$ is injective if and only if B is primitive in L.*

Proof. The commutative diagram with exact rows

$$
\begin{array}{ccccccccc}
0 & \longrightarrow & A \oplus B & \longrightarrow & L & \longrightarrow & H_i & \longrightarrow & 0 \\
  &   & \downarrow &   & \downarrow &   & \downarrow &   & \\
0 & \longrightarrow & A & \longrightarrow & A^{\vee} & \longrightarrow & D_A & \longrightarrow & 0
\end{array}
$$

leads to the following short exact sequence of the kernels:

$$0 \to B \to B^{\perp\perp} \to \ker(p_A) \to 0$$

(note that the kernel of the map $L \to A^{\vee}, v \mapsto \langle v, \cdot \rangle|_A$ is the primitive hull of B). Hence p_A is injective if and only if $B = B^{\perp\perp}$, i.e. B is a primitive sublattice.   □

Example 3.11. We consider the rank 2 lattice L with Gram matrix $\left(\begin{smallmatrix} 2 & 0 \\ 0 & 2 \end{smallmatrix}\right)$; let e_1, e_2 be an orthogonal basis, so that $e_1^2 = e_2^2 = 2$. With $T = \langle 8 \rangle$ having basis $2e_1$ and $K := T^{\perp}$, we get that $H_i \to D_T$ is injective whereas $H_i \to D_K$ is not.

Let $j_T: T \hookrightarrow L$ be a sublattice and $K := T^{\perp}$ its orthogonal complement with embedding $j_K: K \hookrightarrow L$. By Lemma 3.1, the overlattice $i := j_T \oplus j_K: T \oplus K \hookrightarrow L$ corresponds to the isotropic subgroup $H_i \subset D_{T\oplus K}$.

By Lemma 3.10, the map $p_T: H_i \hookrightarrow D_{T\oplus K} \twoheadrightarrow D_T$ is always injective, since $K \subset L$ is an orthogonal complement, hence primitive. The map $p_K: H_i \hookrightarrow D_{T\oplus K} \twoheadrightarrow D_K$ is injective if and only if $T \subset L$ is a primitive sublattice.

If $j_T: T \hookrightarrow L$ is primitive, then $\Gamma_i := p_T(H_i) \subseteq D_T$ is a subgroup such that there is a unique, injective homomorphism $\gamma_i: \Gamma_i \to D_K$. The image of γ_i is $p_K(H_i)$ and its graph is H_i.

For fixed lattices L, K, T, we introduce the following sets:

$$\mathcal{P}(T,L) := \{j_T: T \hookrightarrow L \text{ primitive}\},$$

$$\mathcal{P}(T,K,L) := \left\{ (j_T, j_K) \,\middle|\, \begin{array}{l} j_T \in \mathcal{P}(T,L), j_K \in \mathcal{P}(K,L) \\ j_T \oplus j_K \in \mathcal{E}(T \oplus K, L) \end{array} \right\}.$$

As in the previous section, we can consider various notions of equivalence on the set $\mathcal{P}(T,K,L)$ by considering the action of suitable subgroups of $O(T) \times O(K) \times O(L)$. Since we are only interested in overlattices in this section, we shall assume for the rest of this section that

Assumption 3.12 $\qquad\qquad rk(T) + rk(K) = rk(L).$

In the previous section we said that two embeddings define the same overlattice if they differ by the action of $\{id_T\} \times \{id_K\} \times O(L)$ and accordingly we set

$$\mathcal{O}(T,K,L) = (\{id_T\} \times \{id_K\} \times O(L)) \backslash \mathcal{P}(T,K,L).$$

We now also consider a quadratic form (D,q) which will play the role of the discriminant of the overlattice. Choosing a representative L for each element in $\mathcal{G}(\mathrm{sgn}(T \oplus K), q)$, we also introduce the equivalents of the sets of the previous section:

$$\mathcal{P}(T,K,q) := \left\{ (L, j_T, j_K) \,\middle|\, \begin{array}{l} [L] \in \mathcal{G}(\mathrm{sgn}(T \oplus K), q), \\ (j_T, j_K) \in \mathcal{P}(T,K,L) \end{array} \right\},$$

$$\mathcal{Q}(T,K,q) := \{ H \in \mathcal{Q}(T \oplus K, q) \mid p_T|_H \text{ and } p_K|_H \text{ are injective}\}.$$

Dividing out by the action of the overlattice, we also consider $\mathcal{O}(T,K,q)$. The condition in the definition of $\mathcal{Q}(T,K,q)$ means that H is the graph of an injective group homomorphism $\gamma: \Gamma \hookrightarrow D_K$ with $\Gamma := p_T(H)$ and $\mathrm{im}(\gamma) = p_K(H)$. Note that $q_{T \oplus K}|_H = 0$ is equivalent to $q_K \gamma = -q_T|_\Gamma$.

Evidently, $\mathcal{P}(T,K,q)$, respectively $\mathcal{O}(T,K,q)$, is the disjoint union of $\mathcal{P}(T,K,L)$, respectively $\mathcal{O}(T,K,L)$ over representative lattices L of the genus prescribed by $\mathrm{sgn}(T \oplus K)$ and discriminant form q. The difference between $\mathcal{P}(T,K,q)$ and $\mathcal{P}(T,K,L)$ is that the former set does not specify the overlattice but just its genus and we need $\mathcal{P}(T,K,q)$ because we are interested in describing lattices by discriminant forms, but those forms only see the genus.

Lemma 3.13 *For T, K and q as above, the sets $\mathcal{O}(T,K,q)$ and $\mathcal{Q}(T,K,q)$ are in bijection.*

Proof. The main idea is that the restrictions of H and L to the newly introduced sets factor as follows:

$$\begin{array}{ccc}
\mathcal{O}(T,K,q) \hookrightarrow \mathcal{O}(T \oplus K,q) & \qquad & (L, j_T, j_K) \mapsto (L, j_T \oplus j_K) \\
\end{array}$$

$$\begin{array}{ccc}
\mathcal{O}(T,K,q) \longrightarrow \mathcal{O}(T \oplus K,q) \\
\uparrow \; \downdownarrows \qquad\qquad \mathsf{L} \uparrow\;\; \downarrow \mathsf{H} \\
\mathcal{Q}(T,K,q) \hookrightarrow \mathcal{Q}(T \oplus K,q) & \qquad\qquad H \mapsto H
\end{array}$$

Indeed, the map $\mathsf{H}|_{\mathcal{P}(T,K,q)}$ factors via $\mathcal{Q}(T,K,q)$ in view of Lemma 3.10.

In order to see that $\mathsf{L}|_{\mathcal{Q}(T,K,q)}$ factors over $\mathcal{O}(T,K,q)$, we take an isotropic sub-group $H \subset D_{T \oplus K}$. Then we can form the overlattice $L_H = \pi^{-1}(H)$ of $T \oplus K$. Obviously, this gives embeddings $j_T : T \hookrightarrow L_H$ and $j_K : K \hookrightarrow L_H$. These are primitive since the projections $H \to p_T(H)$ and $H \to p_K(H)$ are isomorphisms. Next, the sublattices are orthogonal to each other: $j_T : T \to T^\vee \oplus K^\vee, v \mapsto (\langle v, \cdot\rangle, 0)$ and $j_K : K \to T^\vee \oplus K^\vee, w \mapsto (0, \langle w, \cdot\rangle)$. Finally, they obviously span L_H over $\mathbb{Q}$. □

Fix a subgroup $G_T \subseteq \mathrm{O}(T)$. Two pairs $(L,i,j), (L',i',j') \in \mathcal{P}(T,K,q)$ are called G_T-*equivalent* if there is an isometry $\varphi \colon L \overset{\sim}{\to} L'$ such that $\varphi(iT) = i'T$ and $\varphi_T := (i')^{-1} \circ \varphi|_{i(T)} \circ i \in G_T$ for the induced isometry of T.

Lemma 3.14 [32, 1.15.1] *Let* $H, H' \in \mathcal{Q}(T,K,q)$. *Then* $\mathsf{L}(H)$ *and* $\mathsf{L}(H')$ *are* G_T-*equivalent if and only if there is* $\psi \in G_T \times \mathrm{O}(K)$ *such that* $D_\psi(H) = H'$.

Proof. First note that the condition $D_\psi(H) = H'$ is equivalent to the one in [32]: there are $\psi_T \in G_T$ and $\psi_K \in \mathrm{O}(K)$ such that $D_{\psi_T}(\Gamma) = \Gamma'$ and $D_{\psi_K}\gamma = \gamma'D_{\psi_T}$ where H and H' are the graphs of $\gamma \colon \Gamma \to D_K$ and $\Gamma' \colon H' \to D_K$, respectively.

Suppose that (L,i,j) and (L',i',j') are G_T-equivalent. Thus there is an isometry $\varphi \colon L \overset{\sim}{\to} L'$ with $\varphi(iT) = i'T$. In particular, $\varphi(i(T)_L^\perp) = i'(T)_L^\perp$; using the isomorphisms j and j', we get an induced isometry $\varphi_K \in \mathrm{O}(K)$. We have established the following commutative diagram with exact rows:

$$\begin{array}{ccccccccc}
0 & \longrightarrow & iT \oplus jK & \longrightarrow & L & \longrightarrow & L/(iT \oplus jK) & \longrightarrow & 0 \\
 & & \downarrow{\scriptstyle \varphi_T \oplus \varphi_K} & & \downarrow{\scriptstyle \varphi} & & \downarrow{\scriptstyle D_\varphi} & & \\
0 & \longrightarrow & i'T \oplus j'K & \longrightarrow & L' & \longrightarrow & L'/(i'T \oplus j'K) & \longrightarrow & 0.
\end{array}$$

Put $\psi := (\varphi_T, \varphi_K) \in G_T \times \mathrm{O}(K)$. Using the identification of $L/(iT \oplus jK)$ with $H \subset D_{T \oplus K}$ obtained from i and j (and analogously for H'), the isomorphism D_φ on discriminants turns into the isomorphism $D_\psi \colon H \overset{\sim}{\to} H'$. Note that by construction $\psi^\vee|_L = \varphi$.

Given $\psi \in G_T \times \mathrm{O}(K)$, consider the induced isomorphism on the dual $\psi^\vee \colon (T \oplus K)^\vee \overset{\sim}{\to} (T \oplus K)^\vee$. By the assumption $D_\psi(H) = H'$, this isomorphism restricts to $\varphi := \psi_T^\vee \oplus \psi_K^\vee|_{L_H} \colon L_H \overset{\sim}{\to} L_{H'}$. Finally, under the embeddings $i_H, j_H, i_{H'}, j_{H'}$ the induced isometries of φ combine to $(\varphi_T, \varphi_K) = \psi$. □

Assumption 3.15 *From now on we suppose that the embedding lattice is uniquely determined by the signature (derived from $T \oplus K$) and the discriminant form q. In other words, we postulate that there is a single lattice L in that genus, i.e. $\mathcal{P}(T,K,L) = \mathcal{P}(T,K,q_L)$.*

We say that two primitive embeddings $(i, j), (i', j') \in \mathcal{P}(T, K, L)$ are G_L-*equivalent* if there is an isometry $f \in G_L$ such that $f(iT \oplus jK) = i'T \oplus j'K$.

Combining the two isometry subgroups $G_T \subset O(T)$ and $G_L \subset O(L)$, we say that $(i, j), (i', j') \in \mathcal{P}(T, K, L)$ are *equivalent up to G_L and G_T* if there is an isometry $f \in G_L$ such that $f(iT \oplus jK) = i'T \oplus j'K$ and $f(iT) = i'T$ and $f_T \in G_T$ for the induced isometry.

For later use, we now present a version of Lemma 3.14 in the presence of a subgroup G_L but with Assumption 3.15.

Lemma 3.16 *Assume that L is an overlattice of $T \oplus K$ which is unique within its genus. Let $H, H' \in \mathcal{Q}(T, K, q_L)$.*

Then $\mathsf{L}(H)$ and $\mathsf{L}(H')$ are equivalent in $\mathcal{P}(T, K, L)$ up to G_L and G_T if and only if there is an isometry $\psi \in G_T \times O(K)$ such that $D_\psi(H) = H'$ and $\psi^\vee|_L \in G_L$.

Proof. Note that the G_L-action is well defined by Assumption 3.15. The proof of the lemma is the same as the one of Lemma 3.14, taking into account the additional assumption. $\qquad\square$

Lemma 3.17 ([27, Lemma 23]) *Let L be an overlattice of $T \oplus K$ such that L is unique in its genus and let K' be a lattice in the genus of K. Then there is a bijection $\mathcal{O}(T, K, L) \overset{1:1}{=} \mathcal{O}(T, K', L)$.*

In particular, there is a primitive embedding $K' \hookrightarrow L$ such that L becomes an overlattice of $T \oplus K'$.

Proof. We observe that the set $\mathcal{Q}(T, K, q) = \mathcal{Q}(T, K, L)$ does not really depend on K but rather just on the discriminant form q_K. Hence from Lemma 3.13 and using Assumption 3.15, we get a chain of bijections

$$\mathcal{O}(T, K, L) \overset{1:1}{=} \mathcal{Q}(T, K, L) \overset{1:1}{=} \mathcal{Q}(T, K', L) \overset{1:1}{=} \mathcal{O}(T, K', L)$$

and hence the claim. $\qquad\square$

The situation is particularly nice for indefinite unimodular overlattices where we recover a result proved by Hosono et al.:

Corollary 3.1 ([22, Theorem 1.4]). *Let $T \oplus K$ be indefinite. Then there is a bijection $G_T \backslash \mathcal{Q}(T, K, 0) \overset{1:1}{=} G_T \times O(K) \backslash O(D_K)$, where G_T acts on D_K via $G_T \hookrightarrow O(T) \to O(D_T) \overset{\sim}{\to} O(D_K)$.*

Proof. We have $D_T \cong D_K$ by the following standard argument: the map $L = L^\vee \to T^\vee$ is surjective with kernel K; hence $L \cong T^\vee \oplus K$, and $T^\vee/T \cong (T^\vee \oplus K)/(T \oplus K) \cong L/(T \oplus K)$, similarly for $K^\vee/K$ by symmetry. Also, the forms on D_T and D_K coincide up to sign: $q_T \cong -q_K$. This also shows that subgroups H of Lemmas 3.14 and 3.16 are graphs of isomorphisms.

Therefore, primitive embeddings $T \hookrightarrow L$ are determined by anti-isometries $\gamma\colon D_T \overset{\sim}{\to} D_K$. If there exists such an embedding (and hence such an anti-isometry), this set is bijective to $O(D_T)$. We deduce the claim from Lemma 3.16. $\qquad\square$

Example 3.18. Note that in the unimodular case ($q = 0$), the prescription of T and of the genus of the overlattice (i.e. just the signature in this case) already settles the genus of K by $q_K = -q_T$ and the signature of K is obviously fixed. This statement is wrong in the non-unimodular case: it can happen that a sublattice has two embeddings with orthogonal complements of different discriminant (so in particular of different genus) as in the following example:

Example 3.19. Let $T := \langle 2 \rangle$ with generator t and $L := U \oplus \langle 2 \rangle$ with generators $e, f \in U$, $x \in \langle 2 \rangle$. Consider the embeddings $\iota_1, \iota_2 \colon T \hookrightarrow L$ given by $\iota_1(t) = e + f$ and $\iota_2(t) = x$. Then, bases for the orthogonal complements are $\{e - f, x\} \subset \iota_1(T)^\perp$ and $\{e, f\} \subset \iota_2(T)^\perp$. Hence $\mathrm{disc}(\iota_1(T)^\perp) = 4$ but $\mathrm{disc}(\iota_1(T)^\perp) = 1$.

4 K3 Surfaces

In this text, a *K3 surface* will mean a smooth compact complex surface which is simply connected and carries a nowhere vanishing 2-form. By classical surface theory, the latter two conditions are equivalent to zero irregularity ($H^1(X, \mathcal{O}_X) = 0$) and trivial canonical bundle ($\Omega_X^2 \cong \mathcal{O}_X$). See [2, VIII] or [4] for details.

We denote the *Picard rank* of a K3 surface X by ρ_X. It is the number of independent line bundles on X. If X is projective, then ρ_X is also the number of independent divisor classes and always positive but not vice versa. The cohomology groups listed below carry lattice structures coming from the cup product on the second cohomology:

$H_X^2 = H^2(X, \mathbb{Z})$	full second cohomology,	$\mathrm{sgn}(H_X^2) = (3, 19)$
T_X	transcendental lattice,	$\mathrm{sgn}(T_X) = (2, 20 - \rho_X)$
NS_X	Néron–Severi lattice,	$\mathrm{sgn}(NS_X) = (1, \rho_X - 1)$

where the signatures in the second and third cases are valid only for X projective. Following usage in algebraic geometry, we will often write $\alpha.\beta = (\alpha, \beta)$ for the pairing. Likewise, we will use the familiar shorthand $L.M$ for the pairing of the first Chern classes $c_1(L).c_1(M)$ of two line bundles L and M.

By Poincaré duality, H_X^2 is a unimodular lattice; it follows from Wu's formula that the pairing is even. Indefinite, even, unimodular lattices are uniquely determined by their signature; we get that H_X^2 is isomorphic to the so-called K3 *lattice* made up from three copies of the hyperbolic plane U and two copies of the negative E_8 lattice:

$$L_{K3} = 3U \oplus 2E_8(-1).$$

The Néron–Severi and transcendental lattices are mutually orthogonal primitive sublattices of H_X^2. In particular, H_X^2 is an overlattice of $T_X \oplus NS_X$.

We denote by ω_X the canonical form on X. It has type $(2, 0)$ and is unique up to scalars, since $H^0(X, \Omega_X^2) = \mathbb{C}$ for a K3 surface. By abuse of notation, we also write ω_X for its cohomology class, so that $\omega_X \in T_X \otimes \mathbb{C}$. In fact, T_X is the smallest primitive submodule of H_X^2 whose complexification contains ω_X.

As X is a complex Kähler manifold, the second cohomology H_X^2 comes equipped with a pure Hodge structure of weight 2: $H_X^2 \otimes \mathbb{C} = H^{2,0}(X) \oplus H^{1,1}(X) \oplus H^{0,2}(X)$. Note that $H^{1,1}(X) = (\mathbb{C}\omega_X + \mathbb{C}\overline{\omega}_X)^\perp$. The transcendental lattice T_X is an irreducible Hodge substructure with unchanged $(2,0)$ and $(0,2)$ components.

A *Hodge isometry* of H_X^2 (or T_X) is an isometry that maps each Hodge summand to itself. As the $(2,0)$-component is one-dimensional, Hodge isometries are just isometries $\varphi \colon H_X^2 \xrightarrow{\sim} H_X^2$ with $\varphi_{\mathbb{C}}(\omega_X) = c\omega_X$ for some $c \in \mathbb{C}^*$ (analogous for Hodge isometries of T_X). If L is a lattice with Hodge structure, we denote the group of Hodge isometries by $\mathrm{O}_H(L)$.

The following two Torelli theorems are basic for all subsequent work. They say that essentially everything about a K3 surface is encoded in its second cohomology group, considered as a lattice with Hodge structure—for both the classical and derived point of view. (We repeat Orlov's result about equivalent surfaces up to derived equivalence.)

Classical Torelli Theorem for K3 surfaces. *Two K3 surfaces X and Y are isomorphic if and only if there is a Hodge isometry between their second cohomology lattices H_X^2 and H_Y^2.*

Derived Torelli Theorem for K3 surfaces (**Orlov**). *Two projective K3 surfaces X and Y are derived equivalent if and only if there is a Hodge isometry between the transcendental lattices T_X and T_Y.*

See [4] or [2, Sect. VIII] for the classical case (the latter reference gives an account of the lengthy history of this result) and [34] or [23, Sect. 10.2] for the derived version.

A *marking* of X is the choice of an isometry $\lambda_X \colon H_X^2 \xrightarrow{\sim} L_{\mathrm{K3}}$. The period domain for K3 surfaces is the following open subset of the projectivised K3 lattice:

$$\Omega_{L_{\mathrm{K3}}} = \{\omega \in \mathbb{P}(L_{\mathrm{K3}} \otimes \mathbb{C}) \mid \omega.\omega = 0, \omega.\overline{\omega} > 0\}.$$

Since L_{K3} has signature $(3, k)$ with $k > 2$, this set is connected. By the surjectivity of the period map [2, VIII.14], each point of $\Omega_{L_{\mathrm{K3}}}$ is obtained from a marked K3 surface. Forgetting the choice of marking by dividing out the isometries of the K3 lattice, we obtain a space $\mathcal{F} = \mathrm{O}(L_{\mathrm{K3}}) \backslash \Omega_{L_{\mathrm{K3}}}$ parametrising all (unmarked) K3 surfaces. As is well known, $\mathcal{F}$ is a 20-dimensional, non-Hausdorff space. In particular, it is not a moduli space in the algebro-geometric sense.

Denote by K3$_{\mathrm{FM}}$ the set of all K3 surfaces up to derived equivalence—two K3 surfaces get identified if and only if they are Fourier–Mukai partners, i.e. if and only if their transcendental lattices are Hodge isometric. Its elements are the sets FM(X) of Fourier–Mukai partners of K3 surfaces X. One cannot expect this set to have a good analytic structure: the fibres of the map $\mathcal{F} \to$ K3$_{\mathrm{FM}}$ can become arbitrarily large (see [33]). On the other hand, any K3 surface has only finitely many FM partners ([8]), so that the fibres are finite at least.

Since the transcendental lattices determine D-equivalence by Orlov's derived Torelli theorem, the Fourier–Mukai partners of a K3 surface X are given by embeddings $T_X \subseteq L_{K3}$, modulo automorphisms of T_X. This can be turned into a precise count:

Theorem (Hosono, Lian, Oguiso, Yau [22, Theorem 2.3]). *The set of Fourier–Mukai partners of a* K3 *surface* X *has the following partition:*

$$\mathsf{FM}(X) = \coprod_{S \in \mathcal{G}(NS_X)} O_H(T_X) \times O(S) \backslash O(D_S)$$

with $O(S)$ *and* $O_H(T_X)$ *acting on* $O(D_S)$ *as in Corollary 3.1 above.*

The special case of a generic projective K3 surface, $\mathrm{rk}(NS_X) = 1$, was treated before, leading to a remarkable formula reminiscent of classical genus theory for quadratic number fields (and proved along these lines):

Theorem (Oguiso [33]). *Let* X *be a projective* K3 *surface with* $\mathrm{Pic}(X)$ *generated by an ample line bundle of self-intersection* $2d$. *Then* X *has* $2^{p(d)-1}$ *FM partners, where* $p(d)$ *is the number of distinct prime factors of* d, *and* $p(1) = 1$.

Oguiso's theorem can also be interpreted as a result about polarised K3 surfaces, which we turn to next: recall that the number $2^{p(d)-1}$ is the order of $O(D_{L_{2d}})/\langle \pm 1 \rangle$, where L_{2d} is the orthogonal complement of a primitive vector of degree $2d$ in the K3 lattice L_{K3}.

5 Polarised K3 Surfaces

A *semi-polarised* K3 *surface* of degree $d > 0$ is a pair (X, h_X) of a K3 surface X together with a class $h_X \in NS_X$ of a nef divisor with $h_X^2 = 2d > 0$. A nef divisor of positive degree is also called *pseudo-ample*. We recall that an effective divisor is nef if and only if it intersects all -2-curves nonnegatively [2, Sect. VIII.3]. We will also assume that h_X is primitive, i.e. not a non-trivial integer multiple of another class.

We speak of a *polarised* K3 *surface* (X, h_X) if h_X is the class of an ample divisor. However, we call h_X the *polarisation*, even if it is just nef and not necessarily ample. For details, see [2, Sect. VIII.22]. The relevant geometric lattice is the complement of the polarisation

$$H_X = (h_X)^{\perp}_{H_X^2} \qquad\qquad \text{non-unimodular of signature } (2, 19).$$

which inherits lattice and Hodge structures from H_X^2.
On the side of abstract lattices, recall that $L_{K3} \cong 3U \oplus 2E_8(-1)$; we denote the three orthogonal copies of U in L_{K3} by $U^{(1)}$, $U^{(2)}$ and $U^{(3)}$. Basis vectors e_i, f_i of $U^{(i)}$, defined by $e_i^2 = f_i^2 = 0$ and $e_i.f_i = 1$, always refer to such a choice. For $h \in L_{K3}$ with $h^2 > 0$, set

$$L_h = h^\perp_{L_{K3}} \qquad\qquad \text{non-unimodular of signature } (2, 19),$$

$$L_{2d} = 2U \oplus \langle -2d \rangle \oplus 2E_8(-1) \qquad \text{the case } h = e_3 + d f_3.$$

The choice of the lattice L_{2d} is motivated as follows: since all primitive vectors of fixed length appear in a single $O(L_{K3})$-orbit by Eichler's criterion (Lemma 2.5), we can assume $h = e_3 + d f_3$. Note that $H_X \cong L_{2d}$ as lattices. Obviously, $D_{L_{2d}}$ is the cyclic group of order $2d$. The non-unimodular summand $\langle -2d \rangle$ of L_{2d} is generated by $e_3 - d f_3$; thus, $D_{L_{2d}}$ is generated by the integer-valued functional $\frac{1}{2d}(e_3 - d f_3, \cdot)$. The quadratic form $D_{L_{2d}} \to \mathbb{Q}/2\mathbb{Z}$ is then given by mapping this generator to the class of $\frac{-2d}{4d^2} = \frac{-1}{2d}$.

There are two relevant groups in this situation: the full isometry group $O(L_{2d})$ and the subgroup $\tilde{O}(L_{2d})$ of *stable* isometries which by definition act trivially on the discriminant $D_{L_{2d}}$. The next lemma gives another description of stable isometries.

Lemma 5.1 *The stable isometry group coincides with the group of L_{K3}-isometries stabilising h, i.e. $\tilde{O}(L_{2d}) = \{g \in O(L_{K3}) \mid g(h) = h\}$.*

Proof. Given $g \in O(L_{K3})$ with $g(h) = h$, we make use of the fact that the discriminant groups of $h^\perp = L_{2d}$ and $\langle 2d \rangle$ (the latter generated by h) are isomorphic and their quadratic forms differ by a sign. This is true because these are complementary lattices in the unimodular L_{K3}; see the proof of Corollary 3.1. The induced maps on discriminants, $D_{g,h^\perp} : h^\perp \xrightarrow{\sim} h^\perp$ and $D_{g,h} : \langle 2d \rangle \xrightarrow{\sim} \langle 2d \rangle$, are the same under the above identification. Since $D_{g,h}$ is the identity by assumption, $D_{g,h^\perp}$ is, too. Hence, $g|_{L_{2d}}$ is stable.

On the other hand, any $f \in O(L_{2d})$ allows defining an isometry $\tilde{f}$ of the lattice $L_{2d} \oplus \mathbb{Z}h$, by mapping h to itself. Note that $L_{2d} \oplus \mathbb{Z}h \subset L_{K3}$ is an overlattice. If f is a stable isometry, i.e. $D_f = \mathrm{id}$, then $\tilde{f}$ extends to an isometry of L_{K3}. (This can be seen by considering a representative of the discriminate of the two summands, i.e. by explicit gluing; cf. Remark 3.2.)   □

An isomorphism of semi-polarised K3 surfaces is an isomorphism of the surfaces respecting the polarisations. Here we recall two Torelli theorems which are essential for the construction of moduli spaces of K3 surfaces.

Strong Torelli Theorem for polarised K3 surfaces. *Given two properly polarised K3 surfaces (X, h_X) and (Y, h_Y), i.e. h_X and h_Y are ample classes, and a Hodge isometry $\varphi : H^2_X \xrightarrow{\sim} H^2_Y$ with $\varphi(h_X) = h_Y$, there is an isomorphism $f : Y \xrightarrow{\sim} X$ such that $\varphi = f^*$.*

This result only holds for polarised K3 surfaces. For semi-polarised K3 surfaces, we have a different result, where we say that (X, h_X) and (Y, h_Y) are *isomorphic* if there is an isomorphism $f : X \xrightarrow{\sim} Y$ such that $f^*(h_Y) = h_X$.

Torelli Theorem for semi-polarised K3 surfaces. *Two semi-polarised K3 surfaces (X, h_X) and (Y, h_Y) are isomorphic if and only if there is a Hodge isometry $\varphi : H^2_X \xrightarrow{\sim} H^2_Y$ with $\varphi(h_X) = h_Y$.*

Proof. Let $\varphi\colon H_X^2 \xrightarrow{\sim} H_Y^2$ be a Hodge isometry with $\varphi(h_X) = h_Y$. Since h_X is not ample, we cannot immediately invoke the strong Torelli theorem. Following [4, p. 151], let Γ be the subgroup of the Weyl group of Y generated by those roots of $H^2(Y,\mathbb{Z})$ which are orthogonal to h_Y. Then Γ acts transitively on the chambers of the positive cone, whose closure contains h_Y. Hence we can find an element $w \in \Gamma$ such that $w(h_Y) = h_Y$ and $w \circ \varphi$ maps the ample cone of X to the ample cone of Y. By the strong Torelli theorem, $w \circ \varphi$ is now induced by an isomorphism $f\colon Y \xrightarrow{\sim} X$. This gives the claim.   □

The counterpart of the unpolarised period domain is the open subset

$$\Omega_{L_{2d}}^{\pm} = \{\omega \in \mathbb{P}(L_{2d} \otimes \mathbb{C}) = \mathbb{P}^{20} \mid (\omega,\omega) = 0, (\omega,\overline{\omega}) > 0\} = \Omega_{L_{K3}} \cap h^{\perp}$$

where we abuse notation to also write $h^{\perp}$ for the projectivised hyperplane. Obviously, both $O(L_{2d})$ and its subgroup $\tilde{O}(L_{2d})$ act on $\Omega_{L_{2d}}^{\pm}$. Since the signature of L_{2d} is $(2,19)$, the action is properly discontinuous.

Furthermore, signature $(2,19)$ also implies that $\Omega_{L_{2d}}^{\pm}$ has two connected components. These are interchanged by the (stable) involution induced by $\mathrm{id}_{U^{(1)}} \oplus (-\mathrm{id}_{U^{(2)}}) \oplus \mathrm{id}_{U^{(3)} \oplus 2E_8(-1)}$. Denote by $\Omega_{L_{2d}}^{+}$ one connected component; this is a type IV domain. Also, let $O^+(L_{2d})$ and $\tilde{O}^+(L_{2d})$ be the subgroups of the (stable) isometry group of L_{2d} fixing the component. They are both arithmetic groups, as they have finite index in $O(L_{2d})$.

Next, let $\Delta \subset L_{2d}$ be the subset of all (-2)-classes, and for $\delta \in \Delta$ denote by $\delta^{\perp} \subset L_{2d} \otimes \mathbb{C}$ the associated hyperplane ("wall"). In analogy to the unpolarised case, we define a parameter space as the quotient by the group action— however, there are certain differences to be explained below: let

$$\mathcal{F}_{2d} = \tilde{O}(L_{2d})\backslash\Omega_{L_{2d}}^{\pm} = \tilde{O}^+(L_{2d})\backslash\Omega_{L_{2d}}^{+}.$$

This space has an analytic structure as the quotient of a type IV domain by a group acting properly discontinuously. Furthermore, $\mathcal{F}_{2d}$ is actually quasi-projective by Baily–Borel [1]. Note that the group actions preserve the collection of walls $\delta^{\perp}$, which by abuse of notation are given the same symbol in the quotient. Hence, the group action also preserves the complement

$$\mathcal{F}_{2d}^{\circ} = \mathcal{F}_{2d} \setminus \bigcup_{\delta \in \Delta} \delta^{\perp}.$$

The definition of $\mathcal{F}_{2d}^{\circ}$ means that -2-classes orthogonal to the polarisation are transcendental. In other words, the polarisation is ample, as it is nef and non-zero on all -2-curves.

The subspace $\mathcal{F}_{2d}^{\circ}$ is the moduli space of pairs (X, h_X) consisting of (isomorphism classes of) a K3 surface X and the class h_X of an ample, primitive line bundle with $h_X^2 = 2d$: given such a pair, we choose a *marking*, i.e. an isometry $\lambda_X\colon H_X^2 \xrightarrow{\sim} L_{K3}$ such that $\lambda_X(h_X) = h$. This induces $\lambda_X|_{H_X}\colon H_X \xrightarrow{\sim} L_{2d}$ and gives the period point $\lambda_X(\omega_X) \in \Omega_{2d}^{\pm}$. Since h_X is ample, the period point avoids the walls.

Conversely, given an $\tilde{O}(L_{2d})$-orbit of a point $[\omega] \in \Omega_{2d}^{\pm}$ not on any wall, we get a pair (X, h_X) by considering $[\omega]$ as period point for the full K3 lattice: this uses the surjectivity of the period map. Now our assumptions on ω imply $h_X^2 = 2d$ and that h_X is ample as ω avoids the walls. Then, the strong Torelli theorem says that both the K3 surface X and the polarisation h_X are unique (up to isomorphism).

Finally, using again the surjectivity of the period map, one can find for every element $[\omega] \in \mathcal{F}_{2d} \setminus \mathcal{F}_{2d}^{\circ}$ a semi-polarised K3 surface (X, h_X) and a marking $\varphi \colon H^2(X, \mathbb{Z}) \to L_{K3}$ with $\varphi(h_X) = h$, $\varphi([\omega_X]) = \omega$. The fact that the points contained in $\mathcal{F}_{2d}$ correspond to isomorphism classes of semi-polarised K3 surfaces of degree $2d$ now follows from the Torelli theorem for semi-polarised K3 surfaces.

Example 5.2. For $d = 1$, the smallest example of a proper semi-polarisation (i.e. nef, not ample) occurs for a generic elliptic K3 surface X with section. Its Néron–Severi lattice will be generated by the section s and a fibre f. The intersection form on $NS(X)$ is $\left(\begin{smallmatrix} -2 & 1 \\ 1 & 0 \end{smallmatrix}\right)$ and we set $D := s + 2f$. This effective divisor is primitive and nef as $D.s = 0$, $D.f = 1$, $D^2 = 2$.

We remark that the lattice L_{2d} is more difficult to work with than L_{K3} as it is not unimodular anymore. On the other hand, the moduli space $\mathcal{F}_{2d}$ of $2d$-polarised K3 surfaces is a quasi-projective variety which is a huge improvement over $\mathcal{F} = O(L_{K3}) \backslash \Omega_{L_{K3}}$.

In the polarised case, another natural quotient appears, taking the full isometry group of the lattice L_{2d}:

$$\hat{\mathcal{F}}_{2d} = O(L_{2d}) \backslash \Omega_{2d}^{\pm} = O^+(L_{2d}) \backslash \Omega_{2d}.$$

(The unpolarised setting has $D_{L_{K3}} = 0$ and hence there is only one natural group to quotient by.)

There is an immediate quotient map $\pi \colon \mathcal{F}_{2d} \to \hat{\mathcal{F}}_{2d}$. It has finite fibres and was investigated by Stellari:

Lemma 5.3 ([42, Lemma 2.3]) *The degree of π is $\deg(\pi) = 2^{p(d)-1}$ where $p(d)$ is the number of distinct primes dividing d (see also [43, Theorem 2.2]).*

Proof. The degree is given by the index $[O(L_{2d}) : \tilde{O}(L_{2d})]$ up to the action of the non-stable isometry $-\text{id}$ which acts trivially on the period domain.

We use the exact sequence $0 \to \tilde{O}(L_{2d}) \to O(L_{2d}) \to O(D_{L_{2d}}) \to 0$ where the right-hand zero follows from [32, Theorem 1.14.2]. The index thus equals the order of the finite group $O(D_{L_{2d}})$. Now $D_{L_{2d}}$ is the cyclic group of order $2d$ and decomposes into the product of various p-groups. Automorphisms of $D_{L_{2d}}$ factorise into automorphisms of the p-groups. However, the only automorphisms of $\mathbb{Z}/p^l$ respecting the quadratic (discriminant) form are those induced by $1 \mapsto \pm 1$. Hence $|O(D_{L_{2d}})| = 2^{p(d)}$. The degree is then $2^{p(d)-1}$, taking the non-stable isometry $-\text{id}$ of L_{2d} into account.

In case $d = 1$, we have $\tilde{O}(L_{2d}) = O(L_{2d})$, fitting with $p(1) = 1$. $\qquad\square$

Points of $\hat{\mathcal{F}}_{2d}$ or rather the fibres of π have the following property (see also [43, Theorem 2.2]):

Lemma 5.4 *Given semi-polarised* K3 *surfaces* (X, h_X) *and* (Y, h_Y) *with* $\pi(X, h_X) = \pi(Y, h_Y)$, *there are Hodge isometries* $H_X \cong H_Y$ *and* $T_X \cong T_Y$. *In particular,* X *and* Y *are FM partners.*

Proof. Fix markings $\lambda_X \colon H_X^2 \xrightarrow{\sim} L_{K3}$ and $\lambda_Y \colon H_Y^2 \xrightarrow{\sim} L_{K3}$ with $\lambda_X(h_X) = \lambda_Y(h_Y) = h$ and $\lambda_X(\omega_X) = \omega$, $\lambda_Y(\omega_Y) = \omega'$. By $\pi(X, h_X) = \pi(Y, h_Y)$, there is some $g \in O(L_{2d})$ such that $g(\lambda_X(\omega_X)) = \lambda_Y(\omega_Y)$. In particular, the primitive lattices generated by ω_X and ω_Y (which are the transcendental lattices T_X and T_Y) get mapped into each other by $\lambda_Y^{-1} \circ g \circ \lambda_X$. Thus, the latter isometry respects the Hodge structures and induces Hodge isometries $T_X \xrightarrow{\sim} T_Y$ and $H_X = h_X^{\perp} \xrightarrow{\sim} H_Y = h_Y^{\perp}$.   □

6 Polarisation and FM Partners

In this section, we want to consider the relationship between polarisations and FM partners. A priori these concepts are very different: the condition that two K3 surfaces are derived equivalent is a property of their transcendental lattices, whereas the existence of polarisations concerns the Néron–Severi group. Indeed, we shall see that there are FM partners where one K3 surface carries a polarisation of given degree but the other does not. On the opposite side, we shall see in the next Sect. 7 that one can count the number of FM partners among polarised K3 surfaces of a given degree.

Introduce the set

$$\text{K3}_{\text{FM}}^{2d} := \left\{ X \,\middle|\, \begin{array}{l} \text{K3 surface admitting a primitive} \\ \text{nef line bundle } L \text{ with } L^2 = 2d \end{array} \right\} / \sim$$

where $X \sim Y$ if and only if $D^b(X) \cong D^b(Y)$.

We shall first discuss two examples which shed light on the relationship between FM partnership and existence of polarisations.

Example 6.1. Derived equivalence does not respect the existence of polarisations of a given degree. To give an example, we use the rank 2 Néron–Severi lattices defined by $\left(\begin{smallmatrix} 0 & -7 \\ -7 & -2 \end{smallmatrix} \right)$ and $\left(\begin{smallmatrix} 0 & -7 \\ -7 & 10 \end{smallmatrix} \right)$ which are related by the rational base change $\frac{1}{3} \left(\begin{smallmatrix} 1 & 0 \\ -2 & 9 \end{smallmatrix} \right)$. The former obviously represents -2, whereas the latter does not. Furthermore, the latter primitively represents 10 via the vector $(0, 1)$ and 6 via the vector $(2, 3)$, whereas the former does not. For example, if we had $10 = -14xy - 2y^2$, then y would have to be one of $1, 2, 5, 10$ up to sign and neither of these eight cases works.

The orthogonal lattices in L_{K3} are isomorphic, as follows from Nikulin's criterion. Denote the common orthogonal complement by T. As in the previous example, we choose a general vector $\omega \in T_{\mathbb{C}}$ with $(\omega, \omega) = 0$, $(\omega, \overline{\omega}) > 0$. We see that T admits primitive embeddings $\iota, \iota' \colon T \hookrightarrow L_{K3}$ such that $\iota(T)^{\perp}$ does not contain any vectors of square $2d$ whereas $\iota'(T)^{\perp}$ does, for $d = 3$ or $d = 5$. Furthermore, $\iota'(T)^{\perp}$

does not contain any -2-classes. The surface X' corresponding to $\iota'(\omega) \in T_{\mathbb{C}}$ can actually be $2d$-polarised since $\iota'(\omega) \in L_h^{\perp} \cong L_{2d}$ and we have $\mathcal{F}_h^{\circ} = \mathcal{F}_{2d}$ by the absence of -2-classes. On the other hand, the surface X corresponding to $\iota(\omega)$ has no $2d$-(semi-)polarisations—there are not even classes in $NS(X)$ of these degrees.

Example 6.2. Let $d > 1$ be an integer, not divisible by 3 such that $2d$ can be represented primitively by the positive definite root lattice A_2 (e.g. $d = 7$). Let $T = 2U \oplus E_8(-1) \oplus E_6(-1) \oplus \langle -2d \rangle$. Following an idea of M. Schütt, we construct two primitive embeddings $\iota, \iota' \colon T \hookrightarrow L_{2d}$. Both of them are the identity on $2U \oplus E_8(-1)$. On the $E_6(-1) \oplus \langle -2d \rangle$ part of T, we use

$$\iota \colon E_6(-1) \hookrightarrow E_8(-1), \langle -2d \rangle \xrightarrow{\sim} \langle -2d \rangle \quad \text{with } \iota(T)_{L_{2d}}^{\perp} \cong A_2(-1)$$
$$\iota' \colon E_6(-1) \oplus \langle -2d \rangle \hookrightarrow E_8(-1) \qquad\qquad \text{with } \iota'(T)_{L_{2d}}^{\perp} \cong \langle -6d \rangle \oplus \langle -2d \rangle.$$

We choose a general point $\omega \in T_{\mathbb{C}}$ with $(\omega, \omega) = 0$ and $(\omega, \bar{\omega}) > 0$. Then the points $\iota(\omega)$ and $\iota'(\omega)$ in $\mathcal{F}_{2d}$ represent (semi-)polarised K3 surfaces (X, h_X) and $(X', h_{X'})$. In the first case h_X is only semi-polarised, as $A_2(-1)$ contains (-2)-vectors; in other words $\iota(\omega) \in \mathcal{F}_{2d} \setminus \mathcal{F}_{2d}^{\circ}$. In the second case $\iota(\omega) \in \mathcal{F}_{2d}^{\circ}$ since the orthogonal complement of $h_{X'}$ in $NS(X')$ equals $\langle -2d \rangle \oplus \langle -6d \rangle$ which does not contain a (-2)-class. This shows that there are examples of polarised and semi-polarised K3 surfaces of the same degree which have the same FM partner.

Incidentally we notice that $NS(X) \cong NS(X') \cong A_2(-1) \oplus \langle 2d \rangle$. This follows from Nikulin's criterion since both lattices have rank 3 and length 1 since we have assumed that $(3, d) = 1$.

We want to construct a map

$$\tau \colon \mathcal{F}_{2d} \to \text{K3}_{\text{FM}}^{2d}.$$

By Lemma 5.4, we have a map $\pi \colon \hat{\mathcal{F}}_{2d} \to \text{K3}_{\text{FM}}^{2d}$; combining it with $\sigma \colon \mathcal{F}_{2d} \to \hat{\mathcal{F}}_{2d}$, we obtain a commutative triangle

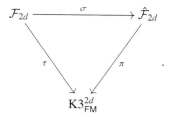

By the counting results of Proposition 7.1, the fibres are finite. Here we give a geometric argument for that fact, following Stellari [43, Lemma 2.5], where we pay special attention to the 'boundary points' of $\mathcal{F}_{2d}$.

Proposition 6.3. *Given a $2d$-(semi-)polarised K3 surface (X, h_X), there are only finitely many $2d$-(semi-)polarised K3 surfaces (Y, h_Y) up to isomorphism with $D^b(X) \cong D^b(Y)$.*

Proof. Disregarding polarisations, there are only finitely many FM partners of X, as X is a smooth projective surface [8]. Given such an FM partner Y, consider the set $A_{Y,2d} = \{c \in C(Y) \mid c^2 = 2d\}$ of elements of length $2d$ in the positive cone of Y. If the divisor c is ample, then $3c$ is very ample, by Saint-Donat's result [40]. By Bertini, there are irreducible divisors $D \in |3c|$. The set $B_{Y,18d} = \{\mathcal{O}_X(D) \mid D^2 = 18d, D \text{ irreducible}\}$ of divisor classes of irreducible divisors of length $18d$ is finite up to automorphisms of Y by Sterk [44]. As $A_{Y,2d} \to B_{Y,18d}, c \mapsto |3c|$ is injective (this uses $H^1(\mathcal{O}_Y) = 0$), this shows that the number of non-isomorphic $2d$-polarisations on Y is finite.

However, there are points (Y, h_Y) where h_Y is only pseudo-ample. Denote the set of pseudo-ample divisors of degree $2d$ by $\overline{A}_{Y,2d} = \{c \in \overline{C}(Y) \mid c^2 = 2d\}$. If we have a non-ample polarisation h_Y, then contracting the finitely many -2-curves which are orthogonal to h_Y produces a projective surface Y' with only ADE singularities, trivial canonical bundle and $H^1(\mathcal{O}_{Y'}) = 0$. Morrison shows that Saint-Donat's result is also true for this surface [28, Sect. 6.1], i.e. $3c$ is again very ample. We can then proceed as above, as the generic divisor in $|3c|$ will be irreducible and avoid the finitely many singularities. Sterk's result on finiteness of $B_{Y',18d}/\mathrm{Aut}(Y')$ still applies as he simply assumes that a linear system is given whose generic member is irreducible.   □

7 Counting FM Partners of Polarised K3 Surfaces in Lattice Terms

Taking our cue from the fact that the fibres of $\mathcal{F} \to \mathcal{F}/\mathrm{FM}$ are just given by FM partners (the unpolarised case), and the latter can be counted in lattice terms, we study the following general setup: let L be an indefinite, even lattice, let T be another lattice, occurring as a sublattice of L, and let finally $G_T \subseteq O(T)$ and $G_L \subseteq O(L)$ be two subgroups, the latter normal. As in Sect. 3, we consider the set $\mathcal{P}(T, L)$ of all primitive embeddings $\iota: T \hookrightarrow L$. This set is partitioned into $\mathcal{P}(T, K, L)$, containing all primitive embeddings $\iota: T \hookrightarrow L$ with $\iota(T)_L^\perp \cong K$.

In the application to geometry, we will have $L = L_{2d} = h^\perp$ the perpendicular lattice of the polarisation inside the K3 lattice, $T = T_X$ the transcendental lattice of a K3 surface X and K the orthogonal complement of h in $NS(X)$ the Néron–Severi lattice of X. By Nikulin's criterion (Lemma 2.4), L_{2d} is unique in its genus, thus fulfilling Assumption 3.15. As to the groups, $G_T = O_H(T)$ is the group of Hodge isometries of T_X, and G_L is either the full or the stable isometry group of L_{2d}.

We recall when two embeddings $\iota_1, \iota_2: T \hookrightarrow L$ are equivalent with respect to G_T and G_L (see page 348): if there are isometries $g \in G_T$ and $\tilde{g} \in G_L$ such that

$$
\begin{array}{ccc}
T & \xrightarrow{\ \iota_1\ } & L \\
\downarrow{g} & & \downarrow{\tilde{g}} \\
T & \xrightarrow{\ \iota_2\ } & L
\end{array}
\ .
$$

This corresponds to orbits of the action $G_T \times G_L \times \mathcal{P}(T, L) \to \mathcal{P}(T, L)$, $(g, \tilde{g}) \cdot \iota = \tilde{g}\iota g^{-1}$.

All of this is essentially the setting of [22]—the novelty is the subgroup G_L, which always was the full orthogonal group in loc. cit.

Proposition 7.1. *For a 2d-polarised K3 surface* (X, h_X), *there are bijections*

$$\sigma^{-1}([X, h_X]) \overset{1:1}{=} O_H(T_X) \times O(H_X) \backslash \mathcal{P}(T_X, H_X),$$

$$\tau^{-1}([X, h_X]) \overset{1:1}{=} O_H(T_X) \times \tilde{O}(H_X) \backslash \mathcal{P}(T_X, H_X).$$

Example 7.1. The unpolarised analogue of the proposition was given in Theorem 2.4 of [22], stating $\mathsf{FM}(X) = O_H(T_X) \times O(H_X^2) \backslash \mathcal{P}(T_X, H_X^2)$.

Proof. The proof proceeds along the lines of [22, Theorem 2.4]. Fix a marking $\lambda_X \colon H_X \overset{\sim}{\to} L_{2d}$ for X. Set $T := \lambda_X(T_X)$. This yields a primitive embedding

$$\iota_0 \colon T \xrightarrow[\lambda_X^{-1}]{\sim} T_X \lhook\joinrel\longrightarrow H_X \xrightarrow[\lambda_X]{\sim} L_{2d} \ .$$

This embedding (or rather the equivalence class of $\iota_0 \lambda_X(\omega_X)$) gives a point in $\mathcal{F}_{2d}$. By definition of $\mathcal{F}_{2d}$, this period point does not depend on the choice of marking.

If (Y, h_Y) belongs to (a period point given by an embedding in) $\mathcal{P}(T, L_{2d})$, then—as the transcendental lattice is the smallest lattice containing the canonical form in its complexification—there is a Hodge isometry $T_X \cong T_Y$; hence $D^b(X) \cong D^b(Y)$ and then $\mathsf{FM}(X) = \mathsf{FM}(Y)$. We therefore get maps

$$\tilde{c} \colon \mathcal{P}(T, L_{2d}) \to \tau^{-1}(X, h_X),$$

$$c \colon \mathcal{P}(T, L_{2d}) \xrightarrow{\tilde{c}} \tau^{-1}(X, h_X) \xrightarrow{\pi} \sigma^{-1}(X, h_X).$$

with the fibre $\tau^{-1}(X, h_X)$ consisting of FM partners of (X, h_X) up to isomorphism.

The map $\tilde{c}$ is surjective (and hence c is, as well): if $(Y, h_Y) \in \mathcal{F}_{2d}$ is an FM partner of X, then we first fix a marking $\lambda_Y \colon H_Y \overset{\sim}{\to} L_{2d}$. By the derived Torelli theorem, there is a Hodge isometry $g \colon T_X \overset{\sim}{\to} T_Y$. Using g and the markings for X and Y, we produce an embedding

$$\iota \colon T_X \xrightarrow[g]{\sim} T_Y \lhook\joinrel\longrightarrow H_Y \xrightarrow[\lambda_Y]{\sim} L_{2d} \xrightarrow[\lambda_X^{-1}]{\sim} H_X \ .$$

This gives a point $\iota \in \mathcal{P}(T_X, H_X)$ and by construction, $\tilde{c}(\iota) = (Y, h_Y) \in \tau^{-1}(X, h_X)$.

For brevity, we temporarily introduce shorthand notation

$$\mathcal{P}^{\mathrm{eq}}(T_X, H_X) := O_H(T_X) \times O(H_X) \backslash \mathcal{P}(T_X, H_X),$$

$$\tilde{\mathcal{P}}^{\mathrm{eq}}(T_X, H_X) := O_H(T_X) \times \tilde{O}(H_X) \backslash \mathcal{P}(T_X, H_X),$$

and the goal is to show $\tilde{\mathcal{P}}^{\mathrm{eq}}(T_X, H_X) = \tau^{-1}([X, h_X])$ and $\mathcal{P}^{\mathrm{eq}}(T_X, H_X) = \sigma^{-1}([X, h_X])$.

Now suppose that two embeddings $\iota, \iota' \colon T \hookrightarrow L_{2d}$ give the same equivalence class in $\tilde{\mathcal{P}}^{\mathrm{eq}}(T, L_{2d})$. This means that there exist isometries $g \in O(T)$ and $\tilde{g} \in \tilde{O}(L_{2d})$ with $\iota' \circ g = \tilde{g} \circ \iota$. Denote the associated polarised K3 surfaces by (Y, h_Y) and $(Y', h_{Y'})$; choose markings λ_Y and $\lambda_{Y'}$ as above. Then we obtain a Hodge isometry

$$
\begin{array}{ccccccc}
H_Y & \xrightarrow[\lambda_Y]{\sim} & L_{2d} & \xrightarrow[\tilde{g}]{\sim} & L_{2d} & \xrightarrow[\lambda_{Y'}^{-1}]{\sim} & H_{Y'} \\
\downarrow & & \uparrow{\scriptstyle\iota} & & \uparrow{\scriptstyle\iota'} & & \downarrow \\
T_Y & \xrightarrow{\sim} & T & \xrightarrow[g]{\sim} & T & \xrightarrow{\sim} & T_{Y'}
\end{array}
$$

and hence (Y, h_Y) and $(Y', h_{Y'})$ define the same point in $\mathcal{F}_{2d}$. Thus the map $\tilde{c}$ factorises over equivalences classes and descends to a surjective map $\tilde{c} \colon \tilde{\mathcal{P}}^{\mathrm{eq}}(T, L_{2d}) \to \tau^{-1}(X, h_X)$.

Analogous reasoning applies if ι and ι' are equivalent in $\mathcal{P}^{\mathrm{eq}}(T, L_{2d})$: we get isometries $g \in O(T)$ and $\hat{g} \in O(L_{2d})$ with $\iota' \circ g = \hat{g} \circ \iota$ and use a diagram similar to the one above. In this case, with the isometry $\hat{g}$ not necessarily stable, we can only derive that the period points coincide in $\hat{\mathcal{F}}_{2d}$; hence $c \colon \mathcal{P}^{\mathrm{eq}}(T, L_{2d}) \to \sigma^{-1}(X, h_X)$.

Finally, we show that these maps are injective, as well. Let $[\iota], [\iota'] \in \tilde{\mathcal{P}}^{\mathrm{eq}}(T, L_{2d})$ be two equivalence classes of embeddings with $\tilde{c}([\iota]) = \tilde{c}([\iota'])$. This implies the existence of a stable isometry in $O(L_{2d})$ mapping $\omega \mapsto \omega'$ where ω and ω' are given by the construction of the map $\tilde{c}$ (they correspond to semi-polarised K3 surfaces (Y, h_Y) and $(Y', h_{Y'})$). Using markings $H_Y \xrightarrow{\sim} L_{2d}$ and $H_{Y'} \xrightarrow{\sim} L_{2d}$, we get an induced Hodge isometry $\varphi \colon H_Y \xrightarrow{\sim} H_{Y'}$ with $\varphi(\omega_Y) = \omega_{Y'}$ and $\varphi(h_Y) = h_{Y'}$. Once more invoking the minimality of transcendental lattices, we also get a Hodge isometry $\varphi_T \colon T_Y \xrightarrow{\sim} T_{Y'}$. These isometries combine to

$$
\begin{array}{ccccccc}
L_{2d} & \xrightarrow[\lambda_Y^{-1}]{\sim} & H_Y & \xrightarrow[\varphi]{\sim} & H_{Y'} & \xrightarrow[\lambda_{Y'}]{\sim} & L_{2d} \\
\uparrow{\scriptstyle\iota} & & \downarrow & & \downarrow & & \uparrow{\scriptstyle\iota'} \\
T & \xrightarrow{\sim} & T_Y & \xrightarrow[\varphi_T]{\sim} & T_{Y'} & \xrightarrow{\sim} & T
\end{array}
$$

the outer square of which demonstrates $[\iota] = [\iota']$.

For $[\iota], [\iota'] \in \mathcal{P}^{\mathrm{eq}}(T, L_{2d})$ with $c([\iota]) = c([\iota'])$, we argue analogously, only now starting with an isometry of L_{2d} mapping $\omega \to \omega'$ which is not necessarily stable. Since periods of the K3 surfaces get identified up to $O(L_{2d})$ in this case, the outer square gives $[\iota] = [\iota']$ up to $O_{\mathsf{H}}(T)$ and $O(L_{2d})$. $\square$

From Lemma 3.16 and Proposition 7.1, we derive the following statement.

Proposition 7.2. *Given an $2d$-polarised K3 surface (X, h_X), there are bijections*

$$
\sigma^{-1}([X, h_X]) \overset{1:1}{=} \bigsqcup_S O_{\mathsf{H}}(T_X) \times O(H_X) \backslash \mathcal{Q}(T_X, S, q_{2d})
$$

$$
\tau^{-1}([X, h_X]) \overset{1:1}{=} \bigsqcup_S O_{\mathsf{H}}(T_X) \times \tilde{O}(H_X) \backslash \mathcal{Q}(T_X, S, q_{2d}),
$$

where the unions run over isomorphism classes of even lattices S which admit an overlattice $S \oplus T_X \hookrightarrow H_X$ such that the induced embedding $S \subseteq H_X$ is primitive. The discriminant q_{2d} on the right-hand sides has $D_{q_{2d}} = \mathbb{Z}/2d$ as abelian group with $q_{2d}(1) = \frac{-1}{2d}$.

Proof. The fibres are obviously partitioned by the orthogonal complements that can occur (this is in general a bigger choice than just of an element in the genus).

Once a complement S is chosen, then the set $\mathcal{P}^{\mathrm{eq}}(T_X, S, H_X)$, i.e. the set of embeddings of T_X into H_X with complement isomorphic to S, up to Hodge isometries of T_X and isometries of H_X, coincides with $(\mathrm{O}(H_X), \mathrm{O_H}(T_X))$-equivalence classes; this follows from Lemma 3.16. Analogous reasoning addresses the σ-fibres. $\square$

Example 7.2. Following Remark 7.1 and Lemma 3.13, we note that the formula for FM partners of an unpolarised K3 surface X from [22] can be written as

$$\mathsf{FM}(X) = \coprod_S \mathrm{O_H}(T_X) \times \mathrm{O}(H_X^2) \backslash \mathcal{Q}(T_X, S, 0)$$

where S now runs through isomorphism classes of lattices admitting an overlattice $S \oplus T_X \hookrightarrow H_X^2$ such that S is primitive in H_X^2. As H_X^2 is unimodular, the genus of S is uniquely determined by that of T_X (see Remark 3.18). One candidate for S is $NS(X)$, and we can describe $\mathsf{FM}(X)$ as the same union, with S running through the genus $\mathcal{G}(NS_X)$. By Lemma 3.17, the sets $\mathcal{Q}(T_X, S, 0)$ are all mutually bijective.

We will temporarily work with the sets $\mathcal{P}(T_X, S, H_X^2)$ instead. On each such set, $\mathrm{O_H}(T_X)$ and $\mathrm{O}(H_X^2)$ act in the natural way. Hence, there are not just bijections $\mathcal{P}(T_X, S, H_X^2) \overset{1:1}{=} \mathcal{P}(T_X, S', H_X^2)$ but also bijections of the quotients by $\mathrm{O_H}(T_X) \times \mathrm{O}(H_X^2)$. We thus get

$$\mathsf{FM}(X) \overset{1:1}{=} \mathcal{G}(NS_X) \times (\,\mathrm{O_H}(T_X) \times \mathrm{O}(H_X^2) \backslash \mathcal{P}(T_X, NS_X, H_X^2))$$
$$\overset{1:1}{=} \mathcal{G}(NS_X) \times (\,\mathrm{O_H}(T_X) \times \mathrm{O}(H_X^2) \backslash \mathcal{Q}(T_X, NS_X, 0)).$$

However, by Corollary 3.1, this implies

$$\mathsf{FM}(X) \overset{1:1}{=} \mathcal{G}(NS_X) \times (\,\mathrm{O_H}(T_X) \times \mathrm{O}(NS_X) \backslash \mathrm{O}(D_{NS_X})).$$

Similar formulae in the polarised (hence non-unimodular) case are generally wrong.

8 Examples

Proposition 7.1 phrases the problem of classifying polarised K3 surfaces up to derived equivalence in lattice terms. Using the results of Sect. 3, this can be rephrased as Proposition 7.2 which clearly makes this a finite problem. Given $h_X \in L_{K3}$ primitive and $T_X \subseteq H_X = h_X^\perp$, or equivalently, $T_X \subset L_{2d}$, one can (in principle) list all potential subgroups H of the discriminant group. This, together with the fact that $\text{Hom}(L_{2d}, H)$ is finite, makes it possible to test all potential overlattice groups.

Picard Rank One

We consider the special case of Picard rank 1. Here, $h_X^\perp = T_X$. Also, any FM partner of a $2d$-polarised K3 surface is again canonically $2d$-polarised (since the orthogonal complement of the transcendental lattice is necessarily of the form $\langle -2d \rangle$). Oguiso showed that the number of non-isomorphic FM partners is $2^{p(d)-1}$ (where $p(d)$ is the number of prime divisors) [33]. This is also half of the order of $O(D_{L_{2d}})$.

Stellari [43, Theorem 2.2] shows that the group $O(D_{L_{2d}})/\{\pm\text{id}\}$ acts simply transitively on the fibre $\tau^{-1}(X, h_X)$. In particular, σ is one to one on these points.

We look at the situation from the point of view of Proposition 7.2. In this case $T_X = H_X$ and $O_H(T_X) \times O(H_X) \backslash \mathcal{P}(T_X, H_X)$ clearly contains only one element, which says that the fibre of σ contains only one element. The situation is different for τ. For this we have to analyse the action of the quotient $O_H(T_X) \times O(H_X)/ O_H(T_X) \times \tilde{O}(H_X) \cong O(D_{L_{2d}})$. We note that $-\text{id}$ is contained in both $O_H(T_X)$ and $O(H_X)$, and the element $(-\text{id}, -\text{id})$ acts trivially on $\mathcal{P}(T_X, H_X)$. On the other hand, since the Picard number of X is 1, it follows that every element in $O_H(T_X)$ extends to an isometry of $H^2(X)$ which maps h to $\pm h$. Hence the group $O(D_{L_{2d}})/\langle \pm 1 \rangle$ acts transitively and freely on the fibre of τ showing again that the number of FM partners equals $2^{p(d)-1}$.

Large Picard Rank

For Picard ranks of 12 or more, derived equivalence implies isomorphism since any Hodge isometry of T_X lifts to an isometry of H_X^2, using [32, 1.14.2]: we have $\ell(NS_X) = \ell(T_X) \leq \text{rk}(T_X) = 22 - \varrho(X)$ for the minimal number of generators of D_{NS_X}; the lifting is possible if $2 + \ell(NS_X) \leq \text{rk}(NS_X) = \varrho(X)$—hence $\ell \leq 10$ and $\varrho \geq 12$. Still, there can be many non-isomorphic polarisations on the same surface. See below for an example where the fibres of τ and σ can become arbitrarily large in the case of $\varrho = 20$ maximal.

Positive Definite Transcendental Lattice

We consider the following candidates for transcendental lattices: $T = \left(\begin{smallmatrix} 2a & 0 \\ 0 & 2b \end{smallmatrix} \right)$ with $a > b > 0$. We denote the standard basis vectors for T by u and v, so that $u^2 = 2a$ and $v^2 = 2b$. Note that the only isometries of this lattice are given by sending the basis vectors u, v to $\pm u, \pm v$. In the lattice $L_{2d} = 2U \oplus 2E_8(-1) \oplus \langle -2d \rangle$, denote by l a generator of the non-unimodular summand $\langle -2d \rangle$.

In this setting, we are looking for embeddings $\iota_1, \iota_2 \colon T \hookrightarrow L_{2d}$ such that $\iota_1(v)$ and $\iota_2(v)$ belong to different $O(L_{2d})$-orbits. This would then immediately imply that the two embeddings cannot be equivalent. In order to show this, we appeal to Eichler's criterion.

Let us restrict to the special case $d = b = p^3$ for a prime p. Recall that the *divisor* of a vector w is the positive generator of the ideal (w, L_{2d}). We want the divisor of the vector v to be p^2. Setting, for $c \in \mathbb{Z}$,

$$v_c := p^2 e_2 + p(1 + c^2) f_2 + cl,$$

we have $v_c^2 = 2p^3$ and $\mathrm{div}(v_c) = (p^2, p(1+c^2), 2cp^3)$. Choosing c with $1 + c^2 \equiv 0 \ (p)$, which enforces $p \equiv 1 \ (4)$, we get $\mathrm{div}(v_c) = p^2$. Now, by Eichler's criterion, the $\tilde{O}(L_{2p^3})$-orbit of v_c is determined by the length $v_c^2 = 2p^3$ and the class $[v_c/\mathrm{div}(v_c)] = [c/p^2] \in D_{L_{2p^3}}$. The latter discriminant group is cyclic of order $2p^3$. Hence, the number of orbits of vectors with length $2p^3$ and divisor p^2 equals the number of solutions of $1 + c^2 \equiv 0 \ (p)$ for $c = 0, \dots p^2 - 1$. The equation $1 + c^2 \equiv 0$ has two solutions in $\mathbb{Z}/p$, as $p \equiv 1 \ (4)$, hence $2p$ solutions in $\mathbb{Z}/p^2$. (The above computation is a very special case of the obvious adaption of [18, Proposition 2.4] to the lattice L_{2d}.)

Together with $u := e_1 + af_1$, we get $2p$ lattices $T_c = \langle u, v_c \rangle$ embedded into L_{2p^3} and such that these embeddings are pairwise nonequivalent under the action of $\tilde{O}(L_{2p^3})$. The discriminant group of L_{2p^3} is $D_{L_{2p^3}} \cong \mathbb{Z}/2p^3$, hence $O(D_{L_{2p^3}}) = \{\pm \mathrm{id}\} \cong \mathbb{Z}/2$. We have to take the action of $O(L_{2p^3})/\tilde{O}(L_{2p^3}) \cong O(D_{L_{2p^3}})$ on the set of $\tilde{O}(L_{2p^3})$-orbits into account. As this is a 2-group, there must at least p orbits under the action of $O(L_{2p^3})$. We remark that $O(D_{L_{2p^3}})$ is a 2-group in greater generality, see [18, Proposition 2.5].

In particular, the number of pairwise nonequivalent embeddings is finite, but unbounded when varying T.

Unimodular $(T)^{\perp}_{L_{2d}}$

We use that there are precisely two inequivalent negative definite unimodular even lattices of rank 16, namely, $2E_8$ and D_{16}^+ (the latter is an extension of the non-unimodular root lattice D_{16}); see [11, Sect. 16.4]. They become equivalent after adding a hyperbolic plane: $2E_8 \oplus U \cong D_{16}^+ \oplus U$, since unimodular even indefinite lattices are determined by rank and signature. Setting $T := 2U \oplus \langle -2d \rangle$, we get

$$2E_8(-1) \oplus T \cong L_{2d} \cong D_{16}^+(-1) \oplus T.$$

Hence, since the orthogonal complements are different, there must at least be two different embeddings of T into L_{2d}. This example allows for arbitrary polarisations (in contrast to the previous one).

Acknowledgements We thank F. Schulze for discussions concerning lattice theory. We are grateful to M. Schütt and to the referee who improved the article considerably. The first author would like to thank the organisers of the Fields Institute Workshop on Arithmetic and Geometry of K3 surfaces and Calabi–Yau threefolds held in August 2011 for a very interesting and stimulating meeting. The second author has been supported by DFG grant Hu 337/6-1 and by the DFG priority programme 1388 "representation theory".

References

1. W.L. Baily Jr., A. Borel, Compactification of arithmetic quotients of bounded symmetric domains. Ann. Math. (2) **84**, 442–528 (1966)
2. W. Barth, K. Hulek, C. Peters, A. Van de Ven, *Compact Complex Surfaces*, 2nd edn. (Springer, Berlin, 2004)
3. C. Bartocci, U. Bruzzo, D. Hernàndez-Ruipérez, *Fourier–Mukai and Nahm Transforms in Geometry and Mathematical Physics* (Birkhäuser, Basel, 2009)
4. A.I. Beauville, J.-P. Bourguignon, M. Demazure (eds.), Géométrie des surfaces K3: modules et périodes, in *Séminaires Palaiseau*. Astérisque, vol. 126. Société Mathématique de France (Paris 1985)
5. A. Beilinson, J. Bernstein, P. Deligne, Faisceaux perverse, in *Astérisque*, vol. 100 Société Mathématique de France (Paris 1980)
6. A.I. Bondal, D.O. Orlov, Semiorthogonal decompositions for algebraic varieties. MPI preprint 15 (1995), also `math.AG/9506012`
7. A.I. Bondal, D.O. Orlov, Reconstruction of a variety from the derived category and groups of autoequivalences. Compos. Math. **125**, 327–344 (2001)
8. T. Bridgeland, A. Maciocia, Complex surfaces with equivalent derived categories. Math. Z. **236**, 677–697 (2001)
9. T. Bridgeland, Flops and derived categories. Invent. Math. **147**, 613–632 (2002)
10. T. Bridgeland, Spaces of stability conditions, in *Algebraic Geometry*, Seattle, 2005. Proceedings of Symposia in Pure Mathematics, vol. 80, Part 1 (American Mathematical Society, Providence, 2009), pp. 1–21
11. J.H. Conway, N.J.A. Sloane, *Sphere Packings, Lattices, and Groups*, 2nd edn. (Springer, Berlin, 1993)
12. D.A. Cox, *Primes of the Form $x^2 + ny^2$* (Wiley, New York, 1989)
13. M. Eichler, *Quadratische Formen und Orthogonale Gruppen* (Springer, Berlin, 1952)
14. W. Fulton, Intersection theory, *Ergebnisse der Mathematik und ihrer Grenzgebiete* (2), vol. 3 (Springer, Berlin, 1998)
15. P. Gabriel, Des catégories abéliennes. Bull. Soc. Math. Fr. **90**, 323–448 (1962)
16. S.I. Gelfand, Y.I. Manin, *Methods of Homological Algebra* (Springer, Berlin, 1997)
17. V. Golyshev, V. Lunts, D.O. Orlov, Mirror symmetry for abelian varieties. J. Algebr. Geom. **10**, 433–496 (2001)
18. V.A. Gritsenko, K. Hulek, G.K. Sankaran, The Kodaira dimension of the moduli of K3 surfaces. Invent. Math. **169**, 519–567 (2007)
19. D. Happel, in *Triangulated Categories in the Representation Theory of Finite Dimensional Algebras*. LMS Lecture Notes Series, vol. 119 (Cambridge University Press, Cambridge, 1988)
20. H. Hartmann, Cusps of the Kähler moduli space and stability conditions on K3 surfaces Math. Ann. doi:10.1007/s00208-011-0719-3 Math. Ann. **354**, 1–42 (2012)
21. R. Hartshorne, in *Residues and Duality*. Lecture Notes in Mathematics, vol. 20 (Springer, Berlin, 1966)
22. S. Hosono, B.H. Lian, K. Oguiso, S.-T. Yau, Fourier–Mukai number of a K3 surface, in *Algebraic Structures and Moduli Spaces*. CRM Proceedings and Lecture Notes, vol. 38 (AMS Providence, 2004), pp. 177–192
23. D. Huybrechts, *Fourier–Mukai Transforms in Algebraic Geometry* (Oxford University Press, Oxford, 2006)
24. B.W. Jones, in *The Arithmetic Theory of Quadratic Forms*. Carus Mathemathical Monographs, vol. 10 (Mathematical Association of America, Buffalo, New York 1950)

25. Y. Kawamata, *D*-equivalence and *K*-equivalence. J. Differ. Geom. **61**, 147–171 (2002)
26. S. Ma, Fourier–Mukai partners of a K3 surface and the cusps of its Kahler moduli. Int. J. Math. **20**, 727–750 (2009)
27. D.R. Morrison, On K3 surfaces with large Picard number. Invent. Math. **75**, 105–121 (1984)
28. D.R. Morrison, in *The Geometry of K3 Surfaces*. Lecture Notes, Cortona, Italy, 1988. http://www.cgtp.duke.edu/ITP99/morrison/cortona.pdf
29. S. Mukai, Duality between $D(X)$ and $D(\hat{X})$ with its application to Picard sheaves. Nagoya Math. J. **81**, 153–175 (1981)
30. S. Mukai, On the moduli space of bundles on K3 surfaces I, in *Vector Bundles on Algebraic Varieties*, vol. 11 (The Tata Institute of Fundamental Research in Mathematics, Bombay, 1984), pp. 341–413
31. Y. Namikawa, Mukai flops and derived categories. J. Reine Angew. Math. **560**, 65–76 (2003)
32. V. Nikulin, Integral symmetric bilinear forms and some of their applications. Math. USSR Izv. **14**, 103–167 (1980)
33. K. Oguiso, K3 surfaces via almost-primes. Math. Res. Lett. **9**, 47–63 (2002)
34. D.O. Orlov, Equivalences of derived categories and K3-surfaces. J. Math. Sci. **84**, 1361–1381 (1997)
35. D.O. Orlov, Derived categories of coherent sheaves on abelian varieties and equivalences between them. Izv. Math. **66**, 569–594 (2002)
36. D.O. Orlov, Triangulated categories of singularities and D-branes in Landau-Ginzburg models. Trudy Steklov Math. Inst. **204**, 240–262 (2004)
37. D. Ploog, Equivariant equivalences for finite group actions. Adv. Math. **216**, 62–74 (2007)
38. A. Polishchuk, Symplectic biextensions and generalizations of the Fourier–Mukai transforms. Math. Res. Lett. **3**, 813–828 (1996)
39. A. Rudakov et al., *Helices and Vector Bundles*. London Mathematical Society Lecture Note Series, vol. 148 (1990)
40. B. Saint-Donat, Projective models of K-3 surfaces. Am. J. Math. **96**, 602–639 (1974)
41. C.L. Siegel, Equivalence of quadratic forms. Am. J. Math. **63**, 658–680 (1941)
42. P. Stellari, Some remarks about the FM-partners of K3 surfaces with small Picard numbers 1 and 2. Geom. Dedicata **108**, 1–13 (2004)
43. P. Stellari, A finite group acting on the moduli space of K3 surfaces. Trans. Am. Math. Soc. **360**, 6631–6642 (2008)
44. H. Sterk, Finiteness results for algebraic K3 surfaces. Math. Z. **189**, 507–513 (1985)
45. J.-L. Verdier, Des catégories dérivées des catégories abéliennes, Ph.D. thesis, 1967; Asterisque, vol. 239, 1996
46. G.L. Watson, *Integral Quadratic Forms* (Cambridge University Press, Cambridge, 1960)
47. C. Weibel, *An Introduction to Homological Algebra* (Cambrige University Press, Cambridge, 1994)
48. D. Zagier, *Zetafunktionen und quadratische Körper* (Springer, Berlin, 1981)

On a Family of K3 Surfaces with $\mathscr{S}_4$ Symmetry

Dagan Karp, Jacob Lewis, Daniel Moore, Dmitri Skjorshammer, and Ursula Whitcher

Abstract The largest group which occurs as the rotational symmetries of a three-dimensional reflexive polytope is S_4. There are three pairs of three- dimensional reflexive polytopes with this symmetry group, up to isomorphism. We identify a natural one-parameter family of K3 surfaces corresponding to each of these pairs, show that S_4 acts symplectically on members of these families, and show that a general K3 surface in each family has Picard rank 19. The properties of two of these families have been analyzed in the literature using other methods. We compute the Picard–Fuchs equation for the third Picard rank 19 family by extending the Griffiths–Dwork technique for computing Picard–Fuchs equations to the case of semi-ample hypersurfaces in toric varieties. The holomorphic solutions to our Picard–Fuchs equation exhibit modularity properties known as "Mirror Moonshine"; we relate these properties to the geometric structure of our family.

Key words: K3 surfaces, Reflexive polytopes, Picard–Fuchs equations, Griffiths–Dwork technique, Mirror Moonshine

Mathematics Subject Classifications (2010): Primary 14J28, 14J33; Secondary 11G42, 14J15, 14G35

D. Karp · D. Moore · D. Skjorshammer
Department of Mathematics, Harvey Mudd College, 301 Platt Boulevard,
Claremont, CA 91711, USA
e-mail: dk@math.hmc.edu; taupin@gmail.com; dmitriskj@gmail.com

J. Lewis
Fakultät für Mathematik, Universität Wien, Garnisongasse 3/14,
1090 Wien, Austria
e-mail: jacobml@u.washington.edu

U. Whitcher (✉)
Department of Mathematics, 508 Hibbard Humanities Hall, University
of Wisconsin–Eau Claire, Eau Claire, WI 54702, USA
e-mail: whitchua@uwec.edu

R. Laza et al. (eds.), *Arithmetic and Geometry of K3 Surfaces and Calabi–Yau Threefolds*,   367
Fields Institute Communications 67, DOI 10.1007/978-1-4614-6403-7_12,
© Springer Science+Business Media New York 2013

1 Introduction

Families of Calabi–Yau varieties with discrete symmetry groups provide a fertile source of examples and conjectures in geometry and theoretical physics. Greene and Plesser's construction of the mirror to a family of Calabi–Yau threefolds relied on the construction of a special pencil of threefolds admitting a discrete group symmetry (see [14]). More recent studies of Calabi–Yau threefolds with discrete symmetry groups include [2, 9, 24].

In the case of K3 surfaces, actions of a finite group of *symplectic* automorphisms, which preserve the holomorphic two-form, are of particular interest. Nikulin classified the finite abelian groups which can act symplectically on K3 surfaces in [27]. Mukai showed in [26] that any finite group G with a symplectic action on a K3 surface is a subgroup of a member of a list of eleven groups, and gave an example of a symplectic action of each of these maximal groups. Xiao and Kondō gave alternate proofs of the classification in [19, 36], respectively; [36, Table 2] includes a complete list of finite groups which admit symplectic group actions on K3 surfaces.

If a K3 surface X admits a symplectic action by a group G, then the Picard group of X must contain a primitive definite sublattice S_G; in [35], the last author gives a procedure for computing the lattice invariants of S_G for any of the groups in [36, Table 2]. The relationship between a symplectic action and the Picard group has been worked out in detail for particular finite groups; cf. [10–13, 15, 28]. Thus, symplectic group actions may be used to identify K3 surfaces with high Picard rank.

Families of K3 surfaces with Picard rank 19 admit a particularly nice construction of the mirror map, which relates the moduli of a family of K3 surfaces to the moduli of the mirror family (see [7]). One may study the mirror map using Picard–Fuchs differential equations. The dissertation [31] uses symplectic group actions to produce pencils of K3 surfaces with Picard rank 19 in projective space $\mathbb{P}^3$ and the weighted projective space $\mathbb{P}(1, 1, 1, 3)$, and computes the associated Picard–Fuchs equations.

In [23], Lian and Yau drew attention to the "Mirror Moonshine" phenomenon, whereby the holomorphic solution to the Picard–Fuchs equation for a family of K3 surfaces becomes a Γ-modular form of weight 2 for some genus 0 modular group $\Gamma \subset PSL_2(\mathbb{R})$. Verrill and Yui studied the mirror map and "Mirror Moonshine" for specific pencils of K3 surfaces in [34]. In [8], Doran uses Picard–Fuchs differential equations to determine when a mirror map for a Picard rank 19 family of K3 surfaces is an automorphic function.

In this paper, we study special one-parameter families of K3 hypersurfaces in toric varieties obtained from three-dimensional reflexive polytopes. The key idea is to use symmetries of the polytopes to identify a group action on a family of hypersurfaces. The largest group which occurs as the rotational symmetries of a three-dimensional reflexive polytope is S_4. Up to automorphism, there are three pairs of three-dimensional reflexive polytopes with this symmetry group. We identify a natural one-parameter family of K3 surfaces corresponding to each of these pairs, show that S_4 acts symplectically on members of these families, and show that a general K3 surface in each family has Picard rank 19.

The Picard–Fuchs equations for two of our families were analyzed in [18, 29, 33]. We compute the Picard–Fuchs equation for the third Picard rank 19 family, using coordinates which arise naturally from the reflexive polytope. In order to do so, we extend the Griffiths–Dwork algorithm to the case of semi-ample hypersurfaces in toric varieties. Our method relies on the theory of residue maps for hypersurfaces in toric varieties developed in [1] and extended in [25]. Our families all exhibit "Mirror Moonshine": we show that the modularity properties are a natural consequence of underlying geometric structures.

2 Toric Varieties and Semiample Hypersurfaces

2.1 Toric Varieties and Reflexive Polytopes

We begin by recalling some standard constructions involving toric varieties. Let N be a lattice isomorphic to $\mathbb{Z}^n$. The dual lattice M of N is given by $\mathrm{Hom}(N, \mathbb{Z})$; it is also isomorphic to $\mathbb{Z}^n$. We write the pairing of $v \in N$ and $w \in M$ as $\langle v, w \rangle$. A *cone* in N is a subset of the real vector space $N_{\mathbb{R}} = N \otimes \mathbb{R}$ generated by nonnegative $\mathbb{R}$-linear combinations of a set of vectors $\{v_1, \ldots, v_m\} \subset N$. We assume that cones are strongly convex, that is, they contain no line through the origin. Note that each face of a cone is a cone.

A *fan* Σ consists of a finite collection of cones such that each face of a cone in the fan is also in the fan, and any pair of cones in the fan intersects in a common face. We say Σ is *simplicial* if the generators of each cone in Σ are linearly independent over $\mathbb{R}$. If every element of $N_{\mathbb{R}}$ belongs to some cone in Σ, we say Σ is *complete*. In the following, we shall restrict our attention to complete fans.

A fan Σ defines a toric variety V_Σ. We may describe V_Σ using homogeneous coordinates, in a process analogous to the construction of $\mathbb{P}^n$ as a quotient space of $(\mathbb{C}^*)^n$. We follow the exposition in [17]. Let $\Sigma(1) = \{\rho_1, \ldots, \rho_q\}$ be the set of one-dimensional cones of Σ. For each $\rho_j \in \Sigma(1)$, let v_j be the unique generator of the semigroup $\rho_j \cap N$.

We construct the toric variety V_Σ as follows. To each edge $\rho_j \in \Sigma(1)$, we associate a coordinate z_j. Let $\mathscr{S}$ denote any subset of $\Sigma(1)$ that does *not* span a cone of Σ. Let $\mathscr{V}(\mathscr{S}) \subseteq \mathbb{C}^q$ be the linear subspace defined by setting $z_j = 0$ for each $\rho_j \in \mathscr{S}$. Let $Z(\Sigma)$ be the union of the spaces $\mathscr{V}(\mathscr{S})$. Note that $(\mathbb{C}^*)^q$ acts on $\mathbb{C}^q - Z(\Sigma)$ by coordinatewise multiplication. Fix a basis for N, and suppose that v_j has coordinates $(v_{j1}, \ldots, v_{jn})$ with respect to this basis. Consider the map $\varphi : (\mathbb{C}^*)^q \to (\mathbb{C}^*)^n$ given by

$$\varphi(t_1, \ldots, t_q) \mapsto \left(\prod_{j=1}^{q} t_j^{v_{j1}}, \ldots, \prod_{j=1}^{q} t_j^{v_{jn}} \right)$$

Then the toric variety V_Σ associated with the fan Σ is given by

$$V_\Sigma = (\mathbb{C}^q - Z(\Sigma))/\operatorname{Ker}(\varphi).$$

Given a lattice polytope Δ in N, we define its *polar polytope* Δ° to be $\Delta^\circ = \{w \in M \mid \langle v, w \rangle \geq -1 \, \forall v \in K\}$. If Δ° is also a lattice polytope, we say that Δ is a reflexive polytope and that Δ and Δ° are a mirror pair.

Example 1. The generalized octahedron in N with vertices at $(\pm 1, 0, \ldots, 0)$, $(0, \pm 1, \ldots, 0), \ldots, (0, 0, \ldots, \pm 1)$ is a reflexive polytope. Its polar polytope is the hypercube with vertices at $(\pm 1, \pm 1, \ldots, \pm 1)$.

A reflexive polytope must contain $\mathbf{0}$; furthermore, $\mathbf{0}$ is the only interior lattice point of the polytope. We may obtain a fan R by taking cones over the faces of Δ. Let Σ be a simplicial refinement of R such that the one-dimensional cones of Σ are generated by the nonzero lattice points v_k, $k = 1, \ldots, q$, of Δ; we call such a refinement a *maximal projective subdivision*. Then the variety V_Σ is an orbifold; if $n = 3$, V_Σ is smooth (see [6]).

Example 2. Let $N \cong \mathbb{Z}^3$, and let Δ be the octahedron with vertices $v_1 = (1, 0, 0)$, $v_2 = (0, 1, 0)$, $v_3 = (0, 0, 1)$, $v_4 = (-1, 0, 0)$, $v_5 = (0, -1, 0)$, and $v_6 = (0, 0, -1)$ (Fig. 1). Then the only lattice points of Δ are the vertices and the origin. Let R be the fan obtained by taking cones over the faces of Δ. Then R defines a toric variety V_R which is isomorphic to $\mathbb{P}^1 \times \mathbb{P}^1 \times \mathbb{P}^1$.

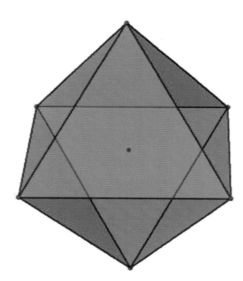

Fig. 1: The octahedron of Example 2

Proof. The vertices of the octahedron $v_1, \ldots, v_6$ generate the one-dimensional cones $\rho_1, \ldots, \rho_6$ of R. The two-element subsets of $\Sigma(1)$ that do not span cones are $\{\rho_1, \rho_4\}$,

$\{\rho_2, \rho_5\}$, and $\{\rho_3, \rho_6\}$; larger subsets of $\Sigma(1)$ that do not span cones contain one of the two-element subsets that do not span cones. Thus, $Z(\Sigma)$ consists of points of the form $(0, z_2, z_3, 0, z_5, z_6)$, $(z_1, 0, z_3, z_4, 0, z_6)$, or $(z_1, z_2, 0, z_4, z_5, 0)$.

The map φ is given by:

$$\varphi(t_1, t_2, t_3, t_4, t_5, t_6) = (t_1 t_4^{-1}, t_2 t_5^{-1}, t_3 t_6^{-1})$$

Then V_R is given by the quotient $\mathbb{C}^6 \backslash Z(\Sigma)/\ker(\varphi)$, where $\ker(\varphi)$ contains points satisfying $t_1 = t_4, t_2 = t_5$, and $t_3 = t_6$. This corresponds to the equivalence relations

$$(z_1, z_2, z_3, z_4, z_5, z_6) \sim (\lambda_1 z_1, z_2, z_3, \lambda_1 z_4, z_5, z_6)$$

$$(z_1, z_2, z_3, z_4, z_5, z_6) \sim (z_1, \lambda_2 z_2, z_3, z_4, \lambda_2 z_5, z_6)$$

$$(z_1, z_2, z_3, z_4, z_5, z_6) \sim (z_1, z_2, \lambda_3 z_3, z_4, z_5, \lambda_3 z_6)$$

where $\lambda_1, \lambda_2, \lambda_3 \in \mathbb{C}^*$. Thus, V_R is isomorphic to the toric variety $\mathbb{P}^1 \times \mathbb{P}^1 \times \mathbb{P}^1$.

Example 3. Let $N \cong \mathbb{Z}^3$, and let Δ be the octahedron with vertices $v_1 = (1, 0, 0)$, $v_2 = (1, 2, 0)$, $v_3 = (1, 0, 2)$, $v_4 = (-1, 0, 0)$, $v_5 = (-1, -2, 0)$, and $v_6 = (-1, 0, -2)$ (Fig. 2). Let R be the fan obtained by taking cones over the faces of Δ. Then R defines a toric variety V_R which is isomorphic to $(\mathbb{P}^1 \times \mathbb{P}^1 \times \mathbb{P}^1)/(\mathbb{Z}_2 \times \mathbb{Z}_2 \times \mathbb{Z}_2)$. If Σ is a simplicial refinement of R such that the one-dimensional cones of Σ are generated by the nonzero lattice points of Δ, then V_Σ is a smooth variety and the map $V_\Sigma \to V_R$ is a resolution of singularities.

Proof. As in Example 2, $Z(\Sigma)$ consists of points of the form $(0, z_2, z_3, 0, z_5, z_6)$, $(z_1, 0, z_3, z_4, 0, z_6)$, or $(z_1, z_2, 0, z_4, z_5, 0)$. The map φ is defined as

$$\varphi(t_1, t_2, t_3, t_4, t_5, t_6) = (t_1 t_2 t_3 t_4^{-1} t_5^{-1} t_6^{-1}, t_2^2 t_5^{-2}, t_3^2 t_6^{-2}).$$

Thus, elements of $\ker(\varphi)$ must satisfy $t_1 t_2 t_3 = t_4 t_5 t_6$, $t_2^2 = t_5^2$, and $t_3^2 = t_6^2$. These equations simplify to $t_1^2 = t_4^2, t_2^2 = t_5^2$ and $t_3^2 = t_6^2$. We obtain the equivalence relations

$$(z_1, z_2, z_3, z_4, z_5, z_6) \sim (\lambda_1 z_1, z_2, z_3, \pm \lambda_1 z_4, z_5, z_6)$$

$$(z_1, z_2, z_3, z_4, z_5, z_6) \sim (z_1, \lambda_2 z_2, z_3, z_4, \pm \lambda_2 z_5, z_6)$$

$$(z_1, z_2, z_3, z_4, z_5, z_6) \sim (z_1, z_2, \lambda_3 z_3, z_4, z_5, \pm \lambda_3 z_6)$$

where $\lambda_1, \lambda_2, \lambda_3 \in \mathbb{C}^*$.

We conclude that V_R is isomorphic to $(\mathbb{P}^1 \times \mathbb{P}^1 \times \mathbb{P}^1)/(\mathbb{Z}_2 \times \mathbb{Z}_2 \times \mathbb{Z}_2)$. Since $n = 3$, the simplicial refinement Σ yields a smooth variety.

2.2 Semiample Hypersurfaces and the Residue Map

In this section, we review properties of hypersurfaces in toric varieties, and give a brief outline of the results of [25] on the residue map in this setting. Let Σ be a

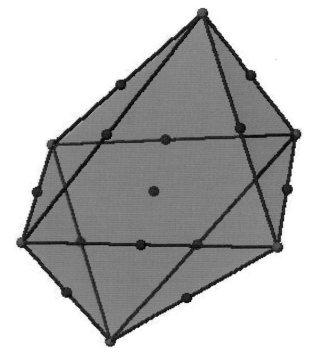

Fig. 2: The octahedron of Example 3

complete, simplicial n-dimensional fan, and let $S = \mathbb{C}[z_1, \ldots, z_q]$ be the homogeneous coordinate ring of the corresponding toric variety V_Σ. Each variable z_i defines an irreducible torus-invariant divisor D_i, given by the points where $z_i = 0$. The homogeneous coordinate ring is graded by the Chow group of V_Σ, according to the rule

$$\deg\left(\prod_{i=1}^{n} x_i^{a_i}\right) = \sum_{i=1}^{n} a_i D_i.$$

A homogeneous polynomial p in S_β defines a hypersurface X in V_Σ.

Definition 1. [1] If the partial derivatives $\partial p / \partial z_i$, $i = 1, \ldots, q$ do not vanish simultaneously on X, we say X is *quasismooth*.

Definition 2. [25] If the partial derivatives $z_i \, \partial p / \partial z_i$, $i = 1, \ldots, q$ do not vanish simultaneously on X, we say X is *regular* and p is *nondegenerate*.

Let R be a fan over the faces of a reflexive polytope, and assume Σ is a refinement of R. We have a proper birational morphism $\pi : V_\Sigma \to V_R$. Let Y be an ample divisor in V_R, and suppose $X = \pi^*(Y)$. Then X is semiample:

Definition 3. [6, Lemma 4.1.2] We say that a Cartier divisor D is *semiample* if D is generated by global sections and the intersection number $D^n > 0$.

Note that if Σ is not identical to R, then X is not ample. If Σ is a maximal projective subdivision of R, then general representatives X of the anticanonical class of V_Σ are Calabi–Yau varieties; if $n = 3$, then the representatives are K3 surfaces. (See [6] and [25, Sect. 1] for a more detailed exposition.)

Now, let us assume that X is a semiample, quasismooth hypersurface defined by a polynomial $p \in S_\beta$. The *residue map* relates the cohomology of $V_\Sigma - X$ to the cohomology of X:

$$\text{Res} : H^n(V_\Sigma - X) \to H^{n-1}(X).$$

In order to give a precise definition of the residue map, let us represent elements of $H^n(V_\Sigma - X)$ using rational forms. Choose an integer basis $m_1, \ldots, m_n$ for the dual lattice M. For any n-element subset $I = \{i_1, \ldots, i_n\}$ of $\{1, \ldots, q\}$, let $\det v_I = \det(\langle m_j, v_{i_k} \rangle_{1 \le j, i_k \le n})$, $dz_I = dz_{i_1} \wedge \cdots \wedge dz_{i_n}$, and $\hat{z}_I = \prod_{i \notin I} z_i$. Let Ω be the n-form on V_Σ given in global homogeneous coordinates by $\sum_{|I|=n} \det v_I \hat{z}_I dz_I$. (Note that if $V_\Sigma = \mathbb{P}^n$, then Ω is the usual holomorphic form on $\mathbb{P}^n$.) Let $\beta_0 = \sum_{i=1}^n \deg(x_i)$, and let $A \in S_{(a+1)\beta - \beta_0}$. Then the rational form $\omega_A := \frac{A\Omega}{p^{a+1}}$ is a class in $H^n(V_\Sigma - X)$. Let γ be any $n - 1$-cycle in X, and let $T(\gamma)$ be the tube over γ in $V_\Sigma - X$. Then the residue of ω_A is the class in $H^{n-1}(X)$ satisfying

$$\int_\gamma \text{Res}\left(\frac{A\Omega}{p^{a+1}}\right) = \int_{T(\gamma)} \frac{A\Omega}{p^{a+1}}. \tag{1}$$

The residue class $\text{Res}(\omega_A)$ lies in $H^{n-1-a,a}(X)$. (See [25, Sect. 3].) We shall have occasion to use the following special case of this construction:

Lemma 1. [25, Sect. 3] *Let X be a quasismooth K3 hypersurface in V_Σ described in global homogeneous coordinates by a polynomial p. Then $\omega := \text{Res}(\Omega/p)$ generates $H^{2,0}(X)$.*

We may use rational forms and the residue map to relate $H^{n-1}(X)$ to certain quotient rings.

Definition 4. [1] Let $p \in S_\beta$. Then the *Jacobian ideal* $J(p)$ is the ideal of S generated by the partial derivatives $\partial p / \partial z_i$, $i = 1, \ldots, q$, and the *Jacobian ring* $R(p)$ is the quotient ring $S/J(p)$. The Jacobian ring inherits a grading from S.

Proposition 1. [25, Sect. 3] *If X is a quasismooth, semiample hypersurface defined by a polynomial p, then the residue map induces a well-defined map of rings*

$$\text{Res}_J : R(p) \to H^{n-1}(X)$$

satisfying $\text{Res}_J([A]_{R(p)}) = \text{Res}(\omega_A)$.

If V_Σ is isomorphic to $\mathbb{P}^n$ or a weighted projective space, then the map Res_J is injective. (See [1, Sect. 11].) One may obtain injective maps for more general ambient spaces by working with a different quotient ring; however, these results only apply when the hypersurface X is regular.

Definition 5. [1] Let $p \in S_\beta$. Then the ideal $J_1(p)$ is the ideal quotient

$$\langle z_1 \partial p / \partial z_1, \ldots, z_q \partial p / \partial z_q \rangle : z_1 \cdots z_q.$$

The ring $R_1(p)$ is the quotient ring $S / J_1(p)$; this ring inherits a grading from S.

Theorem 1. [25, Theorem 4.4] *If X is a regular, semiample hypersurface defined by a polynomial p, then the residue map induces a well-defined, injective map of rings*

$$\mathrm{Res}_{J_1} : R_1(p) \to H^{n-1}(X)$$

satisfying $\mathrm{Res}_{J_1}([A]_{R_1(p)}) = \mathrm{Res}(\omega_A)$.

3 Three Symmetric Families of K3 Surfaces

3.1 Symplectic Group Actions on K3 Surfaces

Let X be a K3 surface and let g be an automorphism of X. We say that g *acts symplectically* if $g^*(\omega) = \omega$, where ω is the unique holomorphic two-form on X. If G is a finite group of automorphisms of X, we say G *acts symplectically* on X if every element of G acts symplectically.

The cup product induces a bilinear form $\langle , \rangle$ on $H^2(X, \mathbb{Z}) \cong H \oplus H \oplus H \oplus E_8 \oplus E_8$. (We take E_8 to be negative definite.) Using this form, we define $S_G = (H^2(X, \mathbb{Z})^G)^\perp$. The Picard group of X, $\mathrm{Pic}(X)$, consists of $H^{1,1}(X) \cap H^2(X, \mathbb{Z})$; the group $\mathscr{T}(X) \subseteq H^2(X, \mathbb{Z})$ of transcendental cycles is defined as $(\mathrm{Pic}(X))^\perp$. Nikulin showed that the groups $\mathrm{Pic}(X)$ and S_G are related:

Proposition 2. [27, Lemma 4.2] $S_G \subseteq \mathrm{Pic}(X)$ *and* $\mathscr{T}(X) \subseteq H^2(X, \mathbb{Z})^G$. *The lattice S_G is nondegenerate and negative definite.*

The rank of the lattice S_G depends only on the group G. [36, Table 2] lists the rank of S_G for each group G which admits a symplectic action on a K3 surface; a discussion of methods for computing lattice invariants of S_G may be found in [35].

Lemma 2. [35, Example 2.1] *Let X be a K3 surface which admits a symplectic action by the permutation group $G = \mathscr{S}_4$. Then $\mathrm{Pic}(X)$ admits a primitive sublattice S_G which has rank 17 and discriminant $d(S_G) = -2^6 \cdot 3^2$.*

3.2 An $\mathscr{S}_4$ Symmetry of Polytopes and Hypersurfaces

Let Δ be a reflexive polytope in a lattice $N \cong \mathbb{Z}^3$, and let Σ be a simplicial refinement of the fan over the faces of Δ. Demazure and Cox showed that the automorphism group A of the toric variety V_Σ is generated by the big torus $T \cong (\mathbb{C}^*)^3$, symmetries of the fan Σ induced by lattice automorphisms, and one-parameter families derived from the "roots" of V_Σ (see [6]). We are interested in finite subgroups of A which act symplectically on K3 hypersurfaces X in V_Σ.

Let us consider the automorphisms of V_Σ induced by symmetries of the fan Σ. Since Σ is a refinement of the fan R consisting of cones over the faces of Δ, the group of symmetries of Σ must be a subgroup H' of the group H of symmetries of Δ (viewed as a lattice polytope). We will identify a family $\mathscr{F}_\Delta$ of K3 surfaces in V_Σ on which H' acts by automorphisms, and then compute the induced action of G on the $(2, 0)$ form of each member of the family.

Let $h \in H'$, and let X be a K3 surface in V_Σ defined by a polynomial p in global homogeneous coordinates. Then h maps lattice points of Δ to lattice points of Δ, so we may view h as a permutation of the global homogeneous coordinates z_i: h is an automorphism of X if $p \circ h = p$. Alternatively, since H is the automorphism group of both Δ and its polar dual polytope Δ°, we may view h as an automorphism of Δ°: from this vantage point, we see that h acts by a permutation of the coefficients c_x of p, where each coefficient c_x corresponds to a point $x \in \Delta^\circ$. Thus, if h is to preserve X, we must have $c_x = c_y$ whenever $h(x) = y$. We may define a family of K3 surfaces fixed by H' by requiring that $c_x = c_y$ for any two lattice points $x, y \in \Delta^\circ$ which lie in the same orbit of H':

Proposition 3. *Let $\mathscr{F}_\Delta$ be the family of K3 surfaces in V_Σ defined by the following family of polynomials in global homogeneous coordinates:*

$$p = \left(\sum_{Q \in \mathcal{O}} c_Q \sum_{x \in Q} \prod_{k=1}^{q} z_k^{\langle v_k, x \rangle + 1} \right) + \prod_{k=1}^{q} z_k,$$

where $\mathcal{O}$ is the set of orbits of nonzero lattice points in Δ° under the action of H'. Then H' acts by automorphisms on each K3 surface X in $\mathscr{F}_\Delta$.

Proposition 4. *Let X be a quasismooth K3 surface in the family $\mathscr{F}_\Delta$, and let $h \in H' \subset \mathbf{GL}(3, \mathbb{Z})$. Then $h^*(\omega) = (\det h)\omega$.*

Proof. Once again, we use the fact that we may view h as either an automorphism of the lattice N which maps Δ to itself, or as an automorphism of the dual lattice M which restricts to an automorphism of Δ°. (If we fix a basis $\{n_1, n_2, n_3\}$ of N, take the dual basis $\{m_1, m_2, m_3\} = \{n_1^*, n_2^*, n_3^*\}$ on M, and treat h as a matrix, then h acts on M by the inverse matrix.) By Proposition 1, each choice of basis for M yields a generator of $H^{3,0}(V)$. Thus, if Ω is the generator of $H^{3,0}(V)$ corresponding to a fixed choice of integer basis m_1, m_2, m_3, we see that we may obtain a new generator Ω' of $H^{3,0}(V)$ by applying the change of basis h^{-1} to M. Recall that $\Omega = \sum_{|I|=3} \det v_I \hat{z}_I dz_I$, where $\det v_I = \det (\langle m_j, v_{i_k} \rangle_{1 \le j, i_k \le 3})$.

We compute:

$$\Omega' = \sum_{|I|=3} \det\left(h^{-1}(v_I)\right)\hat{z}_I dz_I \tag{2}$$

$$= \sum_{|I|=3} \det\left(h^{-1}\right)\det v_I \hat{z}_I dz_I \tag{3}$$

$$= \det h \sum_{|I|=3} \det v_I \hat{z}_I dz_I \tag{4}$$

since $\det h = \pm 1$.

By Proposition 3, $h^*(p) = p$, so $h^*(\omega) = \mathrm{Res}(\Omega'/p) = (\det h)\omega$.

Thus the group G of orientation-preserving automorphisms of Δ which preserve Σ acts symplectically on quasismooth members of $\mathscr{F}_\Delta$.

The largest group which occurs as the orientation-preserving automorphism group of a three-dimensional lattice polytope is S_4. There are three distinct pairs of isomorphism classes of reflexive polytopes which have this symmetry group. In the following examples, we analyze families derived from these pairs of polytopes.

Example 4. Let Δ be the cube with vertices of the form $(\pm 1, \pm 1, \pm 1)$. The dual polytope Δ° is an octahedron, with vertices $\{(\pm 1, 0, 0), (0, \pm 1, 0), (0, 0, \pm 1)\}$. We may choose our fan Σ such that the group of lattice automorphisms of Δ preserves Σ. The group G of orientation-preserving automorphisms of Δ is isomorphic to S_4. $\mathscr{F}_\Delta$ is a one-parameter family, and if X is a quasismooth member of $\mathscr{F}_\Delta$, rank $\mathrm{Pic}(X) \geq 19$.

Proof. The action of G on Δ° has two orbits: the origin, and the vertices of the octahedron. Thus, $\mathscr{F}_\Delta$ is a one-parameter family. Using Lemma 2, we conclude that for any quasismooth member of $\mathscr{F}_\Delta$, rank $S_G = 17$.

Let X be a quasismooth member of $\mathscr{F}_\Delta$. We wish to determine which of the divisors of X inherited from the ambient toric variety V_Σ are in $H^2(X, \mathbb{Z})^G$. The action of G on the lattice points of Δ has four orbits: the origin, the vertices of the cube, the interior points of edges, and interior points of faces. Let $v_1, \ldots, v_8$ be the vertices of the cube and $v_9, \ldots, v_{20}$ be the interior points of edges; let $W_1, \ldots, W_{20}$ be the corresponding torus-invariant divisors of the toric variety V_Σ. Since $v_1, \ldots, v_8$ and $v_9, \ldots, v_{20}$ are orbits of the action of G, $W_1 + \cdots + W_8$ and $W_9 + \cdots + W_{20}$ are elements of $\mathrm{Pic}(V)$ which are fixed by G. These two divisors span a rank-two lattice in $\mathrm{Pic}(V)$. Since there are no lattice points strictly in the interior of the edges of Δ° and none of the points $v_1, \ldots, v_{20}$ lies in the relative interior of a facet of Δ, $W_k \cap X$ is connected and nonempty for $1 \leq k \leq 20$ and the divisors $W_1 \cap X + \cdots + W_8 \cap X$ and $W_9 \cap X + \cdots + W_{20} \cap X$ span a rank-two lattice in $\mathrm{Pic}(X)$. This rank-two lattice is contained in $H^2(X, \mathbb{Z})^G$.

Since S_G is the orthogonal complement of $H^2(X, \mathbb{Z})^G$, rank $\mathrm{Pic}(X) \geq 17+2 = 19$.

Remark 1. This family is analyzed in [18, 29].

Example 5. Let Δ be a three-dimensional reflexive polytope with fourteen vertices and twelve faces. Up to lattice isomorphism, Δ is unique; moreover, Δ has the most vertices of any three-dimensional reflexive polytope. We may choose our fan Σ

such that the group of lattice automorphisms of Δ preserves Σ. The group G of orientation-preserving automorphisms of Δ is isomorphic to S_4, and $\mathscr{F}_\Delta$ is a one-parameter family. If X is a quasismooth member of $\mathscr{F}_\Delta$, rank $\mathrm{Pic}(X) \geq 19$.

Proof. The lattice points of Δ° consist of vertices and the origin, and G acts transitively on the vertices of Δ°, so $\mathscr{F}_\Delta$ is a one-parameter family. As above, Lemma 2 shows that for any quasismooth member of $\mathscr{F}_\Delta$, rank $S_G = 17$.

Let X be a quasismooth member of $\mathscr{F}_\Delta$. Once again, we determine which of the divisors of X inherited from the ambient toric variety V_Σ are in $H^2(X, \mathbb{Z})^G$. The action of G on the lattice points of Δ has three orbits; one orbit contains the origin, another contains eight vertices, and the last contains the remaining six vertices. Let $\{v_1, \ldots, v_8\}$ and $\{v_9, \ldots, v_{14}\}$ be the vertex orbits; let $W_1, \ldots, W_{14}$ be the corresponding torus-invariant divisors of V_Σ. Then $W_1 + \cdots + W_8$ and $W_9 + \cdots + W_{14}$ are elements of $\mathrm{Pic}(V)$ fixed by the action of G; these two divisors span a rank-two lattice in $\mathrm{Pic}(V)$. Since there are no lattice points strictly in the interior of the edges of Δ° and the facets of Δ have no points in their relative interiors, $W_k \cap X$ is connected and nonempty for $1 \leq k \leq 14$ and the divisors $W_1 \cap X + \cdots + W_8 \cap X$ and $W_9 \cap X + \cdots + W_{14} \cap X$ span a rank-two lattice in $\mathrm{Pic}(X)$. This rank-two lattice is contained in $H^2(X, \mathbb{Z})^G$, so rank $\mathrm{Pic}(X) \geq 17 + 2 = 19$.

Remark 2. An explicit analysis of the same family appears in [33].

Example 6. Let Δ be the octahedron with vertices $(1, 1, 1)$, $(-1, -1, 1)$, $(-1, 1, -1)$, $(1, -1, 1)$, $(1, 1, -1)$, and $(-1, -1, -1)$. The polar dual Δ° has vertices $(1, 0, 0)$, $(0, 1, 0)$, $(0, 0, 1)$, $(-1, 1, 1)$, $(1, -1, -1)$, $(0, 0, -1)$, $(0, -1, 0)$, and $(-1, 0, 0)$. We may choose our fan Σ such that the group of lattice automorphisms of Δ preserves Σ. The group G of orientation-preserving automorphisms of Δ is isomorphic to S_4. $\mathscr{F}_\Delta$ is a one-parameter family. If X is a quasismooth member of $\mathscr{F}_\Delta$, rank $\mathrm{Pic}(X) \geq 19$.

Proof. The action of G on Δ° has two orbits, the origin and the polytope's vertices, so $\mathscr{F}_\Delta$ is a one-parameter family. As in the previous example, Lemma 2 shows that for any quasismooth member of $\mathscr{F}_\Delta$, rank $S_G = 17$.

Let X be a quasismooth member of $\mathscr{F}_\Delta$. As before, we determine which of the divisors of X inherited from the ambient toric variety V_Σ are in $H^2(X, \mathbb{Z})^G$. The action of G on the lattice points of Δ has three orbits: the origin, the octahedron's vertices, and the interior points of edges. Let $v_1, \ldots, v_6$ be the vertices and $v_7, \ldots, v_{18}$ be the interior points of edges; let $W_1, \ldots, W_{18}$ be the corresponding torus-invariant divisors of V_Σ. Then $W_1 + \cdots + W_6$ and $W_7 + \cdots + W_{18}$ are elements of $\mathrm{Pic}(V)$ fixed by the action of G. These two divisors span a rank-two lattice in $\mathrm{Pic}(V)$. Since there are no lattice points strictly in the interior of the edges of Δ° and the facets of Δ have no points in their relative interiors, $W_k \cap X$ is connected and nonempty for $1 \leq k \leq 18$ and the divisors $W_1 \cap X + \cdots + W_6 \cap X$ and $W_7 \cap X + \cdots + W_{18} \cap X$ span a rank-two lattice in $\mathrm{Pic}(X)$. This rank-two lattice is contained in $H^2(X, \mathbb{Z})^G$. Thus, rank $\mathrm{Pic}(X) \geq 17 + 2 = 19$.

4 Picard–Fuchs Equations

4.1 The Griffiths–Dwork Technique

A *period* is the integral of a differential form with respect to a specified homology
class. The *Picard–Fuchs differential equation* of a family of varieties is a differential equation which describes the way the value of a period changes as we move
through the family. We may use Picard–Fuchs differential equations for periods of
holomorphic forms to understand the way the complex structure of a family of varieties varies within the family. The *Griffiths–Dwork technique* provides an algorithm
for computing Picard–Fuchs equations for families of hypersurfaces in projective
space. This technique has been generalized to hypersurfaces in weighted projective space and in some toric varieties. Unlike other methods for computing Picard–
Fuchs equations, the Griffiths–Dwork technique allows the study of arbitrary rational parametrizations.

Let us begin by reviewing the Griffiths–Dwork technique for one-parameter fami-
lies of hypersurfaces X_t in $\mathbb{P}^n$ described by homogeneous polynomials p_t of degree ℓ.
We may define a flat family of cycles γ_t. We then differentiate as follows:

$$\frac{d}{dt}\int_{\gamma_t}\mathrm{Res}\left(\frac{A\Omega}{p_t^k}\right)=\int_{\gamma_t}\mathrm{Res}\left(\frac{d}{dt}\left(\frac{A\Omega}{p_t^k}\right)\right) \tag{5}$$

$$=-k\int_{\gamma_t}\mathrm{Res}\left(\frac{(\frac{dp_t}{dt})A\Omega}{p_t^{k+1}}\right).$$

Thus, we may express successive derivatives of the period $\int_{\gamma_t}\frac{\Omega}{p}$ as periods of the
residues of rational forms. If $H^{n-1}(X,\mathbb{C})$ is r-dimensional as a vector space over $\mathbb{C}$,
then at most r residues of rational forms can be linearly independent. Therefore, the
period must satisfy a linear differential equation with coefficients in $\mathbb{C}(t)$ of order
at most r; this linear differential equation is the Picard–Fuchs differential equation
which we seek.

In order to compute the Picard–Fuchs differential equation in practice, we need
a way to compare expressions of the form $\mathrm{Res}\left(\frac{A\Omega}{p_t^k}\right)$ to expressions of the form
$\mathrm{Res}\left(\frac{B\Omega}{p_t^{k+1}}\right)$. Suppose we have an element of $H^{n-1}(X,\mathbb{C})$ of the form $\mathrm{Res}\left(\frac{K\Omega_0}{p^{k+1}}\right)$, where
$K=\sum_i A_i\frac{\partial p}{\partial x_i}$ is a member of the Jacobian ideal, and each A_i is a homogeneous
polynomial of degree $k\cdot\ell-n$. Then the following equation allows us to reduce the
order of the pole:

$$\frac{\Omega_0}{p^{k+1}}\sum_i A_i\frac{\partial p}{\partial x_i}=\frac{1}{k}\frac{\Omega_0}{p^k}\sum_i\frac{\partial A_i}{\partial x_i}+\text{exact terms} \tag{6}$$

We may find the Picard–Fuchs equation by systematically taking derivatives of
$\int_{\gamma_t}\mathrm{Res}\left(\frac{\Omega_0}{p}\right)$ and using (6) to rewrite the results in terms of a standard basis for

$H^{n-1}(X, \mathbb{C})$. This method is known as the *Griffiths–Dwork technique*. Practical implementations of the Griffiths–Dwork technique use the Jacobian ring $J(p)$ and the induced residue map Res_J to transform the problem into a computation suitable for a computer algebra system. (See [5], [6] or [9] for a more detailed discussion of the technique.)

In order to extend the Griffiths–Dwork technique to hypersurfaces in toric varieties, we need two tools: an appropriate version of the residue map, and an analogue of (6) to reduce the order of the poles. In the case of semiample hypersurfaces in toric varieties, we may use the results of [1, 25] described in Sect. 2.2 to define Res. We must be aware, however, that the induced residue map Res_J need not be injective for an arbitrary family of semiample hypersurfaces.

To construct an analogue of (6), we note that the results of [1] apply in the semiample case:

Definition 6 ([1, Definition 9.8]). Let $i \in \{1, \ldots, q\}$. We define the $(n-1)$-form Ω_i on V_Σ as follows:

$$\Omega_i = \sum_{|J|=n-1, i \notin J} \det(v_{\{i\} \cup J}) \hat{z}_{\{i\} \cup J} dz_J.$$

Here we use the convention that i is the first element of $\{i\} \cup J$.

Lemma 3 ([1, Lemma 10.7]). *If* $A \in S_{k\beta - \beta_0 + \beta_i}$, *then*

$$d\left(\frac{A\Omega_i}{p^k}\right) = \frac{\left(p\frac{\partial A}{\partial z_i} - kA\frac{\partial p}{\partial z_i}\right)\Omega_0}{p^{k+1}}.$$

Now, let X be a hypersurface in a toric variety V_Σ described by a homogeneous polynomial $p \in S_\beta$. Suppose we have an element of $H^{n-1}(X, \mathbb{C})$ of the form $\mathrm{Res}\left(\frac{K\Omega_0}{p^{k+1}}\right)$, where $K = \sum_i A_i \frac{\partial p}{\partial x_i}$ is a member of the Jacobian ideal, and $A_i \in S_{k\beta - \beta_0 + \beta_i}$. The following reduction of pole order equation follows immediately:

$$\frac{\Omega_0}{p^{k+1}} \sum_i A_i \frac{\partial p}{\partial x_i} = \frac{1}{k}\frac{\Omega_0}{p^k} \sum_i \frac{\partial A_i}{\partial x_i} + \text{exact terms} \qquad (7)$$

4.2 A Picard–Fuchs Equation

Let Δ be the octahedron with vertices $(1, 1, 1)$, $(-1, -1, 1)$, $(-1, 1, -1)$, $(1, -1, 1)$, $(1, 1, -1)$, and $(-1, -1, -1)$, as in Example 6, and let $\mathscr{F}_\Delta$ be the associated one-parameter family. In this section, we describe the Picard–Fuchs equation for $\mathscr{F}_\Delta$. We use our result to show that the Picard rank of a general member of $\mathscr{F}_\Delta$ is exactly 19.

Doran analyzed the properties of Picard–Fuchs equations for lattice-polarized families of K3 surfaces with Picard rank 19 in [8], and showed that the Picard–Fuchs equations for the K3 surfaces are related to Picard–Fuchs equations for families of elliptic curves.

Proposition 5 ([30, Lemma 3.1.(b)]). *Let $L(y)$ be a homogeneous linear differential polynomial with coefficients in $\mathbb{C}(t)$. Then there exists a homogeneous linear differential equation $M(y) = 0$ with coefficients in $\mathbb{C}(t)$ and solution space the $\mathbb{C}$-span of*

$$\{v_1 v_2 \mid L(v_1) = 0 \text{ and } L(v_2) = 0\}.$$

Definition 7. We call the operator $M(y)$ constructed above the *symmetric square* of L.

The symmetric square of the second-order linear, homogeneous differential equation

$$a_2 \frac{\partial^2 \omega}{\partial t^2} + a_1 \frac{\partial \omega}{\partial t} + a_0 \omega = 0$$

is

$$a_2^2 \frac{\partial^3 \omega}{\partial t^3} + 3a_1 a_2 \frac{\partial^2 \omega}{\partial t^2} + (4a_0 a_2 + 2a_1^2 + a_2 a_1' - a_1 a_2') \frac{\partial \omega}{\partial t}$$
$$+ (4a_0 a_1 + 2a_0' a_2 - 2a_0 a_2') \omega = 0 \tag{8}$$

where primes denote derivatives with respect to t.

Theorem 2. [8, Theorem 5] *The Picard–Fuchs equation of a family of rank-19 lattice-polarized K3 surfaces is a third-order ordinary differential equation which can be written as the symmetric square of a second-order homogeneous linear Fuchsian differential equation.*

Recall that a member of $\mathscr{F}_\Delta$ is described by the polynomial

$$p = \left(c_Q \sum_{x \in Q} \prod_{k=1}^{18} z_k^{\langle v_k, x \rangle + 1} \right) + \prod_{k=1}^{18} z_k, \tag{9}$$

where Q is the orbit consisting of the nonzero lattice points of Δ°. To simplify our computations, we set $t = \frac{1}{c_Q}$ and work with hypersurfaces $X_t \in \mathscr{F}_\Delta$ described by the polynomial

$$f = \left(\sum_{x \in Q} \prod_{k=1}^{18} z_k^{\langle v_k, x \rangle + 1} \right) + t \prod_{k=1}^{18} z_k. \tag{10}$$

Theorem 3. *The Picard–Fuchs equation for $\mathscr{F}_\Delta$ is*

$$\frac{d^3 \omega}{dt^3} + \frac{6(t^2 - 32)}{t(t^2 - 64)} \frac{d^2 \omega}{dt^2} + \frac{7t^2 - 64}{t^2(t^2 - 64)} \frac{d\omega}{dt} + \frac{1}{t(t^2 - 64)} \omega = 0. \tag{11}$$

Proof. We apply the Griffiths–Dwork technique. Let $\omega = \int \mathrm{Res}\left(\frac{\Omega}{f}\right)$ be a period of the holomorphic form. The parameter t only appears in a single term of f, so the derivatives of ω have a particularly nice form:

$$\frac{d^j}{dt^j}\omega = \int (-1)^j j!(z_1 \ldots z_{18})^j \mathrm{Res}\left(\frac{\Omega}{f^{j+1}}\right). \tag{12}$$

Using the computer algebra system MAGMA [3], we find that $(z_1 \ldots z_{18})^3 \in J$. We may now apply (7) to compare $\frac{d^3}{dt^3}\omega$ to lower-order terms. We conclude that ω must satisfy (11).

Corollary 1. *A general member of $\mathscr{F}_\Lambda$ has Picard rank 19.*

Proof. By Example 6, a general member of $\mathscr{F}_\Lambda$ has Picard rank at least 19. Families of K3 surfaces of Picard rank 20 are isotrivial, so if all members of $\mathscr{F}_\Lambda$ had Picard rank 20, ω would be constant. But a constant, non-trivial holomorphic two-form ω cannot satisfy (11).

We now show that (11) is the symmetric square of a second-order differential equation, as predicted by Theorem 2. Multiplying (11) by $t^2(t^2-64)$ and simplifying, we find that ω satisfies

$$t^2(t^2 - 64)\frac{d^3\omega}{dt^3} + 6t(t^2 - 32)(t^2 - 64)\frac{d^2\omega}{dt^2} + (7t^2 - 64)(t^2 - 64)\frac{d\omega}{dt}$$
$$+ t(t^2 - 64)\omega = 0$$

Comparing with (8), we see that the parameters a_2, a_1, and a_0 are given by $a_2 = t(t^2 - 64)$, $a_1 = 2t^2 - 64$ and $a_0 = \frac{t}{4}$. Therefore, the symmetric square root of (11) is

$$\frac{d^2\omega}{dt^2} + \frac{(2t^2 - 64)}{t(t^2 - 64)}\frac{d\omega}{dt} + \frac{1}{4(t^2 - 64)}\omega = 0. \tag{13}$$

The symmetric square root is linear and Fuchsian, as expected.

5 Modularity and Its Geometric Meaning

All three S_4 symmetric families of K3 surfaces exhibit "Mirror Moonshine" [23]: the mirror map is related to a hauptmodul for a genus 0 modular group $\Gamma \subset PSL_2(\mathbb{R})$, which gives a natural identification of the base minus the discriminant locus with $\mathbb{H}/\Gamma$ or a finite cover of $\mathbb{H}/\Gamma$, where $PSL_2(\mathbb{R})$ acts on the upper half-plane $\mathbb{H}$ as linear fractional transformations. Under this identification, the holomorphic solution to the Picard–Fuchs equation becomes a Γ-modular form of weight 2.

In the cases studied in this article, this modularity is not an accident, but rather is a consequence of special geometric properties of the K3 surfaces.

5.1 Elliptic Fibrations on K3 Surfaces

We can determine the geometric structures related to modularity by identifying elliptic fibrations with section on these K3 surfaces. We briefly recall a few facts about elliptic fibrations with section on K3 surfaces.

Definition 8. An *elliptic K3 surface with section* is a triple (X, π, σ) where X is a K3 surface, and $\pi : X \to \mathbb{P}^1$ and $\sigma : \mathbb{P}^1 \to X$ are morphisms with the generic fiber of π an elliptic curve and $\pi \circ \sigma = \mathrm{id}_{\mathbb{P}^1}$.

Any elliptic curve over the complex numbers can be realized as a smooth cubic curve in $\mathbb{P}^2$ in *Weierstrass normal form*

$$y^2 z = 4x^3 - g_2 x z^2 - g_3 z^3 \tag{14}$$

Conversely, (14) defines a smooth elliptic curve provided $\mathrm{Disc}(g_2, g_3) = g_2^3 - 27 g_3^2 \neq 0$.

Similarly, an elliptic K3 surface with section can be embedded into the $\mathbb{P}^2$ bundle $\mathbb{P}(\mathscr{O}_{\mathbb{P}^1} \oplus \mathscr{O}_{\mathbb{P}^1}(4) \oplus \mathscr{O}_{\mathbb{P}^1}(6))$ as a subvariety defined by (14), where now g_2, g_3 are global sections of $\mathscr{O}_{\mathbb{P}^1}(8)$, $\mathscr{O}_{\mathbb{P}^1}(12)$ respectively (i.e. they are homogeneous polynomials of degrees 8 and 12). The singular fibers of π are the roots of the degree 24 homogeneous polynomial $\mathrm{Disc}(g_2, g_3) = g_2^3 - 27 g_3^2 \in H^0(\mathscr{O}_{\mathbb{P}^1}(24))$. Tate's algorithm [32] can be used to determine the type of singular fiber over a root p of $\mathrm{Disc}(g_2, g_3)$ from the orders of vanishing of g_2, g_3 , and $\mathrm{Disc}(g_2, g_3)$ at p.

Proposition 6 ([4, Lemma 3.9]). *A general fiber of π and the image of σ span a copy of H in $\mathrm{Pic}(X)$. Further, the components of the singular fibers of π that do not intersect σ span a sublattice S of $\mathrm{Pic}(X)$ orthogonal to this H, and $\mathrm{Pic}(X)/(H \oplus S)$ is isomorphic to the Mordell–Weil group $MW(X, \pi)$ of sections of π.*

When K3 surfaces are realized as hypersurfaces in toric varieties, one can construct elliptic fibrations combinatorially from the three-dimensional reflexive polytope Δ. As before, let Σ be a refinement of the fan over faces of Δ. Suppose $P \subset N$ is a plane such that $\Delta \cap P$ is a reflexive polygon ∇, let m be a normal vector to P, and let Ξ be the fan over faces of ∇. Then P induces a torus-invariant map $V_\Sigma \to \mathbb{P}^1$ with generic fiber V_Ξ, given in homogeneous coordinates by

$$\pi_P : (z_1, \ldots z_r) \mapsto \left[\prod_{\langle v_i, m \rangle > 0} z_i^{\langle v_i, m \rangle}, \prod_{\langle v_i, m \rangle < 0} z_i^{-\langle v_i, m \rangle} \right] \tag{15}$$

Restricting π_P to an anticanonical K3 surface, we get an elliptic fibration. If ∇ has an edge without interior points, this fibration will have a section as well. See [20] for more details.

Example 7. We can use such an elliptic fibration with section to study the lattice structure of the Picard group of a generic member X of the family defined by (10). The map $\pi : V_\Sigma \to \mathbb{P}^1$ defined by the procedure above with $m = (0, 1, 0)$, is an elliptic fibration with section $\sigma : \mathbb{P}^1 \to X$.

For this particular π, examining the singular fibers gives an embedding of the rank 19 lattice $H \oplus S = H \oplus D_6 \oplus D_6 \oplus A_3 \oplus A_1 \oplus A_1$ into Pic(X). Because this fibration has more than one section, $H \oplus S \neq \mathrm{Pic}(X)$. To determine $MW(X, \sigma) = NS(X)/(H \oplus S)$, we note that the order of this group must divide 16, the square root of the determinant of the intersection matrix of $H \oplus S$. By putting the fibration into the Legendre normal form

$$y^2 z = x(x + z)(x + \frac{stz}{16(1 + t)^2}) \tag{16}$$

one can see immediately that there are three two-torsion sections, namely $[0, 1, 0]$, $[0, 0, 1]$, $[-1, 0, 1]$, and $[-\frac{st}{16(1+t)^2}, 0, 1]$. Applying results of [16] shows there are no four- or eight-torsion sections. Hence $MW(X, \pi) \simeq \mathbb{Z}/2 \times \mathbb{Z}/2$. While this still doesn't completely determine Pic(X), we know now that it a rank 19 lattice of signature $(1, 18)$ with discriminant ± 16 which contains the sublattice $H \oplus D_6 \oplus D_6 \oplus A_3 \oplus A_1 \oplus A_1$.

5.2 Kummer and Shioda–Inose Structures Associated to Products of Elliptic Curves

Let E_1, E_2 be elliptic curves. We think of E_1 and E_2 as quotients $\mathbb{C}/(\mathbb{Z} \oplus \mathbb{Z}\tau_1)$, $\mathbb{C}/(\mathbb{Z} \oplus \mathbb{Z}\tau_2)$ for $\tau_1, \tau_2 \in \mathbb{H}$. The action of $\{\pm 1\}$ on $A = E_1 \times E_2$ has sixteen fixed points, leading to sixteen nodes on the quotient $\bar{A} = A/\{\pm 1\}$. The minimal resolution of $\bar{A}$ is a K3 surface $Km(A)$ called the *Kummer Surface* of A. The Picard group of $Km(A)$ contains a lattice DK of rank 18, generated by the sixteen exceptional curves of the resolution, together with the strict transforms of the images of $E_1 \times \{pt\}$, $\{pt\} \times E_2$. Conversely, any K3 surface X with a primitive embedding $DK \hookrightarrow \mathrm{Pic}(X)$ is isomorphic to $Km(A)$ for some $A = E_1 \times E_2$ [4, Proposition 3.21].

A Kummer surface carries a symplectic involution β, with the minimal resolution of $Km(A)/\beta$ again a K3 surface $SI(A)$, called the *Shioda–Inose surface* of A. [4] shows that $X \simeq SI(A)$ for $A = E_1 \times E_2$ if and only if the rank 18 lattice $H \oplus E_8 \oplus E_8$ embeds primitively into Pic(X). Generically, this will be exactly the Picard lattice, and so the transcendental lattice will be $H \oplus H$.

If E_1, E_2 are n-isogenous, i.e. if there exists a degree n morphism $E_1 \rightarrow E_2$, then the Picard ranks of $Km(A)$ and $SI(A)$ have rank 19, with an extra generator corresponding to the strict transform of the graph of the isogeny. In this case, the Picard lattice of the Shioda–Inose surface will generically be $H \oplus E_8 \oplus E_8 \oplus \langle -2n \rangle$, and the transcendental lattice will be $H \oplus \langle 2n \rangle$.

E_1, E_2 are n-isogenous if and only if, up to the action of $PSL_2(\mathbb{Z})$ on $\mathbb{H}$, $\tau_2 = \frac{-1}{n\tau_1}$. (Note then that the relation is symmetric; given an isogeny $E_1 \rightarrow E_2$, there exists a dual isogeny $E_2 \rightarrow E_1$.) Thus if

$$\Gamma_0(n) = \left\{ \begin{pmatrix} a & b \\ c & d \end{pmatrix} \in PSL_2(\mathbb{Z}) \mid c \equiv 0 \ (\mathrm{mod}\ n) \right\} \tag{17}$$

then the moduli space of ordered pairs of n-isogenous elliptic curves is given by $X_0(n) = \mathbb{H}/\Gamma_0(n)$. To form the moduli space of products of n-isogenous elliptic curves, we need to quotient also by the involution $\tau \mapsto \frac{-1}{n\tau}$ on $X_0(n)$. We call the function $w_h : \mathbb{H} \to \mathbb{H}$ defined by $w_h(\tau) = \frac{-1}{h\tau}$ an *Atkin–Lehner map*. Note that w_h can be represented by the matrix $\begin{pmatrix} 0 & \frac{-1}{\sqrt{h}} \\ \sqrt{h} & 0 \end{pmatrix} \in PSL_2(\mathbb{R})$, and also that if $h|n$, then w_h descends to an involution on $X_0(n)$. We write $\Gamma_0(n) + h$ for the subgroup of $PSL_2(\mathbb{R})$ generated by $\Gamma_0(n)$ and w_h, and $X_0(n) + h$ for the quotient of $X_0(n)$ by w_h (or equivalently for $\mathbb{H}/(\Gamma_0(n) + h)$).

Thus, $X_0(n) + n$ is the moduli space of products of n-isogenous elliptic curves, and hence also of the Kummer surfaces and Shioda–Inose surfaces associated to such products. It is important to note, however, that while the transcendental lattices of $E_1 \times E_2$ and $SI(E_1 \times E_2)$ are isomorphic, the transcendental lattice of $Km(E_1 \times E_2)$ differs from these by scaling by 2.

5.3 Modular Groups Associated to Our Families of K3 Surfaces

For Examples 3.5 and 3.7, Γ is $\Gamma_0(6) + 6$ and $\Gamma_0(6) + 3$ respectively ([18, Proposition 5.4], [33, Theorem 2]).

In these two cases, explicit calculations of Picard lattices in [29, 33] show that the K3 surfaces have Shioda–Inose structures associated to the product of 6- and 3-isogenous elliptic curves respectively. The transcendental lattices of the generic K3's in these pencils are $H \oplus \langle 12 \rangle$ and $H \oplus \langle 6 \rangle$ respectively. The role of $\Gamma_0(6) + 6$ for Example 3.5, then, follows from identifying the base of the family with a compactification of the moduli space $X_0(6) + 6$ of $SI(E_1 \times E_2)$ for E_1, E_2 6-isogenous. Similarly, in the case of Example 3.7, $\Gamma_0(6) + 3 \subset \Gamma_0(3) + 3$, so this example realizes the base of the family as a covering of the moduli space of the Shioda–Inose surface $SI(E_1 \times E_2)$ for E_1, E_2 3-isogenous.

Example 3.9 is somewhat different. In this case, the K3 surfaces are not Shioda–Inose surfaces but Kummer surfaces. To see this, we will use the elliptic fibration of Example 7. Elliptic fibrations on $Km(E_1 \times E_2)$ have been classified by [21], where in particular they show that generically $Km(E_1 \times E_2)$ has a fibration giving lattice $H \oplus D_6 \oplus D_6 \oplus (A_1)^{\oplus 4}$ and Mordell–Weil group $\mathbb{Z}/(2) \oplus \mathbb{Z}/(2)$. If the two elliptic curves are presented in Legendre normal form

$$y^2 = x(x - 1)(x - \lambda_i)$$

then [21] gives the Legendre equation for this fibration as

$$Y^2 = X(X - u(u - 1)(\lambda_2 u - \lambda_1))(X - u(u - \lambda_1)(\lambda_2 u - 1))$$

(where u is an appropriately chosen parameter on the base of the fibration).

Comparing with our fibration, we see that our family then sits inside the family of $Km(E_1 \times E_2)$ as a locus where two of the A_1 singular fibers collide to give an A_3 singular fiber. The only possibilities are for $\lambda_1 = \lambda_2$ or $\lambda_1 = 1/\lambda_2$. In either case, E_1 and E_2 must be isomorphic. Thus, our family is the family of K3 surfaces of the form $Km(E \times E)$.

To determine for what group Γ this family is modular, we consider the symmetric square root of the Picard–Fuchs equation given in (13). By scaling the solutions appropriately, we may put this equation into a projective normal form $\frac{d^2 f}{dt^2} + Q(t)f = 0$, where

$$Q(t) = \frac{\left(t^2 - 8t + 64\right)\left(t^2 + 8t + 64\right)}{4(t-8)^2 t^2 (t+8)^2} \tag{18}$$

Changing variables via $t = \frac{1}{iz}$ and comparing with the table of [22], we see that

$$\Gamma = \Gamma_0(4|2) = \left\{ \begin{pmatrix} a & b/2 \\ 4c & d \end{pmatrix} \in PSL_2(\mathbb{R}) \;\middle|\; a, b, c, d \in \mathbb{Z} \right\} \tag{19}$$

Acknowledgements The second author was supported by the NSF Grant No. OISE-0965183. The third, fourth, and fifth authors were supported in part by No. DMS-0821725. The authors thank Andrey Novoseltsev for thoughtful discussion of computational techniques and Charles Doran for inspirational conversations.

References

1. V. Batyrev, D. Cox, On the Hodge structure of projective hypersurfaces in toric varieties. Duke Math. J. **75**(2), 293–338 (1994)
2. G. Bini, B. van Geemen, T.L. Kelly, Mirror quintics, discrete symmetries and Shioda maps. J. Algebr. Geom. **21**(3), 401–412 (2012)
3. W. Bosma, J. Cannon, C. Playoust, The Magma algebra system, I. The user language. J. Symbolic Comput. **24**(3–4), 235–265 (1997)
4. A. Clingher, C. Doran, Modular invariants for lattice polarized K3 surfaces. Mich. Math. J. **55**(2), 355–393 (2007)
5. A. Clingher, C. Doran, J. Lewis, U. Whitcher, Normal forms, K3 surface moduli, and modular parametrizations, in *CRM Proceedings and Lecture Notes*, vol. 47 Amer. Math. Soc., Providence, RI, (2009), pp. 81–98
6. D. Cox, S. Katz, *Mirror Symmetry and Algebraic Geometry* (American Mathematical Society, Providence, 1999)
7. I.V. Dolgachev, Mirror symmetry for lattice polarized $K3$ surfaces. Algebraic geometry, 4. J. Math. Sci. **81**(3), 2599–2630 (1996)
8. C. Doran, Picard-Fuchs uniformization and modularity of the mirror map. Comm. Math. Phys. **212**(3), 625–647 (2000)
9. C. Doran, B. Greene, S. Judes, Families of quintic Calabi-Yau 3-folds with discrete symmetries. Comm. Math. Phys. **280**(3), 675–725 (2008)
10. A. Garbagnati, Elliptic K3 surfaces with abelian and dihedral groups of symplectic automorphisms (2009) [arXiv:0904.1519]
11. A. Garbagnati, Symplectic automorphisms on Kummer surfaces. Geometriae Dedicata **145**, 219–232 (2010)

12. A. Garbagnati, The Dihedral group $\mathscr{D}_5$ as group of symplectic automorphisms on K3 surfaces. Proc. Am. Math. Soc. **139**(6), 2045–2055 (2011)

13. A. Garbagnati, A. Sarti, Symplectic automorphisms of prime order on $K3$ surfaces. J. Algebra **318**(1), 323–350 (2007)

14. B. Greene, M. Plesser, Duality in Calabi-Yau moduli space. Nucl. Phy. B. **338**(1), 15–37 (1990)

15. K. Hashimoto, Period map of a certain K3 family with an $\mathscr{S}_5$-action. J. Reine Angew. Math. **652**, 1–65 (2011)

16. G.H. Hitching, Quartic equations and 2-division on elliptic curves (2007) [arXiv:0706.4379]

17. K. Hori, S. Katz, A. Klemm, R. Pandharipande, R. Thomas, C. Vafa, R. Vakil, E. Zaslow, in *Mirror Symmetry*. Clay Mathematics Monographs, vol. 1 (American Mathematical Society/-Clay Mathematics Institute, Providence/Cambridge, 2003)

18. S. Hosono, B.H. Lian, K. Oguiso, S.-T. Yau, Autoequivalences of derived category of a K3 surface and monodromy transformations. J. Algebr. Geom. **13**(3), 513–545 (2004)

19. S. Kondō, Niemeier lattices, Mathieu groups, and finite groups of symplectic automorphisms of $K3$ surfaces. With an appendix by Shigeru Mukai. Duke Math. J. **92**(3), 593–603 (1998)

20. M. Kreuzer, Calabi-Yau 4-folds and toric fibrations. J. Geom. Phys. **26**, 272–290 (1998)

21. M. Kuwata, T. Shioda, Elliptic parameters and defining equations for elliptic fibrations on a Kummer surface, in *Algebraic Geometry in East Asia – Hanoi 2005*. Advanced Studies in Pure Mathematics, vol. 50 (Mathematical Society of Japan, Tokyo, 2008), pp. 177–215

22. B.H. Lian, J.L. Wiczer, Genus zero modular functions (2006), http://people.brandeis.edu/~lian/Schiff.pdf. Accessed March 2013

23. B.H. Lian, S.-T. Yau, Mirror maps, modular relations and hypergeometric series I. Appeared as "Integrality of certain exponential series", in *Lectures in Algebra and Geometry*, ed. by M.-C. Kang. Proceedings of the International Conference on Algebra and Geometry, Taipei, 1995 (International Press, Cambridge, 1998), pp. 215–227

24. C. Luhn, P. Ramond, Quintics with finite simple symmetries. J. Math. Phys. **49**(5) 053525, 14 (2008)

25. A. Mavlyutov, Semiample hypersurfaces in toric varieties. Duke Math. J. **101**(1), 85–116 (2000)

26. S. Mukai, Finite groups of automorphisms and the Mathieu group. Invent. Math. **94**(1), 183–221 (1988)

27. V. Nikulin, Finite automorphism groups of Kähler K3 surfaces. Trans. Mosc. Math. Soc. **38**(2), 71–135 (1980)

28. K. Oguiso, D.-Q. Zhang, The simple group of order 168 and $K3$ surfaces, in *Complex Geometry*, Göttingen, 2000 (Springer, Berlin, 2002)

29. C. Peters, J. Stienstra, A pencil of $K3$-surfaces related to Apéry's recurrence for $\zeta(3)$ and Fermi surfaces for potential zero, in *Arithmetic of Complex Manifolds*, Erlangen, 1988 (Springer, Berlin, 1989)

30. M. Singer, Algebraic relations among solutions of linear differential equations: Fano's theorem. Am. J. Math. **110**, 115–143 (1988)

31. J.P. Smith, *Picard-Fuchs Differential Equations for Families of K3 Surfaces*, University of Warwick, 2006 [arXiv:0705.3658v1] (2007)

32. J. Tate, Algorithm for determining the type of a singular fiber in an elliptic pencil, in *Modular Functions of One Variable IV*. Lecture Notes in Mathematics, vol. 476 (Springer, Berlin, 1975), pp. 33–52

33. H. Verrill, Root lattices and pencils of varieties. J. Math. Kyoto Univ. **36**(2), 423–446 (1996)

34. H. Verrill, N. Yui, Thompson series, and the mirror maps of pencils of $K3$ surfaces, in *The Arithmetic and Geometry of Algebraic Cycles*, Banff, AB, 1998. CRM Proceedings Lecture Notes, vol. 24 (AMS, Providence, 2000), pp. 399–432

35. U. Whitcher, Symplectic automorphisms and the Picard group of a K3 surface. Comm. Algebra **39**(4), 1427–1440 (2011)

36. G. Xiao, Galois covers between K3 surfaces. Ann. l'Institut Fourier **46**(1), 73–88 (1996)

K_1^{ind} of Elliptically Fibered $K3$ Surfaces: A Tale of Two Cycles

Matt Kerr

Abstract We discuss two approaches to the computation of transcendental invariants of indecomposable algebraic K_1 classes. Both the construction of the classes and the evaluation of the regulator map are based on the elliptic fibration structure on the family of $K3$ surfaces. The first computation involves a Tauberian lemma, while the second produces a "Maass form with two poles".

Key words: Regulator map, Indecomposable K-theory, $K3$ surfaces, Tauberian theory, Higher Green's function

Mathematics Subject Classifications (2010): Primary 14C25; Secondary 14C30, 14J28

1 Introduction

By a seminal result of Chen and Lewis [8], one already knows that (for fixed lattice L) on a very general L-polarized $K3$ surface X, the indecomposable K_1-classes proliferate like loaves and fishes to span $H_{tr}^{1,1}(X, \mathbb{R})$ under the real regulator map. However, things in general are not settled for ever: the literature lacks a large class of concrete, nontrivial examples occurring in modular families (with the possible exception of Collino's examples obtained by degenerating the Ceresa cycle [11][1]). The most natural source of such examples should be cycles supported on singular fibers of Kodaira type $I_{n \geq 1}$ in torically-induced or Weierstrass-type internal fibrations. In this paper we consider two families of higher Chow cycles of this type, and

M. Kerr (✉)
Department of Mathematics, Campus Box 1146, Washington University in St. Louis, St. Louis, MO 63130, USA
e-mail: matkerr@math.wustl.edu

[1] Though presented on Jacobians of genus 2 curves, these can be transferred (using a correspondence) to the corresponding family of Kummer $K3$ surfaces.

R. Laza et al. (eds.), *Arithmetic and Geometry of K3 Surfaces and Calabi–Yau Threefolds*,   387
Fields Institute Communications 67, DOI 10.1007/978-1-4614-6403-7_13,
© Springer Science+Business Media New York 2013

investigate properties of the transcendental functions produced by the real regulator map (and a variant reviewed in Sect. 2).

The first cycle is on the family of $H \oplus E_8 \oplus E_8 \oplus \langle -12 \rangle$-polarized Kummer $K3$'s studied by Beukers, Peters and Stienstra [5, 20, 21], which is parametrized by $\Gamma_1(6)^{+6} \backslash \mathfrak{H}$. By representing it as a family of toric hypersurfaces, one may produce an elliptic structure by restricting a fibration of the ambient toric Fano threefold constructed by appropriately "slicing" its reflexive polytope [2, 22]. In the spirit of mirror symmetry, we perform a power series computation of the transcendental regulator for our cycle (Sect. 3, with a technical detail resolved in Sect. 5). For our second example, we revisit the computation of [9] for the Clingher–Doran $M :=$ $H \oplus E_8 \oplus E_8$-polarized two-parameter family of $K3$'s, and prove the results (partially described there) relating the real regulator to higher Green's functions and the thesis of A. Mellit [18].

While neither wise nor foolish, nor meriting any superlative degree of comparison, we hope these constructions lead to something far, far better (or just more general). The author would also like to thank Chuck Doran, James Lewis, Greg Pearlstein, Duco van Straten, and Stefan Müller–Stach for discussions related to this paper, and to acknowledge partial support from NSF Standard Grant DMS-1068974. We are especially grateful to Adrian Clingher for supplying Remark 4.

2 Real and Transcendental Regulators

We shall introduce only the groups and maps we require; for a more general treatment of cycle maps see Lewis's lectures in this volume [17] (or Sect. 1 of [13]). Let X be a smooth $K3$ surface over $\mathbb{C}$, and consider the abelian group of "empty rational equivalences"

$$\tilde{K}_1(X) := \frac{\left\{ \overset{\text{(finite sums)}}{\sum} q_j.(f_j, D_j) \, \middle| \, \begin{array}{l} q_j \in \mathbb{Q}, \, D_j \subset X \text{ curves, } f_j \in \mathbb{C}(\tilde{D}_j)^*; \\ \text{ and } \sum q_j (\iota_j)_*((f_j)) = 0 \end{array} \right\}}{\langle (f, D) + (g, D) - (fg, D) \rangle}$$

where $\iota_j : \tilde{D}_j \to X$ is the composition of the inclusion of the curve with its desingularization. Algebraic K_1 is the quotient by Tame symbols

$$K_1(X) := \tilde{K}_1(X) \big/ \text{Tame}\{K_2(\mathbb{C}(X))\},$$

with $\mathbb{Q}$-coefficients understood (here and throughout). There is a "formal" (but always zero) fundamental class map

$$cl : K_1(X) \to Hg^{2,1}(X) := F^2 H^3(X, \mathbb{C}) \cap H^3(X, \mathbb{Q}) = \{0\}$$

which is "computed" by sending

$$Z = \sum q_j.(f_j, D_j) \longmapsto \begin{cases} \Omega_Z := \frac{1}{2\pi i} \sum q_j(\iota_j)_* \frac{df_j}{f_j} \in F^2 \mathcal{D}^3(X) \\ T_Z := \sum q_j(\iota_j)_*(f_j^{-1}(\mathbb{R}^-)) \in Z^3_{\text{top}}(X) \end{cases}.$$

Vanishing of $Hg^{2,1}(X)$ implies the existence of a $(2,0)$ current and piecewise $\mathcal{C}^\infty$ chain

$$\left. \begin{array}{l} \Xi \in F^2 \mathcal{D}^2(X) \\ \Gamma \in C^2_{\text{top}}(X) \end{array} \right\} \text{ such that } \begin{cases} \Omega_Z = d\Xi \\ T_Z = \partial\Gamma \end{cases}.$$

The Abel–Jacobi map

$$AJ : K_1(X) \to J^{2,1}(X) := \frac{H^2(X, \mathbb{C})}{H^{2,0}(X, \mathbb{C}) \oplus H^2(X, \mathbb{Q})} \cong \frac{\{F^1 H^2(X, \mathbb{C})\}^\vee}{H_2(X, \mathbb{Q})}$$

is the basic invariant, and a special case of the arithmetic Bloch–Beilinson conjecture says it should be injective. Writing $\log^-(\cdot)$ for the (discontinuous) branch with imaginary part $\in (-\pi, \pi]$ (thought of as a 0-current), and $\delta_{(\cdot)}$ for the current of integration over a chain, AJ is induced by

$$Z \longmapsto \tilde{R}_Z := \underbrace{\frac{1}{2\pi i} \sum q_j(\iota_j)_* \log^-(f_j)}_{R_Z} - \Xi + \delta_\Gamma \in \mathcal{D}^2(X).$$

To spell this out, evaluating the $\tilde{R}_Z$ against a d-closed smooth test form $\omega \in F^1 A^2(X)$ gives

$$AJ(Z)(\omega) = \frac{1}{2\pi i} \sum q_j \int_{\tilde{D}_j} (\log^-(f_j)) \iota_j^* \omega + \int_\Gamma \omega,$$

where Γ is defined "up to a cycle".

Now the group which interests us is the indecomposables

$$K_1^{\text{ind}}(X) := K_1(X) / \text{image} \left(\mathbb{C}^* \otimes \text{Div}(X) \right),$$

and it is conjecturally detected by

$$\overline{AJ} : K_1^{\text{ind}}(X) \to \left\{ F^1 H^2_{tr}(X, \mathbb{C}) \right\}^\vee \Big/ H^{tr}_2(X, \mathbb{Q}).$$

Since $\overline{AJ}$ is hard to compute, one tends instead to compute one of two "quotients". The so-called **transcendental regulator**

$$\Psi : K_1^{\text{ind}}(X) \to \left\{ \Omega^2(X) \right\}^\vee \Big/ \text{image} \left\{ H^{tr}_2(X, \mathbb{Q}) \right\}$$

is given (on $\omega^{2,0} \in \Omega^2(X)$) by

$$\Psi(Z)(\omega^{2,0}) = \int_\Gamma \omega^{2,0}.$$

Since image$\{H_2^{tr}(X, \mathbb{Q})\}$ is intractable for fixed X (except for Picard rank 20), this is primarily of use variationally: if X_t is a family of $K3$ surfaces over a Zariski open $U \subset \mathbb{P}^1$, carrying

- An algebraic family of cycles $Z_t \in K_1^{ind}(X_t)$.
- A smoothly varying (but possibly multivalued) family of chains Γ_t as above.
- An algebraically family of holomorphic forms $\omega_t \in \Omega^2(X_t)$ with Picard–Fuchs operator D_{PF}^ω annihilating its periods,

then

$$D_{PF}^\omega \int_{\Gamma_t} \omega_t \in \mathbb{C}(t)$$

is an invariant of $\{Z_t\}$ [12, Theorem 3.2]. We will compute this "inhomogeneous term" for the cycle in Sect. 2, with a small caveat (cf. Remark 3).

For the **real regulator**

$$r : K_1^{ind}(X) \to \left\{ H_{tr}^{1,1}(X, \mathbb{R}) \right\}^\vee,$$

which is really the *imaginary* part of $\overline{AJ}$, the main difficulty is in producing appropriate test forms. It is defined by

$$r(Z)(\omega_\mathbb{R}) = \Re \left\{ 2\pi i \int \tilde{R}_Z \wedge \omega_\mathbb{R} \right\} = \sum_i q_i \int_{\tilde{D}_j} \log |f_j| \iota_j^* \omega_\mathbb{R},$$

on 2-forms $\omega_\mathbb{R}$ which must be *smooth, real, d-closed, of pure type* (1, 1)*, and orthogonal to* $H_{alg}^{1,1}$. This approach is applied to a family of cycles in Sect. 3.

In both cases, the cycles of interest arise from an elliptic fibration of X

$$
\begin{array}{ccc}
 & p & \leftarrow 0, \infty \\
X \supset & D & \leftarrow \mathbb{P}_z^1 \\
\downarrow & \downarrow & \\
\mathbb{P}^1 \ni & \{t_0\} &
\end{array}
$$

with a nodal rational (Kodaira type I_1) fiber. The class of $(z^a, D) \in K_1^{ind}(X)$ is independent of how we scale the coordinate z; it depends only on a. The **primitive class** associated to such a fiber, defined up to sign, is the one with $|a| = 1$. Note that its construction requires normalizing D, which can have implications for its minimal field of definition (or its monodromy). It has been known for a long time that similar constructions on I_n fibers in modular elliptic surfaces have trivial class in K_1^{ind}, being in the Tame image of Beilinson's Eisenstein symbols [4]. More recent work of Asakura showed that this is not so on elliptic "Tate surfaces" [1], but did not compute the regulator. We defer to [9] for further discussion of the context for these computations; our personal interest lies in the novel relationships between geometry and arithmetic they uncover through transcendental means.

3 The Apéry Family and an Inhomogeneous Picard–Fuchs Equation

Our first "mathematical short story" begins with the Laurent polynomial

$$\phi(u, v, w) := \frac{(u-1)(v-1)(w-1)(1-u-v+uv-uvw)}{uvw}$$

and the toric threefold $\mathbb{P}_\Delta$ attached to its Newton polytope

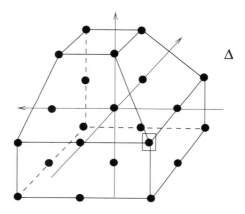

Δ

(whose singularity corresponding to the ⊡ shall not trouble us). The minimal resolution X_t of the Zariski closure of $\{1 - t\phi = 0\}$ in $\mathbb{P}_\Delta$ defines a $K3$ surface for $t \notin \{0, (\sqrt{2} \pm 1)^4, \infty\} =: \mathcal{L}$, which has Picard rank 19 for general t and is birational to the family considered in [20] (cf. [13]). We shall work with $t \neq 0$ small, for which the singular fibers of the internal elliptic fibration[2]

$$\pi : X_t \to \mathbb{P}^1$$
$$(u,v,w) \mapsto w$$

have Kodaira types

$w =$	0	1	∞	$w(t)$	3 more near ∞
type	I_1^*	I_5	I_8	I_1	3 more I_1's

More precisely, a computation shows that the I_1 fibers occur at the solutions of

$$0 = (t^3)w^4 + (3t^2 - 2t^3)w^3 + (t^3 + 5t^2 + 3t)w^2 + (-8t^2 - 20t + 1)w + (16t), \quad (1)$$

all of which but

$$w(t) = -16tH(t) := -16t\{1 + 20t + 456t^2 + 11280t^3 + \cdots\}$$

---

[2] In the setup of [2], π is induced by slicing Δ horizontally. This suggests a significant generalization of the computation carried out in this section. Also note that this particular π has Mordell–Weil rank 1.

are large (and asymptotic to $-t^{-1} + 3e^{\frac{2\pi i}{3}j}t^{-\frac{2}{3}} + \cdots$ for $j = 0, 1, 2$). With our running assumption of "t small", w is of course 1-to-1.

Remark 1. Globally speaking, (1) tells us how singular fibers swap and collide; taking the fixed fibers into account, a resultant calculation shows that all collisions occur for $t \in \mathcal{L}' := \mathcal{L} \cup \left\{1, \frac{27}{40}\right\}$.

On the family of $K3$ surfaces $\{X_t\}$, let φ_t represent a family of topological two-cycles with class in $H^2_{tr}(X_t)$ and vanishing in homology at $t = 0$, where X_t degenerates[3] to $X_0 = \mathbb{P}_{\Delta} \backslash (\mathbb{C}^*)^3$. Its class is invariant about $t = 0$ and unique up to scale; we fix this by saying that its image under the map $Tube : H_2(X_t) \to H_3(\mathbb{P}_{\Delta} \backslash X_t)$ is the class of the torus $|u| = |v| = |w| = 1$. We shall also need relative vanishing cycles for the internal fibration. Similarly, for w close to 1, we let $\varphi_{t,w}$ denote a family of 1-cycles on the elliptic curves $X_{t,w} := \pi^{-1}(w)$ vanishing in $H_1(X_{t,1})$; and let it also denote the multivalued family resulting from their topological continuation. The link between these cycles is via the Lefschetz thimble

$$\Phi_{t,w_0} := \bigcup_{w \in \overrightarrow{1.w_0}} \varphi_{t,w} \in C^2_{\mathrm{top}}(X_t),$$

which has monodromy $\Phi_{t,w_0} \mapsto \Phi_{t,w_0} + \varphi_t$ as w_0 goes about the unit circle counterclockwise. (Here, "φ_t" is to be understood up to coboundary.) That w_0 is going around both 0 and $\mathsf{w}(t)$ is what is important here; that $X_{t,1}$ is singular is not an issue. The monodromy is the same on circles of radius less than 1 and $\geq \mathsf{w}(t)$.

Let $Z_t \in K_1^{\mathrm{ind}}(X_t)$ be the primitive class supported on $X_{t,\mathsf{w}(t)}$, and note that $\Gamma_t := \Phi_{t,\mathsf{w}(t)}$ bounds on T_{Z_t}. Over $\mathbb{P}^1 \backslash \mathcal{L}'$, the continuation of Z_t has significant monodromy, which can be eliminated by lifting to (the preimage of $\mathbb{P}^1 \backslash \mathcal{L}'$ in) a double-cover of the curve (1). However, as long as t remains small, we need only that Z_t has no monodromy about $t = 0$; one way to see this is by a limiting argument, cf. Remark 2 below. For the family of holomorphic 2-forms, take

$$\omega_t := \frac{1}{2\pi i} Res_{X_t} \hat{\omega}_t := \frac{1}{2\pi i} Res_{X_t} \left\{ \frac{\frac{du}{u} \wedge \frac{dv}{v} \wedge \frac{dw}{w}}{1 - t\phi(u,v,w)} \right\}$$

and write also

$$\omega_{t,w} := \frac{1}{2\pi i} Res_{X_{t,w}} \hat{\omega}_{t,w} := \frac{1}{2\pi i} Res_{X_{t,w}} \left\{ \frac{\frac{du}{u} \wedge \frac{dv}{v}}{1 - t\phi(u,v,w)} \right\}.$$

Then we are aiming to compute

$$\Psi(Z_t)(\omega_t) = \int_{\Phi_{t,\mathsf{w}(t)}} \omega_t = \frac{1}{2\pi i} \int_1^{\mathsf{w}(t)} \left(\int_{|u|=|v|=1} \frac{\frac{du}{u} \wedge \frac{dv}{v}}{1 - t\phi(u,v,w)} \right) \frac{dw}{w}$$

in terms of power series in t.

---

[3] This degeneration is not semistable, which can be fixed by blowing up the components of X_0 at a few points; this need not trouble us.

Fix real numbers $0 < \eta < \alpha \ll 1$. We will study the behavior of the (singular, multivalued) functions

$$v(t, t_0) := \int_{\Phi_{t, \mathsf{w}(t_0)}} \omega_t , \qquad \tilde{v}(t, t_0) := v(t, t_0) - \frac{\log(t_0)}{2\pi i} \int_{\varphi_t} \omega_t$$

on the set

$$\mathcal{S} := \{|t| \le \alpha + \eta\} \times \{\alpha - \eta \le |t_0| \le \alpha + \eta\} \subset \mathbb{C}^2.$$

For fixed t, the previous remarks on Φ_{t, w_0} imply that $\tilde{v}$ has no monodromy in t_0 about a circle of radius $\ge |t|$. For fixed t_0, $\tilde{v}$ (or v) has no monodromy in t about circles of radius $\le |t_0|$, while remaining finite as $t \to 0$; and so we may write (uniquely)

$$\tilde{v}(t, t_0) = \sum_{n=0}^{\infty} \mathfrak{A}_n(t_0) t^n \quad \text{for } |t| < |t_0|.$$

As $\mathsf{w} \to \mathsf{w}(t)$, $\int_{\varphi_{t,\mathsf{w}}} \omega_{t,\mathsf{w}}$ is asymptotic to (a constant multiple of) $\log(\mathsf{w} - \mathsf{w}(t))$, which translates to $(\mathsf{w}_0 - \mathsf{w}(t)) \log(\mathsf{w}_0 - \mathsf{w}(t))$-type behavior for $\int_{\Phi_{t,\mathsf{w}_0}} \omega_t$ and thence to $(t_0 - t)$ $\log(t_0 - t)$ for $\tilde{v}$ (or v). More precisely, we must have on $\mathcal{S}$

$$\tilde{v}(t, t_0) = \left\{ (t - t_0) \log\left(\frac{t}{t_0} - 1\right) \right\} F_0(t, t_0) + G_0(t, t_0) \tag{2}$$

and (therefore)

$$\delta_t \tilde{v}(t, t_0) = \log\left(\frac{t}{t_0} - 1\right) F(t, t_0) + G(t, t_0) \tag{3}$$

where $F, G, F_0, G_0 \in \mathcal{O}(\mathcal{S})$ and $\delta_t := t\frac{\partial}{\partial t}$.

Clearly, the function we must compute is $v(t) := v(t, t)$. By the above formula, at least on the annulus $\mathcal{A} := \{\alpha - \eta \le |t_0| \le \alpha + \eta\}$, $\tilde{v}(t) := \tilde{v}(t, t)$ is monodromy-free about 0.

Lemma 1. $\tilde{v}(t)$ extends to a holomorphic function on the disk $D := \{|t| < \alpha + \eta\}$, and so is representable by power series on $\mathcal{A}$, viz.

$$\tilde{v}(t) = \sum_{m=0}^{\infty} v_m t^m.$$

Proof. Since the family $\{Z_t\}$ extends to a (global algebraic) higher Chow cycle on a cover of the total space $\cup_{t \in \mathbb{P}^1 \setminus \mathcal{L}'} X_t$, the associated higher normal function is admissible on $D \setminus \{0\}$. (See for example [7].) One easily deduces (as in the proof of Proposition 5.28 in [24]) that its period $v(t)$, and hence $\tilde{v}(t)$, is of the form $\sum_{a,q} f_{a,q}(t) t^a \log^q(t)$ on $D \setminus \{0\}$, where $a \in \mathbb{Q} \cap [0, 1)$, $q \in \mathbb{Z}_{\ge 0}$, and $f_{a,q} \in \mathcal{O}(D)$. Any function of this form with no monodromy is in $\mathcal{O}(D)$.

The proof of the following key "Tauberian lemma" is deferred to Sect. 5:

Lemma 2. $\sum_{n=0}^{\infty} \mathfrak{A}_n(t) t^n$ converges uniformly on $\{|t| = \alpha\}$, to $\tilde{v}(t)$.

The computations below will show (without using Lemma 2!) that the $\mathfrak{A}_n$ are given by Laurent series on $\mathcal{A}$ with poles of order n. Assuming this, we may apply Cauchy's theorem and Lemma 2 to obtain

$$\nu_m = \frac{1}{2\pi} \int_{|t|=\alpha} \frac{\tilde{v}(t)}{t^{m+1}} dt = \lim_{N\to\infty} \int_{|t|=\alpha} \frac{\sum_{n=0}^{N} \mathfrak{A}_n(t)t^n}{t^{m+1}} dt$$

$$= \lim_{N\to\infty} \sum_{n=0}^{N} [\mathfrak{A}_n(t)t^n]_m = \sum_{n=0}^{\infty} [\mathfrak{A}_n(t)t^n]_m$$

$$= \sum_{n=0}^{\infty} [\mathfrak{A}_n(t)]_{m-n}, \tag{4}$$

where $[\cdot]_m$ takes the mth power series coefficient. (Notice that a corollary here is that the last sum itself is convergent.) This will justify the rearrangements we perform below.

Fix $w_0 \in \bar{D}_1^*$ (i.e., $0 < |w_0| \le 1$), and assume $t \ne 0$ is "sufficiently small". Then we have

$$\int_{\Phi_{t,w_0}} \omega_t$$

$$= 2\pi i \int_1^{w_0} \left(\frac{1}{(2\pi i)^2} \int_{|u|=|v|=1} \frac{d\log u \wedge d\log v}{1 - t\phi(u,v,w)} \right) \frac{dw}{w}$$

$$= 2\pi i \int_1^{w_0} \sum_{n\ge 0} \frac{t^n(w-1)^n}{w^n} \left[\frac{(u-1)^n(v-1)^n(1-u-v+uv-uvw)^n}{u^n v^n} \right]_{(0,0)} \frac{dw}{w}$$

where $[\cdot]_{(0)}$ takes the coefficient of $u^0 v^0$, which in this case equals

$$\sum_{k=0}^{n} (-w)^{n-k} \binom{n}{k} \binom{n+k}{n}^2.$$

With this substitution, the above integral

$$= 2\pi i \sum_{n=0}^{\infty} \sum_{k=0}^{n} (-1)^{n-k} \binom{n}{k} \binom{n+k}{k}^2 t^n \int_1^{w_0} \frac{(w-1)^n}{w^{k+1}} dw$$

$$= 2\pi i \sum_{n=0}^{\infty} t^n \sum_{\substack{k,\ell=0 \\ k\ne\ell}}^{n} (-1)^{\ell-k} \binom{n}{k}\binom{n}{\ell}\binom{n+k}{n}^2 \frac{w_0^{\ell-k}-1}{\ell-k}$$

$$+ 2\pi i \log w_0 \sum_{n=0}^{\infty} t^n \sum_{k=0}^{n} \binom{n}{k}^2 \binom{n+k}{n}^2$$

in which we recognize the sum in the second term as $\sum_{n=0}^{\infty} t^n [\phi^n]_{(0)} = \frac{1}{(2\pi i)^2} \int_{\varphi_\lambda} \omega_t$.

Having carried out this calculation, the result can be continued in t to $|t| < |w^{-1}(w_0)|$ as $\int_{\Phi_{t,w_0}} \omega_t$ is holomorphic there. We conclude that for $(t, t_0) \in S$ with $|t| < |t_0|$,

$$
\begin{aligned}
\frac{\tilde{v}(t, t_0)}{2\pi i} &= \frac{1}{2\pi i} \int_{\Phi_{t,w(t_0)}} \omega_t - \frac{\log t_0}{(2\pi i)^2} \int_{\varphi_t} \omega_t \\
&= \sum_{n=0}^{\infty} \left\{ \begin{array}{l} \sum_{\substack{k,\ell=0 \\ k \neq \ell}}^{n} \binom{n}{k}\binom{n}{\ell}\binom{n+k}{n}^2 \frac{16^{\ell-k} t_0^{\ell-k} H(t_0)^{\ell-k} - (-1)^{\ell-k}}{\ell - k} \\ + \sum_{k=0}^{n} \binom{n}{k}^2 \binom{n+k}{n}^2 \log(-16H(t_0)) \end{array} \right\} t^n,
\end{aligned}
$$

and the term in braces is our $\frac{1}{2\pi i}\mathfrak{A}_n(t_0)$ from above. Interpreting powers and log of $H(t_0)$ as power series in t_0, the claim below Lemma 2 is now verified. We may summarize what has been proved by saying that $\tilde{v}(t)$ may be computed *by substituting $t_0 = t$ in the last sum and rearranging by power of t.* Each coefficient becomes an infinite series (due to the terms with $k > \ell$) whose convergence is nontrivial and guaranteed by the preceding argument, as is the convergence of the resulting power series for small t. See Remark 3 below for the precise domain of convergence.

Performing this computation—that is, applying (4)—we find the first few power series coefficients:

$$
\frac{v_0}{2\pi i} = \log 16 - \sum_{n \geq 1} \frac{\binom{2n}{n}^2}{16^n n},
$$

$$
\frac{v_1}{2\pi i} = 22 + 5\log 16 - 20 \sum_{n \geq 2} \frac{\binom{2n}{n}^2}{16^n(n-1)},
$$

$$
\frac{v_2}{2\pi i} = \frac{1703}{4} + 73\log 16 - 8 \sum_{n \geq 3} \frac{\binom{2n}{n}^2}{16^n} \frac{259n^2 - 258n + 64}{(n-2)(2n-1)^2}.
$$

In particular, we recognize[4] the first of these as $\frac{8}{\pi}G$, where G is Catalan's constant $\sum_{k \geq 0} \frac{(-1)^k}{(2k+1)^2}$; one naturally wonders if the others hold arithmetic interest. The sought-for function is, of course,

$$
\frac{v(t)}{2\pi i} = \sum_{m \geq 0} \frac{v_m}{2\pi i} t^m + \frac{\log t}{(2\pi i)^2} \int_{\varphi_t} \omega_t.
$$

The log term can be removed by tweaking our choice of Γ by a cycle (and hence v by a period); this will simplify the computation with D_{PF}. Using results from [20], one can compute the periods of integral cycles φ, ξ, η (sent by monodromy about $t = 0$ to $\varphi, \xi - 12\varphi, \eta + \xi - 6\varphi$ resp.) to be

---

[4] Cf. (for example) [13, (6.15)ff].

$$\int_{\varphi_t} \omega_t = (2\pi i)^2 \{1 + 5t + 73t^2 + 1445t^3 + \cdots\},$$

$$\int_{\xi_t} \omega_t = -12\frac{\log t}{2\pi i} \int_{\varphi_t} \omega_t + (2\pi i)\{-144t - 2520t^2 - \cdots\},$$

$$\int_{\eta_t} \omega_t = 6\frac{\log^2 t}{(2\pi i)^2} \int_{\varphi_t} \omega_t + \frac{\log t}{2\pi i} \int_{\varphi_\xi} \omega_t - 864t^2 - 25920t^3 - \cdots,$$

the coefficients in the first of which are just $[\phi^n]_{(0)}$. Replacing Γ_t by $\hat{\Gamma}_t := \Gamma_t + \frac{1}{12}\xi_t$, changes $\frac{\nu}{2\pi i}$ to

$$\frac{\hat{\nu}(t)}{2\pi i} = \frac{\nu(t)}{2\pi i} + \frac{1}{12}\frac{1}{2\pi i} \int_{\xi_t} \omega_t = A + Bt + \cdots$$

where $A = \frac{\nu_0}{2\pi i}$, $B = \frac{\nu_1}{2\pi i} - 12$.

Using [op. cit.], one finds that the Picard–Fuchs operator killing periods of ω_t is

$$D_{\mathrm{PF}}^\omega = (t^2 - 34t + 1)\delta_t^3 + 2t(t - 17)\delta_t^2 + 3t(t - 9)\delta_t + t(t - 5).$$

Applied to our "higher normal function", this gives

$$\left(D_{\mathrm{PF}}^\omega \frac{\nu}{2\pi i} =\right) D_{\mathrm{PF}}^\omega \frac{\hat{\nu}}{2\pi i} = (B - 5A)t + \text{h.o.t.}$$

where

$$B - 5A = 10 + 5\sum_{n\geq 1} \frac{\binom{2n}{n}^2}{16^n n} - 20\sum_{n\geq 2} \frac{\binom{2n}{n}^2}{16^n(n - 1)}$$

$$= 10 + 5\sum_{n\geq 1} \frac{\binom{2n}{n}^2}{16^n n(2n + 2)} > 0.$$

So $\hat{\nu}$ is not a period, and we conclude

Theorem 1. *For very general t, (the continuation of) Z_t has nontrivial class in $K_1^{\mathrm{ind}}(X_t)$, detected by the transcendental regulator.*

Remark 2. The I_1 fiber $X_{t,w(t)}$ supporting Z_t limits, as $t \to 0$, to the nodal rational curve $Y = \{16uv = (u - 1)^2(v - 1)^2, w = 0\} \subset X_0$. It follows that $X_{t,w(t)}$ admits a normalization over $\mathbb{C}[[t]]$, justifying the statement that Z_t has no monodromy about 0.

In fact, Z_t itself limits to a class $Z_0 \in H_{\mathcal{M}}^2(X_0, \mathbb{Q}(3))$ in motivic cohomology, and one can use this give an alternative proof of Z_t's nontriviality. More precisely, in the sense of [13, Sect. 6] Z_0 belongs to $W_{-2}H_{\mathcal{M}}^2(X_0, \mathbb{Q}(3)) \cong CH^2(\mathrm{Spec}(\mathbb{C}), 3) \cong K_3^{\mathrm{ind}}(\mathbb{C})$ with AJ map to $\mathbb{C}/\mathbb{Q}(2)$. Now the tangent vectors of Y at the singularity $(u, v) = (-1, -1)$ have slopes $\pm i$, which implies that Y admits a normalization— and hence that Z_0 is defined—over $\mathbb{Q}(i)$ (but not $\mathbb{Q}$). By Beilinson's variant of Borel's theorem (cf. [3, 19], and especially [13, Proposition 6.2]) AJ of any cycle in $K_3^{\mathrm{ind}}(\mathbb{Q}(i))$ has imaginary part a rational multiple of G.

With our choices above, $AJ(Z_0)$ must match up with $\lim_{t\to 0} \hat{v}(t)$ (cf. [13, Proposition 6.3]). In a different guise, $AJ(Z_0)$ has been computed in [15, Sect. 4][5] and comes out to exactly $16iG$. This proves immediately that $\hat{v}$ cannot have been a period, and satisfyingly explains the presence of $\frac{8}{\pi}G = \frac{16iG}{2\pi i}$ as the leading coefficient above.

Note that the computational method really requires little more than knowing ϕ, $\mathrm{w}(t)$, and D_{PF}, and is likely to work in greater generality than the approach outlined in the last remark. For instance, uncovering Z_0 in general could require a nontrivial moving lemma calculation, and even here we did not discover Z_0 until the presence of G in $\hat{v}_0$ suggested it.

Remark 3. Because there are no collisions of internal singular fibers until $t_0 = (\sqrt{2}-1)^4$, Z_t remains well-defined and $\hat{v}(t) = \Psi(Z_t)(\omega_t)$ monodromy- and pole-free on D_{t_0}. Since Z_t (hence $\hat{v}$) has monodromy about t_0, this is precisely the radius of convergence of $\sum v_m t^m$.

The monodromy of $\{Z_t\}$ means that the "inhomogeneous term" $D_{PF}^\omega \hat{v}$ is algebraic rather than rational in t. It becomes single-valued upon pullback to the double cover of (1) which makes $\{Z_t\}$ globally well-defined (so that monodromy of the pullback of $\hat{v}$ is by periods alone). It was pointed out by D. van Straten that the curve (1) is in fact rational; it is not known whether this is so for the double cover.

4 M-Polarized $K3$ Surfaces and a Higher Green's Function

The cycle whose real regulator we shall study appeared in Sect. 6 of [9], and we shall preface our second tale with a review of that construction, starting with a brief summary of material from [10]. Let E_λ, for each $\lambda \in \mathbb{P}^1 \setminus \{0, 1, \infty\}$, denote the Legendre elliptic curve $\{y^2 = x(x-1)(x-\lambda)\}$. Given $(a,b) \in \mathbb{C}^2$, the minimal resolution of

$$\left\{ Y^2 Z - (4u^3 - 3au - b)W^2 Z - \frac{1}{2}Z^2 W - \frac{1}{2}W^3 = 0 \right\} \subset \mathbb{P}^2_{[Y:Z:W]} \times \mathbb{P}^1_u$$

defines a $K3$ surface $X_{a,b}$ of *Shioda–Inose type*: that is, its Hodge structure $H^2_{tr}(X_{a,b})$ is integrally isomorphic to $H^2_{tr}(E_{\lambda_1} \times E_{\lambda_2})$ for certain Legendre parameters λ_1, λ_2. It turns out that these must satisfy $j(\lambda_1)j(\lambda_2) = a^3$, $j(\lambda_1) + j(\lambda_2) = a^3 - b^2 + 1$. The natural Weierstrass fibration $\theta : X_{a,b} \to \mathbb{P}^1_u$ has an I_{12}^* singular fiber over $u = \infty$, which gives the generic Picard rank 18. However, for a $K3$ we must have $\deg(\theta_* \omega_{X_{a,b}/\mathbb{P}^1}) = 2$, which implies the presence (for generic a,b) of 6 additional singular fibers, each of type I_1. Our cycle $Z_{a,b} \in K_1^{ind}(X_{a,b})$ will be the primitive class supported on one of these, ignoring for the moment *which* one, as well as issues of sign and monodromy.

One of the achievements of [10] was an explicit correspondence inducing the isomorphism of Hodge structures above. To produce this, notice that the I_{12}^* embeds a D_{16}^+ lattice in $H^2(X_{a,b}, \mathbb{Z})$. This implies the existence of two sections of θ with

---

[5] Where G is incorrectly identified as a transcendental number; that is the conjecture, but its irrationality is still unproven. This has no bearing on nontriviality of $16iG$ modulo $\mathbb{Q}(2)$.

2-torsion difference, translating by which gives a Nikulin involution $\mathcal{N}$. This involution has a fixed point (the node) on each I_1 and two fixed points on the I_{12}^*, from which one deduces that the minimal resolution of $X_{a,b}/\mathcal{N}$ has one I_6^* and 6 I_2 fibers. This Kummer surface, which we denote $\mathcal{K}_{\lambda_1,\lambda_2}$, fits into a diagram of the form

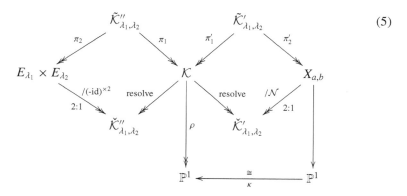

$$(5)$$

where for the moment we think of (a,b) and (λ_1,λ_2) as fixed and very general. Explicitly, $\mathcal{K}_{\lambda_1,\lambda_2}$ can be given as the minimal resolution of $\check{\mathcal{K}}_{\lambda_1,\lambda_2} :=$

$$\left\{ U^2 X_1 X_2 = (X_1 - V)(X_1 - \lambda_1 V)(X_2 - V)(X_2 - \lambda_2 V) \right\} \subset \mathbb{P}^3_{[X_1:X_2:U:V]},$$

and the elliptic fibration ρ by $\left[\sum_{i=1}^2 \left(-\frac{X_i^2}{\lambda_i} + \frac{\lambda_i+1}{\lambda_i} X_i V \right) - V^2 : X_1 X_2 \right]$, in terms of which the I_2 fibers lie over $1, \frac{1}{\lambda_1}, \frac{1}{\lambda_2}, \frac{1}{\lambda_1\lambda_2}, \frac{\lambda_1\lambda_2+1}{\lambda_1\lambda_2}, \frac{\lambda_1+\lambda_2}{\lambda_1\lambda_2}$ and the I_6^* fiber over ∞.

The cycle $Z_{a,b}$ is taken to lie on the I_1 fiber over $\kappa^{-1}(1)$. With the aid of the diagram

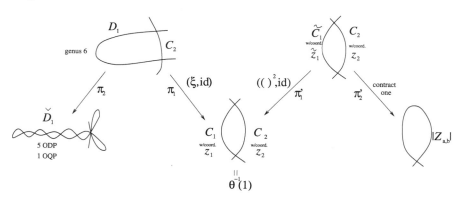

of curves (rational except for D_1 and $\check{D}_1$) in the top half of (5), we may construct classes in K_1^{ind} of $\mathcal{K}_{\lambda_1,\lambda_2}$ and $E_{\lambda_1} \times E_{\lambda_2}$ with the same real regulator class as $Z_{a,b}$. Indeed, noting that

$$(\pi_2')_* \left\{ (\tilde{C}_1, \tilde{z}_1) + (C_2, z_2) \right\} = Z_{a,b},$$

we can set

$$\Im_{\lambda_1,\lambda_2} := (\pi_1')_* \left\{ (\tilde{C}_1, \tilde{z}_1) + (C_2, z_2) \right\} \equiv \frac{1}{2} (\pi_1)_* \left\{ (D_1, z_1 \circ \xi) + (C_2, z_2^2) \right\}$$

and $\mathfrak{W}_{\lambda_1,\lambda_2} := \frac{1}{2}(\pi_2)_* \left\{ (D_1, z_1 \circ \xi) + (C_2, z_2^2) \right\}$. Explicit normalization of C_1 (or rather its image $\check{C}_1$ in $\check{\mathcal{K}}_{\lambda_1,\lambda_2}$) shows that $\Im_{\lambda_1,\lambda_2}$ has $\pm$ monodromy in accordance with $\sqrt{\frac{\lambda_1(\lambda_2-1)}{(\lambda_1-1)\lambda_2}}$. Note that this function is constant on the diagonal.

For our test form, we now let[6]

$$\omega_{\mathbb{R},\underline{\lambda}} := \Re \left\{ \frac{dx_1}{y_1} \wedge \overline{\left(\frac{dx_2}{y_2} \right)} \right\} \in A^{1,1}(E_{\lambda_1} \times E_{\lambda_2}).$$

The maps π_i, π_i' are isomorphisms on $H^2_{tr,\mathbb{Q}}$, and we denote also by $[\omega_{\mathbb{R}}]_{\mathcal{K}}$, $[\omega_{\mathbb{R}}]_X$ two more classes, in $H^{1,1}_{tr,\mathbb{R}}$ of $\mathcal{K}_{\lambda_1,\lambda_2}$ resp. $X_{a,b}$, such that all three agree under pullback to $\tilde{\mathcal{K}}_{\lambda_1,\lambda_2}, \check{\mathcal{K}}'_{\lambda_1,\lambda_2}$. Since AJ commutes with pushforward, we have $r(Z_{a,b})\{[\omega_{\mathbb{R}}]_X\} = r(\Im_{\lambda_1,\lambda_2})\{[\omega_{\mathbb{R}}]_{\mathcal{K}}\} = r(\mathfrak{W}_{\lambda_1,\lambda_2})(\omega_{\mathbb{R}}) =$

$$\mathcal{R}(\lambda_1, \lambda_2) := \frac{1}{2} \int_{D_1} (\log|z_1 \circ \xi|) \pi_2^* \iota_{D_1}^* \omega_{\mathbb{R}} = \frac{1}{2} \int_{C_1} \log|z_1| \underbrace{\xi_* \{\iota_{D_1}^* \pi_2^* \omega_{\mathbb{R}}\}}_{\in \mathcal{D}^{1,1}(C_1)}. \tag{6}$$

At this point, it is convenient to specialize to the Picard rank 19 locus $\lambda_1 = \lambda_2 =: \lambda$, along which the collisions of singular fibers do not affect $\rho^{-1}(1)$; this eliminates the monodromy in $\Im$, hence that in $\mathcal{R}$. Writing $z = x + iy = re^{i\phi}$, the computation of (6) carried out in [9] specializes to

$$\mathcal{R}(\lambda) := \mathcal{R}(\lambda, \lambda) = -4|\lambda + 1|\Re \int_{\mathbb{P}^1} z \log \left| \frac{z+i}{z-i} \right| \frac{P_\lambda(z)\overline{Q_\lambda(z)}}{|S_\lambda(z)|} dz \wedge d\bar{z} \tag{7}$$

where

$$\begin{cases} P_\lambda(z) := (\lambda^2 - \lambda - 1)z^4 + 2z^2 + (\lambda^3 - \lambda^2 - 2\lambda + 1) \\ Q_\lambda(z) := (\lambda^3 - \lambda^2 - 2\lambda + 1)z^4 + 2z^2 + (\lambda^2 - \lambda - 1) \\ S_\lambda(z) := (z^2 - \lambda)(1 - \lambda z^2)(z^2 + 1)\left(z^2 - (1 + \lambda - \lambda^2)\right) \\ \qquad\qquad \times \left((1 + \lambda - \lambda^2)z^2 - 1\right)\left(z^4 + (\lambda^3 - 3\lambda)z^2 + 1\right). \end{cases}$$

An analytic argument [op. cit.] is required to show that $\lim_{\lambda \to 1} \mathcal{R}(\lambda, \lambda)$ agrees with[7]

$$\mathcal{R}(1) = -16 \int_{\mathbb{P}^1} \frac{\log \left| \frac{z+i}{z-i} \right| r \sin \phi}{|z^2 + 1||z^2 - 1|^2} dx\,dy < 0, \tag{8}$$

whereupon we have

---

[6] We will usually drop the subscript $\underline{\lambda}$.

[7] Since $\left| \log \left| \frac{z+i}{z-i} \right| \right| < C \left| z^2 - 1 \right|$ for z near ± 1, this clearly converges.

Theorem 2. *For very general* (a, b) *resp.* (λ_1, λ_2), *(the continuation of)* $Z_{a,b}$ *resp.* $\mathfrak{Z}_{\lambda_1,\lambda_2}$, $\mathfrak{W}_{\lambda_1,\lambda_2}$ *is real-regulator indecomposable.*

We now turn to the magnificent properties of the function $\mathcal{R}$. More precisely, writing

$$f(\lambda) := i \int_{E_\lambda} \frac{dx}{y} \wedge \overline{\left(\frac{dx}{y}\right)}, \qquad \eta_\lambda := \frac{\omega_{\mathbb{R},\lambda}}{f(\lambda)} \in A^{1,1}_{\mathbb{R}}(E_\lambda \times E_\lambda),$$

and $\lambda : \Gamma(2) \backslash \mathfrak{H} \stackrel{\cong}{\to} \mathbb{P}^1 \backslash \{0, 1, \infty\}$ for the classical elliptic modular function, we will study

$$\Psi(\tau) := \frac{\mathcal{R}(\lambda(\tau))}{f(\lambda(\tau))} = r\left(\mathfrak{W}_{\lambda(\tau)}\right)\left(\eta_{\lambda(\tau)}\right)$$

for $\tau \in \mathfrak{H}$. As pointed out to the author by C. Doran, some of the general results below have also appeared in A. Mellit's thesis [18]; we expect that a simple exposition of these matters is nevertheless of value.

Denote by $\mathcal{E} \stackrel{\pi}{\to} \mathfrak{H}$ the family of elliptic curves with fibers $\pi^{-1}(\tau) = \mathbb{C}/\mathbb{Z}\langle 1, \tau\rangle$, by $\mathcal{E}^{(2)} \stackrel{\pi^{(2)}}{\to} \mathfrak{H}$ its fiber-product with itself, and by $\mathcal{E}_U$ resp. $\mathcal{E}^{(2)}_U$ the restrictions to an analytic open neighborhood U in any fundamental domain for a congruence subgroup $\Gamma \subset SL_2(\mathbb{Z})$. Let $\mathfrak{Y}$ denote a (complex-) analytic family of $\tilde{K}_1$-cycles on the fibres of $\pi^{(2)}$. We may regard this as an "analytic higher Chow cycle" on $\mathcal{E}^{(2)}_U$— that is, as a formal sum $\sum q_i.(F_i, \mathcal{S}_i)$ of surfaces paired with meromorphic functions F_i on their analytic desingularizations, with sum of divisors $\sum q_i(F_i) = 0$ in $\mathcal{E}^{(2)}_U$. Therefore, $r_{\mathfrak{Y}} := \sum q_i \log|F_i|\delta_{\mathcal{S}_i}$ makes sense as a $(1, 1)$ normal current on $\mathcal{E}^{(2)}_U$, of intersection type with respect to the fibers; likewise for the closed $(2, 1)$ current $\Omega'_{\mathfrak{Y}} := (2\pi i \Omega_{\mathfrak{Y}} =) \sum q_i \frac{dF_i}{F_i} \delta_{\mathcal{S}_i}$.

Write $\tau = X + iY$. Let $z = x + iy$ resp. z_1, z_2 be the usual coordinates (modulo $\mathbb{Z}\langle 1, \tau\rangle$) on fibers of π resp. $\pi^{(2)}$. By abuse of notation, we have

$$\eta_\tau := \frac{dz_1 \wedge d\bar{z}_2 + d\bar{z}_1 \wedge dz_2}{4Y} \qquad (9)$$

which is in general Γ-invariant, and in case $\Gamma = \Gamma(2)$ matches up with the form $\eta_{\lambda(\tau)}$ under the isomorphism $(\pi^{(2)})^{-1}(\tau) \cong (E_{\lambda(\tau)})^{\times 2}$. We shall denote by $\mathcal{H}^k_{\pi^{(2)}}$, $\mathcal{H}^{p,q}_{\pi^{(2)}}$ the C^∞ relative cohomology sheaves on U, and by $\mathcal{L}^\bullet$ the Leray filtration on C^∞ forms $A^k(\mathcal{E}^{(2)}_U)$. Calling $\alpha \in \mathcal{L}^a A^k(\mathcal{E}^{(2)}_U)$ $\pi^{(2)}$-closed if $d\alpha \in \mathcal{L}^{a+1}$, we have natural maps

$$[\quad]^{\{a\}}_U : \mathcal{L}^a A^m(\mathcal{E}^{(2)}_U)_{\pi^{(2)}\text{-cl}} \longrightarrow A^a(U; \mathcal{H}^{m-a}_{\pi^{(2)}})$$

to cohomology-sheaf valued forms.

Lemma 3. *There exists a smooth form* $\tilde{\eta} \in A^{1,1}(\mathcal{E}^{(2)}_U)$ *pulling back to* η_τ *on fibers, and satisfying:*

(i) $[\partial\tilde{\eta}]^{\{1\}}_U \in A^{0,1}(U; \mathcal{H}^{2,0}_{\pi^{(2)}})$;

(ii) $[\bar{\partial}\tilde{\eta}]^{\{1\}}_U \in A^{1,0}(U; \mathcal{H}^{0,2}_{\pi^{(2)}})$; *and*

(iii) $\bar{\partial}\partial\tilde{\eta} = \frac{1}{2Y^2}\tilde{\eta} \wedge d\tau \wedge d\bar{\tau}$.

Proof. Consider the C^∞ uniformization $F : U \times (\mathbb{C}/\mathbb{Z}\langle 1, i\rangle) \xrightarrow{\cong} \mathcal{E}_U$ given by $F(\tau, w) := (\tau, \Re(w) + \tau\Im(w))$. According to the easy pullback computation

$$F^*\left(dz - \frac{y}{\Im(\tau)}d\tau\right) = \Re(dw) + \tau\Im(dw),$$

$\widetilde{dz} := dz - \frac{y}{\Im(\tau)}d\tau \in A^{1,0}(\mathcal{E}_U)$ is smooth and well-defined on $\mathcal{E}_U$ (whereas "dz" is not), while pulling back to dz on fibers. We compute

$$d(\widetilde{dz}) = \frac{d\bar{z} - dz}{\tau - \bar\tau} \wedge d\tau + \frac{z - \bar{z}}{(\tau - \bar\tau)^2}d\tau \wedge d\bar\tau,$$

from which it follows that

$$\partial(\widetilde{dz}) = \frac{\widetilde{dz} \wedge d\tau}{\bar\tau - \tau}, \quad \bar\partial(\widetilde{dz}) = \frac{d\bar{z} \wedge d\tau}{\tau - \bar\tau} + \frac{z - \bar{z}}{(\tau - \bar\tau)^2}d\tau \wedge d\bar\tau = \frac{\overline{\widetilde{dz}} \wedge d\tau}{\tau - \bar\tau},$$

and then (by conjugation) $\partial(\overline{\widetilde{dz}}) = \frac{\overline{\widetilde{dz}} \wedge d\bar\tau}{\bar\tau - \tau}$.

Swtiching to $\mathcal{E}_U^{(2)}$, since

$$\tilde\eta := \frac{i}{2}\frac{\widetilde{dz_1} \wedge \overline{\widetilde{dz_2}} + \overline{\widetilde{dz_1}} \wedge \widetilde{dz_2}}{\tau - \bar\tau}$$

pulls back to η_τ on fibers, it is vertically closed. More concretely, we easily compute

$$\partial\tilde\eta = i\frac{\widetilde{dz_1} \wedge \widetilde{dz_2}}{(\tau - \bar\tau)^2} \wedge d\bar\tau, \quad \bar\partial\tilde\eta = -i\frac{\overline{\widetilde{dz_1}} \wedge \overline{\widetilde{dz_2}}}{(\tau - \bar\tau)^2} \wedge d\tau,$$

which gives *(i)–(ii)*. At this point, *(iii)* is easy and left to the reader.

The next few Lemmas deduce properties of the function

$$\Upsilon(\tau) := \mathcal{R}(\mathfrak{Y}_\tau)(\eta_\tau) = \pi_*^{(2)}(r_\mathfrak{Y} \wedge \tilde\eta)$$

on U; of course, we have the case $\mathfrak{Y} = \mathfrak{W}|_U$ and $\Upsilon = \Psi|_U$ (and $\Gamma = \Gamma(2)$) in mind.

Lemma 4. *We have $\Delta_{hyp}\Upsilon = -2\Upsilon$, where*

$$\Delta_{hyp} := -Y^2\Delta = -4Y^2\frac{\partial}{\partial\bar\tau}\frac{\partial}{\partial\tau}$$

is the hyperbolic Laplacian.

Proof. We shall use the fact that the pairing

$$\pi_*^{(2)} : \mathcal{D}^{1,1}(\mathcal{E}_U^{(2)})_{\pi^{(2)}\text{-cl}} \otimes \mathcal{L}^1 A^{1,2}(\mathcal{E}_U^{(2)})_{\pi^{(2)}\text{-cl}} \longrightarrow \mathcal{D}^{0,1}(U)$$

factors, via $[\ ]_U^{\{0\}} \otimes [\ ]_U^{\{1\}}$, through $\mathcal{D}^0(U; \mathcal{H}_{\pi^{(2)}}^{1,1}) \otimes A^{0,1}(U; \mathcal{H}_{\pi^{(2)}}^{1,1})$. In particular, any components of type $A^{1,0}(U; \mathcal{H}_{\pi^{(2)}}^{0,2})$ in the right-hand factor are killed. (A similar observation applies to $\mathcal{L}^1 A^{2,1}(\mathcal{E}_U^{(2)})_{\pi^{(2)}\text{-cl}}$.) Moreover, $r_{\mathfrak{Y}}$ belongs to the left-hand factor, with $d[r_{\mathfrak{Y}}] = \frac{1}{2}\Omega_{\mathfrak{Y}}' + \frac{1}{2}\overline{\Omega_{\mathfrak{Y}}'}$.

From Lemma 3(ii), we have

$$\bar{\partial}\Upsilon = \bar{\partial}\pi_*^{(2)}(r_{\mathfrak{Y}} \wedge \tilde{\eta}) = \pi_*^{(2)}\left(\frac{1}{2}\overline{\Omega_{\mathfrak{Y}}'} \wedge \tilde{\eta}\right) + \pi_*^{(2)}\left(r_{\mathfrak{Y}} \wedge \bar{\partial}\tilde{\eta}\right) = \frac{1}{2}\pi_*^{(2)}\left(\overline{\Omega_{\mathfrak{Y}}'} \wedge \tilde{\eta}\right).$$

Since $\partial[\overline{\Omega_{\mathfrak{Y}}'}] = 0$,

$$\partial\bar{\partial}\Upsilon = -\frac{1}{2}\pi_*^{(2)}\left(\overline{\Omega_{\mathfrak{Y}}'} \wedge \partial\tilde{\eta}\right) = \pi_*^{(2)}\left(r_{\mathfrak{Y}} \wedge \bar{\partial}\partial\tilde{\eta}\right) - \pi_*^{(2)}\left(\bar{\partial}[r_{\mathfrak{Y}} \wedge \partial\tilde{\eta}]\right)$$

which by Lemma 3(i),(iii)

$$= \pi_*^{(2)}\left(r_{\mathfrak{Y}} \wedge \bar{\partial}\partial\tilde{\eta}\right) = \frac{1}{2Y^2}\pi_*^{(2)}\left(r_{\mathfrak{Y}} \wedge \tilde{\eta}\right)d\tau \wedge d\bar{\tau} = \frac{\Upsilon}{2Y^2}d\tau \wedge d\bar{\tau}.$$

Lemma 5. *Let* $\tau_0 \in U$ *be a CM point (i.e. quadratic irrationality), so that* $\pi^{-1}(\tau_0)$ *is a CM elliptic curve. Assume that* $\mathfrak{Y}_{\tau_0}$ *is defined over* $\bar{\mathbb{Q}}$. *Then* $\Upsilon(\tau_0)$ *is of the form* $\sum \bar{\mathbb{Q}}\log\bar{\mathbb{Q}}$ *(i.e., is a sum of algebraic multiples of logarithms of algebraic numbers).*

Proof. On $(\pi^{(2)})^{-1}(\tau_0) =: E_0 \times E_0$, write $\mathfrak{D}_1 = E_0 \times \{0\}$, $\mathfrak{D}_2 = \{0\} \times E_0$, $\mathfrak{D}_3 = \Delta_{E_0}$, and $\mathfrak{D}_4$ for the graph of multiplication by τ_0. The presence of $\mathfrak{D}_4$ makes $H^{1,1}(E_0 \times E_0)$, and thus $[\eta_{\tau_0}]$, algebraic. In fact, a simple computation shows that

$$\eta_{\tau_0} \equiv \alpha_1 \delta_{\mathfrak{D}_1} + \alpha_2 \delta_{\mathfrak{D}_2} + \alpha_3 \delta_{\mathfrak{D}_3} + \alpha_4 \delta_{\mathfrak{D}_4}$$

in $H^{1,1}$, with $\alpha_1 := \frac{1-X_0}{2Y_0}$, $\alpha_2 := \frac{|\tau_0|^2 - X_0}{2Y_0}$, $\alpha_3 := \frac{X_0}{2Y_0}$, $\alpha_4 := -\frac{1}{2}$ (all obviously in $\mathbb{Q}(\Im(\tau_0))$). We may assume (by Bloch's moving lemma [6]) that $\mathfrak{Y}_{\tau_0} = \sum_i(g_i, D_i)$ with D_i and g_i $\bar{\mathbb{Q}}$-rational, and such that D_i intersects $\mathfrak{D}_j$ properly (with multiplicities all 1) away from $|(g_i)|$. This yields immediately $\Upsilon(\tau_0) =$

$$r(\mathfrak{Y}_{\tau_0})(\eta_{\tau_0}) = \sum_i \sum_{j=1}^4 \alpha_j \int_{D_i} \log|g_i|\delta_{\mathfrak{D}_j}$$

$$= \sum_{i,j} \sum_{p \in D_i \cap \mathfrak{D}_j} \alpha_j \log|g_i(p)|,$$

with $g_i(p) \in \bar{\mathbb{Q}}$.

Next, we allow $\mathfrak{Y}$ to fail to be a cycle over a point $\hat{\tau} \in U$; that is, suppose that the 1-cycle $\mathcal{C} := \sum q_i(F_i)$ is supported on $\left(\pi^{(2)}\right)^{-1}(\hat{\tau})$. We say $\mathfrak{Y}$ is singular at $\hat{\tau}$.

Lemma 6. *If* $\hat{\tau}$ *is a CM point, then either (i)* Υ *is smooth at* $\hat{\tau}$ *or (ii)* $\Upsilon \sim c\log|\tau - \hat{\tau}|$ *as* $\tau \to \hat{\tau}$, *for some* $c \in \mathbb{Q}(\Im(\hat{\tau}))^*$. *If* $\hat{\tau}$ *is not a CM point, then the singularity is apparent; that is,* Υ *remains smooth at* $\hat{\tau}$.

Proof. Since $\sum q_i \int_{S_{i,\tau}} |\eta_\tau|$ is bounded by a constant and $F_i|_\tau$ depends algebraically on τ, $\Upsilon(\tau) = \sum q_i \int_{S_{i,\tau}} (\log |F_i|_\tau|)\eta_\tau$ (resp. $\frac{\partial \Upsilon}{\partial \bar{\tau}}$) is bounded by a multiple of $\log |\tau - \hat{\tau}|$ (resp. $\frac{1}{\tau - \bar{\tau}}$). As in the proof of Lemma 4, we have $\bar{\partial}\Upsilon = \frac{1}{2}\pi_*^{(2)}\left(\overline{\Omega_{\mathfrak{Y}}'} \wedge \tilde{\eta}\right)$, but $\partial[\overline{\Omega_{\mathfrak{Y}}}] = -2\pi i \delta_{\mathcal{C}}$ instead of 0. Hence

$$\partial\bar{\partial}\Upsilon = -\frac{1}{2}\pi_*^{(2)}\left(\overline{\Omega_{\mathfrak{Y}}'} \wedge \partial\tilde{\eta}\right) - \frac{2\pi i}{2}\pi_*^{(2)}\left(\delta_C \wedge \tilde{\eta}\right)$$

$$= \frac{\Upsilon}{2Y^2}d\tau \wedge d\bar{\tau} - \pi i c \delta_{\{\hat{\tau}\}},$$

where $c := \int_{\mathcal{C}} \eta_{\hat{\tau}}$ belongs to $\mathbb{Q}(\Im(\hat{\tau}))$ by the proof of Lemma 5 if $\hat{\tau}$ is CM. From $\Delta_{\mathrm{hyp}}(\Upsilon - c \log |\tau - \hat{\tau}|) = -2\Upsilon$ it now follows that $\frac{\Upsilon - c \log |\tau - \hat{\tau}|}{\log |\tau - \hat{\tau}|} \to 0$.

If $\hat{\tau}$ is not CM, or if $c = 0$ above, then $[\mathcal{C}]$ extends to a section of $\mathcal{H}^{1,1}_{\pi^{(2)},\mathrm{alg}}$; indeed, $\mathcal{C} = \mathcal{M} \cdot (\pi^{(2)})^{-1}(\hat{\tau})$ for some surface $\mathcal{M} = \sum q_i' \mathcal{M}_i \subset \mathcal{E}_U^{(2)}$. Subtracting the decomposable cycle $\sum q_i'.(\tau - \hat{\tau}, \mathcal{M}_i)$ from $\mathfrak{Y}$ (and applying Bloch's moving lemma to make it properly intersect the fiber $(\pi^{(2)})^{-1}(\hat{\tau})$) removes the singularity without affecting Υ.

Finally, let $\bar{\mathcal{E}}_\Gamma^{(2)} \xrightarrow{\bar{\pi}} X(\Gamma)$ be Shokurov's smooth compactification [23] of the Kuga modular variety $\Gamma \backslash \mathcal{E}^{(2)} \to Y(\Gamma)$. (In case $\Gamma = \Gamma(2)$, it has fibers $E_\lambda \times E_\lambda$ for $\lambda \in Y(\Gamma)$.) Consider a (higher Chow) precycle $\mathfrak{Y}_\Gamma = \sum q_i.(F_i, S_i)$ on $\bar{\mathcal{E}}_\Gamma^{(2)}$, with "boundary" $\sum q_i(F_i)$ supported on $\bar{\pi}^{-1}(\Xi)$ for some finite set $\Xi \subset X(\Gamma)$. Let η_Γ be the (nonholomorphic, real) section of the logarithmically extended Hodge bundle $\mathcal{H}^{1,1}_{\bar{\pi},e} \to X(\Gamma)$ provided by (9). Asymptotics of $\Upsilon_\Gamma(x) := r(\mathfrak{Y}_{\Gamma,x})(\eta_{\Gamma,x})$ at points in $\Xi \cap Y(\Gamma)$ are clear from Lemma 6, so let $\kappa \in X(\Gamma) \backslash Y(\Gamma)$ be a cusp (with local holomorphic coordinate q).

Lemma 7. *(a) $|\Upsilon_\Gamma|$ is bounded by a constant near κ. (b) If $\kappa \notin \Xi$, then this bound is improved to a constant multiple of $\frac{1}{\log |q|}$.*

Proof. Assume for simplicity κ is unipotent, so that $\bar{\mathcal{E}}_\Gamma^{(1)}$ has a Néron N-gon over $q = 0$. Writing ω_q for a local generator of the extended relative canonical sheaf, $\log |q| \eta_{\Gamma,q} = \Re(\omega_{q,1} \wedge \bar{\omega}_{q,2})$ limits to a nonzero homology class on $\bar{\pi}^{-1}(\kappa)$. If $\kappa \notin \Xi$, then $r(\mathfrak{Y}_\Gamma)$ restricts to a cohomology class on $\bar{\pi}^{-1}(\kappa)$, which pairs with the former to give a (finite) number. Dividing by $\log |q|$ gives (b). For (a), the beginning of the proof of Lemma 6 shows that when $x \in \Xi$, the bound is worse than that in (b) by a factor of $\log |q|$.

For our purposes, a *higher Green's function* $G(\tau)$ of weight $2k$ and level Γ on $\mathfrak{H}$ will be defined by the following properties:

- G is smooth and real-valued on $\mathfrak{H}^\circ := \mathfrak{H} \backslash \{\Gamma.\hat{\tau}\}$ for some $\hat{\tau} \in \mathfrak{H}$.
- G is Γ-invariant.
- $\Delta_{\mathrm{hyp}}G = k(1 - k)G$ (on $\mathfrak{H}^\circ$).
- G tends to zero at all cusps.
- $G(\tau) \sim c \log |\tau - \hat{\tau}|$ (as $\tau \to \hat{\tau}$) for some $c \in \overline{\mathbb{Q}}^*$.

Uniqueness is clear given $c, k, \hat{\tau}, \Gamma$: the difference of two distinct such functions would be a Maass form with eigenvalue -2, which is impossible since Δ_{hyp} is a positive definite operator. Existence is explained in [18].

Under the conditions that $\hat{\tau}$ is CM and $S_{2k}(\Gamma) = \{0\}$, Gross and Zagier [14] conjectured (roughly) that

$$\text{for any CM point } \tau_0, \ G(\tau_0) \text{ is of the form } \sum \overline{\mathbb{Q}} \log \overline{\mathbb{Q}}. \tag{10}$$

(Clearly, its validity is independent of $c \in \overline{\mathbb{Q}}^*$.) Mellit was able to prove this for the case $k = 2$, $\hat{\tau} = i$, $\Gamma = PSL_2(\mathbb{Z})$ using the above ideas together with an explicit family of cycles. Noting that $S_4(\Gamma(2)) = \{0\}$, $\lambda(1 + i) = -1$ and $\lambda(\frac{1+i}{2}) = 2$, our cycle leads to another case:

Theorem 3. Ψ is a $\overline{\mathbb{Q}}$-linear combination of two higher Green's function of weight 4 and level $\Gamma(2)$ with $\hat{\tau} = 1 + i$ and $\frac{1+i}{2}$; moreover, it verifies conjecture (10).

Proof. This follows from Lemmas 4–7, once we verify that $\mathfrak{W}$ extends to a cycle on $\bar{\mathcal{E}}_{\Gamma}^{(2)} \backslash \bar{\pi}^{-1}(\{-1, 2\}) \xrightarrow{\bar{\pi}} X(\Gamma) \backslash \{-1, 2\}$.[8] Equivalently, we may check this for $\mathfrak{Z}$ on the (1-parameter) Kummer family, for which the following analysis on the singular model $\check{\mathcal{K}}_{\lambda,\lambda}$ will suffice. Referring to (6) and (7), the function on the nodal rational curve $\check{C}_1 \subset \check{\mathcal{K}}_{\lambda,\lambda} \subset \mathbb{P}^3$ whose zero and pole cancel at the node is $z_1 = \frac{z+i}{z-i}$. By a computation in [9], $\check{C}_1$ is a double cover of a rational curve with parameter

$$u = \frac{1 - \lambda z^2}{z^2 - \lambda}, \tag{11}$$

in terms of which its equation is

$$U^2(u^2 + \lambda u + 1)^2 = V^2(u + 1)^2(u + \lambda)(\lambda u + 1). \tag{12}$$

We need to determine the values of λ for which (12) acquires a component where $z_1 \equiv 0$ or ∞ ($\Leftrightarrow z^2 \equiv -1$), i.e. where $\mathfrak{Z} := (z_1, \check{C}_1)$ has boundary.

Inverting the parameter (11) to $z^2 = \frac{1+\lambda u}{u+\lambda}$, this happens when $\lambda = -1$, and also if (12) has a component with $u \equiv -1$, which occurs when $\lambda = 2$. (In spite of singular fiber collisions or degeneration of the K3 at $\lambda = 0, 1, \infty$, the cycle extends.) It follows that $\mathfrak{Z}$ has boundary only at $\lambda = -1, 2$. In fact, z_1 limits to both 0 and ∞ on components of $\check{C}_1$ in each case, and so the boundary cannot be corrected by adding a decomposable cycle.

So for example this shows that $\Psi(\zeta_6) = \mathcal{R}(\zeta_6)/f(\zeta_6)$ (made quite explicit by (7)) satisfies (10). More generally, one might optimistically view (10) as predicting (for each $k, \hat{\tau}, \Gamma$ as above) the existence of a family of indecomposable K_1-classes. The moral of this story is perhaps that (generalized) algebraic cycles are far more

---

[8] It is possible, but tedious, to instead check the asymptotics for Ψ at $0, \pm 1, 2, \infty$ directly from the formula (7), cf. the appendix to Sect. 6 of [9].

ubiquitous, and useful, than the Hodge or Bloch–Beilinson conjectures would suggest on their own.

Remark 4. In fact, we can say precisely what the linear combination in Theorem 3 is. A computation by A. Clingher (private communication) shows that there is a rational involution of the family $\{\mathcal{K}_{\lambda,\lambda}\}_{\lambda \in \mathbb{P}^1}$ over $\lambda \mapsto 1 - \lambda$ sending the alternate fibration $\rho \mapsto \frac{1-2\lambda+\lambda^2\rho}{(1-\lambda)^2}$ (hence $\rho^{-1}(1) \to \rho^{-1}(1)$) and restricting to the identity on $\mathcal{K}_{\frac{1}{2},\frac{1}{2}}$. Since the cycle family $\{\mathfrak{Z}_{\lambda,\lambda}\}$ is preserved by this involution, Ψ is invariant under $\lambda \leftrightarrow 1 - \lambda$, and so we get that the $\bar{\mathbb{Q}}$-coefficients of $\log|\lambda + 1|$ and $\log|\lambda - 2|$ in Ψ are equal.

5 Proof of the Tauberian Lemma 2

Though we could not find this result in the literature, what follows makes substantial use of ideas from [16]. We will give a fairly detailed proof, since those working in cycles may not be familiar with this part of complex analysis. We retain the notation of Sect. 3, with α fixed throughout.

Since F, G, F_0, G_0 are holomorphic on $\mathcal{S}$, they are uniformly continuous there, hence also in

$$S := \{(t, t_0) \in \mathcal{S} \mid |t_0| = \alpha, \, |t| \leq \alpha\}.$$

It is clear from Sect. 3 that $\tilde{v}$ is (uniformly) continuous on S. Defining also

$$S_0 := \left\{(t, t_0) \in \mathcal{S} \,\middle|\, |t_0| = \alpha, \, \frac{t}{t_0} \in [0, 1]\right\},$$

we have $S_0 \subset S \subset \mathcal{S}$.

We work first on S_0, writing $t = \beta t_0$ ($\beta \in [0, 1]$) with $t_0 = \alpha e^{i\lambda_0}$. Set

$$V(\beta, \lambda_0) := \tilde{v}\left(\beta \alpha e^{i\lambda_0}, \alpha e^{i\lambda_0}\right),$$

$a_n(\lambda_0) := \mathfrak{A}_n(\alpha e^{i\lambda_0}) \alpha^n e^{in\lambda_0}$, and $s_N(\lambda_0) := \sum_{n=0}^N a_n(\lambda_0)$, so that $V(\beta, \lambda_0) = \sum_{n=0}^\infty a_n(\lambda_0)\beta^n$ for $\beta \in [0, 1)$. We will show that as $N \to \infty$,

$$s_N(\lambda_0) \text{ converges uniformly to } V(1, \lambda_0) \tag{13}$$

in λ_0. On $\{|t| = \alpha\}$, $\tilde{v}(t) = V(1, \lambda_0)$ and $\sum_{n=0}^N \mathfrak{A}_n(t)t^n = s_N(\lambda_0)$, so (13) is equivalent to Lemma 2.

The first step is to break this problem into three pieces ((*i*)–(*iii*) below). Using[9] $1 - \beta^n \leq n(1 - \beta)$ for $\beta \in [0, 1]$,

---

[9] To see this, examine the function $(n - 1) - n\beta + \beta^n$.

$$|s_N(\lambda_0) - V(\beta, \lambda_0)| = \left| \sum_{n=1}^{N} a_n(\lambda_0)(1 - \beta^n) - \sum_{n=N+1}^{\infty} a_n(\lambda_0)\beta^n \right|$$

$$\leq \sum_{n=1}^{N} n(1 - \beta) |a_n(\lambda_0)| + \frac{1}{N} \sum_{n=N+1}^{\infty} n |a_n(\lambda_0)| \beta^n$$

$$\leq (1 - \beta) \sum_{n=1}^{N} |na_n(\lambda_0)| + \frac{1}{N(1 - \beta)} \sup_{n>N} |na_n(\lambda_0)|,$$

and so

$$\left| s_N(\lambda_0) - V\left(1 - \frac{1}{N}, \lambda_0\right) \right| \leq \frac{1}{N} \sum_{n=1}^{N} |na_n(\lambda_0)| + \sup_{n>N} |na_n(\lambda_0)|.$$

Noting $V(1, \lambda_0) = \tilde{v}\left(\alpha e^{i\lambda_0}, \alpha e^{i\lambda_0}\right)$, we therefore have the bound

$$|s_N(\lambda_0) - \tilde{v}(t_0, t_0)|$$

$$\leq \left| s_N(\lambda_0) - V\left(1 - \frac{1}{N}, \lambda_0\right) \right| + \left| V\left(1 - \frac{1}{N}, \lambda_0\right) - V(1 - \lambda_0) \right|$$

$$\leq \underbrace{\frac{1}{N} \sum_{n=1}^{N} |na_n(\lambda_0)|}_{(i)} + \underbrace{\sup_{n>N} |na_n(\lambda_0)|}_{(ii)} + \underbrace{\left| V\left(1 - \frac{1}{N}, \lambda_0\right) - V(1, \lambda_0) \right|}_{(iii)}. \qquad (14)$$

To prove (13), we need to bound (i)–(iii) uniformly in λ_0 (by taking N sufficiently large). In fact, (iii) is obvious by uniform continuity of V on S_0, and so we turn to (ii).

Now $\delta_t \tilde{v}(t, t_0) = \sum_{n=0}^{\infty} n\mathfrak{A}_n(t_0)t^n$ in $\mathcal{S}$ for $|t| < |t_0|$; moreover, for *fixed* t_0 with $|t_0| = \alpha$, the function $\delta_t \tilde{v}(t, t_0)$ on $\{|t| = \alpha\}$ is both L^1 and L^2 (as log, $\log^2$ are integrable). Working on S, with $t = \gamma t_0 = \beta t_0 e^{i\lambda} = \beta \alpha e^{i(\lambda + \lambda_0)}$ ($|\gamma| \leq 1$), we now define

$$V_t(\gamma, \lambda_0) := (\delta_t \tilde{v})(\gamma \alpha e^{i\lambda_0}, \alpha e^{i\lambda_0}),$$

which for $|\gamma| < 1$

$$= \sum_{n=0}^{\infty} na_n(\lambda_0)\gamma^n.$$

By the Cauchy integral formula, for $0 < |\beta| < 1$,

$$n\mathfrak{A}_n(t_0) = \frac{1}{2\pi i} \int_{|t|=\alpha\beta} \frac{(\delta_t \tilde{v})(t, t_0)}{t^{n+1}} dt$$

$$= \frac{1}{2\pi i} \int_{-\pi}^{\pi} \frac{(\delta_t \tilde{v})(\beta t_0 e^{i\lambda}, t_0)}{(\beta \alpha e^{i(\lambda + \lambda_0)})^{n+1}} i\beta \alpha e^{i(\lambda + \lambda_0)} d\lambda$$

$$= \frac{1}{2\pi \alpha^n \beta^n e^{in\lambda_0}} \int_{-\pi}^{\pi} V_t\left(\beta e^{i\lambda}, \lambda_0\right) e^{-in\lambda} d\lambda. \qquad (15)$$

For fixed (small) $\epsilon > 0$, the reader will readily verify that

$$\lim_{\beta \to 1^-} \int_{-\epsilon}^{\epsilon} \left| \log(e^{i\lambda} - 1) - \log(\beta e^{i\lambda} - 1) \right| d\lambda = 0.$$

In conjunction with (3), and the uniform continuity of $V_t(\gamma, \lambda_0)$ (in γ) on $\{|\gamma| \le 1\}\backslash\{\arg(\gamma) \in (-\epsilon, \epsilon)\}$, this implies

$$\lim_{\beta \to 1^-} \int_{-\pi}^{\pi} \left| V_t(e^{i\lambda}, \lambda_0) - V_\lambda(\beta e^{i\lambda}, \lambda_0) \right| d\lambda = 0.$$

Therefore, taking the limit of (15) as $\beta \to 1$, we obtain

$$na_n(\lambda_0) = n\mathfrak{A}_n(t_0)\alpha^n e^{in\lambda_0} = \frac{1}{2\pi} \int_{-\pi}^{\pi} V_t\left(e^{i\lambda}, \lambda_0\right) e^{-in\lambda} d\lambda \tag{16}$$

and then

$$na_n(\lambda_0 + \delta) - na_n(\lambda_0)$$

$$= \frac{1}{2\pi} \int_{-\pi}^{\pi} \left\{ V_t\left(e^{i\lambda}, \lambda_0 + \delta\right) - V_t\left(e^{i\lambda}, \lambda_0\right) \right\} e^{-in\lambda} d\lambda$$

$$= \frac{1}{2\pi} \int_{-\pi}^{\pi} \left\{ \begin{array}{l} (\delta_t\tilde{v})(e^{i\lambda} \cdot \alpha e^{i(\lambda_0+\delta)}, \alpha e^{i(\lambda_0+\delta)}) \\ -(\delta_t\tilde{v})\left(e^{i\lambda} \cdot \alpha e^{i\lambda_0}, \alpha e^{i\lambda_0}\right) \end{array} \right\} e^{-in\lambda} d\lambda$$

$$= \frac{1}{2\pi} \int_{-\pi}^{\pi} \left\{ \begin{array}{l} \log(e^{i\lambda} - 1)\left[F\left(\alpha e^{i(\lambda_0+\lambda+\delta)}, \alpha e^{i(\lambda_0+\delta)}\right) - F\left(\alpha e^{i(\lambda_0+\lambda)}, \alpha e^{i\lambda_0}\right) \right] \\ + \left[G\left(\alpha e^{i(\lambda_0+\lambda+\delta)}, \alpha e^{i(\lambda_0+\delta)}\right) - G\left(\alpha e^{i(\lambda_0+\lambda)}, \alpha e^{i\lambda_0}\right) \right] \end{array} \right\}$$
$$\times\, e^{-in\lambda} d\lambda.$$

By uniform continuity of F and G, the differences in square brackets can be bounded $< \varepsilon$ by taking δ sufficiently small. Together with L^1 integrability of $\log(e^{i\lambda} - 1)$, this gives (uniform) continuity of $a_n(\lambda_0)$. Similar reasoning shows that $\int_{-\pi}^{\pi} \left| V_t(e^{i\lambda}, \lambda_0) \right|^2 d\lambda$ is (uniformly) continuous in λ_0.

As $V_t(e^{i\lambda}, \lambda_0)$ is L^2, Parseval gives

$$\sum_{n=0}^{\infty} |na_n(\lambda_0)|^2 = \frac{1}{2\pi} \int_{-\pi}^{\pi} \left| V_t(e^{i\lambda}, \lambda_0) \right|^2 d\lambda.$$

The right-hand side minus the Nth partial sums of the left yields a decreasing sequence of continuous, non-negative functions limiting to 0 pointwise. A standard argument using compactness of the circle shows this limit must be uniform. This proves that

$$|na_n(\lambda_0)| \to 0 \text{ uniformly in } \lambda_0,$$

which takes care of (14)(ii).

To treat $(14)(i)$, let $\epsilon > 0$ be given, and let $N \in \mathbb{N}$ be such that $n \geq N \implies$ $|na_n(\lambda_0)| < \frac{\epsilon}{2}$ $(\forall \lambda_0)$. For all $n \leq N$ (and hence for all n), there exists $M \in \mathbb{N}$ such that $|na_n(\lambda_0)| \leq M$. Now, taking $m \geq \frac{2NM}{\epsilon}$, we have

$$\frac{1}{m} \sum_{n=0}^{m} |na_n(\lambda_0)| \leq \frac{\epsilon}{2} \cdot \frac{1}{NM} \sum_{n=0}^{N} |na_n(\lambda_0)| + \frac{1}{m} \sum_{n=N+1}^{m} |na_n(\lambda_0)|$$
$$< \frac{\epsilon}{2} + \frac{\epsilon}{2} = \epsilon,$$

uniformly in λ_0, which completes the proof.

References

1. M. Asakura, *On dlog image of K_2 of elliptic surface minus singular fibers*, Preprint, Available at arXiv:math/0511190v4
2. A. Avram, M. Kreuzer, M. Mendelberg, H. Skarke, Searching for $K3$ fibrations. Nucl. Phys. **B494**, 567–589 (1997)
3. A. Beilinson, Higher regulators and values of L-functions. J. Sov. Math. **30**, 2036–2070 (1985)
4. A. Beilinson, Higher regulators of modular curves, in *Applications of Algebraic K-Theory to Algebraic Geometry and Number Theory*, Boulder, CO, 1983. Contemporary Mathematics, vol. 55 (AMS, Providence, 1986), pp. 1–34
5. F. Beukers, C. Peters, A family of $K3$ surfaces and $\zeta(3)$. J. Reine Angew. Math. **351**, 42–54 (1984)
6. S. Bloch, The moving lemma for higher Chow groups. J. Algebr. Geom. **3**(3), 537–568 (1993)
7. P. Brosnan, G. Pearlstein, C. Schnell, The locus of Hodge classes in an admissible variation of mixed Hodge structure. C. R. Math. Acad. Sci. Paris **348**(11–12), 657–660 (2010)
8. X. Chen, J. Lewis, The Hodge-D-conjecture for K3 and Abelian surfaces. J. Algebr. Geom. **14**(2), 213–240 (2005)
9. X. Chen, C. Doran, M. Kerr, J. Lewis, Normal functions, Picard-Fuchs equations, and elliptic fibrations on $K3$ surfaces, Preprint, Available at arXiv:1108.2223v2
10. A. Clingher, C. Doran, Modular invariants for lattice polarized $K3$ surfaces. Mich. Math. J. **55**(2), 355–393 (2007)
11. A. Collino, Griffiths's infinitesimal invariant and higher K-theory on hyperelliptic Jacobians. J. Algebr. Geom. **6**(3), 393–415 (1997)
12. P. del Angel, S. Müller-Stach, Differential equations associated to families of algebraic cycles. Ann. l'institut Fourier **58**(6), 2075–2085 (2008)
13. C. Doran, M. Kerr, Algebraic K-theory of toric hypersurfaces. CNTP **5**(2), 397–600 (2011)
14. B. Gross, D. Zagier, Heegner points and derivatives of L-series. Invent. Math. **84**(2), 225–320 (1986)
15. M. Kerr, A regulator formula for Milnor K-groups. K-Theory **29**(3), 175–210 (2003)
16. J. Korevaar, Tauberian theory: a century of developments (Springer, Berlin, 2004)
17. J. Lewis, Transcendental Methods in the Study of Algebraic Cycles with a Special Emphasis on Calabi-Yau Varieties, in *Arithmetic and Geometry of K3 surfaces and Calabi-Yau three-folds* (this volume), (Springer, 2013)
18. A. Mellit, Higher Green's functions for modular forms. The University of Bonn, Ph.D. Thesis, 2008, Available at arXiv:0804.3184
19. J. Neukirch, The Beilinson conjecture for algebraic number fields, in *Beilinson's Conjectures on Special Values of L-Functions*, ed. by Rapoport et al. (Academic, New York, 1988)
20. C. Peters, Monodromy and Picard-Fuchs equations for families of $K3$ surfaces and elliptic curves. Ann. ENS **19**, 583–607 (1986)

21. C. Peters, J. Stienstra, A pencil of $K3$ surfaces related to Apéry's recurrence for $\zeta(3)$ and Fermi surfaces for potential zero, in *Arithmetic of Complex Manifolds*, Erlangen, 1988. Springer Lecture Notes in Mathematics, vol. 1399 (1989), pp. 110–127
22. F. Rohsiepe, *Fibration structure in toric Calabi-Yau fourfolds* [aXiv:hep-th/0502138]
23. S. Shokurov, Holomorphic forms of highest degree on Kuga's modular varieties (Russian) Mat. Sb. (N.S.) **101** (143), (1), 131–157, 160 (1976)
24. J. Steenbrink, S. Zucker, Variation of mixed Hodge structure, I. Invent. Math. **80**, 489–542 (1985)

A Note About Special Cycles on Moduli Spaces of K3 Surfaces

Stephen Kudla

Abstract We describe the application of the results of Kudla–Millson on the modularity of generating series for cohomology classes of special cycles to the case of lattice polarized K3 surfaces. In this case, the special cycles can be interpreted as higher Noether–Lefschetz loci. These generating series can be paired with the cohomology classes of complete subvarieties of the moduli space to give classical Siegel modular forms with higher Noether–Lefschetz numbers as Fourier coefficients. Examples of such complete families associated to quadratic spaces over totally real number fields are constructed.

Key words: K3 surface, Shimura varieties, Modular forms

Mathematics Subject Classifications: Primary 14J28; Secondary 11F41, 11F46

1 Introduction

This article contains a short survey of some results about special cycles on certain Shimura varieties that occur as moduli spaces of lattice polarized K3 surfaces. The two points that may be of interest are the Siegel modular forms arising as generating series for higher Noether–Lefschetz numbers and the description of some complete subvarieties in certain of these moduli spaces. The families of K3 surfaces parametrized by these subvarieties ought to have particularly nice properties and I do not know to what extent they already occur implicitly or explicitly in the literature. Finally, I ask the indulgence of the experts in this area for my very naive treatment of things that may be very well known to them.

S. Kudla (✉)
Department of Mathematics, University of Toronto, Toronto, ON M5S 2E4, Canada,
e-mail: skudla@math.toronto.edu

R. Laza et al. (eds.), *Arithmetic and Geometry of K3 Surfaces and Calabi–Yau Threefolds*,   411
Fields Institute Communications 67, DOI 10.1007/978-1-4614-6403-7_14,
© Springer Science+Business Media New York 2013

2 Special Cycles for Orthogonal Groups

We begin by reviewing a very special case of old joint work with John Millson
[13–15].

2.1 Arithmetic Quotients

Let L, $(\ ,\ )$ be a lattice with a $\mathbb{Z}$-valued symmetric bilinear form of signature $(2, n)$.
In particular, for $V = L_{\mathbb{Q}} = L \otimes_{\mathbb{Z}} \mathbb{Q}$, the dual lattice

$$L^{\vee} = \{\, x \in V \mid (x, L) \subset \mathbb{Z} \,\}$$

contains L. Let

$$D = D(L) = \{\, w \in V_{\mathbb{C}} \mid (w, w) = 0,\ (w, \bar{w}) > 0 \,\}/\mathbb{C}^{\times}$$
$$\simeq \{\text{oriented positive 2-planes in } V_{\mathbb{R}}\}$$
$$\simeq SO(V_{\mathbb{R}})/K$$

be the associated symmetric space, where $K \simeq SO(2) \times SO(n)$ is the stabilizer of
an oriented positive 2-plane. Here $V_{\mathbb{C}} = V \otimes_{\mathbb{Q}} \mathbb{C}$ (resp. $V_{\mathbb{R}} = V \otimes_{\mathbb{Q}} \mathbb{R}$), and we
sometimes write $\underline{w}$ for the image of $w \in V_{\mathbb{C}}$ in D. Let $\Gamma_L \subset SO(V)$ be the isometry
group of L and let $\Gamma \subset \Gamma_L$ be a subgroup of finite index. Then

$$M_{\Gamma} = \Gamma \backslash D(L), \qquad \dim_{\mathbb{C}} M_{\Gamma} = n$$

is (isomorphic to) a quasi-projective variety. This variety is a connected component
of a Shimura variety[1] and has a model defined over a cyclotomic field. For small
values of n, this space can be the moduli space of polarized K3 surfaces and, for
smaller values, of abelian varieties.

2.2 Special Cycles

To define special cycles, suppose that $x \in L$ is a vector with $(x, x) < 0$, and let

$$D_x = \{\underline{w} \in D \mid (x, \underline{w}) = 0\} = D(L \cap x^{\perp}).$$

Thus D_x has codimension 1 in D and gives rise to a divisor

$$Z(x) : \Gamma_x \backslash D_x \longrightarrow \Gamma \backslash D = M_{\Gamma},$$

---

[1] For $n \leq 2$, we need to add the condition that Γ be a congruence subgroup; this is automatic for
$n \geq 3$. In general, M_{Γ} can have two components, since $D(L)$ does.

in M_Γ, where Γ_x is the stabilizer of x in Γ. We call such a divisor, which depends only on the Γ-orbit of x, a special divisor.

Composite divisors are defined as follows. Let $\Gamma_L^o \subset \Gamma_L$ be the subgroup of isometries that act trivially on $L^\vee/L$, and suppose that $\Gamma \subset \Gamma_L^o$. For $t \in \mathbb{Z}_{>0}$ and $h \in L^\vee$ a coset representative for $L^\vee/L$, let

$$L(t, h) = \{x \in L + h \mid (x, x) = -t\}.$$

There is an associated special divisor

$$Z(t, h) = \sum_{\substack{x \in L(t,h) \\ \bmod \Gamma}} Z(x).$$

Remark. For the following observation, see [32], Lemma 1.7. The space D parametrizes polarized Hodge structures of weight 2 on the rational vector space V with $\dim_\mathbb{C} V_\mathbb{C}^{2,0} = 1$. Such a HS is *simple* if it does not contain any proper rational Hodge substructure. Then, in fact, the polarized HS corresponding to $\underline{w} \in D$ is simple if and only if $\underline{w} \notin D_x$ for any nonzero vector $x \in V$. Thus the set

$$D - \bigcup_{x \in V, \, x \neq 0} D_x$$

parametrizes the simple HS's of this type.

More generally, for $1 \leq r \leq n$, consider an r-tuple of vectors

$$\mathbf{x} = [x_1, \ldots, x_r], \qquad x_i \in V,$$

and suppose that

$$T(\mathbf{x}) = -(\mathbf{x}, \mathbf{x}) = -((x_i, x_j)) > 0.$$

Let

$$D_\mathbf{x} = \{\underline{w} \in D \mid (\mathbf{x}, \underline{w}) = 0\},$$
$$\Gamma_\mathbf{x} = \text{stablizer of } \mathbf{x} \text{ in } \Gamma,$$

and

$$Z(\mathbf{x}) : \Gamma_\mathbf{x} \backslash D_\mathbf{x} \longrightarrow \Gamma \backslash D = M_\Gamma.$$

The condition $T(\mathbf{x}) > 0$ implies that $D(\mathbf{x})$ has codimension r in D so that the special cycle $Z(\mathbf{x})$ has codimension r in M_Γ. Again, this cycle depends only on the Γ-orbit of $\mathbf{x}$.

Again there is a composite version. For

$$T \in \text{Sym}_r(\mathbb{Z})_{>0}, \qquad \mathbf{h} \in (L^\vee)^r,$$

let

$$L(T, \mathbf{h}) = \{\mathbf{x} \in L^r + \mathbf{h} \mid T(\mathbf{x}) = T\},$$

and define the cycle

$$Z(T, \mathbf{h}) = \sum_{\substack{\mathbf{x} \in L(T, \mathbf{h}) \\ \bmod \Gamma}} Z(\mathbf{x}).$$

These are the special cycles in question.

Remarks. (a) In the rest of this note, we will suppress the coset parameter h, although it plays an evident and important role in many places in the literature.

(b) We can make the same construction with any $T \in \mathrm{Sym}_r(\mathbb{Z})_{\geq 0}$ except that we require that for $\mathbf{x} \in L(T, \mathbf{h})$ the subspace spanned by the components of $\mathbf{x}$ has dimension equal to the rank of T. We write $Z^{\mathrm{naive}}(T, \mathbf{h})$ for the resulting cycle. It has codimension equal to the rank of T. For example, for $\mathbf{h} = 0$,

$$Z^{\mathrm{naive}}(0, \mathbf{h}) = M_\Gamma,$$

while, for $\mathbf{h} \neq 0$, $Z^{\mathrm{naive}}(0, \mathbf{h})$ is empty.

3 Modular Generating Series

First, for $\tau = u + iv \in \mathfrak{H}$, the upper half-plane, consider the series

$$\phi_1(\tau) = [Z(0)] + \sum_{t \in \mathbb{Z}_{>0}} [Z(t)] \, q^t, \qquad [Z(t)] \in H^2(M_\Gamma).$$

Here $[Z(0)] = [\omega] = c_1(\mathcal{L})$ is the Chern class of the tautological line bundle

$$\mathcal{L} \longrightarrow M_\Gamma,$$

defined by:

$$\mathcal{L} = b^*(\mathcal{O}(-1)), \qquad b : D(L) \longrightarrow D(L)^\vee,$$

where

$$D^\vee(L) = \{\, w \in V_\mathbb{C} \mid (w, w) = 0 \,\}/\mathbb{C}^\times \subset \mathbb{P}(V_\mathbb{C})$$

is the compact dual of $D(L)$. Here $H^r(M_\Gamma)$ is the usual Betti cohomology group of the (quasi-projective) variety M_Γ with complex coefficients.[2]

Theorem 1 ([15]) $\varphi_1(\tau)$ *is an elliptic modular form of weight* $\frac{n}{2} + 1$ *and level determined*[3] *by* L, *valued in* $H^2(M_\Gamma)$.

---

[2] If Γ has fixed points on $D(L)$, M_Γ is viewed as an orbifold and $H^r(M_\Gamma)$ is the space of Γ/Γ_1-invariants in $H^r(M_{\Gamma_1})$ where $\Gamma_1 \subset \Gamma$ is a normal subgroup of finite index which acts freely on $D(L)$.

[3] This means that the components of $\phi_1(\tau)$ with respect to any basis of the finite dimensional space $H^2(M_\Gamma)$ are scalar valued modular forms of the given weight and level. The level divides $4|L^\vee/L|$ and is determined by the usual recipe for theta functions, cf., for example, [29], Sect. 2.

The analogous generating series with values in the first Chow group $\mathrm{CH}^1(M_\Gamma)\otimes_\mathbb{Z}$ $\mathbb{C}$ of M_Γ was considered by Borcherds[4]

Theorem 2 ([2])

$$\phi_1^{CH}(\tau) = \{Z(0)\} + \sum_{t\in\mathbb{Z}_{>0}} \{Z(t)\}\, q^t, \qquad \{Z(t)\} \in \mathrm{CH}^1(M_\Gamma) \otimes \mathbb{C}$$

is an elliptic modular form of weight $\frac{n}{2} + 1$, etc. Its image under the cycle class map

$$\mathrm{CH}^1(M_\Gamma) \otimes \mathbb{C} \longrightarrow H^2(M_\Gamma)$$

is $\phi_1(\tau)$.

More generally, for any r, $1 \le r \le n$, and for $\tau = u + iv \in \mathfrak{H}_r$, the Siegel space of genus r, we can define a generating series

$$\phi_r(\tau) = \sum_{\substack{T\in\mathrm{Sym}_r(\mathbb{Z}) \\ T\ge 0}} [Z^{\mathrm{naive}}(T)] \cup [\omega]^{r-\mathrm{rank}T}\, q^T,$$

$$= [\omega]^r + \sum_{\substack{\mathrm{rank}T < r \\ T\ne 0}} [Z^{\mathrm{naive}}(T)] \cup [\omega]^{r-\mathrm{rank}T}\, q^T + \sum_{T\in\mathrm{Sym}_r(\mathbb{Z})_{>0}} [Z(T)]\, q^T.$$

The point is that we have shifted the classes $[Z^{\mathrm{naive}}(T)]$ by a suitable power of $[\omega]$ so that all of the coefficients lie in $H^{2r}(M_\Gamma)$.

Theorem 3 ([15]) (contd.) *$\phi_r(\tau)$ is a Siegel modular form of weight $\frac{n}{2} + 1$ and level determined by L, valued in $H^{2r}(M_\Gamma)$.*

Remark 1. (1) In [10] I asked whether the Chow group version $\phi_r^{CH}(\tau)$ is a Siegel modular form valued in $\mathrm{CH}^r(M_\Gamma)$. Using Borcherds' result mentioned above and an inductive argument based on Fourier–Jacobi expansions, Wei Zhang proved this in his Columbia thesis [33], conditionally on some finiteness/convergence assumption.
(2) A key point is that the pairing of $\phi_r(\tau)$ with any compactly supported class $c \in H_c^{2n-2r}(M_\Gamma)$ defines a classical Siegel modular form[5] of the same weight and level as $\phi_r(\tau)$.
(3) It is worth noting that the modularity of the $\phi_r(\tau)$ is proved by constructing a theta function $\theta(\tau, \varphi_{KM}^{(r)})$ valued in $A^{(r,r)}(M_\Gamma)$ the space of smooth differential forms of type (r, r) on M_Γ. This theta function is a non-holomorphic modular form of weight $\frac{n}{2} + 1$ in τ, analogous to the classical Siegel theta function for an indefinite quadratic form. Moreover, it is a closed (r, r)-form and its cohomology class $[\theta(\tau, \varphi_{KM}^{(r)})] \in H^{(r,r)}(M_\Gamma)$ is $\phi_r(\tau)$. In particular, if the class

---

[4] More precisely, he views M_Γ as an orbifold/stack and defines a group by generators and relations that maps to the usual Chow group, at least after tensoring with $\mathbb{C}$.

[5] This form will be vector-valued if we keep track of the parameter $\mathbf{h} \in L^\vee/L$.

$c \in H_c^{2n-2r}(M_\Gamma)$ is the class of an algebraic cycle $c = [S]$, as will be the case in several examples below, the pairing of c and $\phi_r(\tau)$ is given by an integral

$$c \cdot \phi_r(\tau) = \int_S \theta(\tau, \varphi_{KM}^{(r)}).$$

(4) The extension of such integrals to classes c that are not compactly supported has been studied in important work of Funke and Millson, [5–7], and involves interesting correction terms related to the boundary of the Borel–Serre compactification of M_Γ.

4 The Case of K3 Surfaces

In some cases, the M_Γ's and the $Z(T)$'s can be interpreted in terms of moduli spaces of lattice polarized[6] K3 surfaces. Here is an amateur's version of this. Start from the K3 lattice:

$$K = H^3 \oplus E_8(-1)^2, \qquad \mathrm{sig}(K) = (3, 19),$$

where H is the unimodular hyperbolic plane, and, letting $K_\mathbb{Q} = K \otimes_\mathbb{Z} \mathbb{Q}$, choose an orthogonal decomposition

$$K_\mathbb{Q} = V \oplus V'$$

with

$$\mathrm{sig}(V') = (1, 19 - n), \qquad \mathrm{sig}(V) = (2, n),$$

for some n, $0 \le n \le 19$. Let

$$L = K \cap V, \qquad L' = K \cap V',$$

be the corresponding primitive sublattices in K. Note that they are both even integral.

A marked L'-polarized K3 surface is a collection (X, u, λ) where X is an algebraic K3 surface over $\mathbb{C}$,

$$u : H^2(X, \mathbb{Z}) \xrightarrow{\sim} K$$

is an isometry for the intersection form on $H^2(X, \mathbb{Z})$, a marking, and

$$\lambda : L' \hookrightarrow \mathrm{Pic}(X) \subset H^2(X, \mathbb{Z})$$

is an embedding such that

$$u \circ \lambda : L' \hookrightarrow K$$

is the given inclusion. Moreover, λ is required to satisfy the "ample cone" condition,[7] cf. Sect. 10 of [4].

---

[6] Good references for lattice polarized K3 surfaces are [3, 4].

[7] More precisely, let $(L')_{-2}$ be the set of vectors x in L' with $(x, x) = -2$, and let $\mathcal{V}(L')^0$ be one component of the set of vectors $x \in L'_\mathbb{R}$ with $(x, x) > 0$, i.e., a choice of positive light cone. Finally,

These conditions imply that

(a) The embedding $\lambda : L' \hookrightarrow Pic(X)$ is primitive and isometric.
(b) The period point of (X, u) lies in $D(V)$.

Recall that the period point of (X, u) is the complex line $u_{\mathbb{C}}(H^{2,0}(X))$, where $u_{\mathbb{C}} : H^2(X, \mathbb{Z}) \otimes_{\mathbb{Z}} \mathbb{C} \longrightarrow K_{\mathbb{C}}$ is the complex linear extension of u. The moduli space of such gadgets (X, λ), obtained by eliminating the marking, is the quotient

$$M_\Gamma = \Gamma \backslash D(V),$$
$$\Gamma = \Gamma_L^0 = \{\gamma \in \Gamma_L \mid \gamma|_{L^\vee/L} = 1\}.$$

This gives a moduli interpretation of M_Γ. Details of this construction can be found in [4], Sect. 10.

Remark 2. (1) It is easy to check that, for $n \leq 17$, any *rational* quadratic space of signature $(2, n)$ can occur as V. If $n = 18$ or 19, there are restrictions on V, since then $L'_{\mathbb{Q}}$ has rank 2 or 1.

(2) In certain cases, a precise description of what lattices L' and L can occur was given by Nikulin [25]. As summarized in [24], Sect. 2, the result is the following. If $0 \leq n \leq 9$, then any even integral lattice L of signature $(2, n)$ can occur, and, if $n < 9$, the primitive embedding $L \hookrightarrow K$ is unique up to an isometry of K. Similarly, if $0 \leq n' \leq 10$, then any even integral lattice L' of signature $(1, n')$ can occur, and, if $n' < 10$, the embedding $L' \hookrightarrow K$, is unique up to an isometry of K.

(3) In such a family the generic element has Picard number $\rho(X) = rank(L') = 20 - n$,

For small values of n, the resulting M_Γ's are familiar classical objects. Here is a table:

n	ρ	$G = SO(V)$	M_Γ classically	Accidental iso.
0	20	$SO(2)$	$U(1)$	CM
1	19	$SO(2, 1)$	$SL(2, \mathbb{R})$	Shimura curves
2	18	$SO(2, 2)$	$SL(2, \mathbb{R}) \times SL(2, \mathbb{R})$	Hilbert modular surfaces
3	17	$SO(2, 3)$	$Sp(2, \mathbb{R})$	Siegel threefolds
4	16	$SO(2, 4)$	$SU(2, 2)$	Unitary Shimura fourfolds
—	—	—	—	—
19	1	$SO(2, 19)$	—	Moduli of polarized K3's

let $C(L')^+$ be a component of

$$\mathcal{V}(L')^0 - \bigcup_{x \in (L')_{-2}} x^\perp.$$

Then the ample cone condition is that

$$\lambda_{\mathbb{R}}(C(L')^+) \cap \mathcal{K}_X \neq \emptyset$$

where $\mathcal{K}_X$ is the closure of the ample cone in $Pic(X)_{\mathbb{R}}$.

For example, for $n = 19$, we have $L' = (2d)$, and we get the moduli space of polarized K3's of polarization degree $2d$. Note that K is an even lattice and represents every positive even integer.

At the other extreme, for $n = 0$, we have $\text{sig}(L) = (2, 0)$, rank $L' = 20$, and X is a singular K3 surface.

4.1 Modular Interpretation of the Special Cycles

In the case of families of lattice polarized K3 surfaces M_Γ for a lattice L of signature $(2, n)$ as described in the previous section, vectors in L correspond to additional elements of $\text{Pic}(X)$. Let $N = \mathbb{Z}^r$, and, for $T \in \text{Sym}_r(\mathbb{Z})_{>0}$, let $N = N_T$ be the quadratic lattice of signature $(0, r)$ defined by $-T$.

Proposition 1 *The codimension r cycle $Z(T)$ can be identified with the locus of objects (X, λ, j) where*

$$j : N_T \hookrightarrow \text{Pic}(X)$$

is a quadratic embedding with

$$j(N_T) \cdot \lambda(L') = 0.$$

Here, if (X, λ, u) is a marked object, then

$$u \circ j : \mathbb{Z}^r = N_T \hookrightarrow L, \qquad e_i \mapsto x_i,$$

determines an r-tuple $\mathbf{x} \in L^r$ with $T(\mathbf{x}) = -T$ and the period point of (X, u) lies in $D_\mathbf{x} \subset D(L)$.

Remark 3. (1) In this construction, we have fixed the basis $\mathbb{Z}^r \simeq N$. A change in this basis corresponds to a right multiplication of the row vectors $\mathbf{x}$ by an element of $\text{GL}_r(\mathbb{Z})$.

(2) For $r = 1$, we are imposing a single additional class in $\text{Pic}(X)$ and the $Z(t)$'s are essentially the Noether–Lefschetz divisors in M_Γ, cf. [8, 23].

4.2 Some Applications

I. Suppose that, for a smooth projective curve C,

$$i_\pi : C \longrightarrow M_\Gamma$$

is a morphism corresponding to a family

$$\pi : X \longrightarrow C$$

of L'-polarized K3 surfaces. Recall that $Z(t) \to M_\Gamma$ is the locus of collections (X, λ, j), $j \cdot j = -t$, and consider the fiber product

$$
\begin{array}{ccc}
Z(t) \times_{M_\Gamma} C & \longrightarrow & Z(t) \\
\downarrow & & \downarrow \\
C & \longrightarrow & M_\Gamma.
\end{array}
$$

Then, for $[C] \in H_c^{2n-2}(M_\Gamma)$,

$$
\begin{aligned}
[Z(t)] \cdot [C] &= \deg \mathcal{O}_{Z(t)}|_C. \\
&= \sum_{z \in C} \#\{(X_z, \lambda_z, j) \mid j \cdot j = -t, \ j \cdot \lambda = 0\} \qquad \text{(generically)} \\
&=: m(t, X/C) = \text{Noether–Lefschetz number.}
\end{aligned}
$$

For example, if the Picard number of a generic member of the family is $20 - n$, then C is not contained in any of the $Z(t)$'s and the loci in question are all finite sets of points.

Corollary 1

$$
\phi_1(\tau) \cdot [C] = \deg \mathcal{L}|_C + \sum_{t \in \mathbb{Z}_{>0}} m(t, X/C)\, q^t,
$$

is an elliptic modular form of weight $\frac{n}{2} + 1$ and level determined by L. In particular, the numbers $m(t, X/C)$ are the Fourier coefficients of this form.

This result is due to Maulik–Pandharipande [23], where it is derived from Borcherds' Theorem and its significance in Gromov–Witten theory is explained.

II. Similarly, suppose that

$$
\pi : X \longrightarrow S, \qquad i_\pi : S \longrightarrow M_\Gamma
$$

is a family of L'-polarized K3 surfaces where S is a projective surface. For $T \in \mathrm{Sym}_2(\mathbb{Z})_{>0}$, $Z(T) \to M_\Gamma$ is the locus of collections $(X, \lambda, \mathbf{j})$, where $\mathbf{j} = [j_1, j_2]$ is a pair of classes in $\mathrm{Pic}(X)$ orthogonal to $\lambda(L')$ and with matrix of inner products $\mathbf{j} \cdot \mathbf{j} = -T$. Consider

$$
\begin{array}{ccc}
Z(T) \times_{M_\Gamma} S & \longrightarrow & Z(T) \\
\downarrow & & \downarrow \\
S & \longrightarrow & M_\Gamma.
\end{array}
$$

Then, for the cohomology class $[S] \in H_c^{2n-4}(M_\Gamma)$, we can define

$$
[Z(T)] \cdot [S] = \sum_{z \in S} \#\{(X_z, \lambda_z, \mathbf{j}) \mid \mathbf{j} \cdot \mathbf{j} = -T, \ \mathbf{j} \cdot \lambda = 0\} \qquad \text{(generically)}
$$

$$
=: m(T, X/S) = \text{a higher Noether–Lefschetz number.}
$$

Here there can be curves in $Z(T) \cap S$ and, in this case, more care must be taken to interpret the intersection number $[Z(T)] \cdot [S]$. By the modularity results above, the $m(T, X/S)$ are Fourier coefficients of a Siegel modular form of genus 2 and weight $\frac{n}{2} + 1$. It would be interesting to compute these Siegel modular forms for specific families $X \to S$, e.g., ones coming from explicit classical geometry.

III. Suppose that L has signature $(2, 2)$ and is anisotropic. Then we have a family where $S = M_\Gamma$ is itself a projective surface, so that

$$\phi_2(\tau, M_\Gamma) \in H^4(M_\Gamma) \xrightarrow{\text{deg}} \mathbb{C}.$$

The following result from [10] is obtained by combining the results of [15] with the extended Siegel–Weil formula of [16].

Theorem 4 *Assume that condition (1) below holds. Then*

$$\deg \phi_2(\tau, M_\Gamma) = E\left(\tau, \frac{1}{2}, L\right)$$

where $E(\tau, s, L)$ is a Siegel Eisenstein series of genus 2 and weight 2 associated to L, evaluated at the Siegel–Weil critical point $s = s_0 = \frac{1}{2}$.

Note that, for $T > 0$, $Z(T)$ is a 0-cycle, and, when $T \geq 0$ has rank 1, then $Z(T)^{\text{naive}}$ is a curve on M_Γ. Thus

$$\deg \phi_2(\tau, M_\Gamma) = \text{vol}(M_\Gamma, \omega^2) + \sum_{\text{rank}(T)=1} \text{vol}(Z(T)^{\text{naive}}, \omega) \, q^T$$

$$+ \sum_{T>0} \deg(Z(T)) \, q^T.$$

Here, the Siegel–Einstein series is defined by

$$E(\tau, s, L) = \sum_{\gamma \in \Gamma'_\infty \backslash \Gamma'} \det(c\tau + d)^{\frac{n}{2}+1} |\det(c\tau + d)|^{s-s_0} \det(v)^{\frac{1}{2}s-s_0} \, \Phi(\gamma, L),$$

where

$$\gamma = \begin{pmatrix} a & b \\ c & d \end{pmatrix} \in \Gamma' = \text{Sp}_2(\mathbb{Z}).$$

and $\Phi(\gamma, L)$ is a generalized Gauss sum attached to γ and L, [12, 30]. This series is termwise absolutely convergent for $\text{Re}(s) > \frac{3}{2}$. Its value at $s_0 = \frac{1}{2}$ is defined by analytic continuation, [16]. The main point behind Theorem 4 is that, as explained in Remark 1(3), there is a genus 2 theta function $\theta(\tau, \varphi_{KM}^{(2)})$ valued in $A^{(2,2)}(M_\Gamma)$, i.e., in top degree forms, and the degree generating series $\deg \phi_2(\tau, M_\Gamma)$ is obtained by integrating this form over M_Γ. Let $H = O(V)$ be the orthogonal group of V and assume that there is an open compact subgroup $K \subset H(\mathbb{A}_f)$, the group of finite adèle points of H, such that

$$H(\mathbb{A}) = H(\mathbb{Q})H(\mathbb{R})K, \qquad \Gamma = H(\mathbb{Q}) \cap K. \tag{1}$$

Then the geometric integral of $\theta(\tau, \varphi_{KM}^{(2)})$ over

$$M_\Gamma \simeq H(\mathbb{Q})\backslash H(\mathbb{A})/K_\infty K$$

coincides with the integral of a scalar valued theta function on $H(\mathbb{A})$ over the adelic quotient $H(\mathbb{Q})\backslash H(\mathbb{A})$—cf. [11], Sect. 4, for a more detailed discussion. The extended Siegel–Weil formula, [12, 16], identifies the result as a special value at $s = s_0$, perhaps outside the range of absolute convergence, of a certain Eisenstein series attached to L.

IV. Here is an amusing example. Let $F = \mathbb{Q}(\sqrt{d})$, $d \in \mathbb{Z}_{>0}$ square free, be a real quadratic field with ring of integers O_F. Let M be a projective O_F-lattice with a symmetric O_F-bilinear form $(\,,\,)_M$, and suppose that the signature of M is given by

$$\mathrm{sig}(M) = ((2, m), (0, m + 2)).$$

Let L be M, viewed as a $\mathbb{Z}$-module, with bilinear form $(\,,\,)_L$ given by

$$(x, y)_L = \mathrm{tr}_{F/\mathbb{Q}}(x, y)_M.$$

The signature of L is $(2, 2m + 2)$. Let $V = L \otimes_\mathbb{Z} \mathbb{Q}$ and note that

$$V \otimes_\mathbb{Q} \mathbb{R} = (V \otimes_F F) \otimes_\mathbb{Q} \mathbb{R} = (V \otimes_{F,\sigma_1} \mathbb{R}) \times (V \otimes_{F,\sigma_2} \mathbb{R}),$$

where σ_1 and σ_2 are the two real embeddings of F. The two factors on the right have signatures $(2, m)$ and $(0, m + 2)$ respectively. Then there is an embedding

$$D(M) = D(V \otimes_{F,\sigma_1} \mathbb{R}) \hookrightarrow D(V_\mathbb{R}) = D(L).$$

Let $\Gamma_M \subset \Gamma$ be the subgroup of O_F-linear isometries of in Γ. Then

$$\Gamma_M \backslash D(M) \longrightarrow \Gamma \backslash D(L),$$

is an algebraic cycle of codimension $m + 2$, and, since the quotient $\Gamma_M \backslash D(M)$ is compact, this cycle is projective.

For example, when $m = 2$, we get a projective surface

$$S \longrightarrow M_\Gamma = M_\Gamma(2, 6), \qquad [S] \in H_c^8(M_\Gamma).$$

It would be nice to have a concrete description of the corresponding family of K3 surfaces $X \to S$. The Siegel modular generating function for higher Noether–Lefschetz numbers for this family is

$$\phi_2(\tau, M_\Gamma) \cdot [S] = i^* E_F\left(\tau, \frac{1}{2}, \Phi\right), \qquad i : SL_2/\mathbb{Q} \longrightarrow SL_2/F,$$

the pullback of the special value of a genus 2 Hilbert–Siegel Eisenstein series $E_F(\tau, \frac{1}{2}, \Phi)$ of weight $(2, 2)$ over F to a weight 4 modular Siegel modular form over $\mathbb{Q}$! This is again a consequence of the extended Siegel–Weil formula together with a seesaw identity, [9]. Note that, in general, such a pullback will now have lots of cuspidal components and so the Fourier coefficients of such a form are essentially more complicated than those of an Eisenstein series like that occurring in Theorem 4.

Remark 4. A more or less explicit geometric construction of an example for $m = 1$ is described in [32], Sect. 3. In this case, $\text{sig}(M) = ((2, 1), (0, 3))$ and the base of the family will be a Shimura curve C over F. In this case, the generating series for the Noether–Lefschetz numbers will be an elliptic modular form of weight 3 arising as the pullback for a Hilbert modular Eisenstein series of weight $(3/2, 3/2)$.

V. The previous example can be further generalized. Suppose that F is a totally real number field with ring of integers O_F, $|F : \mathbb{Q}| = d > 1$, and real embeddings $\sigma_i : F \hookrightarrow \mathbb{R}$, $1 \leq i \leq d$. Let $M, (\,,\,)_M$ be a quadratic O_F-lattice of rank $m + 2$ over O_F and with

$$\text{sig}(M) = ((2, m), (0, m + 2)^{d-1}).$$

Define a quadratic lattice L as in example **IV**, so that

$$\text{sig}(L) = (2, d(m + 2) - 2).$$

Again setting $V = L \otimes_{\mathbb{Z}} \mathbb{Q}$, we have

$$V \otimes_{\mathbb{Q}} \mathbb{R} = (V \otimes_{F, \sigma_1} \mathbb{R}) \times (V \otimes_{F, \sigma_2} \mathbb{R}) \times \cdots \times (V \otimes_{F, \sigma_d} \mathbb{R}),$$

an embedding

$$D(M) = D(V \otimes_{F, \sigma_1} \mathbb{R}) \hookrightarrow D(V_{\mathbb{R}}) = D(L),$$

and a projective algebraic cycle

$$Y = \Gamma_M \backslash D(M) \longrightarrow \Gamma \backslash D(L) = M_\Gamma,$$

of codimension $(d - 1)(m + 2)$ and dimension m. For this construction to fall into the world of K3 moduli, we must require that

$$2 \leq d(m + 2) \leq 21,$$

and hence we have the following table of the possibilities: (Note that $N = \dim M_\Gamma = d(m + 2) - 2$.)

d	2	3	4	5	6	7	8	9	10
$m = \dim Y$	$0 \leq m \leq 8$	$0 \leq m \leq 5$	$0 \leq m \leq 3$	$0 \leq m \leq 2$	0, 1	0, 1	0	0	0
$N = \dim M_\Gamma$	$2 \leq N \leq 18$	$4 \leq N \leq 19$	$6 \leq N \leq 19$	$8 \leq N \leq 18$	10, 16	12, 19	14	16	18

It would be interesting to give an account of the families of K3 surfaces over the projective varieties Y occurring here.

One nice case is the following. Let $k = \mathbb{Q}(\zeta_{13})$ be the 13th cyclotomic field and let $F = \mathbb{Q}(\zeta_{13} + \zeta_{13}^{-1})$, so that $|F : \mathbb{Q}| = 6$. Let $M \otimes_{O_F} F$ to be the space of elements of trace 0 in a quaternion algebra B/F with $B \otimes_{F,\sigma_1} \mathbb{R} = M_2(\mathbb{R})$ and $B \otimes_{F,\sigma_i} \mathbb{R} \simeq \mathbb{H}$, the Hamiltonian quaternions, for $i > 1$. To specify B, we need to choose an additional set $\Sigma_f(B)$ of finite places of F with $|\Sigma_f(B)|$ odd. The algebra B is determined by the condition that, for a finite place v of F, $B \otimes_F F_v$ is a division algebra if and only if $v \in \Sigma_f(B)$. The simplest choice would be $\Sigma_f(B) = \{v_{13}\}$, where v_{13} is the unique place above 13. We fix a maximal order O_B in B and let M be the set of trace 0 elements in it. As quadratic form on M, we take $(x, y) = \text{tr}(xy^t)$. In this case, $V = L \otimes_{\mathbb{Z}} \mathbb{Q}$ has signature $(2, 16)$, and, as noted in Remark 2(1), the rational quadratic space V occurs as a summand of $K_\mathbb{Q}$. I have not checked[8] if the lattice L just defined can occur as $V \cap K$. In any case, we obtain a Shimura curve C embedded in the 16-dimensional moduli space M_Γ, and the generating series for the Noether–Lefschetz numbers for the associated family of K3 surfaces will be an elliptic modular form of weight 9 arising as the pullback of a Hilbert modular Eisenstein series of weight $(3/2, \ldots, 3/2)$ for F.

5 Kuga–Satake Abelian Varieties and Special Endomorphisms

In this last section, we give another interpretation of the special cycles $Z(T)$. For convenience, we change the sign and consider rational quadratic forms of signature $(n, 2)$.

5.1 The Kuga–Satake Construction

The moduli spaces M_Γ of lattice polarized K3 surfaces carry families of abelian varieties arising from (a slight variant of) the Kuga–Satake construction, which we now review.

For a quadratic lattice L of signature $(n, 2)$ and for $V = L_\mathbb{Q}$, let

$$C = C(V) = C^+(V) \oplus C^-(V)$$

be the Clifford algebra of V. For a basis $v_1, \ldots, v_{n+2}$ of V, let $\delta = v_1 \cdots v_{n+2} \in C(V)$, and let $k = \mathbb{Q}(\delta)$ be the discriminant field. For n even, $\delta \in C^+(V)$, k is the center of $C^+(V)$, and δ anticommutes with elements of $C^-(V)$. For n is odd, $\delta \in C^-(V)$ and k is central in all of $C(V)$. Now

$$O_C = C(L) = \text{Clifford algebra of } L$$

---

[8] If it differs from the primitive lattice $V \cap K$, then some "level structure" will have to be introduced.

is an order in C and $O_k = k \cap C(L)$ is an order in k. We obtain a real torus

$$A^{\text{top}} = C(L_{\mathbb{R}})/C(L)$$

of dimension 2^{n+2} with an action

$$\iota : C(L) \otimes_{\mathbb{Z}} O_k \longrightarrow \text{End}(A^{\text{top}}), \qquad \iota(c \otimes \alpha) : x \longmapsto cx\alpha,$$

via left-right multiplication. Let

$$G = \text{GSpin}(V) = \{g \in C^+(V)^{\times} \mid gVg^{-1} = V\}$$

and, for an element $a \in C^+(L) \cap C(V)^{\times}$ with $a^t = -a$, define an alternating form on C by

$$\langle x, y \rangle = \langle x, y \rangle_a = \text{tr}(axy^t).$$

Here $x \mapsto x^t$ is the main involution on $C(V)$, determined by the conditions $(xy)^t = y^t x^t$ for all $x, y \in C$, and $v^t = v$ for all $v \in V \subset C^-(V)$. Note that for $g \in G$,

$$\langle xg, yg \rangle = \nu(g) \langle x, y \rangle, \qquad \nu(g) = gg^t = \text{spinor norm}.$$

A variation of complex structures on A^{top} is defined as follows. For an oriented negative 2-plane $z \in D(L)$ in $V_{\mathbb{R}}$, let z_1, z_2 be a properly oriented orthonormal basis and let $j_z = z_1 z_2 \in C^+(V_{\mathbb{R}})$. This element is independent of the choice of basis. Note that $j_z^2 = -1$, so that right multiplication by j_z defines a complex structure on $C(V_{\mathbb{R}})$, and hence we have a complex torus $A_z = (A^{\text{top}}, j_z)$ for each $z \in D(L)$. Since the action of $C(L) \otimes O_k$ commutes with the right multiplication by j_z, we have

$$\iota : O_C \otimes_{\mathbb{Z}} O_k \longrightarrow \text{End}(A_z) \tag{1}$$

with

$$\iota(c \otimes \alpha)^* = \iota(c^* \otimes \alpha^t), \qquad c^* = ac^t a^{-1},$$

for the Rosati involution determined by $\langle\ ,\ \rangle$. If $\gamma \in \Gamma \subset \Gamma_L$, and[9] if $\tilde{\gamma} \in O_C^{\times}$ is an element mapping to γ under the natural homomorphism $\text{GSpin}(V) \to \text{SO}(V)$, then right multiplication by $\tilde{\gamma}$ on $C(L_{\mathbb{R}})$ induces an isomorphism of the complex tori $A_{\gamma(z)}$ and A_z, equivariant for the action of $O_C \otimes O_k$. Finally, fix a rational splitting $V = V^+ + V^-$ of signature $(n, 0) + (0, 2)$ and let a_1, a_2 be a $\mathbb{Z}$-basis for the negative definite lattice $L^- = L \cap V^-$. Let $a = a_1 a_2$. With this choice of a, the form $\langle\ ,\ \rangle_a$ is $\mathbb{Z}$-valued on $C(L)$ and defines a Riemann form on each complex torus A_z, i.e., the form $\langle xj_z, y \rangle_a$ is symmetric and definite (hence positive definite on one of the connected components of $D(L)$). Thus, M_{Γ} carries a family of polarized abelian varieties:

$$\pi : \text{KS}(L) \longrightarrow M_{\Gamma}, \qquad (A_z, \iota, \lambda), \qquad \dim A_z = 2^{n+1}$$

---

[9] In general, there is an orbifold issue here that can be eliminated by introducing a suitable level structure.

with endomorphisms (1). Note that these abelian varieties are $\mathbb{Z}_2$-graded,

$$A_z = A_z^+ \oplus A_z^-,$$

since the construction respects the decomposition $C(L) = C^+(L) \oplus C^-(L)$.

Remark 5. (1) For $n \leq 4$, M_Γ is the moduli space for such PE type abelian varieties and is a Shimura variety of PEL type.[10]

(2) For general $n \geq 5$, the abelian varieties in the family are not characterized by the given PE; more Hodge classes are required, and M_Γ is a Shimura variety of Hodge type. The family π corresponds to a morphism

$$i_\pi : M_\Gamma \longrightarrow A_g, \qquad g = 2^{n+1}.$$

5.2 Special Cycles and Special Endomorphisms

Special endomorphisms arise as follows. For $x \in L$, let

$$r_x : A^{\text{top}} \longrightarrow A^{\text{top}} = C(L_\mathbb{R})/C(L),$$

be the endomorphism induced by right multiplication by x. Note that this endomorphism has degree 1 with respect to the $\mathbb{Z}_2$-grading, and that, since $x^2 = Q(x)$ in $C(L)$,

$$r_x^2 = [Q(x)],$$

is just multiplication by the integer $Q(x) = (x, x) \in \mathbb{Z}$. Moreover,

$$r_x \circ \iota(c \otimes b) = \begin{cases} \iota(c \otimes b) \circ r_x & \text{for } n \text{ odd} \\ \iota(c \otimes b^\sigma) \circ r_x & \text{for } n \text{ even.} \end{cases} \tag{2}$$

and the adjoint of r_x with respect to $\langle \ , \ \rangle$ is

$$(r_x)^* = r_{x^\iota} = r_x. \tag{3}$$

Definition 1. For a given $z \in D(L)$, the endomorphism r_x is said to be a special endomorphism of the abelian variety A_z when it is holomorphic, i.e., when $r_x \in \text{End}(A_z)$.

Lemma 1. *Let* $L(A_z)$ *be the space of special endomorphisms of* A_z, *with quadratic form defined by* $j^2 = Q(j)$. *Then*

$$L(A_z) = L \cap z^\perp$$

is an isometry.

---

[10] We could add a level structure here.

Note that, r_x is a special endomorphism of A_z precisely when x commutes with j_z. Since conjugation by j_z induces the endomorphism of V that is -1 on z and $+1$ on $z^\perp$, it follows that r_x is a special endomorphism precisely when $x \in z^\perp$, i.e., $z \in D_x$.

Corollary 2 *For* $T \in \mathrm{Sym}_r(\mathbb{Z})_{>0}$, *the special cycle* $Z(T)$ *is the image in* M_Γ *of*

$$\tilde{Z}(T) = \{(A_z, \mathbf{j}) \mid \mathbf{j} \in L(A_z)^r, \ (\mathbf{j}, \mathbf{j}) = T\}.$$

In fact, to give the right definition of special endomorphism, one should start with an object $(A, \iota, \lambda, \epsilon, \xi)$ where A is an abelian variety (scheme) with $O_C \otimes O_{\mathbf{k}}$-action ι, a $\mathbb{Z}/2\mathbb{Z}$-grading ϵ and additional Hodge tensors ξ. A special endomorphism is then an element $j \in \mathrm{End}(A)$ of degree 1 with respect to the $\mathbb{Z}/2\mathbb{Z}$-grading satisfying conditions (2) and (3), with an additional compatibility with respect to the Hodge tensors ξ. For $0 \le n \le 3$, where the classes ξ can be expressed in terms of endomorphisms, such a definition of special endomorphism was given in [19] $(n = 0)$, [20] $(n = 1)$, [17] $(n = 2)$, and [18] $(n = 3)$ and used to define special cycles in integral models of the associated PEL type Shimura varieties. The results of these papers concerning the intersection number of integral special cycles and Fourier coefficients of modular forms should have some consequences in the arithmetic theory of K3 surfaces. To reveal it one would have to relate these integral models to the integral models of moduli spaces of K3 surfaces developed by Rizov, [26, 27]. The extension to the case of general n, i.e., to integral models of Shimura varieties attached to $\mathrm{GSpin}(n, 2)$ is the subject of ongoing work of Andreatta–Goren [1] and of Howard–Madapusi (private communication).

Remark 6. In the references that follow, I have made no attempt at completeness and apologize in advance for the many omitted citations.

References

1. F. Andreatta, E. Goren, Research report on a conjecture of Bruinier-Yang, Oberwolfach report, July 2012
2. R.E. Borcherds, The Gross-Kohnen-Zagier theorem in higher dimensions. Duke Math. J. **97**, 219–233 (1999)
3. I. Dolgachev, Mirror symmetry for lattice polarized K3 surfaces. J. Math. Sci. **81**, 2599–2630 (1996)
4. I. Dolgachev, S. Kondo, Moduli spaces of K3 surfaces and complex ball quotients, in *Arithmetic and Geometry Around Hypergeometric Functions*. Progress in Mathematics, vol. 260 (Birkhäuser, Basel, 2007), pp. 43–100
5. J. Funke, J. Millson, Cycles in hyperbolic manifolds of non-compact type and Fourier coefficients of Siegel modular forms. Manuscripta Math. **107**, 409–444 (2002)
6. J. Funke, J. Millson, Cycles with local coefficients for orthogonal groups and vector valued Siegel modular forms. Am. J. Math. **128**, 899–948 (2006)
7. J. Funke, J. Millson, Spectacle cycles with coefficients and modular forms of half-integral weight. Advanced Lectures in Mathematics (ALM), vol. 19 (International Press, Somerville, 2011)
8. A. Klemm, D. Maulik, R. Pandharipande, E. Scheidegger, Noether-Lefschetz theory and the Yau-Zaslow conjecture. J. AMS **23**, 1013–1040 (2010)

9. S. Kudla, Seesaw dual reductive pairs, in *Automorphic Forms of Several Variables*. Taniguchi Symposium, Katata 1983. Progress in Mathematics, vol. 46 (Birkhäuser, Boston, 1984), pp. 244–268

10. S. Kudla, Algebraic cycles on Shimura varieties of orthogonal type. Duke Math. J. **86**(1), 39–78 (1997)

11. S. Kudla, Integrals of Borcherds forms. Compos. Math. **137**, 293–349 (2003)

12. S. Kudla, On some extensions of the Siegel-Weil formula, in *Eisenstein Series and Applications*, ed. by W.-T. Gan, S. Kudla, Y. Tschinkel. Progress in Mathematics, vol. 258 (Birkhäuser, Boston, 2008)

13. S. Kudla, J. Millson, The theta correspondence and harmonic forms I. Math. Ann. **274**, 353–378 (1986)

14. S. Kudla, J. Millson, The theta correspondence and harmonic forms II. Math. Ann. **277**, 267–314 (1987)

15. S. Kudla, J. Millson, Intersection numbers of cycles on locally symmetric spaces and Fourier coefficients of holomorphic modular forms in several complex variables. Publ. Math. IHES **71**, 121–172 (1990)

16. S. Kudla, S. Rallis, On the Weil-Siegel formula. Crelle **387**, 1–68 (1988)

17. S. Kudla, M. Rapoport, Arithmetic Hirzebruch–Zagier cycles. J. Reine Angew. Math. **515**, 155–244 (1999)

18. S. Kudla, M. Rapoport, Cycles on Siegel 3-folds and derivatives of Eisenstein series. Ann. Sci. École. Norm. Sup. **33**, 695–756 (2000)

19. S. Kudla, M. Rapoport, T. Yang, On the derivative of an Eisenstein series of weight 1. Int. Math. Res. Not. **7**, 347–385 (1999)

20. S. Kudla, M. Rapoport, T. Yang, in *Modular Forms and Special Cycles on Shimura Curves*. Annals of Mathematics Studies, vol. 161 (Princeton University Press, Princeton, 2006)

21. M. Kuga, I. Satake, Abelian varieties attached to polarized K3 surfaces. Math. Ann. **109**, 239–242 (1967)

22. K. Madapusi, Integral canonical models of Spin Shimura varieties, (2012) [arXiv:1212.1243]

23. D. Maulik, R. Pandharipande, Gromov-Witten theory and Noether-Lefschetz theory (2010) [arXiv:0705.1653v2]

24. D. Morrison, On K3 surfaces with large Picard number. Invent. Math. **75**, 105–121 (1984)

25. V. Nikulin, Integral symmetric bilinear forms and some of their applications. Izv. Akad. Nauk. SSSR **43**, 111–177 (1979); Maht. USSR Izvestija **14**, 103–167 (1980)

26. J. Rizov, Moduli stacks of polarized K3 surfaces in mixed characteristic. Serdica Math. J. **32**, 131–178 (2006)

27. J. Rizov, Kuga-Satake abelian varieties in positive characteristic. J. Reine Angew. Math. **648**, 13–67 (2010)

28. I. Satake, Clifford algebras and families of abelian varieties. Nagoya Math. J. **27**, 435–446 (1966)

29. G. Shimura, On modular forms of half-integral weight. Ann. Math. (2) **97**, 440–481 (1973)

30. C.L. Siegel, *Lecture on the Analytic Theory of Quadratic Forms, Buchhandlung Robert Peppmüller* (Göttingen, Germany, 1963)

31. B. van Geemen, Kuga-Satake varieties and the Hodge conjecture, in *The Arithmetic and Geometry of Algebraic Cycles*, ed. by B.B. Gordon et al. (Kluwer, Dordrecht, 2000), pp. 51–82

32. B. van Geemen, Real multiplication on K3 surfaces and Kuga Satake varieties. Mich. Math. J. **56**, 375–399 (2008)

33. W. Zhang, Modularity of generating functions of special cycles on Shimura varieties, Ph.D. thesis, Columbia University, 2009

Enriques Surfaces of Hutchinson–Göpel Type and Mathieu Automorphisms

Shigeru Mukai and Hisanori Ohashi

Abstract We study a class of Enriques surfaces called of Hutchinson–Göpel type. Starting with the projective geometry of Jacobian Kummer surfaces, we present the Enriques' sextic expression of these surfaces and their intrinsic symmetry by $G = C_2^3$. We show that this G is of Mathieu type and conversely, that these surfaces are characterized among Enriques surfaces by the group action by C_2^3 with prescribed topological type of fixed point loci. As an application, we construct Mathieu type actions by the groups $C_2 \times \mathfrak{A}_4$ and $C_2 \times C_4$. Two introductory sections are also included.

Key words: Enriques surfaces, Mathieu groups

Mathematics Subject Classifications (2010): Primary 14J28; Secondary 14E07, 14J50

1 Introduction

From a curve C of genus two and its Göpel subgroup $H \subset (\mathrm{Jac}\ C)_{(2)}$, we can construct an Enriques surface $(\mathrm{Km}\ C)/\varepsilon_H$, which we call *of Hutchinson–Göpel type*. We may say that surfaces of this type among all Enriques surfaces occupy an equally important place as Jacobian Kummer surfaces $\mathrm{Km}\ C$, or $\mathrm{Km}\,(\mathrm{Jac}\ C)$, do among all $K3$ surfaces. In [9] we characterized these Enriques surfaces as those which have numerically reflective involutions.

S. Mukai
Research Institute for Mathematical Sciences, Kyoto University, Kyoto 606-8502, Japan
e-mail: mukai@kurims.kyoto-u.ac.jp

H. Ohashi (✉)
Faculty of Science and Technology, Department of Mathematics, Tokyo University of Science, 2641 Yamazaki, Noda, Chiba 278-8510, Japan
e-mail: ohashi@ma.noda.tus.ac.jp

R. Laza et al. (eds.), *Arithmetic and Geometry of K3 Surfaces and Calabi–Yau Threefolds*,   429
Fields Institute Communications 67, DOI 10.1007/978-1-4614-6403-7_15,
© Springer Science+Business Media New York 2013

In this paper, we will study the group action of Mathieu type on these Enriques surfaces of Hutchinson–Göpel type. In particular we will characterize them by using a special sort of action of Mathieu type by the elementary abelian group C_2^3. As a byproduct, we will also give examples of actions of Mathieu type by the groups $C_2 \times \mathfrak{A}_4$ (of order 24) and $C_2 \times C_4$ (of order 8). These constructions are crucial in the study of automorphisms of Mathieu type on Enriques surfaces; in particular they answer the conjecture we posed in the lecture notes [11].

Our starting point is the fact that the Kummer surface Km C is the $(2, 2, 2)$-Kummer covering[1] of the projective plane $\mathbb{P}^2$,

$$\text{Km } C \xrightarrow{C_2^3} \mathbb{P}^2,$$

whose equation can be written in the form

$$u^2 = q_1(x, y, z),\ v^2 = q_2(x, y, z),\ w^2 = q_3(x, y, z).$$

All branch curves $\{(x, y, z) \in \mathbb{P}^2 \mid q_i(x, y, z) = 0\}$ $(i = 1, 2, 3)$ are reducible conics and our Enriques surface S of Hutchinson–Göpel type sits in between this covering as the quotient of Km C by the free involution

$$(u, v, w) \mapsto (-u, -v, -w).$$

By computing invariants, we will see that S is the normalization of the singular sextic surface

$$x^2 + y^2 + z^2 + t^2 + \left(\frac{a}{x^2} + \frac{b}{y^2} + \frac{c}{z^2} + \frac{d}{t^2} \right) xyzt = 0 \tag{1}$$

in $\mathbb{P}^3$, where $a, b, c, d \in \mathbb{C}^*$ are constants. They satisfy the condition $abcd = 1$ corresponding to the Cremona invariance of the six lines $\{q_1 q_2 q_3 = 0\} \subset \mathbb{P}^2$.

In general, an involution σ acting on an Enriques surface is said to be *Mathieu* or *of Mathieu type* if its Lefschetz number $\chi_{\text{top}}(\text{Fix } \sigma)$ equals four,[2] [12]. This is equivalent to saying that the Euler characteristic of the fixed curves $\text{Fix}^-(\sigma)$ is equal to 0 (see the beginning of Sect. 7 for this notation). We have the following classification of $\text{Fix}^-(\sigma)$ according to its topological types.

(M0) $\text{Fix}^-(\sigma) = \emptyset$, namely σ is a *small* involution.
(M1) $\text{Fix}^-(\sigma)$ is a single elliptic curve.
(M2) $\text{Fix}^-(\sigma)$ is a disjoint union of two elliptic curves.

---

[1] This octic model of Km C is different from the standard nonsingular octic model given by the smooth complete intersection of three diagonal quadrics. See (⋆2) of Sect. 5.

[2] This number is exactly the number of fixed points of non-free involutions in the small Mathieu group M_{12}, which implies that the character of Mathieu involutions on $H^*(S, \mathbb{Q})$ coincides with that of involutions in M_{11}. This is the origin of the terminology. See also [11].

(M3) Fix$^-(\sigma)$ is a disjoint union of a genus $g \geq 2$ curve and $(g-1)$ smooth rational curves.[3]

Our motivation comes from the following observation.

Observation 1. The Enriques surface $S = (\mathrm{Km}\ C)/\varepsilon_H$ of Hutchinson–Göpel type has an action of Mathieu type[4] by the elementary abelian group $G = C_2^3$ with the following properties. Let h be the polarization of degree 6 given by (1) above.

(1) The group G preserves the polarization h up to torsion.
(2) There exists a subgroup G_0 of index two, which preserves the polarization h while the coset $G \setminus G_0$ sends h to $h + K_S$.
(3) All involutions in G_0 are of type (M2) above.
(4) All involutions in $G \setminus G_0$ are of type (M0) above.

These are the properties of Mathieu type actions by which we characterize Enriques surfaces of Hutchinson–Göpel type.

Theorem 1. *Let S be an Enriques surface with a group action of Mathieu type by $G = C_2^3$ which satisfies the properties (3) and (4) in Observation 1 for a subgroup G_0 of index two. Then S is isomorphic to an Enriques surface of Hutchinson–Göpel type.*

Our proof of Theorem 1 (Sect. 7) exhibits the effective divisor h of Observation 1 in terms of the fixed curves of the group action. In particular we can reconstruct the sextic equation (1) of S. In this way, we see that the group action perfectly characterizes Enriques surfaces of Hutchinson–Göpel type and all parts of Observation 1 hold true.

The sextic equation (1) also has the following application to our study of Mathieu automorphisms.

Theorem 2. *Among those Enriques surfaces of Hutchinson–Göpel type (1), there exists a 1-dimensional subfamily whose members are acted on by the group $C_2 \times \mathfrak{A}_4$ of Mathieu type. Similarly there exists another one-dimensional subfamily whose members are acted on by the group $C_2 \times C_4$ of Mathieu type.*

The paper is organized as follows. Sections 2 and 3 give an introduction to Enriques surfaces. In Sect. 2 we explain the constructions of Enriques surfaces from rational surfaces, while in Sect. 3 we focus on the quotients of Kummer surfaces. In Sect. 4, we introduce a larger family of sextic Enriques surfaces which we call of diagonal type. They contain our Enriques surfaces of Hutchinson–Göpel type as a subfamily of codimension one. We derive the sextic equation by computing the invariants from a $K3$ surface which is a degree 8 cover of the projective plane $\mathbb{P}^2$. In Sect. 5 we restrict the family to the Hutchinson–Göpel case. We give a discussion of the related isogenies between Kummer surfaces and also give the definition of the

---

[3] In fact only $g = 2$ is possible.

[4] This means that every involution is Mathieu.

group action by $G = C_2^3$. In Sect. 6 we use the sextic equation to study their singularities and give a precise computation for the group actions. Theorem 2 is proved here. In Sect. 7 we prove Theorem 1.

Throughout the paper, we work over the field $\mathbb{C}$ of complex numbers.

2 Rational Surfaces and Enriques Surfaces

An algebraic surface is *rational* if it is birationally equivalent to the projective plane $\mathbb{P}^2$. It is easy to see that a rational surface has vanishing geometric genus $p_g = 0$ and irregularity $q = 0$. At the beginning of the history of algebraic surfaces the converse problem was regarded as important.

Problem 1. Is an algebraic surface with $p_g = q = 0$ rational?

Enriques surfaces were discovered by Enriques as counterexamples to this problem. They have Kodaira dimension $\kappa = 0$. Nowadays we know that even some algebraic surfaces of general type have also $p_g = q = 0$, the Godeaux surfaces for example.

Definition 1. An algebraic surface S is an *Enriques surface* if it satisfies $p_g = 0, q = 0$ and $2K_S \sim 0$.

By the adjunction formula, a nonsingular rational curve $C \subset S$ satisfies $(C^2) = -2$, hence there are no exceptional curves of the first kind on S. This means that S is *minimal* in its birational equivalence class.

If a $K3$ surface X admits a fixed-point-free involution ε, then the quotient surface X/ε is an Enriques surface. Conversely, for an Enriques surface S the canonical double cover

$$X = \operatorname{Spec}_S(\mathcal{O}_S \oplus \mathcal{O}_S(K_S))$$

turns out to be a $K3$ surface and is called the $K3$-*cover* of S. Since a $K3$ surface is simply connected, π is the same as the universal covering of S. In this way, an Enriques surface is nothing but a $K3$ surface modded out by a fixed-point-free involution ε.

Example 1. Let X be a smooth complete intersection of three quadrics in $\mathbb{P}^5$, defined by the equations

$$q_1(x) + r_1(y) = q_2(x) + r_2(y) = q_3(x) + r_3(y) = 0,$$

where $(x : y) = (x_0 : x_1 : x_2 : y_0 : y_1 : y_2) \in \mathbb{P}^5$ are homogeneous coordinates of $\mathbb{P}^5$. If the quadratic equations q_i, r_i ($i = 1, 2, 3$) are general so that the intersections $q_1 = q_2 = q_3 = 0$ and $r_1 = r_2 = r_3 = 0$ considered in $\mathbb{P}^2$ are both empty, then the involution

$$\varepsilon: (x : y) \mapsto (x : -y)$$

is fixed-point-free and we obtain an Enriques surface $S = X/\varepsilon$.

As we mentioned, an Enriques surface appeared as a counterexample to Problem 1. Even though it is not a rational surface, it is closely related to them; plenty of examples of Enriques surfaces are available by the quadratic twist construction as follows.

Let us consider a rational surface R and a divisor B belonging to the linear system $|-2K_R|$. The double cover of R branched along B,

$$X = \operatorname{Spec}_R(\mathcal{O}_R \oplus \mathcal{O}_R(-K_R)) \to R,$$

gives a $K3$ surface if B is nonsingular. More generally, if B has at most simple singularities, then X has at most rational double points and its minimal desingularization $\tilde{X}$ is a $K3$ surface.

Example 2. Well-known examples are given by sextic curves in $R = \mathbb{P}^2$ or curves of bidegree $(4, 4)$ in $R = \mathbb{P}^1 \times \mathbb{P}^1$.

Let us assume that the surface R admits an involution $e \colon R \to R$ which is *small*, namely with at most finitely many fixed points over R. Further, let us assume that the curve B is invariant under e, $e(B) = B$. Then we can lift e to involutions of X. There are two lifts, one of which acts symplectically on X (namely acts on the space $H^0(\Omega_X^2)$ trivially) and the other anti-symplectically (namely acts by (-1) on the space $H^0(\Omega_X^2)$). We denote the latter by ε. (The former is exactly the composition of ε and the covering transformation.) We can see that ε acts on X freely and the quotient X/ε gives an Enriques surface if B is disjoint from the fixed points of e. We call this Enriques surface the *quadratic twist* of R by (e, B).

Example 3. Let e_0 be an arbitrary involution of $\mathbb{P}^1$ and consider the small involution $e = (e_0, e_0)$ acting on $R = \mathbb{P}^1 \times \mathbb{P}^1$. According to our recipe, we can construct an Enriques surface S which is the quadratic twist of R obtained from e and an e-stable divisor B of bidegree $(4, 4)$.

Example 4. We consider the Cremona transformation

$$e \colon (x : y : z) \mapsto (1/x : 1/y : 1/z)$$

of $\mathbb{P}^2$, where $(x : y : z)$ are the homogeneous coordinates of $\mathbb{P}^2$. Let B be a sextic curve with nodes or cusps at three points $(1 : 0 : 0), (0 : 1 : 0), (0 : 0 : 1)$ and such that $e(B) = B$. (More generally, the singularities at the three points can be any simple singularities of curves.) Then we can construct the quadratic twist S of $\mathbb{P}^2$ by (e, B).

In this example, it might be easier to consider the surface R obtained by blowing the three points up. The Cremona transformation e induces a biregular automorphism of R and the strict transform $\overline{C}$ of C belongs to the linear system $|-2K_R|$. The Enriques surface S is nothing but the quadratic twist of the surface R by $(e, \overline{C})$.

We borrowed the terminology from the following example.

Example 5 (Kondo [8], Hulek–Schütt [4]). Let $f\colon R \to \mathbb{P}^1$ be a rational elliptic surface with the zero-section and a 2-torsion section. Let e be the translation by the 2-torsion section, which we assume to be small. Let B be a sum of two fibers of f. Then B belongs to $|-2K_R|$ and is obviously stable under e. Thus we obtain an Enriques surface from the quadratic twist construction. In this case the Enriques surface naturally has an elliptic fibration $S \to \mathbb{P}^1$. In the theory of elliptic curves this is called the quadratic twist of f.

We remark that the Enriques surface obtained as the quadratic twist of a rational surface always admits a nontrivial involution. In general any involution σ of an Enriques surface admits two lifts to the $K3$-cover X, one of which is symplectic and the other non-symplectic. We denote the former by σ_K and the latter by σ_R. With one exception, the quotient X/σ_R becomes a rational surface.

This operation can be seen as the converse construction of the quadratic twist. The exception appears in the case where σ_R is also a fixed-point-free involution, in which case the quotient X/σ_R is again an Enriques surface.

3 Abelian Surfaces and Enriques Surfaces

A two-dimensional torus $T = \mathbb{C}^2/\Gamma$, where $\Gamma \simeq \mathbb{Z}^4$ is a full lattice in $\mathbb{C}^2$, is acted on by the involution $(-1)_T$. It has 16 fixed points which are exactly the 2-torsion points $T_{(2)}$ of T. The *Kummer surface* $\mathrm{Km}\,T$ is obtained as the minimal desingularization of the quotient surface $\overline{\mathrm{Km}}\,T = T/(-1)_T$. This is known to be a $K3$ surface, equipped with 16 exceptional (-2)-curves.

When T is isomorphic to the direct product $E_1 \times E_2$ of elliptic curves, the Kummer surface $\mathrm{Km}(E_1 \times E_2)$ is the same as the desingularized double cover of $\mathbb{P}^1 \times \mathbb{P}^1$ defined by

$$\overline{\mathrm{Km}}(E_1 \times E_2)\colon w^2 = x(x-1)(x-\lambda)y(y-1)(y-\mu), \tag{2}$$

where $\lambda, \mu \in \mathbb{C} - \{1, 0\}$ are constants and x, y are inhomogeneous coordinates of $\mathbb{P}^1$. The strict transforms of the eight divisors on $\mathbb{P}^1 \times \mathbb{P}^1$ defined by

$$x = 0, 1, \infty, \lambda \text{ and } y = 0, 1, \infty, \mu \tag{3}$$

give eight smooth rational curves on $\mathrm{Km}(E_1 \times E_2)$. In this product case, together with 16 exceptional curves, it has 24 smooth rational curves with the following configuration (called the *double Kummer configuration*) (Fig. 1).

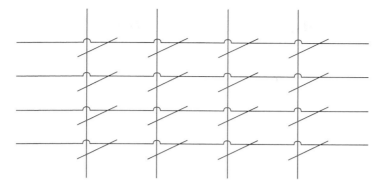

Fig. 1: The double Kummer configuration

There are many studies on Km T when T is a principally polarized abelian surface, too. In this case, using the theta divisor Θ, the linear system $|2\Theta|$ gives an embedding of the singular surface $T/(-1)_T$ into $\mathbb{P}^3$ as a quartic surface

$$x^4 + y^4 + z^4 + t^4 + A(x^2t^2 + y^2z^2) + B(y^2t^2 + x^2z^2) + C(z^2t^2 + x^2y^2) + Dxyzt = 0,$$
$$A, B, C, D \in \mathbb{C}$$

which is stable under the Heisenberg group action.

Let us consider the following question: How many Enriques surfaces are there whose universal covering is one of these Kummer surfaces Km T? The easiest example is given by the following.

Example 6 (Lieberman). On the Kummer surface $\mathrm{Km}(E_1 \times E_2)$ of product type (2), we have the involutive action

$$\varepsilon \colon (x, y, w) \mapsto \left(\frac{\lambda}{x}, \frac{\mu}{y}, \frac{\lambda\mu w}{x^2 y^2} \right).$$

We can see easily that ε is fixed-point-free. Hence $\mathrm{Km}(E_1 \times E_2)/\varepsilon$ is an Enriques surface, which is the quadratic twist of $\mathbb{P}^1 \times \mathbb{P}^1$ by $e \colon (x, y) \mapsto (\lambda/x, \mu/y)$ and the branch divisor (3).

The surface $\mathrm{Km}(E_1 \times E_2)$ is equivalently the desingularized double cover of $\mathbb{P}^2$ branched along six lines

$$x = 0, 1, \lambda, \quad y = 0, 1, \mu.$$

(See Fig. 2.) The involution above is given as the lift of the Cremona involution $(x, y) \mapsto (\frac{\lambda}{x}, \frac{\mu}{y})$, which exhibits the Enriques surface $\mathrm{Km}(E_1 \times E_2)/\varepsilon$ as the quadratic twist of blown up $\mathbb{P}^2$. Another Enriques surface can be obtained from the surface

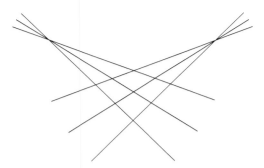

Fig. 2: Product Kummer as a double plane

$Km(E_1 \times E_2)$ as follows when $\lambda \neq \mu$. We note that under this condition, the three lines passing through two of the points $(0, 0), (1, 1), (\lambda, \mu)$ can be given by

$$x - y, \quad \mu x - \lambda y, \quad (\mu - 1)(x - 1) - (\lambda - 1)(y - 1).$$

We make the coordinate change

$$X = \frac{\mu x - \lambda y}{x - y}, Y = \frac{(\mu - 1)(x - 1) - (\lambda - 1)(y - 1)}{x - y}.$$

The six branch lines then become

$$X = \lambda, \mu$$
$$Y = \lambda - 1, \mu - 1$$
$$X/Y = \lambda/(\lambda - 1), \mu/(\mu - 1).$$

These six lines are preserved by the Cremona transformation

$$(X, Y) \mapsto \left(\frac{\lambda\mu}{X}, \frac{(\lambda - 1)(\mu - 1)}{Y} \right).$$

Hence the Kummer surface

$Km(E_1 \times E_2)$:
$$w^2 = (X - \lambda)(X - \mu)(Y - \lambda + 1)(Y - \mu + 1)(\lambda Y - (\lambda - 1)X)(\mu Y - (\mu - 1)X)$$

has the automorphism

$$\varepsilon : (X, Y, w) \mapsto \left(\frac{\lambda\mu}{X}, \frac{(\lambda - 1)(\mu - 1)}{Y}, \frac{\lambda(\lambda - 1)\mu(\mu - 1)w}{X^2 Y^2} \right)$$

whenever $\lambda \neq \mu$. Moreover, this automorphism has no fixed points; hence we obtain the Enriques surface $Km(E_1 \times E_2)/\varepsilon$. This Enriques surface with $\lambda = \mu = 1^{1/3}$

was found by Kondo and constructed in full generality by Mukai [9]. It is called an Enriques surface *of Kondo–Mukai type.*

Remark 1. It is interesting to find the limit of the above Enriques surface $\mathrm{Km}(E_1 \times E_2)/\varepsilon$ when λ goes to μ. The limit is no longer an Enriques surface but a rational surface with quotient singularities of type $\frac{1}{4}(1, 1)$. A more precise description is the following: Let R be the minimal resolution of the double cover of $\mathbb{P}^2$ branched along the union of four tangent lines

$$x = 0, \; x - 2y + z = 0, \; x - 2\lambda y + \lambda^2 z = 0, \; z = 0$$

of the conic $xz = y^2$. The pullback of the conic splits into two smooth rational curves C_1 and C_2 in R. Let R' be the blow-up of R at the four points $C_1 \cap C_2$. Then the strict transforms of C_1 and C_2 become (-4)-$\mathbb{P}^1$'s. The limit of the Enriques surface $\mathrm{Km}(E_1 \times E_2)/\varepsilon$ is the rational surface R' contracted along these two (-4)-$\mathbb{P}^1$'s.

Remark 2 (Ohashi [13]). When E_1 and E_2 are taken generically, these two surfaces are the only Enriques surfaces (up to isomorphism) whose universal covering is the surface $\mathrm{Km}(E_1 \times E_2)$.

Let us proceed to the study of $\mathrm{Km}(A)$, where (A, Θ) is a principally polarized abelian surface. In this case, there are three Enriques surfaces known whose universal coverings are isomorphic to $\mathrm{Km}\,A$ [10, 14]. Here we introduce the surface obtained from a Göpel subgroup $H \subset A_{(2)}$. The next observation is fundamental.

Lemma 1. *Suppose that we are given six distinct lines $l_1, \ldots, l_6$ in the projective plane, whose three intersection points $p_1 = l_1 \cap l_4, p_2 = l_2 \cap l_5, p_3 = l_3 \cap l_6$ are not collinear and such that the lines $\overline{p_i p_j}$ are different from l_i. Then the following conditions are equivalent.*

(1) A suitable quadratic Cremona transformation with center p_1, p_2, p_3 sends l_1, l_2, l_3 to l_4, l_5, l_6 respectively.
(2) All $l_1, \ldots, l_6$ are tangent to a smooth conic or both l_1, l_2, l_3 and l_4, l_5, l_6 are concurrent (after suitable renumberings $2 \leftrightarrow 5$ or $3 \leftrightarrow 6$).

Proof. This is an extended version of [10, Proposition 5.1]. We sketch the proof. Let us choose linear coordinates $(x : y : z)$ such that p_1, p_2, p_3 are the vertices of the coordinate triangle $xyz = 0$. Then the six lines are given by

$$l_i : y = \alpha_i x \; (i = 1, 4), \; l_j : z = \alpha_j y \; (j = 2, 5), \; l_k : x = \alpha_k z \; (k = 3, 6)$$

for $\alpha_1, \ldots, \alpha_6 \in \mathbb{C}^*$. We easily see that the condition (1) is equivalent to $\prod_{i=1}^{6} \alpha_i = 1$. Let us consider a conic in the *dual* projective plane

$$Q : ax^2 + by^2 + cz^2 + dyz + ezx + fxy = 0.$$

For Q to contain the six points q_i corresponding to l_i, we have the following conditions:

$$\alpha_1 \alpha_4 = \frac{b}{a}, \quad \alpha_2 \alpha_5 = \frac{c}{b}, \quad \alpha_3 \alpha_6 = \frac{a}{c},$$

$$\alpha_1 + \alpha_4 = \frac{f}{a}, \quad \alpha_2 + \alpha_5 = \frac{d}{b}, \quad \alpha_3 + \alpha_6 = \frac{e}{c}.$$

Thus $\prod_{i=1}^{6} \alpha_i = 1$ is equivalent to the existence of such Q. If Q is smooth, then the former condition of (2) is satisfied by taking the dual of Q. If Q is a union of two distinct lines, then the points q_i and q_{i+3} must lie on different components for $i = 1, 2, 3$, hence the latter configuration of (2) occurs. (For the same reason, Q cannot be a double line.)

We have already encountered the latter configuration of lines in Fig. 2; in this case the double cover of $\mathbb{P}^2$ branched along $\sum l_i$ is birational to $\mathrm{Km}(E_1 \times E_2)$. Even in the former case of (2) of Lemma 1, the lift of the Cremona involution to the double cover gives an automorphism of $\mathrm{Km}(A)$ without fixed points. Hence we obtain an Enriques surface $\mathrm{Km}(A)/\varepsilon$.

This Enriques surface is described in the following way (and characterized by the presence of a numerically reflective involution) by Mukai [10]. Let $H \subset A_{(2)}$ be a Göpel subgroup; namely, H is a subgroup consisting of four elements and the Weil pairing with respect to 2Θ,

$$A_{(2)} \times A_{(2)} \to \mu_2,$$

is trivial on $H \times H$. There are 15 such subgroups. One such H defines four nodes of the Kummer quartic surface in $\mathbb{P}^3$, and if we take the homogeneous coordinates $(x : y : z : t)$ of $\mathbb{P}^3$ so that the coordinate points coincide with the four nodes, then the Kummer quartic surface has the equation

$$q(xt + yz, yt + xz, zt + xy) + (\text{const.})xyzt = 0. \tag{4}$$

(We assume that the four nodes are not coplanar.) This equation is invariant under the standard Cremona transformation

$$(x : y : z : t) \mapsto \left(\frac{1}{x} : \frac{1}{y} : \frac{1}{z} : \frac{1}{t} \right).$$

Moreover, this involutive automorphism is free from fixed points over the Kummer quartic surface. Let us denote by ε_H this free involution of $\mathrm{Km}(A)$. The Enriques surface $\mathrm{Km}(A)/\varepsilon_H$ is thus determined by the principally polarized abelian surface (A, Θ) and the Göpel subgroup H. We call this surface *the Enriques surface of Hutchinson–Göpel type* since the expression (4) was first found by Hutchinson [5] using theta functions. (See also Keum [7, Sect. 3].)

Remark 3. The limit of the Enriques surface $\mathrm{Km}(A)/\varepsilon_H$ when H becomes coplanar is also a rational surface with two quotient singular points of type $\frac{1}{4}(1, 1)$ as in Remark 1.

4 Sextic Enriques Surfaces of Diagonal Type

Now we consider the Kummer $(2, 2, 2)$-covering of the projective plane $\mathbb{P}^2$ with coordinates $x = (x_1 : x_2 : x_3)$ branched along three conics $q_i(x) = 0$, $i = 1, 2, 3$:

$$\overline{X}: w_1^2 = q_1(x), \ w_2^2 = q_2(x), \ w_3^2 = q_3(x).$$

These equations define a $(2, 2, 2)$ complete intersection in $\mathbb{P}^5$ with homogeneous coordinates $(w_1 : w_2 : w_3 : x_1 : x_2 : x_3)$. Hence the minimal desingularization X of $\overline{X}$ is a $K3$ surface if it has at most rational double points. It has the action by C_2^3 arising from covering transformations. Among them, we focus on the involution

$$\varepsilon: (w_1 : w_2 : w_3 : x_1 : x_2 : x_3) \mapsto (-w_1 : -w_2 : -w_3 : x_1 : x_2 : x_3).$$

It is free of fixed points on X if and only if the locus $q_1(x) = q_2(x) = q_3(x) = 0$ is empty in $\mathbb{P}^2$. In this way we obtain the Enriques surface $S = X/\varepsilon$.

Let us specialize to the case where all $q_i(x)$ are reducible conics. More precisely our assumption is as follows.

($\star$) The conic $\{q_i = 0\}$ is the sum of two lines l_i, l_{i+3} ($i = 1, 2, 3$) for six distinct lines $l_1, \ldots, l_6$. The three points $l_1 \cap l_4, l_2 \cap l_5, l_3 \cap l_6$ are also distinct.

Under assumption ($\star$), the $(2, 2, 2)$-covering $\overline{X}$ has at most rational double points and we obtain the minimal desingularization X and the quotient Enriques surface S. The singularities of $\overline{X}$ consists of 12 nodes located above the three points $l_1 \cap l_4, l_2 \cap l_5, l_3 \cap l_6$. (It follows that the Enriques surface S contains six disjoint smooth rational curves as images of the exceptional curves.)

Remark 4. The quotient surface $\overline{X}/\varepsilon$ is nothing but the normalization of the surface

$$u_1^2 = q_1 q_3, \ u_2^2 = q_2 q_3$$

which is the covering of $\mathbb{P}^2$ of degree 4.

The projection of $\mathbb{P}^2$ from the singular point of q_i defines a rational map to the projective line, which in turn defines an elliptic fibration on X and on S. We denote by G_0 the Galois group of $S \to \mathbb{P}^2$. Each nontrivial element $g \in G_0$ corresponds to and defines the double covering of the rational elliptic surface branched along two smooth fibers. Hence Fix(g) has two smooth elliptic curves as its 1-dimensional components. This shows

Proposition 1. *Under assumption* ($\star$), *the action of $G_0 \simeq C_2^2$ on the Enriques surface S is of Mathieu type and every nontrivial element has (M2) type.*

For later use, we give the sextic equation of the Enriques surface S under the condition ($\star$). Here we additionally assume that the three points Sing(q_i) ($i = 1, 2, 3$)

are not collinear. (See also Remark 7.) Then we can choose homogeneous coordinates of $\mathbb{P}^2$ so that the three points are the coordinate points $(x_1 : x_2 : x_3) = (0 : 0 : 1), (0 : 1 : 0), (1 : 0 : 0)$. The degree 8 cover X over $\mathbb{P}^2$ has the form

$$w_i^2 = \frac{x_{i+1} - \alpha_i x_{i+2}}{x_{i+1} - \beta_i x_{i+2}}, (i = 1, 2, 3 \in \mathbb{Z}/3), \tag{5}$$

hence it has the following field of rational functions

$$\mathbb{C}\left(\frac{x_1}{x_2}, \frac{x_2}{x_3}, \sqrt{\frac{x_2 - \alpha_1 x_3}{x_2 - \beta_1 x_3}}, \sqrt{\frac{x_3 - \alpha_2 x_1}{x_3 - \beta_2 x_1}}, \sqrt{\frac{x_1 - \alpha_3 x_2}{x_1 - \beta_3 x_2}}\right).$$

Here we put $q_i(x) = (\text{const.})(x_{i+1} - \alpha_i x_{i+2})(x_{i+1} - \beta_i x_{i+2})$. X is exactly the minimal model of this field of algebraic functions in two variables. Since we have the relations

$$\frac{x_{i+1}}{x_{i+2}} = \frac{\beta_i w_i^2 - \alpha_i}{w_i^2 - 1}, i = 1, 2, 3,$$

by multiplying them, X is also the minimal desingularization of the $(2, 2, 2)$ divisor

$$(\star\star) \quad (\beta_1 w_1^2 - \alpha_1)(\beta_2 w_2^2 - \alpha_2)(\beta_3 w_3^2 - \alpha_3) = (w_1^2 - 1)(w_2^2 - 1)(w_3^2 - 1)$$

in $\mathbb{P}^1 \times \mathbb{P}^1 \times \mathbb{P}^1$. Here we consider $w_i(i = 1, 2, 3)$ as inhomogeneous coordinates of $\mathbb{P}^1 \times \mathbb{P}^1 \times \mathbb{P}^1$.

Proposition 2. *Assume that the three reducible conics $q_1 = 0, q_2 = 0, q_3 = 0$ satisfy $(\star)$ and the three points $\mathrm{Sing}\, q_1, \mathrm{Sing}\, q_2, \mathrm{Sing}\, q_3$ are not collinear. Then the Enriques surface $S \to \mathbb{P}^2$ is isomorphic to the minimal desingularization of the sextic surface in $\mathbb{P}^3$ defined by*

$$(\star \star \star) \quad a_0 x_0^2 + a_1 x_1^2 + a_2 x_2^2 + a_3 x_3^2 = \left(\frac{b_0}{x_0^2} + \frac{b_1}{x_1^2} + \frac{b_2}{x_2^2} + \frac{b_3}{x_3^2}\right) x_0 x_1 x_2 x_3,$$

where we put

$$a_0 = \alpha_1 \alpha_2 \alpha_3 - 1, a_1 = \alpha_1 \beta_2 \beta_3 - 1,$$
$$a_2 = \beta_1 \alpha_2 \beta_3 - 1, a_3 = \beta_1 \beta_2 \alpha_3 - 1,$$
$$b_0 = \beta_1 \beta_2 \beta_3 - 1, b_1 = \beta_1 \alpha_2 \alpha_3 - 1,$$
$$b_2 = \alpha_1 \beta_2 \alpha_3 - 1, b_3 = \alpha_1 \alpha_2 \beta_3 - 1$$

Proof. The Enriques surface is the quotient of the $(2, 2, 2)$ surface $(\star\star)$ by the involution

$$(w_1, w_2, w_3) \mapsto (-w_1, -w_2, -w_3)$$

followed by the minimal desingularization. We focus on the ambient spaces and construct a birational map between $\mathbb{P}^3$ and the quotient of $\mathbb{P}^1 \times \mathbb{P}^1 \times \mathbb{P}^1$ by the involution above.

We consider a rational map

$$\mathbb{P}^1 \times \mathbb{P}^1 \times \mathbb{P}^1 \dashrightarrow \mathbb{P}^3$$

defined by $(w_1, w_2, w_3) \mapsto (x_0 : x_1 : x_2 : x_3) = (1 : w_2 w_3 : w_1 w_3 : w_1 w_2)$. It has four points of indeterminacy $(\infty, \infty, \infty), (\infty, 0, 0), (0, \infty, 0), (0, 0, \infty)$. In other words, the rational map is the projection of the Segre variety $\mathbb{P}^1 \times \mathbb{P}^1 \times \mathbb{P}^1 \subset \mathbb{P}^7$ from the 3-space spanned by the four points. The indeterminacy is resolved by blowings up and we obtain a morphism

$$\mathrm{Bl}_{4\text{-pts}}(\mathbb{P}^1 \times \mathbb{P}^1 \times \mathbb{P}^1) \to \mathbb{P}^3.$$

This morphism factors through the double cover

$$Y: w^2 = x_0 x_1 x_2 x_3$$

of $\mathbb{P}^3$ branched along the tetrahedron and $\mathrm{Bl}_{4\text{-pts}}(\mathbb{P}^1 \times \mathbb{P}^1 \times \mathbb{P}^1) \to Y$ is a birational morphism which contracts six quadric surfaces

$$w_1 = 0, \infty, \quad w_2 = 0, \infty, \quad w_3 = 0, \infty,$$

into six edges. Since $(\star\star)$ is an irreducible surface which does not contain any of these six quadric surfaces, by multiplying $w_1^2 w_2^2 w_3^2$,

$$(\beta_1 x_2 x_3 - \alpha_1 x_0 x_1)(\beta_2 x_1 x_3 - \alpha_2 x_0 x_2)(\beta_3 x_1 x_2 - \alpha_3 x_0 x_3)$$
$$= (x_2 x_3 - x_0 x_1)(x_1 x_3 - x_0 x_2)(x_1 x_2 - x_0 x_3)$$

defines the sextic surface which is birational to the Enriques surface. By reducing coefficients, we obtain $(\star \star \star)$.

Remark 5. In the proof we have used the four invariants $1, w_2 w_3, w_1 w_3, w_1 w_2$. Instead, we could use the anti-invariants $w_1 w_2 w_3, w_1, w_2, w_3$ to obtain another sextic equation. In this case the indeterminacies are given by

$$(0, 0, 0), (0, \infty, \infty), (\infty, 0, \infty), (\infty, \infty, 0)$$

and the computation results in the sextic surface

$$(\star \star \star'): \sum_{i=0}^{3} b_i x_i^2 = x_0 x_1 x_2 x_3 \sum_{i=0}^{3} \frac{a_i}{x_i^2}.$$

This is nothing but the surface obtained from $(\star \star \star)$ by applying the standard Cremona transformation $(x_i) \mapsto (1/x_i)$.

Remark 6. More generally, a $(2, 2, 2)$ $K3$ surface in $\mathbb{P}^1 \times \mathbb{P}^1 \times \mathbb{P}^1$ which is invariant under the involution

$$(w_1, w_2, w_3) \mapsto (-w_1, -w_2, -w_3)$$

is mapped to the sextic Enriques surface

$$q(x_0, x_1, x_2, x_3) = x_0 x_1 x_2 x_3 \sum_{i=0}^{3} \frac{b_i}{x_i^2},$$

not necessarily of diagonal type. The proof is the same as above.

As is well-known, these sextic surfaces have double lines along the six edges of the tetrahedron $x_0 x_1 x_2 x_3 = 0$.

5 Action of C_2^3 of Mathieu Type on Enriques Surfaces of Hutchinson–Göpel Type

In this section we study Enriques surfaces of Hutchinson–Göpel type explained in Sect. 3. We show that they are $(2, 2)$-covers of the projective plane $\mathbb{P}^2$ branched along three reducible conics and extend the action of $G_0 \simeq C_2^2$ to an action of C_2^3, which is still of Mathieu type.

Let us begin with the configuration of six distinct lines $l_1, \ldots, l_6$ in $\mathbb{P}^2$. We recall that there exists uniquely a C_2^5-cover of $\mathbb{P}^2$ branched along these lines; it is represented by the diagonal complete intersection surface in $\mathbb{P}^5$ as

$$W: \sum_{i=1}^{6} a_i x_i^2 = \sum_{i=1}^{6} b_i x_i^2 = \sum_{i=1}^{6} c_i x_i^2 = 0,$$

where $(x_1 : \cdots : x_6)$ are the homogeneous coordinates of $\mathbb{P}^5$.

We restrict ourselves to the case

($\star$0) All $l_1, \ldots, l_6$ are tangent to a smooth conic $Q \subset \mathbb{P}^2$.

More concretely, we have a nonsingular curve B of genus two

$$(\star 1)\ w^2 = \prod_{i=1}^{6} (x - \lambda_i),\ \lambda_i \in \mathbb{C}$$

and the quadratic Veronese embedding $v_2 \colon \mathbb{P}^1 \to \mathbb{P}^2$ whose image is $Q = v_2(\mathbb{P}^1)$ so that the lines $l_1, \cdots, l_6$ are nothing but the tangent lines to Q at $v_2(\lambda_i)$. By an easy computation (e.g. [10, Sect. 5]), the desingularized double cover of $\mathbb{P}^2$ branched along the sum $\sum_{i=1}^{6} l_i$ is isomorphic to the Jacobian Kummer surface Km B of the curve B. The C_2^5-cover branched along six lines in this case is given by the equation

$$W: \sum_{i=1}^{6} x_i^2 = \sum_{i=1}^{6} \lambda_i x_i^2 = \sum_{i=1}^{6} \lambda_i^2 x_i^2 = 0.$$

The morphism from W to the double plane branches only along the 15 exceptional curves of Km B corresponding to 15 nonzero 2-torsions of $J(B)$, hence the induced map $W \dashrightarrow \mathrm{Km}\ B$ is the same as induced from the multiplication morphism $x \mapsto 2x$ of $J(B)$. In particular we see that W is isomorphic to $\mathrm{Km}\ B$. (See [15, Theorem 2.5] for the alternative proof using the traditional quadric line complex.)

We take the subgroup H_0 of $J(B)$ consisting of 2-torsions $p_1 - p_4, p_2 - p_5, p_3 - p_6$ and the zero element. Here p_i are the Weierstrass points corresponding to $\lambda_i \in \mathbb{C}$. This H_0 is a Göpel subgroup of $J(B)$ and the quotient abelian surface $J(B)/H_0$ again has a principal polarization. There are two cases:

1. The quotient surface $J(B)/H_0$ is isomorphic to the Jacobian $J(C)$ of a curve C of genus two.
2. The surface $J(B)/H_0$ is isomorphic to a product $E_1 \times E_2$ of two elliptic curves.

The group H_0 acts on the Kummer surface $W \simeq \mathrm{Km}\ B$ by the formulas

$$(x_1 : x_2 : x_3 : x_4 : x_5 : x_6) \mapsto (-x_1 : x_2 : x_3 : -x_4 : x_5 : x_6), \text{ and}$$
$$(x_1 : x_2 : x_3 : x_4 : x_5 : x_6) \mapsto (x_1 : -x_2 : x_3 : x_4 : -x_5 : x_6).$$

Hence the quotient $\mathrm{Km}\ B/H_0$ is a C_2^3-cover of $\mathbb{P}^2$ branched along the three reducible conics

$$(\star 2) \quad l_1 + l_4 : q_1 = 0, \ l_2 + l_5 : q_2 = 0, \ l_3 + l_6 : q_3 = 0.$$

Proposition 3. *Assume in* $(\star 2)$ *that the three points* $\mathrm{Sing}\ q_i$ $(i = 1, 2, 3)$ *are not collinear. Then the minimal resolution of the quotient surface* $\mathrm{Km}\ B/H_0$ *is isomorphic to the Jacobian Kummer surface* $\mathrm{Km}\ C$ *of* C *and the involution*

$$(w_1, w_2, w_3) \mapsto (-w_1, -w_2, -w_3)$$

of $\mathrm{Km}\ C$ *coincides with the Hutchinson–Göpel involution* ε_H *associated to the Göpel subgroup* $H := J(B)_{(2)}/H_0$ *of* $J(C)$ [10]. *In particular, the Enriques cover* $S \to \mathbb{P}^2$ *of degree 4 with branch curve* $(\star 2)$ *is an Enriques surface of Hutchinson–Göpel type.*

Proof. We consider the polar m_i of Q at the point $\mathrm{Sing}\ q_i = l_i \cap l_{i+3}$, namely the line connecting $v_2(\lambda_i)$ and $v_2(\lambda_{i+3})$. Since $\mathrm{Sing}\ q_i$ are not collinear, m_1, m_2, m_3 are not concurrent.

We introduce homogeneous coordinates $(x_1 : x_2 : x_3)$ such that m_1, m_2, m_3 are defined by x_1, x_2, x_3. Let $q(x_1, x_2, x_3)$ be the defining equation of Q. Replacing q, m_1, m_2, m_3 by suitable constant multiplications, we can put the defining equations of the conics $l_i + l_{i+3} : q_i = 0$ as $-q + x_i^2$. Now the $K3$ surface $\overline{X}$ is defined by the equations

$$w_i^2 = -q(x_1, x_2, x_3) + x_i^2 \ (i = 1, 2, 3).$$

In particular we see that $\overline{X}$ is contained in the $(2, 2)$ complete intersection

$$V : w_1^2 - x_1^2 = w_2^2 - x_2^2 = w_3^2 - x_3^2$$

in $\mathbb{P}^5$. This (quartic del Pezzo) threefold V is nothing but the image of the rational map

($\star$3) $\mathbb{P}^3 \dashrightarrow \mathbb{P}^5$

$$(x : y : z : t) \mapsto (x_1 : x_2 : x_3 : w_1 : w_2 : w_3)$$
$$= (xt + yz : yt + xz : zt + xy : xt - yz : yt - xz : zt - xy)$$

More precisely, V is isomorphic to the $\mathbb{P}^3$ first blown up at four coordinate points and then contracted along the six (-2) smooth rational curves which are strict transforms of the six edges of the tetrahedron $xyzt = 0$. The rational map ($\star$3) induces a birational equivalence between $\overline{X}$ and the quartic surface

($\star$4) $q(xt + yz, yt + xz, zt + xy) = 4xyzt.$

Under ($\star$3), the involution $(\underline{x} : \underline{w}) \mapsto (\underline{x} : -\underline{w})$ corresponds to the Cremona involution

$$(x : y : z : t) \mapsto \left(\frac{1}{x} : \frac{1}{y} : \frac{1}{z} : \frac{1}{t} \right).$$

Hence S is of Hutchinson–Göpel type (see Sect. 3).

Remark 7. The collinearity property of the three points $\text{Sing } q_i$ ($i = 1, 2, 3$) is equivalent to the three quadratic equations $(x - \lambda_i)(x - \lambda_{i+3})$ being linearly dependent. In this case, there exists an involution σ of $\mathbb{P}^1$ which sends λ_i to λ_{i+3} for $i = 1, 2, 3$. This involution σ lifts to an involution $\tilde{\sigma}$ of the curve B in ($\star$1) and the quotient $B/\tilde{\sigma}$ becomes an elliptic curve. We call such pair (B, H_0) *bielliptic*. In this case the quotient $J(B)/H_0$ is isomorphic to the product of two elliptic curves as principally polarized abelian surfaces.

Corollary 1. *Assume that the pair (C, H) is not bielliptic. Then the Enriques surface* Km C/ε_H *obtained from the curve C of genus two and the Göpel subgroup $H \subset J(C)_{(2)}$ is isomorphic to the desingularization of the $(2, 2)$-cover of the projective plane $\mathbb{P}^2$ branched along three reducible conics ($\star$2) satisfying the condition ($\star$0).*

Proof. The quotient abelian surface $J(C)/H$ has a principal polarization which is not reducible. Hence it is isomorphic to the Jacobian $J(B)$ of some curve B of genus two. Also the quotient $H_0 = J(C)_{(2)}/H$ gives a Göpel subgroup of $J(B)$. The pair (B, H_0) is not bielliptic, hence the three points $\text{Sing } q_i$ ($i = 1, 2, 3$) are not collinear. By the proposition, Km C/ε_H is isomorphic to the Enriques surface which is the $(2, 2)$-covering of the projective plane.

By Lemma 1, we have a Cremona involution σ which exchanges l_1, l_2, l_3 with l_4, l_5, l_6 respectively. This involution σ lifts to Km C, hence we obtain an action by C_2^4 on Km C and on the Enriques surface S we get the extension of $G_0 \simeq C_2^2$ to the group $G \simeq C_2^3$. The Cremona involution σ has only four isolated fixed points. Hence the lift of σ as an anti-symplectic involution of Km C has no fixed points. This together with Proposition 1 proves the following.

Proposition 4. *The Enriques surface* Km C/ε_H *of Hutchinson–Göpel type has an action of Mathieu type by the elementary abelian group $G \simeq C_2^3$.*

In fact, every involution in the coset $G \setminus G_0$ has type (M0). Although we can prove this from geometric consideration so far, we postpone it until Theorem 3 where a straightforward computation of the fixed locus is given.

Remark 8. The image T of the rational map $(\star 3)$ is the octahedral toric threefold and its automorphism group is isomorphic to the semi-direct product $(\mathbb{C}^*)^3.(\mathfrak{S}_4 \times \mathfrak{S}_2)$. The obvious C_2^3 of Aut(Km C) is induced from the Klein's four-group in $\mathfrak{S}_4$ and the Cremona involution, the generator of $\mathfrak{S}_2$. But any lift of the Cremona involution σ does not come from Aut T.

Let us study the symmetry of the sextic surface

$$(\star 5) \quad \sum_{i=0}^{3} a_i x_i^2 = \left(\sum_{i=0}^{3} \frac{b_i}{x_i^2} \right) x_0 x_1 x_2 x_3.$$

The group $G_0 \simeq C_2^2$ acts by the simultaneous change of signs of two coordinates. The coefficients a_i, b_i ($i = 0, \ldots, 3$) are given as in Proposition 2. When we are treating Enriques surfaces of Hutchinson–Göpel type, since the six lines satisfy the condition $(\star 0)$, we have

$$\prod_{i=1}^{3} \alpha_i \prod_{i=1}^{3} \beta_i = 1.$$

By the identity

$$\prod_{i=0}^{3} a_i - \prod_{i=0}^{3} b_i = \left(\prod_{i=1}^{3} \alpha_i \prod_{i=1}^{3} \beta_i - 1 \right) \prod_{i=1}^{3} (\alpha_i - \beta_i),$$

we obtain $\prod_{i=0}^{3} a_i = \prod_{i=0}^{3} b_i$. By choosing the constants appropriately, the sextic surface $(\star 5)$ acquires the action of the standard Cremona involution

$$(x_0 : x_1 : x_2 : x_3) \mapsto \left(\frac{(\text{const.})}{x_0} : \frac{(\text{const.})}{x_1} : \frac{(\text{const.})}{x_2} : \frac{(\text{const.})}{x_3} \right). \tag{6}$$

This action together with G_0 gives us the action of $G \simeq C_2^3$.

Remark 9. (1) When a principally polarized abelian surface A is the product $E_1 \times E_2$, then the morphism $\Phi_{|2\Theta|} : A \to \mathbb{P}^3$ is of degree 2 onto a smooth quadric. The limit of Enriques surfaces of Hutchinson–Göpel type, when (Jac C, H) becomes $(E_1 \times E_2, H_0)$, is the Enriques surface Km$(E_1 \times E_2)/\varepsilon$ of Lieberman type (Example 6) or Kondo–Mukai type according as the Göpel subgroup H_0 is a product or not. Km$(E_1 \times E_2)$ is also a $(2, 2, 2)$-cover of $\mathbb{P}^2$ branched along three reducible quadrics $(\star 2)$. In the latter case they satisfy $(\star 0)$ and Sing(q_i) ($i = 1, 2, 3$) are collinear.
(2) When the three points Sing(q_i) are collinear, there exists an involution of $\mathbb{P}^2$ which exchanges l_i with l_{i+3} for $i = 1, 2, 3$. Thus we have an extension of the

group action of G_0 to a group C_2^3 in this case, too. However this action is not of Mathieu type. In this case the Enriques surface coincides with the one in [9] and the coset $C_2^3 \setminus G_0$ contains a numerically trivial involution.

Further discussions on these topics will be pursued elsewhere.

6 Examples of Mathieu Actions by Large Groups

In this section we treat more directly the sextic Enriques surfaces of Hutchinson–Göpel type (Sect. 5, ($\star$ 5)). We start by studying the singularities of sextic Enriques surfaces and then as an application we give explicit examples of Enriques surfaces of Hutchinson–Göpel type which are acted on by the groups $C_2 \times \mathfrak{A}_4$ and $C_2 \times C_4$ of Mathieu type [12].

We recall the sextic equation of an Enriques surface of diagonal type from Proposition 2,

$$F(x_0, x_1, x_2, x_3) =$$
$$(a_0 x_0^2 + a_1 x_1^2 + a_2 x_2^2 + a_3 x_3^2)x_0 x_1 x_2 x_3 + \left(\frac{b_0}{x_0^2} + \frac{b_1}{x_1^2} + \frac{b_2}{x_2^2} + \frac{b_3}{x_3^2}\right) x_0^2 x_1^2 x_2^2 x_3^2, \quad (7)$$

where $\prod_i a_i \prod_i b_i \neq 0$. Let $\overline{S}$ be the singular surface defined by F. By Bertini's theorem every general element in this linear system is smooth outside the coordinate tetrahedron $\Delta = \{x_0 x_1 x_2 x_3 = 0\}$, whereas along the intersection $\Delta \cap \overline{S}$ it always has singularities.

At each coordinate point, say at $P = (0 : 0 : 0 : 1)$, F is expanded to

$$(a_3 x_3^3)x_0 x_1 x_2 + (\text{higher terms in } x_0, x_1, x_2)$$

as a polynomial in the variables x_0, x_1, x_2. This shows that $\overline{S}$ has an ordinary triple point at P. It can be resolved by the normalization $\pi \colon S \to \overline{S}$ and $\pi^{-1}(P)$ consists of three points. These three points correspond to the three components $\{x_i = 0\}(i = 0, 1, 2)$ of the resolution of the triple point $\{x_0 x_1 x_2 = 0\}$, so it may be natural to denote them by

$$\pi^{-1}(P) = \{(\overline{0} : 0 : 0 : 1), (0 : \overline{0} : 0 : 1), (0 : 0 : \overline{0} : 1)\}.$$

Along each edges of the tetrahedron Δ, say along $l = \{x_0 = x_1 = 0\}$, F is expanded to

$$x_2 x_3((b_0 x_2 x_3)x_1^2 + (a_2 x_2^2 + a_3 x_3^2)x_1 x_0 + (b_1 x_2 x_3)x_0^2) + (\text{higher terms in } x_0, x_1)$$

as a polynomial in the variables x_0, x_1. Therefore $\overline{S}$ has the singularity of ordinary double lines at $(0 : 0 : x_2 : x_3)$ if $g(T) = (b_0 x_2 x_3)T^2 + (a_2 x_2^2 + a_3 x_3^2)T + b_1 x_2 x_3$ has only simple roots; if $g(T)$ has multiple roots, it becomes a pinch point (also called a

Whitney umbrella singularity). We see that both of these singularities are resolved by the normalization π. The double cover $\tilde{l} := \pi_*^{-1}(l) \to l$ branches at the pinch points. Since the discriminant condition of $g(T)$, $\begin{vmatrix} 2b_0 x_2 x_3 & a_2 x_2^2 + a_3 x_3^2 \\ a_2 x_2^2 + a_3 x_3^2 & 2b_1 x_2 x_3 \end{vmatrix} = 0$, gives in general four pinch points, the curve $\tilde{l}$ is an elliptic curve. At each coordinate point, say at $(0:0:0:1)$, we see that $\tilde{l}$ contains exactly the two points $(\bar{0}:0:0:1)$ and $(0:\bar{0}:0:1)$. We denote by $\tilde{l}_{ij}$ the strict transform of the edge $l_{ij} = \{x_i = x_j = 0\}$.

As is proved in Sect. 5, the Enriques surface $\overline{S}$ of Hutchinson–Göpel type satisfies $\prod_i a_i = \prod_i b_i$ in (7). By a suitable scalar multiplication of coordinates, the equation of $\overline{S}$ is normalized into

$$(x_0^2 + x_1^2 + x_2^2 + x_3^2) + \left(\frac{b_0}{x_0^2} + \frac{b_1}{x_1^2} + \frac{b_2}{x_2^2} + \frac{b_3}{x_3^2}\right) x_0 x_1 x_2 x_3, \quad \prod_{i=0}^{3} b_i = 1. \qquad (8)$$

To make use of Cremona transformations, we work also with $B_i = \sqrt{b_i}$, $\prod_i B_i = 1$. With the previous notation we can give a full statement of Proposition 4.

Theorem 3. *The Enriques surface of Hutchinson–Göpel type (8) has the automorphisms*

$$s_1: (x_0:x_1:x_2:x_3) \mapsto (-x_0:-x_1:x_2:x_3),$$
$$s_2: (x_0:x_1:x_2:x_3) \mapsto (-x_0:x_1:-x_2:x_3),$$
$$\sigma: (x_0:x_1:x_2:x_3) \mapsto \left(\frac{B_0}{x_0}:\frac{B_1}{x_1}:\frac{B_2}{x_2}:\frac{B_3}{x_3}\right).$$

The involutions s_1, s_2 generate the group $G_0 \simeq C_2^2$ and s_1, s_2, σ generate the group $G \simeq C_2^3$. Their types as regards the fixed locus are as follows.

(1) Every non-identity element of G_0 has type (M2).
(2) Every element of the coset $G \setminus G_0$ has type (M0).

Proof. Let $\pi: S \to \overline{S}$ be the normalization. Then S is the smooth minimal model and the actions extend. It is easy to see that

$$\mathrm{Fix}(s_1) = \tilde{l}_{01} \cup \tilde{l}_{23} \cup \{(0:0:\bar{0}:1), (0:0:1:\bar{0}), (\bar{0}:1:0:0), (1:\bar{0}:0:0)\}$$

and similarly for $s_2, s_3 = s_1 s_2$. This shows the assertion about fixed points of s_i. As for σ and σs_i, first we note that they exchange the three pairs of opposite edges of the tetrahedron Δ. Hence their fixed loci exist only in the complement of the coordinate hyperplanes, on which the whole group acts biregularly. In fact we see that, for example, the fixed points of σ consist of the four points of the form

$$(e_0 \sqrt{B_0} : e_1 \sqrt{B_1} : e_2 \sqrt{B_2} : e_3 \sqrt{B_3}),$$

where $e_i \in \{\pm 1\}$ satisfy $\prod e_i = -1$ and we fix once and for all $\sqrt{B_i}$ for which $\prod \sqrt{B_i} = 1$. Thus the fixed points of Cremona transformations are of type (M0).

We remark that if the pair (C, H) (or (B, H_0)) admits a special automorphism, it induces a further automorphism of the Enriques surface $S = (\mathrm{Km}\, C)/\varepsilon_H$. Theorem 4 below is an example of this general idea. Recall that a finite group of semi-symplectic automorphisms[5] of an Enriques surface is called *of Mathieu type* if every element g of order 2 or 4 acts with $\chi_{\mathrm{top}}(\mathrm{Fix}(g)) = 4$ (see [12]). Our theorem provides two examples of large group actions of Mathieu type on Enriques surfaces.

Theorem 4. *Let S be the Enriques surface of Hutchinson–Göpel type in Sect. 5.*

(1) If B admits an automorphism ψ of order 3, then B has a Göpel subgroup H_0 preserved by ψ, and S acquires an action of Mathieu type by the group $C_2 \times \mathfrak{A}_4$.

(2) If B has an action by the dihedral group D_8 of order 8, then B has a Göpel subgroup H_0 preserved by D_8, and S acquires an action by the group $C_2^3 \rtimes C_2^2$. The restriction to a certain subgroup isomorphic to $C_2 \times C_4$ is of Mathieu type.

Proof. Consult, e.g., [2] or [6, Sect. 8] for automorphisms of curves of genus two.

(1) We may assume that B is defined by

$$w^2 = (x^3 - \lambda^3)(x^3 - \lambda^{-3}), \lambda \in \mathbb{C}^*.$$

Then the curve B has the automorphism $\psi \colon (x, w) \mapsto (\zeta_3 x, w)$ where ζ_3 is the primitive cube root of unity. We label the six branch points as

$$\lambda_i = \zeta_3^{i-1}\lambda \ (i = 1, 2, 3), \quad \lambda_i = \zeta_3^{i-1}\lambda^{-1} \ (i = 4, 5, 6)$$

Then the automorphism ψ acts on the Göpel subgroup $H_0 \subset J(B)_{(2)}$ of Sect. 5 as follows.

$$p_1 - p_4 \mapsto p_2 - p_5 \mapsto p_3 - p_6 \mapsto p_1 - p_4.$$

The induced automorphism on S is denoted by the same letter ψ. More explicitly, by (5) and Proposition 2, we see that ψ permutes the coordinates w_1, w_2, w_3 and x_1, x_2, x_3. This symmetry has the effect on (8) of $\overline{S}$ that $b_1 = b_2 = b_3 =: A^2$, hence we get the family

$$(x_0^2 + x_1^2 + x_2^2 + x_3^2) + \left(\frac{1}{A^6 x_0^2} + \frac{A^2}{x_1^2} + \frac{A^2}{x_2^2} + \frac{A^2}{x_3^2}\right) x_0 x_1 x_2 x_3 = 0. \quad (9)$$

The action of ψ is given by $(x_0 : x_1 : x_2 : x_3) \mapsto (x_0 : x_3 : x_1 : x_2)$ and it extends the group $G = C_2^3$ to $C_2 \times \mathfrak{A}_4$. Theorem 3 shows that its unique 2-Sylow subgroup acts with Mathieu character, hence this is an example of a family with a group action of Mathieu type.

---

[5] An automorphism is semi-symplectic if it acts on the space $H^0(S, \mathcal{O}_S(2K_S))$ trivially.

(2) We may assume that B is defined by

$$w^2 = x(x^2 - \lambda^2)\left(x^2 - \frac{1}{\lambda^2}\right), \quad \lambda \in \mathbb{C}^*,$$

and has the automorphisms

$$\psi\colon (x, w) \mapsto (-x, \sqrt{-1}w), \quad \varphi\colon (x, w) \mapsto \left(\frac{1}{x}, \frac{w}{x^3}\right).$$

These ψ, φ generate a group isomorphic to D_8. Here we label the six branch points as

$$\lambda_1 = 0, \ \lambda_4 = \infty; \ \lambda_2 = \lambda, \ \lambda_5 = -\lambda; \ \lambda_3 = 1/\lambda, \ \lambda_6 = -1/\lambda.$$

Then the Göpel subgroup H_0 is preserved by ψ, φ. This extends the group G of automorphisms of the Enriques surface S in Theorem 3 to $C_2^3 \rtimes C_2^2$ (the index 4 of the extention from G corresponds to the order of the reduced automorphism group $\langle \psi, \varphi \rangle / \psi^2$ of B).

Their action on the equation is as follows. As before, we use the notation of (5) and Proposition 2. First the action of φ on w_i is given by

$$w_1 \mapsto w_1^{-1}, \ w_2 \leftrightarrow w_3$$

and we have $(1 : w_2 w_3 : w_3 w_1 : w_1 w_2) \mapsto (w_1 : w_1 w_2 w_3 : w_2 : w_3)$. In view of Remark 5, this is the same as a Cremona transformation followed by a permutation. Next, the action of ψ on w_i is given by

$$w_1 \mapsto w_1, \ w_2 \mapsto w_2^{-1}, \ w_3 \mapsto w_3^{-1}$$

and we have $(x_0 : x_1 : x_2 : x_3) \mapsto (x_1 : x_0 : x_3 : x_2)$. This implies that our Enriques surface has the sextic equation

$$(x_0^2 + x_1^2 + x_2^2 + x_3^2) + \sqrt{-1}\left(\frac{A^2}{x_0^2} + \frac{A^2}{x_1^2} + \frac{1}{A^2 x_2^2} + \frac{1}{A^2 x_3^2}\right)x_0 x_1 x_2 x_3 = 0. \qquad (10)$$

(See Remark 10.) Although S has a group of automorphisms of order 32, most of them do not satisfy the Mathieu condition. However, we claim that there are some that do.

Claim. The action

$$g\colon (x_0 : x_1 : x_2 : x_3) \mapsto \left(\frac{A}{x_1} : \frac{A}{x_0} : \frac{1}{A x_3} : -\frac{1}{A x_2}\right)$$

is of Mathieu type of order 4.

Proof. We show this by the computation of the fixed locus. The action of g on the edges of the tetrahedron Δ is as follows: it exchanges l_{01} and l_{23}, while it stabilizes the other four edges. For each intersecting pair of stable edges, we have an isolated fixed point. Hence there are four isolated fixed points. These are exactly the isolated points of $\mathrm{Fix}(s_1)$, in view of the relation $g^2 = s_1$.

In this way, we find that the family (10) has the Mathieu type actions generated by g and

$$h: (x_0 : x_1 : x_2 : x_3) \mapsto \left(\frac{A}{x_1} : -\frac{A}{x_0} : \frac{1}{Ax_3} : \frac{1}{Ax_2}\right).$$

The relations $g^2 = h^2 = s_1, gh = hg$ show that they in fact generate the group $C_2 \times C_4$.

Remark 10. The coefficient $\sqrt{-1}$ is a kind of subtlety of Enriques surfaces, which makes (10) irreducible. Note that without this adjustment, we obtain the reducible equation

$$(x_0^2 + x_1^2 + x_2^2 + x_3^2) + \left(\frac{A^2}{x_0^2} + \frac{A^2}{x_1^2} + \frac{1}{A^2 x_2^2} + \frac{1}{A^2 x_3^2}\right) x_0 x_1 x_2 x_3$$

$$= \left(x_0^2 + x_1^2 + \frac{x_0 x_1}{A^2 x_2 x_3}(x_2^2 + x_3^2)\right)\left(1 + A^2 \frac{x_2 x_3}{x_0 x_1}\right).$$

The octahedral Enriques surface. A careful look at (9) and (10) shows that they have a member $\overline{S}_{oct}$ in common,

$$\overline{S}_{oct}: (x_0^2 + x_1^2 + x_2^2 + x_3^2) + \sqrt{-1}\left(\frac{1}{x_0^2} + \frac{1}{x_1^2} + \frac{1}{x_2^2} + \frac{1}{x_3^2}\right) x_0 x_1 x_2 x_3 = 0.$$

This surface is associated to the curve B (($\star 1$), Sect. 5) which is ramified over the six vertices of the regular octahedron inscribed in $\mathbb{P}^1$. Hence we call the desingularization S_{oct} of this surface the *octahedral Enriques surface.*

The additional automorphisms are quite visible on $\overline{S}_{oct}$; it is generated by the symmetric group $\mathfrak{S}_4$ acting on the coordinates $\{x_i\}$ and three involutions $\beta_j: (x_i) \mapsto (\epsilon_{ij}/x_i)$ ($j = 1, 2, 3$) where ϵ_{ij} takes value -1 if $i = j$ and 1 otherwise. Thus S_{oct} is acted on by the group $C_2^3 \mathfrak{S}_4$ of order 192. For convenience, we give here a table of topological structures of the fixed loci of these automorphisms, sorted by the conjugacy classes in $C_2^3 \mathfrak{S}_4$.

Representative	Length	Order	Fixed loci
id	1	1	S_{oct}
β_1	4	2	{4pts.}
$\beta_1\beta_2$	3	2	(Two elliptic curves) + {4pts.}
$\beta_1(x_0x_1)(x_2x_3)$	12	4	{4pts.}
$\beta_1\beta_2(x_0x_1)(x_2x_3)$	6	2	(Two rational curves) + {4pts.}
$(x_0x_1)(x_2x_3)$	6	2	{4pts.}
$\beta_2(x_0x_1)$	12	2	(A rational curve) + {4pts.}
(x_0x_1)	12	2	(A genus-two curve) + {4pts.}
$\beta_1(x_0x_1)$	12	4	{2pts.}
$\beta_1\beta_2(x_0x_1)$	12	4	{2pts.}
$\beta_1(x_1x_2x_3)$	32	6	{1pt.}
$(x_1x_2x_3)$	32	3	{3pts.}
$\beta_1(x_0x_1x_2x_3)$	24	4	{4pts.}
$(x_0x_1x_2x_3)$	24	4	{2pts.}

We remark that, as the specialization of the families (9) and (10), the group $C_2 \times \mathfrak{A}_4$ is generated by $\beta_2\beta_3, \beta_3\beta_1, \beta_1\beta_2, (x_1x_2x_3)$ and the group $C_2 \times C_4$ is generated by $\beta_1(x_0x_1)(x_2x_3), \beta_3(x_0x_1)(x_2x_3)$.

7 The Characterization

In this section we prove a converse of Theorem 3 (resp. Proposition 1), stating that sextic Enriques surfaces of Hutchinson–Göpel type (resp. of diagonal type) are characterized by the group actions by G (resp. G_0).

To begin with, let us recall the study of involutions of Mathieu type. Every involution s on an Enriques surface S acts on the space $H^0(S, \mathcal{O}_S(2K_S))$ trivially. This means that at a fixed point P of s, the derivative of s satisfies $\det(ds)_P = \pm 1$. The fixed point P is called *symplectic* (resp. *anti-symplectic*) according to the value $\det(ds)_P = +1$ (resp. -1). The set of symplectic (resp. anti-symplectic) fixed points is denoted by $\mathrm{Fix}^+(s)$ (resp. $\mathrm{Fix}^-(s)$). Geometrically, $\mathrm{Fix}^+(s)$ is exactly the set of isolated fixed points and $\mathrm{Fix}^-(s)$ is the set of fixed curves since s has order two. By topological and holomorphic Lefschetz formulas, we see always $\#\mathrm{Fix}^+(s) = 4$ and the Mathieu condition is equivalent to $\chi_{top}(\mathrm{Fix}^-(s)) = 0$. A more precise argument shows that there are only four types (M0)-(M3) mentioned in the introduction.

Lemma 2. *Let s be an involution of (M2) type on an Enriques surface S; we denote the two elliptic curves of $\mathrm{Fix}^-(s)$ by E, F. Then there exists an elliptic fibration $S \to \mathbb{P}^1$ in which $2E$ and $2F$ are multiple fibers.*

Proof. It is well-known that the linear system of some multiple of E gives an elliptic fibration $f: S \to \mathbb{P}^1$, [1, Chap. VIII]. Since s fixes the fibers E, F and those which contain the four points of $\mathrm{Fix}^+(s)$, s acts on the base trivially. Thus s is induced from an automorphism s_0 of the Jacobian fibration $J(f)$. Since s does not have fixed horizontal curves, s_0 acts as a fiberwise translation. Hence E and F must be multiple fibers.

Here we first give the characterization of Enriques surfaces of diagonal type.

Proposition 5. *Let S be an Enriques surface with an action of Mathieu type by the group $G_0 := C_2^2$ such that every nontrivial element is of (M2) type. Then S is birationally equivalent to the sextic Enriques surface of diagonal type, Proposition 2, $(\star \star \star)$.*

Proof. We let the group $G_0 = \{1, s_1, s_2, s_3\}$ and let E_i, F_i be the two elliptic curves in the fixed locus $\mathrm{Fix}(s_i)$ respectively for $i = 1, 2, 3$. Lemma 2 shows that the divisor class of $2E_i \sim 2F_i$ defines an elliptic fibration $f_i \colon S \to \mathbb{P}^1$. Moreover, since f_i has exactly two multiple fibers, for $j \neq i$ the curves E_j and F_j are horizontal in the fibration f_i. In particular the intersections

$$E_i \cap E_j, E_i \cap F_j, F_i \cap E_j, F_i \cap F_j \tag{11}$$

are all nonempty.

On the other hand, each of the four intersections of (11) defines an isolated fixed point of s_k because $s_i s_j = s_k$, where k is taken as the element in $\{1, 2, 3\} \setminus \{i, j\}$. These isolated fixed points belong to the set $\mathrm{Fix}^+(s_k)$, which consists of four points. Hence we see that the intersections of (11) all are transversal and consists of one point.

Next let us consider the linear system $\mathcal{L} := |E_1 + E_2 + E_3|$ with $\mathcal{L}^2 = 6$. This is a nef and big divisor, hence it maps S into $\mathbb{P}^3$. Note that the relation

$$E_1 + E_2 + E_3 \sim E_1 + F_2 + F_3 \sim F_1 + E_2 + F_3 \sim F_1 + F_2 + E_3 \tag{12}$$

shows that $\mathcal{L}$ is base-point-free. Then we can use [3, Remark 7.9] to see that at least either $\mathcal{L}$ or $\mathcal{L} + K_S$ gives a birational morphism onto a sextic surface. Noting that $\mathcal{L} + K_S$ is nothing but the system $|E_1 + E_2 + F_3|$, exchanging E_3 and F_3 if necessary, we can assume that $\mathcal{L}$ gives a birational morphism φ onto a sextic surface $\overline{S} \subset \mathbb{P}^3$. As is known, $\overline{S}$ becomes a sextic surface with double lines along edges of a tetrahedron Δ. In our case the edges of Δ consist of the images of the six elliptic curves $E_1, \cdots, F_3$.

We denote by x_0, x_1, x_2, x_3 the respective global sections of $\mathcal{O}_S(E_1 + E_2 + E_3)$ corresponding to the divisors (12). In these coordinates Δ is nothing but the coordinate tetrahedron $\Delta = \{x_0 x_1 x_2 x_3 = 0\}$. Thus our surface $\overline{S}$ belongs to the following linear system of sextics,

$$q(x_0, x_1, x_2, x_3) + x_0 x_1 x_2 x_3 (b_0/x_0^2 + b_1/x_1^2 + b_2/x_2^2 + b_3/x_3^2),$$

where q is a quadric and $b_0, \cdots, b_3 \in \mathbb{C}$ are constants.

The involution s_i induces a linear transformation of the ambient $\mathbb{P}^3$. More precisely, since s_i stabilizes each divisor in (12), $s_i(x_j)$ is just a scalar multiple of x_j for any i, j. By considering their fixed locus, we easily deduce that this action is given by changing the signs of two coordinates. Since $\overline{S}$ is invariant under this change of signs, we have $q(x, y, z, t) = a_0 x_0^2 + a_1 x_1^2 + a_2 x_2^2 + a_3 x_3^2$. Therefore, S is birationally equivalent to a sextic Enriques surface of diagonal type.

Proof of Theorem 1. We identify the subgroup G_0 with the one in the previous proposition and keep the same notation. Recall from Sect. 5 that the sextic surface ($\star 5$) is an Enriques surface of Hutchinson–Göpel type exactly when $\prod_i a_i = \prod_i b_i$. This is the case when there exists an action of standard Cremona transformation (6). Let $\sigma \in G - G_0$. We claim that σ exchanges E_i and F_i for any i.

Suppose σ preserves E_1 and F_1. Then we would obtain an effective action of $\langle \sigma, s_2 \rangle \simeq C_2^2$ on both E_1 and F_1. Since s_2 has fixed points on them, it negates the periods. It follows that the elements σ and σs_2, both in the set $G \setminus G_0$, cannot act on E_1 freely, so that for example it would happen that σ has four fixed points on E_1 and σs_2 has four fixed points on F_1. But this is not possible, since on the $K3$ cover X, the symplectic lift $\tilde{\sigma}$ has eight fixed points inside the inverse image of E_1 which is an irreducible elliptic curve (since E_1 is a double fiber). Thus we have proved that σ exchanges E_1 and F_1. The same applies to E_i and F_i for $i = 2, 3$.

Thus σ sends $\sum E_i$ to $\sum F_i$. It follows that σ transforms the sextic model defined by $\mathcal{L}$ to a sextic model defined by $\mathcal{L} + K_S$. As was noticed in Remark 5 (or [11, Remark 4.2]), these two models are related via the Cremona transformation. Thus the Enriques surface is of Hutchinson–Göpel type.

Acknowledgements We are grateful to the organizers of the interesting Workshop on Arithmetic and Geometry of $K3$ surfaces and Calabi–Yau threefolds. The second author is grateful to Professor Shigeyuki Kondo for discussions and encouragement.

This work is supported in part by the JSPS Grant-in-Aid for Scientific Research (B) 17340006, (S) 19104001, (S) 22224001, (A) 22244003, for Exploratory Research 20654004 and for Young Scientists (B) 23740010.

References

1. W. Barth, K. Hulek, C. Peters, A. Van de Ven, in *Compact Complex Surfaces*, 2nd enlarged edn. Erg. der Math. und ihrer Grenzgebiete, 3. Folge, Band 4 (Springer, Berlin, 2004)
2. O. Bolza, On binary sextics with linear transformations into themselves. Am. J. Math. **10**, 47–70 (1888)
3. F.R. Cossec, Projective models of Enriques surfaces. Math. Ann. **265**, 283–334 (1983)
4. K. Hulek, M. Schütt, Enriques surfaces and Jacobian elliptic $K3$ surfaces. Math. Z. **268**, 1025–1056 (2011)
5. J.I. Hutchinson, On some birational transfromations of the Kummer surface into itself. Bull. Am. Math. Soc. **7**, 211–217 (1901)
6. J. Igusa, Arithmetic variety of moduli for genus two. Ann. Math. **72**, 612–649 (1960)
7. J.H. Keum, Every algebraic Kummer surface is the K3-cover of an Enriques surface. Nagoya Math. J. **118**, 99–110 (1990)
8. S. Kondo, Enriques surfaces with finite automorphism groups. Jpn. J. Math. **12**, 191–282 (1986)

9. S. Mukai, Numerically trivial involutions of Kummer type of an Enriques surface. Kyoto J. Math. **50**, 889–902 (2010)
10. S. Mukai, Kummer's quartics and numerically reflective involutions of Enriques surfaces. J. Math. Soc. Jpn. **64**, 231–246 (2012)
11. S. Mukai, Lecture notes on $K3$ and Enriques surfaces (Notes by S. Rams), in *Contributions to Algebraic Geometry*. IMPANGA Lecture Notes (European Mathematical Society Publishing House) Zurich, 2012
12. S. Mukai, H. Ohashi, Finite groups of automorphisms of Enriques surfaces and the Mathieu group M_{12} (in preparation)
13. H. Ohashi, On the number of Enriques quotients of a $K3$ surface. Publ. Res. Inst. Math. Sci. **43**, 181–200 (2007)
14. H. Ohashi, Enriques surfaces covered by Jacobian Kummer surfaces. Nagoya Math. J. **195**, 165–186 (2009)
15. T. Shioda, Some results on unirationality of algebraic surfaces. Math. Ann. **230**, 153–168 (1992)

Quartic K3 Surfaces and Cremona Transformations

Keiji Oguiso

Dedicated to Professor Doctor Klaus Hulek on the occasion of his sixtieth birthday

Abstract We prove that there is a smooth quartic K3 surface automorphism that is not derived from the Cremona transformation of the ambient three-dimensional projective space. This gives a negative answer to a question of Professor Marat Gizatullin.

Key words: Automorphisms, Quartic K3 surfaces, Cremona transformations

Mathematics Subject Classifications (2010): Primary 14J28; Secondary 14E07, 14J50

1 Introduction

Throughout this note, we work over the complex number field $\mathbf{C}$.

In his lecture "Quartic surfaces and Cremona transformations" [2] in the workshop on Arithmetic and Geometry of K3 surfaces and Calabi–Yau threefolds held at the Fields Institute (August 16–25, 2011), Professor Igor Dolgachev discussed the following question with several beautiful examples supporting it:

Question 1. Let $S \subset \mathbf{P}^3$ be a smooth quartic K3 surface. Is any biregular automorphism g of S (as abstract variety) derived from a Cremona transformation of the ambient space $\mathbf{P}^3$? More precisely, is there a birational automorphism $\tilde{g}$ of $\mathbf{P}^3$ such that $\tilde{g}_*(S) = S$ and $g = \tilde{g}|S$? Here $\tilde{g}_*(S)$ is the proper transform of S, and $\tilde{g}|S$ is the necessarily biregular, birational automorphism of S then induced by $\tilde{g}$.

K. Oguiso (✉)
Department of Mathematics, Osaka University, Toyonaka 560-0043 Osaka, Japan

Korea Institute for Advanced Study, 560-0043 Hoegiro 87, Seoul 130-722, Korea
e-mail: oguiso@math.sci.osaka-u.ac.jp

R. Laza et al. (eds.), *Arithmetic and Geometry of K3 Surfaces and Calabi–Yau Threefolds*,   455
Fields Institute Communications 67, DOI 10.1007/978-1-4614-6403-7_16,
© Springer Science+Business Media New York 2013

Later, Dolgachev pointed out to me that, to his best knowledge, Gizatullin was the first who asked this question. The aim of this short note is to give a negative answer to the question:

Theorem 1. *(1) There exists a smooth quartic K3 surface* $S \subset \mathbf{P}^3$ *of Picard number 2 such that* $\mathrm{Pic}\,(S) = \mathbf{Z}h_1 \oplus \mathbf{Z}h_2$ *with intersection form:*

$$((h_i.h_j)) = \begin{pmatrix} 4 & 20 \\ 20 & 4 \end{pmatrix}.$$

(2) Let S be as above. Then $\mathrm{Aut}\,(S)$ *has an element g such that it is of infinite order and* $g^*(h) \neq h$. *Here* $\mathrm{Aut}\,(S)$ *is the group of biregular automorphisms of S as an abstract variety, and* $h \in \mathrm{Pic}\,(S)$ *is the hyperplane section class.*

(3) Let S and g be as above. Then there is no element $\tilde{g}$ *of* $\mathrm{Bir}\,(\mathbf{P}^3)$ *such that* $\tilde{g}_*(S) = S$ *and* $g = \tilde{g}|S$. *Here* $\mathrm{Bir}\,(\mathbf{P}^3)$ *is the Cremona group of* $\mathbf{P}^3$, *i.e., the group of birational automorphisms of* $\mathbf{P}^3$.

Our proof is based on a result of Takahashi concerning the log Sarkisov program [7], which we quote as Theorem 2, and standard argument concerning K3 surfaces.

Remark 1. (1) Let $C \subset \mathbf{P}^2$ be a smooth cubic curve, i.e., a smooth curve of genus 1. It is classical that any element of $\mathrm{Aut}\,(C)$ is derived from a Cremona transformation of the ambient space $\mathbf{P}^2$. In fact, this follows from the fact that any smooth cubic curve is written in Weierstrass form after a linear change of coordinates and the explicit form of the group law in terms of the coordinates.

(2) Let n be an integer such that $n \geq 3$ and $Y \subset \mathbf{P}^{n+1}$ be a smooth hypersurface of degree $n + 2$. Then Y is an n-dimensional Calabi–Yau manifold. It is well known that $\mathrm{Bir}\,(Y) = \mathrm{Aut}\,(Y)$, it is a finite group, and any element of $\mathrm{Aut}\,(Y)$ is derived from a biregular automorphism of the ambient space $\mathbf{P}^{n+1}$. In fact, the statement follows from $K_Y = 0$ in $\mathrm{Pic}\,(Y)$ (adjunction formula), $H^0(T_Y) = 0$ (by $T_Y \simeq \Omega_Y^{n-1}$ together with Hodge symmetry), and $\mathrm{Pic}\,(Y) = \mathbf{Z}h$, where h is the hyperplane class (Lefschetz hyperplane section theorem). We note that $K_Y = 0$ implies that any birational automorphism of Y is an isomorphism in codimension one, so that for any birational automorphism g of Y, we have a well-defined group isomorphism g^* on $\mathrm{Pic}\,(Y)$. Then $g^*h = h$. This implies that g is biregular and it is derived from an element of $\mathrm{Aut}\,(\mathbf{P}^{n+1}) = \mathrm{PGL}\,(\mathbf{P}^{n+1})$.

2 Proof of Theorem 1(1)(2)

In this section, we shall prove Theorem 1(1)(2) by dividing it into several steps. The last lemma (Lemma 5) will be used also in the proof of Theorem 1(3).

Lemma 1. *There is a projective K3 surface such that* $\mathrm{Pic}\,(S) = \mathbf{Z}h_1 \oplus \mathbf{Z}h_2$ *with*

$$((h_i.h_j)) = \begin{pmatrix} 4 & 20 \\ 20 & 4 \end{pmatrix}.$$

Proof. Note that the abstract lattice given by the symmetric matrix above is an even lattice of rank 2 with signature $(1, 1)$. Hence the result follows from [4], Corollary (2.9), which is based on the surjectivity of the period map for K3 surfaces (see, e.g., [1, Page 338, Theorem 14.1]) and Nikulin's theory [5] of integral bilinear forms.

From now on, S is a K3 surface in Lemma 1.

Note that the cycle map $c_1 : \mathrm{Pic}\,(S) \to \mathrm{NS}\,(S)$ is an isomorphism for a K3 surface. So, we identify these two spaces. $\mathrm{NS}\,(S)_{\mathbf{R}}$ is $\mathrm{NS}\,(S) \otimes_{\mathbf{Z}} \mathbf{R}$. The positive cone $P(S)$ of S is the connected component of the set

$$\{x \in \mathrm{NS}\,(S)_{\mathbf{R}} \mid (x^2)_S > 0\} \ ,$$

containing the ample classes. The ample cone $\mathrm{Amp}\,(S) \subset \mathrm{NS}\,(S)_{\mathbf{R}}$ of S is the open convex cone generated by the ample classes.

Lemma 2. *NS (S) represents neither 0 nor -2. In particular, S has no smooth rational curve and no smooth elliptic curve and $(C^2)_S > 0$ for all nonzero effective curves C in S. In particular, the positive cone of S coincides with the ample cone of S.*

Proof. We have $((xh_1 + yh_2)^2)_S = 4(x^2 + 10xy + y^2)$. Hence there is no $(x, y) \in \mathbf{Z}^2$ such that $((xh_1 + yh_2)^2)_S \in \{-2, 0\}$.

Lemma 3. *After replacing h_1 by $-h_1$, the line bundle h_1 is very ample. In particular, $\Phi_{|h_1|} : S \to \mathbf{P}^3$ is an isomorphism onto a smooth quartic surface.*

Proof. h_1 is non-divisible in $\mathrm{Pic}\,(S)$ by construction. It follows from Lemma 2 and $(h_1^2)_S = 4 > 0$ that one of $\pm h_1$ is ample with no fixed component. By replacing h_1 by $-h_1$, we may assume that it is h_1. Then, by Saint-Donat [6], Theorem 6.1, h_1 is a very ample line bundle with the last assertion.

By Lemma 3, *we may and will assume that $S \subset \mathbf{P}^3$ and denote this inclusion by ι and a general hyperplane section by h. That is, $h = H \cap S$ for a general hyperplane $H \subset \mathbf{P}^3$, from now on.* Note that $h = h_1$ in $\mathrm{Pic}\,(S)$.

Lemma 4. *There is an automorphism g of S such that g is of infinite order and $g^*(h) \neq h$ in $\mathrm{Pic}\,(S)$.*

There are several ways to prove this fact. The following simpler proof was suggested by the referee.

Proof. Let us consider the following orthogonal transformation σ of NS (S)):

$$\sigma(h_1) = 10h_1 - h_2 \ , \ \sigma(h_2) = h_1 \ .$$

It is straightforward to see that σ is certainly an element of $O(\mathrm{NS}\,(S))$ and preserves the positive cone of S, which is also an ample cone of S by Lemma 2. Note also that σ is of infinite order, because one of the eigenvalues is $5 + 4\sqrt{6} > 1$.

Let n be a positive integer such that $\sigma^n = id$ on the discriminant group $(\mathrm{NS}\,(S))^*/\mathrm{NS}\,(S)$. Such an n exists as $(\mathrm{NS}\,(S))^*/\mathrm{NS}\,(S)$ is a finite set. Let $T(S)$

be the transcendental lattice of S, i.e., the orthogonal complement of NS (S) in $H^2(S, \mathbf{Z})$. Then, by [5], Proposition 1.6.1, the isometry $(\sigma^n, id_{T(S)})$ of O(NS $(S)) \times$ O($T(S)$) extends to an isometry τ of $H^2(S, \mathbf{Z})$. Since τ also preserves the Hodge decomposition and the ample cone, there is then an automorphism g of S such that $g^* = \tau$ by the global Torelli theorem for K3 surfaces (see, e.g., [1], Chap. VIII). This g satisfies the requirement.

Let g be as in Lemma 4. Then the pair $(S \subset \mathbf{P}^3, g)$ satisfies all the requirements of Theorem 1(1)(2).

Lemma 5. *Let* $(S \subset \mathbf{P}^3, g)$ *be as in Theorem 1(1)(2). Let* $C \subset S$ *be a nonzero effective curve of degree* < 16, *i.e.,*

$$(C \cdot h)_S = (C \cdot H)_{\mathbf{P}^3} < 16 .$$

Then $C = S \cap T$ *for some hypersurface* T *in* $\mathbf{P}^3$.

Proof. Recall that $h = h_1$ in Pic (S). There are $m, n \in \mathbf{Z}$ such that $C = mh_1 + nh_2$ in Pic (S). Then

$$(C \cdot h)_S = 4(m + 5n) > 0 , \quad (C^2)_S = 4(n^2 + 10mn + m^2) > 0 .$$

Here the last inequality follows from Lemma 2. Thus, if $(C \cdot h)_S < 16$, then $m + 5n$ is either 1, 2, or 3 by $m, n \in \mathbf{Z}$. Hence we have either one of

$$m = 1 - 5n , \quad , m = 2 - 5n , \quad m = 3 - 5n .$$

Substituting into $n^2 + 10mn + m^2 > 0$, we obtain one of either

$$1 - 24n^2 > 0 , \quad 4 - 24n^2 > 0 , \quad 9 - 24n^2 > 0 .$$

Since $n \in \mathbf{Z}$, it follows that $n = 0$ in each case. Therefore, in Pic (S), we have $C = mh$ for some $m \in \mathbf{Z}$. Since $H^1(\mathbf{P}^3, \mathcal{O}_{\mathbf{P}^3}(\ell)) = 0$ for all $\ell \in \mathbf{Z}$, the natural restriction map

$$\iota^* : H^0(\mathbf{P}^3, \mathcal{O}_{\mathbf{P}^3}(m)) \to H^0(S, \mathcal{O}_S(m))$$

is surjective for all $m \in \mathbf{Z}$. This implies the result.

3 Proof of Theorem 1(3)

In his paper [7], Theorem 2.3 and Remark 2.4, N. Takahashi proved the following remarkable theorem as a nice application of the log Sarkisov program (for terminologies, we refer to [3]):

Theorem 2. *Let* X *be a Fano manifold of dimension* ≥ 3 *with Picard number 1, $S \in |-K_X|$ be a smooth hypersurface. Let* $\Phi : X \cdots \to X'$ *be a birational map to a*

Q-*factorial terminal variety* X' *with Picard number* 1, *which is not an isomorphism, and* $S' := \Phi_* S$. *Then:*

(1) If $\mathrm{Pic}\,(X) \to \mathrm{Pic}\,(S)$ *is surjective, then* $K_{X'} + S'$ *is ample.*
(2) Let $X = \mathbf{P}^3$ *and* H *be a hyperplane of* $\mathbf{P}^3$. *Note that then* S *is a smooth quartic* $K3$ *surface. Assume that any irreducible reduced curve* $C \subset S$ *such that* $(C \cdot H)_{\mathbf{P}^3} < 16$ *is of the form* $C = S \cap T$ *for some hypersurface* $T \subset \mathbf{P}^3$. *Then* $K_{X'} + S'$ *is ample.*

Applying Theorem 2(2), we shall complete the proof of Theorem 1(3) in the following slightly generalized form:

Theorem 3. *Let* $S \subset \mathbf{P}^3$ *be a smooth quartic* $K3$ *surface. Then:*

(1) Any automorphism g *of* S *of infinite order is not the restriction of a biregular automorphism of the ambient space* $\mathbf{P}^3$, *i.e., the restriction of an element of* $\mathrm{PGL}(\mathbf{P}^3)$.
(2) Assume further that S *contains no curves of degree* < 16 *which are not cut out by a hypersurface. Then, any automorphism* g *of* S *of infinite order is not the restriction of a Cremona transformation of the ambient space* $\mathbf{P}^3$.

Recalling Lemma 5, we see that the pair $(S \subset \mathbf{P}^3, g)$ in Theorem 1(1)(2) satisfies all the requirements of Theorem 3(2). So, Theorem 1(3) follows from Theorem 3(2). We prove Theorem 3.

Proof. Let us first show (1). Consider the group $G := \{g \in \mathrm{PGL}(\mathbf{P}^3) \mid g(S) = S\}$. Let H be the connected component of $\mathrm{Hilb}\,(\mathbf{P}^3)$ containing S. Then G is the stabilizer group of the point $[S] \in H$ under the natural action of $\mathrm{PGL}(\mathbf{P}^3)$ on H. In particular, G is a Zariski closed subset of the affine variety $\mathrm{PGL}(\mathbf{P}^3)$. In particular, G has only finitely many irreducible components. Note that the natural map $G \to \mathrm{Aut}\,(S)$ is injective and $H^0(S, T_S) = 0$. Thus dim $G = 0$. Hence G is a finite set.

Let $g \in \mathrm{Aut}\,(S)$. If there is an element $\tilde{g} \in \mathrm{PGL}\,(\mathbf{P}^3)$ such that $g = \tilde{g}|S$, then $g \in G$, and therefore g is of finite order. This proves (1).

Let us show (2). We argue by contradiction, i.e., *assuming to the contrary that there would be a birational map* $\tilde{g} : \mathbf{P}^3 \cdots \to \mathbf{P}^3$ *such that* $\tilde{g}_*(S) = S$ *and that* $g = \tilde{g}|S$, *we shall derive a contradiction.*

We shall divide it into two cases:
(i) $\tilde{g}$ is an isomorphism, and (ii) $\tilde{g}$ is not an isomorphism.

Case (i). By (1), g would be of finite order, a contradiction.
Case (ii). By the case assumption, our S would satisfy all the conditions of Theorem 2(2). Recall also that $\tilde{g}_* S = S$. However, then, by Theorem 2(2), $K_{\mathbf{P}^3} + S$ would be ample, a contradiction to $K_{\mathbf{P}^3} + S = 0$ in $\mathrm{Pic}\,(\mathbf{P}^3)$.

This completes the proof.

Acknowledgements This research was supported by JSPS Gran-in-Aid (B) No 22340009, by JSPS Grant-in-Aid (S), No 22224001, and by KIAS Scholar Program.

I would like to express my thanks to Professor Igor Dolgachev for his kind explanation on this question with history and the organizers of the workshop "Arithmetic and Geometry of K3 surfaces and Calabi–Yau threefolds" held at the Fields Institute (August 16–25, 2011) for an invitation with full financial support. I would like to express my thanks to the referee for many valuable comments including the simplification of the proof of Lemma 4.

References

1. W. Barth, K. Hulek, C. Peters, A. Van de Ven, *Compact Complex Surfaces*, 2nd enlarged edn. (Springer, Berlin, 2004)
2. I. Dolgachev, in *Quartic Surfaces and Cremona Transformations*. A Lecture in the Workshop on Arithmetic and Geometry of K3 Surfaces and Calabi-Yau Threefolds Held at the University of Toronto and the Fields Institute, Math. Soc. Japan, Tokyo, 16–25 August, 2011
3. Y. Kawamata, K. Matsuda, K. Matsuki, Introduction to the minimal model problem, in *Algebraic Geometry*, Sendai, 1985. Advanced Studies in Pure Mathematics, Math. Soc. Japan, Tokyo, vol. 10 (1987), pp. 283–360
4. D. Morrison, On K3 surfaces with large Picard number. Invent. Math. **75**, 105–121 (1984)
5. V.V. Nikulin, Integer symmetric bilinear forms and some of their geometric applications. Izv. Akad. Nauk SSSR Ser. Mat. **43**, 111–177 (1979)
6. B. Saint-Donat, Projective models of K3 surfaces. Am. J. Math. **96**, 602–639 (1974)
7. N. Takahashi, An application of Noether-Fano inequalities. Math. Zeit. **228**, 1–9 (1998)

Invariants of Regular Models of the Product of Two Elliptic Curves at a Place of Multiplicative Reduction

Chad Schoen

Abstract The divisor class group, (co)homology, and Picard group of the closed fibers of various regular proper models of the product of two elliptic curves at a place of multiplicative reduction are computed. The variation of the isomorphism class of the closed fiber with the variation of the elliptic curves is discussed. The higher direct images of the sheaf, $\mathbb{Z}/n$, are computed when n is prime to the residue characteristic.

Key words: Degenerations of surfaces, Degenerations of abelian varieties, Regular models

Mathematics Subject Classifications (2010): Primary 14D06; Secondary 11G10

1 Introduction

The purpose of this note is to compute invariants of the degeneration of a product of two elliptic curves at a place where both have split multiplicative reduction. It is natural to focus on such degenerations, since the reduction of an elliptic curve at a given place becomes either good or split multiplicative after an appropriate finite base change (cf. [23, VII.5.4]). The Weil divisor class group, the (co)homology, and the Picard group of the closed fiber of the canonical regular model will be computed. Write V for the closed fiber, $CH^1(V)$ for the Weil divisor class group, I_m and $I_{m'}$ for the Kodaira types of the reductions of the two elliptic curves. Set $m'' = gcd(m, m')$. Our first result is

Theorem 1. $CH^1(V) \simeq \mathbb{Z}^{2mm'+2} \oplus \mathbb{Z}/m''$.

C. Schoen (✉)
Department of Mathematics, Duke University, Box 90320, Durham, NC 27708-0320, USA
e-mail: schoen@math.duke.edu

R. Laza et al. (eds.), *Arithmetic and Geometry of K3 Surfaces and Calabi–Yau Threefolds*, 461
Fields Institute Communications 67, DOI 10.1007/978-1-4614-6403-7_17,
© Springer Science+Business Media New York 2013

It is interesting to note that the torsion may be non-trivial. The same phenomenon occurs for the (co)homology with $\mathbb{Z}_l$ coefficients (see Proposition 1). Torsion is not detected by the method commonly applied to analyze the cohomology of degenerate fibers, the local invariant cycle theorem [8, 3.11], as this requires cohomology with $\mathbb{Q}_l$ coefficients as well as additional hypotheses on the degeneration. Some related techniques have been applied to Chow groups tensored with $\mathbb{R}$ [4, 4.4], but again the torsion goes undetected.

In the notation introduced above, the main result about the Picard group is the following:

Theorem 2. $Pic(V) \simeq (k^*)^2 \times \mathbb{Z}^{2mm'+2-\kappa}$, where $\kappa \in \{0, 1\}$. If k is contained in the algebraic closure of a finite field, then $\kappa = 0$. Otherwise both values of κ can occur.

It is intriguing that $Pic(V)$ is not always determined by the reduction type. This points to the perhaps surprising fact that the isomorphism class of the special fiber, V, may vary as the elliptic curves change, even as the reduction type is held fixed. The Chow groups and (co)homology are insensitive to this variation, but the Picard group registers it faintly in the variation of κ.

The first part of the paper is organized as follows: The first section serves to set notation. In Sect. 3 Theorem 1 is proved. Brief treatments of the homology and cohomology of $V_{\bar{k}}$ and of the cycle class map, $CH^1(V) \to H_2(V_{\bar{k}}, \mathbb{Z}_l(-1))$, are given in Sect. 4. The variation of the closed fiber V is the subject of Sect. 5. Theorem 2 is proved in Sect. 6.

As mentioned above, there is a canonical choice of regular, projective model for a product of two elliptic curves at a place where both have multiplicative reduction. Unfortunately, this canonical model has the defect of never being semi-stable. Semi-stable degenerations are often the simplest to work with and play a distinguished role in the theory of degenerations [8, Sect. 3]. In Sect. 7 it is noted that semi-stable models exist if $m > 1$ and $m' > 1$ and that similar results to those proved for the closed fiber of the canonical model hold for the closed fibers of the semi-stable models. It is noteworthy that if a semi-stable model exists, there are usually several different ones and the various closed fibers are not generally isomorphic. Nonetheless their Weil divisor class groups, (co)homology, and Picard groups are isomorphic.

The final section of the paper is devoted to another important invariant of the degeneration, the higher direct image sheaves, $R^i f_* \mathbb{Z}/n$, where f is the structure morphism of the canonical regular, projective model and n is not divisible by the characteristic of the residue field. The stalk at the geometric closed point is the cohomology of the closed fiber described in Corollary 1. The stalk at a geometric generic point is the cohomology of the product of two elliptic curves. Thus the task is to describe the cospecialization map between the stalks. This is accomplished with the help of the semi-stable models introduced in Sect. 7 and the vanishing cycle spectral sequence.

One hopes that local results about degenerations will have global applications. The global objects most closely related to the degenerations treated here are the projective threefolds obtained by desingularizing the fiber product of two projective elliptic surfaces over a common base curve. In fact this is an important class of threefolds which have found diverse applications, many closely related to the themes of this conference [14, Chap. 2], [5, 18, 19]. The main result of Sect. 8 finds application in [22] where it plays an important role in the proof of a formula for the torsion in the third and fourth cohomology of these desingularized fiber products. A formula of this type for any class of projective threefolds appears to be new.

2 Notations

We collect notations in this section for easy reference.

2.1 Basic Notations

K = a discretely valued field.
$\mathfrak{o}$ = the corresponding valuation ring.
$T = Spec(\mathfrak{o})$.
k = the residue field, assumed perfect.
$0 = Spec(k)$.
E = elliptic curve over K with split multiplicative reduction of type I_m [13, 10.2.2].
E' = an elliptic curve over K with split multiplicative reduction of type $I_{m'}$.
$\pi : \mathcal{E} \to T$ the relatively minimal model of E [13, Sect. 9.3.3].
$\pi' : \mathcal{E}' \to T$ the relatively minimal model of E'.
$\overline{W} = \mathcal{E} \times_T \mathcal{E}'$, the fiber product.
$\overline{f} : \overline{W} \to T$ the canonical morphism.
$\overline{V} = \overline{f}^{-1}(0)$, the closed fiber of the fiber product.
$\sigma : W \to \overline{W}$ the blow-up of the ideal sheaf of the non-regular locus, $\overline{W}_{sing}$.
$f = \overline{f} \circ \sigma : W \to T$, the canonical, regular, projective model.
$V = f^{-1}(0)$, the closed fiber of the canonical, regular, projective model.
$m'' = gcd(m, m')$.
$\overline{k}$ = an algebraic closure of k.

2.2 Notations Related to the Closed Fiber, F, of $\pi : \mathcal{E} \to T$

$F = \pi^{-1}(0)$. $F' = (\pi')^{-1}(0)$.
F_i with $i \in \mathbb{Z}/m$ denote the irreducible components of F indexed so that F_0 meets the identity section of $\mathcal{E}$ and F_i meets F_{i-1} and F_{i+1}.
F'_j with $j \in \mathbb{Z}/m'$ denote the irreducible components of F' with analogous indexing.
F_{sing} = the singular locus of F. F'_{sing} = the singular locus of F'.
q_i with $i \in \mathbb{Z}/m$ denote the points of F_{sing} indexed so that $q_i \in F_i \cap F_{i+1}$
q'_j with $j \in \mathbb{Z}/m'$ denote the points of F'_{sing} indexed analogously.

Completion gives an isomorphism, $\widehat{\mathcal{O}_{\mathcal{E},q_i}} \simeq \widehat{\mathfrak{o}}[[x,y]]/(xy-u)$, where $u \in \mathfrak{o}$ is a uniformizing parameter [13, 10.3.20]. It follows that the point $(q_i, q_j') \in F_{sing} \times F_{sing}' \subset F \times F' \simeq \overline{V} \subset \overline{W}$ is an isolated ordinary double point of $\overline{W}$. The projectivized tangent cone to $\overline{W}$, Q_{ij}, is isomorphic to $\mathbb{P}^1 \times_k \mathbb{P}^1$, while the projectivized tangent cone of $\overline{V}$, $\Xi_{ij} \subset Q_{ij}$, may be identified with the configuration of four rulings,

$$\mathbb{P}^1 \times \{0, \infty\} \cup \{0, \infty\} \times \mathbb{P}^1 \subset \mathbb{P}^1 \times_k \mathbb{P}^1.$$

Identify Q_{ij} with the fiber of the blow-up, $\sigma : W \to \overline{W}$, over (q_i, q_j').

2.3 Notation Related to Components of V and Their Intersections

$L = V_{red}$, the reduced subscheme of V (cf. Lemma 5).
$P_{ij} \subset W$ denotes the strict transform of $F_i \times F_j \subset \overline{V} \subset \overline{W}$.
$Q := \coprod_{i,j} Q_{ij}$, the disjoint union.
$P := \coprod_{i,j} P_{ij}$, the disjoint union.
$a_{ij} = q_i \times F_{j+1}' \subset P_{i,j+1} \cap P_{i+1,j+1} \subset W$.
$b_{ij} = F_{i+1} \times q_j' \subset P_{i+1,j} \cap P_{i+1,j+1} \subset W$.

When $m > 1$ and $m' > 1$ define
$c_{ij} = Q_{ij} \cap P_{i+1,j+1}$,
$d_{ij} = Q_{ij} \cap P_{i,j+1}$,
$e_{ij} = Q_{ij} \cap P_{i+1,j}$,
$f_{ij} = Q_{ij} \cap P_{ij}$.
It will be necessary to define $c_{ij}, d_{ij}, e_{ij}, f_{ij}$ even when $m = 1$ or $m' = 1$. If $m = m' = 1$ the tangent cone to F_0 at q_0 (respectively to F_0' at q_0') consists of two lines denoted, TC_+ and TC_- (respectively TC_+' and TC_-'). Define $c_{00} \subset Q_{00}$ (respectively d_{00}, e_{00}, f_{00}) to be the projectivization of $TC_+ \times TC_+'$ (respectively $TC_- \times TC_+'$, $TC_+ \times TC_-'$, $TC_- \times TC_-'$). If $m = 1$ but $m' > 1$ write $T_{q_j} F_j'$ for the tangent space to F_j' at q_j' and define c_{0j} (respectively d_{0j}, e_{0j}, f_{0j}) to be the projectivization of $TC_+ \times T_{q_j} F_{j+1}'$ (respectively of $TC_- \times T_{q_j} F_{j+1}'$, $TC_+ \times T_{q_j} F_j'$, $TC_- \times T_{q_j} F_j'$) in Q_{0j}. Similarly if $m > 1$ but $m' = 1$, let c_{i0} (respectively d_{i0}, e_{i0}, f_{i0}) denote the projectivization of $T_{q_i} F_{i+1} \times TC_+'$ (respectively $T_{q_i} F_i \times TC_+'$, $T_{q_i} F_{i+1} \times TC_-'$, $T_{q_i} F_i \times TC_-'$) in Q_{i0}.
In any case $c_{ij} \simeq d_{ij} \simeq e_{ij} \simeq f_{ij} \simeq \mathbb{P}_k^1$ as abstract varieties. Together with a_{ij} and b_{ij} these are the irreducible components of the singular locus of L. Figure 1 should help visualize $L = V_{red}$.

3 The Weil Divisor Class Group of V

The purpose of this section is to prove Theorem 1. The main tool in the computation of $CH^1(V)$ is the (exact) localization sequence [6, 1.8],

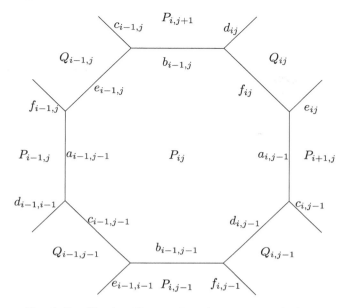

Fig. 1: $L = V_{red}$ locally near P_{ij} when $m > 1$ and $m' > 1$

$$CH^1(V - Q, 1) \xrightarrow{\partial} CH^1(Q) \to CH^1(V) \to CH^1(V - Q) \to 0,$$

where for any reduced, separated, finite type k-scheme, Z,

$$CH^1(Z, 1) := \text{Ker } k(Z)^* \xrightarrow{div} Div(Z).$$

To define the boundary map, view $f \in CH^1(V - Q, 1)$ as a rational function on the closure, $\check{V}$, of $V - Q$ in V. Extend to an element $f \in k(\check{V})^* \times k(Q)^* \simeq k(V)^*$ by means of an arbitrary factor in $k(Q)^*$. Then $div(f)$ has support in Q and defines an element $\partial(f) \in CH^1(Q)$ independent of choices.

We remark in passing that $CH^1(Z, 1)$ is canonically isomorphic to the Bloch higher Chow group designated by the same symbol [17, 1.2]. However familiarity with higher Chow groups is not needed to understand the arguments which follow.

The group $CH^1(Z, 1)$ is seldom finitely generated, but may be replaced by a more manageable group as follows: Let $CH^1(Z, 1)_0 \subset CH^1(Z, 1)$, denote the image of the obvious map,

$$CH^0(Z) \otimes CH^1(Spec(k), 1) \to CH^1(Z, 1).$$

Define $CH^1(Z, 1)_1 := CH^1(Z, 1)/CH^1(Z, 1)_0$. Now $CH^1(V - Q, 1)_0 \subset \text{Ker}(\partial)$ and $CH^1(V - Q, 1)_1$ is a finitely generated free abelian group. We want a basis. For this we introduce some notation:

$g_i \in k(F_i)^*$ satisfies $div(g_i) = q_i - q_{i-1}$. ($i \in \mathbb{Z}/m$.)
$g'_j \in k(F'_j)^*$ satisfies $div(g'_j) = q'_j - q'_{j-1}$. ($j \in \mathbb{Z}/m'$.)

$\overline{V}_{ss} := (\overline{V}_{sing})_{sing}.$

For each $j \in \mathbb{Z}/m'$ define an element

$$\eta'_j \in CH^1(\overline{V}, 1)_1 \simeq CH^1(\overline{V} - \overline{V}_{ss}, 1)_1 \simeq CH^1(V - Q, 1)_1,$$

by pulling back $\prod_{i \in \mathbb{Z}/m} g_i \in CH^1(F, 1)$ to $CH^1(F \times F'_j, 1)$ and then extending by 1 on the other irreducible components of $\overline{V}$. Similarly for each $i \in \mathbb{Z}/m$ pull back $\prod_{j \in \mathbb{Z}/m'} g'_j \in CH^1(F', 1)$ to $CH^1(F_i \times F', 1)$ and extend by 1 to define $\eta_i \in CH^1(V - Q, 1)_1$.

Lemma 1. $\{\eta_0, \ldots, \eta_{m-1}, \eta'_0, \ldots, \eta'_{m'-1}\}$ *is a basis for* $CH^1(V - Q, 1)_1$.

Proof. For linear independence note that the product

$$\eta_0^{c_0} \cdots \eta_{m-1}^{c_{m-1}} \eta_0'^{c'_0} \cdots \eta_{m'-1}'^{c'_{m'-1}}$$

restricts to $g_i^{c'_j} g_j'^{c_i}$ on P_{ij}. Thus the product is constant on each irreducible component if and only if each c_i and c'_i is zero.

An arbitrary element of $CH^1(V - Q, 1)_1$ may be represented by a rational function, $f \in k(V - Q)^* \simeq k(\overline{V} - \overline{V}_{ss})^*$ whose restriction to each $F_i \times F'_j$ has the form $g_i^{c'_j} g_j'^{c_i}$. Fix $i_1 \in \mathbb{Z}/m$. The condition $div(f) = 0$ forces the exponent of g_i to be independent of $i \in \mathbb{Z}/m$ for fixed j. Similarly the exponent of g'_j is independent of $j \in \mathbb{Z}/m'$ for fixed i. Thus the given set generates. $\square$

For the purpose of describing the image of the map, ∂, it is convenient to work with a different basis of $CH^1(V - Q, 1)_1$. Writing the group law additively we define

$$D_0 := \sum_{i \in \mathbb{Z}/m} [i]_m \eta_i - \sum_{j \in \mathbb{Z}/m'} [j]_{m'} \eta'_j \in CH^1(V - Q, 1)_1$$

where $[\ ]_m : \mathbb{Z}/m \to \mathbb{Z}$ is the section of the quotient map which takes values in $[0, m - 1]$ and $[\ ]_{m'} : \mathbb{Z}/m' \to \mathbb{Z}$ is defined analogously. Set

$A := \{\sum_{i \in \mathbb{Z}/m} \eta_i, \ \sum_{i' \in \mathbb{Z}/m'} \eta'_{i'}\},$
$B := \{\eta_1, \ldots, \eta_{m-1}\},$
$C := \{\eta'_2, \ldots, \eta'_{m'-1}\},$
$D := \{D_0\}.$

Lemma 2. *A basis of* $CH^1(\overline{V}, 1)_1$ *is given by*

(i) $A \cup B \cup C \cup D$ *when* $m > 1$ *and* $m' > 1$.
(ii) A *when* $m = m' = 1$.
(iii) $A \cup B$ *when* $m > 1$ *and* $m' = 1$.

Proof. Straightforward from Lemma 1. $\square$

In Lemma 2 we have ignored the case $m = 1$, $m' > 1$ as this becomes case (iii) after interchanging the factors in the fiber product.

Define $\delta_{ij} = c_{ij} - d_{ij} \in CH^1(Q_{ij})$ and observe that $\{c_{ij}\} \cup \{\delta_{ij}\}$ is a basis of $CH^1(Q)$.

Lemma 3. *(i)* $\partial \eta_i = \sum_{j \in \mathbb{Z}/m'} (\delta_{ij} - \delta_{i-1,j})$.
(ii) $\partial \eta'_j = \sum_{i \in \mathbb{Z}/m} (\delta_{ij} - \delta_{i,j-1})$.
(iii) ∂ *maps* $Span(B \cup C)$ *isomorphically to a direct summand,* $\Sigma \subset CH^1(Q)$.
(iv) *Assume* $m > 1$ *and* $m' > 1$. *Then* $\frac{1}{m''} \partial D_0 \in CH^1(Q)/\Sigma$ *is indivisible.*
(v) $A = \text{Ker}(\partial)$.

Proof. (i) Regard g'_j as element of $k(P_{ij})^*$. From Fig. 1

$$div(g'_j) = b_{i-1,j} + e_{i-1,j} + f_{ij} - (b_{i-1,j-1} + c_{i-1,j-1} + d_{i,j-1}).$$

Now $\partial \eta_i = \sum_{j \in \mathbb{Z}/m'} div(g'_j)$ and the assertion follows from $\delta_{ij} = f_{ij} - d_{ij}$ and $\delta_{i-1,j} = c_{i-1,j} - e_{i-1,j}$.

(ii) The proof is analogous to that of (i).

(iii) Form a matrix with columns indexed by $\eta \in B \cup C$ and rows indexed by $(i, j) \in \mathbb{Z}/m \times \mathbb{Z}/m'$. The (i, j)-th entry in column η is the coefficient of δ_{ij} in $\partial \eta$. Restricting (i, j) to lie in the set,

$$S := \{(0, 0), (1, 0), \ldots, (m - 2, 0)\} \cup \{(0, 1), \ldots, (0, m' - 2)\},$$

gives an $(m + m' - 3) \times (m + m' - 3)$ square matrix in which all entries above the diagonal are zero and diagonal entries are -1.

(iv) The coefficient of δ_{ij} in ∂D_0 is

$$[i]_m - [i + 1]_m - [j]_{m'} + [j + 1]_{m'} = \begin{cases} 0, & \text{if } i \neq m - 1, j \neq m' - 1 \\ m, & \text{if } i = m - 1, j \neq m' - 1 \\ -m', & \text{if } i \neq m - 1, j = m' - 1 \\ m - m', & \text{if } i = m - 1, j = m' - 1 \end{cases}.$$

For $(i, j) \in S$ the coefficient of δ_{ij} in ∂D_0 is zero. Since the coefficients of $\delta_{0,m'-1}$ $(-m'/m'')$ and $\delta_{m-1,0}$ (m/m'') in $\partial D_0/m''$ are coprime, the assertion follows.

(v) By (i) and (ii) $A \subset \text{Ker}(\partial)$. Equality follows from (iii) and (iv). $\square$

Lemma 4. *(i)* $CH^1(V - Q) \simeq CH^1(\overline{V}) \simeq \mathbb{Z}^{m+m'}$.
(ii) $CH^1(V) \simeq \mathbb{Z}^{2mm'+2} \oplus \mathbb{Z}/m''$.
(iii) *If* $CH^1(V)_{tors} \neq 0$, *for any* $\mathbf{i}, \mathbf{j}$, $\tau_{\mathbf{i}\mathbf{j}} := \frac{m}{m''} \sum_j \delta_{\mathbf{i},j} - \frac{m'}{m''} \sum_i \delta_{i,\mathbf{j}}$ *is a generator.*

Proof. (i) The first isomorphism follows from $\overline{V} - \overline{V}_{ss} \simeq V - Q$ and the fact that $\overline{V}_{ss} \subset \overline{V}$ has codimension 2. To compute $CH^1(\overline{V})$ observe that $CH^1(F_i \times F'_j)$ is generated by $q_i \times F'_j$ and $F_i \times q'_j$. In $CH^1(F_i \times F'_j)$ we have the relations,

$$q_{i-1} \times F'_j = q_i \times F'_j \quad \text{and} \quad F_i \times q'_j = F_i \times q'_{j-1}.$$

Thus $q_i \times F'_j \in CH^1(\overline{V})$ is independent of i and $F_i \times q'_j \in CH^1(\overline{V})$ is independent of j. Thus there is a surjective map, $\mathbb{Z}^{m+m'} \to CH^1(\overline{V})$. To show injectivity intersect these divisors with the pullbacks to $\overline{V}$ of the standard generators of the Néron–Severi groups, $N.S.(F) := Pic(F)(k)/Pic^0(F)(k) \simeq \mathbb{Z}^m$ and $N.S.(F') \simeq \mathbb{Z}^{m'}$.

(ii) Since $CH^1(V - Q)$ is a free abelian group, $CH^1(V) \simeq Coker(\partial) \oplus CH^1(V - Q)$. The first summand is easily computed from the previous lemma and the basis of $CH^1(Q) \simeq \mathbb{Z}^{2mm'}$ given above.

(iii) By Lemma 3(iv) a generator of $Coker(\partial)_{tors}$ is given by $\tau_{m-1,m'-1}$. In fact $\tau_{i,j} = \tau_{m-1,m'-1} \in Coker(\partial)$ for all $\mathbf{i} \in \mathbb{Z}/m, \mathbf{j} \in \mathbb{Z}/m'$ by Lemma 3(i)–(ii).   □

Remark 1. There are interesting extensions of this material to the case that E and E' are allowed to have genus > 1 which I hope to discuss in a sequel to this paper.

4 The Homology and Cohomology of $V_{\bar{k}}$

In this section the residue field, k, is assumed to be algebraically closed. The goal is to compute the étale homology of V [12]. When $k = \mathbb{C}$ this may be identified with the Borel–Moore homology of the associated analytic space [25, 2.8.4]. If the analytic space is compact, this is isomorphic to the singular homology. Fix a prime $l \neq char(k)$ and let l'' denote the largest power of l which divides m''.

Proposition 1. *The étale homology is given by*

(i) $H_0(V, \mathbb{Z}_l) \simeq \mathbb{Z}_l$.
(ii) $H_1(V, \mathbb{Z}_l) \simeq \mathbb{Z}_l^2$.
(iii) $H_2(V, \mathbb{Z}_l) \simeq \mathbb{Z}_l^{2mm'+2}(1) \oplus \mathbb{Z}/l''(1) \oplus \mathbb{Z}_l$.
(iv) $H_3(V, \mathbb{Z}_l) \simeq \mathbb{Z}_l^2(1)$.
(v) $H_4(V, \mathbb{Z}_l) \simeq \mathbb{Z}_l^{2mm'}(2)$.

The first step towards a proof is the following:

Proposition 2. *There is a commutative diagram with exact rows,*

$$
\begin{array}{ccccccccc}
CH^1(\overline{V}, 1)_1 \otimes \mathbb{Z}_l & \xrightarrow{\bar{\partial}} & CH^1(Q) \otimes \mathbb{Z}_l & \longrightarrow & CH^1(V) \otimes \mathbb{Z}_l & \longrightarrow & CH^1(\overline{V}) \otimes \mathbb{Z}_l & \longrightarrow & 0 \\
\alpha \downarrow \wr & & \beta \downarrow \wr & & \gamma \downarrow & & \upsilon \downarrow & & \\
H_3(\overline{V}) & \xrightarrow{\bar{\partial}_H} & H_2(Q) & \longrightarrow & H_2(V) & \xrightarrow{\epsilon} & H_2(\overline{V}) & \longrightarrow & 0.
\end{array}
$$

where $\bar{\partial}$ is induced from the map ∂ of Sect. 3 and υ maps $CH^1(\overline{V}) \otimes \mathbb{Z}_l$ isomorphically to the Künneth component,

$$H_2(\overline{V})_0 := H_2(F) \otimes H_0(F', \mathbb{Z}_l) \oplus H_0(F, \mathbb{Z}_l) \otimes H_2(F'),$$

of $H_2(\overline{V})$ and where the omitted coefficients in the homology are $\mathbb{Z}_l(-1)$.

Proof of Proposition 1 assuming Proposition 2. Define $H_2(V)_0 = \epsilon^{-1}(H_2(\overline{V})_0)$. When the subscript $_0$ is placed on the last two terms in the bottom row of the diagram all vertical maps become isomorphisms. Use the isomorphism, $H_1(F, \mathbb{Z}_l) \otimes H_1(F', \mathbb{Z}_l) \simeq \mathbb{Z}_l$, the Künneth formula [12, 4.1], and Lemma 4(ii) to conclude

$$H_2(V, \mathbb{Z}_l(-1)) \simeq \mathbb{Z}_l^{2mm'+2} \oplus \mathbb{Z}/l'' \oplus \mathbb{Z}_l(-1).$$

The bottom row of the diagram in Proposition 2 is part of a longer exact sequence,

$$0 \to H_4(Q) \to H_4(V) \to H_4(\overline{V}) \to H_3(Q) \to H_3(V) \to H_3(\overline{V}) \xrightarrow{\overline{\partial}_H} H_2(Q) \to H_2(V) \to \ldots$$

This comes from the exact étale homology sequence of the pair (V, Q) [3, 1.1.1, 1.2.3, 2.1] and the observation that $H_i(V - Q) \simeq H_i(\overline{V})$ for $i > 1$ since

$$V - Q \simeq \overline{V} - \overline{V}_{ss} \qquad \text{and} \qquad H_i(\overline{V} - \overline{V}_{ss}) \simeq H_i(\overline{V})$$

for $i > 1$ by the exact sequence for the pair $(\overline{V}, \overline{V}_{ss})$. Now

$$H_3(V) \simeq \mathrm{Ker}(\overline{\partial}_H) \simeq \mathrm{Ker}(\overline{\partial}) \simeq \mathbb{Z}_l^2$$

by Lemma 3(v). $H_4(V)$ is the free module on the irreducible components. The rank of $H_1(V)$ may be computed from the Euler characteristic, $e(V) = 4mm'$. To show $H_1(V)_{tors} = 0$ use the Leray spectral sequence for the map, $V \to \overline{V}$, to conclude that $H^2(V, \mathbb{Z}_l)_{tors} = 0$ and apply the isomorphism [12, 2.2]:

$$H_i(V, \mathbb{Z}_l) \simeq Hom(H^i(V, \mathbb{Z}_l), \mathbb{Z}_l) \oplus Hom(H^{i+1}(V, \mathbb{Z}_l)_{tors}, \mathbb{Q}_l/\mathbb{Z}_l). \qquad (1)$$

Proof of Proposition 2. The top row in the diagram is obtained from the localization sequence in Sect. 3 by applying $\otimes \mathbb{Z}_l$ and recalling the isomorphisms $CH^1(\overline{V})_1 \simeq CH^1(V - Q, 1)_1$ and $CH^1(\overline{V}) \simeq CH^1(V - Q)$ from Sect. 3. The maps β, γ, υ are the usual functorial cycle class maps [12, Sect. 6]. That υ gives an isomorphism, $CH^1(\overline{V}) \otimes \mathbb{Z}_l \simeq H_2(\overline{V})_0$, is immediate from the explicit descriptions of these two modules (cf. the proof of Lemma 4(i)).

It remains only to describe the map α and verify the commutativity of the first square in the diagram. This may be done using the classical cycle class map for divisors on a non-singular surface and some diagram chasing. First some notation:

$\check{V}$ = the closure of $V - Q$ in V.
$\dot{V} = \check{V} - \check{V}_{sing}$.
$S = (V - Q)_{sing}$
$\ddot{V} = (V - Q) - S$.
$Z = \dot{V} - \ddot{V}$.
$\nu : \widetilde{V} \to V - Q$, the normalization. ($\widetilde{V}$ is non-singular.)
$\widetilde{S} = \widetilde{V} - \ddot{V}$, a double cover of S.
Now use the commutative diagram

$$CH^1(\overline{V},1) \xrightarrow{\sim} CH^1(V-Q,1)$$

$$\downarrow a$$

$$H^0(\dot{V},\mathbb{G}_m) \longrightarrow H^0(\ddot{V},\mathbb{G}_m) \xrightarrow{\delta} H^1_Z(\dot{V},\mathbb{G}_m)$$

$$\downarrow b \qquad\qquad \downarrow d$$

$$H^1(\ddot{V},\mu_{l^n}) \xrightarrow{\delta_l} H^2_Z(\dot{V},\mu_{l^n})$$

$$\downarrow c\,\wr \qquad\qquad \downarrow e\,\wr$$

$$H_3(\ddot{V},\mathbb{Z}/l^n(-1)) \xrightarrow{\delta_H} H_2(Z,\mathbb{Z}/l^n(-1)) \xrightarrow{\zeta} H_2(Q,\mathbb{Z}/l^n(-1)).$$

Here the second row comes from the exact sequence of the pair, $\ddot{V} \subset \dot{V}$. The last term is canonically identified with $H^0_Z(\dot{V},\underline{Div})$ so that the map δ identifies with $f \mapsto div(f)$ [15, VI.6]. The middle set of vertical maps come from the Kummer sequence. To define ζ replace Z by its closure in Q, this doesn't change H_2, and push forward. Thus $\beta \circ \bar{\partial}$ in Proposition 2 comes from $\zeta \circ e \circ d \circ \delta \circ a$ by passing to the inverse limit over n.

To finish the argument we show that $c \circ b \circ a$ may be viewed as a map to $H_3(\overline{V},\mathbb{Z}/l^n(-1))$. Then α in Proposition 2 may be constructed from $\varprojlim_n(c \circ b \circ a)$. The point is that $H_3(\overline{V},\mathbb{Z}/l^n(-1))$ is canonically identified with $H_3(V-Q,\mathbb{Z}/l^n(-1))$ and the latter is canonically identified with a subgroup of $H_3(\dot{V},\mathbb{Z}/l^n(-1))$ via the exact sequence of the pair, $\ddot{V} \subset V - Q$, as in the final row of the commutative diagram,

$$H^0(\widetilde{V},\mathbb{G}_m) \longrightarrow H^0(\ddot{V},\mathbb{G}_m) \xrightarrow{\delta_{\mathbb{G}}} H^1_{\widetilde{S}}(\widetilde{V},\mathbb{G}_m)$$

$$\downarrow \qquad\qquad \downarrow b \qquad\qquad \downarrow g$$

$$H^1(\widetilde{V},\mu_{l^n}) \longrightarrow H^1(\ddot{V},\mu_{l^n}) \longrightarrow H^2_{\widetilde{S}}(\widetilde{V},\mu_{l^n})$$

$$\downarrow \wr \qquad\qquad \downarrow c\,\wr \qquad\qquad \downarrow h\,\wr$$

$$H_3(\widetilde{V},\mathbb{Z}/l^n(-1)) \longrightarrow H_3(\ddot{V},\mathbb{Z}/l^n(-1)) \longrightarrow H_2(\widetilde{S},\mathbb{Z}/l^n(-1))$$

$$\downarrow \qquad\qquad \| \qquad\qquad \downarrow j$$

$$H_3(V-Q,\mathbb{Z}/l^n(-1)) \longrightarrow H_3(\ddot{V},\mathbb{Z}/l^n(-1)) \longrightarrow H_2(S,\mathbb{Z}/l^n(-1)).$$

As above, the map, $h \circ g \circ \delta_{\mathbb{G}}$, sends a rational function to the homology class of its divisor. For $f \in CH^1(V-Q,1)_1$, $div(f) = 0 \in Div(V-Q)$, so $h \circ g \circ \delta_{\mathbb{G}} \circ a(f) \in \mathrm{Ker}(j)$. Thus the image of $c \circ b \circ a$ lies in $H_3(V-Q,\mathbb{Z}/l^n(-1))$ as desired.

Finally α in Proposition 2 sends the basis in Lemma 1 of $CH^1(V-Q,1)_1$ to the standard basis of $H_3(\overline{V},\mathbb{Z}_l(-1))$ coming from the Künneth decomposition and is thus an isomorphism. $\square$

The cohomology, $H^\bullet(V,\mathbb{Z}_l)$, may be computed from Proposition 1 and (1). A standard exact sequence then computes $H^\bullet(V,\mathbb{Z}/l^r)$ [15, V.1.11]. This determines $H^\bullet(V,\mathbb{Z}/n)$ for n not divisible by $char(k)$. For use in Sect. 8 we record the result.

Corollary 1. *Assume char(k) does not divide n. Set $n'' = \gcd(n, m'')$. Then*

(i) $H^0(V, \mathbb{Z}/n) \simeq \mathbb{Z}/n\mathbb{Z}$.

(ii) $H^1(V, \mathbb{Z}/n) \simeq (\mathbb{Z}/n\mathbb{Z})^2$.

(iii) $H^2(V, \mathbb{Z}/n) \simeq \mathbb{Z}/n\mathbb{Z} \oplus (\mathbb{Z}/n\mathbb{Z})^{2mm'+2}(-1) \oplus \mathbb{Z}/n''\mathbb{Z}(-1)$.

(iv) $H^3(V, \mathbb{Z}/n) \simeq (\mathbb{Z}/n\mathbb{Z})^2(-1) \oplus \mathbb{Z}/n''\mathbb{Z}(-1)$.

(v) $H^4(V, \mathbb{Z}/n) \simeq (\mathbb{Z}/n\mathbb{Z})^{2mm'}(-2)$.

Remark 2.   (i) Corollary 1(iv) was obtained previously by a less natural method [20, 12.1].

(ii) Presumably the map α in Proposition 2 is a particular case of a general cycle class map for higher Chow groups, cf. [2, Sect. 4].

5 Variation of the Isomorphism Class of V

The purpose of this section is to show that the isomorphism class of V may vary as the elliptic curves E and E' vary even as m and m' remain fixed. This is perhaps surprising as the closed fiber F (respectively F') does not change as E (respectively E') varies over elliptic curves with split I_m (respectively $I_{m'}$) reduction. Thus $\overline{V} = F \times F'$ also does not vary. Furthermore $L := V_{red}$ is the union of the reduced exceptional divisor, Q, and the blow-up, $V^!$, of $\overline{V} = F \times F'$ along $F_{sing} \times F'_{sing}$. The isomorphism classes of Q and $V^!$ also do not change as E and E' vary. To show that L varies we examen how Q is attached to $V^!$.

For concreteness we assume $m = m' = 1$. This implies in particular that $\mathcal{E}$ and $\mathcal{E}'$ are minimal Weierstrass models [24, IV.9]. Thus $\mathcal{E}$ is determined by a Weierstrass equation,

$$Y^2 + a_1 XY + a_3 Y - (X^3 + a_2 X^2 + a_4 X + a_6) = 0,$$

where we may assume that a_3, a_4, a_6 lie in the maximal ideal, $\mathfrak{m}$ of $\mathfrak{o}$, $b_2 := a_1^2 + 4a_2 \in \mathfrak{o}^*$ and $a_6 \notin \mathfrak{m}^2$ [24, p. 370]. The closed fiber is given by

$$Y^2 + a_1 XY - a_2 X^2 = X^3.$$

Since the reduction is split, the left hand side of the equation factors in $\mathfrak{o}/\mathfrak{m}[X, Y]$ as the product of distinct linear factors. Assume that $\mathfrak{o}$ is henselian, so that this factorization lifts to $\mathfrak{o}[X, Y]$. After a linear change of variables the equation for $\mathcal{E}$ becomes,

$$xy - g_1(x, y) - u - uh_1(x, y) = 0,$$

where $g_1(x, y)$ is a homogeneous cubic, $u \in \mathfrak{o}$ is a uniformizer, and $h_1(x, y) \in (x, y) \subset \mathfrak{o}[x, y]$. Assume that $\mathcal{E}'$ is given by

$$ax'y' - ag_2(x', y') - u - uh_2(x', y') = 0,$$

where $a \in \mathfrak{o}^*$, g_2 is a homogeneous cubic and $h_2(x', y') \in (x', y')$. Write $\mathcal{E}'_a$ for $\mathcal{E}'$ to emphasize the dependence on a. Define $\overline{W}_a := \mathcal{E} \times_T \mathcal{E}'_a$ and let W_a denote the blow-up along $(\overline{W}_a)_{sing}$.

Proposition 3. *Given $a, a' \in \mathfrak{o}^*$, the reduced closed fibers $L_a \subset W_a$ and $L_{a'} \subset W_{a'}$ are isomorphic if and only if a and a' have the same image in k^*.*

Proof. An affine open subscheme of the fiber product $\mathcal{E} \times_T \mathcal{E}'_a$ is given by Spec of

$$\mathfrak{o}[x, y, x', y']/(xy - g_1(x, y) - u - uh_1(x, y), \ ax'y' - ag_2(x', y') - u - uh_2(x', y')).$$

The corresponding equation for the closed fiber $\overline{V}$ ($u = 0$) is independent of a:

$$xy - g_1(x, y) = 0 = x'y' - g_2(x', y').$$

The tangent space to the fiber product at the singular point is given by Spec of $\mathfrak{o}[x, y, x', y']/(u) \simeq k[x, y, x', y']$. The projectivized tangent cone of the fiber product is the quadric surface in $Proj(k[x, y, x', y'])$,

$$Q_a : \quad xy - ax'y' = 0.$$

Let $V^! \to \overline{V} = F \times F'$ denote the blow-up at the point $F_{sing} \times F'_{sing}$. The reduced closed fiber, L_a, in the desingularized fiber product is obtained by gluing Q_a to $V^!$ along the configuration of four lines, Ξ, in $Proj(k[x, y, x', y'])$ defined by the homogeneous ideal $(xy, x'y')$. The identification depends only on the image of a in k^*. To show that $L_a \neq L_{a'}$ it suffices to show that there is no isomorphism, $\varphi : (Q_a, \Xi) \to (Q_{a'}, \Xi)$, such that $\varphi|_\Xi = \psi|_\Xi$ for some $\psi \in Aut(V^!)$. Now

$$Aut(Q_a) \simeq Aut(\mathbb{P}^1 \times \mathbb{P}^1) \simeq (PGL_2 \times PGL_2) \rtimes \langle \tau \rangle,$$

is a projective orthogonal group; the involution τ interchanges the factors in the product.

Every isomorphism, $\varphi : Q_a \to Q_{a'}$, is the restriction of an automorphism of the ambient projective space. If φ stabilizes Ξ then, after composing with τ if necessary, it fixes Ξ_{sing} pointwise. In x, y, x', y' coordinates Ξ_{sing} consists of the four points $(1 : 0 : 0 : 0),\ldots,(0 : 0 : 0 : 1)$. Thus φ acts on the coordinates x, y, x', y' by a diagonal matrix, $diag(\alpha, \beta, \alpha', \beta')$. Since $\varphi(Q_a) = Q_{a'}$, $\alpha'\beta'\alpha^{-1}\beta^{-1} = a'/a \in k^*$.

View $\overline{V}$ as the self-product of $\mathbb{P}^1$ with 0 and ∞ identified. The action of $\alpha \in \mathbb{G}_m$ on $(\mathbb{P}^1, \{0, \infty\})$ gives an action on the tangent cone of $\mathbb{P}^1/\{0, \infty\}$ at the singular point. The coordinates x, y (respectively x', y') above are naturally coordinates on the tangent cone of F (respectively F') at the singular point. The action of $\alpha \in \mathbb{G}_m$ on these coordinates, which may be identified with uniformizers on $\mathbb{P}^1$ at 0 and ∞, is by α, α^{-1}. Since the normalization of $\overline{V}$ is isomorphic to $\mathbb{P}^1 \times \mathbb{P}^1$ and the inverse image of $\overline{V}_{sing}$ is a configuration of four lines which may be identified with Ξ we find

$$Aut(V^!) \simeq Aut(\overline{V}) \simeq Aut(\mathbb{P}^1 \times \mathbb{P}^1, \Xi) \simeq (\mathbb{G}_m \times \mathbb{G}_m) \rtimes \langle \tau \rangle.$$

Consequently, the restriction of $Aut(V^!)$ to $Aut(\varXi)$ is represented by matrices of the form, $diag(\alpha, \alpha^{-1}, \alpha', (\alpha')^{-1})$, times a power of τ. One checks that the composition,

$$diag(\alpha, \beta, \alpha', \beta') \rightarrow Aut(\mathbb{P}^3, \varXi) \rightarrow Aut(\varXi),$$

is injective. Thus if $a \neq a' \in k^*$, there is no $\psi \in Aut(V^!)$ with $\varphi|_{\varXi} = \psi|_{\varXi}$.   □

Remark 3. Even when $(m, m') \neq (1, 1)$ the isomorphism class of the reduced closed fiber of the desingularized fiber product may vary. Replacing E and E_a' with appropriate isogenous elliptic curves leads to finite étale $\mathbb{Z}/m \times \mathbb{Z}/m'$-covers of L_a. As L_a varies in moduli, not all of its $\mathbb{Z}/m \times \mathbb{Z}/m'$-étale covers can be isomorphic if k is an infinite field.

6 The Picard Group of V

In this section we prove Theorem 2 of the introduction.

Proposition 4. $Pic(V) \simeq (k^*)^2 \oplus \mathbb{Z}^{2mm'+2-\kappa}$, where $\kappa \in \{0, 1\}$.

The precise value of κ will be discussed later in this section. The proof of Proposition 4 will be given in several steps.

Lemma 5. *The exceptional divisor in the blow-up,* $\sigma : W \rightarrow \overline{W}$, *appears with multiplicity two in the closed fiber,* $V \subset W$.

Proof. Up to étale morphisms we may replace $\mathcal{E}$ with $Spec(\mathfrak{o}[x, y]/(xy - u))$ and $\overline{W}$ with $Spec(\mathfrak{o}[x, y, x', y']/(xy - u, x'y' - u))$, where $u \in \mathfrak{o}$ is a uniformizer. The maximal ideal at the non-regular point is (x, y, x', y'). An affine chart in the blow-up is given by applying $Spec$ to the $\mathfrak{o}$-algebra homomorphism,

$$\mathfrak{o}[x, y, x', y']/(xy - u, x'y' - u) \quad \rightarrow \quad \mathfrak{o}[x, Y, X', Y']/(x^2 Y - u, x^2 X' Y' - u),$$
$$x, y, x', y' \quad \mapsto \quad x, xY, xX', xY'.$$

The assertion follows since $u \in (x)^2$, $u \notin (x)^3$, and (x) defines the exceptional locus.   □

Lemma 6. $Pic(V) \simeq Pic(V_{red})$.

Proof. Recall $L := V_{red}$. The ideal sheaf, $\mathcal{I}_L$, of L in V is isomorphic to $\mathcal{O}_Q(1)$, which is componentwise the restriction of $\mathcal{O}_{\mathbb{P}^3}(1)$ to a quadric hypersurface. The lemma follows from the exact sequence,

$$1 \rightarrow 1 + \mathcal{I}_L \rightarrow \mathcal{O}_V^* \rightarrow \mathcal{O}_L^* \rightarrow 1,$$

and $H^i(Q, \mathcal{O}_Q(1)) = 0$ for $i \in \{1, 2\}$.   □

Write $L_{sing} \subset L$ for the singular locus with its reduced scheme structure. Let $v_0 : L^{[0]} \to L$ and $v_1 : L^{[1]} \to L_{sing}$ denote the normalizations. Let $v_2 : L^{[2]} := (L_{sing})_{sing} \to L_{sing}$ denote the inclusion.

Proposition 5. *There is an exact sequence of étale sheaves on L,*

$$1 \to \mathbb{G}_{mL} \xrightarrow{v_0^*} v_{0*}\mathbb{G}_{mL^{[0]}} \xrightarrow{\beta^*} v_{1*}\mathbb{G}_{mL^{[1]}} \xrightarrow{\alpha^*} v_{2*}\mathbb{G}_{mL^{[2]}} \to 1.$$

Proof. The main point is to specify liftings, $\alpha_i : L^{[2]} \to L^{[1]}$ (respectively $\beta_j : L^{[1]} \to L^{[0]}$) of the inclusion, $v_2 : L^{[2]} \to L_{sing}$ (respectively of $v_1 : L^{[1]} \to L$) for $i \in \{0, 1, 2\}$ (respectively for $j \in \{0, 1\}$) in such a way that the restrictions of the α_i's (respectively the β_j's) to each component of $L^{[2]}$ (respectively $L^{[1]}$) give all possible liftings and such that the identities

$$\beta_0 \circ \alpha_1 = \beta_0 \circ \alpha_0, \qquad \beta_0 \circ \alpha_2 = \beta_1 \circ \alpha_0, \qquad \beta_1 \circ \alpha_2 = \beta_1 \circ \alpha_1, \qquad (2)$$

hold. Define $\alpha^* = \alpha_0^* \cdot (\alpha_1^*)^{-1} \cdot \alpha_2^*$ and $\beta^* = \beta_0^* \cdot (\beta_1^*)^{-1}$. The identities imply that $\alpha^* \circ \beta^* = 1$. Furthermore $\beta^* \circ v_0^* = 1$. That the natural maps, $im(v_0^*) \to \text{Ker}(\beta^*)$ and $im(\beta^*) \to \text{Ker}(\alpha^*)$, are isomorphisms and that α^* is surjective may be checked on stalks using the fact that L is a local normal crossing surface.

To describe one choice of α_i's and β_j's which fulfill the conditions when $m = m' = 1$ consider the following picture of the two components, P and Q, of $L^{[0]}$ and the curves on these components which get identified to form L_{sing}:   □

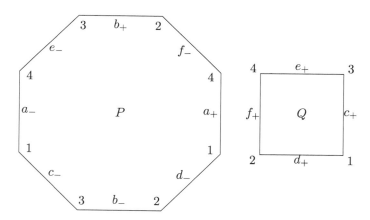

Fig. 2: $L^{[0]}$ when $m = m' = 1$

The curves are labeled a, b, c, etc. as in Fig. 1. The + and − copies of each curve are identified by v_0 to form L_{sing} (e.g., c_+ and c_- are identified to the component c of L_{sing}). The vertex labels $1, 2, 3, 4$ correspond to the singular points of L_{sing}. That is the three vertecies labeled 1 represent the inverse image of $1 \in (L_{sing})_{sing}$ in $L^{[0]}$ and similarly for 2, etc.

Now we fix the lifting β_0 of v_1 on the components of $L^{[1]}$. This determines β_1:

$$\beta_0 : \quad a, \ b, \ c, \ d, \ e, \ f \ \mapsto \ a_-, \ b_-, \ c_+, \ d_+, \ e_+, \ f_+$$
$$\beta_1 : \quad a, \ b, \ c, \ d, \ e, \ f \ \mapsto \ a_+, \ b_+, \ c_-, \ d_-, \ e_-, \ f_-$$

The required identities uniquely determine the liftings α_i of v_2:

$$\alpha_0 : \quad 1 \quad 2 \quad 3 \quad 4 \ \mapsto \ c, \quad d, \quad c, \quad e$$
$$\alpha_1 : \quad 1 \quad 2 \quad 3 \quad 4 \ \mapsto \ a, \quad b, \quad b, \quad a$$
$$\alpha_2 : \quad 1 \quad 2 \quad 3 \quad 4 \ \mapsto \ d, \quad f, \quad e, \quad f.$$

To treat the case that m and m' are arbitrary, write $L_{m,m'}$ for V_{red}. To verify exactness we are free to replace the base ring $\mathfrak{o}$ by a strict henselization. Now the m-torsion group scheme $\mathcal{E}[m] \simeq (\mathbb{Z}/m)^2$. Similarly $\mathcal{E}'[m'] \simeq (\mathbb{Z}/m')^2$. Translation by appropriate torsion points gives rise to a free action of $\mathbb{Z}/m \times \mathbb{Z}/m'$ on $L_{m,m'}$ and on $L_{m,m'}^{[i]}$ for $i \in \{0, 1, 2\}$. Write $L_{m,m'} \to L_{1,1}$ for the quotient map. There is a commutative diagram of finite, étale $\mathbb{Z}/m \times \mathbb{Z}/m'$-covers:

$$
\begin{array}{ccccccc}
L_{m,m'}^{[2]} & \xrightarrow{\ \widetilde{\alpha}_i\ } & L_{m,m'}^{[1]} & \xrightarrow{\ \widetilde{\beta}_j\ } & L_{m,m'}^{[0]} & \xrightarrow{\ \widetilde{v}_0\ } & L_{m,m'} \\
\downarrow & & \downarrow & & \downarrow{\scriptstyle\gamma} & & \downarrow \\
L_{1,1}^{[2]} & \xrightarrow{\ \alpha_i\ } & L_{1,1}^{[1]} & \xrightarrow{\ \beta_j\ } & L_{1,1}^{[0]} & \xrightarrow{\ v_0\ } & L_{1,1}.
\end{array}
$$

Here $\widetilde{v}_0$ is the normalization. The map $\widetilde{\beta}_j$ is uniquely determined by specifying that the middle square commute and that $\widetilde{v}_0 \circ \widetilde{\beta}_j : L_{m,m'}^{[1]} \to (L_{m,m'})_{sing}$ is the normalization. Furthermore $\widetilde{\alpha}_i$ is determined by specifying that the left square commute and that

$$\widetilde{v}_0 \circ \widetilde{\beta}_0 \circ \widetilde{\alpha}_i : L_{m,m'}^{[2]} \to ((L_{m,m'})_{sing})_{sing}$$

is the canonical identification. Now (2) holds with $\widetilde{\alpha}_i$ replacing α_i and $\widetilde{\beta}_j$ replacing β_j, because a point in $L_{m,m'}^{[0]}$ whose image in $L_{m,m'}$ is fixed is specified by its image under γ. $\qquad\square$

Lemma 7. *There is an exact sequence,* $1 \to (k^*)^2 \to Pic(L) \to M \to 0$, *where* M *is a free abelian group of rank at most* $2mm' + 2$.

Proof. Since $\mathbb{G}_{mL}$ is quasi-isomorphic to the complex,

$$v_{0*}\mathbb{G}_{mL^{[0]}} \xrightarrow{\ \beta^*\ } v_{1*}\mathbb{G}_{mL^{[1]}} \xrightarrow{\ \alpha^*\ } v_{2*}\mathbb{G}_{mL^{[2]}}$$

we may compute $Pic(V) \simeq H^1(L, \mathbb{G}_{mL})$ using the spectral sequence for hypercohomology. This yields the exact sequence,

$$1 \to E_\infty^{10} \to H^1(L, \mathbb{G}_m) \to E_\infty^{01} \to 1,$$

where $E_\infty^{10} \simeq E_2^{10}$ is computed as H^1 of the dual graph of L with k^*-coefficients [16, p. 105]. Since the dual graph triangulates a torus $E_2^{10} \simeq (k^*)^2$. As all the components of $L^{[0]}$ are rational surfaces and those of $L^{[1]}$ are isomorphic to $\mathbb{P}^1$, the differential,

$$d_1^{01} : E_1^{01} \simeq H^1(L^{[0]}, \mathbb{G}_m) \overset{H^1(\beta^*)}{\longrightarrow} H^1(L^{[1]}, \mathbb{G}_m) \simeq E_1^{11},$$

after tensoring with $\mathbb{Z}_l$ may be identified with the map on cohomology, $H^2(\beta^*, \mathbb{Z}_l)$. The kernel is isomorphic to $\mathbb{Z}_l^{mm'+2}$ [20, Sect. 13]. Thus $E_2^{01} \simeq \mathbb{Z}^{2mm'+2}$. Since

$$M = E_\infty^{01} \simeq E_3^{01} \simeq \operatorname{Ker} d_2^{01} : E_2^{01} \to E_2^{20},$$

the lemma follows.   □

Corollary 2. *Suppose k is a subfield of the algebraic closure of a finite field. Then* $\operatorname{rank}(M) = 2mm' + 2$.

Proof. Observe that E_2^{20} is a torsion group since it is a quotient of the torsion group,

$$E_1^{20} \simeq H^0(L^{[2]}, \mathbb{G}_m) \simeq (k^*)^{4mm'}.$$

Since the image of d_2^{01} is torsion, $M \simeq \operatorname{Ker}(d_2^{01}) \simeq \mathbb{Z}^{2mm'+2}$.   □

Remark 4. In fact $E_2^{20} \simeq H^2(S^1 \times S^1, k^*) \simeq k^*$ since the dual graph of L triangulates the torus, $S^1 \times S^1$ [16, p. 105].

Now Proposition 4 is an immediate consequence of part (i) of the next proposition.

Proposition 6. *(i) $\operatorname{rank}(M) \geq 2mm' + 1$.*
(ii) If E and E' are isogenous over K, then $\operatorname{rank}(M) = 2mm' + 2$.

Proof. Consider the composition, ξ,

$$\xi : Pic(W) \overset{i_V^*}{\to} Pic(V) \to M \overset{cl}{\to} H^2(V, \mathbb{Q}_l(1)) \simeq R^2 f_* \mathbb{Q}_l(1)|_0 \to (j_* j^* R^2 f_* \mathbb{Q}_l(1))|_0,$$

where cl is the cycle class map, the isomorphism is the proper base change map [15, VI.2], and the final term may be decomposed by applying the Künneth theorem to the cohomology of $(E \times E')_{\bar{K}}$. Now the closures in W of $E \times e'$, $e \times E'$ (and if E is isogenous to E', of the graph of an isogeny) give rise to elements of $Pic(W)$ whose images under ξ are non-zero and lie in distinct Künneth components. Hence they are linearly independent. Since ξ maps the image of $i_{V*} : CH^0(V) \to Pic(W)$ to zero, the proposition will follow from the next lemma.

Lemma 8. $\operatorname{rank}(\operatorname{im} i_{V*} i_V^* CH^0(V) \to Pic(V)) = 2mm' - 1$.

Proof. This is a consequence of the known rank of the intersection matrix of the fiber components of a fibered surface. For the convenience of the reader we give the argument. Since $i_{V*}([V]) = 0 \in Pic(W)$, $\operatorname{rank}(i_{V*} i_V^*(CH^0(V))) \leq 2mm' - 1$. To show the opposite inequality let $S \subset W$ be a regular, relatively very ample hypersurface with the property that each irreducible component of V meets S in a prime divisor. Now the intersection matrix between the irreducible components of the closed fibers of W and S,

$$(V_i \cdot S_j)_W = (S_i \cdot S_j)_S,$$

has rank $2mm' - 1$ by [13, Theorem 9.1.23]. □
□

In order to establish Theorem 2 of the Introduction it remains to show that κ in the statement of Proposition 4. can take the value 1 when k is not contained in the algebraic closure of a finite field. The computation will be done in the concrete setting of Sect. 5 where $m = m' = 1$. Let V_a denote the closed fiber of W_a, defined in Sect. 5. Set $L_a = V_{a \, red}$. Since a will be fixed, we often drop it from the notation. Define κ as in Proposition 4.

Proposition 7. *If $a \in k^*$ is not a root of unity, then $\kappa = 1$.*

Proof. By the formula for M in the proof of Lemma 7 $\kappa = 1$, if there is an element $\Delta \in E_2^{01}$ such that $d_2^{01}(\Delta) \in E_2^{20}$ has infinite order. By Remark 4 $E_2^{20} \simeq k^*$. The idea is to choose a cover $\mathfrak{U}$ of L, to represent Δ by a Cech cocycle with respect to $\mathfrak{U}$ and to compute $d_2^{01}(\Delta)$ as in [7, p. 446] using the diagram,

$$
\begin{array}{ccccc}
C^1(\mathfrak{U}, \nu_{0*}\mathbb{G}_m) & \xrightarrow{\beta^*} & C^1(\mathfrak{U}, \nu_{1*}\mathbb{G}_m) & \xrightarrow{\beta^*} & C^1(\mathfrak{U}, \nu_{2*}\mathbb{G}_m) \\
\uparrow{\delta_0} & & \uparrow{\delta_1} & & \uparrow{\delta_2} \\
C^0(\mathfrak{U}, \nu_{0*}\mathbb{G}_m) & \xrightarrow{\beta^*} & C^0(\mathfrak{U}, \nu_{1*}\mathbb{G}_m) & \xrightarrow{\alpha^*} & C^0(\mathfrak{U}, \nu_{2*}\mathbb{G}_m),
\end{array}
$$

where α^* and β^* are defined in Proposition 5 and the vertical maps are the usual Cech coboundaries. The cocycle Δ will come from a Cartier divisor, Δ, on $L^{[0]}$ which is the sum of two rational curves, $\Delta_P + \Delta_Q$. The cover $\mathfrak{U}$ will be by Zariski open subsets and chosen as simply as possible to represent Δ by functions on open subsets.

Recall that $L^{[0]}$ is the disjoint union of the two components P and Q pictured in Fig. 2. One may view P as the blow-up of $\mathbb{P}^1 \times \mathbb{P}^1$ at the four points

$$((1:0),(1:0)), \quad ((1:0),(0:1)), \quad ((0:1),(1:0)), \quad ((0:1),(0:1)),$$

where the exceptional curves c_-, d_-, f_-, e_- contract to these points in the order indicated. Let $\Delta_P \subset P$ denote the strict transform of the diagonal of $\mathbb{P}^1 \times \mathbb{P}^1$ in P. The inclusion of the line Δ_Q in the surface $Q = Q_a$ is defined by the inclusion of ideals $(x - ax', y - y') \supset (xy - ax'y')$ using the homogeneous coordinates of Proposition 3. The lines on Q defined by the ideals $(x, x'), (x', y), (x, y'), (y, y')$ are labeled respectively c_+, d_+, e_+, f_+ in Fig. 2. The cohomology class $\Delta \in H^1(L^{[0]}, \mathbb{G}_m)$ associated to the Cartier divisor, $\Delta = \Delta_P + \Delta_Q$, lies in $E_2^{01} = \operatorname{Ker}(d_1^{01}) = \operatorname{Ker}(H^1(\beta^*))$ because $\mathcal{O}_P(\Delta_P)$ and $\mathcal{O}_Q(\Delta_Q)$ have isomorphic restrictions to each component of $L^{[1]}$, namely $\mathcal{O}_{\mathbb{P}^1}(1)$ for c and f and $\mathcal{O}_{\mathbb{P}^1}$ for d and e.

Choose a Zariski open cover, $\mathfrak{U} = \{U_0, U_1, U_2, \ldots\}$ of L and functions, $h_i \in v_{0*}\mathcal{O}_{L^{[0]}}(U_i)$, to represent the effective Cartier divisor, $\varDelta$, as follows:

$U_0 : v_0^{-1}(U_0)$ does not meet the support of $\varDelta$. It contains all labeled curves in Fig. 2 except c_-, f_-, c_+, f_+. Furthermore $h_0 = 1 \in v_{0*}\mathcal{O}_{L^{[0]}}(U_0)$.

U_1: The only labeled curves in Fig. 2 meeting $v_0^{-1}(U_1)$ are c_- and c_+. Choose $h_1 \in v_{0*}\mathcal{O}_{L^{[0]}}(U_1)$ such that $div(h_1) = \varDelta|_{v_0^{-1}(U_1)}$ and

$$\beta^*(h_1) = h_1 \circ \beta_0 / h_1 \circ \beta_1 = 1|_{U_1 \cap L_{sing}}.$$

This is possible because the Cartier divisors $\varDelta_P$ and $\varDelta_Q$ both restrict to give the same divisor, the point $(0 : 0 : 1 : 1)$ in the above coordinate system, on c. Finally U_0, U_1 restrict to give an open cover of c.

U_2: The only labeled curves in Fig. 2 meeting $v_0^{-1}(U_2)$ are f_+ and f_-. U_0, U_2 restrict to give an open cover of f. Choose $h_2 \in v_{0*}\mathcal{O}_{L^{[0]}}(U_2)$ such that $div(h_2) = \varDelta|_{v_0^{-1}(U_2)}$ and

$$\beta^*(h_2) = h_2 \circ \beta_0 / h_2 \circ \beta_1 = (x - x')/(x - ax')|_{V_2},$$

where $V_2 = U_2 \cap L_{sing}$. This is possible because $\varDelta_P$ restricts to the point $(1 : 1 : 0 : 0)$ on f while $\varDelta_Q$ restricts to $(a : 1 : 0 : 0)$.

$U_i, i > 2$: No labeled curves in Fig. 2 meet $v_0^{-1}(U_i)$; $div(h_i) = \varDelta|_{v_0^{-1}(U_i)}$.

Define $\partial\varDelta \in C^1(\mathfrak{U}, v_{0*}\mathbb{G}_m)$ by $\partial\varDelta(U_i \times_L U_j) = h_j/h_i \in v_{0*}\mathcal{O}_{L^{[0]}}(U_i \times_L U_j)$. The Cech coboundary of $\partial\varDelta$ vanishes. Thus $\partial\varDelta$ gives rise to an element in $H^1(\mathfrak{U}, v_{0*}\mathbb{G}_m)$, whose image in $H^1(L, v_{0*}\mathbb{G}_m) \simeq E_1^{01}$ is denoted $\varDelta$. As noted above, $\varDelta \in \mathrm{Ker}(d_1^{01}) = E_2^{01}$. To compute $d_2^{01}(\varDelta)$ it suffices to find $\omega \in C^0(\mathfrak{U}, v_{1*}\mathbb{G}_m)$ with $\delta_1(\omega) = \beta^*(\partial\varDelta)$ and to then evaluate $\alpha^*(\omega) \in \mathrm{Ker}(\delta_2)$.

Set $r = \beta^*(\partial\varDelta)$. Now $r_{ij} \in v_{1*}\mathbb{G}_m(U_i \times_L U_j)$ is given by

$$r_{20} = (x - x')/(x - ax'), \qquad r_{ij} = 1 \quad \text{if} \quad (i, j) \notin \{(0, 2), (2, 0)\}.$$

Define $\omega \in C^0(\mathfrak{U}, v_{1*}\mathbb{G}_m)$ by defining $\omega_0 \in \mathbb{G}_m(v_1^{-1}(U_0))$ by

$$\omega_0|_{f \cap v_1^{-1}(U_0)} := (x - x')/(x - ax') \quad \text{and} \quad \omega_0|_{(a \cup b \cup c \cup d \cup e) \cap v_1^{-1}(U_0))} = 1,$$

and $\omega_i = 1 \in \mathbb{G}_{mL_i}(v_1^{-1}(U_i))$ for $i \geq 1$. Now $\delta_1(\omega) = r$.

Observe that $C^0(\mathfrak{U}, v_{2*}\mathbb{G}_{mL^{[2]}}) \simeq \mathrm{Ker}(\delta_2) \simeq \mathbb{G}_m(L^{[2]})$ as $U_0 \cap (L_{sing})_{sing} = (L_{sing})_{sing}$ and $U_i \cap (L_{sing})_{sing} = \emptyset$ for $i > 0$. Under this identification $\alpha^*(\omega)$ becomes the function on $L^{[2]}$ whose value at $4 \in L^{[2]}$ (see Fig. 2) is a^{-1} and whose value elsewhere is 1. The natural composition,

$$\mathbb{G}_m(L^{[2]}) \simeq \mathrm{Ker}(\delta_2) \to E_1^{02} \to E_1^{02}/im(d_1^{01}) \simeq E_2^{02} \simeq k^*,$$

takes a function on $L^{[2]}$ to the product of its values. Thus $\alpha^*(\omega_2)$ maps to an element of infinite order if $a \in k^*$ has infinite order. $\qquad\square$

Remark 5. One may rephrase the results about $Pic(V)$ in the language of Picard schemes. Write $P(V)$ for the Picard scheme of V [10, 4.18.3], $P^0(V)$ for the connected component of the identity [10, 5.1] and define $N.S.(V) := P(V)(k)/P^0(V)(k)$. Then $P^0(V) \simeq \mathbb{G}_m^2$ and $N.S.(V) \simeq \mathbb{Z}^{2mm'+2-\kappa}$.

7 Semi-stable Models and Small Resolutions

Semi-stable degenerations play a distinguished role in the theory of degenerations because of their simplicity and the techniques available for their construction [9, Chap. II]. Recall that semi-stable models are regular and proper and the closed fiber is a reduced normal crossing divisor [8, Sect. 3.1]. The canonical model, W, is never semi-stable by Lemma 5. In this section we recall the use of small resolutions to create semi-stable models when $m > 1$ and $m' > 1$ and various variants of this result which apply even when $m = 1$ or $m' = 1$. Then we compare the invariants of the closed fiber of the resulting models with the invariants of the closed fiber of the canonical model.

   Suppose that $m > 1$ and $m' > 1$. Then it is possible to desingularize the scheme $\overline{W}$ without introducing exceptional divisors. Note that the hypothesis $m > 1$ and $m' > 1$ implies that the irreducible components of $\overline{V}$ are non-singular. They are Weil divisors which are not Cartier, because they are non-singular at singular points of $\overline{W}$. To desingularize $\overline{W}$ choose an order on the irreducible components of $\overline{V}$. Blow up the first component giving a scheme, $\overline{W}_1$. Now blow up $\overline{W}_1$ along the strict transform of the second irreducible component of $\overline{V}$. Continue until a strict transform of every irreducible component of $\overline{V}$ has been blown up. The composition of blow-ups is a projective morphism, $\gamma : \widehat{W} \to \overline{W}$. An easy local computation [1, pp. 177–178], [19, Sect. 1] at $\overline{W}_{sing}$ shows that $\widehat{W}$ is regular, $\widehat{f} := \overline{f} \circ \gamma : \widehat{W} \to T$ is semi-stable, and the positive dimensional fibers of γ, are all isomorphic to $\mathbb{P}^1$. Blowing up $\widehat{W}$ up along these $\mathbb{P}^1$'s recovers W. The map $\gamma : \widehat{W} \to \overline{W}$ is called a small resolution because the exceptional locus consists of curves.

   When $m = 1$ or $m' = 1$ each component of $\overline{V}$ is a Cartier divisor on $\overline{W}$, since it is the pull back of a fiber component, $F_i \subset \mathcal{E}$ or $F'_{i'} \subset \mathcal{E}'$, via projection on one of the factors. Note that F_i and $F'_{i'}$ are Cartier divisors since $\mathcal{E}$ and $\mathcal{E}'$ are regular. Blowing up a Cartier divisor has no effect on the ambient space. If E and E' are not isogenous, the semi-local ring, $\mathcal{O}_{\overline{W},\overline{W}_{sing}}$, is factorial and no small resolution of $\overline{W}$ exists in the category of schemes [1, p. 179], [19, Proof of 3.1(ii–iii)], [21, 10.1]. On the other hand if $\mathfrak{o}$ is the local ring at a point on a smooth algebraic curve over $\mathbb{C}$, then one can associate to the scheme $\overline{W}$ the germ of a complex analytic space along $\overline{V}$, which we denote $\overline{W}^{an}$. An irreducible component of $\overline{V}$ will have two (four if $m = m' = 1$) branches passing through each point of $\overline{W}^{an}_{sing}$ which it contains. In the complex analytic category we may pick one of these branches and blow it up obtaining a small resolution. If this is done at each point in $\overline{W}^{an}_{sing}$ we obtain the germ, $\widehat{W}^{an}$, of a complex manifold along the normal crossing closed fiber, $\widehat{V}^{an}$. This gives a semi-stable model in the complex analytic sense.

An analogous construction is possible in the category of algebraic spaces [1, pp. 177–185] for arbitrary m and m'. The closed fiber $\widehat{V} \subset \widehat{W}$ is always a proper scheme, since its normalization is projective [11, Corollary 48].

A disadvantage of working with $\widehat{W}$ in place of W, is that the isomorphism class of the former depends on the order in which the blow-ups (or modifications of algebraic spaces) are performed. The same is generally true for $\widehat{V}$. Nonetheless many of the invariants of $\widehat{V}$ are independent of these choices as the following proposition shows:

Proposition 8. (i) $CH^1(\widehat{V}) \simeq (\mathbb{Z})^{mm'+2} \oplus \mathbb{Z}/m''$.

(ii) *Suppose that char(k) does not divide n. Then* $H^0(\widehat{V}_{\bar{k}}, \mathbb{Z}/n) \simeq \mathbb{Z}/n$,
$$H^1(\widehat{V}_{\bar{k}}, \mathbb{Z}/n) \simeq (\mathbb{Z}/n)^2,$$
$$H^2(\widehat{V}_{\bar{k}}, \mathbb{Z}/n) \simeq \mathbb{Z}/n \oplus (\mathbb{Z}/n)^{mm'+2}(-1) \oplus \mathbb{Z}/n''(-1),$$
$$H^3(\widehat{V}_{\bar{k}}, \mathbb{Z}/n) \simeq (\mathbb{Z}/n)^2(-1) \oplus \mathbb{Z}/n''(-1),$$
and $H^4(\widehat{V}_{\bar{k}}, \mathbb{Z}/n) \simeq (\mathbb{Z}/n)^{mm'}(-2)$.

(iii) *There is a short exact sequence,*
$$1 \to Pic(\widehat{V}) \xrightarrow{\rho^*} Pic(V) \to \mathbb{Z}^{mm'} \to 0.$$

(iv) *If* $\kappa = 1$ *and either* $m = 1$ *or* $m' = 1$, *then* $\widehat{V}$ *is not projective.*

Proof. (i) The natural morphism $\rho : V \to \widehat{V}$ collapses each Q_{ij} to a $\mathbb{P}^1$ and exactly one of the curves, c_{ij} or d_{ij}, to a point. Write t_{ij} for the contracted curve and define $\mathcal{T} = Span(\{t_{ij}\})$. Recall that $\{t_{ij}\} \cup \{\delta_{ij}\}$ is a basis of $CH^1(Q)$. By Lemma 3(i)–(ii) the natural map, $\mathcal{T} \to CH^1(Q)/im(\partial)$ is a split injection. Thus the following exact sequence splits:
$$0 \to \mathcal{T} \to CH^1(V) \to CH^1(\widehat{V}) \to 0.$$

(ii) In the Leray spectral sequence associated to $\rho : V \to \widehat{V}$, $E_3 \simeq E_2$. The only possible non-zero differential in E_3 is the map,
$$d_3^{02} : (\mathbb{Z}/n)^{mm'} \simeq H^0(\widehat{V}_{\bar{k}}, R^2\rho_*\mathbb{Z}/n) \simeq E_3^{02} \to E_3^{30} \simeq H^3(\widehat{V}_{\bar{k}}, \mathbb{Z}/n),$$
whose kernel has the form,
$$Ker(d_3^{02}) \simeq H^2(V_{\bar{k}}, \mathbb{Z}/n)/H^2(\widehat{V}_{\bar{k}}, \mathbb{Z}/n).$$

The term on the right contains a submodule isomorphic to $(\mathbb{Z}/n)^{mm'}$ generated by the chern classes of the normal bundles, $\mathcal{N}_{Q_{ij}/W}$. Thus $d_3^{02} = 0$, the spectral sequence degenerates at E_2 and the assertion follows from Corollary 1.

(iii) Apply the Leray spectral sequence for the morphism, $\rho_{red} : V_{red} \to \widehat{V}$, to the sheaf $\mathbb{G}_m$. We have $\rho_{red*}\mathbb{G}_m \simeq \mathbb{G}_m$ and $R^1\rho_{red*}\mathbb{G}_m \simeq \upsilon_*\mathbb{Z}$, where υ is the inclusion of the positive dimensional fibers of $\widehat{V} \to \widehat{V}$. The subgroup of $H^1(V, \mathbb{G}_m)$ generated by isomorphism classes of sheaves of the form, $\mathcal{O}_W(Q_{ij})|_V$, is free of rank mm'. Under the natural map, $H^1(V, \mathbb{G}_m) \to H^0(\widehat{V}, R^1\rho_{red*}\mathbb{G}_m)$, it is mapped isomorphically to the target. The assertion follows since $Pic(V) \simeq Pic(V_{red})$ by Lemma 6.

(iv) Assume $m = 1$. By Proposition 4 and (iii) $N.S.(\widehat{V}) \simeq \mathbb{Z}^{m'+1}$. Let $e \subset \mathcal{E}$ (respectively $e' \subset \mathcal{E}'$) denote the identity section. Let $\mathcal{P} \subset Pic(\widehat{V})$ denote the subgroup generated by restricting invertible sheaves associated to the following Cartier divisors on $\overline{W}$ to $\overline{V}$:

$$\mathcal{E} \times_X e', \qquad e \times_X \mathcal{E}', \qquad \text{and} \qquad \text{irreducible components of } \overline{V}.$$

The latter are Cartier divisors, since they are pulled back from the components of F' on the regular surface, $\mathcal{E}'$. Now $\mathcal{P} \simeq \mathbb{Z}^{m'+1}$ and pullback induces an injection, $\mathcal{P} \to N.S.(\widehat{V})$. Thus the cokernel of the pullback, $\gamma^* : Pic(\overline{V}) \to Pic(\widehat{V})$, is finite. Consequently there is no invertible sheaf on $\widehat{V}$ whose restriction to the curves on $\widehat{V}$ which are collapsed by γ is ample. Thus $\widehat{V}$ has no ample invertible sheaf. $\qquad\qquad\square$

8 The Sheaves $R^i f_* \mathbb{Z}/n$

For the purpose of computing cohomology by the Leray spectral sequence it is important to describe the sheaves $R^i f_* \mathbb{Z}/n$. This will be accomplished in this section under the assumption that $char(k) \nmid n$ and the discrete valuation ring $\mathfrak{o}$ is strictly henselian. Tate twists will be ignored.

8.1 Notations

$n'' = gcd(m'', n)$.
$t = Spec(K)$.
$j : t \to T$, the inclusion of the generic point.
$j_0 : 0 \to T$, the inclusion of the closed point.

To begin observe that the espace étalé of the sheaf $R^1 \pi_* \mu_n$ may be identified with the n-torsion sub group scheme of the Néron model of E which meets only the identity component of the closed fiber. In particular, the stalk at 0 is isomorphic to $\mathbb{Z}/n$, the stalk at t to $(\mathbb{Z}/n)^2$ and the cospecialization map [15, II.3.16] is injective.

Lemma 9. (i) $f_* \mathbb{Z}/n \simeq \mathbb{Z}/n$.
(ii) $R^1 f_* \mathbb{Z}/n \simeq R^1 \pi_* \mathbb{Z}/n \oplus R^1 \pi'_* \mathbb{Z}/n$.
(iii) $R^3 f_* \mathbb{Z}/n \simeq R^1 \pi_* \mathbb{Z}/n \oplus R^1 \pi'_* \mathbb{Z}/n \oplus j_{0*} \mathbb{Z}/n''$.
(iv) $R^4 f_* \mathbb{Z}/n \simeq \mathbb{Z}/n \oplus j_{0*}(\mathbb{Z}/n)^{2mm'-1}$.

Proof. (ii) By the Leray spectral sequence for the composition, $f = \overline{f} \circ \sigma$, $R^1 \overline{f}_* \mathbb{Z}/n \simeq R^1 f_* \mathbb{Z}/n$. The assertion follows from the Künneth formula [15, VI.8.5].

(iii) The Leray spectral sequence for $f = \overline{f} \circ \sigma$ gives an exact sequence,

$$\overline{f}_* R^2 \sigma_* \mathbb{Z}/n \to R^3 \overline{f}_* \mathbb{Z}/n \overset{\sigma^*}{\to} R^3 f_* \mathbb{Z}/n \to 0,$$

where the support of the first term is 0. By the Künneth formula [15, VI.8.5]

$$R^3 \overline{f}_* \mathbb{Z}/n \simeq R^1 \pi_* \mathbb{Z}/n \otimes R^2 \pi'_* \mathbb{Z}/n \oplus R^2 \pi_* \mathbb{Z}/n \otimes R^1 \pi'_* \mathbb{Z}/n.$$

Now

$$R^2\pi_*\mu_n \simeq \mathbb{Z}/n \oplus j_{0*}(\mathbb{Z}/n)^{m-1} \quad \text{and} \quad R^2\pi'_*\mu_n \simeq \mathbb{Z}/n \oplus j_{0*}(\mathbb{Z}/n)^{m'-1},$$

where the summands with support 0 are generated by the cohomology classes of the fiber components $F_1,\ldots,F_{m-1}$ and $F'_1,\ldots,F'_{m'-1}$. This establishes the desired isomorphism up to a summand supported at 0. The latter is determined by Corollary 1(iv), since $R^3 f_*\mathbb{Z}/n|_0 \simeq H^3(V,\mathbb{Z}/n)$.

(iv) The Leray spectral sequence for $f = \overline{f} \circ \sigma$ gives a split exact sequence,

$$0 \to R^4\overline{f}_*\mathbb{Z}/n \to R^4 f_*\mathbb{Z}/n \to \overline{f}_*R^4\sigma_*\mathbb{Z}/n \to 0.$$

Now use the isomorphisms, $\overline{f}_*R^4\sigma_*\mathbb{Z}/n \simeq j_{0*}(\mathbb{Z}/n)^{mm'}$ and

$$R^4\overline{f}_*\mathbb{Z}/n \simeq R^2\pi_*\mathbb{Z}/n \otimes R^2\pi'_*\mathbb{Z}/n \simeq \mathbb{Z}/n \oplus j_{0*}(\mathbb{Z}/n)^{mm'-1}.$$

□

In order to describe the sheaf $R^2 f_*\mathbb{Z}/n$, consider the composition of canonical maps,

$$R^2 f_*\mathbb{Z}/n \xrightarrow{r} j_*j^*R^2 f_*\mathbb{Z}/n \xrightarrow{j_*\circ pr} j_*j^*(R^1\pi_*\mathbb{Z}/n \otimes R^1\pi'_*\mathbb{Z}/n), \tag{3}$$

where pr is the Künneth projection on the stalk at the generic point. Denote the image of this composition by $\mathcal{G}_n$. Since the other Künneth components of $j_*j^*R^2 f_*\mathbb{Z}/n$ are Tate twists of constant sheaves corresponding to the cohomology classes of the two factors in the product, we have, $im(r) \simeq (\mathbb{Z}/n)^2 \oplus \mathcal{G}_n$, ignoring Tate twists.

Proposition 9. $R^2 f_*\mathbb{Z}/n \simeq j_{0*}(\mathbb{Z}/n\mathbb{Z})^{2mm'-1} \oplus (\mathbb{Z}/n\mathbb{Z})^2 \oplus \mathcal{G}_n$.

To prepare for the proof of Proposition 9 it is convenient to recall a tool for working in the category, $\mathfrak{S}$, of étale sheaves of $\mathbb{Z}/n$ modules on T. This is the category, $\mathfrak{M}$, the objects of which are triples, (N_0, N, φ), where N is a $\mathbb{Z}/n\mathbb{Z}$-module with an action of $I := Gal(\overline{K}/K)$, N_0 is a $\mathbb{Z}/n\mathbb{Z}$-module, and $\varphi : N_0 \to N^I$ is a group homomorphism. A morphism in $\mathfrak{M}$, $\Psi : (N_0, N, \varphi) \to (N'_0, N', \varphi')$, is given by a I-module homomorphism, $\psi : N \to N'$, and a group homomorphism, $\psi_0 : N_0 \to N'_0$, such that $\varphi' \circ \psi_0 = \psi \circ \varphi$.

Lemma 10. *An equivalence of categories,* $\mathfrak{S} \to \mathfrak{M}$, *is defined by*

$$\mathcal{F} \to (j_0^*\mathcal{F}, j^*\mathcal{F}|_{\overline{K}}, \varphi),$$

where the cospecialization map, φ, *corresponds to the natural map,* $j_0^*\mathcal{F} \to j_0^*j_*j^*\mathcal{F}$.

Proof. [15, II.3.10–3.12].

Proof of Proposition 9. It is convenient to apply the vanishing cycle spectral sequence [8, Sect. 3],

$$E_2^{pq} = H^p(\widehat{V}_{\bar{k}}, R^q \Psi \mathbb{Z}/n) \Rightarrow H^{p+q}(E \times E'_{\bar{R}}, \mathbb{Z}/n).$$

So that the sheaves $R^q \Psi \mathbb{Z}/n$ may be identified with more familiar sheaves on the closed fiber, we work with a semi-stable model in the sense of [8, 3.1]. As noted in the previous section this precludes working with W. We assume until further notice that $m > 1$ and $m' > 1$ and that the semi-stable model, $\widehat{W}$, is constructed from $\overline{W}$ by successively blowing up fiber components as described in Sect. 7. Then there are canonical identifications of sheaves on $\widehat{V}$ [8, Théorème 3.2(c)]:

$$R^0 \Psi \mathbb{Z}/n \simeq \mathbb{Z}/n, \qquad R^1 \Psi \mathbb{Z}/n \simeq (\nu_* \mathbb{Z}/n)/(\mathbb{Z}/n),$$

where $\nu : \widehat{V}^{[0]} \to \widehat{V}$ is the normalization. The cospecialization map may be expressed in terms of the map $\widehat{\varphi}$,

$$R^2 \widehat{f}_* \mathbb{Z}/n|_0 \simeq H^2(\widehat{V}_{\bar{k}}, \mathbb{Z}/n) \simeq H^2(\widehat{V}_{\bar{k}}, R^0 \Psi \mathbb{Z}/n) \xrightarrow{\widehat{\varphi}} H^2(E \times E'_{\bar{R}}, \mathbb{Z}/n) \simeq R^2 \widehat{f}_* \mathbb{Z}/n|_{\overline{K}}.$$

which appears in the five term exact sequence associated to the spectral sequence,

$$0 \to H^1(\widehat{V}_{\bar{k}}, \mathbb{Z}/n) \to H^1(E \times E'_{\bar{R}}, \mathbb{Z}/n) \to H^0(\widehat{V}_{\bar{k}}, R^1 \Psi \mathbb{Z}/n) \xrightarrow{d_2^{01}}$$

$$H^2(\widehat{V}_{\bar{k}}, \mathbb{Z}/n) \xrightarrow{\widehat{\varphi}} H^2(E \times E'_{\bar{R}}, \mathbb{Z}/n). \tag{4}$$

The first two terms in (4) are free $\mathbb{Z}/n$-modules of ranks 2 and 4 respectively. The third term is free of rank $mm' + 1$ as one sees from the exact sequence of sheaves on $\widehat{V}$:

$$0 \longrightarrow \mathbb{Z}/n \longrightarrow \nu_* \mathbb{Z}/n \longrightarrow R^1 \Psi \mathbb{Z}/n \longrightarrow 0,$$

h^0	1	mm'
h^1	2	0

where the numbers give the ranks of cohomology groups which are free $\mathbb{Z}/n$-modules. It follows that $\mathrm{Ker}(\widehat{\varphi})$ is a free $\mathbb{Z}/n$-module of rank $mm' - 1$.

The Leray spectral sequence for the morphism, $\rho : W \to \widehat{W}$, which contracts each Q_{ij} to a $\mathbb{P}^1$ gives an exact sequence

$$0 \to R^2 \widehat{f}_* \mathbb{Z}/n \to R^2 f_* \mathbb{Z}/n \xrightarrow{\xi} \widehat{f}_* R^2 \rho_* \mathbb{Z}/n.$$

The sheaf on the right is isomorphic to $j_{0*}(\mathbb{Z}/n)^{mm'}$. The map ξ has a left inverse given by taking the cohomology classs of the Q_{ij}'s in W. In the notation of (3) it follows that $\mathrm{Ker}(r) \simeq j_{0*}(\mathbb{Z}/n)^{2mm'-1}$. Using Lemma 10 and the fact that $\mathbb{Z}/n$ is an injective $\mathbb{Z}/n$-module one verifies that $\text{Ker}(r)$ is a direct summand of $R^2 f_* \mathbb{Z}/n$ [22, 2.4(iii)]. This proves Proposition 9 when $m > 1$ and $m' > 1$.

The case where $m = 1$ or $m' = 1$ may be reduced to the situation considered above. Fix $m_0 \in \mathbb{N}$ prime to nmm'. As $\mathfrak{o}$ is strictly henselian the Néron models of E and E' have unique constant subgroup schemes over T which may be identified with $\mathbb{Z}/m_0$. Let $\tilde{E} \to E$ (respectively $\tilde{E}' \to E'$) denote the isogeny which is dual to the isogeny obtained by taking the quotient by this subgroup scheme. Extending the isogeny to the relative minimal model, $\widetilde{\mathcal{E}}$ of $\tilde{E}$ (respectively $\widetilde{\mathcal{E}}'$ of $\tilde{E}'$) yields a finite étale $\mathbb{Z}/m_0$ cover $\widetilde{\mathcal{E}} \to \mathcal{E}$ (respectively $\widetilde{\mathcal{E}}' \to \mathcal{E}'$). Take the fiber product and blow up the singular locus to get a finite étale cover, $\widetilde{W} \to W$, which is Galois with Galois group, $H := (\mathbb{Z}/m_0)^2$. Consider the natural map,

$$R^2 f_* \mathbb{Z}/n \to R^2 \widetilde{f}_* \mathbb{Z}/n.$$

There is a corresponding map in the category, $\mathfrak{M}$,

$$\Psi : (H^2(V, \mathbb{Z}/n),\ H^2(E \times E'_{\overline{K}}, \mathbb{Z}/n),\ \varphi)$$
$$\to (H^2(\tilde{V}, \mathbb{Z}/n),\ H^2(\tilde{E} \times \tilde{E}'_{\overline{K}}, \mathbb{Z}/n),\ \tilde{\varphi}),$$

where $\widetilde{V} \subset \widetilde{W}$ denotes the closed fiber. The action of H on $H^2(\tilde{E} \times \tilde{E}'_{\overline{K}}, \mathbb{Z}/n)$ is trivial, since translation by torsion points acts trivially on cohomology. By Corollary 1(iii) and the Hochschild–Serre spectral sequence,

$$(\mathbb{Z}/n)^{2mm'+3} \simeq H^2(V, \mathbb{Z}/n) \simeq H^2(\widetilde{V}, \mathbb{Z}/n)^H.$$

Since $\mathbb{Z}/n$-free $\mathbb{Z}/n[H]$-modules are semi-simple, the isomorphism,

$$\psi : H^2(E \times E'_{\overline{K}}, \mathbb{Z}/n) \to H^2(\tilde{E} \times \tilde{E}'_{\overline{K}}, \mathbb{Z}/n),$$

maps the image of φ isomorphically to the image of $\widetilde{\varphi}$. From the analysis of the case $m > 1$ and $m' > 1$ above, $H^2(\widetilde{V}, \mathbb{Z}/n) \simeq (\mathbb{Z}/n)^{mm'+3}$ (as $n'' = 1$) and $\text{Ker}(\widetilde{\varphi}) \simeq (\mathbb{Z}/n)^{mm'-1}$ in (4) so

$$im(\widetilde{\varphi}) \simeq im(\widetilde{\varphi}) \simeq (\mathbb{Z}/n)^4.$$

Thus $im(\varphi) \simeq (\mathbb{Z}/n)^4$. Apply Corollary 1(iii) to V to conclude that $\text{Ker}(\varphi) \simeq (\mathbb{Z}/n)^{2mm'-1}$ as desired. $\qquad\square$

To describe the structure of the sheaf $\mathcal{G}_n$ it is convenient to introduce the notation (M_0, M, φ) (respectively (M'_0, M', φ')) for the object of $\mathcal{M}$ corresponding to the sheaf $R^1 \pi_* \mathbb{Z}/n$ (respectively $R^1 \pi'_* \mathbb{Z}/n$). Since φ and φ' are injective, we may identify M_0 with a submodule of M^I and M'_0 with a submodule of $(M')^I$. Define

$$M_* := (M \otimes M')^I \cap (M \otimes M'_0 + M_0 \otimes M') \subset M \otimes M'.$$

Proposition 10. *(i)* $\mathcal{G}_n|_0 \simeq (\mathbb{Z}/n\mathbb{Z})^2 \oplus \mathbb{Z}/n''$.
(ii) There is a short exact sequence,

$$0 \to R^1\pi_*\mathbb{Z}/n \otimes R^1\pi'_*\mathbb{Z}/n \to \mathcal{G}_n \to j_{0*}(\mathbb{Z}/n \oplus \mathbb{Z}/n'') \to 0.$$

(iii) $M_* \simeq (\mathbb{Z}/n)^2 \oplus \mathbb{Z}/n''$.

(iv) The object $(M_*, M \otimes M', i)$ of $\mathcal{M}$ corresponds to $\mathcal{G}_n$. Here $i : M_* \to M \otimes M'$ is the inclusion.

Proof. (i) The structure of the stalk of $\mathcal{G}_n$ at 0 follows from Corollary 1(iii) and Proposition 9.

(ii) This follows from the Künneth formula applied to $R^2\bar{f}_*\mathbb{Z}/n$, the injectivity of $R^2\bar{f}_*\mathbb{Z}/n \to R^2 f_*\mathbb{Z}/n$, the isomorphism, $(R^1\pi_*\mathbb{Z}/n \otimes R^1\pi'_*\mathbb{Z}/n)|_0 \simeq \mathbb{Z}/n$, and (i).

(iii) The inertia group acts through its tame quotient which is topologically generated by a single element, τ. By [24, V.5 Exercise 13] one may choose a basis e_0 of M_0 (respectively e'_0 of M'_0) and extend to a basis $\{e_0, e_1\}$ of M (respectively $\{e'_0, e'_1\}$ of M') so that τ acts by

$$e_0 \mapsto e_0, \qquad e_1 \mapsto e_1 + me_0, \qquad e'_0 \mapsto e'_0, \qquad e'_1 \mapsto e'_1 + m'e'_0.$$

Define

$$\widetilde{M} := Span\left\{e_0 \otimes e'_0, \; \frac{m'}{m''}e_1 \otimes e'_0 - \frac{m}{m''}e_0 \otimes e'_1\right\} \subset M \otimes M'.$$

Reducing the matrix for $\tau - Id$ acting on $M \otimes M'$ to Smith normal form gives $diag(m'', m'', 0, 0)$. Thus

$$(M \otimes M')^I = \widetilde{M} + \frac{n}{n''}(M \otimes M').$$

As the second term is a free $\mathbb{Z}/n''$-module, $M_* \simeq \widetilde{M} \oplus \mathbb{Z}/n'' \simeq (\mathbb{Z}/n)^2 \oplus \mathbb{Z}/n''$.

(iv) The cup product pairing on $H^2(E \times E'_{\bar{K}}, \mathbb{Z}/n)$ gives rise to pairing on $M \otimes M'$ with respect to which,

$$(e_0 \otimes e'_0)^\perp = M \otimes M'_0 + M_0 \otimes M'.$$

To complete the proof of (iv) it suffices to check that $\mathcal{G}_n|_0$ is orthogonal to $e_0 \otimes e'_0$.

A generator, $\omega \in H^1(F, \mathbb{Z}/n) \otimes H^1(F', \mathbb{Z}/n)$, gives by the Künneth formula an element of $H^2(\overline{V}, \mathbb{Z}/n)$. By the Leray spectral sequence this group maps injectively to $H^2(V, \mathbb{Z}/n)$ which by the proper base change theorem may be identified with the stalk, $R^2 f_*\mathbb{Z}/n|_0$. After multiplying ω by an element in $(\mathbb{Z}/n)^*$ if necessary, the element, $r(\omega)$, in the following commutative diagram involving stalks of sheaves at 0, may be identified with $e_0 \otimes e'_0$.

$$R^2 f_* \mathbb{Z}/n|_0 \xrightarrow{\ r\ } (j_* j^* R^2 f_* \mathbb{Z}/n)_0 \xrightarrow{j_* \circ pr} [j_* j^* (R^1 \pi_* \mathbb{Z}/n \otimes R^1 \pi'_* \mathbb{Z}/n)]_0$$

$$\downarrow \omega\cup \qquad\qquad\qquad \downarrow r(\omega)\cup \qquad\qquad\qquad\qquad \downarrow r(\omega)\cup$$

$$R^4 f_* \mathbb{Z}/n|_0 \longrightarrow (j_* j^* R^4 f_* \mathbb{Z}/n)_0 =\!\!=\!\!=\!\!= (j_* j^* R^4 f_* \mathbb{Z}/n)_0.$$

The orthogonality of $\mathcal{G}_n|_0$ and $e_0 \otimes e'_0$ is equivalent to the composition of maps from the upper left along the top row then down the right hand side being zero. To check this it suffices show that the vertical map on the left is zero:

Lemma 11. *The cup product map,* $H^2(V, \mathbb{Z}/n) \overset{\omega\cup}{\to} H^4(V, \mathbb{Z}/n)$, *is zero.*

Proof. We may replace V by $L := V_{red}$ [15, I.3.23]. There is an exact sequence of sheaves on L,

$$0 \to \mathbb{Z}/n \overset{\nu_0^*}{\to} \nu_{0*}\mathbb{Z}/n \overset{\beta^*}{\to} \nu_{1*}\mathbb{Z}/n \overset{\alpha^*}{\to} \nu_{2*}\mathbb{Z}/n \to 0,$$

whose derivation is similar to Proposition 5 and a corresponding spectral sequence, $E_1^{pq} \simeq H^q(L^{[p]}, \mathbb{Z}/n) \Rightarrow H^{p+q}(L, \mathbb{Z}/n)$. Since the dual graph of L triangulates a topological two torus, one gets an isomorphism (cf. [16, p. 105])

$$\mathbb{Z}/n \simeq E_2^{20} \simeq \mathrm{Ker}\,[\,H^2(L, \mathbb{Z}/n) \overset{H^2(\nu_0^*)}{\longrightarrow} H^2(L^{[0]}, \mathbb{Z}/n)\,].$$

As $H^2(\nu_0^*)(\omega) = 0$, $\mathrm{Ker}(H^2(\nu_0^*)) \simeq (\mathbb{Z}/n)\omega$. Furthermore $\omega \cup \omega = 0$, since the map,

$$H^1(F) \otimes H^1(F) \overset{\cup}{\to} H^2(F),$$

is zero. The lemma now follows from the commutativity of the diagram,

$$
\begin{array}{ccc}
H^2(L, \mathbb{Z}/n) & \xrightarrow{\ \omega\cup\ } & H^4(L, \mathbb{Z}/n) \\
{\scriptstyle H^2(\nu_0^*)}\big\downarrow & & \big\downarrow{\scriptstyle H^4(\nu_0^*)}\ \wr \\
H^2(L^{[0]}, \mathbb{Z}/n) & \xrightarrow{H^2(\nu_0^*)(\omega)\cup} & H^4(L^{[0]}, \mathbb{Z}/n).
\end{array}
$$

□

□

Acknowledgements The hospitality of the Isaac Newton Institute for Mathematical Sciences and of the Max-Planck-Institut für Mathematik, Bonn where some of this work was done is gratefully acknowledged as is the partial support of the NSF, DMS 99-70500, and the NSA, H98230-08-1-0027. Thanks to A. Weisse for help with the figures and to the referee for comments which led to an improvement of the exposition and to Debbie Iscoe at the Fields Institute for translating AMSTeX to LaTeX.

References

1. M. Artin, *Théorèmes de représentabilité pour les espaces algébriques* (Presses de l'université de Montréal, Montreal, Quebec, 1973)
2. S. Bloch, Algebraic cycles and the Beilinson conjectures, in *The Lefschetz Centennial Conference, Part I*, Mexico City, 1984. Contemporary Mathematics, vol. 58 (American Mathematical Society, Providence, 1986), pp. 65–79
3. S. Bloch, A. Ogus, Gersten's conjecture and the homology of schemes. Ann. Sci. Éx. Norm. Sup. 4e Sér. t. **7**, 181–202 (1974)
4. S. Bloch, H. Gillet, Ch. Soulé, Algebraic cycles on degenerate fibers, in *Arithmetic Geometry*, Cortona, 1994. Sympos. Math., vol. XXXVII (Cambridge University Press, Cambridge, 1997), pp. 45–69
5. S. Cynk, M. Schütt, Non-liftable Calabi-Yau spaces. Ark. Mat. **50**, 23–40 (2012)
6. W. Fulton, *Intersection Theory* (Springer, New York, 1984)
7. P. Griffiths, J. Harris, *Principles of Algebraic Geometry* (Wiley, New York, 1978)
8. L. Illusie, Autour du théorème de monodromie locale. Astérisque **223**, 9–57 (1994)
9. G. Kempf, F. Knudsen, D. Mumford, B. Saint-Donat, in *Toroidal Embeddings I*. Lecture Notes in Mathematics, vol. 339 (Springer, New York, 1973)
10. S. Kleiman, The Picard scheme, in *Fundamental Algebraic Geometry*. Mathematical Surveys and Monographs, vol. 123 (American Mathematical Society, Providence, 2005), pp. 235–321
11. J. Kollar, Quotients by finite equivalence relations [arXiv:0812.3608]
12. G. Laumon, Homologie étale, Exposé VIII in *Séminaire de géometrie analytique*, ed. by A. Douady, J.-L. Verdier. Astérisque, vol. 36–37 (1976), pp. 163–188
13. Q. Liu, *Algebraic Geometry and Arithmetic Curves* (Oxford University Press, Oxford, 2002)
14. Ch. Meyer, in *Modular Calabi-Yau Threefolds*. Fields Institute Monograph, vol. 22 (American Mathematical Society, Providence, 2005)
15. J. Milne, *Étale Cohomology* (Princeton University Press, Princeton, 1980)
16. D. Morrison, The Clemens-Schimd exact sequence and applications, in *Topics in Transcendental Algebraic Geometry*, ed. by P. Griffiths (Princeton University Press, Princeton, 1984), pp. 101–119
17. E. Nart, The Bloch complex in codimension one and arithmetic duality. J. Number Theor. **32**, 321–331 (1989)
18. C. Schoen, Complex multiplication cycles on elliptic modular threefolds. Duke Math. J. **53**, 771–794 (1986)
19. C. Schoen, On fiber products of rational elliptic surfaces with section. Math. Z. **197**, 177–199 (1988)
20. C. Schoen, Complex varieties for which the Chow group mod n is not finite. J. Algebr. Geom. **11**, 41–100 (2002)
21. C. Schoen, Desingularized fiber products of semi-stable elliptic surfaces with vanishing third Betti number. Compos. Math. **145**, 89–111 (2009)
22. C. Schoen, Torsion in the cohomology of desingularized fiber products of elliptic surfaces. Mich. Math. J. (to appear)
23. J. Silverman, *The Arithmetic of Elliptic Curves* (Springer, New York, 1986)
24. J. Silverman, *Advanced Topics in the Arithmetic of Elliptic Curves* (Springer, New York, 1994)
25. J.-L. Verdier, Classe d'homologie associeé à un cycle, Exposé VI, in *Séminaire de Géometrie Analytique*, ed. by A. Douady, J.-L. Verdier. Astérisque, vol. 36–37 (Soc. Math. France, Paris, France, 1976), pp. 101–151

Part III
Research Articles: Arithmetic and Geometry of Calabi-Yau Threefolds and Higher Dimentional Varieties

Dynamics of Special Points on Intermediate Jacobians

Xi Chen and James D. Lewis

Abstract We prove some general density statements about the subgroup of invertible points on intermediate jacobians; namely those points in the Abel–Jacobi image of nullhomologous algebraic cycles on projective algebraic manifolds.

Key words: Abel–Jacobi map, Intermediate jacobian, Normal function, Chow group

Mathematics Subject Classifications (2010): Primary 14C25; Secondary 14C30, 14C35

1 Introduction

Let $X/\mathbb{C}$ be a projective algebraic manifold, $\mathrm{CH}^r(X)$ the Chow group of codimension r algebraic cycles on X (with respect to the equivalence relation of rational equivalence), and $\mathrm{CH}^r_{\mathrm{hom}}(X)$ the subgroup of cycles that are nullhomologous under the cycle class map to singular cohomology with $\mathbb{Z}$-coefficients. Largely in relation to the celebrated Hodge conjecture, as well as with regard to equivalence relations on algebraic cycles, the Griffiths Abel–Jacobi map

$$\Phi_r : \mathrm{CH}^r_{\mathrm{hom}}(X) \to J^r(X) \overset{\text{Carlson}}{\simeq} \mathrm{Ext}^1_{\mathrm{MHS}}(\mathbb{Z}(0), H^{2r-1}(X, \mathbb{Z}(r))),$$

has been a focus of attention for the past 60 years. The role of the Abel–Jacobi map in connection to the celebrated Hodge conjecture began with the work of Lefschetz in his proof of his famous "Lefschetz $(1, 1)$ theorem"; and which inspired Griffiths to develop his program of updating Lefschetz's ideas as a general line of attack on

X. Chen (✉) · J.D. Lewis
632 Central Academic Building, University of Alberta, Edmonton, AB, Canada T6G 2G1
e-mail: xichen@math.ualberta.ca; lewisjd@ualberta.ca

R. Laza et al. (eds.), *Arithmetic and Geometry of K3 Surfaces and Calabi–Yau Threefolds*, 491
Fields Institute Communications 67, DOI 10.1007/978-1-4614-6403-7_18,
© Springer Science+Business Media New York 2013

the Hodge conjecture (see [8], as well as [7, Lec. 6, 12, 14]). To this day, a *precise* statement about what the image of Φ_r is in general seems rather elusive. What we do know is that there are examples where the image of Φ_r can be a countable set (even infinite dimensional over $\mathbb{Q}$) [1, 3], or completely torsion [2]. One can ask whether the image of Φ_r is always dense in $J^r(X)$, but even that is unlikely to be true in light of some results in the literature inspired by some of the conjectures in [4]. In this paper, we seek to come up with a general statement about the image of Φ_r, which however is modest, is indeed is better than no statement at all. There are two key ideas exploited in this paper, viz., the business of Lefschetz pencils and associated normal functions, and the classical Kronecker's theorem (see [5] (Chap. XXIII)), which we state in the following form:

Theorem 1. *Let $A = \mathbb{R}^n/\mathbb{Z}^n$ be a compact real torus of dimension n. For a point $p = (x_1, x_2, \ldots, x_n) \in A$, $\mathbb{Z}p = \{kp : k \in \mathbb{Z}\}$ is dense in A if and only if $1, x_1, x_2, \ldots, x_n$ are linearly independent over $\mathbb{Q}$. In particular, the set*

$$\{p \in A : \mathbb{Z}p \text{ is not dense in } A\} \tag{1}$$

is of the first Baire category.

The main results are stated in Theorem 2 and Corollaries 1 and 2 below.

We are grateful to our colleagues Matt Kerr, Phillip Griffiths and Chuck Doran for enlightening discussions, as well as the referee for suggesting improvements in presentation.

2 Some Preliminaries

All integral cohomology is intended modulo torsion. Let $X/\mathbb{C}$ be a projective algebraic manifold of dimension $2m$ and $\{X_t\}_{t \in \mathbb{P}^1}$ a Lefschetz pencil of hyperplane sections of X arising from a given polarization on X. Let $D := \bigcap_{t \in \mathbb{P}^1} X_t$ be the (smooth) base locus and $\overline{X} = B_D(X)$, the blow-up. One has a diagram:

$$\overline{X}_U \hookrightarrow \overline{X}$$

$$\rho_U \downarrow \qquad \downarrow \rho \tag{2}$$

$$U \xrightarrow{j} \mathbb{P}^1,$$

where $\Sigma := \mathbb{P}^1 \setminus U = \{t_1, \ldots, t_M\}$ is the singular set, viz., where the fibers are singular Lefschetz hyperplane sections. One has a short exact sequence of sheaves

$$0 \to j_* R^{2m-1} \rho_{U,*} \mathbb{Z} \to \overline{\mathcal{F}}^{m,*} \to \overline{\mathcal{J}} \to 0, \tag{3}$$

where

$$\overline{\mathcal{F}}^{m,*} = \mathcal{O}_{\mathbb{P}^1}\left(\coprod_{t\in\mathbb{P}^1} \frac{H^{2m-1}(X_t,\mathbb{C})}{F^m H^{2m-1}(X_t,\mathbb{C})}\right) \quad \text{(canonical extension)},$$

and where the cokernel sheaf $\overline{\mathcal{J}}$ is the sheaf of germs of normal functions. The canonical (sometimes called the privileged) extension $\overline{\mathcal{F}}^{m,*}$ of the vector bundle

$$\mathcal{F}^{m,*} := \mathcal{O}_U\left(\coprod_{t\in U} \frac{H^{2m-1}(X_t,\mathbb{C})}{F^m H^{2m-1}(X_t,\mathbb{C})}\right),$$

is introduced in [8] (as well as in the references cited there). It plays a role in the required limiting behaviour of the group $H^0(\mathbb{P}^1,\overline{\mathcal{J}})$ of normal functions "at the boundary", viz., at Σ. Roughly speaking then, a normal function $\nu \in H^0(\mathbb{P}^1,\overline{\mathcal{J}})$ is a holomorphic cross-section,

$$\nu : \mathbb{P}^1 \to \coprod_{t\in\mathbb{P}^1} J^m(X_t),$$

where for $t \in \Sigma$, $J^m(X_t)$ are certain "generalized" intermediate jacobians, and where ν is locally liftable to a section of $\overline{\mathcal{F}}^{m,*}$.[1] The results in [8, Corollary 4.52] show that (3) induces a short exact sequence:

$$0 \to J^m(X) \to H^0(\mathbb{P}^1,\overline{\mathcal{J}}) \xrightarrow{\delta} H^1(\mathbb{P}^1, j_* R^{2m-1}\rho_{U,*}\mathbb{Z})^{(m,m)} \to 0, \tag{4}$$

where it is also shown that with respect to the aforementioned polarization of X defining primitive cohomology,

$$H^1(\mathbb{P}^1, j_* R^{2m-1}\rho_{U,*}\mathbb{C}) \simeq \mathrm{Prim}^{2m}(X,\mathbb{C}) \bigoplus H_\nu^{2m-2}(D,\mathbb{C}), \tag{5}$$

where $H_\nu^{2m-2}(D,\mathbb{C}) = \ker(H^{2m-2}(D,\mathbb{C}) \to H^{2m+2}(X,\mathbb{C}))$, (induced by the inclusion $D \hookrightarrow X$, and where $H^1(\mathbb{P}^1, j_* R^{2m-1}\rho_{U,*}\mathbb{Z})^{(m,m)}$ are the integral classes of Hodge type (m,m) in $H^1(\mathbb{P}^1, j_* R^{2m-1}\rho_{U,*}\mathbb{Z})$, and the fixed part $J^m(X)$ is the Griffiths intermediate jacobian of X. It should be pointed out that there is an intrinsically defined Hodge structure on the space $H^1(\mathbb{P}^1, j_* R^{2m-1}\rho_{U,*}\mathbb{C})$ and that (5) is an isomorphism of Hodge structures [9]. For $t \in U$, the Lefschetz theory guarantees an orthogonal decomposition

$$H^{2m-1}(X_t,\mathbb{C}) = H^{2m-1}(X,\mathbb{C}) \oplus H_\nu^{2m-1}(X_t,\mathbb{C}),$$

where by the weak Lefschetz theorem, $H^{2m-1}(X,\mathbb{C})$ is identified with its image $H^{2m-1}(X,\mathbb{C}) \hookrightarrow H^{2m-1}(X_t,\mathbb{C})$ and integrally speaking, $H_\nu^{2m-1}(X_t,\mathbb{Z})$ is the space

---

[1] There is also a horizontality condition attached to the definition of normal functions of families of projective algebraic manifolds, which automatically holds in the Lefschetz pencil situation (see [8, Theorem 4.57]).

generated by the vanishing cocycles $\{\delta_1, \ldots, \delta_M\}$ (cf. [7, Lec. 6, p. 71]). For fixed $t \in U$, we put

$$J_v^m(X_t) = \text{Ext}^1_{\text{MHS}}(\mathbb{Z}(0), H_v^{2m-1}(X_t, \mathbb{Z}(m))).$$

For each $t_i \in \Sigma$, we recall the Picard–Lefschetz transformation T_i, and formula $T_i(\gamma) = \gamma + (-1)^m(\gamma, \delta_i)\delta_i$, where $(\delta_i, \delta_j) := (\delta_i, \delta_j)_{X_t} \in \mathbb{Z}$ is the cup product on X_t (followed by the trace). Note that a lattice in $H_v^{2m-1}(X_t, \mathbb{Z})$ (i.e. defining a basis of $H_v^{2m-1}(X_t, \mathbb{Q})$), is given (up to relabelling) by a suitable subset $\{\delta_1, \ldots, \delta_{2g}\}$, $(2g \leq M)$, of vanishing cocycles. However we are going to choose our lattice generators $\{\delta_1, \ldots, \delta_{2g}\}$ more carefully as follows:

- Given δ_1, choose δ_2 such that $(\delta_1, \delta_2) \neq 0$. Since $(\delta_j^2) = 0 \; \forall \; j = 1, \ldots, M$, it follows that $\{\delta_1, \delta_2\}$ are $\mathbb{Q}$-independent.
- Next, we argue inductively on k with $1 \leq k \leq 2g - 1$, that

$$\{\delta_1, \ldots, \delta_k, \delta_{k+1}\} \text{ are } \mathbb{Q}\text{-independent and} \tag{6}$$

$$(\delta_\ell, \delta_{k+1}) \neq 0 \text{ for some } \ell \in \{1, \ldots, k\}.$$

Indeed if (6) failed to hold for any or all such k, then in light of the Picard–Lefschetz formula, $\{\delta_1, \ldots, \delta_M\}$ would not be conjugate under the monodromy group action [7, Lec. 6, p. 71].

3 Main Results

The class $\delta(v) \in H^1(\mathbb{P}^1, j_* R^{2m-1} \rho_{U,*} \mathbb{Z})^{(m,m)}$ is called the topological invariant or the cohomology class of the normal function v.

Theorem 2. *Let $v \in H^0(\mathbb{P}^1, \overline{\mathcal{J}})$ be a normal function with nontrivial cohomology class, i.e., satisfying $\delta(v) \neq 0$. Then for very general $t \in U$, the subgroup $\langle v(t) \rangle \subset J_v^m(X_t)$ generated by $v(t)$, is dense in the strong topology. In particular, the family of rational curves in the manifold (see [8], Proposition 2.9):*

$$\mathbf{J} := \coprod_{t \in \mathbb{P}^1} J_v^m(X_t),$$

(viz., the images of non-constant holomorphic maps $\mathbb{P}^1 \to \mathbf{J}$), is dense in the strong topology.

Proof. From the Picard–Lefschetz formula,

$$N_i = \log T_i = (T_i - I), \text{ using } (T_i - I)^2 = 0.$$

Now let $v \in H^0(\mathbb{P}^1, \overline{\mathcal{J}})$ and $\omega \in H^0(\mathbb{P}^1, \overline{\mathcal{F}}^m)$ be given. Note that

$$v : \mathbb{P}^1 \to \mathbf{J},$$

defines a rational curve on $\mathbf{J}$. Next, the images

$$\{[\delta_1], \ldots, [\delta_{2g}]\} \text{ in } F^{m,*} H_v^{2m-1}(X_t, \mathbb{C}) := H_v^{2m-1}(X_t, \mathbb{C})/F^m H_v^{2m-1}(X_t, \mathbb{C}),$$

define a lattice. In terms of this lattice and modulo the fixed part $J^m(X)$, a local lifting of v is given by $\sum_{j=1}^{2g} x_j(t)[\delta_j]$, for suitable real-valued functions $\{x_j(t)\}$, multivalued on U. Let $T_i v(\omega(t))$ be the result of analytic continuation of $v(\omega(t))$ counterclockwise in $\mathbb{P}^1$ about t_i and $N_i v(\omega(t)) = T_i v(\omega(t)) - v(\omega(t))$. About t_i, we pick up a period

$$N_i v(\omega(t)) = c_i \int_{\delta_i} \omega(t), \text{ for some } c_i \in \mathbb{Z},$$

dependent only on v (not on ω), where we identify δ_i with its corresponding homology vanishing cycle via Poincaré duality. Likewise in terms of the lattice description,

$$N_i v(\omega(t)) = \sum_{j=1}^{2g} T_i(x_j(t)) \int_{\delta_j + (-1)^m (\delta_j, \delta_i) \delta_i} \omega(t) - \sum_{j=1}^{2g} x_j(t) \int_{\delta_j} \omega(t)$$

$$= \sum_{j=1}^{2g} N_i(x_j(t)) \int_{\delta_j} \omega(t) + (-1)^m \left(\sum_{j=1}^{2g} T_i(x_j(t))(\delta_j, \delta_i) \right) \cdot \int_{\delta_i} \omega(t).$$

Thus

$$c_i = N_i(x_i(t)) + (-1)^m \sum_{j=1}^{2g} T_i(x_j(t))(\delta_j, \delta_i), \tag{7}$$

and

$$N_i(x_j(t)) = 0 \text{ for all } i \neq j. \tag{8}$$

Hence $T_i(x_j(t)) = x_j(t)$ for all $i \neq j$ and further, using $(\delta_i, \delta_i) = 0$, we can rewrite (7) as:

$$c_i = N_i(x_i(t)) + (-1)^m \sum_{j=1}^{2g} x_j(t)(\delta_j, \delta_i). \tag{9}$$

Note that if $N_i(x_i(t)) = 0$ for all i, then from the linear system in (9), $x_i(t) \in \mathbb{Q}$ for all i, and so $\delta(v) = 0 \in H^1(\mathbb{P}^1, j_* R^{2m-1} \rho_{U,*} \mathbb{Q})$. Now suppose that we have a nontrivial relation:

$$\sum_{j=1}^{2g} \lambda_j x_j(t) = \lambda_0, \text{ for some } \lambda_i \in \mathbb{Q}, \forall i, t \in U. \tag{10}$$

Then by (8) and (10) we have

$$\lambda_i N_i(x_i(t)) = \sum_{j=1}^{2g} \lambda_j N_i(x_j(t)) = 0.$$

So $\lambda_i \neq 0 \Rightarrow N_i(x_i(t)) = 0$. Let us assume for the moment that $\lambda_1 \neq 0$. Then $N_1(x_1(t)) = 0$, hence from (9):

$$(-1)^m c_1 = (\delta_2, \delta_1)x_2(t) + (\delta_3, \delta_1)x_3(t) + \cdots + (\delta_{2g}, \delta_1)x_{2g}(t), \qquad (11)$$

and applying N_2 and (6) we arrive at

$$0 = N_2(c_1) = (\delta_2, \delta_1)N_2(x_2(t)) \Rightarrow N_2(x_2(t)) = 0.$$

Hence again from (9):

$$(-1)^m c_2 = (\delta_1, \delta_2)x_1(t) + (\delta_3, \delta_2)x_3(t) + \cdots + (\delta_{2g}, \delta_2)x_{2g}(t). \qquad (12)$$

Applying N_3 to both (11) and (12), and (6) we arrive at

$$(0, 0) = (N_3(c_1), N_3(c_2)) = ((\delta_3, \delta_1), (\delta_3, \delta_2)) \cdot N_3(x_3(t)) \Rightarrow N_3(x_3(t)) = 0,$$

and so on. Now it may happen that $\lambda_1 = 0$. Since $(\lambda_1, \ldots, \lambda_{2g}) \neq (0, \ldots, 0)$ we can assume that $\lambda_{\ell_1} \neq 0$ for some $1 \leq \ell_1 \leq 2g$. Thus by (10), $N_{\ell_1}(x_{\ell_1}(t)) = 0$ and accordingly by (9):

$$(-1)^m c_{\ell_1} = \sum_{j=1}^{2g} (\delta_j, \delta_{\ell_1})x_j(t). \qquad (13)$$

By (6), $(\delta_{\ell_2}, \delta_{\ell_1}) \neq 0$ for some $1 \leq \ell_2 < \ell_1$ (assuming $\ell_1 > 1$). Applying N_{ℓ_2} to (13), we arrive at $N_{\ell_2}(x_{\ell_2}(t)) = 0$, and hence again by (9):

$$(-1)^m c_{\ell_2} = \sum_{j=1}^{2g} (\delta_j, \delta_{\ell_2})x_j(t).$$

Again by (6), $(\delta_{\ell_3}, \delta_{\ell_2}) \neq 0$ for some $1 \leq \ell_3 < \ell_2$ (assuming $\ell_2 > 1$), and thus we can repeat this process until we get $N_1(x_1(t)) = 0$. This puts in the situation of (11), where the same arguments imply that $N_i(x_i(t)) = 0$ for all $i = 1, \ldots, 2g$.

Corollary 1. *Let V be a general quintic threefold. Then the image of the Abel–Jacobi map $AJ : CH^2_{\mathrm{hom}}(V) \to J^2(V)$ is a countable dense subset of $J^2(V)$.*

Proof. Let $X \subset \mathbb{P}^5$ be the Fermat quintic fourfold, and $\{X_t\}_{t \in \mathbb{P}^1}$ a Lefschetz pencil of hyperplane sections of X. We will assume the notation given in diagram (2). For the Fermat quintic, it is easy to check that $H^1(\mathbb{P}^1, R^3\rho_{U,*}\mathbb{Q})^{(2,2)} \neq 0$, so by the sequence in (4), there exists $v \in H^0(\mathbb{P}^1, \overline{\mathcal{J}})$ such that $\delta(v) \neq 0 \in H^1(\mathbb{P}^1, R^3\rho_{U,*}\mathbb{Q})$ (this being related to Griffiths' famous example [3]). Thus by Theorem 2 and for general $t \in \mathbb{P}^1$, the Abel–Jacobi image is dense in $J^2(X_t)$. But it is well known that the lines in X_t for general $t \in \mathbb{P}^1$, deform in the universal family of quintic threefolds in $\mathbb{P}^4$. The corollary follows from this.

Remark 1. In light of the conjectures in [4], Corollary 1 most likely does not generalize to higher degree general hypersurface threefolds. However there is a different kind of generalization that probably holds. Namely, let S be the universal family of smooth threefolds $\{V_t\}_{t \in S}$ of degree d say in $\mathbb{P}^4$. Put

$$\mathbf{J}_S := \coprod_{t \in S} J^2(V_t),$$

and

$$\mathbf{J}^2_{S,\mathrm{inv}} := \mathrm{Image}\Big(\coprod_{t \in S} \mathrm{CH}^2_{\mathrm{hom}}(V_t) \xrightarrow{\text{Abel--Jacobi}} \mathbf{J}_S \Big).$$

Then in the strong topology, we anticipate that $\mathbf{J}^2_{S,\mathrm{inv}} \subset \mathbf{J}_S$ is a dense subset.

In this direction, we have the following general result.

Corollary 2. *Let $\coprod_{\lambda \in S_0} W_\lambda \to S_0$ be a smooth proper family of $2m$-dimensional projective varieties in some $\mathbb{P}^N$ with the following property:*

There exists a dense subset $\Sigma \subset S_0$ such that $\lambda \in \Sigma \Rightarrow \mathrm{Prim}^{m,m}_{\mathrm{alg}}(W_\lambda, \mathbb{Q}) \neq 0$, where Prim is primitive cohomology with respect to the embedding $W_\lambda \subset \mathbb{P}^N$. Further, let us assume that $H^{2m-1}(W_\lambda, \mathbb{Q})^{\pi_1(S_0)} = H^{2m-1}(W_\lambda, \mathbb{Q})$ and let

$$T := \{t := (c, \lambda) \in \mathbb{P}^{N,*} \times S_0 \mid V_t := \mathbb{P}^{N-1}_c \cap W_\lambda \text{ smooth, } \& \dim V_t = 2m - 1\},$$

with corresponding $\mathbf{J}^m_{T,\mathrm{inv}} \subset \mathbf{J}^m_T$ (where this jacobian space only involves the orthogonal complement of the fixed part of a corresponding variation of Hodge structure). Then in the strong topology $\mathbf{J}^m_{T,\mathrm{inv}}$ is dense in $\mathbf{J}^m_T$.

Proof. This easily follows from the techniques of this section and is left to the reader.

Remark 2. (i) The following is obvious, but certainly merits mentioning: *Let us assume given the setting and assumptions in Corollary 2, and further assume that for all $\lambda \in \Sigma$ and general c with $t = (c, \lambda) \in T$, the m-th $\mathbb{Q}$-Griffiths group $\{\mathrm{CH}^m_{\mathrm{hom}}(V_t)/\mathrm{CH}^m_{\mathrm{alg}}(V_t)\} \otimes \mathbb{Q} = 0$. Then $\mathbf{J}^m_T$ is a family of Abelian varieties.* (The general Hodge conjecture would imply in this situation that $\mathbf{J}^m_{T,\mathrm{inv}} = \mathbf{J}^m_T$, but we don't yet know this.)

(ii) Our results say nothing about the *arithmetic* nature of the invertible points on the jacobians. Matt Kerr pointed out to us Proposition 124 in [6, p. 92], which appears to be related to our results, and may have some potential in this direction; albeit it is unclear how to move forward with this.

Acknowledgements Both authors partially supported by a grant from the Natural Sciences and Engineering Research Council of Canada.

References

1. C.H. Clemens, Homological equivalence, modulo algebraic equivalence, is not finitely generated. Publ. I.H.E.S. **58**, 19–38 (1983)
2. M. Green, Griffiths' infinitesimal invariant and the Abel-Jacobi map. J. Differ. Geom. **29**, 545–555 (1989)
3. P.A. Griffiths, On the periods of certain rational integrals: I and II. Ann. Math. Second Series **90**(3), 460–541 (1969)
4. P.A. Griffiths, J. Harris, On the Noether-Lefschetz theorem and some remarks on codimension two cycles. Math. Ann. **271**(1), 31–51 (1985)
5. G.H. Hardy, E.M. Wright, *An Introduction to the Theory of Numbers*, 5th edn. (The Clarendon Press/Oxford University Press, New York, 1979), pp. xvi+426 [ISBN: 0-19-853170-2; 0-19-853171-0]
6. M. Kerr, G. Pearlstein, An exponential history of functions with logarithmic growth, in *Topology of Stratified Spaces*. MSRI Pub., vol. 58 (Cambridge University Press, New York, 2010)
7. J.D. Lewis, in *A Survey of the Hodge Conjecture*, 2nd edn. CRM Monograph Series, vol. 10 (AMS, Providence, 1999)
8. S. Zucker, Generalized intermediate jacobians and the theorem on normal functions. Invent. Math. **33**, 185–222 (1976)
9. S. Zucker, Hodge theory with degenerating coefficients: L_2 cohomology in the Poincaré metric. Ann. Math. **109**(3), 415–476 (1979)

Calabi–Yau Conifold Expansions

Slawomir Cynk and Duco van Straten

Abstract We describe examples of computations of Picard–Fuchs operators for families of Calabi–Yau manifolds based on the expansion of a period near a conifold point. We find examples of operators without a point of maximal unipotent monodromy, thus answering a question posed by J. Rohde.

Key words: Calabi–Yau threefolds, Picard–Fuchs operator, Maximal unipotent monodromy, Conifold point

Mathematics Subject Classifications (2010): Primary 14J32; Secondary 14Qxx, 32S40, 34M15

1 Introduction

The computation of the instanton numbers n_d for the quintic $X \subset \mathbb{P}^4$ using the period of the quintic mirror Y by P. Candelas, X. de la Ossa and co-workers [10] marked the beginning of intense mathematical interest in the mechanism of mirror symmetry that continues to the present day. On a superficial and purely computational level the calculation runs as follows: one considers the hypergeometric differential operator

$$\mathcal{P} = \theta^4 - 5^5 t(\theta + \tfrac{1}{5})(\theta + \tfrac{2}{5})(\theta + \tfrac{3}{5})(\theta + \tfrac{4}{5})$$

D. van Straten (✉)
Institut für Mathematik, Johannes Gutenberg University, 55099 Mainz, Germany
e-mail: straten@mathematik.uni-mainz.de

S. Cynk
Institut of Mathematics, Jagiellonian University, Łojasiewicza 6, 30-348 Kraków, Poland
e-mail: slawomir.cynk@uj.edu.pl

R. Laza et al. (eds.), *Arithmetic and Geometry of K3 Surfaces and Calabi–Yau Threefolds*,   499
Fields Institute Communications 67, DOI 10.1007/978-1-4614-6403-7_19,
© Springer Science+Business Media New York 2013

where $\theta = t\frac{d}{dt}$ denotes the logarithmic derivation. The power series

$$\varphi(t) = \sum \frac{(5n)!}{(n!)^5} t^n$$

is the unique holomorphic solution $\varphi(t) = 1 + \dots$ to the differential equation

$$\mathcal{P}\varphi = 0.$$

There is a unique second solution ψ that contains a log:

$$\psi(t) = \log(t)\varphi(t) + \rho(t)$$

where $\rho \in t\mathbb{Q}[[t]]$. We now define

$$q := e^{\psi/\varphi} = te^{\rho/\varphi} = t + 770t^2 + \dots.$$

We can use q as a new coordinate, and as such it can be used to bring the operator $\mathcal{P}$ into the local normal form

$$\mathcal{P} = D^2 \frac{5}{K(q)} D^2$$

where $D = q\frac{d}{dq}$ and $K(q)$ is a power series. When we write this series $K(q)$ in the form of a Lambert series

$$K(q) = 5 + \sum_{d=1}^{\infty} n_d d^3 \frac{q^d}{1 - q^d},$$

one can read off the numbers

$$n_1 = 2875, \quad n_2 = 609250, \quad n_3 = 317206375, \dots.$$

The data in the calculation are tied to two Calabi–Yau threefolds:

A. The quintic threefold $X \subset \mathbb{P}^4$ ($h^{11} = 1, h^{12} = 101$). The n_d have the interpretation of number of rational degree d curves on X, counted in the Gromov–Witten sense (see [11, 15]).

B. The quintic mirror Y ($h^{11} = 101, h^{12} = 1$). Y is member of a pencil $\mathcal{Y} \longrightarrow \mathbb{P}^1$, and $\mathcal{P}$ is Picard–Fuchs operator of this family. The series φ is the power-series expansion of a special period near the point 0, which is a point of maximal unipotent monodromy, a so-called MUM-point.

As one can see, the whole calculation depends only on the differential operator $\mathcal{P}$ or its holomorphic solution φ and never uses any further geometrical properties of X or Y, except maybe for choice of 5, which is the degree of X.

In [1] this computation was taken as the starting point to investigate so-called *CY3-operators*, which are Fuchsian differential operators $\mathcal{P} \in \mathbb{Q}(t, \theta)$ of order four with the following properties:

1. The operator has the form

$$P = \theta^4 + tP_1(\theta) + \ldots + t^r P_r(\theta)$$

where the P_i are polynomials of degree at most four. This implies in particular that 0 is a MUM-point.
2. The operator $\mathcal{P}$ is *symplectic*. This means the $\mathcal{P}$ leaves invariant a symplectic form in the solution space. The operator than is formally self-adjoint, which can be expressed by a simple condition on the coefficients [1, 7].
3. The holomorphic solution $\varphi(t)$ is in $\mathbb{Z}[[t]]$.
4. Further integrality properties: the expansion of the q-coordinate has integral coefficients, and the instanton numbers are integral (possibly up to a common denominator) [26, 29].

There is an ever-growing list of operators satisfying the first three and probably the last conditions [2]. It starts with the above operator and continues with 13 further hypergeometric cases, which are related to Calabi–Yau threefolds that are complete intersections in weighted projective spaces. Recently, M. Bogner and S. Reiter [7, 8] have classified and constructed the symplectically rigid Calabi–Yau operators, thus providing a solid understanding for the beginning of the list.

Another nice example is operator no. 25 from the list:

$$\mathcal{P} = \theta^4 - 4t(2\theta + 1)^2(11\theta^2 + 11\theta + 3) - 16t^2(2\theta + 1)^2(2\theta + 3)^2.$$

The holomorphic solution of the operator is $\varphi(t) = \sum A_n t^n$ where

$$A_n := \binom{2n}{n}^2 \sum_{k=1}^{n} \binom{n}{k}^2 \binom{n+k}{k}.$$

This operator was obtained in [6] as follows: one considers the Grassmannian $Z := G(2,5)$, a Fano manifold of dimension 6, with $Pic(Z) \approx \mathbb{Z}$, with ample generator h, the class of a hyperplane section in the Plücker embedding. As the canonical class of Z is $-5h$, the complete intersection $X := X(1,2,2)$ by hypersurfaces of degree $1, 2, 2$ is a Calabi–Yau threefold with $h^{11} = 1, h^{12} = 61$. The small quantum cohomology of Z is known, so that one can compute its quantum D-module. The quantum Lefschetz theorem then produces the above operator nr. 25 which thus provides the numbers n_d for X:

$$n_1 = 400, \quad n_2 = 5540, \quad n_3 = 164400, \ldots$$

Also, a mirror manifold $Y = Y_t$ was described as (the resolution of the toric closure of) a hypersurface in the torus $(\mathbb{C}^*)^4$ given by a Laurent polynomial.

The question arises which operators in the list are related in a similar way to a mirror pair (X, Y) of Calabi–Yau threefolds with $h^{11}(X) = h^{12}(Y) = 1$. This is certainly not to be expected for all operators, but it suggests the following attractive problem.

Problem. A. Construct examples of Calabi–Yau threefolds X with $h^{11} = 1$ and try
to identify the associated quantum differential equation.
  B. Construct examples of pencils of Calabi–Yau threefolds $\mathcal{Y} \longrightarrow \mathbb{P}^1$ with
$h^{12}(Y_t) = 1$ and try to compute the associated Picard–Fuchs equation.

It has been shown that in many cases one can predict from the operator $\mathcal{P}$ alone
topological invariants of X like $(h^3, c_2(X)h, c_3(X))$ [27] and the zeta function of Y_t
[23, 30]. In either case we see that the operators of the list provide predictions
for the existence of Calabi–Yau threefolds with quite precise properties. Recently,
A. Kanazawa [18] has used weighted Pfaffians to construct some Calabi–Yau threefolds X whose existence were predicted in [27]. In this note we report on work in
progress to compute the Picard–Fuchs equation for a large number of families of
Calabi–Yau threefolds with $h^{12} = 1$.

2 How to Compute Picard–Fuchs Operators

2.1 The Method of Griffiths–Dwork

For a smooth hypersurface $Y \subset \mathbb{P}^n$ defined by a polynomial $F \in \mathbb{C}[x_0, \ldots, x_n]$ of
degree d, one has a useful representation of (the primitive part of) the middle cohomology $H^{n-1}_{prim}(Y)$ using residues of differential forms on the complement $U := \mathbb{P}^n \setminus Y$.
One can work with the complex of differential forms with poles along Y and com-
pute modulo exact forms. Although this method was used in the nineteenth century
by mathematicians like Picard and Poincaré, it was first developed in full generality
by P. Griffiths [16] and B. Dwork [13] in the sixties of the last century.

The *Griffiths' isomorphism* identifies the Hodge space $H^{p,q}_{prim}$ with a graded piece
of the Jacobian algebra

$$R := \mathbb{C}[x_0, \ldots, x_n]/(\partial_0 F, \partial_1 F, \ldots, \partial_n F).$$

More precisely one has

$$R_{d(k+1)-(n+1)} \xrightarrow{\approx} H^{n-1-k,k}_{prim}(Y)$$

$$P \mapsto Res(\tfrac{P\Omega}{F^{k+1}})$$

where $\Omega := \iota_E(dx_0 \wedge dx_1 \wedge \ldots dx_n)$ and $E = \sum x_i \partial/\partial x_i$ is the Euler vector field. This
enables us to find an explicit basis.

If the polynomial F depends on a parameter t, we obtain a pencil $\mathcal{Y} \longrightarrow \mathbb{P}^1$
of hypersurfaces, which can be seen as a smooth hypersurface Y_t over the function
field $K := \mathbb{C}(t)$, and the above method provides a basis $\omega_1, \ldots, \omega_r$ of differential
forms over K. We now can differentiate the differential forms ω_i with respect to t

and express the result in the basis. This step involves a Gröbner-basis calculation. As a result we obtain an $r \times r$ matrix $A(t)$ with entries in K such that

$$\frac{d}{dt}\begin{pmatrix} \omega_1 \\ \omega_2 \\ \ldots \\ \omega_r \end{pmatrix} = A(t)\begin{pmatrix} \omega_1 \\ \omega_2 \\ \ldots \\ \omega_r \end{pmatrix}.$$

The choice of a *cyclic vector* for this differential system then provides a differential operator $\mathcal{P} \in \mathbb{C}(t, \theta)$ that annihilates all period integrals $\int_\gamma \omega$. In the situation of Calabi–Yau manifolds there is always a natural vector obtained from the holomorphic differential. For details we refer to the literature, for example [11].

This methods works very well in simple examples and has been used by many authors. It can be generalised to the case of (quasi-)smooth hypersurfaces in weighted projective spaces and more generally complete intersections in toric varieties [4]. Also, it is possible to handle families depending on more than one parameter. A closely related method for tame polynomials in affine space has been implemented by M. Schulze [25] and H. Movasati [21] in SINGULAR. The ultimate generalisation of the method would be an implementation of the *direct image functor in the category of D-modules*, which in principle can be achieved by Gröbner-basis calculations in the Weyl algebra.

The GRIFFITHS–DWORK method however also has some drawbacks:

- In many situations the varieties one is interested in have singularities. For the simplest types of singularities, it is still possible to adapt the method to take the singularities into account, but the procedure becomes increasingly cumbersome for more complicated singularities.
- In many situations the variety under consideration is given by some geometrical construction, and a description with equations seems less appropriate.

In some important situations the following *alternative method* can be used with great success.

2.2 Method of Period Expansion

In order to find Picard–Fuchs operator for a family $\mathcal{Y} \longrightarrow \mathbb{P}^1$, one does the following:

- Find the explicit power-series expansion of a single period

$$\varphi(t) = \int_{\gamma_t} \omega_t = \sum_{n=0}^{\infty} A_n t^n.$$

- Find a differential operator

$$\mathcal{P} = P_0(\theta) + tP_1(\theta) + \ldots + t^r P_r(\theta)$$

that annihilates φ by solving the linear recursion

$$\sum_{i=0}^{r} P_i(n)A_{n-i} = 0$$

on the coefficients. Here the P_i are polynomials in θ of a certain degree d. As $\mathcal{P}$ contains $(d + 1)(r + 1)$ coefficients, we need the expansion of φ only up to sufficiently high order to find it.

This quick-and-dirty method surely is very old and goes back to the time of EULER. And of course, many important issues arise like: *To what order do we need to compute our period?* For this one needs a priori estimates for d and r, which might not be available. Or *Is the operator $\mathcal{P}$ really the Picard–Fuchs operator of the family?* We will not discuss these issues here in detail, as they are not so important in practice: one expands until one finds an operator, and if the monodromy representation is irreducible, the operator obtained is necessarily the Picard–Fuchs operator.

However, it is obvious that the method stands or falls with our ability to find such an explicit period expansion. It appears that the *critical points* of our family provide the clue.

Principle
If one can identify explicitly a vanishing cycle, then its period can be computed "algebraically".

If our family $\mathcal{Y} \longrightarrow \mathbb{P}^1$ is defined over $\mathbb{Q}$, or more generally over a number field, then it is known that such expansions are G-functions and thus have very strong arithmetical properties [3].

Rather than trying to prove here a general statement in this direction, we will illustrate the principle in two simple examples. The appendix contains a general statement that covers the case of a variety acquiring an ordinary double point.

I. Let us look at the LEGENDRE family of elliptic curves given by the equation

$$y^2 = x(t - x)(1 - x).$$

If the parameter t is a small positive real number, the real curve contains a cycle γ_t that runs from 0 to t and back. If we let t go to zero, this loop shrinks to a point and the curve acquires an A_1 singularity. The period of the holomorphic differential $\omega = dx/y$ along this loop is

$$\varphi(t) = \int_{\gamma_t} \omega = 2F(t)$$

where

$$F(t) := \int_0^t \frac{dx}{\sqrt{(x(t-x)(1-x)}}.$$

By the substitution $x \mapsto tx$ we get

$$F(t) = \int_0^1 \frac{1}{\sqrt{(1-xt)}} \frac{dx}{\sqrt{x(1-x)}}.$$

The first square root expands as

$$\frac{1}{\sqrt{(1-xt)}} = \sum_{n=0}^{\infty} \binom{2n}{n} \left(\frac{xt}{4}\right)^n$$

so that

$$F(t) = \sum_{n=0}^{\infty} \binom{2n}{n} \left(\int_0^1 \frac{x^n}{\sqrt{x(1-x)}} dx\right) t^n.$$

The appearing integral is well known since the work of WALLIS and is a special case of EULERS beta integral.

$$\int_0^1 \frac{x^n}{\sqrt{(x(1-x)}} dx = \pi \binom{2n}{n} \frac{1}{4^n}.$$

So the final result is the beautiful series

$$F(t) = \pi \sum_{n=0}^{\infty} \binom{2n}{n}^2 \left(\frac{t}{16}\right)^n$$

$$= \pi \left(1 + \left(\frac{1}{2}\right)^2 t + \left(\frac{1 \cdot 3}{2 \cdot 4}\right)^2 t^2 + \left(\frac{1 \cdot 3 \cdot 5}{2 \cdot 4 \cdot 6}\right)^2 t^3 + \dots\right).$$

From this series it is easy to see that the second-order operator with $F(t)$ as solution is

$$4\theta^2 - t(2\theta + 1)^2.$$

In fact, the first six coefficients suffice to find the operator.
This should be compared to the GRIFFITHS–DWORK method, which would consist of considering the basis

$$\omega_1 = dx/y, \omega_2 = xdx/y$$

of differential forms on E_t and expressing the derivative

$$\partial_t \omega_1 = -\frac{x(1-x)dx}{(x(t-x)(1-x))^{3/2}}$$

in terms of ω_1, ω_2 modulo exact forms.

II. In mirror symmetry one often encounters families of Calabi–Yau manifolds that arise from a Laurent polynomial

$$f \in \mathbb{Z}[x_1, x_1^{-1}, x_2, x_2^{-1}, \ldots, x_n, x_n^{-1}].$$

Such a Laurent polynomial f determines a family of hypersurfaces in a torus given by

$$V_t := \{1 - tf(x_1, \ldots, x_n) = 0\} \subset (\mathbb{C}^*)^n.$$

In case the Newton polyhedron $N(f)$ of f is *reflexive*, a crepant resolution of the closure of V_t in the toric manifold determined by $N(f)$ will be a Calabi–Yau manifold Y_t. To compute its Picard–Fuchs operator, the GRIFFITHS–DWORK method is usually cumbersome.

The holomorphic $n - 1$-form on Y_t is given on V_t

$$\omega_t := Res_{V_t}\left(\frac{1}{1 - tf} \frac{dx_1}{x_1} \frac{dx_2}{x_2} \cdots \frac{dx_n}{x_n}\right).$$

There is an $n - 1$-cycle γ_t on V_t whose LERAY coboundary is homologous to $T := T_\epsilon := \{|x_i| = \epsilon\} \subset (\mathbb{C}^*)^n$. The so-called *principal period* is

$$\varphi(t) = \int_{\gamma_t} \omega_t = \frac{1}{(2\pi i)^n} \int_T \frac{1}{1 - tf} \frac{dx_1}{x_1} \frac{dx_2}{x_2} \cdots \frac{dx_n}{x_n} = \sum_{n=0}^{\infty} [f^n]_0 t^n$$

where $[g]_0$ denotes the *constant term* of the Laurent series g. For this reason, the series $\varphi(t)$ is sometimes called the *constant term series* of the Laurent polynomial. This method was used in [5] to determine the Picard–Fuchs operator for certain families Y_t and has been popular ever since. A fast implementation for the computation of $[g]_0$ was realised by P. Metelitsyn [19].

3 Double Octics

One of the simplest types of Calabi–Yau threefolds is the so-called *double octic* , which is a double cover Y of $\mathbb{P}^3$ ramified over a surface of degree 8. It can be given by an equation of the form

$$u^2 = f_8(x, y, z, w)$$

and thus can be seen as a hypersurface in weighted projective space $\mathbb{P}(1^4, 4)$. For a general choice of f_8 the variety Y is smooth and has Hodge numbers $h^{11} = 1$, $h^{12} = 149$. A nice subclass of such double octics consists of those for which f_8 is a product of eight planes. In that case Y has singularities at the intersections of the planes. In the generic such situation Y is singular along $8.7/2 = 28$ lines, and by blowing up these lines (in any order), we obtain a smooth Calabi–Yau manifold $\tilde{Y}$

with $h^{11} = 29, h^{12} = 9$. By taking the eight planes in special positions, the double cover Y acquires other singularities, and a myriad of different Calabi–Yau threefolds with various Hodge numbers appear as crepant resolutions $\tilde{Y}$. In [20], 11 configurations leading to rigid Calabi–Yau varieties were identified. Furthermore, C. Meyer listed 63 one-parameter families which thus give 63 special one-parameter families of Calabi– Yau threefolds $\tilde{Y}_t$, and it is for these that we want to compute the associated Picard–Fuchs equation. Due to the singularities of f_8, a Griffiths–Dwork approach is cumbersome, if not impossible. So we resort to the *period expansion method*.

In many of the 63 cases one can identify a *vanishing tetrahedron*: for a special value of the parameter one of the eight planes passes through a triple point of intersection, caused by three other planes. In appropriate coordinates we can write our affine equation as

$$u^2 = xyz(t - x - y - z)P_t(x, y, z)$$

where P_t is the product of the other four planes and we assume $P_0(0, 0, 0) \neq 0$. Analogous to the above calculation with the elliptic curve we now "see" a cycle γ_t, which consists of two copies of the real tetrahedron T_t bounded by the plane $x = 0$, $y = 0, z = 0, x + y + z = t$. For $t = 0$ the tetrahedron shrinks to a point. So we have

$$\varphi(t) = \int_{\gamma_t} \omega = 2F(t),$$

where

$$F(t) = \int_{T_t} \frac{dxdydz}{\sqrt{xyz(t - x - y - z)P_t(x, y, z)}}.$$

Proposition 1. *The period $\varphi(t)$ expands in a series of the form*

$$\varphi(t) = \pi^2 t(A_0 + A_1 t + A_2 t^2 + \ldots)$$

with $A_i \in \mathbb{Q}$ if $P_t(x, y, z) \in \mathbb{Q}[x, y, z, t], P_0(0, 0, 0) \neq 0$.

Proof. When we replace x, y, z by tx, ty, tz, respectively, we obtain an integral over the standard tetrahedron $T := T_1$:

$$F(t) = t \int_T \frac{dxdydz}{\sqrt{xyz(1 - x - y - z)}} \frac{1}{\sqrt{P_t(tx, ty, tz)}}.$$

We can expand the last square root in a power series

$$\frac{1}{\sqrt{P_t(tx, ty, tz)}} = \sum_{iklm} C_{iklm} x^k y^l z^m t^i$$

and thus find $F(t)$ as a series

$$F(t) = t \sum_{i,k,l,m} \int_T \frac{x^k y^l z^m dxdydz}{\sqrt{xyz(1 - x - y - z)}} C_{iklm} t^i.$$

The integrals appearing in this sum can be evaluated easily in terms of the *generalised beta integral*

$$\int_T x_1^{\alpha_1-1} x_2^{\alpha_2-1} \ldots x_n^{\alpha_n-1} (1 - x_1 - \ldots - x_n)^{\alpha_{n+1}-1} dx_1 dx_2 \ldots dx_n$$
$$= \Gamma(\alpha_1)\Gamma(\alpha_2)\ldots\Gamma(\alpha_{n+1})/\Gamma(\alpha_1 + \alpha_2 + \ldots + \alpha_{n+1}).$$

In particular we get

$$\int_T \frac{x^k y^l z^m dxdydz}{\sqrt{xyz(1 - x - y - z)}} = \frac{\Gamma(k + 1/2)\Gamma(l + 1/2)\Gamma(m + 1/2)\Gamma(1/2)}{\Gamma(k + l + m + 2)}$$
$$= \pi^2 \frac{(2k)!(2l)!(2m)!}{4^{k+l+m}k!l!m!(k + l + m + 1)!} \in \pi^2 \mathbb{Q}$$

and thus we get an expansion of the form

$$F(t) = \pi^2 t(A_0 + A_1 t + A_2 t^2 + A_3 t^3 + \ldots)$$

where $A_i \in \mathbb{Q}$ when $P_t(x, y, z) \in \mathbb{Q}[x, y, z, t]$.   □

Example 1. Configuration no. 36 of C. Meyer ([20], p. 57) is equivalent to the double octic with equation

$$u^2 = xyz(t - x - y - z)(1 - x)(1 - z)(1 - x - y)(1 + (t - 2)x - y - z).$$

A smooth model has $h^{11} = 49, h^{12} = 1$. For $t = 0$ the resolution is a rigid Calabi–Yau with $h^{11} = 50, h^{12} = 0$, corresponding to arrangement no. 32. The expansion of the tetrahedral integral around $t = 0$ reads

$$F(t) = \pi^2 t(1 + t + \frac{43}{48} t^2 + \frac{19}{24} t^3 + \frac{10811}{15360} t^4 + \frac{9713}{15360} t^5 + \ldots).$$

The operator is determined by the first 34 terms of the expansion and reads

$$32\,\theta(\theta - 2)(\theta - 1)^2 - 16\,t\theta(\theta - 1)\left(9\,\theta^2 - 13\,\theta + 8\right)$$
$$+ 8\,t^2\theta\left(33\,\theta^3 - 32\,\theta^2 + 38\,\theta - 10\right) - t^3(252\,\theta^4 + 104\,\theta^3 + 304\,\theta^2 + 76\,\theta + 20)$$
$$+ t^4(132\,\theta^4 + 224\,\theta^3 + 292\,\theta^2 + 160\,\theta + 38)$$
$$- t^5(36\,\theta^4 + 104\,\theta^3 + 140\,\theta^2 + 88\,\theta + 21) + 4\,t^6\,(\theta + 1)^4.$$

The Riemann symbol of this operator is

$$\left\{\begin{array}{cccc} 0 & 1 & 2 & \infty \\ \hline 0 & 0 & 0 & 1 \\ 1 & 0 & 0 & 1 \\ 1 & 0 & 2 & 1 \\ 2 & 0 & 2 & 1 \end{array}\right\}.$$

At 0 we have indeed a "conifold point" with its characteristic exponents 0, 1, 1, 2. At $t = 1$ and $t = \infty$ we find MUM-points. M. Bogner has shown that via a quadratic transformation this operator can be transformed to operator number 10* from the AESZ list, which has Riemann symbol

$$\left\{ \begin{matrix} 0 & 1/256 & \infty \\ 0 & 0 & 1/2 \\ 0 & 0 & 1 \\ 0 & 1 & 1 \\ 0 & 1 & 3/2 \end{matrix} \right\}$$

which is symplectically rigid [8]. So the family of double octics provides a clean B-interpretation for this operator.

Example 2. Configuration no. 70 of Meyer is isomorphic to

$$u^2 = xyz(x + y + z - t)(1 - x)(1 - z)(x + y + z - 1)(x/2 + y/2 + z/2 - 1).$$

Again, for general t we obtain a Calabi–Yau threefold with $h^{11} = 49, h^{12} = 1$ and for $t = 0$ we have $h^{11} = 50, h^{12} = 0$, corresponding to the rigid Calabi–Yau of configuration no. 69 of [20]. The tetrahedral integral expands as

$$F(t) = \pi^2 t \left(1 + \frac{13}{16} t + \frac{485}{768} t^2 + \frac{12299}{24576} t^3 + \frac{534433}{1310720} t^4 + \frac{21458473}{62914560} t^5 + \dots \right)$$

and is annihilated by the operator

$$16 \theta (\theta - 2)(\theta - 1)^2 - 2 t\theta (\theta - 1)\left(24 \theta^2 - 24 \theta + 13 \right)$$
$$+ t^2 \theta^2 \left(52 \theta^2 + 25 \right) - 2 t^3 \left(3 \theta^2 + 3 \theta + 2 \right)(2 \theta + 1)^2$$
$$+ t^4 (2 \theta + 1)(\theta + 1)^2 (2 \theta + 3).$$

The Riemann symbol of this operator is:

$$\left\{ \begin{matrix} 0 & 1 & 2 & \infty \\ 0 & 0 & 0 & 1/2 \\ 1 & 0 & 0 & 1 \\ 1 & 1 & 1 & 1 \\ 2 & 1 & 1 & 3/2 \end{matrix} \right\}$$

so we see that it has *no point of maximal unipotent monodromy*!

The first examples of families Calabi–Yau manifolds without MUM-point were described by J. Rohde [22] and studied further by A. Garbagnati and B. van Geemen [14]. It should be pointed out that in those cases the associated Picard–Fuchs operator was of second order, contrary to the above fourth-order operator. M. Bogner has checked that this operator has $Sp_4(\mathbb{C})$ as differential Galois group. It is probably one of the simplest examples of this sort. J. Hofmann has calculated with his package [17] the integral monodromy of the operator. In an appropriate basis it reads

$$T_2 = \begin{pmatrix} 2 & -7 & 1 & 0 \\ 0 & 1 & 0 & 1 \\ -1 & 7 & 0 & 7 \\ 0 & 0 & 0 & 1 \end{pmatrix}, \quad T_1 = \begin{pmatrix} -1 & -2 & 0 & 0 \\ 2 & 3 & 0 & 0 \\ 11 & 7 & 2 & 1 \\ -3 & 1 & -1 & 0 \end{pmatrix},$$

$$T_0 = \begin{pmatrix} 1 & 0 & 0 & 0 \\ 0 & 1 & 0 & 0 \\ 0 & 0 & 1 & 0 \\ 8 & -16 & 4 & 1 \end{pmatrix}, \quad T_\infty = \begin{pmatrix} 0 & 23 & -3 & -2 \\ 0 & -15 & 2 & 1 \\ 0 & -84 & 11 & 6 \\ 1 & -75 & 11 & 4 \end{pmatrix}$$

with $T_2 T_1 T_0 T_\infty = id$.

As Calabi–Yau operators in the sense of [1] need a to have a MUM, W. Zudilin has suggested to call an operator without such a point of maximal unipotent monodromy an *orphan*.

Example 3. Configuration no. 254 of C. Meyer gives a family of Calabi–Yau three-folds with $h^{11} = 37, h^{12} = 1$:

$$u^2 = xyz(t - x - y - z)P_t(x, y, z)$$

with

$$P_t(x, y, z) = (1 - 3z + t - t^2x + tz - tx - 2y)(1 - z + tx - 2x)$$
$$\cdot (1 - tx + z)(1 + t - t^2x + tz - 5tx + z - 2y - 4x).$$

For $t = 0$ we obtain the rigid configuration no. 241 with $h^{11} = 40, h^{12} = 0$. The tetrahedral integral expands as

$$F(t) = \pi^2 t \left(1 + \frac{1}{2}t + \frac{37}{24}t^2 + \frac{41}{16}t^3 + \frac{13477}{1920}t^4 + \frac{14597}{768}t^5 + \ldots \right).$$

The operator is very complicated and has the following Riemann symbol:

$$\left\{ \begin{array}{ccccccccc} \alpha_1 & \alpha_2 & 0 & \rho_1 & \rho_2 & \rho_3 & -1 & 1 & \infty \\ \hline 0 & 0 & 0 & 0 & 0 & 0 & 0 & 0 & 3/2 \\ 1 & 1 & 1 & 1 & 1 & 1 & 0 & 0 & 3/2 \\ 1 & 1 & 1 & 3 & 3 & 3 & 0 & 0 & 3/2 \\ 2 & 2 & 2 & 4 & 4 & 4 & 0 & 0 & 3/2 \end{array} \right\}$$

where at 0 and $\alpha_{1,2} = -2 \pm \sqrt{5}$ we find conifold points, at the $\rho_{1,2,3}$, roots of the cubic equation $2t^3 - t^2 - 3t + 4 = 0$ we have apparent singularities and at $-1, 1$ we find point of maximal unipotent monodromy, which we also find at ∞, after taking a square root. This operator was not known before.

These three examples illustrate the current *win–win–win* aspect of these calculations. It can happen that the operator is known, in which case we get a nice geometric incarnation of the differential equation. It can happen that the operator does not have a MUM-point, in which case we have found a further example of family of Calabi–Yau threefolds without a MUM-point. From the point of mirror symmetry these cases are of special importance, as the torus for the SYZ fibration, which in the ordinary cases vanishes at the MUM-point, is not in sight. Or it can happen that we find a *new* operator with a MUM-point, thus extending the AESZ table [2].

Many more examples have been computed, in particular also for other types of families, like fibre products of rational elliptic surfaces of the type considered by C. Schoen [24]. The first example of $Sp_4(\mathbb{C})$-operators without MUM-point was found among these [28]. A paper collecting our results on periods of double octics and fibre products is in preparation [12].

4 An Algorithm

Let $\mathcal{Y}$ be a smooth variety of dimension n and $f : \mathcal{Y} \longrightarrow \mathbb{P}^1$ a nonconstant map to $\mathbb{P}^1$ and let $P \in \mathcal{Y}$ be a critical point. In order to analyse the local behaviour of periods of cycles vanishing at P, we replace $\mathcal{Y}$ by an affine part, on which we have a function $f : \mathcal{Y} \longrightarrow \mathbb{A}^1$, with $f(P) = 0$. An n-form

$$\omega \in \Omega^n_{\mathcal{Y},P}$$

gives rise to a family of differential forms on the fibres of f:

$$\omega_t := Res_{Y_t}\left(\frac{\omega}{f - t}\right).$$

The period integrals

$$\int_{\gamma_t} \omega_t$$

over cycles γ_t vanishing at P only depend on the class of ω in the *Brieskorn module* at P, which is defined as

$$\mathcal{H}_P := \Omega^n_{\mathcal{Y},P}/df \wedge d\Omega^{n-2}_{\mathcal{Y},P}.$$

If P is an isolated critical point, it was shown in [9] that the completion $\widehat{\mathcal{H}_P}$ is a (free) $\mathbb{C}[[t]]$-module of rank $\mu(f, P)$, the Milnor number of f at P. In particular, if f has an A_1-singularity at P, we have $\mu(f, P) = 1$, and the image of the class of ω under the isomorphism $\widehat{\mathcal{H}_P} \longrightarrow \mathbb{C}[[t]]$ is, up to a factor, just the expansion of the integral of the vanishing cycle. We will now show how one can calculate this with a simple algorithm.

Proposition 2. *If $f : \mathcal{Y} \longrightarrow \mathbb{A}^1$ and the critical point P of type A_1. If $f : \mathcal{Y} \longrightarrow \mathbb{A}^1$, P and $\omega \in \Omega_{\mathcal{Y},P}$ are defined over $\mathbb{Q}$, then the period integral over the vanishing cycle $\gamma(t)$*

$$\varphi(t) = \int_{\gamma(t)} \omega_t$$

has an expansion of the form

$$\varphi(t) = ct^{n/2-1}(1 + A_1 t + A_2 t^2 + \ldots)$$

where

$$c = d\frac{n}{2}\frac{\Gamma(1/2)^n}{\Gamma(\frac{n}{2} + 1)}$$

where $d^2 \in \mathbb{Q}$ and the $A_i \in \mathbb{Q}$ can be computed via a simple algorithm.

Proof. As P and f are defined over $\mathbb{Q}$, we may assume that in appropriate formal coordinates x_i on $\mathcal{Y}$, we have $P = 0$, $f(P) = 0$, and the map is represented by a series

$$f = f_2 + f_3 + f_4 + \ldots$$

where f_2 is a nondegenerate quadratic form and the $f_d \in \mathbb{Q}[x_1, \ldots, x_n]$ are homogeneous polynomials of degree d. After a linear coordinate transformation (which may involve a quadratic field extension), we may and will assume that

$$f_2 = x_1^2 + x_2^2 + \ldots + x_n^2.$$

For $t > 0$ small enough, the part of solution set $\{(x_1, x_2, \ldots, x_n) \in \mathbb{R}^n \mid f = t\}$ near 0 looks like a slightly bumped sphere $\gamma(t)$ and is close to standard sphere $\{(x_1, x_2, \ldots, x_n) \in \mathbb{R}^n \mid f_2 = t\}$. This is the vanishing cycle we want to integrate $\omega_t = Res(\Omega/(f - t))$ over. Note that

$$\int_0^t \int_{\gamma(t)} \omega_t = \int_{\Gamma(t)} \omega$$

where

$$\Gamma(t) = \cup_{s \in [0,t]} \gamma(s) = \{(x_1, x_2, \ldots, x_n) \in \mathbb{R}^n \mid f \le t\}$$

is the Lefschetz thimble, which is a slightly bumped ball, that is near to the standard ball

$$B(t) := \{(x_1, x_2, \ldots, x_n) \in \mathbb{R}^n \mid f_2 \le t\}.$$

The idea is now to change to coordinates that map f into its quadratic part f_2. An automorphism $\varphi : x_i \mapsto y_i$ of the local ring $R := \mathbb{Q}[[x_1, x_2, \ldots, x_n]]$ is given by n-tuples of series $(y_1, y_2, \ldots, y_n)$ with the property that

$$\left|\frac{\partial y}{\partial x}\right| = \begin{vmatrix} \frac{\partial y_1}{\partial x_1} & \cdots & \frac{\partial y_1}{\partial x_n} \\ \cdots & \cdots & \cdots \\ \frac{\partial y_n}{\partial x_1} & \cdots & \frac{\partial y_n}{\partial x_n} \end{vmatrix} \notin \mathcal{M} := (x_1, x_2, \ldots, x_n) \subset R.$$

One has the following Formal Morse Lemma: there exist an automorphism φ of R such that

$$\varphi(f) = f_2.$$

Such a φ is obtained by an iteration: if

$$f = f_2 + f_k + f_{k+1} + \ldots,$$

then we can find an automorphism φ_k such that

$$\varphi_k(f) = f_2 + \tilde{f}_{k+1} + \ldots.$$

To find φ_k it is sufficient to write $f_k = \sum a_i \partial f / \partial x_i$ and set $\varphi_k(x_i) = x_i - a_i$.

Alternatively, we may say that one can find formal coordinates $y_i = \varphi(x_i)$ such that

$$f_2 + f_3 + \ldots = y_1^2 + y_2^2 + \ldots + y_n^2.$$

By the transformation formula for integrals we get

$$\int_{\Gamma(t)} \omega = \int_{B(t)} \varphi^*(\omega).$$

When we write

$$\omega := A(x) dx_1 dx_2 \ldots dx_n,$$

then

$$\varphi^*(\omega) = A(x(y)) \left| \frac{\partial x(y)}{\partial y} \right| dy_1 dy_2 \ldots dy_n$$

which can be expanded in a series in the coordinates y_i as

$$\varphi^*(\omega) = \sum_\alpha J_\alpha y^\alpha dy_1 dy_2 \ldots dy_n$$

where the $J_\alpha \in \mathbb{Q}$. So we get

$$\int_{\Gamma(t)} \omega = \int_{B(t)} \varphi^*(\omega) = \sum_\alpha J_\alpha \int_{B(t)} y^\alpha dy_1 dy_2 \ldots dy_n.$$

The integrals

$$I(\alpha) := \int_{B(t)} y^\alpha dy_1 dy_2 \ldots dy_n$$

can be reduced to the generalised beta integral, and one has

Lemma 1. *(i)*

$$I(\alpha_1, \alpha_2, \ldots, \alpha_n) = 0$$

when some α_i is odd. (ii)

$$I(2k_1, 2k_2, \ldots, 2k_n)$$
$$= \frac{\Gamma(k_1 + 1/2)\Gamma(k_2 + 1/2) \ldots \Gamma(k_n + 1/2)}{\Gamma(k_1 + k_2 + \ldots + n/2 + 1)} t^{k_1 + k_2 \ldots + k_n + n/2}.$$

As a consequence we have

$$\int_{\Gamma(t)} = \sum J_\alpha I(\alpha) t^{k_1 + k_2 + \ldots + k_n + n/2}$$

$$= I(0) t^{n/2} (1 + a_1 t + a_2 t^2 + \ldots).$$

The coefficient

$$I(0) = \frac{\Gamma(1/2)^n}{\Gamma(n/2 + 1)}$$

is the volume of the n-dimensional unit ball. As $I(\alpha)/I(0, 0, \ldots, 0) \in \mathbb{Q}$, we see that the a_i are also in $\mathbb{Q}$.

So we see that the period integral

$$\varphi(t) = \frac{d}{dt} \int_{\Gamma(t)} \omega$$

has, up to a prefactor, a series expansion with rational coefficients that can be computed algebraically be a very simple although memory-consuming algorithm. Pavel Metelitsyn is currently working on an implementation.

Acknowledgements We would like to thank the organisers for inviting us to the *Workshop on Arithmetic and Geometry of K3 surfaces and Calabi–Yau threefolds* held in the period 16–25 August 2011 at the Fields Institute. We also thank M. Bogner and J. Hofmann for help with the analysis of the examples. Furthermore, I thank G. Almkvist and W. Zudilin for continued interest in this crazy project. Part of this research was done during the stay of the first named author as a guest professor at the *Schwerpunkt Polen* of the Johannes Gutenberg–Universität in Mainz.

References

1. G. Almkvist, W. Zudilin, Differential equations, mirror maps and zeta values, in *Mirror Symmetry. V.* AMS/IP Studies in Advanced Mathematics, vol. 38 (American Mathematical Society, Providence, 2006), pp. 481–515
2. G. Almkvist, C. van Enckevort, D. van Straten, W. Zudilin, Tables of Calabi–Yau equations, arXiv:math/0507430 [math.AG]
3. Y. André, *G*-functions and geometry, in *Aspects of Mathematics*, E13, Friedr (Vieweg & Sohn, Braunschweig, 1989)
4. V. Batyrev, D. Cox, On the Hodge structure of projective hypersurfaces in toric varieties. Duke Math. J. **75**(2), 293–338 (1994)
5. V. Batyrev, D. van Straten, Generalized hypergeometric functions and rational curves on Calabi–Yau complete intersections in toric varieties. Comm. Math. Phys. **168**(3), 493–533 (1995)
6. V. Batyrev, I. Ciocan-Fontanine, B. Kim, D. van Straten, Conifold transitions and mirror symmetry for Calabi–Yau complete intersections in Grassmannians. Nucl. Phys. B **514**(3), 640–666 (1998)
7. M. Bogner, On differential operators of Calabi–Yau type, Thesis, Mainz, 2012
8. M. Bogner, S. Reiter, On symplectically rigid local systems of rank four and Calabi-Yau operators. J. Symbolic Comput. **48**, 64–100 (2013)

9. E. Brieskorn, Die Monodromie der isolierten singularitäten von Hyperflächen. Manuscripta Math. **2**, 103–161 (1970)
10. P. Candelas, X. de la Ossa, P. Green, L. Parkes, An exactly soluble superconformal theory from a mirror pair of Calabi–Yau manifolds. Phys. Lett. B **258**(1–2), 118–126 (1991)
11. D. Cox, S. Katz, in *Mirror Symmetry and Algebraic Geometry*. Mathematical Surveys and Monographs, vol. 68 (American Mathematical Society, Providence, 1999)
12. S. Cynk, D. van Straten, Picard-Fuchs equations for double octics and fibre products (in preparation)
13. B. Dwork, On the zeta function of a hypersurface, III. Ann. Math. (2) **83**, 457–519 (1966)
14. A. Garbagnati, B. van Geemen, The Picard–Fuchs equation of a family of Calabi–Yau threefolds without maximal unipotent monodromy. Int. Math. Res. Not. IMRN, 16 (2010), pp. 3134–3143
15. A. Givental, The mirror formula for quintic threefolds, in *Northern California Symplectic Geometry Seminar*. American Mathematical Society Translations Series 2, vol. 196 (American Mathematical Society, Providence, 1999), pp. 49–62
16. P. Griffiths, On the periods of certain rational integrals I, II. Ann. Math. (2) **90**, 460–495 (1969); Ann. Math. (2) **90**, 496–541 (1969)
17. J. Hofmann, A Maple package for the monodromy calculations (in preparation)
18. A. Kanazawa, Pfaffian Calabi–Yau Threefolds and Mirror Symmetry arXiv: 1006.0223 [math.AG]
19. P. Metelitsyn, How to compute the constant term of a power of a Laurent polynomial efficiently arXiv:1211.3959 [cs.SC]
20. C. Meyer, in *Modular Calabi–Yau Threefolds*. Fields Institute Monographs, vol. 22 (American Mathematical Society, Providence, 2005)
21. H. Movesati, Calculation of mixed Hodge structures, Gauss–Manin connections and Picard–Fuchs equations, in *Real and Complex Singularities*. Trends in Mathematics (Birkhäuser, Basel, 2007), pp. 247–262
22. J. Rohde, Maximal automorphisms of Calabi–Yau manifolds versus maximally unipotent monodromy. Manuscripta Math. **131**(3–4), 459–474 (2010)
23. K. Samol, D. van Straten, Frobenius polynomials for Calabi–Yau equations. Comm. Number Theor. Phys. **2**(3), 537–561 (2008)
24. C. Schoen, On fiber products of rational elliptic surfaces with section. Math. Z. **197**(2), 177–199 (1988)
25. M. Schulze, Good bases for tame polynomials. J. Symbolic Comput. **39**(1), 103–126 (2005)
26. A. Schwarz, V. Vologodsky, Integrality theorems in the theory of topological strings. Nucl. Phys. B **821**(3), 506–534 (2009)
27. C. van Enckevort, D. van Straten, Monodromy calculations of fourth order equations of Calabi–Yau type, in *Mirror Symmetry. V*. AMS/IP Studies in Advanced Mathematics, vol. 38 (American Mathematical Society, Providence, 2006), pp. 539–559
28. D. van Straten, Conifold period expansion. Oberwolfach Reports No. 23/2012
29. V. Vologodsky, Integrality of instanton numbers, arXiv:0707.4617 [math.AG]
30. J.D. Yu, Notes on Calabi–Yau ordinary differential equations. Comm. Number Theor. Phys. **3**(3), 475–493 (2009)

Quadratic Twists of Rigid Calabi–Yau Threefolds Over $\mathbb{Q}$

Fernando Q. Gouvêa, Ian Kiming, and Noriko Yui

Abstract We consider rigid Calabi–Yau threefolds defined over $\mathbb{Q}$ and the question of whether they admit quadratic twists. We give a precise geometric definition of the notion of a quadratic twists in this setting. Every rigid Calabi–Yau threefold over $\mathbb{Q}$ is modular so there is attached to it a certain newform of weight 4 on some $\Gamma_0(N)$. We show that quadratic twisting of a threefold corresponds to twisting the attached newform by quadratic characters and illustrate with a number of obvious and not so obvious examples. The question is motivated by the deeper question of which newforms of weight 4 on some $\Gamma_0(N)$ and integral Fourier coefficients arise from rigid Calabi–Yau threefolds defined over $\mathbb{Q}$ (a geometric realization problem).

Key words: Rigid Calabi–Yau threefolds over $\mathbb{Q}$, Twists, Modular forms, Holomorphic 3-forms

Mathematics Subject Classifications (2010): Primary 14J32; Secondary 11F80, 11F11

F. Q. Gouvêa
Department of Mathematics, Colby College, Waterville, ME 04901, USA
e-mail: fqgouvea@colby.edu

I. Kiming (✉)
Department of Mathematics, University of Copenhagen, Universitetsparken 5,
DK 2100 Ø, Copenhagen, Denmark
e-mail: kiming@math.ku.dk

N. Yui
Department of Mathematics and Statistics, Queen's University, Kingston,
ON, Canada K7L 3N6
e-mail: yui@mast.queensu.ca

R. Laza et al. (eds.), *Arithmetic and Geometry of K3 Surfaces and Calabi–Yau Threefolds*,   517
Fields Institute Communications 67, DOI 10.1007/978-1-4614-6403-7_20,
© Springer Science+Business Media New York 2013

1 Introduction

Suppose X is a rigid Calabi–Yau threefold defined over $\mathbb{Q}$. As Gouvêa and Yui observe in [8] (see also [4, 5]), it follows from work of Khare and Winterberger that X is modular: The L-series of X coincides with the L-series of a certain newform f of weight 4 on some $\Gamma_0(N)$. Alternatively, there is a newform f with integer coefficients such that, for any prime ℓ, the ℓ-adic representation of $G_\mathbb{Q} = \mathrm{Gal}(\overline{\mathbb{Q}}/\mathbb{Q})$ on $H^3(\bar{X}, \mathbb{Q}_\ell)$ is isomorphic to the ℓ-adic representation of $G_\mathbb{Q}$ attached to f.

Very little seems to be known about the form f. Notably, the relation between the conductor N and the geometry of X still seems to be poorly understood. (See the discussions and conjectures of Sect. 6.4 of [12], as well as the paper [3].)

Another unresolved and probably very hard question is the following. The form f above obviously has integral Fourier coefficients. Can one conversely characterize the newforms of weight 4 on some $\Gamma_0(N)$ with integral coefficients that arise from rigid Calabi–Yau threefolds over $\mathbb{Q}$? Do all such forms arise from Calabi–Yau threefolds? (This is a kind of the geometric realization problem. See [6] for the case of "singular" K3 surfaces and forms of weight 3.)

A very weak version of this question is the topic of this paper: Given a rigid Calabi–Yau threefold X with form f as above, for any non-square rational number d there is a twist f_d of f by the quadratic character corresponding to the quadratic extension $K = \mathbb{Q}(\sqrt{d})$ over $\mathbb{Q}$. This f_d is again of the above form and so we can ask whether f_d arises from a rigid Calabi–Yau threefold X_d over $\mathbb{Q}$. This will be the case whenever X admits a quadratic twist by d in the sense we discuss next.

2 Quadratic Twists of Rigid Calabi–Yau Threefolds

Let X be a rigid Calabi–Yau threefold defined over $\mathbb{Q}$. Suppose that $d \in \mathbb{Q}^\times$ is a squarefree integer, let $K := \mathbb{Q}(\sqrt{d})$, and let σ be the non-trivial automorphism of $K/\mathbb{Q}$. We say that a rigid Calabi–Yau threefold X_d defined over $\mathbb{Q}$ is a *twist of X by d* if there exist:

- An involution ι of $X_\mathbb{Q}$ that acts as -1 on $H^3(\bar{X}, \mathbb{Q}_\ell)$ for some prime ℓ.
- An isomorphism $\theta \colon (X_d)_K \xrightarrow{\cong} X_K$ defined over K

such that:

$$\theta^\sigma \circ \theta^{-1} = \iota.$$

Notice that the condition $\iota = \theta^\sigma \circ \theta^{-1}$ necessarily implies that the involution ι satisfies $\iota^\sigma = \iota$, i.e., that ι is defined over $\mathbb{Q}$. Conversely, given an involution ι on $X_\mathbb{Q}$, one can always find the isomorphism θ. One takes the quotient of $X_K = X_\mathbb{Q} \otimes K$ by $\iota \otimes \sigma$, checks that it is defined over $\mathbb{Q}$ and that it is the twist X_d as above. This will become clear in the examples below: whenever we can find the appropriate involution we can also construct a twist.

Since X is a rigid Calabi–Yau threefold, $H^{3,0}(X)$ is one-dimensional so there is a unique (up to scalar) holomorphic 3-form Ω on X. The involution ι should act on Ω non-symplectically, sending it to $-\Omega$. Conversely, since X is rigid we have $h^{2,1}(X) = 0$, so if ι sends Ω to $-\Omega$ we see that ι acts as -1 on all of $H^3(\bar{X}, \mathbb{Q}_\ell)$. (Here the rigidity of X is used in an essential way.)

This is the method that we will primarily employ in the examples below to ensure this part of the condition on the involution ι.

One could envision relaxing the above definition in the direction of just requiring the existence of an algebraic correspondence between $(X_d)_K$ and X_K and still retain (a somewhat stronger version of) the theorem below. However, in the examples that we will give, we actually find isomorphisms in all cases and have hence chosen to work with the above definition.

The principles of proof of the following theorem should be well-known, but we provide the details because of lack of a precise reference.

Recall that, given a newform f of some weight and a non-square $d \in \mathbb{Q}$ there is a twist f_d of f by d which is again a newform of the same weight as f (but potentially at another level) and whose attached ℓ-adic Galois representation (for some prime ℓ and hence for all primes ℓ) is isomorphic to the ℓ-adic representation attached to f twisted by the quadratic character χ corresponding to $K/\mathbb{Q}$. If the Fourier coefficients of f and f_d are a_n and b_n, respectively, we have the relation $b_p = \chi(p)a_p$ for almost all primes p. In particular, since χ is quadratic, if f has coefficients in $\mathbb{Z}$ then so does f_d.

Theorem 1. *In the above setting, suppose that the newform (of weight 4) attached to X is f. Then, if X_d is a twist by d of X the newform attached to X_d is f_d, the twist of f by the Dirichlet character χ corresponding to the quadratic extension $K = \mathbb{Q}(\sqrt{d})$ of $\mathbb{Q}$.*

If we keep all hypotheses above except possibly that ι acts as -1 on $H^3(\bar{X}, \mathbb{Q}_\ell)$, we can still deduce that the newform attached to X_d is either f or f_d.

Proof. Fix a prime number ℓ, and consider the ℓ-adic Galois representations ρ and ρ_d attached to X and X_d, respectively: these are given by the action of $G_{\mathbb{Q}} = \mathrm{Gal}(\bar{\mathbb{Q}}/\mathbb{Q})$ on the two-dimensional $\mathbb{Q}_\ell$-vector spaces $V := H^3(\bar{X}, \mathbb{Q}_\ell)$ and $V_d := H^3(\bar{X}_d, \mathbb{Q}_\ell)$, respectively.

Now, the newform f attached to X is determined uniquely by the requirement that its attached ℓ-adic representation be isomorphic to ρ. Similarly, the newform attached to X_d is determined by the requirement that its attached ℓ-adic representation be isomorphic to ρ_d.

Since the ℓ-adic representation attached to f_d is isomorphic to the twist by χ of the one attached to f, we see that what we have to prove boils down to:

$$\rho_d \cong \rho \otimes \chi.$$

Put $N := G_K = \mathrm{Gal}(\bar{\mathbb{Q}}/K)$ so that N is a normal subgroup of $G_{\mathbb{Q}}$. The existence of the isomorphism $\theta: (X_d)_{/K} \cong X_{/K}$ defined over K translates into the existence of a $\mathbb{Q}_\ell$-linear isomorphism $V_d \to V$ commuting with the action of G_K. That is, in matrix terms we have an invertible matrix A with:

$$\rho(n)A = A\rho_d(n) \quad \text{for all } n \in N.$$

In matrix terms the conjugate isomorphism θ^σ of V_d onto V is then given by the matrix

$$\rho(\sigma)A\rho_d(\sigma)^{-1};$$

notice that we have here viewed σ as an element of $G_\mathbb{Q}$ via choice of a representative; the expression $\rho(\sigma)A\rho_d(\sigma)^{-1}$ does not depend on this choice.

If now ι acts as -1 on $H^3(\bar{X}, \mathbb{Q}_\ell)$ we can deduce that the matrix

$$\rho(\sigma)A\rho_d(\sigma)^{-1}A^{-1}$$

is a non-trivial involution.

Define the representation ρ' of $G_\mathbb{Q}$ by $\rho' := A^{-1}\rho A$ so that $\rho'(n) = \rho_d(n)$ for $n \in N$. Then, for arbitrary $g \in G_\mathbb{Q}$ and $n \in N$ we have

$$\rho_d(g)\rho'(n)\rho_d(g)^{-1} = \rho_d(g)\rho_d(n)\rho_d(g)^{-1}$$
$$= \rho_d(gng^{-1}) = \rho'(gng^{-1}) = \rho'(g)\rho'(n)\rho'(g)^{-1}$$

so that

$$\rho'(g)^{-1}\rho_d(g)\rho'(n) = \rho'(n)\rho'(g)^{-1}\rho_d(g)$$

i.e., for any $g \in G_\mathbb{Q}$ the matrix $\rho'(g)^{-1}\rho_d(g)$ commutes with all matrices $\rho'(n), n \in N$.

Now, suppose first that ρ (and hence ρ') is absolutely irreducible when restricted to N. In that case we deduce that $\rho'(g)^{-1}\rho_d(g)$ is a scalar matrix, say with diagonal entry $\mu(g)$. We have $\mu(n) = 1$ for $n \in N$ and see that $g \mapsto \mu(g)$ is in fact a character of $G_\mathbb{Q}$ factoring through $N = G_K$. So, either $\mu = 1$ or $\mu = \chi$.

If we had $\mu = 1$ we would have $A^{-1}\rho(g)A = \rho'(g) = \rho_d(g)$ for all $g \in G_\mathbb{Q}$ and so in particular the matrix

$$\rho(\sigma)A\rho_d(\sigma)^{-1}A^{-1}$$

would be trivial. As we noted above, this can not happen if ι acts as -1 on $H^3(\bar{X}, \mathbb{Q}_\ell)$. Hence, in that case we must have $\mu = \chi$ and so $\rho_d = \rho' \otimes \chi \cong \rho \otimes \chi$, as desired.

Suppose now that ρ is not absolutely irreducible when restricted to G_K. The same is then true of ρ' and ρ_d. In this case it is known, cf. (4.4), (4.5) of [15], that ρ' is induced from the ℓ-adic representation ψ attached to a Grössencharacter over K: $\rho' = \mathrm{Ind}_{K/\mathbb{Q}}(\psi)$, and $\rho'_{|G_K}$ splits up as the sum of the two characters ψ and ψ^σ. Notice that $\mathrm{Ind}_{K/\mathbb{Q}}(\psi) = \mathrm{Ind}_{K/\mathbb{Q}}(\psi^\sigma)$. Since ρ' and ρ_d have the same restriction to G_K we may then conclude that in fact $\rho' = \rho_d$ as representations of $G_\mathbb{Q}$, and hence that the newform attached to X_d is f. Furthermore, the matrix $\rho(\sigma)A\rho_d(\sigma)^{-1}A^{-1}$ must then be trivial, and so we see that this case in fact does not materialize if ι acts as -1 on $H^3(\bar{X}, \mathbb{Q}_\ell)$.

Remark 1. What we have proved, in fact, is that if ι is nontrivial the Galois representation on the middle cohomology of X_d is isomorphic to the tensor product of the representation on the middle cohomology of X and the one-dimensional Galois representation corresponding to $K = \mathbb{Q}(\sqrt{d})$. For rigid Calabi–Yau manifolds, we know these representations correspond to modular forms, but the question can, of course, be asked without knowing anything about modularity. We are grateful to the referee for pointing this out to us.

2.1 Easy Examples of Twists

The standard, simple example of twisting is of course for an elliptic curve E over $\mathbb{Q}$, say given by a Weierstrass equation $y^2 = x^3 + ax^2 + bx + c$. The twisted curve E_d is then given by the equation $dy^2 = x^3 + ax^2 + bx + c$ with the isomorphism $\theta : E_d \to E$ defined over $K = \mathbb{Q}(\sqrt{d})$ by $\theta(x, y) = (x, \sqrt{d}y)$. The corresponding involution ι is $(x, y) \mapsto (x, -y)$. It is clear that ι sends the holomorphic 1-form $\Omega = \dfrac{dx}{y}$ to $-\Omega$.

For a number of rigid Calabi–Yau threefolds over $\mathbb{Q}$ we can display twists by essentially the same method: Consider for examples the various cases of double octic Calabi–Yau threefolds over $\mathbb{Q}$ (see [12] for instance for a good overview). They are defined as hypersurfaces of the form

$$y^2 = f_8(x_1, x_2, x_3, x_4)$$

where f_8 is a degree 8 homogeneous polynomial. As in the case of elliptic curves, we have an obvious twist given by

$$dy^2 = f_8(x_1, x_2, x_3, x_4).$$

The corresponding involution is of course again given by $y \mapsto -y$. Again it is clear that ι sends the holomorphic 3-form

$$\Omega = \frac{\sum_{i=1}^{4}(-1)^i x_i \, dx_1 \wedge \cdots \wedge \widehat{dx_i} \wedge \cdots \wedge dx_4}{y}$$

to $-\Omega$.

This is also completely analogous to the case certain modular double sextic $K3$ surfaces, see [13]. These have form

$$w^2 = f_6(x, y, z)$$

where f_6 is a projective smooth curve of degree 6. As above, we get a twist of this surface (in the sense analogous to Theorem 1) via the twisted equation

$$dw^2 = f_6(x, y, z)$$

for a non-square rational number d. The involution is again given by $w \mapsto -w$. The holomorphic 2-form

$$\Omega = \frac{z\,dx \wedge dy - x\,dy \wedge dz + y\,dx \wedge dz}{w}$$

is sent by ι to $-\Omega$.

2.2 Self-fiber Products of Rational Elliptic Surfaces with Section and Their Twists

Slightly more complicated examples arise in connection with the rigid Calabi–Yau threefolds studied by H. Verrill in the appendix to [21]: She determined the L-series via the point counting method for the six isomorphism classes of rigid Calabi–Yau threefolds constructed as self-fiber products of rational elliptic surfaces with section by Schoen [17]. Along the way, she discussed twists by quadratic characters.

These six rigid Calabi–Yau threefolds over $\mathbb{Q}$ are defined as follows: Start with semi-stable families of elliptic curves $\pi : \mathcal{Y} \to \mathbb{P}^1$, i.e., $\mathcal{Y}$ is a smooth surface and the singular fibers have type I_m. Beauville [1] gave a complete list of these families. These are realized as the resolutions of singular surfaces $\bar{\mathcal{Y}} \subset \mathbb{P}^2 \times \mathbb{P}^1$ given by the following equations:

#	Equation for $\bar{\mathcal{Y}}$	
I	$(x^3 + y^3 + z^3)\mu$	$= \lambda xyz$
II	$x(x^2 + z^2 + 2zy)\mu$	$= \lambda(x^2 - y^2)z$
III	$x(x - z)(y - z)\mu$	$= \lambda(x - y)yz$
IV	$(x + y + z)(xy + yz + zx)\mu$	$= \lambda xyz$
V	$(x + y)(xy - z^2)\mu$	$= \lambda xyz$
VI	$(x^2y + y^2z + z^2x)\mu$	$= \lambda xyz$

The fibration $\bar{\pi} : \bar{\mathcal{Y}} \to \mathbb{P}^1$ is given by projecting to $\mathbb{P}^1$, and $\mathcal{Y}$ is obtained by resolving $\bar{\mathcal{Y}}$. Now take the self-fiber product $\mathcal{Y} \times_{\mathbb{P}^1} \mathcal{Y}$. Schoen [17] shows that a small resolution exists and that the resulting smooth variety X is a rigid Calabi–Yau threefold defined over $\mathbb{Q}$. Thus, in each case there is a newform of weight 4 attached to f. In each case, the form was identified by Verrill via determination of the L-series of X (point counting.) Here is the table of newforms from Verrill.

#	Newform	Modular group	Level
I	$\eta(q^3)^8$	$\Gamma(3)$	9
II	$\eta(q^2)^4\eta(q^4)^4$	$\Gamma_1(4) \cap \Gamma(2)$	8
III	$\eta(q)^4\eta(q^5)^4$	$\Gamma_1(5)$	5
IV	$\eta(q)^2\eta(q^2)^2\eta(q^3)^2\eta(q^6)^2$	$\Gamma_1(6)$	6
V	$\eta(q^4)^{16}\eta(q^8)^{-4}\eta(q^2)^{-4}$	$\Gamma_0(8) \cap \Gamma_1(4)$	16
VI	$\eta(q^3)^8$	$\Gamma_0(9) \cap \Gamma_1(3)$	9

In each case, one can display a twist X_d of X so that X_d corresponds to twisting the newform by the quadratic character belonging to $\mathbb{Q}(\sqrt{d})/\mathbb{Q}$. Consider for instance type V above. Given a non-square $d \in \mathbb{Q}^\times$ let X_d be the variety arising from the equation

$$(x + y)(xy - dz^2)\mu = \lambda xyz$$

by a process analogous to the one leading to X above.

Then we have an isomorphism $\theta \colon X_d \to X$ defined over $\mathbb{Q}(\sqrt{d})$ and given by

$$((x : y : z), (\mu : \lambda)) \mapsto ((\sqrt{d}x : \sqrt{d}y : z), (\mu : \sqrt{d}\lambda)).$$

In the setup of Theorem 1, the involution ι is given by

$$\iota((x : y : z), (\mu : \lambda)) = ((-x : -y : z), (\mu : -\lambda)).$$

That X_d is a genuine twist of X, i.e., that the attached newform is f_d rather than f can be ascertained via point counting, cf. appendix in [21].

The other examples can be dealt with in similar fashions.

2.3 The Schoen Quintic and Its Quadratic Twists

As a more interesting test case, we consider the Schoen quintic

$$x_0^5 + x_1^5 + x_2^5 + x_3^5 + x_4^5 = 5x_0x_1x_2x_3x_4.$$

We write

$$f = f(x_0, x_1, x_2, x_3, x_4) = x_0^5 + x_1^5 + x_2^5 + x_3^5 + x_4^5 - 5x_0x_1x_2x_3x_4 = 0.$$

This is a singular threefold with 125 nodes (ordinary double points) as only singularities, and a small resolution of singularities produces a rigid Calabi–Yau threefold X that is known, cf. [16], to be associated to a newform of weight 4 and level 25 (the modular form $25k4A1$); see also [12], Sect. 3.1.

We seek an involution ι of X that acts on $H^{3,0}(X)$ as multiplication by -1. Since $H^{3,0}(X)$ is generated by a unique holomorphic 3-form Ω (up to scalar), ι should send Ω to $-\Omega$.

To determine the action of ι on Ω we can use either of the following two arguments:

According to Cox and Katz [2], especially Sect. 2.3 and the formula (2.7) therein, Ω can be computed on the smooth part as

$$\Omega = \mathrm{Res}(\frac{\omega}{f})$$

where

$$\omega = \sum_{i=0}^{4} (-1)^i x_i \, dx_0 \cdots \wedge \widehat{dx_i} \wedge \cdots dx_4$$

and where 'Res' denotes Poincaré residue.

Alternatively, it follows from Lemma 1 below that we have a holomorphic 3-form

$$\frac{dx_0 \wedge dx_1 \wedge dx_2}{\partial f / \partial x_3}.$$

on the Zariski open set where x_4 and $\partial f / \partial x_3$ are both non-vanishing, and that this extends to the Calabi–Yau threefold X.

Can one construct the requisite quadratic twists of the Schoen quintic? In Gouvêa and Yui [8] it was briefly asserted that quadratic twist indeed exists for the Schoen quintic. We now discuss details of this claim.

Proposition 1. *For any non-square $d \in \mathbb{Q}^\times$ the Schoen quintic has a twist by d. The corresponding involution ι defined over $\mathbb{Q}$ is given explicitly on the coordinates by*

$$\iota : (x_0, x_1, x_2, x_3, x_4) \mapsto (x_1, x_0, x_2, x_3, x_4),$$

and sends $\Omega \in H^{3,0}(X)$ to $-\Omega$.

Proof. First, it is plain that ι sends the above Ω to $-\Omega$ (using any of the two descriptions of Ω.)

Put $u = x_0 + x_1$ and $v = x_0 - x_1$. Then the equation for the quintic equation can be written as a polynomial in u and v^2 as follows:

$$u^5 + 10u^3 v^2 + 5uv^4 + 16(x_2^5 + x_3^5 + x_4^5) - 20(u^2 - v^2)x_2 x_3 x_4 = 0.$$

Replacing v by $\sqrt{d}v$, we obtain the following quintic equation:

$$(*) \qquad u^5 + 10du^3 v^2 + 5d^2 uv^4 + 16(x_2^5 + x_3^5 + x_4^5) - 20(u^2 - dv^2)x_2 x_3 x_4 = 0,$$

and we see how to apply Theorem 1: The equation $(*)$ gives rise to a rigid Calabi–Yau threefold X_d defined over $\mathbb{Q}$. Then we have an isomorphism $\theta \colon X_d \to X$ defined over $\mathbb{Q}(\sqrt{d})$ and given by

$$(u, v, x_2, x_3, x_4) \mapsto (u, \sqrt{d}v, x_2, x_3, x_4)$$

so that $\theta^\sigma \circ \theta^{-1}$ is the involution given by $(u, v, x_2, x_3, x_4) = (u, -v, x_2, x_3, x_4)$. This is precisely the involution ι so the existence of the twist follows from Theorem 1.

2.4 Explicit Description for a Holomorphic 3-Form for a Complete Intersection Calabi–Yau Threefold

Before we go into further examples, we give an explicit description of a holomorphic 3-form for a complete intersection Calabi–Yau threefold, by the Griffiths residue theorem or its generalized version. We are grateful to Bert van Geemen for communicating to us the following lemma as well as its proof.

Lemma 1. *Let $Y = V(f_1, \cdots, f_k)$ be a complete intersection in $\mathbb{P}^n$ of dimension $d =: n - k$ where $f_1, \ldots, f_k$ are homogeneous equations in the homogeneous variables $x_0, \ldots, x_n$. Assume that Y is a normal crossings divisor.*

Let $i_0 \in \{0, \ldots, n\}$, let $I \subseteq \{0, \ldots, n\} \setminus \{i_0\}$ have cardinality k, and consider

$$D_I := \det\left(\frac{\partial f_i}{\partial x_j}\right)_{\substack{1 \le i \le k \\ j \in I}}$$

Then, on the Zariski open set where x_{i_0} and D_I are both non-vanishing, a holomorphic d-form is given by

$$\Omega = \frac{\bigwedge_{j \in \{0, \ldots, n\} \setminus (\{i_0\} \cup I)} dx_j}{D_I}$$

If additionally Y has a crepant resolution X that is Calabi–Yau variety of dimension $\dim X \le 3$, then Ω extends to all of X.

Let us remind that a "crepant resolution" is one that does not change the canonical class, cf. [14], Sect. 2.

The Lemma applies to the Schoen quintic as well as the threefolds that we shall consider below because the singularities involved are ordinary double point in all cases.

The proof is given below in the appendix.

2.5 Two Rigid Calabi–Yau Threefolds of Werner and van Geemen

Werner and van Geemen [20] constructed a number of examples of rigid Calabi–Yau threefolds over $\mathbb{Q}$. They are complete intersection Calabi–Yau threefolds.

We consider two of them. First, the rigid Calabi–Yau threefold denoted by $\tilde{V}_{33}$: Let $V_{33} \subset \mathbb{P}^5$ be the threefold defined by the system of equations

$$\begin{aligned} x_0^3 + x_1^3 + x_2^3 + x_3^3 &= 0 \\ x_2^3 + x_3^3 + x_4^3 + x_5^3 &= 0. \end{aligned}$$

V_{33} has 9 singularities, and let $\tilde{V}_{33}$ be the blow up of V_{33} along its singular locus (big resolution). Then $\tilde{V}_{33}$ is a rigid Calabi–Yau threefold over $\mathbb{Q}$, and it is modular with a corresponding newform f of weight 4 on $\Gamma_0(9)$:

$$f(q) = \eta(q^3)^8.$$

It is shown by Kimura [11] that if $E \subset \mathbb{P}^2$ is the curve defined by $x_0^3 + x_1^3 + x_2^3 = 0$, then there is a dominant rational map from E^3 to V_{33} of degree 3. Consequently, the L-series coincide. By Lemma 1, a holomorphic 3-form of V_{33} is given in affine coordinates by

$$\Omega = \frac{dx_2 \wedge dx_3 \wedge dx_4}{x_1^2 x_5^2}.$$

Proposition 2. *For any non-square $d \in \mathbb{Q}^\times$ the rigid Calabi–Yau threefold $\tilde{V}_{33}$ has a twist $\tilde{V}_{33,d}$ by d. The corresponding involution ι is defined by permuting x_2 and x_3.*

Proof. Put $u = x_2 + x_3$, $v = x_2 - x_3$. Then the equation for V_{33} can be expressed in terms of x_0, x_1, x_4, x_5 and u and v^2:

$$4x_0^3 + 4x_1^3 + u^3 + 3uv^2 \qquad\qquad = 0$$
$$u^3 + 3uv^2 + 4x_4^3 + 4x_5^3 = 0.$$

Replacing v by $\sqrt{d}v$ in this system we obtain a system of equations that gives rise to $\tilde{V}_{33,d}$. Applying Theorem 1 shows that $\tilde{V}_{33,d}$ is twist by d of $\tilde{V}_{33}$ with the corresponding involution given by $v \mapsto -v$, i.e., by $(x_2, x_3) \mapsto (x_3, x_2)$.

The holomorphic 3-form Ω above clearly changes sign when x_2 and x_3 are interchanged.

Secondly, we can consider the rigid Calabi–Yau threefold denoted by $\tilde{V}_{24}$: Let $V_{24} \subset \mathbb{P}^5$ be the threefold defined by the equations:

$$x_0^2 + x_1^2 + x_2^2 - x_3^2 - x_4^2 - x_5^2 = 0$$
$$x_0^4 + x_1^4 + x_2^4 - x_3^4 - x_4^4 - x_5^4 = 0.$$

Then V_{24} has 122 nodes (ordinary double points) as only singularities. Let $\tilde{V}_{24}$ be the blow up of V_{24} along its singular locus (small resolution). Then $\tilde{V}_{24}$ is a rigid Calabi–Yau threefold over $\mathbb{Q}$, and it is modular with a corresponding newform g which is the newform of weight 4 on $\Gamma_0(12)$.

Proposition 3. *For any non-square $d \in \mathbb{Q}^\times$ the rigid Calabi–Yau threefold $\tilde{V}_{24}$ has a twist $\tilde{V}_{24,d}$ by d. The corresponding involution ι is given by:*

$$\iota : x_1 \mapsto -x_1 \quad (or \; x_2 \mapsto -x_2)$$

and all other coordinates fixed with $x_0 \neq 0$.

Proof. Replacing x_1^2 and x_1^4 in the defining equations for $\tilde{V}_{24}$ by dx_1^2 and $d^2x_1^4$, respectively, we get a system of equations that give rise to $\tilde{V}_{24,d}$ isomorphic to $\tilde{V}_{24}$ over

$\mathbb{Q}(\sqrt{d})$. The corresponding involution ι is clearly as stated. A holomorphic 3-form on V_{24} is given by

$$\Omega = \frac{dx_3 \wedge dx_4 \wedge dx_5}{8x_1 x_2^3 - 8x_1^3 x_2}$$

and under the involution $x_1 \mapsto -x_1$, Ω is mapped to $-\Omega$.

2.6 The Rigid Calabi–Yau Threefold of van Geemen and Nygaard

Another interesting example is the case of the rigid Calabi–Yau threefold of van Geemen and Nygaard. In [18], van Geemen and Nygaard gave an example of a rigid Calabi–Yau threefold defined over $\mathbb{Q}$: Let $Y \subset \mathbb{P}^7$ be the complete intersection of the four quadrics:

$$\begin{aligned}
y_0^2 &= x_0^2 + x_1^2 + x_2^2 + x_3^2 \\
y_1^2 &= x_0^2 - x_1^2 + x_2^2 - x_3^2 \\
y_2^2 &= x_0^2 + x_1^2 - x_2^2 - x_3^2 \\
y_3^2 &= x_0^2 - x_1^2 - x_2^2 + x_3^2 .
\end{aligned}$$

The variety Y has 96 isolated singularities, which are ordinary double points. Let X be a (small) blow-up of Y along its singular locus. (A recent article of Freitag and Salvati-Manni [7] asserts that Y admits a resolution that is a projective Calabi–Yau threefold, X.) Then X is a rigid Calabi–Yau threefold over $\mathbb{Q}$. Its attached newform is the unique newform of weight 4 on $\Gamma_0(8)$, cf. [18], Theorem 2.4. Notice that there is a misprint in the equations on p. 56 of that paper: In the second equation x_3^2 should occur with a minus sign as above rather than a plus sign as in [18], p. 56. This is evident from the theta relations on p. 54 of [18].

Again, we instantly see the existence of twists of X via replacing x_0^2 by dx_0^2 in the above equations. Thus:

Proposition 4. *For any non-square $d \in \mathbb{Q}^\times$ the above rigid Calabi–Yau threefold X has a twist X_d by d. The corresponding involution ι is given by*

$$x_0 \mapsto -x_0$$

and all other coordinates fixed.

Proof. The only thing we need to check is whether a holomorphic 3-form is send by ι to $-\Omega$. Let

$$f_0 := y_0^2 - (x_0^2 + x_1^2 + x_2^2 + x_3^2),$$

and similarly, define f_1, f_2 and f_3 by the second, third and the fourth equation, respectively. Then Ω may be given by

$$\Omega = \frac{dx_0 \wedge dx_2 \wedge dx_3}{D}$$

where

$$D = \det \left(\frac{\partial f_i}{\partial y_j} \right)_{0 \le i,j \le 3} = 2^4 y_0 y_1 y_2 y_3.$$

Thus the involution given by $\iota : x_0 \mapsto -x_0$ and fixing all other coordinates will send Ω to $-\Omega$.

3 Remarks on the Levels of Twists

Suppose that $d \in \mathbb{Z}$ is squarefree and suppose that f is a newform of level N. One may ask about the level of the twisted newform f_d. Viewed from the Galois representation side, this amounts to asking for the conductor of $\rho \otimes \chi$ where ρ has conductor N and χ is a (quadratic) character of conductor D, say (so D divides $4d$ in the above setup).

As is well-known, the answer is ND^2 if $(N, D) = 1$. (This can be proved either via basic theory of conductors, or, alternatively, more directly via the theory of modular forms.) When $(N, D) > 1$, however, the question has no simple answer: the level of the twisted representation will depend heavily on the behavior of the representation ρ at inertia groups over common prime divisors of N and D. Nevertheless, see [9, 10] for the cases where ρ has a "small" image.

For the concrete examples of twisting that we have discussed in this paper, a more modest question can be asked, namely whether the newform that we start with is "twist minimal" or not, i.e., whether it has the lowest level among all of its quadratic twists. This question can be easily answered with a little computation.

Let us for example consider the case of the Schoen quintic. By Proposition 5.3 of [16] (and its proof) one has that the attached newform f is of weight 4 on $\Gamma_0(25)$ given as an explicit linear combination of certain η-products. From this explicit description of f one computes that the coefficient of q^2 in its q-expansion is -84.

On the other hand, there is a unique newform f_0 of weight 4 on $\Gamma_0(5)$, namely $f_0(z) := \eta(z)^4 \eta(5z)^4$. We compute that the coefficient of q^2 in the q-expansion of f_0 is -4.

Since $-4 \ne \pm(-84)$ we can deduce that f_0 is not a twist of f and hence that f is twist minimal. In particular, the unique newform f_0 of level 4 is not the form attached to a twist of the Schoen quintic. Is it attached to any Calabi–Yau threefold?

4 Final Remarks

4.1 An Explicit Unresolved Case

As we implied in the introduction, the main contribution of this paper is to put focus on the question whether any rigid Calabi–Yau threefold over $\mathbb{Q}$ has a twist by d

for any non-square $d \in \mathbb{Q}$. In contrast with the classical situation involving elliptic curves, the question for rigid Calabi–Yau threefolds over $\mathbb{Q}$ seems genuinely more difficult. One difference is that one does not know in general the automorphism group of a rigid Calabi–Yau threefold over $\mathbb{Q}$. Another point is the poor understanding of the conductor of a rigid Calabi–Yau threefold over $\mathbb{Q}$, i.e., the level of its associated newform.

For many other cases than the ones we have considered here, the existence of quadratic twists of a given Calabi–Yau threefold over $\mathbb{Q}$ can be shown along the same lines as above, i.e., inspection of the defining equation(s) combined with an application of Theorem 1. For instance, one can try to show the existence of quadratic twists for many rigid Calabi–Yau threefolds over $\mathbb{Q}$ discussed in Meyer [12].

However, in some cases the question does not seem as easy. For instance, does the rigid Calabi–Yau threefold of Hirzebruch (Theorem 5.11 in Yui [21]) admit quadratic twists? Let X_0 be the quintic threefold defined over $\mathbb{Q}$ by the equation $F(x, y) - F(u, w) = 0$ where

$$F(x, y) = \left(x + \frac{1}{2}\right)\left(y^4 - y^2(2x^2 - 2x + 1) + \frac{1}{5}(x^2 + x - 1)^2\right).$$

Then X_0 has 126 nodes (ordinary double points) as only singularities. Let X be the blow up of X_0 along its singular locus. Then X is a rigid Calabi–Yau threefold defined over $\mathbb{Q}$ with the Euler characteristic 306. The map sending y to $-y$ gives rise to an involution on X and this raises the obvious question of whether this induces a nontrivial involution on H^3 so that we get a quadratic twist of X by replacing y^2 by dy^2.

4.2 The Question About Existence of Geometric Twists

Does there exist a rigid Calabi–Yau threefold X defined over $\mathbb{Q}$ and a non-square rational d such that X does not have a quadratic twist X_d by d?

Perhaps, in order to approach this question, one needs to loosen the definition of "quadratic twist by d" so that the existence of X_d becomes *equivalent to* (rather than just implying) the existence of a rigid Calabi–Yau threefold over $\mathbb{Q}$ whose attached ℓ-adic Galois representation (for some prime ℓ) is the twist by the quadratic character of $\mathbb{Q}(\sqrt{d})/\mathbb{Q}$ of the ℓ-adic representation attached to the original threefold. Maybe this is possible by considering, more generally, algebraic correspondences rather than isomorphisms in the setting of Theorem 1.

Calabi–Yau threefolds, even rigid ones, in general, may not have involutions; even when they do, it might be rather difficult to find one. So geometric realization of modular forms of weight 4 on some $\Gamma_0(N)$ with integral Fourier coefficients along our proposed approach may in fact not be overly promising. But this remark once again raises the question of when the kind of twisting that we have discussed in this paper is possible.

4.3 The Fixed Point Set of the Involution ι

We should include here a description of the fixed point set of the involution ι acting on a rigid Calabi–Yau threefold (though we did not make use of it in the examples.) The following observation is due to B. van Geemen.

Let X be a Calabi–Yau threefold over $\mathbb{Q}$. Let ι be an involution acting on X. Then the fixed point set of ι on X is determined as follows.

Suppose that p is a fixed point of ι, then in suitable local coordinates $z_i, i = 1, 2, 3$, $z_i(p) = 0$, and

$$\iota(z_1, z_2, z_3) = (e_1 z_1, e_2 z_2, e_3 z_3) \quad \text{where } e_i = \pm 1,$$

and the dimension of the fixed point set is thus the number of e_i's which are equal to $+1$ (so if $\iota \neq 1$, then at least one e_i must be -1.)

If Ω is the nowhere vanishing holomorphic 3-form in $H^{3,0}(X)$, then in these coordinates,

$$\Omega = f(z_1, z_2, z_3) dz_1 \wedge dz_2 \wedge dz_3 \quad \text{and } f(0,0,0) \neq 0.$$

But $f(0,0,0) \neq 0$ forces that

$$f(e_1 z_1, e_2 z_2, e_3 z_3) = +f(z_1, z_2, z_3),$$

and hence

$$\iota^* \Omega = e_1 e_2 e_3 \Omega.$$

From this we see that if the fixed point locus consists of isolated points and divisors, then all or exactly one of the e_i are -1, else two of the $e_i = +1$ and the one -1.

Appendix

Proof (Proof of Lemma 1). Permuting variables if necessary we may, and will, assume $i_0 = 0$ and $I = \{1, \ldots, k\}$. We put $D := D_I$ with this particular I.

Let us first recall some general facts about the Griffiths residue map: Suppose that V is a smooth projective variety of dimension n and that W is a smooth codimension 1 subvariety, or, more generally, a normal crossings divisor. The Griffiths residue map (see for instance the proof of Proposition 8.32 in [19]) is a surjective homomorphism

$$\text{Res} : \ \Omega_V^n(\log W) \to \Omega_W^{n-1}$$

defined as follows: If $(U, (z_1, \ldots, z_n))$ is a complex chart of V such that W is given locally by the equation $z_1 = 0$ then $\Omega_V^n(\log W)_{|U}$ is the free sheaf of $\mathcal{O}_U$-modules generated by

$$\alpha := \frac{dz_1}{z_1} \wedge dz_2 \wedge \ldots \wedge dz_n,$$

and Res is then defined locally by

$$\operatorname{Res}(g\alpha) := (g dz_2 \wedge \ldots \wedge dz_n)_{U \cap W}.$$

Notice that $(U \cap W, (z_1, \ldots, z_{n-1})_U)$ is a complex chart of W.

Now, for the proof of the Lemma, let us first consider the case $k = 1$ where we specialize the above to the situation $V = \mathbb{P}^n$ and $W = Y$ the hypersurface given by the equation $f_1 = 0$. Consider the open set U_0 where $x_0 \neq 0$, let $z_i := x_i/x_0$ on U_0, and let $F := f_1/x_0^t$ where t is the degree of f_1. Then $\omega_n := dz_1 \wedge \ldots \wedge dz_n$ is a generator of $\Omega^n_{U_0}$.

The open subset U_0' of U_0 where $\partial F/\partial z_1 \neq 0$ coincides with the open subset where $\partial f_1/\partial x_1 \neq 0$. On U_0' we have local coordinates $F, z_2, \ldots, z_n$. Since

$$dF = \sum_{i=1}^{n} (\partial F/\partial z_i) dz_i$$

we find

$$dF \wedge dz_2 \wedge \ldots \wedge dz_n = \left(\frac{\partial F}{\partial z_1}\right) \omega_n$$

so that

$$\operatorname{Res}(\frac{\omega_n}{F})_{|U_0'} = \operatorname{Res}(\frac{dF \wedge dz_2 \wedge \ldots \wedge dz_n}{F\left(\frac{\partial F}{\partial z_1}\right)})_{|U_0'} = \left(\frac{dz_2 \wedge \ldots \wedge dz_n}{\left(\frac{\partial F}{\partial z_1}\right)}\right)_{|U_0'}$$

which coincides up to a power of x_0 with

$$\frac{dx_2 \wedge \ldots \wedge dx_n}{\left(\frac{\partial f_1}{\partial x_1}\right)}$$

on U_0'.

Since the residue is holomorphic on all of Y, this differential form on U_0' will extend holomorphically to all of Y.

For the general case, one can argue inductively with respect to k: We see Y as the end of a chain $Y := Y_k \subseteq \ldots \subseteq Y_1 \subseteq \mathbb{P}^n$ where each Y_i is a codimension 1 subvariety of the Y_{i-1} defined by the equation $f_i = 0$. Dividing by suitable powers of x_0 to define F_i from f_i as above and retaining local coordinates $z_i := x_i/x_0$ for $i > 0$, the conclusion is now that we get a holomorphic 3-form on U_0 (where $x_0 \neq 0$) by taking the residue of the form $\omega_n/(F_1 \cdots F_k)$.

If we define $\mathcal{D} := \det\left(\frac{\partial F_i}{\partial z_j}\right)_{1 \leq i,j \leq k}$ then

$$dF_1 \wedge \ldots dF_k \wedge dz_{k+1} \wedge \ldots \wedge dz_n = \mathcal{D} \cdot dz_1 \wedge \ldots \wedge dz_n$$

as $dF_j = \sum_{i=1}^{n} (\partial F_j/\partial z_i) dz_i$, and by the definition of the determinant and alternating property of wedge products. Redefining U_0' as the open subset of U_0 where $\mathcal{D} \neq 0$

then U_0' coincides with the open subset of U_0 where $D \neq 0$ as D differs from $\mathcal{D}$ by a power of x_0.

Thus, on U_0' we can compute the above residue:

$$\mathrm{Res}(\frac{\omega_n}{F_1 \cdots F_k})_{|U_0'} = \mathrm{Res}(\frac{dF_1 \wedge \ldots dF_k \wedge dz_{k+1} \wedge \ldots \wedge dz_n}{F_1 \cdots F_k \mathcal{D}})_{|U_0'}$$

$$= \left(\frac{dz_{k+1} \wedge \ldots \wedge dz_n}{\mathcal{D}}\right)_{|U_0'}$$

which coincides with

$$\frac{dx_{k+1} \wedge \ldots dx_n}{D}$$

up to a power of x_0 on U_0'.

Again this form extends to all of Y for the same reasons as in the case $k = 1$.

Now suppose that Y has a crepant resolution X that is Calabi–Yau variety of dimension $\dim X \leq 3$. There is then a surjective map $\Omega_X^3 \to \Omega_Y^3$. Hence the holomorphic 3-form on Y that we constructed above on Y extends to a holomorphic 3-form on X.

Acknowledgements We would like to thank Bert van Geemen, Ken-Ichiro Kimura, James D. Lewis, and Matthias Schütt for helpful discussions on this topic.

We would also like to thank the referee for going meticulously through the manuscript and pointing out a number of inaccuracies in the previous version.

The first author thanks Queen's University for its hospitality when this work was done and Colby College for research funding. The second author thanks The Danish Council for Independent Research for support. The third author thanks the Natural Sciences and Engineering Research Council of Canada (NSERC) for their support.

References

1. A. Beauville, Le familles stables de courbes elliptiques sur $\mathbb{P}^1$ admettant quatre fibres singulières. C. R. Acad. Sci. Paris Sér. I Math. **294**, 657–660 (1982)
2. D.A. Cox, S. Katz, in *Mirror Symmetry and Algebraic Geometry*. Mathematical Surveys and Monographs, vol. 68 (American Mathematical Society, Providence, 1999)
3. L. Dieulefait, Computing the level of a modular rigid Calabi–Yau threefold. Exp. Math. **13**, 165–169 (2004)
4. L. Dieulefait, On the modularity of rigid Calabi-Yau threefolds: Epilogue. Zap. Nauchn. Sem. S.-Peterburg. Otdel. Mat. Inst. Steklov. (POMI) **377** (2010); Issledovaniya po Teorii Chisel. **10**, 44–49, 241; translation in J. Math. Sci. (N.Y.) **171**(6), 725–727 (2010) [arXiv:0908.1210v2]
5. L. Dieulefait, J. Manoharmayum, Modularity of rigid Calabi–Yau threefolds over $\mathbb{Q}$, in *Calabi-Yau Varieties and Mirror Symmetry*. Fields Institute Communications, vol. 38 (American Mathematical Society, Providence, 2003), pp. 159–166
6. N.D. Elkies, M. Schütt, Modular Forms and K3 Surfaces, Preprint [arXiv:0809.0830v2]
7. E. Freitag, R. Salvati-Manni, On Siegel threefolds with a projective Calabi–Yau model. Comm. Number Theor. Phys. **5**(3), 713–750 (2011)

8. F. Gouvêa, N. Yui, Rigid Calabi–yau threefolds over $\mathbb{Q}$ are modular. Expo. Math. **29**, 142–149 (2011)

9. I. Kiming, On the liftings of 2-dimensional projective Galois representations over $\mathbb{Q}$. J. Number Theor. **56**, 12–35 (1996)

10. I. Kiming, H. Verrill, On modular mod ℓ representations with exceptional images. J. Number Theor. **110**, 236–266 (2005)

11. K. Kimura, A rational map between two threefolds, in *Mirror Symmetry V* (International Press/American Mathematical Society, Providence, 2006), pp. 87–88

12. C. Meyer, in *Modular Calabi–Yau Threefolds* (American Mathematical Society, Providence, 2005)

13. U. Persson, Double sextics and singular $K3$ surfaces. Lect. Notes Math. **1124**, 262–328 (1985)

14. M. Reid, Canonical 3-folds, in *Journées de Géometrie Algébrique d'Angers, Juillet 1979/ Algebraic Geometry*, ed. by A. Beauville, Angers, 1979 (Sijthoff & Noordhoff, Alphen aan den Rijn, Netherlands, 1980), pp. 273–310

15. K.A. Ribet, Galois representations attached to eigenforms with nebentypus, in *Modular Functions of One Variable V*, ed. by J.-P. Serre, D.B. Zagier. Lecture Notes in Mathematics, vol. 601 (Springer, Berlin, 1977), pp. 17–52

16. C. Schoen, On the geometry of a special determinantal hypersurface associated to the Mumford-Horrocks vector bundle. J. Reine Angew. Math. **364**, 85–111 (1986)

17. C. Schoen, On fiber products of rational elliptic surfaces with section. Math. Z. **197**, 177–199 (1988)

18. B. van Geemen, N.O. Nygaard, On the geometry and arithmetic of some Siegel modular threefolds. J. Number Theor. **53**, 45–87 (1995)

19. C. Voisin, *Hodge Theory and Complex Algebraic Geometry I* (Cambridge University Press, Cambridge, 2002)

20. J. Werner, B. van Geemen, New examples of threefolds with $c_1 = 0$. Math. Z. **203**, 211–225 (1990)

21. N. Yui, Update on the modularity of Calabi-Yau varieties, with an appendix by Helena Verrill, in *Calabi–Yau Varieties and Mirror Symmetry* (American Mathematical Society, Providence, 2003), pp. 307–362

Counting Sheaves on Calabi–Yau and Abelian Threefolds

Martin G. Gulbrandsen

Abstract We survey the foundations for Donaldson–Thomas invariants for stable sheaves on algebraic threefolds with trivial canonical bundle, with emphasis on the case of abelian threefolds.

Key words: Donaldson–Thomas invariants, Calabi–Yau threefolds, Abelian three-folds, Moduli of sheaves, Virtual fundamental classes

Mathematics Subject Classifications (2010): Primary 14N35; Secondary 14K05, 14D20

Let X be a Calabi–Yau threefold, in the weak sense that the canonical sheaf ω_X is trivial. The aim of Donaldson–Thomas theory is to make sense of "counting" the number of stable sheaves on X.

This text consists of two parts: in the first part, Sect. 1, we give an informal and somewhat simplified introduction to the foundations for Donaldson–Thomas invariants, following Behrend–Fantechi [2, 3], Li–Tian [12], Siebert [17], Thomas [19], Huybrechts–Thomas [7], Behrend [1], and Joyce–Song [8]. We put some emphasis on the possibility of having nontrivial $H^1(\mathcal{O}_X)$, so that line bundles may deform, as we have the abelian situation in mind: in the second part, Sect. 2, we discuss recent work by the author [5], where we modify the standard setup surveyed in the first part, to obtain nontrivial Donaldson–Thomas invariants for abelian threefolds X.

M.G. Gulbrandsen (✉)
Stord/Haugesund University College, Haugesund, Norway
e-mail: martin.gulbrandsen@hsh.no

R. Laza et al. (eds.), *Arithmetic and Geometry of K3 Surfaces and Calabi–Yau Threefolds*,   535
Fields Institute Communications 67, DOI 10.1007/978-1-4614-6403-7_21,
© Springer Science+Business Media New York 2013

1 Virtual Counts

We work over $\mathbb{C}$ for simplicity. Fix a polarization H on the Calabi–Yau threefold X, and let M denote the Simpson moduli space [6, 18] of H-stable coherent sheaves on X, with fixed Chern character ch $\in \bigoplus_p H^{2p}(X, \mathbb{Q})$. We assume that M is compact, for instance by choosing the Chern classes such that strictly semistable sheaves are excluded for numerical reasons. For simplicity we shall also assume that there is a universal family, denoted $\mathscr{F}$, on $X \times M$.

The *virtual dimension* is the guess at dim M one obtains from deformation theory; at any point $p \in M$ it is

$$d^{\mathrm{vir}} = \dim \underbrace{\mathrm{Ext}^1(\mathscr{F}_p, \mathscr{F}_p)}_{\text{tangents}} - \dim \underbrace{\mathrm{Ext}^2(\mathscr{F}_p, \mathscr{F}_p)}_{\text{obstructions}} = 0$$

by Serre duality. Our aim is to "count M", even if the prediction fails, so that M has positive dimension. The number arrived at is the Donaldson–Thomas invariant $\mathrm{DT}(M)$.

1.1 Deformation Invariance

Here is a thought model: if M fails to be finite, suppose we can deform X to a new Calabi–Yau threefold X' such that the corresponding moduli space M' of sheaves on X' is finite. Then we want to declare that M should have had the same number of points as M', but for some reason M came out oversized. So we define the virtual count $\mathrm{DT}(M)$ as the number of points in M'. Of course we ask whether this count is independent of the chosen deformation and, if it is, whether it can be phrased intrinsically on M. The answer is affirmative, and this intrinsically defined invariant is the Donaldson–Thomas invariant.

This presentation is misleading in that we rarely can find a finite M', but it motivates the following demands: the virtual count should be such that

- If M is finite and reduced, then $\mathrm{DT}(M)$ is its number of points.
- $\mathrm{DT}(M_t)$ is constant in (smooth) families X_t.

Note that the topological Euler characteristic does specialize to the number of points when M is finite and reduced, but it is certainly not invariant under deformation of X. We return to Euler characteristics in Sect. 1.4.

1.2 Virtual Fundamental Class

The Donaldson–Thomas invariant is defined as the degree of the *virtual fundamental class* $[M]^{\mathrm{vir}}$, which is a Chow class on M of dimension $d^{\mathrm{vir}} = 0$. It is in some sense a characteristic class attached to an obstruction theory on M. The same construction underlies Gromov–Witten invariants, where M is instead a moduli space for stable maps. This machinery was first developed by Li–Tian [12]; our presentation follows Behrend–Fantechi [2], and is also influenced by Siebert [17].

Here is a toy model (cf. [19, Sect. 3]), which serves as a guide for the actual construction. Suppose the moduli space comes out naturally as the zero locus $M = Z(s)$ of a section $s \in \Gamma(V, E)$ of a vector bundle on a smooth variety V. Then the expected dimension of M is

$$d^{\text{vir}} = \dim V - \text{rk}E$$

and if this is indeed the dimension of M, then its fundamental class is $c_{\text{top}}(E)$. But in any case, there is the localized top Chern class $\mathbb{Z}(s)$ [4, Sect. 14.1], which is a degree d^{vir} class in the Chow group of M. This should be our $[M]^{\text{vir}}$.

The section s embeds M into E; recall that deformation to the normal cone in this context [4, Remark 5.1.1] says that as $\lambda \to \infty$, the locus $\lambda s(M) \subset E$ becomes the normal cone $C_{M/V} \subset E|_M$. The localized top Chern class $\mathbb{Z}(s)$ is the (refined) intersection of $C_{M/V}$ with the zero section of $E|_M$. Thus we can forget about V: all we need to be able to write down our toy virtual fundamental class $[M]^{\text{vir}} = \mathbb{Z}(s)$ is a cone ($C_{M/V}$) in a vector bundle ($E|_M$) on M. It turns out that the normal cone $C_{M/V}$, or at least an essential part of it, is in some sense intrinsic to M, whereas an embedding into a vector bundle is a (perfect) obstruction theory on M.

Now return to the actual moduli space M: choose an embedding $M \subset V$ into a smooth variety V (our M is projective, so we may take $V = \mathbb{P}^n$). Let $\mathscr{I} \subset \mathscr{O}_V$ be the ideal of the embedding. The natural map

$$L_M \colon \mathscr{I}/\mathscr{I}^2 \to \Omega_V|_M, \tag{1}$$

considered as a complex with objects in degrees -1 and 0, is the truncated cotangent complex for M. We define

$$T_V = \text{SpecSym}\Omega_V,$$
$$N_{M/V} = \text{SpecSym}\mathscr{I}/\mathscr{I}^2,$$
$$C_{M/V} = \text{Spec} \bigoplus \mathscr{I}^d/\mathscr{I}^{d+1}$$

(the first two are vector space fibrations, with possibly varying fibre dimensions, and the last one is a cone fibration) so (1) gives a map

$$T_V|_M \to N_{M/V} \tag{2}$$

and there is an embedding $C_{M/V} \subseteq N_{M/V}$. Now Behrend–Fantechi define stack quotients

$$N_M = [N_{M/V}/T_V|_M], \qquad\qquad C_M = [C_{M/V}/T_V|_M]$$

and prove that they are independent of V. They are the *intrinsic normal space* and *intrinsic normal cone*. The reader not comfortable with stacks can safely view N_M as the map (2) between vector space fibrations, modulo some equivalence relation, and similarly for C_M (this viewpoint is carried further by Siebert [17]). But it is also useful for intuition to think of them as somewhat weird vector space and cone fibrations over M. For instance, the fibre of N_M over a smooth point $p \in M$ is the trivial vector space together with the stabilizer group $T_M(p)$.

So we have the intrinsically defined (stacky) normal cone C_M on M, embedded into N_M. But the latter is a (stacky) vector space fibration, which is not necessarily locally free. So the lacking piece of data is an embedding of N_M into a vector bundle. We enlarge our notion of vector bundles to allow stack quotients $E = [E^1/E^0]$, where $E^0 \to E^1$ is a linear map of vector bundles on M. Ad hoc, we define a map $f: N_M \subset E$ to be something induced by a commutative square

$$\begin{array}{ccc} T_M|_V & \longrightarrow & N_{M/V} \\ {\scriptstyle f^0}\downarrow & & \downarrow{\scriptstyle f^1} \\ E^0 & \longrightarrow & E^1 \end{array} \qquad (3)$$

and to count as an embedding, the map f^1 should take distinct $T_M|_V$-orbits in $N_{M/V}$ to distinct E^0-orbits in E^1, i.e. the induced map on cokernels should be a *monomorphism*. Furthermore we require that the induced map on kernels should be an *isomorphism*, so that stabilizer groups in N_M and in E agree.

In summary, we want to equip the scheme M with a (stacky) vector bundle E and an embedding $N_M \subset E$ of the intrinsic normal space. Then we define the virtual fundamental class $[M]^{\mathrm{vir}}$ as the intersection of the cone $C_M \subset E$ with the zero section in E. (This does make sense on stacks [10].)

1.3 Obstruction Theory

Obstruction theory is a systematic answer to the problem of extending a morphism $f: T \to M$ to an infinitesimal thickening $T \subset \overline{T}$, where the ideal $\mathscr{I} \subset \mathscr{O}_{\overline{T}}$ of T has square zero. We may and will assume that T and $\overline{T}$ are affine. To each such situation, there is a canonical obstruction class [2, Sect. 4]

$$\omega \in \mathrm{Ext}_T^1(f^*L_M, \mathscr{I})$$

whose vanishing is equivalent to the existence of a morphism $\overline{f}: \overline{T} \to M$ extending f. Moreover, when $\omega = 0$, the set of such extensions $\overline{f}$ forms in a natural way a torsor under $\mathrm{Hom}_T(f^*L_M, \mathscr{I})$. These statements, if unfamiliar to the reader, may be taken on trust for the purposes of this text.

Obstruction theory is connected with the construction of virtual fundamental classes as follows: diagram (3) is obtained by applying $\mathrm{SpecSym}(-)$ to the dual diagram of coherent sheaves

$$
\begin{array}{ccc}
\Omega_M|_V & \longleftarrow & \mathcal{I}/\mathcal{I}^2 \\
\phi^0 \uparrow & & \phi^{-1} \uparrow \\
\mathcal{E}^0 & \longleftarrow & \mathcal{E}^{-1}
\end{array}
$$

i.e. a morphism of complexes

$$\phi \colon \mathcal{E} \to L_M$$

(to be precise, this happens in the derived category, so quasi-isomorphisms are inverted).

Theorem 1 (Behrend–Fantechi [2]). *Let $\mathcal{E}$ be a complex of locally free sheaves concentrated in nonpositive degrees and let $\phi \colon \mathcal{E} \to L_M$ be a morphism (in the derived category). Then the following are equivalent:*

(i) *For each deformation situation $T \subset \overline{T}$, $f \colon T \to M$, there is a morphism $\overline{f} \colon \overline{T} \to M$ extending f if and only if*

$$\phi^*(\omega) \in \mathrm{Ext}^1_T(f^* \mathcal{E}, \mathcal{I})$$

vanishes; furthermore when $\phi^*(\omega) = 0$, the set of such extensions $\overline{f}$ forms a torsor under $\mathrm{Hom}_T(f^* \mathcal{E}, \mathcal{I})$.

(ii) *The morphism ϕ induces an isomorphism in degree 0 and an epimorphism in degree -1.*

The condition that f in Diagram (3) induces an isomorphism on kernels and a monomorphism on cokernels translates precisely to the condition on ϕ in (ii) in the theorem. Thus an embedding $N_M \subset E$ is equivalent to an obstruction theory of a particular kind:

Definition 1 (Behrend–Fantechi [2]). A *perfect obstruction theory* on M is a two term complex of locally free sheaves $\mathcal{E}$ together with a morphism $\phi \colon \mathcal{E} \to L_M$, such that ϕ induces an isomorphism in degree 0 and an epimorphism in degree -1.

The point is, of course, that our moduli space M, or rather the subspace $M(\mathcal{L}) \subset M$ of sheaves with fixed determinant line bundle $\mathcal{L}$, carries a natural perfect obstruction theory. To construct the obstruction theory, we assume the rank r is nonzero, and consider the trace map

$$\mathrm{tr} \colon \mathcal{H}om(\mathcal{F}, \mathcal{F}) \to \mathcal{O}_{X \times M}$$

(do this in the derived category, so $\mathcal{H}om$ means derived $\mathcal{H}om$; if $\mathcal{F}$ is locally free it doesn't matter, of course). Let $\mathcal{F}_T$ be the sheaf on $T \times X$ obtained by pulling back the universal family $\mathcal{F}$ along $f \colon T \to M(\mathcal{L})$. To extend f to $\overline{T}$ is the same as to extend $\mathscr{F}_T$ to a $\overline{T}$-flat family on $\overline{T} \times X$ with constant determinant. The trace map induces

$$\mathrm{tr}^i \colon \mathrm{Ext}^i(\mathscr{F}_T, \mathscr{F}_T \otimes_{\mathscr{O}_T} \mathscr{I}) \to H^i(\mathscr{I}).$$

We will use a subscript 0 on $\mathscr{H}om$ and Ext^i to indicate the kernels of tr and tr^i. By reasonably elementary arguments (see e.g. Thomas [19]), the existence of such an extension is equivalent to the vanishing of a certain class

$$\omega \in \mathrm{Ext}^2_0(\mathscr{F}_T, \mathscr{F}_T \otimes_{\mathscr{O}_T} \mathscr{I}). \tag{4}$$

Moreover, when $\omega = 0$, the set of extensions of $\mathscr{F}_T$ form a torsor under

$$\mathrm{Ext}^1_0(\mathscr{F}_T, \mathscr{F}_T \otimes_{\mathscr{O}_T} \mathscr{I}). \tag{5}$$

This elementary obstruction theory can be lifted to a Behrend–Fantechi type theory, and for this step we will be brief: the diagonal map $\mathscr{O}_{X \times M} \to \mathscr{H}om(\mathscr{F}, \mathscr{F})$ composed with the trace map is multiplication by the rank r, hence there is a splitting

$$\mathscr{H}om(\mathscr{F}, \mathscr{F}) = \mathscr{H}om_0(\mathscr{F}, \mathscr{F}) \oplus \mathscr{O}_{X \times M}. \tag{6}$$

There is a natural morphism (essentially the Atiyah class of $\mathscr{F}$ [7])

$$\phi \colon \mathscr{E} = (p_{2*}\mathscr{H}om_0(\mathscr{F}, \mathscr{F}))^{\vee}[-1] \to L_M \tag{7}$$

(again, derived functors), whose restriction to $M(\mathscr{L})$ is a perfect obstruction theory. In fact, there are $\mathscr{O}_T$-linear isomorphisms

$$\mathrm{Ext}^i(\mathscr{F}_T, \mathscr{F}_T \otimes_{\mathscr{O}_T} \mathscr{I}) \cong \mathrm{Ext}^{i-1}(f^*\mathscr{E}, \mathscr{I})$$

such that the obstruction class in (4) agrees with the one in Theorem 1 (i), and the torsor structures are the same [7, Theorem 4.1].

If we used the full $\mathscr{H}om$ instead of the trace free $\mathscr{H}om_0$ in (7), the complex $\mathscr{E}$ would be too big in two ways: firstly, it would not be concentrated in degrees $[-1, 0]$; secondly and more seriously, even if we truncate it, it would contain a trivial summand by (6), causing the virtual fundamental class to be zero (just as, in our toy model in Sect. 1.2, the top Chern class of a vector bundle with a trivial summand is zero). This is why we are led to fixing the determinant, as the trace free part of $\mathscr{H}om(\mathscr{F}, \mathscr{F})$ is precisely what controls the deformation theory for sheaves with fixed determinant.

Definition 2 (Thomas [19], Huybrechts–Thomas [7]). The Donaldson–Thomas invariant $\mathrm{DT}(M(\mathscr{L}))$ of $M(\mathscr{L})$ is the degree of the virtual fundamental class $[M(\mathscr{L})]^{\mathrm{vir}}$ associated to the canonical perfect obstruction theory (7).

The Donaldson–Thomas invariant does fulfill our two requirements from Sect. 1.1. Firstly, if $M(\mathscr{L})$ happens to be finite and reduced, and somewhat more generally: finite and a complete intersection, then the truncated cotangent complex is itself a perfect obstruction theory, and the associated virtual fundamental class is the usual fundamental class, hence its degree is the length of $M(\mathscr{L})$ as a finite scheme.

Secondly, the obstruction theory we have sketched above generalizes to the relative situation of a moduli space $M \to S$ for sheaves on the fibres of a family $X \to S$ of Calabi–Yau threefolds. The relative obstruction theory gives rise to a virtual fundamental class on the whole family, which restricts to the fibrewise virtual fundamental class. Consequently, the degree of the virtual fundamental class is constant among the fibres.

1.4 Behrend's Weighted Euler Characteristic

A priori, the Donaldson–Thomas invariant defined above may depend on the choice of obstruction theory on $M(\mathscr{L})$. But in fact, the invariant can be rephrased entirely in terms of the intrinsic geometry of $M(\mathscr{L})$: the decisive property of the obstruction theory (7) is that it is not only of virtual dimension 0, but it is *symmetric* [3, Definition 1.10]. Roughly speaking, symmetry is a refinement of the property that

$$\mathrm{Ext}^1(\mathscr{F}_p, \mathscr{F}_p) \quad \text{and} \quad \mathrm{Ext}^2(\mathscr{F}_p, \mathscr{F}_p)$$

are Serre dual (and so are the trace free versions).

Now, for any scheme Y, Behrend defines an integral invariant

$$\nu \colon Y \to \mathbb{Z}$$

with the properties (among others) that $\nu(p)$ only depends on an étale neighbourhood of $p \in Y$, at smooth points $\nu = (-1)^{\dim_p M}$, and $\nu^{-1}(n)$ is a constructible subset for all $n \in \mathbb{Z}$.

Theorem 2 (Behrend [1]). *If Y is a compact scheme with a perfect symmetric obstruction theory, then the degree of the associated virtual fundamental class equals the ν-weighted Euler characteristic*

$$\tilde{\chi}(Y) = \sum_{n \in \mathbb{Z}} n \chi(\nu^{-1}(n)).$$

The theorem has at least three important consequences: Firstly, as promised, the Donaldson–Thomas invariant is an intrinsic invariant of $M(\mathscr{L})$. Secondly, the weighted Euler characteristic is directly accessible for computation in examples [3]. But deformation invariance does not follow from the weighted Euler characteristic formulation; this is a consequence of the virtual fundamental class machinery. So, for instance, in the presence of strictly semi-stable sheaves, one might attempt to define generalized Donaldson–Thomas invariants as the weighted Euler characteristic of either the non-compact moduli space $M(\mathscr{L})$ or the compactified moduli space $\overline{M}(\mathscr{L})$ for semi-stable sheaves, but these numbers would not be deformation invariant. Still, and this is the third consequence, Behrend's weighted Euler characteristic is the starting point for Joyce–Song's [8] correct (i.e. deformation invariant) and somewhat mysterious way of counting strictly semi-stable sheaves (see also Kontsevich–Soibelman [9]). In a nontrivial manner, these generalized Donaldson–Thomas invariants take into account all ways of putting together stable (Jordan–Hölder) factors to form semi-stable sheaves as iterated extensions. At present, this theory only covers the situation where $H^1(\mathscr{O}_X) = 0$, so that line bundles do not deform.

2 Abelian Threefolds

Let X be an abelian threefold. The theory outlined in the first part of this text applies, but almost always results in vanishing Donaldson–Thomas invariants. We will investigate why this is so, and how the setup can be adjusted to give nontrivial invariants.

The Chern character of the sheaves parametrized by M will be written

$$\text{ch} = r + c_1 + \gamma + \chi \tag{8}$$

where r and χ are integers, c_1 is a divisor class and γ is a curve class.

2.1 Determinants

Let us apply Behrend's weighted Euler characteristic to see, in concrete terms, why the Donaldson–Thomas invariant of the full moduli space M is zero. Afterwards, we shall see that restricting to $M(\mathscr{L})$ does not help.

Assume the rank r is nonzero. Let $\delta \colon M \to \text{Pic}^{c_1}(X) \cong \widehat{X}$ be the morphism that sends a sheaf $\mathscr{F}$ to its determinant line bundle $\det(\mathscr{F})$. Then δ is surjective, in fact all fibres $M(\mathscr{L}) = \delta^{-1}(\mathscr{L})$ are isomorphic. This can be seen by letting $\widehat{X} = \text{Pic}^0(X)$ act on M by twist:

$$\widehat{X} \times M \to M, \quad (\xi, \mathscr{F}) \mapsto \mathscr{F} \otimes \mathscr{P}_\xi$$

where we write $\mathscr{P}_\xi$ for the invertible sheaf corresponding to $\xi \in \widehat{X}$. Since

$$\det(\mathscr{F} \otimes \mathscr{P}_\xi) = \det(\mathscr{F}) \otimes \mathscr{P}_{r\xi},$$

it follows that every orbit in M surjects onto $\text{Pic}^{c_1}(X)$, and every fibre of δ can be moved to any other fibre by the action of some element $\xi \in \widehat{X}$.

The topological Euler characteristic of M thus equals the product of the Euler characteristics of a fibre $M(\mathscr{L})$ and the base $\widehat{X}$, but the latter has Euler characteristic zero. Thus $\chi(M) = 0$. Via a stratification, the same argument works for Behrend's weighted Euler characteristic: write $M = \bigcup_n M_n$ where $M_n \subset M$ is the constructible subset $\nu^{-1}(n)$. Each M_n is invariant under the $\widehat{X}$-action, so for all n, $\chi(M_n)$ equals the product of the Euler characteristic of a fibre $M_n \cap M(\mathscr{L})$ and the base $\widehat{X}$, hence is zero. Thus

$$\tilde{\chi}(M) = \sum_n n\chi(M_n) = 0.$$

This argument applies to any, not necessarily abelian, X. But in the abelian case, the weighted Euler characteristic of $M(\mathscr{L})$ is usually zero, too. We look at an example before handling the general situation.

Example 1. Let $\mathrm{Hilb}^n(X)$ be the Hilbert scheme of finite subschemes $Z \subset X$ of length n. By associating with Z its ideal $\mathscr{I}_Z$, we view the Hilbert scheme as a moduli space for rank 1 sheaves. These sheaves may be deformed either by moving Z around, or by twisting with invertible sheaves in $\mathrm{Pic}^0(X)$, so the full moduli space is $M = \widehat{X} \times \mathrm{Hilb}^n(X)$, first projection is the determinant map $M \to \widehat{X}$, and

$$M(\mathscr{O}_X) = \mathrm{Hilb}^n(X)$$

is a moduli space for rank 1 sheaves with fixed determinant $\mathscr{O}_X$. Writing $\sum Z$ for the sum under the group law on X, of the zero cycle underlying Z, we find a second fibration:

$$\mathrm{Hilb}^n(X) \to X, \quad Z \mapsto \textstyle\sum Z \tag{9}$$

By translation with elements $x \in X$, any fibre can be moved to any other fibre, so that by repeating the argument above, we conclude that Behrend's weighted Euler characteristic of $\mathrm{Hilb}^n(X)$ is zero.

The "second fibration" (9) on the Hilbert scheme generalizes as follows: let $\widehat{\delta}\colon M \to X$ be the morphism that takes a sheaf $\mathscr{F}$ to the determinant of its Fourier–Mukai transform $\widehat{\mathscr{F}}$ (we should warn the reader that in the literature, the notation $\widehat{\mathscr{F}}$ is usually reserved for WIT-sheaves [14, Definition 2.3]; our $\widehat{\mathscr{F}}$ may well be a complex). The invertible sheaf $\det(\widehat{\mathscr{F}})$ will be called the *codeterminant* of $\mathscr{F}$. It belongs to some component of $\mathrm{Pic}(\widehat{X})$, which we identify with $\mathrm{Pic}^0(\widehat{X}) = X$. In general, there is no relation between the determinant and the codeterminant. More precisely, let $X \times \widehat{X}$ act on M by translation and twist:

$$(X \times \widehat{X}) \times M \to M, \quad (x, \xi; \mathscr{F}) \mapsto T^*_{-x}\mathscr{F} \otimes \mathscr{P}_\xi$$

Write $\phi_{c_1}\colon X \to \widehat{X}$ for the homomorphism $x \mapsto \mathscr{O}_X(T^*_x D - D)$, for any divisor D representing c_1. Via Poincaré duality, the curve class γ in (8) corresponds to a divisor class on $\widehat{X}$ [15, Proposition 1.17]; we shall write $\psi_\gamma\colon \widehat{X} \to \widehat{\widehat{X}} = X$ for the associated homomorphism. With this notation, the action of $X \times \widehat{X}$ on a fixed sheaf $\mathscr{F} \in M$, composed with the determinant/codeterminant $(\widehat{\delta}, \delta)\colon M \to X \times \widehat{X}$ is easily computed [5, Proposition 2.2]: it is

$$\begin{pmatrix} \chi & -\psi_\gamma \\ -\phi_{c_1} & r \end{pmatrix} \in \mathrm{End}(X \times \widehat{X}). \tag{10}$$

Thus the condition that the rank r is nonzero, that we used to ensure that δ was a fibration, is now replaced by the condition that *this matrix is an isogeny.* Then $X \times \widehat{X}$-orbits in M surjects onto $X \times \widehat{X}$ via $(\widehat{\delta}, \delta)$, all fibres are isomorphic and the weighted Euler characteristic is zero. Fixing just one determinant does not help: we need to fix both to obtain a nontrivial invariant.

Our object is thus the fibre $M(\mathscr{L}', \mathscr{L}) \subset M$ of $(\widehat{\delta}, \delta)$, parametrizing sheaves with determinant $\mathscr{L}$ and codeterminant $\mathscr{L}'$. Examples indicate that its weighted Euler characteristic is nonzero in general, so there are no further fibrations, which is a good thing, since we have run out of group actions on M.

Proposition 1. *Suppose X has Picard number* 1. *Then the matrix* (10) *is an isogeny if and only if*

$$3r\chi \neq c_1\gamma.$$

See [5, Lemma 2.3] for a statement without the Picard number restriction, and proof.

The proposition shows that our isogeny condition is satisfied for almost all choices of Chern classes. If the condition does fail, as it does for instance for rank 2 vector bundles $\mathscr{F}$ with $c_1(\mathscr{F}) = 0$, we may replace $\mathscr{F}$ with $\mathscr{F}(H)$ and try again. One can show [5, Proposition 3.5] that by such tricks (to be made precise in 2.2), the inequality $3r\chi \neq c_1\gamma$ can always be forced to hold, with the sole exception of Mukai's semi-homogeneous sheaves, whose moduli spaces are fully understood anyway [13].

This observation points to an arbitrariness in the definition of the determinant/codeterminant map, which changes when $\mathscr{F}$ is replaced by $\mathscr{F}(H)$, since the Fourier–Mukai transform does not preserve tensor product. We will fix this arbitrariness in Sect. 2.2. This can be contrasted with the situation for abelian surfaces [20], where the determinant/codeterminant pair is just the Albanese map of the moduli space M, and hence is entirely intrinsic. For abelian threefolds, the moduli space M is not, in general, fibred over its Albanese:

Example 2. Let $C \subset X$ be a non hyperelliptic genus 3 curve, embedded into its Jacobian by an Abel–Jacobi map. Any deformation of C is a translate $T_x(C)$ by some point $x \in X$, and in fact the Hilbert scheme component containing C is isomorphic to X [11]. Now consider the Hilbert scheme component H parametrizing translations of C together with a possibly embedded point. As in Example 1, H can be viewed as a moduli space for rank 1 sheaves with fixed determinant. There is a map

$$H \to X^2$$

sending a point $T_x(C) \cup y$ in H to the pair (x, y). This is clearly the Albanese map, and it is generically bijective. However, the fibre over a pair (x, y) for which $y \in T_x(C)$, consists of all embedded points in $T_x(C)$ supported at y, hence is a $\mathbb{P}^1$. In particular, the Albanese fibres of H are not isomorphic.

2.2 Translation and Twist

We assume that the matrix (10) is an isogeny throughout this section.

The fibres $M(\mathscr{L}', \mathscr{L})$ of $(\widehat{\delta}, \delta)$ intersect each $X \times \widehat{X}$-orbit in finitely many points. The canonical object lurking here is the quotient space

$$K = M/X \times \widehat{X} = M(\mathscr{L}', \mathscr{L})/G$$

(which we consider as a Deligne–Mumford stack), where G is the (finite) kernel of the isogeny (10). The point is that the "tricks" we alluded to above, such as twisting with a divisor, preserves the $X \times \widehat{X}$-action on M, so that K does not change. More generally, suppose Y is a second abelian threefold and there exists a derived equivalence $F: D(X) \xrightarrow{\sim} D(Y)$. Then we may equally well consider M as a moduli space for sheaves $\mathscr{F}$ on X or as a moduli space for $F(\mathscr{F})$ on Y (this may be a complex, and not a sheaf, and for this reason we work with complexes from the start in [5]). Orlov shows that there is an induced isomorphism $X \times \widehat{X} \cong Y \times \widehat{Y}$ such that the two actions on M are compatible [16, Corollary 2.13], so $M/X \times \widehat{X} \cong M/Y \times \widehat{Y}$. In this sense, the space K is invariant under derived equivalence, although $M(\mathscr{L}', \mathscr{L})$ is not.

Although weighted Euler characteristics of $M(\mathscr{L}', \mathscr{L})$ and K make sense, it is not clear that they are of interest (in particular, whether they are invariant under deformation of X) unless there is an underlying perfect obstruction theory.

Theorem 3. *There is a perfect symmetric obstruction theory on* $M(\mathscr{L}', \mathscr{L})$.

Proof. We sketch the main points; the details can be found in [5]. Consider the problem of extending $f: T \to M(\mathscr{L}', \mathscr{L})$ over $T \subset \overline{T}$. The trace map on $\widehat{X} \times M$,

$$\widehat{\mathrm{tr}}: \mathscr{H}om(\widehat{\mathscr{F}}, \widehat{\mathscr{F}}) \to \mathcal{O}_{\widehat{X} \times M}$$

together with the Fourier–Mukai-induced isomorphism

$$p_{2*}\mathscr{H}om(\mathscr{F}, \mathscr{F}) \cong p_{2*}\mathscr{H}om(\widehat{\mathscr{F}}, \widehat{\mathscr{F}}) \tag{11}$$

(in the derived category) give new trace maps

$$\widehat{\mathrm{tr}}^i: \mathrm{Ext}^i(\mathscr{F}_T, \mathscr{F}_T \otimes_{\mathcal{O}_T} \mathscr{I}) \to H^i(\mathscr{I}).$$

Switching back and forth between X and $\widehat{X}$ we thus see that the obstruction class ω for extending f to $\overline{T}$ is a class in $\ker(\mathrm{tr}^2) \cap \ker(\widehat{\mathrm{tr}}^2)$, and when $\omega = 0$, the set of such extensions, with fixed determinant and codeterminant, is a torsor under $\ker(\mathrm{tr}^1) \cap \ker(\widehat{\mathrm{tr}}^1)$.

Again the setup can be lifted to a Behrend–Fantechi obstruction theory: the two trace maps taken together and pushed down to M

$$p_{2*}\mathscr{H}om(\mathscr{F}, \mathscr{F}) \to p_{2*}\mathcal{O}_{\widehat{X} \times M} \oplus p_{2*}\mathcal{O}_{X \times M} \tag{12}$$

(derived functors) is a split epimorphism in degrees 1 and 2: in fact, the two diagonal maps

$$\mathscr{O}_{X \times M} \rightarrow \mathscr{H}om(\mathscr{F}, \mathscr{F})$$
$$\mathscr{O}_{\widehat{X} \times M} \rightarrow \mathscr{H}om(\widehat{\mathscr{F}}, \widehat{\mathscr{F}})$$

give, via (11), a map

$$p_{2*} \mathscr{O}_{\widehat{X} \times M} \oplus p_{2*} \mathscr{O}_{X \times M} \rightarrow p_{2*} \mathscr{H}om(\mathscr{F}, \mathscr{F}). \tag{13}$$

The composition of (13) with (12) realizes the splitting: in degree 1, it is an endomorphism of $H^1(\mathscr{O}_{\widehat{X}}) \oplus H^1(\mathscr{O}_X)$, which is nothing but the derivative of the isogeny (10) at $(0, 0) \in X \times \widehat{X}$, hence an isomorphism. By duality, the degree 2 part is an isomorphism, too. Thus, writing $\tau^{[1,2]}$ for the truncation of a complex to degrees $[1, 2]$, we have produced a splitting

$$\tau^{[1,2]} p_{2*} \mathscr{H}om(\mathscr{F}, \mathscr{F}) = \mathscr{E} \oplus \tau^{[1,2]}(p_{2*} \mathscr{O}_{\widehat{X} \times M} \oplus p_{2*} \mathscr{O}_{X \times M}). \tag{14}$$

The morphism (7) induces

$$\phi \colon \mathscr{E}^{\vee}[-1] \rightarrow L_M$$

whose restriction to $M(\mathscr{L}', \mathscr{L})$ is a perfect symmetric (via duality) obstruction theory.

The two trivial summands in (14) shows, in terms of the virtual fundamental class machinery, why it is not enough to fix one determinant, as this kills just one of the summands.

Instead of worrying about whether the obstruction theory on $M(\mathscr{L}', \mathscr{L})$ descends to K, we define the Donaldson–Thomas invariant directly:

Definition 3. Assume (10) is an isogeny, and let G be its (finite) kernel. Then the Donaldson–Thomas invariant of K is

$$DT(K) = \frac{1}{|G|} \deg[M(\mathscr{L}', \mathscr{L})]^{\text{vir}}.$$

Theorem 3 generalizes to the relative situation, so that deformation invariance for the virtual fundamental class of $M(\mathscr{L}', \mathscr{L})$ holds. Since the kernel G of the isogeny (10) has constant order in families, the Donaldson–Thomas invariant $DT(K)$ is invariant under deformations of X. Moreover, it agrees with Behrend's weighted Euler characteristic, hence is an intrinsic invariant of K.

Example 3. We return to the Hilbert scheme of points in Example 1. The summation map $\mathrm{Hilb}^n(X) \rightarrow X$ agrees, up to sign, with the codeterminant map (use that, modulo short exact sequences, $\mathscr{O}_Z$ is equivalent to a sum of skyscrapers $k(z)$, with $z \in Z$ repeated according to multiplicity, and $\widehat{k(z)} = \mathscr{P}_z$).

The moduli space $M(\mathscr{O}_{\widehat{X}}, \mathscr{O}_X)$ for ideals with trivial determinant and codeterminant, is thus nothing but the locus

$$K^n(X) = \{Z \in \mathrm{Hilb}^n(X) \mid \Sigma Z = 0\}.$$

(For abelian surfaces X, this locus $K^n(X)$ is the generalized Kummer variety of Beauville.) The kernel G of the isogeny (10) is in this case the group of n-torsion points X_n in $X \times 0 \subset X \times \widehat{X}$, so

$$K = K^n(X)/X_n.$$

Behrend–Fantechi [3] found that Behrend's weighted Euler characteristic of the Hilbert scheme of n points on any threefold agrees, up to sign, with the usual Euler characteristic. Their argument can be adapted to $K^n(X)$, showing that its weighted Euler characteristic is $(-1)^{n+1}\chi(K^n(X))$, and so

$$DT(K) = \frac{1}{|X_n|}\tilde{\chi}(K^n(X)) = \frac{(-1)^{n+1}}{n^6}\chi(K^n(X)).$$

See [5, Sect. 4.2] for a conjectural explicit formula for the Euler characteristic of $K^n(X)$.

References

1. K. Behrend, Donaldson-Thomas type invariants via microlocal geometry. Ann. Math. (2) **170**(3), 1307–1338 (2009)
2. K. Behrend, B. Fantechi, The intrinsic normal cone. Invent. Math. **128**(1), 45–88 (1997)
3. K. Behrend, B. Fantechi, Symmetric obstruction theories and Hilbert schemes of points on threefolds. Algebra Number Theor. **2**(3), 313–345 (2008)
4. W. Fulton, in *Intersection theory*, 2nd edn. Ergebnisse der Mathematik und ihrer Grenzgebiete. 3. Folge. A Series of Modern Surveys in Mathematics [Results in Mathematics and Related Areas. 3rd Series. A Series of Modern Surveys in Mathematics], vol. 2 (Springer, Berlin, 1998)
5. M.G. Gulbrandsen, Donaldson–Thomas invariants for complexes on abelian threefolds. Math. Z. **273**(1–2), 219–238 (2013)
6. D. Huybrechts, M. Lehn, The geometry of moduli spaces of sheaves, in *Aspects of Mathematics*, E31, Friedr (Vieweg & Sohn, Braunschweig, 1997)
7. D. Huybrechts, R.P. Thomas, Deformation-obstruction theory for complexes via Atiyah and Kodaira-Spencer classes. Math. Ann. **346**(3), 545–569 (2010)
8. D. Joyce, Y. Song, A theory of generalized Donaldson–Thomas invariants. Mem. Amer. Math. Soc. 217, no. 1020, v+199 (2012)
9. M. Kontsevi ch, Y. Soibelman, Stability structures, motivic Donaldson-Thomas invariants and cluster transformations (2008) arXiv:0811.2435v1 [math.AG]
10. A. Kresch, Cycle groups for Artin stacks. Invent. Math. **138**(3), 495–536 (1999)
11. H. Lange, E. Sernesi, On the Hilbert scheme of a Prym variety. Ann. Mat. Pura Appl. (4) **183**(3), 375–386 (2004)
12. J. Li, G. Tian, Virtual moduli cycles and Gromov-Witten invariants of algebraic varieties. J. Am. Math. Soc. **11**(1), 119–174 (1998)
13. S. Mukai, Semi-homogeneous vector bundles on an Abelian variety. J. Math. Kyoto Univ. **18**(2), 239–272 (1978)
14. S. Mukai, Duality between $D(X)$ and $D(\hat{X})$ with its application to Picard sheaves. Nagoya Math. J. **81**, 153–175 (1981)
15. S. Mukai, Fourier functor and its application to the moduli of bundles on an abelian variety, in *Algebraic Geometry*, Sendai, 1985. Advanced Studies in Pure Mathematics, vol. 10 (North-Holland, Amsterdam, 1987), pp. 515–550

16. D.O. Orlov, Derived categories of coherent sheaves on abelian varieties and equivalences between them. Izv. Ross. Akad. Nauk Ser. Mat. **66**(3), 131–158 (2002)
17. B. Siebert, Virtual fundamental classes, global normal cones and Fulton's canonical classes, in *Frobenius Manifolds*. Aspects Mathematics, E36 (Vieweg, Wiesbaden, 2004), pp. 341–358 (corrected version: arXiv:math/0509076v1 [math.AG])
18. C.T. Simpson, Moduli of representations of the fundamental group of a smooth projective variety, I. Inst. Hautes Études Sci. Publ. Math. **79**, 47–129 (1994)
19. R.P. Thomas, A holomorphic Casson invariant for Calabi-Yau 3-folds, and bundles on $K3$ fibrations. J. Differ. Geom. **54**(2), 367–438 (2000)
20. K. Yoshioka, Moduli spaces of stable sheaves on abelian surfaces. Math. Ann. **321**(4), 817–884 (2001)

The Segre Cubic and Borcherds Products

Shigeyuki Kondō

Abstract We shall construct a five-dimensional linear system of holomorphic automorphic forms on a three-dimensional complex ball by applying Borcherds theory of automorphic forms. We shall show that this linear system gives the dual map from the Segre cubic threefold to the Igusa quartic threefold.

Key words: Segre cubic threefold, Igusa quartic threefold, Ball quotient, Automorphic form

Mathematics Subject Classifications (2010): Primary 14J15; Secondary 11F55, 14J28, 32N15

1 Introduction

The main purpose of this note is to give an application of the theory of automorphic forms on bounded symmetric domains of type IV due to Borcherds [4, 5]. We consider the Segre cubic threefold X, which is a hypersurface of $\mathbf{P}^4$ of degree 3 with ten nodes. The symmetry group $\mathfrak{S}_6$ of degree 6 acts on X as projective transformations. It is known that the Segre cubic X is isomorphic to the Satake–Baily–Borel compactification of an arithmetic quotient of a three-dimensional complex ball $\mathcal{B}$ associated to a Hermitian form of signature $(1,3)$ defined over the Eisenstein integers [9–11]. The complex ball $\mathcal{B}$ can be embedded into a bounded symmetric domain $\mathcal{D}$ of type IV and of dimension 6. By applying Borcherds' theory of automorphic forms on bounded symmetric domains of type IV, we can construct a five-dimensional linear system of holomorphic automorphic forms of weight 6. We shall show that this linear system gives the dual map from the Segre cubic X to its dual Igusa quartic threefold.

S. Kondō (✉)
Graduate School of Mathematics, Nagoya University, Nagoya 464-8602, Japan
e-mail: kondo@math.nagoya-u.ac.jp

R. Laza et al. (eds.), *Arithmetic and Geometry of K3 Surfaces and Calabi–Yau Threefolds*, Fields Institute Communications 67, DOI 10.1007/978-1-4614-6403-7_22, © Springer Science+Business Media New York 2013

B. van Geemen [15] and B. Hunt [9] observed that both the Segre cubic threefold and the Igusa quartic threefold are birational to the moduli space of ordered six points on the projective line. By taking the triple cover of $\mathbf{P}^1$ branched along six points we get a curve of genus 4 with an automorphism of order 3. One can consider the Segre cubic as a compactification of the moduli space of such curves. On the other hand, by taking the double cover of $\mathbf{P}^1$ branched along 6 points, we get a hyperelliptic curve of genus 2. The Igusa quartic is the Satake compactification of an arithmetic quotient of the Siegel space of degree 2 (Igusa [12], page 397). It is classically known that the dual of the Segre cubic is isomorphic to the Igusa quartic (Baker [3], Chap. V). We will give an interpretation of this ball quotient as the moduli space of some $K3$ surfaces with an automorphism of order 3 which are obtained from 6 points on the projective line.

We use an idea of Allcock and Freitag [1] to construct a linear system of automorphic forms. In [1], they consider a four-dimensional complex ball defined over the Eisenstein integers and construct a ten-dimensional linear system of automorphic forms. An arithmetic quotient of the four-dimensional complex ball is birational to the moduli space of marked cubic surfaces. Our complex ball $\mathcal{B}$ appears as a sub-complex ball of Allcock and Freitag's one, and hence one can restrict Allcock and Freitag's linear system to $\mathcal{B}$. However in this note, instead of using their linear system, we apply Borcherds' theory directly to our situation and get a linear system on $\mathcal{B}$.

The plan of this note is as follows. In Sect. 1, we recall the Segre cubic X and some divisors on X. In Sect. 2, we mention the complex ball $\mathcal{B}$, the bounded symmetric domain $\mathcal{D}$ and Heegner divisors on them. In Sect. 3, we recall the Weil representation and calculate its character. In Sect. 4, we shall show that there exist holomorphic automorphic forms on the complex ball $\mathcal{B}$ of weight 45, 5 with known zeros. These forms will be used to determine the zeros of a member of a five-dimensional linear system of automorphic forms on $\mathcal{B}$. In Sect. 5, we construct a five-dimensional linear system of automorphic forms and show that this linear system gives the dual map of the Segre cubic.

2 The Segre Cubic Threefold

In this note we consider the variety X called the *Segre cubic* which is defined by

$$X : \sum_{i=1}^{6} x_i = 0, \quad \sum_{i=1}^{6} x_i^3 = 0$$

in $\mathbf{P}^5$. Obviously the symmetric group $\mathfrak{S}_6$ of degree 6 acts on X projectively. X has ten nodes which are $\mathfrak{S}_6$-orbits of $(1 : 1 : 1 : -1 : -1 : -1)$. A linear section of X given by $x_i + x_j = 0$ is the union of three projective planes given by

$$x_i + x_j = 0, \quad x_k + x_l = 0, \quad x_m + x_n = 0;$$

$$x_i + x_j = 0, \ x_k + x_m = 0, \ x_l + x_n = 0;$$
$$x_i + x_j = 0, \ x_k + x_n = 0, \ x_l + x_m = 0,$$

respectively, where $\{i, j, k, l, m, n\} = \{1, 2, 3, 4, 5, 6\}$. On the other hand a linear section of X given by $x_i - x_j = 0$ is an irreducible cubic surface containing four nodes of X. This irreducible cubic surface with four nodes is projectively unique and is called the *Cayley cubic surface*. Thus we have 15 Cayley cubics and 15 planes on the Segre cubic X. It is known that the dual of X is a quartic threefold Y in $\mathbf{P}^4$, called the *Igusa quartic* [12]. The dual map $d : X \to Y$ is defined on X except at the ten nodes and is birational. It is given by a linear system of quadrics through ten nodes. For more details of these facts, we refer the reader to [10], Chap. 3.

3 A Complex Ball Quotient

It is known that the Segre cubic X is isomorphic to the Satake–Baily–Borel compactification of an arithmetic quotient of a three-dimensional complex ball by a certain arithmetic subgroup ([11], Theorem 1; [10], Chap. 3, 3.2.3). In this section we recall this fact.

3.1 A Complex Ball

Let

$$\mathcal{E} = \mathbf{Z}[\omega], \ \omega = \frac{-1 + \sqrt{-3}}{2}$$

be the ring of Eisenstein integers. Consider the Hermitian lattice

$$\Lambda = \mathcal{E}^{1,3} = \mathcal{E} \oplus \mathcal{E} \oplus \mathcal{E} \oplus \mathcal{E}$$

with the Hermitian form

$$h(x, y) = x_0 \bar{y}_0 - x_1 \bar{y}_1 - x_2 \bar{y}_2 - x_3 \bar{y}_3.$$

We denote by $\mathcal{B}$ the complex ball of dimension 3 defined by

$$\mathcal{B} = \{x \in \mathbf{P}(\Lambda \otimes_{\mathcal{E}} \mathbf{C}) \ : \ h(x, x) > 0\}.$$

Let $\Gamma = \mathrm{Aut}(\Lambda)$ be an arithmetic subgroup of the unitary group $U(3, 1; \mathbf{Q}(\sqrt{-3}))$ with respect to the Hermitian form $h(\, , \,)$. Obviously Γ naturally acts on $\mathcal{B}$. Under the isomorphism

$$\mathcal{E} / \sqrt{-3}\mathcal{E} \simeq \mathbf{F}_3,$$

the Hermitian form h induces a quadratic form q on $\Lambda/\sqrt{-3}\Lambda$ over $\mathbf{F}_3$. We define a subgroup $\Gamma(\sqrt{-3})$ of Γ by

$$\Gamma(\sqrt{-3}) = \mathrm{Ker}\{\Gamma \longrightarrow O(q)\}.$$

Let L be the real lattice corresponding to Λ with the symmetric bilinear form

$$\langle x, y \rangle = h(x, y) + h(y, x).$$

Then $L \cong A_2 \oplus A_2(-1)^3$ where A_2 is a root lattice of rank 2, that is, a positive definite lattice with Gram matrix $\begin{pmatrix} 2 & -1 \\ -1 & 2 \end{pmatrix}$ and $A_2(-1)$ is a negative definite lattice with Gram matrix $\begin{pmatrix} -2 & 1 \\ 1 & -2 \end{pmatrix}$. The action of ω on Λ induces an isometry ι of L of order 3 without non-zero fixed points. We denote by L^* the dual of L: $L^* = \mathrm{Hom}(L, \mathbf{Z})$. Note that $A_2^*/A_2 \cong \mathbf{F}_3$. Let $A_L = L^*/L \cong (\mathbf{F}_3)^4$ and let $q_L : A_L \to \mathbf{Q}/2\mathbf{Z}$ be the discriminant quadratic form of L defined by $q_L(x + L) = \langle x, x \rangle + 2\mathbf{Z}$. The form q_L coincides with q, up to scale, under the isomorphism

$$\Lambda/\sqrt{-3}\Lambda \cong L^*/L.$$

We denote by $O(L)$ the orthogonal group of L and by $O(q_L)$ the group of automorphisms of A_L preserving q_L. Let $\tilde{O}(L)$ be the kernel of the natural map $O(L) \to O(q_L)$. Then the group Γ is naturally isomorphic to the subgroup $O(L, \iota)$ of $O(L)$ consisting of isometries commuting with ι. Under this isomorphism the subgroup $\Gamma(\sqrt{-3})$ corresponds to $O(L, \iota) \cap \tilde{O}(L)$.

Conversely, first consider the lattice L with an automorphism ι of order 3 without non-zero fixed points. Then we can consider L as a $\mathbf{Z}[\omega]$-module by the action $\omega \cdot x = \iota(x)$. The Hermitian form h is given by

$$h(x, y) = \frac{1}{2}\{\frac{\sqrt{-3}}{3}\langle 2\iota(x) + x, y \rangle + \langle x, y \rangle\}.$$

Define

$$\mathcal{D} = \{v \in \mathbf{P}(L \otimes \mathbf{C}) : \langle v, v \rangle = 0, \langle v, \bar{v} \rangle > 0\}.$$

Then $\mathcal{D}$ is a disjoint union of two copies of a bounded symmetric domain of type IV and of dimension 6. Consider the action of ι on $L \otimes \mathbf{C}$. Since ι is defined over $\mathbf{Z}$ and has no non-zero fixed vectors, the eigenspaces $V_\omega, V_{\bar{\omega}}$ with the eigenvalues $\omega, \bar{\omega}$ respectively are isomorphic to $\mathbf{C}^4$. Moreover the restriction $\langle v, \bar{v} \rangle$ to V_ω is a Hermitian form of signature $(1, 3)$. Note that $\langle v, v \rangle = 0$ for any vector v in V_ω or $V_{\bar{\omega}}$ because $\langle v, v \rangle = \langle \iota(v), \iota(v) \rangle = \langle \omega v, \omega v \rangle$ or $\langle v, v \rangle = \langle \bar{\omega} v, \bar{\omega} v \rangle$. Let $i : \Lambda \to \Lambda \otimes_{\mathbf{Z}} \mathbf{C} = L \otimes \mathbf{C}$ be the inclusion map and let $p : L \otimes \mathbf{C} \to V_\omega$ be the projection. For any $\xi \in L$, write $\xi = \xi_\omega + \xi_{\bar{\omega}}$ as an element in $V_\omega \oplus V_{\bar{\omega}}$. Then we can easily see that

$$h(\xi, \xi) = \langle \xi_\omega, \bar{\xi}_\omega \rangle.$$

Hence the map $p \circ i : \Lambda \to V_\omega$ is an isometry which induces an isomorphism from $\mathcal{B}$ to the subdomain

$$\mathcal{D} \cap \mathbf{P}(V_\omega) = \{v \in \mathbf{P}(V_\omega) \ : \ \langle v, \bar{v} \rangle > 0\}$$

of $\mathcal{D}$. Thus the complex ball $\mathcal{B}$ can be embedded into $\mathcal{D}$.

3.2 Roots and Reflections

Following to [1], we recall roots and reflections of the Hermitian lattice Λ. A vector $a \in \Lambda$ is called a *short root* (resp. *long root*) if $h(a, a) = -1$ (resp. $h(a, a) = -2$). For a short root or long root a, consider the following isometry $r_{a,\zeta}$ of Λ with respect to h:

$$r_{a,\zeta} : x \to x - (1 - \zeta) \frac{h(a, v)}{h(a, a)} a.$$

If a is a short root and ζ is a primitive third root of unity ω, $r_{a,\omega}$ is an isometry of Λ of order three sending a to ωa. We call $r_{a,\omega}$ a *trireflection*. If a is a short root or long root, and $\zeta = -1$, then $r_{a,-1}$ is a *reflection* in Γ which is an isometry of order two sending a to $-a$.

For a short root a in Λ, denote by r the corresponding (-2)-vector in L. Then the trireflection $r_{a,\omega}$ induces an isometry

$$s_r \circ s_{\iota(r)} : x \to x + \langle x, r \rangle r + \langle x, \iota(r) \rangle r + \langle x, \iota(r) \rangle \iota(r)$$

of L, where $s_r : x \to x + \langle x, r \rangle r$ is the reflection associated to r. On the other hand, $r_{a,-1}$ induces an isometry of L:

$$x \to x + 2 \langle \frac{r + 2\iota(r)}{3}, x \rangle \iota(r) + 2 \langle \frac{2r + \iota(r)}{3}, x \rangle r.$$

For a long root $a \in \Lambda$, denote by r the corresponding (-4)-vector in L. Then $r_{a,-1}$ induces an isometry of L:

$$x \to x + \langle \frac{r + 2\iota(r)}{3}, x \rangle \iota(r) + \langle \frac{2r + \iota(r)}{3}, x \rangle r.$$

For $a \in \Lambda$, we denote by $\bar{a}$ the image of a in $\Lambda / \sqrt{-3}\Lambda$. We call the images of short roots (resp. long roots) in $\Lambda / \sqrt{-3}\Lambda$ the short roots (resp. long roots), too. We also denote by $\bar{r}_{a,\zeta}$ the isometry on $\Lambda / \sqrt{-3}\Lambda$ induced by $r_{a,\zeta}$. Note that if a is a short root, then $r_{a,\omega}$ is contained in $\Gamma(\sqrt{-3})$, that is, $\bar{r}_{a,\omega}$ acts trivially on $\Lambda / \sqrt{-3}\Lambda$. On the other hand, $\bar{r}_{a,-1}$ acts on $\Lambda / \sqrt{-3}\Lambda$ as a reflection associated to $\bar{a}$.

Lemma 1. (1) *The group Γ acts transitively on the primitive isotropic vectors, on the short roots and on the long roots, respectively.*

(2) *Let a_1, a_2 be two isotropic vectors, or two short roots, or two long roots. Then a_1 and a_2 are equivalent under $\Gamma(\sqrt{-3})$ if and only if their images in $\Lambda/\sqrt{-3}\Lambda$ coincide.*

(3) *The number of non-zero isotropic vectors, short roots and long roots in $\Lambda/\sqrt{-3}\Lambda$ is 20, 30 and 30 respectively.*

(4) *The map $\Gamma \to O(q)$ is surjective and $\Gamma/\Gamma(\sqrt{-3}) \simeq O(q) \simeq \mathfrak{S}_6 \times \mathbf{Z}/2\mathbf{Z}$.*

Proof. In the case of the Hermitian lattice $\mathcal{E}^{1,4}$, Allcock, Carlson and Toledo proved the same assertion (1) ([2], Theorems 7.21, 11.13), and Allcock and Freitag ([1], Proposition 2.1) proved the assertions (2), (3). The same proof works in our case $\mathcal{E}^{1,3}$. The last assertion is well known. For example, see [6], page 4.   □

3.3 Ball Quotient and Heegner Divisors

We denote a vector $\alpha \in A_L = (\mathbf{F}_3)^4$ by $\alpha = (x_1, x_2, x_3, x_4)$ where $x_i \in \mathbf{F}_3$ is in the i-th factor of $L^*/L = A_2^*/A_2 \oplus (A_2(-1)^*/A_2(-1))^{\oplus 3}$. Then an elementary calculation shows the following:

Lemma 2. *The group $O(q_L)$ has four orbits O_i ($i = 1, \ldots, 4$) on A_L :*

$$O_1 = \{0\}, \quad O_2 = \{\alpha \in A_L \ : \ \alpha \neq 0, q_L(\alpha) = 0\},$$
$$O_3 = \{\alpha \in A_L \ : \ q_L(\alpha) = -4/3\}, \quad O_4 = \{\alpha \in A_L \ : \ q_L(\alpha) = -2/3\}.$$

We call $\alpha \in A_L$ a vector of type (00), (0), (1) or (2) if $\alpha \in O_1, O_2, O_3$ or O_4 respectively. We can easily see that $|O_1| = 1, |O_2| = 20, |O_3| = 30, |O_4| = 30$, and

$$O_2 = \{(\pm1, \pm1, 0, 0), (\pm1, 0, \pm1, 0), (\pm1, 0, 0, \pm1), (0, \pm1, \pm1, \pm1)\},$$
$$O_3 = \{(\pm1, 0, 0, 0), (\pm1, \pm1, \pm1, \pm1), (0, \pm1, \pm1, 0), (0, \pm1, 0, \pm1), (0, 0, \pm1, \pm1)\},$$
$$O_4 = \{(0, \pm1, 0, 0), (0, 0, \pm1, 0), (0, 0, 0, \pm1), (\pm1, \pm1, \pm1, 0), (\pm1, \pm1, 0, \pm1),$$
$$(\pm1, 0, \pm1, \pm1)\}.$$

Lemma 3. *Under the canonical isomorphism $\Lambda/\sqrt{-3}\Lambda \cong A_L = L^*/L$, the set of short roots (resp. long roots) in $\Lambda/\sqrt{-3}\Lambda$ corresponds to the set of vectors of norm $-2/3$ (resp. vectors of norm $-4/3$) in A_L. Also the set of isotropic vectors in $\Lambda/\sqrt{-3}\Lambda$ corresponds to the set of isotropic vectors in A_L.*

Let α be a non-isotropic vector in A_L. For a given $n \in \mathbf{Z}$, $n < 0$, we consider a Heegner divisor $\mathcal{D}_{\alpha,n}$ which is the union of the orthogonal complements $r^{\perp}$ in $\mathcal{D}$, where r varies over the vectors in L^* satisfying $\langle r, r \rangle = n$ and $r \bmod L = \alpha$. Obviously $r^{\perp}$ is a bounded symmetric domain of type IV and of dimension 5. In the case $q_L(\alpha) = -2/3$ (resp. $q_L(\alpha) = -4/3$) and $n = -2/3$ (resp. $n = -4/3$), we denote $\mathcal{D}_{\alpha,-2/3}$ (resp. $\mathcal{D}_{\alpha,-4/3}$) by $\mathcal{D}_{\alpha}$ for simplicity and call it a $(-2/3)$-*Heegner divisor* (resp. $(-4/3)$-*Heegner divisor*).

Proposition 1. *The Segre cubic X is isomorphic to the Satake–Baily–Borel compactification $\bar{\mathcal{B}}/\Gamma(\sqrt{-3})$ of the quotient $\mathcal{B}/\Gamma(\sqrt{-3})$ which is, set theoretically, the union of $\mathcal{B}/\Gamma(\sqrt{-3})$ and ten cusps corresponding to ten non-zero isotropic vectors in $A_L/\{\pm1\}$. These ten cusps correspond to ten nodes of the Segre cubic X.*

Proof. The assertion follows from [10], Sect. 3.2.   □

Also, $\bar{\mathcal{B}}/\Gamma(\sqrt{-3})$ contains some divisors called *Heegner divisors*. Let α be a short root in $\Lambda/\sqrt{-3}\Lambda$. Let a be a short root in Λ with a mod $\sqrt{-3}\Lambda = \alpha$. We denote by $a^\perp$ the orthogonal complement of a in $\mathcal{B}$ which is a complex ball of dimension 2. Let

$$\mathcal{H}_\alpha = \bigcup_a a^\perp$$

where a ranges over the set of all short roots satisfying a mod $\sqrt{-3}\Lambda = \alpha$. The image of $\mathcal{H}_\alpha$ in $\bar{\mathcal{B}}/\Gamma(\sqrt{-3})$ is denoted by $\bar{\mathcal{H}}_\alpha$ and is called a (-1)-*Heegner divisor*. There exist 15 (-1)-Heegner divisors $\bar{\mathcal{H}}_\alpha$ corresponding to 15 short roots $\alpha \in (\Lambda/\sqrt{-3}\Lambda)/\{\pm 1\}$, $q(\alpha) = -1$.

Similarly we can define 15 (-2)-Heegner divisors $\bar{\mathcal{H}}_\alpha$ corresponding to 15 long roots $\alpha \in (\Lambda/\sqrt{-3}\Lambda)/\{\pm 1\}$, $q(\alpha) = -2$.

Finally we compare the Heegner divisors in $\mathcal{D}$ and in $\mathcal{B}$. Let $r \in L^*$ be a $(-2/3)$- or $(-4/3)$-vector. Then both $\iota(r)$ and $\iota^2(r)$ are $(-2/3)$- or $(-4/3)$-vectors and r mod $L = \iota(r)$ mod $L = \iota^2(r)$ mod L. Note that $r^\perp$, $\iota(r)^\perp$ and $\iota^2(r)^\perp$ in $\mathcal{D}$ are different, but their restrictions to $\mathcal{B}$ are the same. Thus we have

Lemma 4.

$$\mathcal{D}_\alpha \cap \mathcal{B} = 3\mathcal{H}_\alpha$$

where we identify $(-2/3)$- *(resp.* $(-4/3)$-*) vectors in A_L and short roots (resp. long roots) in $\Lambda/\sqrt{-3}\Lambda$.*

3.4 Interpretation via K3 Surfaces

The complex ball quotient $\mathcal{B}/\Gamma(\sqrt{-3})$ can be considered as the moduli space of lattice polarized $K3$ surfaces. The following is essentially given in [7]. Let $N = U \oplus E_6(-1) \oplus A_2(-1)^{\oplus 3}$. Then N can be primitively embedded into the $K3$ lattice $M = U^{\oplus 3} \oplus E_8(-1)^{\oplus 2}$ whose orthogonal complement is isomorphic to $L = A_2 \oplus A_2(-1)^{\oplus 3}$. Here U is an even lattice with Gram matrix $\begin{pmatrix} 0 & 1 \\ 1 & 0 \end{pmatrix}$ and A_m, E_k are positive definite root lattices defined by the Cartan matrix of type A_m, E_k, and for a lattice $(K, \langle, \rangle)$ we denote by $K(-1)$ the lattice $(K, -\langle, \rangle)$. In the following we consider N and L as sublattices of M. The isometry ι of L of order 3 acts trivially on L^*/L and hence it can be extended to an isometry $\tilde{\iota}$ of M acting trivially on N. Let $\omega \in \mathcal{B}$ with the property $\omega^\perp \cap M = N$. Let S be a $K3$ surface and let $\alpha_S : H^2(S, \mathbf{Z}) \to M$ be an isometry satisfying $(\alpha_S \otimes \mathbf{C})(\omega_S) = \omega$, where ω_S is a holomorphic 2-form on S. By definition the Picard lattice of S is isomorphic to N. Note that the isometry $\tilde{\iota}$ preserves ω_S and acts trivially on the Picard lattice. It now follows from the Torelli type theorem for $K3$ surface [14] that $\tilde{\iota}$ can be represented by an automorphism σ on S of order 3. Thus an open set of $\mathcal{B}/\Gamma(\sqrt{-3})$ is the moduli space of such pairs (S, σ) of $K3$ surfaces S with an automorphism σ of order 3.

In the following, we shall show that S is canonically obtained from six points on $\mathbf{P}^1$. Let $Q = \mathbf{P}^1 \times \mathbf{P}^1$. Let $(u_0 : u_1), (v_0 : v_1)$ be homogeneous coordinates of the first and the second factor of Q. Let $p_1, \ldots, p_6$ be six distinct points on $\mathbf{P}^1$. Consider the divisors on Q defined by

$$L_i = \mathbf{P}^1 \times \{p_i\} \,(1 \le i \le 6), \; D_0 = \{0\} \times \mathbf{P}^1, \; D_1 = \{1\} \times \mathbf{P}^1, \; D_\infty = \{\infty\} \times \mathbf{P}^1.$$

Let $\tilde{Q} \to Q$ be the blow-ups of the 18 points on Q which are the intersection of $L_1, \ldots, L_6$ and D_0, D_1, D_∞. We denote by the $\tilde{L}_1, \ldots, \tilde{L}_6, \tilde{D}_0, \tilde{D}_1$ or $\tilde{D}_\infty$ the strict transform of $L_1, \ldots, L_6, D_0, D_1$ or D_∞ respectively. Let $\pi : \tilde{X} \to \tilde{Q}$ be the triple covering of $\tilde{Q}$ branched along $\tilde{L}_1 + \cdots + \tilde{L}_6 + \tilde{D}_0 + \tilde{D}_1 + \tilde{D}_\infty$. Then $\pi^{-1}(\tilde{L}_i)$ is a (-1)-curve. Let $\tilde{X} \to S$ be the contraction of $\pi^{-1}(\tilde{L}_i)$ to the points q_i. We can easily see that S is a $K3$ surface. The projection from Q to the second factor $\mathbf{P}^1$ induces an elliptic fibration $p : S \to \mathbf{P}^1$ which has six singular fibers of type IV in the notation of Kodaira and three sections. Here three components of the singular fiber of type IV over p_i correspond to three exceptional curves over the three intersection points of L_i and D_0, D_1, D_∞ and three sections correspond to D_0, D_1, D_∞. The classes of components of fibers and a section generate a sublattice of the Picard lattice Pic(S) isomorphic to $U \oplus A_2(-1)^{\oplus 6}$. By adding other two sections, we have a sublattice in Pic(S) isomorphic to $N = U \oplus E_6(-1) \oplus A_2(-1)^{\oplus 3}$. The covering transformation of $\tilde{S} \to \tilde{Q}$ induces an automorphism σ of S of order 3. Note that the set of fixed points of σ consists of six isolated points $q_1, \ldots, q_6$ and three sections. Since σ has a fixed curve as its fixed points, $\sigma^*(\omega_S) = \zeta_3 \omega_S$ where ω_S is a non-zero holomorphic 2-form on S and ζ_3 is a primitive cube root of unity. Thus we have a pair (S, σ) of a $K3$ surface and an automorphism of order 3. This $K3$ surface appears as a degeneration of $K3$ surfaces associated to a smooth cubic surface given in [7].

Next we consider the case that two points among the 6 points coincide. In this case, similarly, we have an elliptic $K3$ surface S' with one singular fiber of type VI*, four singular fibers of type VI and three sections. The Picard lattice of S' is isomorphic to $U \oplus E_6(-1)^2 \oplus A_2(-1)$ and its transcendental lattice is isomorphic to $A_2 \oplus A_2(-1)^{\oplus 2}$. Thus the period domain of $K3$ surfaces S' is a subdomain of $\mathcal{B}$ the orthogonal complement of $A_2(-1)$, that is, a (-1)-Heegner divisor. Thus we have

Proposition 2. 15 (-1)-*Heegner divisors bijectively correspond to* 15 *planes on the Segre cubic* X.

Proof. It is known that 15 planes on the Segre cubic correspond to the moduli of 6 points on the projective line in which two points coincide ([10], Proposition 3.2.7). Hence we have the assertion. □

Lemma 5. *Let* $\bar{\mathcal{H}}_\alpha, \bar{\mathcal{H}}_\beta$ *be* (-1)-*Heegner divisors. Then* $\bar{\mathcal{H}}_\alpha$ *and* $\bar{\mathcal{H}}_\beta$ *meet along a line if and only if* α *and* β *are orthogonal.*

Proof. Note that if α and β are orthogonal, then $\bar{\mathcal{H}}_\alpha \cap \bar{\mathcal{H}}_\beta$ is 1-dimensional, and otherwise $\bar{\mathcal{H}}_\alpha$ and $\bar{\mathcal{H}}_\beta$ meet only at cusps. Thus the assertion follows from the incidence relation between 15 planes on the Segre cubic. □

We shall show that 15 (-2)-Heegner divisors correspond to 15 Cayley cubics on the Segre cubic X (see Lemma 11).

4 Weil Representation

In this section, we recall a representation of $SL(2, \mathbf{Z})$ on the group ring $\mathbf{C}[A_L]$ called the Weil representation. In the following Table 1, for each vector $u \in A_L$ of given type, m_j is the number of vectors v of the same type with $\langle u, v \rangle = 2j/3$.

Table 1: A relation of elements in A_L

u	00	00	00	00	0	00	0	1	11	1	2	22	2
v	00	0	1	2	00	0 1	2	00	0 1	2	00	0 1	2
m_0	1	20	30	30	1	2 12	12	1	8 12	6	1	8 6	12
m_1	0	0	0	0	0	0 9	9	9	0 6	9	12 0	6 12	9
m_2	0	0	0	0	0	0 9	9	9	0 6	9	12 0	6 12	9

Let $T = \begin{pmatrix} 1 & 1 \\ 0 & 1 \end{pmatrix}$, $S = \begin{pmatrix} 0 & -1 \\ 1 & 0 \end{pmatrix}$ be generators of $SL(2, \mathbf{Z})$. Let ρ_L be the Weil representation of $SL(2, \mathbf{Z})$ on $\mathbf{C}[A_L]$ defined by:

$$\rho_L(T)(e_\alpha) = exp(\langle \alpha, \alpha \rangle / 2)e_\alpha, \quad \rho_L(S)(e_\alpha) = \frac{-1}{\sqrt{|A_L|}} \sum_\delta exp(-\langle \delta, \alpha \rangle)e_\delta.$$

Note that the action of the group $O(q_L)$ on $\mathbf{C}[A_L]$ commutes with ρ_L. The action ρ_L factorizes the action of $SL(2, \mathbf{Z}/3\mathbf{Z})$ which is denoted by the same symbol ρ_L. The conjugacy classes of $SL(2, \mathbf{Z}/3\mathbf{Z})$ consist of $\pm E, S, \pm ST, \pm ST^2$. Let χ_i ($1 \le i \le 7$) be the characters of irreducible representations of $SL(2, \mathbf{Z}/3\mathbf{Z})$. The following Table 2 is the character table of $SL(2, \mathbf{Z}/3\mathbf{Z})$. Here $\omega = \frac{-1+\sqrt{-3}}{2}$ and the last line means the number of elements in a given conjugate class.

Table 2: The character table of $SL(2, \mathbf{Z}/3\mathbf{Z})$

	E	$-E$	S	ST^2	$-ST^2$	ST	$-ST$
χ_1	1	1	1	1	1	1	1
χ_2	3	3	-1	0	0	0	0
χ_3	1	1	1	ω^2	ω^2	ω	ω
χ_4	1	1	1	ω	ω	ω^2	ω^2
χ_5	2	-2	0	$-\omega$	ω	ω^2	$-\omega^2$
χ_6	2	-2	0	-1	1	1	-1
χ_7	2	-2	0	$-\omega^2$	ω^2	ω	$-\omega$
	1	1	6	4	4	4	4

Lemma 6. *Let χ be the character of the representation ρ_L of $SL(2, \mathbf{Z}/3\mathbf{Z})$ on $\mathbf{C}[A_L]$. Let $\chi = \sum_i m_i \chi_i$ be the decomposition into irreducible characters. Then $m_1 = 1, m_2 = 10, m_3 = m_4 = 5, m_5 = 5, m_6 = 10, m_7 = 5$.*

Proof. By definition of ρ_L and the Table 1, we can easily see that trace$(E) = 3^4$, trace$(-E) = 1$, trace$(S) = 1$, trace$(ST^2) = -9$, trace$(-ST^2) = 1$, trace$(ST) = 1$, trace$(-ST) = -9$. The assertion now follows from Table 2. □

Definition 1. Let W be the five-dimensional subspace of $\mathbf{C}[A_L]$ which is the direct sum of irreducible representations of SL(2, $\mathbf{Z}/3\mathbf{Z}$) with the character χ_3 in Lemma 6.

Note that O(q_L) naturally acts on W because the actions of O(q_L) and ρ_L commute. In Sect. 6, we associate a five-dimensional space of automorphic forms on $\mathcal{B}$ to W. We remark that there is an another five-dimensional subspace in $\mathbf{C}[A_L]$ which is a direct sum of irreducible representations of SL(2, $\mathbf{Z}/3\mathbf{Z}$) with the character χ_4. The author does not know whether this subspace corresponds to an interesting linear system of automorphic forms on $\mathcal{B}$.

5 Borcherds Products

In this section and the next we shall show the existence of some automorphic forms, called *Borcherds products* and *Gritsenko–Borcherds liftings*, on the bounded symmetric domain $\mathcal{D}$ of type IV. By restricting such forms to the complex ball $\mathcal{B}(\subset \mathcal{D})$, we obtain automorphic forms on $\mathcal{B}$. First we start by introducing the notion of vector-valued modular forms. Let

$$\rho : \mathrm{SL}(2, \mathbf{Z}) \to \mathrm{GL}(V)$$

be a finite-dimensional representation on a complex vector space V. We assume that ρ factors through SL(2, $\mathbf{Z}/N\mathbf{Z}$) for a suitable natural number N. A holomorphic map

$$f : H^+ \to V$$

is called a *vector-valued modular form of weight k and of type ρ* if

$$f(M\tau) = \rho(M)(c\tau + d)^k f(\tau)$$

for any $M = \begin{pmatrix} a & b \\ c & d \end{pmatrix} \in \mathrm{SL}(2, \mathbf{Z})$ and f is meromorphic at cusps. Borcherds products are automorphic forms on $\mathcal{D}$ whose zeros and poles lie on Heegner divisors. Borcherds [4] gave a systematic method to construct such automorphic forms associated with suitable vector-valued modular forms of type ρ_L. Here we employ Borcherds and Freitag's observation in [5, 8] to show the existence of such forms. To do this we introduce the *obstruction space* consisting of all vector-valued modular forms $\{f_\alpha\}_{\alpha \in A_L}$ of weight $(2+6)/2 = 4$ and with respect to the dual representation ρ_L^* of ρ_L:

$$f_\alpha(\tau + 1) = e^{-\pi\sqrt{-1}\,\langle\alpha,\alpha\rangle} f_\alpha(\tau), \quad f_\alpha(-1/\tau) = -\frac{\tau^4}{9} \sum_\beta e^{2\pi\sqrt{-1}\,\langle\alpha,\beta\rangle} f_\beta(\tau).$$

Theorem 1. (Borcherds [5], Freitag [8], Theorem 5.2) *A linear combination*

$$\sum_{\alpha \in A_L, n<0} c_{\alpha,n} \mathcal{D}_{\alpha,n}, \ c_{\alpha,n} \in \mathbf{Z}$$

of Heegner divisors is the divisor of an automorphic form on $\mathcal{D}$ of weight k if for every cusp form

$$f = \{f_\alpha(\tau)\}_{\alpha \in A_L}, \ f_\alpha(\tau) = \sum_{n \in \mathbf{Q}} a_{\alpha,n} e^{2\pi \sqrt{-1} n\tau}$$

in the obstruction space, the relation

$$\sum_{\alpha \in A_L, n<0} a_{\alpha,-n/2} c_{\alpha,n} = 0$$

holds. In this case the weight k is given by

$$k = \sum_{\alpha \in A_L, n \in \mathbf{Z}} b_{\alpha,n/2} c_{\alpha,-n}$$

where $b_{\alpha,n}$ are the Fourier coefficients of the Eisenstein series in the obstruction space with the constant term $b_{0,0} = -1/2$ and $b_{\alpha,0} = 0$ for $\alpha \neq 0$.

In the following we shall study the divisors $\sum_{\alpha \in A_L, n<0} c_{\alpha,n} \mathcal{D}_{\alpha,n}$ where $c_{\alpha,n}$ depends only on the type of α. Recall that $O(q_L)$ has four orbits O_i $(i = 1, \ldots, 4)$ (see Lemma 2). We denote the vector-valued modular form $(f_\alpha)_{\alpha \in A_L}$ of type ρ_L^* by

$$(f_{00}, \ f_0, \ f_1, \ f_2)$$

where each f_t is the sum of the f_α as α varies over the elements of A_L of type t. Then ρ_L^* induces a four-dimensional representation $\bar{\rho}^*$ of $SL(2, \mathbf{Z})$ on $V = \oplus_t \mathbf{C} f_t$. A calculation shows that the generators S, T of $SL(2, \mathbf{Z})$ with respect to this basis is given by

$$\bar{\rho}^*(T) = \begin{pmatrix} 1 & 0 & 0 & 0 \\ 0 & 1 & 0 & 0 \\ 0 & 0 & \omega^2 & 0 \\ 0 & 0 & 0 & \omega \end{pmatrix}, \quad \bar{\rho}^*(S) = \frac{-1}{9} \begin{pmatrix} 1 & 1 & 1 & 1 \\ 20 & -7 & 2 & 2 \\ 30 & 3 & 3 & -6 \\ 30 & 3 & -6 & 3 \end{pmatrix}.$$

Lemma 7. *The dimension of the space of modular forms of weight $4 = (2 + 6)/2$ and of type $\bar{\rho}^*$ is 2. The dimension of the space of Eisenstein forms of weight 4 and of type $\bar{\rho}^*$ is also 2.*

Proof. The dimension is given by

$$d + dk/12 - \alpha(e^{\pi \sqrt{-1}k/2}\bar{\rho}^*(S)) - \alpha((e^{\pi \sqrt{-1}k/3}\bar{\rho}^*(ST))^{-1}) - \alpha(\bar{\rho}^*(T))$$

([5], Sect. 4, [8], Proposition 2.1). Here $k = 4$ is the weight,

$$d = \dim\{x \in V : \bar{\rho}^*(-E)x = (-1)^k x\} = 4$$

and

$$\alpha(A) = \sum_\lambda \alpha$$

where λ runs through all eigenvalues of A and $\lambda = e^{2\pi\sqrt{-1}\alpha}$, $0 \le \alpha < 1$. A direct calculation shows that

$$\alpha(e^{\pi\sqrt{-1}k/2}\bar{\rho}^*(S)) = 1, \ \alpha((e^{\pi\sqrt{-1}k/3}\bar{\rho}^*(ST))^{-1}) = 4/3 \text{ and } \alpha(\bar{\rho}^*(T)) = 1.$$

On the other hand, the space of Eisenstein series is isomorphic to the subspace of V given by

$$\bar{\rho}^*(T)(x) = x, \ \bar{\rho}^*(-E)(x) = (-1)^k x$$

(see Remark 2.2 in [8]). Thus we have the assertion.   □

Next we shall calculate a basis of Eisenstein forms of weight 4 and of type $\bar{\rho}^*$. Let

$$E_1 = G_4(\tau, 0, 1; 3), \ E_2 = G_4(\tau, 1, 0; 3), \ E_3 = G_4(\tau, 1, 1; 3), \ E_4 = G_4(\tau, 1, 2; 3)$$

be Eisenstein series of weight 4 and level 3 (see [8]). Then the action of S, T is as follows:

$$T : E_2 \to E_3 \to E_4 \to E_2,$$

T fixes E_1, and S switches E_1 and E_2, E_3 and E_4 respectively.

Now we can easily see that a basis of Eisenstein forms of weight 4 and of type $\bar{\rho}^*$ is given by

$$f_{00} = aE_1 + b(E_2 + E_3 + E_4),$$

$$f_0 = (-a - 9b)E_1 + (-3a - 7b)(E_2 + E_3 + E_4),$$

$$f_1 = (-3a + 3b)(E_2 + \omega E_3 + \omega^2 E_4),$$

$$f_2 = (-3a + 3b)(E_2 + \omega^2 E_3 + \omega E_4),$$

where a, b are parameters. The Fourier expansions of E_i are given as follows (see [8]):

$$E_1 = \frac{(2\pi)^4}{2 \cdot 3^6} + c(-3^3 q + \cdots),$$

$$E_2 = c(q^{1/3} + (2^3 + 1)q^{2/3} + 3^3 q + \cdots),$$

$$E_3 = c(\omega q^{1/3} + (2^3 + 1)\omega^2 q^{2/3} + 3^3 q + \cdots),$$

$$E_4 = c(\omega^2 q^{1/3} + (2^3 + 1)\omega q^{2/3} + 3^3 q + \cdots),$$

where $c = \frac{(-2\pi\sqrt{-1})^4}{3^4 3!} = \frac{(2\pi)^4}{2 \cdot 3^5}$. Put $a = -\frac{3^6}{(2\pi)^4}$ and $a = -9b$. Then

$$f_{00} = -1/2 + 2 \cdot 3^3 q + \cdots,$$
$$f_0 = 10 \cdot 3^3 q + \cdots,$$
$$f_1 = 135 q^{2/3} + \cdots,$$
$$f_2 = 15 q^{1/3} + \cdots.$$

It follows from Lemma 7 that there are no non-zero cusp forms in the obstruction space. Hence Theorem 1 implies that

Theorem 2. *There exist automorphic forms on $\mathcal{D}$ of weight* 135, 15 *with some character whose zero divisors are the* $(-4/3)$-*Heegner divisor and the* $(-2/3)$-*Heegner divisor, respectively.*

Note that $(-4/3)$-, $(-2/3)$-Heegner divisors meet the complex ball with multiplicity 3 (Lemma 4). Since $\mathcal{B}$ is simply connected, we can take the cube root of these automorphic forms. Thus we have:

Corollary 1. *There exist automorphic forms* Φ_{45}, Φ_5 *on the complex ball $\mathcal{B}$ of weight* 45, 5 *whose zero divisors are the* (-2)-*Heegner divisor and the* (-1)-*Heegner divisor, respectively.*

6 Gritsenko–Borcherds Liftings

In this section, by applying the theory of liftings [4], we construct a linear system of automorphic forms on $\mathcal{B}$ with respect to $\Gamma(\sqrt{-3})$ which gives a birational map from the Segre cubic to the Igusa quartic.

Let ρ_L be the Weil representation given in Sect. 4. We shall construct a five-dimensional space of vector-valued modular forms of weight 4 and of type ρ_L.

Let W be the five-dimensional subspace of $\mathbf{C}[A_L]$ in Definition 1. Recall that $O(q_L)$ naturally acts on W. First we shall consider the following special vectors v_{α_0} in W (We remark that the following definition of v_{α_0} is similar to the one given in Allcock–Freitag [1] to construct liftings in their case). Let $\alpha_0, \alpha_1, \alpha_2, \alpha_3$ be an orthogonal basis of A_L with $q_L(\alpha_0) = -4/3, q_L(\alpha_1) = q_L(\alpha_2) = q_L(\alpha_3) = -2/3$. If we are given α_0, then such a basis is uniquely determined up to sign. For each such basis, we define a vector $v_{\alpha_0} = (c_\alpha)_{\alpha \in A_L}$ in $\mathbf{C}[A_L]$ as follows:

$$c_\alpha = 1, 0, -1$$

according to

$$\prod_i \langle \alpha, \alpha_i \rangle = 1, 0, -1 \in \mathbf{F}_3.$$

For example, assume $\alpha_0 = (1,0,0,0)$, $\alpha_1 = (0,1,0,0)$, $\alpha_2 = (0,0,1,0)$, $\alpha_3 = (0,0,0,1) \in A_L = (\mathbf{F}_3)^4$. Then $c_\alpha \neq 0$ if and only if $\alpha \in \{(\pm 1, \pm 1, \pm 1, \pm 1)\}$.

Lemma 8. *Let* $v_{\alpha_0} = (c_\alpha)$ *be as above. Then*

$$\rho_L(S)(v_{\alpha_0}) = v_{\alpha_0}, \quad \rho_L(T)(v_{\alpha_0}) = \omega v_{\alpha_0}.$$

Moreover $r_{\alpha_i}(v_{\alpha_0}) = -v_{\alpha_0}$ *for the reflection* r_{α_i} *associated with* α_i.

Proof. It suffices to prove the case $\alpha_0 = (1,0,0,0), \alpha_1 = (0,1,0,0), \alpha_2 = (0,0,1,0),$ $\alpha_3 = (0,0,0,1)$. Let $M = \{(\pm 1, \pm 1, \pm 1, \pm 1)\}$. Then $v_{\alpha_0} = \sum_{\alpha \in M} c_\alpha e_\alpha$. If $c_\alpha \neq 0$, then $q_L(\alpha) = -4/3$. Hence $\rho_L(T)(v_{\alpha_0}) = \omega v_{\alpha_0}$. Next consider

$$\rho_L(S)v_{\alpha_0} = -\frac{1}{9} \sum_{\beta \in A_L} \left(\sum_{\alpha \in M} c_\alpha e^{-2\pi \sqrt{-1}\langle \alpha, \beta \rangle} \right) e_\beta.$$

A direct calculation shows that the coefficient

$$\sum_{\alpha \in M} c_\alpha e^{-2\pi \sqrt{-1}\langle \alpha, \beta \rangle}$$

of e_β is 0 if $\beta \notin M$, 9 if $\beta \in M$, $c_\beta = -1$, and -9 if $\beta \in M$, $c_\beta = 1$. The last assertion follows from the definition of v_{α_0}. □

It follows from Lemma 8 that v_{α_0} is contained in W. Thus we have fifteen elements v_{α_0} in W where α_0 is fifteen $(-4/3)$-vectors in $A_L/\{\pm 1\}$ (see Lemma 2).

Lemma 9. *As a* $O(q_L)$ *module,* W *is irreducible.*

Proof. If V is an irreducible representation of $\mathfrak{S}_6$ and $\dim V \geq 2$, then $\dim V \geq 5$. Hence it suffices to see that there are no 1-dimensional subspaces invariant under the action of $\mathfrak{S}_6$. If such a 1-dimensional subspace exists, then all vectors in W are invariant under the action of $\mathfrak{S}_6$. However, any special vector v_{α_0} as above is not invariant under the action of $\mathfrak{S}_6$. This is a contradiction. □

Let $\eta(\tau)$ be the Dedekind eta function. Then

$$\eta(\tau + 1)^8 = \omega \cdot \eta(\tau)^8,$$

$$\eta(-1/\tau)^8 = \tau^4 \cdot \eta(\tau).$$

Therefore, for $v \in W$, $\eta(\tau)^8 \cdot v = (\eta(\tau)^8 \cdot c_\alpha)_{\alpha \in A_L}$ is a vector-valued modular form of weight 4 and of type ρ_L. By applying the Gritsenko–Borcherds lifting ([4], Theorem 14.3), we have

Lemma 10. *There is a five-dimensional space of holomorphic automorphic forms of weight 6 on* $\mathcal{D}$ *with respect to* $\tilde{O}(L)$ *on which* $O(q_L)$ *acts irreducibly.*

Proof. It suffices to see that the lifting of $\eta(\tau)^8 v$ is non-zero. Then the assertion follows from Schur's lemma. We use Theorem 14.3 in [4]. We first note that $L = A_2 \oplus A_2(-1)^{\oplus 3}$ is isomorphic to $U \oplus U(3) \oplus A_2(-1)^{\oplus 2}$ where U (resp. $U(3)$) is an even lattice with Gram matrix $\begin{pmatrix} 0 & 1 \\ 1 & 0 \end{pmatrix}$ (resp. $\begin{pmatrix} 0 & 3 \\ 3 & 0 \end{pmatrix}$). This follows from

[13], Theorem 1.14.2. Let z, z' be a basis of U with $z^2 = z'^2 = 0, \langle z, z' \rangle = 1$, and let $K = z^\perp / \mathbf{Z}z = U(3) \oplus A_2(-1)^{\oplus 2} \subset L$. Let e, f be a basis of $U(3)$ with $e^2 = f^2 = 0, \langle e, f \rangle = 3$. We consider the Fourier expansion around z. Since $\eta(\tau)^8 = q^{1/3} + \cdots$, the initial term of

$$\eta(\tau)^8 v = \sum_{\alpha \in A_L} e_\alpha \sum_{n \in \mathbf{Q}} c_\alpha(n) e^{2\pi \sqrt{-1} n \tau}$$

is

$$\sum_{\alpha \in A_L, \alpha^2 = 2/3} c_\alpha q^{1/3}.$$

If we take $\lambda = (e + f)/3$, then $\langle \lambda, \lambda \rangle = 2/3 > 0$ and hence λ has positive inner products with all elements in the interior of the Weyl chamber. Also note that $L^*/L = K^*/K$. We choose $v = (c_\alpha) \in V$ satisfying $c_\lambda \neq 0$. Now it follows from [4], Theorem 14.3 that the Fourier coefficient of $e^{2\pi \sqrt{-1} \langle \lambda, Z \rangle}$ in the lifting of $\eta(\tau)^8 v$ is equal to

$$c_\lambda(\lambda^2/2) \cdot e^{2\pi \sqrt{-1} \langle \lambda, z' \rangle} = c_\lambda(1/3) = c_\lambda.$$

Hence the lifting of $\eta(\tau)^8 v$ is non-zero. □

Let α_0 be a $(-4/3)$-vector in A_L and let v_{α_0} be the element in W as above. Let F_{α_0} be the restriction of the Gritsenko–Borcherds lifting of $\eta(\tau)^8 \cdot v_{\alpha_0}$ to the complex ball $\mathcal{B}$. Then

Theorem 3. F_{α_0} *is a holomorphic automorphic form of weight 6 on $\mathcal{B}$ with respect to $\Gamma(\sqrt{-3})$ which vanishes exactly along the (-2)-Heegner divisor $\mathcal{H}_{\alpha_0}$ with multiplicity one and the (-1)-Heegner divisors $\mathcal{H}_{\alpha_1}, \mathcal{H}_{\alpha_2}, \mathcal{H}_{\alpha_3}$ with multiplicity three.*

Proof. First recall that the reflection r_α is induced from the reflection $r_{a,-1}$ of Λ where $a \in \Lambda$ is a short or long root with $a \mod \sqrt{-3}\Lambda = \alpha$. It follows from Lemma 8 and the $O(q_L)$-equivariance of the lifting that F_{α_0} vanishes along $\mathcal{H}_{\alpha_i}$ ($i = 0, 1, 2, 3$). Moreover, since the trireflection $r_{a,\omega}$ associated to a short root a is contained in $\Gamma(\sqrt{-3})$, F_{α_0} vanishes along $\mathcal{H}_{\alpha_i}$ ($i = 1, 2, 3$) with multiplicity 3. Then the product of 15 F_{α_0} has weight 90 and vanishes along (-2)-Heegner divisors with at least multiplicity one and along (-1)-Heegner divisors with at least multiplicity $\frac{3 \cdot 15 \cdot 3}{15} = 9$. On the other hand $\Phi_{45} \cdot \Phi_5^9$ has weight 90 and vanishes along (-2)-Heegner divisors with exactly multiplicity one and along (-1)-Heegner divisors with multiplicity 9 (Corollary 1). Then the ratio $\prod_v F_v / \Phi_{45} \cdot \Phi_5^9$ has weight zero and holomorphic, and hence it is constant by the Koecher principle. □

Since trireflections are contained in $\Gamma(\sqrt{-3})$, the covering $\mathcal{B} \to \mathcal{B}/\Gamma(\sqrt{-3})$ is ramified along $(-2/3)$-Heegner divisors. Hence we have

Theorem 4. *The zero divisor (F_{α_0}) on $\bar{\mathcal{B}}/\Gamma(\sqrt{-3})$ is $\bar{\mathcal{H}}_{\alpha_0} + \bar{\mathcal{H}}_{\alpha_1} + \bar{\mathcal{H}}_{\alpha_2} + \bar{\mathcal{H}}_{\alpha_3}$.*

Lemma 11. *Let $\alpha \in A_L$ with $q_L(\alpha) = -4/3$. Then the Heegner divisor $\bar{\mathcal{H}}_\alpha$ coincides with a Cayley cubic on X.*

Proof. The zero divisor of $\varPhi_{45}$ on $\bar{\mathcal{B}}/\Gamma(\sqrt{-3})$ is the union of $\bar{\mathcal{H}}_\alpha$ where α varies over 15 $(-4/3)$-vectors in $A_L/\{\pm 1\}$. On the other hand, as mentioned as above, the covering $\mathcal{B} \to \mathcal{B}/\Gamma(\sqrt{-3})$ is ramified along (-1)-Heegner divisors. Hence the zero divisors of $\varPhi_5^3$ on $\bar{\mathcal{B}}/\Gamma(\sqrt{-3})$ is the union of $\bar{\mathcal{H}}_\alpha$, where α varies over 15 $(-2/3)$-vectors in $A_L/\{\pm 1\}$. Recall that For a $(-2/3)$-vector α in A_L, the Heegner divisor $\bar{\mathcal{H}}_\alpha$ is a plane on X (Proposition 2). Moreover, if $\alpha_1, \alpha_2, \alpha_3$ are mutually orthogonal $(-2/3)$-vectors, then the union of three planes $\bar{\mathcal{H}}_{\alpha_1} + \bar{\mathcal{H}}_{\alpha_2} + \bar{\mathcal{H}}_{\alpha_3}$ is a linear section of X in $\mathbf{P}^4$ (Lemma 5). By comparing the weights of $\varPhi_{45}$ and $\varPhi_5^3$, we can see that each $\bar{\mathcal{H}}_\alpha$ with $q_L(\alpha) = -4/3$ is also a linear section, that is, a cubic surface. Since $\bar{\mathcal{H}}_\alpha$ with $q_L(\alpha) = -4/3$ contains four cusps, it should be isomorphic to a Cayley cubic on X.   □

Hence we conclude:

Lemma 12. *The divisor* $(F_{\alpha_0}) = \bar{\mathcal{H}}_{\alpha_0} + \bar{\mathcal{H}}_{\alpha_1} + \bar{\mathcal{H}}_{\alpha_2} + \bar{\mathcal{H}}_{\alpha_3}$ *is a quadric section of the Segre cubic* $X \subset \mathbf{P}^4$ *where* $\bar{\mathcal{H}}_{\alpha_0}$ *is a Cayley cubic and* $\bar{\mathcal{H}}_{\alpha_i}$ $(i = 1, 2, 3)$ *are planes.*

For an orthogonal basis $\{\alpha_0, \alpha_1, \alpha_2, \alpha_3\}$ of A_L, we can easily see that any isotropic vector in A_L is perpendicular to α_i for some i (see Lemma 2). Hence the divisor $\bar{\mathcal{H}}_{\alpha_0} + \bar{\mathcal{H}}_{\alpha_1} + \bar{\mathcal{H}}_{\alpha_2} + \bar{\mathcal{H}}_{\alpha_3}$ contains 10 nodes of X. Thus each (F_{α_0}) passes through the ten nodes of X, and hence the five-dimensional linear system of automorphic forms has the ten nodes as base points. The linear system defines a rational map $\varphi : X \to \mathbf{P}^5$.

Theorem 5. *The image of* φ *is the Igusa quartic that is the dual of* X.

Proof. Recall that the dual map from the Segre cubic to the Igusa quartic is given by a linear system of quadrics through the ten nodes ([10], Theorem 3.3.12). The assertion now follows from Lemma 12.   □

Acknowledgements The author thanks the referee for his useful suggestions. The author was supported in part by JSPS Grant-in-Aid (S), No 22224001, No 19104001.

References

1. D. Allcock, E. Freitag, Cubic surfaces and Borcherds products. Comm. Math. Helv. **77**, 270–296 (2002)
2. D. Allcock, J.A. Carlson, D. Toledo, The complex hyperbolic geometry of the moduli space of cubic surfaces. J. Algebr. Geom. **11**, 659–724 (2002)
3. H.F. Baker, *Principles of Geometry*, vol. IV (Cambridge University Press, Cambridge, 1925)
4. R. Borcherds, Automorphic forms with singularities on Grassmannians. Invent. Math. **132**, 491–562 (1998)
5. R. Borcherds, The Gross-Kohnen-Zagier theorem in higher dimensions. Duke Math. J. **97**, 219–233 (1999)
6. J.H. Conway et al., *Atlas of Finite Groups* (Oxford University Press, Oxford, 1985)
7. I. Dolgachev, B. van Geemen, S. Kondō, A complex ball uniformization of the moduli space of cubic surfaces via periods of $K3$ surfaces. J. Reine Angew. Math. **588**, 99–148 (2005)

8. E. Freitag, Some modular forms related to cubic surfaces. Kyungpook Math. J. **43**, 433–462 (2003)
9. B. Hunt, A Siegel modular 3-fold that is a Picard modular 3-fold. Comp. Math. **76**, 203–242 (1990)
10. B. Hunt, in *The Geometry of Some Special Arithmetic Quotients*. Lecture Notes in Mathematics, vol. 1637 (Springer, Berlin, 1996)
11. B. Hunt, S.T. Weintraub, Janus-like algebraic varieties. J. Differ. Geom. **39**, 509–557 (1994)
12. J. Igusa, On Siegel modular forms of genus two. Am. J. Math. **86**, 219–245 (1964); Am. J. Math. **88**, 392–412 (1964)
13. V.V. Nikulin, Integral symmetric bilinear formsand its applications. Math. USSR Izv. **14**, 103–167 (1980)
14. I. Piatetskii-Shapiro, I.R. Shafarevich, A Torelli theorem for algebraic surfaces of type $K3$. USSR Izv. **35**, 530–572 (1971)
15. B. van Geemen, Projective models of Picard modular varieties, in *Classification of Irregular Varieties*, ed. by E. Ballico. Proceedings, Trento, 1990. Lecture Notes in Mathematics, vol. 1515 (Springer, Berlin, 1992), pp. 68–99

Quasi-modular Forms Attached to Hodge Structures

Hossein Movasati

Abstract The space D of Hodge structures on a fixed polarized lattice is known as Griffiths period domain, and its quotient by the isometry group of the lattice is the moduli of polarized Hodge structures of a fixed type. When D is a Hermitian symmetric domain, then we have automorphic forms on D, which according to Baily–Borel theorem, they give an algebraic structure to the mentioned moduli space. In this article we slightly modify this picture by considering the space U of polarized lattices in a fixed complex vector space with a fixed Hodge filtration and polarization. It turns out that the isometry group of the filtration and polarization, which is an algebraic group, acts on U and the quotient is again the moduli of polarized Hodge structures. This formulation leads us to a notion of quasi-automorphic forms which generalizes quasi-modular forms attached to elliptic curves.

Key words: Polarized Hodge structure, Period map, Algebraic de Rham cohomology

Mathematics Subject Classifications (2010): 32G20, 11F46, 14C30, 14J15

1 Introduction

In 1970 Griffiths in his article [5] introduced the period domain D and described a project to enlarge D to a moduli space of degenerating polarized Hodge structures. He also asked for the existence of a certain automorphic form theory for D,

H. Movasati (✉)
Instituto de Matemática Pura e Aplicada, IMPA, Estrada Dona Castorina, 110,
22460-320, Rio de Janeiro, RJ, Brazil
e-mail: hossein@impa.br

R. Laza et al. (eds.), *Arithmetic and Geometry of K3 Surfaces and Calabi–Yau Threefolds*,   567
Fields Institute Communications 67, DOI 10.1007/978-1-4614-6403-7_23,
© Springer Science+Business Media New York 2013

generalizing the usual notion of automorphic forms on Hermitian symmetric domains. Since then there have been much effort made on the first part of Griffiths's project (see [8, 13] and the references there). For the second part Griffiths himself introduced the theory of automorphic cohomology; however, the generating function role of automorphic forms is somewhat lacking in this theory.

Some years ago, I was looking for some analytic spaces over D for which one may state the Baily–Borel theorem on the unique algebraic structure of quotients of Hermitian symmetric domains by discrete arithmetic groups. I realized that even in the simplest case of Hodge structures, namely, $h^{01} = h^{10} = 1$, such spaces are not well studied. This led me to the definition of a class of holomorphic functions on the Poincaré upper half plane which generalize the classical modular forms (see [14]). Since a differential operator acts on them, I called them differential modular forms. Soon after I realized that such functions play a central role in mathematical physics and, in particular, in mirror symmetry (see [11] and the references therein). Inspired by this special case of Hodge structures with its fruitful applications, I felt the necessity to develop as much as possible similar theories for an arbitrary type of Hodge structure.

In this note we construct an analytic variety U and an action of an algebraic group G_0 on U from the right such that U/G_0 is the moduli space of polarized Hodge structures of a fixed type. We may pose the following algebraization problem for U, in parallel to the Baily–Borel theorem in [1]: construct functions on U which have some automorphic properties with respect to the action of G_0 and have some finite growth when a Hodge structure degenerates. There must be enough of them in order to enhance U with a canonical structure of an algebraic variety such that the action of G_0 is algebraic. In the case for which the Griffiths period domain is Hermitian symmetric, for instance, for the Siegel upper half plane, this problem seems to be promising but needs a reasonable amount of work if one wants to construct such functions through the inverse of the generalized period maps (see Sect. 5.1). Among them are calculating explicit affine coordinates in certain moduli spaces and calculating Gauss–Manin connections. Some main ingredients of such a study for K3 surfaces endowed with polarizations is already done by many authors; see, for instance, [2] and the references therein. For the case in which the Griffiths period domain is not Hermitian symmetric, we reformulate the algebraization problem further (see Sect. 4.3), and we solve it for the Hodge numbers $h^{30} = h^{21} = h^{12} = h^{03} = 1$ (see Sect. 5.2 and [15]). This gives us a first example of quasi-automorphic forms theory attached to a period domain which is not Hermitian symmetric.

The realization of the algebraization problem in the case of elliptic curves and the corresponding Hodge numbers $h^{10} = h^{01} = 1$ clarifies many details of the previous paragraph; therefore, I explain it here (for more details, see [14]). In this case, $U = \mathrm{SL}(2, \mathbb{Z}) \backslash P$, where

$$P := \{ \begin{pmatrix} x_1 & x_2 \\ x_3 & x_4 \end{pmatrix} \in \mathrm{SL}(2, \mathbb{C}) \mid \mathrm{Im}(x_1 \overline{x_3}) > 0 \}.$$

In order to find an algebraic structure on U we work with the following family of elliptic curves:

$$E_t : y^2 - 4(x - t_1)^3 + t_2(x - t_1) + t_3 = 0,$$

where the parameter $t = (t_1, t_2, t_3)$ is a point of the affine variety

$$T := \{(t_1, t_2, t_3) \in \mathbb{C}^3 \mid 27t_3^2 - t_2^3 \neq 0\}.$$

The generalized period map

$$\mathsf{pm} : T \to U, \tag{1}$$

$$t \mapsto \left[\frac{1}{\sqrt{2\pi i}} \begin{pmatrix} \int_{\delta_1} \frac{dx}{y} & \int_{\delta_1} \frac{x dx}{y} \\ \int_{\delta_2} \frac{dx}{y} & \int_{\delta_2} \frac{x dx}{y} \end{pmatrix} \right]$$

is in fact a biholomorphism. Here, $[\cdot]$ means the equivalence class and $\{\delta_1, \delta_2\}$ is a basis of the $\mathbb{Z}$-module $H_1(E_t, \mathbb{Z})$ with $\langle \delta_1, \delta_2 \rangle = -1$. The algebraic group

$$G_0 = \{ \begin{pmatrix} k & k' \\ 0 & k^{-1} \end{pmatrix} \mid k, k' \in \mathbb{C}, \ k \neq 0 \}$$

acts from the right on U by the usual multiplication of matrices. Under pm the action of G_0 is given by

$$t \bullet g = (t_1 k^{-2} + k' k^{-1}, t_2 k^{-4}, t_3 k^{-6}),$$

$$t = (t_1, t_2, t_3) \in \mathbb{C}^3, \ g = \begin{pmatrix} k & k' \\ 0 & k^{-1} \end{pmatrix} \in G_0.$$

In fact, T is the moduli space of pairs $(E, \{\omega_1, \omega_2\})$, where E is an elliptic curve and $\{\omega_1, \omega_2\}$ is a basis of $H_{dR}^1(E)$ such that ω_1 is represented by a differential form of the first kind and $\frac{1}{2\pi i} \int_E \omega_1 \cup \omega_2 = 1$.

The algebra of quasi-modular forms arises in the following way: we consider the composition of maps

$$\mathbb{H} \overset{i}{\hookrightarrow} P \to U \overset{\mathsf{pm}^{-1}}{\to} T \hookrightarrow \tilde{T}, \tag{2}$$

where $\mathbb{H} = \{\tau \in \mathbb{C} \mid \mathrm{Im}(\tau) > 0\}$ is the upper half plane,

$$i : \mathbb{H} \to P, \ i(\tau) = \begin{pmatrix} \tau & -1 \\ 1 & 0 \end{pmatrix},$$

$P \to U$ is the quotient map and $\tilde{T} = \mathbb{C}^3$ is the underlying complex manifold of the affine variety $\mathrm{Spec}(\mathbb{C}[t_1, t_2, t_3])$. The pullback of the function ring $\mathbb{C}[t_1, t_2, t_3]$ of $\tilde{T}$ by the composition $\mathbb{H} \to \tilde{T}$ is a $\mathbb{C}$-algebra which we call the $\mathbb{C}$-algebra of quasi-modular forms for $SL(2, \mathbb{Z})$. Three Eisenstein series

$$g_i(\tau) = a_k\Big(1 + b_k \sum_{d=1}^{\infty} d^{2k-1} \frac{e^{2\pi i d\tau}}{1 - e^{2\pi i d\tau}}\Big), \quad k = 1, 2, 3, \tag{3}$$

where

$$(b_1, b_2, b_3) = (-24, 240, -504), \quad (a_1, a_2, a_3) = (\frac{2\pi i}{12}, 12(\frac{2\pi i}{12})^2, 8(\frac{2\pi i}{12})^3)$$

are obtained by taking the pullback of the t_i's. Our reformulation of the algebraization problem is based on (2) and the pullback argument; see Sect. 4.3.

We fix some notations from linear algebra. For a basis $\omega_1, \omega_2, \ldots, \omega_h$ of a vector space, we denote by ω an $h \times 1$ matrix whose entries are the ω_i's. In this way we also say that ω is a basis of the vector space. If there is no danger of confusion we also use ω to denote an element of the vector space. We use A^t to denote the transpose of the matrix A. Recall that if δ and ω are two bases of a vector space, $\delta = p\omega$ for some $p \in GL(h, \mathbb{C})$ and a bilinear form on V_0 in the basis δ (resp. ω) has the matrix form A (resp. B) then $pBp^t = A$. By $[a_{ij}]_{h \times h}$ we mean an $h \times h$ matrix whose (i, j) entry is a_{ij}.

2 Moduli of Polarized Hodge Structures

In this section we define the generalized period domain U, and we explain its comparison with the classical Griffiths period domain.

2.1 The Space of Polarized Lattices

We fix a $\mathbb{C}$-vector space V_0 of dimension h, a natural number $m \in \mathbb{N}$ and a $h \times h$ integer-valued matrix Ψ_0 such that the associated bilinear form

$$\mathbb{Z}^h \times \mathbb{Z}^h \to \mathbb{Z}, \ (a, b) \to a^t \Psi_0 b$$

is non-degenerate, symmetric if m is even, and skew if m is odd. Note that, in the case of $\mathbb{Z}$-modules, by non-degenerate we mean that the associated morphism

$$\mathbb{Z}^h \to (\mathbb{Z}^h)^\vee, \ a \to (b \to a^t \Psi_0 b)$$

is an isomorphism, where $\vee$ means the dual of a $\mathbb{Z}$-module.

A lattice $V_\mathbb{Z}$ in V_0 is a $\mathbb{Z}$-module generated by a basis of V_0. A polarized lattice $(V_\mathbb{Z}, \psi_\mathbb{Z})$ of type Ψ_0 is a lattice $V_\mathbb{Z}$ together with a bilinear map $\psi_\mathbb{Z} : V_\mathbb{Z} \times V_\mathbb{Z} \to \mathbb{Z}$ such that in a $\mathbb{Z}$-basis of $V_\mathbb{Z}$, $\psi_\mathbb{Z}$ has the form Ψ_0.

Let $\mathcal{L}$ be the set of polarized lattices of type Ψ_0 in V_0. It has a canonical structure of a complex manifold of dimension $\dim_\mathbb{C}(V_0)^2$. One can take a local chart around $(V_\mathbb{Z}, \psi_\mathbb{Z})$ by fixing a basis of the $\mathbb{Z}$-module $V_\mathbb{Z}$. Usually, we denote an element of

$\mathcal{L}$ by $x, y, \ldots$ and the associated lattice (resp. bilinear form) by $V_{\mathbb{Z}}(x), V_{\mathbb{Z}}(y), \ldots$ (resp. $\psi_{\mathbb{Z}}(x), \psi_{\mathbb{Z}}(y), \ldots$). Let R be any subring of $\mathbb{C}$. For instance, R can be $\mathbb{Q}, \mathbb{R}, \mathbb{C}, \mathbb{Z}$. We define

$$V_R(x) := V_{\mathbb{Z}}(x) \otimes_{\mathbb{Z}} R \text{ and } \psi_R(x) : V_R(x) \times V_R(x) \to R \text{ the induced map.}$$

Conjugation with respect to $x \in \mathcal{L}$ of an element $\omega = \sum_{i=1}^{h} a_i \delta_i \in V_0$, where $V_{\mathbb{Z}}(x) = \sum_{i=1}^{h} \mathbb{Z}\delta_i$, is defined by

$$\overline{\omega}^x := \sum_{i=1}^{h} \overline{a}_i \delta_i,$$

where $\overline{s}$, $s \in \mathbb{C}$ is the usual conjugation of complex numbers.

2.2 Hodge Filtration

We fix Hodge numbers

$$h^{i,m-i} \in \mathbb{N} \cup \{0\}, \ h^i := \sum_{j=i}^{m} h^{j,m-j}, \ i = 0, 1, \ldots, m, \ h^0 = h$$

a filtration

$$F_0^{\bullet} : \{0\} = F_0^{m+1} \subset F_0^m \subset \cdots \subset F_0^1 \subset F_0^0 = V_0, \ dim(F_0^i) = h^i \qquad (4)$$

on V_0 and a bilinear form

$$\psi_0 : V_0 \times V_0 \to \mathbb{C}$$

such that in a basis of V_0, its matrix is Ψ_0 and it satisfies

$$\psi_0(F_0^i, F_0^j) = 0, \ \forall i, j, \ i + j > m. \qquad (5)$$

A basis $\omega_i, i = 1, 2, \ldots, h$ of V_0 is compatible with the filtration $F_0^{\bullet}$ if $\omega_i, i = 1, 2, \ldots, h^i$ is a basis of F_0^i for all i. It is sometimes convenient to fix a basis $\omega_i, i = 1, 2, \ldots, h$ of V_0 which is compatible with the filtration $F_0^{\bullet}$ and such that the polarization matrix $[\psi_0(\omega_i, \omega_j)]$ is a fixed matrix Φ_0:

$$[\psi_0(\omega_i, \omega_j)] = \Phi_0.$$

The matrices Ψ_0 and Φ_0 are not necessarily the same. For any $x \in \mathcal{L}$ we define

$$H^{i,m-i}(x) := F_0^i \cap \overline{F_0^{m-i}}^x$$

and the following properties for $x \in \mathcal{L}$:

1. $\psi_{\mathbb{C}}(x) = \psi_0$.
2. $V_0 = \oplus_{i=0}^{m} H^{i,m-i}(x)$.
3. $(-1)^{\frac{m(m-1)}{2}+i}(\sqrt{-1})^{-m}\psi_{\mathbb{C}}(x)(\omega, \overline{\omega}^x) > 0, \ \forall \omega \in H^{i,m-i}(x), \ \omega \neq 0$.

Throughout the text we call these properties P1, P2, and P3. Fix a polarized lattice $x \in \mathcal{L}$. P1 implies that

$$\psi_0(H^{i,m-i}(x), H^{j,m-j}(x)) = 0 \text{ except for } i + j = m.$$

This is because if $i + j > m$, then $\psi_0(F_0^i, F_0^j) = 0$ and if $i + j < m$, then $\psi_0(\overline{F_0^i}^x, \overline{F_0^j}^x) = 0$. We have also $\sum_i H^{i,m-i}(x) = \oplus_i H^{i,m-i}(x)$ if and only if

$$F_0^i \cap \overline{F_0^j}^x = 0, \ \forall \ i + j > m. \tag{6}$$

If $a_{m-k,k} + \cdots + a_{0,m} = 0$, $a_{i,m-i} \in H^{i,m-i}(x)$ for some $0 \leq k \leq m$ with $a_{m-k,k} \neq 0$, then

$$-a_{m-k,k} = a_{m-k-1,k+1} + \cdots + a_{0,m} \in F_0^{m-k} \cap \overline{F_0^{k+1}}^x \Rightarrow a_{k,m-k} = 0$$

which is a contradiction. The proof in the other direction is a consequence of

$$F_0^i \cap \overline{F_0^j}^x = H^{i,m-i}(x) \cap H^{m-j,j}(x), \ i + j > m.$$

2.3 Period Domain U

Define

$$X := \{x \in \mathcal{L} \mid x \text{ satisfies P1 }\},$$

$$U := \{x \in \mathcal{L} \mid x \text{ satisfies P1,P2, P3 }\}.$$

Proposition 1. *The set X is an analytic subset of $\mathcal{L}$ and U is an open subset of X.*

Proof. Take a basis ω_i, $i = 1, 2, \ldots, h$ of V_0 compatible with the Hodge filtration. The property (5) is given by

$$\psi_{\mathbb{C}}(x)(\omega_r, \omega_s) = 0, \ r \leq h^i, \ s \leq h^j, \ i + j > m$$

and so X is an analytic subset of $\mathcal{L}$.

Now choose a basis δ of $V_{\mathbb{Z}}(x)$ and write $\delta = p\omega$. Using ω we may assume that $V_0 = \mathbb{C}^h$ and δ is constituted by the rows of p. We have

$$\omega = p^{-1}\delta \Longrightarrow \overline{\omega}^x = \overline{p}^{-1}\delta = \overline{p}^{-1}p\omega.$$

Therefore, the rows of $\overline{p}^{-1}p$ are complex conjugates of the entries of ω. Now it is easy to verify that if the property (6), $\dim(H^{i,m-i}(x)) = h^{i,m-i}$ and P3 are valid for one x, then they are valid for all points in a small neighborhood of x (for P3 we may first restrict ψ_0 to the product of sphere of radius 1 and center $0 \in \mathbb{C}^h$).

2.4 An Algebraic Group

Let G_0 be the algebraic group

$$G_0 := \mathrm{Aut}(F_0^\bullet, \psi_0) :=$$

$$\{g : V_0 \to V_0 \text{ linear} \mid g(F_0^i) = F_0^i, \ \psi_0(g(\omega_1), g(\omega_2)) = \psi_0(\omega_1, \omega_2), \omega_1, \omega_2 \in V_0\}.$$

It acts from the right on $\mathcal{L}$ in a canonical way:

$$xg := g^{-1}(x), \ \psi_\mathbb{Z}(xg)(\cdot, \cdot) := \psi_\mathbb{Z}(g(\cdot), g(\cdot)), \ g \in G_0, \ x \in \mathcal{L}.$$

One can easily see that for all $\omega \in V_0$, $x \in \mathcal{L}$, and $g \in G$, we have

$$\overline{\omega}^{xg} = g^{-1}\overline{g(\omega)}^x.$$

Proposition 2. *The properties P1, P2, and P3 are invariant under the action of G_0.*

Proof. The property P1 for xg follows from the definition. Let $x \in \mathcal{L}$, $g \in G_0$ and $\omega \in V_0$. We have

$$H^{i,m-i}(xg) = F_0^i \cap \overline{F_0^{m-i}}^{xg} = F_0^i \cap g^{-1}\overline{g(F_0^{m-i})}^x = F_0^i \cap g^{-1}(\overline{F_0^{m-i}}^x)$$
$$= g^{-1}(F_0^i \cap \overline{F_0^{m-i}}^x) = g^{-1}(H^{i,m-i}(x))$$

and

$$\psi_\mathbb{C}(xg)(\omega, \overline{\omega}^{xg}) = \psi_\mathbb{C}(x)(g(\omega), gg^{-1}\overline{g(\omega)}^x) = \psi_\mathbb{C}(x)(g(\omega), \overline{g(\omega)}^x).$$

These equalities prove the proposition.

The above proposition implies that G_0 acts from the right on U. We fix a basis $\omega_i, i = 1, 2, \ldots, h$, of V_0 compatible with the Hodge filtration $F_0^\bullet$, and if there is no danger of confusion, we identify each $g \in G_0$ with the $h \times h$ matrix $\tilde{g}$ given by

$$[g^{-1}(\omega_1), g^{-1}(\omega_2), \ldots, g^{-1}(\omega_h)] = [\omega_1, \omega_2, \ldots, \omega_h]\tilde{g}. \tag{7}$$

2.5 Griffiths Period Domain

In this section we give the classical approach to the moduli of polarized Hodge structures due to P. Griffiths. The reader is referred to [8, 9] for more developments in this direction.

Let us fix the $\mathbb{C}$-vector space V_0 and the Hodge numbers as in Sect. 2.2. Let also F be the space of filtrations (4) in V_0. In fact, F has a natural structure of a compact smooth projective variety. We fix the polarized lattice $x_0 \in \mathcal{L}$ and define the Griffiths domain

$$D := \{F^\bullet \in \mathrm{F} \mid (V_{\mathbb{Z}}(x_0), \psi_{\mathbb{Z}}(x_0), F^\bullet) \text{ is a polarized Hodge structure }\}.$$

The group

$$\Gamma_{\mathbb{Z}} := \mathrm{Aut}(V_{\mathbb{Z}}(x_0), \psi_{\mathbb{Z}}(x_0))$$

acts on V_0 from the right in the usual way and this gives us an action of $\Gamma_{\mathbb{Z}}$ on D. The space $\Gamma_{\mathbb{Z}} \backslash D$ is the moduli space of polarized Hodge structures.

Proposition 3. *There is a canonical isomorphism*

$$\beta : U/G_0 \xrightarrow{\sim} \Gamma_{\mathbb{Z}} \backslash D.$$

Proof. We take $x \in U$ and an isomorphism

$$\gamma : (V_{\mathbb{Z}}(x), \psi_{\mathbb{Z}}(x)) \xrightarrow{\sim} (V_{\mathbb{Z}}(x_0), \psi_{\mathbb{Z}}(x_0)).$$

The pushforward of the Hodge filtration $F_0^\bullet$ under this isomorphism gives us a Hodge filtration on V_0 with respect to the lattice $V_{\mathbb{Z}}(x_0)$ and so it gives us a point $\beta(x) \in D$. Different choices of γ leads us to the action of $\Gamma_{\mathbb{Z}}$ on $\beta(x)$. Therefore, we have a well-defined map

$$\beta : U \to \Gamma_{\mathbb{Z}} \backslash D.$$

Since $G_0 = \mathrm{Aut}(V_0, F_0^\bullet, \psi_0)$, β induces the desired isomorphism (it is surjective because for any polarized Hodge structure $(V_{\mathbb{Z}}(x_0), \psi_{\mathbb{Z}}(x_0), F^\bullet)$, we have $V_{\mathbb{Z}}(x_0) = V_0$, $\psi_{\mathbb{C}}(x_0) = \psi_0$, and $F^\bullet = g(F_0^\bullet)$ for some $g \in G_0$).

The Griffiths domain is the moduli space of polarized Hodge structures of a fixed type and with a $\mathbb{Z}$-basis in which the polarization has a fixed matrix form. Our domain U is the moduli space of polarized Hodge structures of a fixed type and with a $\mathbb{C}$-basis compatible with the Hodge filtration and for which the polarization has a fixed matrix form.

3 Period Map

In this section we introduce Poincaré duals, period matrices, and Gauss-Manin connections in the framework of polarized Hodge structures.

3.1 Poincaré Dual

In this section we explain the notion of Poincaré dual. Let $(V_{\mathbb{Z}}(x), \psi_{\mathbb{Z}}(x))$ be a polarized lattice and $\delta \in V_{\mathbb{Z}}(x)^{\vee}$, where $\vee$ means the dual of a $\mathbb{Z}$-module. We will use the symbolic integral notation

$$\int_{\delta} \omega := \delta(\omega), \ \forall \omega \in V_0.$$

The equality

$$\int_{\delta} \overline{\omega}^x = \overline{\int_{\delta} \omega}, \ \forall \omega \in V_0, \ \delta \in V_{\mathbb{Z}}(x)^{\vee} \tag{8}$$

follows directly from the definition. The Poincaré dual of $\delta \in V_{\mathbb{Z}}(x)^{\vee}$ is an element $\delta^{\mathsf{pd}} \in V_{\mathbb{Z}}(x)$ with the property

$$\int_{\delta} \omega = \psi_{\mathbb{Z}}(x)(\omega, \delta^{\mathsf{pd}}), \ \forall \omega \in V_{\mathbb{Z}}(x).$$

It exists and is unique because $\psi_{\mathbb{Z}}$ is non-degenerate. Using the Poincaré duality one defines the dual polarization

$$\psi_{\mathbb{Z}}(x)^{\vee}(\delta_i, \delta_j) := \psi_{\mathbb{Z}}(x)(\delta_i^{\mathsf{pd}}, \delta_j^{\mathsf{pd}}), \ \delta_i, \delta_j \in V_{\mathbb{Z}}(x)^{\vee}.$$

We have

$$(A^{\vee}\delta)^{\mathsf{pd}} = A^{-1}\delta^{\mathsf{pd}}, \ \forall A \in \Gamma_{\mathbb{Z}}, \ \delta \in V_{\mathbb{Z}}(x_0)^{\vee},$$

where $A^{\vee} : V_{\mathbb{Z}}(x_0)^{\vee} \to V_{\mathbb{Z}}(x_0)^{\vee}$ is the induced dual map. This follows from:

$$\int_{A^{\vee}\delta} \omega = \int_{\delta} A\omega = \psi_{\mathbb{Z}}(x_0)(A\omega, \delta^{\mathsf{pd}}) = \psi_{\mathbb{Z}}(x_0)(\omega, A^{-1}\delta^{\mathsf{pd}}), \ \forall \omega \in V_0.$$

We define

$$\Gamma_{\mathbb{Z}}^{\vee} := \mathrm{Aut}(V_{\mathbb{Z}}(x_0)^{\vee}, \psi_{\mathbb{Z}}(x_0)^{\vee}).$$

It follows that $\Gamma_{\mathbb{Z}} \to \Gamma_{\mathbb{Z}}^{\vee}$, $A \mapsto A^{\vee}$ is an isomorphism of groups.

3.2 Period Matrix

Let $\omega_i, i = 1, 2, \ldots, h$ be a $\mathbb{C}$-basis of V_0 compatible with $F_0^\bullet$. Recall that ω means the $h \times 1$ matrix with entries ω_i. For $x \in U$, we take a $\mathbb{Z}$-basis δ_i, $i = 1, 2, \ldots, h$ of $V_{\mathbb{Z}}(x)^\vee$ such that the matrix of $\psi_{\mathbb{Z}}(x)^\vee$ in the basis δ is Ψ_0. We define the abstract period matrix/period map in the following way:

$$\mathsf{pm} = \mathsf{pm}(x) = [\int_{\delta_i} \omega_j]_{h \times h} := \begin{pmatrix} \int_{\delta_1} \omega_1 & \int_{\delta_1} \omega_2 & \cdots & \int_{\delta_1} \omega_h \\ \int_{\delta_2} \omega_1 & \int_{\delta_2} \omega_2 & \cdots & \int_{\delta_2} \omega_h \\ \vdots & \vdots & \vdots & \vdots \\ \int_{\delta_h} \omega_1 & \int_{\delta_h} \omega_2 & \cdots & \int_{\delta_h} \omega_h \end{pmatrix}.$$

Instead of the period matrix it is useful to use the matrix

$$\mathsf{q} = \mathsf{q}(x), \quad \text{where } \delta^{\mathsf{pd}} = \mathsf{q}\omega.$$

Then we have

$$\Psi_0^{\mathsf{t}} = \mathsf{pm} \cdot \mathsf{q}^{\mathsf{t}}.$$

If we identify V_0 with $\mathbb{C}^h$ through the basis ω, then q is a matrix whose rows are the entries of δ. We define P to be the set of period matrices pm. We write an element A of $\Gamma_{\mathbb{Z}}$ in a basis of $V_{\mathbb{Z}}(x_0)$ and redefine $\Gamma_{\mathbb{Z}}$:

$$\Gamma_{\mathbb{Z}} := \{A \in \mathrm{GL}(h, \mathbb{Z}) \mid A\Psi_0 A^{\mathsf{t}} = \Psi_0\}.$$

The group $\Gamma_{\mathbb{Z}}$ acts on P from the left by the usual multiplication of matrices and

$$U = \Gamma_{\mathbb{Z}} \backslash P.$$

In a similar way, if we identity each element g of G_0 with the matrix $\tilde{g}$ in (7), then G_0 acts from the right on P by the usual multiplication of matrices.

3.3 A Canonical Connection on $\mathcal{L}$

We consider the trivial bundle $\mathcal{H} = \mathcal{L} \times V_0$ on $\mathcal{L}$. On $\mathcal{H}$ we have a well-defined integrable connection

$$\nabla : \mathcal{H} \to \Omega_{\mathcal{L}}^1 \otimes_{\mathcal{O}_{\mathcal{L}}} \mathcal{H}$$

such that a section s of $\mathcal{H}$ in a small open set $V \subset \mathcal{L}$ with the property

$$s(x) \in \{x\} \times V_{\mathbb{Z}}(x), \quad x \in V$$

is flat. Let $\omega_1, \omega_2, \ldots, \omega_h$ be a basis of V_0 compatible with the Hodge filtration $F_0^\bullet$. We can consider ω_i as a global section of $\mathcal{H}$ and so we have

$$\nabla\omega = A \otimes \omega, \ A = \begin{pmatrix} \omega_{11} & \omega_{12} & \cdots & \omega_{1h} \\ \omega_{21} & \omega_{22} & \cdots & \omega_{2h} \\ \vdots & \vdots & \ddots & \vdots \\ \omega_{h1} & \omega_{h2} & \cdots & \omega_{hh} \end{pmatrix}, \ \omega_{ij} \in H^0(\mathcal{L}, \Omega^1_{\mathcal{L}}). \tag{9}$$

A is called the connection matrix of ∇ in the basis ω. The connection ∇ is integrable and so $dA = A \wedge A$:

$$d\omega_{ij} = \sum_{k=1}^{h} \omega_{ik} \wedge \omega_{kj}, \ i, j = 1, 2, \ldots, h. \tag{10}$$

Let δ be a basis of flat sections. Write $\delta = \mathsf{q}\omega$. We have

$$\omega = \mathsf{q}^{-1}\delta \Rightarrow \nabla(\omega) = d(\mathsf{q}^{-1})\mathsf{q}\omega \Rightarrow$$

$$A = d\mathsf{q}^{-1} \cdot \mathsf{q} = d(\mathsf{pm}^t \cdot \varPsi_0^{-t}) \cdot (\varPsi_0^t \cdot \mathsf{pm}^{-t}) = d(\mathsf{pm}^t) \cdot \mathsf{pm}^{-t}.$$

and so

$$A = d(\mathsf{pm}^t) \cdot \mathsf{pm}^{-t}. \tag{11}$$

where pm is the abstract period map. We have used the equality $\varPsi_0 = \mathsf{pm} \cdot \mathsf{q}^t$. Note that the entries of A are holomorphic 1-forms on $\mathcal{L}$ and a fundamental system for the linear differential equation $dY = A \cdot Y$ in $\mathcal{L}$ is given by $Y = \mathsf{pm}^t$:

$$d\mathsf{pm}^t = A \cdot \mathsf{pm}^t.$$

We define the Griffiths transversality distribution by:

$$\mathcal{F}_{gr} : \omega_{ij} = 0, \ i \le h^{m-x}, \ j > h^{m-x-1}, \ x = 0, 1, \ldots, m-2. \tag{12}$$

A holomorphic map $f : V \to U$, where V is an analytic variety, is called a period map if it is tangent to the Griffiths transversality distribution, that is, for all ω_{ij} as in (12), we have $f^{-1}\omega_{ij} = 0$.

3.4 Some Functions on $\mathcal{L}$

For two vectors $\omega_1, \omega_2 \in V_0$, we have the following holomorphic function on $\mathcal{L}$:

$$\mathcal{L} \to \mathbb{C}, \ x \mapsto \psi_{\mathbb{C}}(x)(\omega_1, \omega_2).$$

We choose a basis ω of V_0 and δ of $V_{\mathbb{Z}}(x)^{\vee}$ for $x \in \mathcal{L}$ and write $\delta^{\mathsf{pd}} = \mathsf{q} \cdot \omega$. Then

$$F := [\psi_{\mathbb{C}}(x)(\omega_i, \omega_j)] = (\mathsf{q}^{-1})\varPsi_0 \mathsf{q}^{-t} = \mathsf{pm}^t \varPsi_0^{-t}\mathsf{pm} \tag{13}$$

(we have used the identity $\Psi_0^t = \mathsf{pm} \cdot \mathsf{q}^t$). The matrix F satisfies the differential equation

$$dF = A \cdot F + F \cdot A^t, \tag{14}$$

where A is the connection matrix. The proof is a straightforward consequence of (13) and (11):

$$
\begin{aligned}
dF &= d(\mathsf{pm}^t \Psi_0^{-t} \mathsf{pm}) \\
&= (d\mathsf{pm}^t)\Psi_0^{-t}\mathsf{pm} + \mathsf{pm}^t \Psi_0^{-t}(d\mathsf{pm}) \\
&= A \cdot F + F \cdot A^t
\end{aligned}
$$

It is easy to check that every solution of the differential equation (14) is of the form $\mathsf{pm}^t \cdot C \cdot \mathsf{pm}$ for some constant $h \times h$ matrix C with entries in $\mathbb{C}$ (if F is a solution of (14), then $F \cdot \mathsf{pm}^{-1}$ is a solution of $dY = A \cdot Y$). We restrict $F, A,$ and pm to U and we conclude that

$$\Phi_0 = \mathsf{pm}^t \Psi_0^{-t} \mathsf{pm} \tag{15}$$

$$A \cdot \Phi_0 = -\Phi_0 \cdot A^t,$$

where by definition $F|_U$ is the constant matrix Φ_0.

We have a plenty of non-holomorphic functions on $\mathcal{L}$. For two elements $\omega_1, \omega_2 \in V_0$, we define

$$\mathcal{L} \to \mathbb{C}, \; x \mapsto \psi_{\mathbb{C}}(x)(\omega_1, \overline{\omega_2}^x).$$

Let ω and δ be as before. We write $\delta^{\mathsf{pd}} = \overline{\mathsf{q}} \cdot \overline{\omega}^x$ and we have

$$G := [\psi_{\mathbb{C}}(x)(\omega_i, \bar{\omega}_j^x)] = \mathsf{pm}^t \Psi_0^{-t} \overline{\mathsf{pm}} = (\mathsf{q}^{-1}) \Psi_0 \overline{\mathsf{q}}^{-t} \tag{16}$$

The matrix G satisfies the differential equation

$$dG = A \cdot G + G \cdot \overline{A}^t, \tag{17}$$

where A is the connection matrix.

4 Quasi-modular Forms Attached to Hodge Structures

In this section we explain what is a quasi-modular form attached to a given fixed data of Hodge structures and a full family of enhanced projective varieties.

4.1 Enhanced Projective Varieties

Let X be a complex smooth projective variety of a fixed topological type. This means that we fix a C^∞ manifold X_0 and assume that X as a C^∞-manifold is isomorphic to

X_0 (we do not fix the isomorphism). Let n be the complex dimension of X and let m be an integer with $1 \leq m \leq n$. We fix an element $\theta \in H^{2n-2m}(X, \mathbb{Z}) \cap H^{n-m,n-m}(X)$. By $H^i(X, \mathbb{Z})$ we mean its image in $H^i(X, \mathbb{C}) = H^i_{\mathrm{dR}}(X)$; therefore, we have killed the torsion. We consider the bilinear map

$$\langle \cdot, \cdot \rangle_{\mathbb{C}} : H^m_{\mathrm{dR}}(X) \times H^m_{\mathrm{dR}}(X) \to \mathbb{C}, \quad \langle \omega, \alpha \rangle = \frac{1}{(2\pi i)^m} \int_X \omega \cup \alpha \cup \theta.$$

The $(2\pi i)^{-m}$ factor in the above definition ensures us that the bilinear map $\langle \cdot, \cdot \rangle_{\mathbb{C}}$ is defined for the algebraic de Rham cohomology (see, for instance, Deligne's lecture in [3]). We assume that it is non-degenerate. The cohomology $H^m_{\mathrm{dR}}(X)$ is equipped with the so-called Hodge filtration $F^\bullet$. We assume that the Hodge numbers $h^{i,m-i}$, $i = 0, 1, 2, \ldots, m$ coincide with those fixed in this article. We consider Hodge structures with an isomorphism

$$(H^m_{\mathrm{dR}}(X), F^\bullet, \langle \cdot, \cdot \rangle_{\mathbb{C}}) \cong (V_0, F_0^\bullet, \psi_0).$$

From now on, by an enhanced projective variety, we mean all the data described in the previous paragraph.

We also need to introduce families of enhanced projective varieties. Let V be an irreducible affine variety and $\mathcal{O}_V$ be the ring of regular functions on V. By definition V is the underlying complex space of $\mathrm{Spec}(\mathcal{O}_V)$ and $\mathcal{O}_V$ is a finitely generated reduced $\mathbb{C}$-algebra without zero divisors. Also, let $X \to V$ be a family of smooth projective varieties as in the previous paragraph. We will also use the notations $\{X_t\}_{t \in V}$ or X/V to denote $X \to V$. The de Rham cohomology $H^m_{\mathrm{dR}}(X/V)$ and its Hodge filtration $F^\bullet H^m_{\mathrm{dR}}(X/V)$ are $\mathcal{O}_V$-modules (see, for instance, [7]), and in a similar way we have $\langle \cdot, \cdot \rangle_{\mathcal{O}_V} : H^m_{\mathrm{dR}}(X/V) \times H^m_{\mathrm{dR}}(X/V) \to \mathcal{O}_V$. Note that we fix an element $\theta \in F^{n-m} H^{2n-2m}_{\mathrm{dR}}(X/V)$ and assume that it induces in each fiber X_t an element in $H^{2n-2m}(X_t, \mathbb{Z})$. We say that the family is enhanced if we have an isomorphism

$$\left(H^m_{\mathrm{dR}}(X/V), F^\bullet H^m_{\mathrm{dR}}(X/V), \langle \cdot, \cdot \rangle_{\mathcal{O}_V} \right) \cong \left(V_0 \otimes_{\mathbb{C}} \mathcal{O}_V, F_0^\bullet \otimes_{\mathbb{C}} \mathcal{O}_V, \psi_0 \otimes_{\mathbb{C}} \mathcal{O}_V \right). \quad (18)$$

We fix a basis ω_i, $i = 1, 2, \ldots, h$ of V_0 compatible with the filtration $F_0^\bullet$. Under the above isomorphism we get a basis $\tilde{\omega}_i$, $i = 1, 2, \ldots, h$ of the $\mathcal{O}_V$-module $H^m_{\mathrm{dR}}(X/V)$ which is compatible with the Hodge filtration and the bilinear map $\langle \cdot, \cdot \rangle_{\mathcal{O}_V}$ written in this basis is a constant matrix. This gives us another formulation of an enhanced family of projective varieties. An enhanced family of projective varieties $\{X_t\}_{t \in V}$ is full if we have an algebraic action of G_0 (defined in Sect. 2.4) from the right on V (and hence on $\mathcal{O}_V$) such that it is compatible with the isomorphism (18). This is equivalent to saying that for X_t and $\tilde{\omega}_i$, $i = 1, 2, \ldots, h$ as above, we have an isomorphism

$$(X_{tg}, [\tilde{\omega}_1, \tilde{\omega}_2, \ldots, \tilde{\omega}_h]) \cong (X_t, [\tilde{\omega}_1, \tilde{\omega}_2, \ldots, \tilde{\omega}_h]g), \quad t \in V, \ g \in G_0,$$

(recall the matrix form of $g \in G_0$ in (7)). A morphism $Y/W \to X/V$ of two families of enhanced projective varieties is a commutative diagram

$$
\begin{array}{ccc}
Y & \to & X \\
\downarrow & & \downarrow \\
W & \to & V
\end{array}
$$

such that

$$
\begin{array}{ccc}
H^m(X/V) & \to & H^m(Y/W) \\
\downarrow & & \downarrow \\
V_0 \otimes_{\mathbb{C}} \mathcal{O}_V & \to & V_0 \otimes_{\mathbb{C}} \mathcal{O}_W
\end{array}
$$

is also commutative.

4.2 Period Map

For an enhanced projective variety X, we consider the image of $H^m(X, \mathbb{Z})$ in $H^m(X, \mathbb{C}) \cong H^m_{\mathrm{dR}}(X) \cong V_0$, and hence we obtain a unique point in U. Note that by this process we kill torsion elements in $H^m(X, \mathbb{Z})$. We fix bases ω_i and $\tilde{\omega}_i$ as in Sect. 4.1 and a basis δ_i, $i = 1, 2, \ldots, h$ of $H_m(X, \mathbb{Z}) = H^m(X, \mathbb{Z})^{\vee}$ with $[\langle \delta_i, \delta_j \rangle] = \Psi_0$, and we see that the corresponding point in $U := \Gamma_{\mathbb{Z}} \backslash P$ is given by the equivalence class of the geometric period matrix $[\int_{\delta_i} \tilde{\omega}_j]$.

For any family of enhanced projective varieties $\{X_t\}_{t \in V}$, we get

$$
\mathsf{pm} : V \to U
$$

which is holomorphic. It satisfies the so-called Griffiths transversality, that is, it is tangent to the Griffiths transversality distribution. It is called a geometric period map. The pullback of the connection ∇ constructed in Sect. 3.3 by the period map pm is the Gauss–Manin connection of the family $\{X_t\}_{t \in V}$. If the family is full, then the geometric period map commutes with the action of G_0:

$$
\mathsf{pm}(tg) = \mathsf{pm}(t)g, \ g \in G_0, \ t \in V.
$$

4.3 Quasi-modular Forms

Let M be the set of enhanced projective varieties with the fixed topological data explained in Sect. 4.1. We would like to prove that M is in fact an affine variety. The first step in developing a quasi-modular form theory attached to enhanced projective varieties is to solve the following conjectures.

Conjecture 1. There is an affine variety T and a full family X/T of enhanced projective varieties which is universal in the following sense: for any family of enhanced projective varieties Y/S, we have a unique morphism of $Y/S \to X/T$ of enhanced projective varieties.

We would also like to find a universal family which describes the degeneration of projective varieties:

Conjecture 2. There is an affine variety $\tilde{T} \supset T$ of the same dimension as T and with the following property: for any family $f : Y \to S$ of projective varieties with fixed prescribed topological data, but not necessarily enhanced and smooth, and with the discriminant variety $\Delta \subset S$, the map $Y \backslash f^{-1}(\Delta) \to S \backslash \Delta$ is an underlying morphism of an enhanced family, and hence, we have the map $S \backslash \Delta \to T$ which extends to $S \to \tilde{T}$. The conjecture is about the existence of $\tilde{T}$ with such an extension property.

Similar to Shimura varieties, we expect that T and $\tilde{T}$ are affine varieties defined over $\bar{\mathbb{Q}}$. Both conjectures are true in the case of elliptic curves (see the discussion in the Introduction). In this case, the function ring of T (resp. $\tilde{T}$) is $\mathbb{C}[t_1, t_2, t_3, \frac{1}{27t_3^2 - t_2^3}]$ (resp. $\mathbb{C}[t_1, t_2, t_3]$). We have also verified the conjectures for a particular class of Calabi–Yau varieties (see Sect. 5.2 and [15]).

Now, consider the case in which both conjectures are true. We are going to explain the rough idea of the algebra of quasi-modular forms attached to all fixed data that we had. It is the pullback of the $\mathbb{C}$-algebra of regular functions in $\tilde{T}$ by the composition

$$\mathbb{H} \overset{i}{\hookrightarrow} P|_{\mathrm{Im(pm)}} \to U|_{\mathrm{Im(pm)}} \overset{\mathsf{pm}^{-1}}{\to} T \hookrightarrow \tilde{T}. \tag{19}$$

Here pm is the geometric period map. We need that the period map is locally injective (local Torelli problem) and hence pm^{-1} is a local inverse map. The set $\mathbb{H}$ is a subset of the set of period matrices P, and it will play the role of the Poincaré upper half plane. If the Griffiths period domain D is Hermitian symmetric, then it is biholomorphic to D (see Sect. 5.1); however, in other cases, it depends on the universal period map $T \to U$ and its dimension is the dimension of the deformation space of the projective variety. In this case we do not need to define $\mathbb{H}$ explicitly (see Sect. 5.2). More details of this discussion will be explained by two examples of the next section.

5 Examples

In this section we discuss two examples of Hodge structures and the corresponding quasi-modular form algebras: those attached to mirror quintic Calabi–Yau varieties and principally polarized Abelian varieties. The details of the first case are done in [15, 16] and we will sketch the results which are related to the main thread of the present text. For the second case there is much work that has been done and I only sketch some ideas. Much of the work for K3 surfaces endowed with polarizations has been already done by many authors; see [2] and the references therein. The generalization of such results to Siegel quasi-modular forms is work for the future.

5.1 Siegel Quasi-modular Forms

We consider the case in which the weight m is equal to 1 and the polarization matrix is:

$$\Psi_0 = \begin{pmatrix} 0 & -I_g \\ I_g & 0 \end{pmatrix},$$

where I_g is the $g \times g$ identity matrix. In this case $g := h^{10} = h^{01}$ and $h = 2g$. We take a basis ω_i, $i = 1, 2, \ldots, 2g$, of V_0 compatible with $F_0^\bullet$, that is, the first g elements form a basis of F_0^1. We further assume that the polarization $\psi_0 : V_0 \times V_0 \to \mathbb{C}$ in the basis ω has the form $\Phi_0 := \Psi_0$. Because of the particular format of Ψ_0, both these assumptions do not contradict each other. We take a basis δ of $V_{\mathbb{Z}}(x)^\vee$ such that the intersection form in this basis is of the form Ψ_0 and we write the associated period matrix in the form

$$[\int_{\delta_i} \omega_j] = \begin{pmatrix} x_1 & x_2 \\ x_3 & x_4 \end{pmatrix},$$

where x_i, $i = 1, \ldots, 4$, are $g \times g$ matrices. Since $\Psi_0^{-t} = \Psi_0$, we have

$$\begin{pmatrix} 0 & -I_g \\ I_g & 0 \end{pmatrix} = \begin{pmatrix} x_1^t & x_3^t \\ x_2^t & x_4^t \end{pmatrix} \begin{pmatrix} 0 & -I_g \\ I_g & 0 \end{pmatrix} \begin{pmatrix} x_1 & x_2 \\ x_3 & x_4 \end{pmatrix}$$

$$= \begin{pmatrix} x_3^t x_1 - x_1^t x_3 & x_3^t x_2 - x_1^t x_4 \\ x_4^t x_1 - x_2^t x_3 & x_4^t x_2 - x_2^t x_4 \end{pmatrix}$$

and

$$[\langle \omega_i, \bar{\omega}_j^x \rangle] = \begin{pmatrix} x_1^t & x_3^t \\ x_2^t & x_4^t \end{pmatrix} \begin{pmatrix} 0 & -I_g \\ I_g & 0 \end{pmatrix} \begin{pmatrix} \bar{x}_1 & \bar{x}_2 \\ \bar{x}_3 & \bar{x}_4 \end{pmatrix}$$

$$= \begin{pmatrix} x_3^t \bar{x}_1 - x_1^t \bar{x}_3 & x_3^t \bar{x}_2 - x_1^t \bar{x}_4 \\ x_4^t \bar{x}_1 - x_2^t \bar{x}_3 & x_4^t \bar{x}_2 - x_2^t \bar{x}_4 \end{pmatrix}.$$

The properties P1, P2, and P3 are summarized in the properties

$$x_3^t x_1 = x_1^t x_3, \quad x_3^t x_2 - x_1^t x_4 = -I_g,$$

$$x_1, x_2 \in \mathrm{GL}(g, \mathbb{C}),$$

$$\sqrt{-1}(x_3^t \bar{x}_1 - x_1^t \bar{x}_3) \text{ is a positive matrix.}$$

By definition P is the set of all $2g \times 2g$ matrices $\begin{pmatrix} x_1 & x_2 \\ x_3 & x_4 \end{pmatrix}$ satisfying the above properties: the matrix $x := x_1 x_3^{-1}$ is well defined and invertible and satisfies the well-known
Riemann relations:

$$x^t = x, \ \mathrm{Im}(x) \text{ is a positive matrix.}$$

The set of matrices $x \in \mathrm{Mat}^{g \times g}(\mathbb{C})$ with the above properties is called the Siegel
upper half space and is denoted by $\mathbb{H}$. We have $U = \Gamma_{\mathbb{Z}} \backslash P$, where

$$\Gamma_{\mathbb{Z}} = \mathrm{Sp}(2g, \mathbb{Z}) = \{\begin{pmatrix} a & b \\ c & d \end{pmatrix} \in \mathrm{GL}(2g, \mathbb{Z}) \mid ab^t = ba^t, \ cd^t = dc^t, \ ad^t - bc^t = I_g\}.$$

We have also

$$G_0 = \{\begin{pmatrix} k & k' \\ 0 & k^{-t} \end{pmatrix} \in \mathrm{GL}(2g, \mathbb{C}) \mid kk'^t = k'k^t\}$$

which acts on P from the right. The group $Sp(2g, \mathbb{Z})$ acts on $\mathbb{H}$ by

$$\begin{pmatrix} a & b \\ c & d \end{pmatrix} \cdot x = (ax + b)(cx + d)^{-1}, \ \begin{pmatrix} a & b \\ c & d \end{pmatrix} \in Sp(2g, \mathbb{Z}), \ x \in \mathbb{H},$$

and we have the isomorphism

$$U/G_0 \to Sp(2g, \mathbb{Z}) \backslash \mathbb{H},$$

given by

$$\begin{pmatrix} x_1 & x_2 \\ x_3 & x_4 \end{pmatrix} \to x_1 x_3^{-1}.$$

To each point x of P we associate a triple $(A_x, \theta_x, \alpha_x)$ as follows: we have $A_x :=
\mathbb{C}^g/\Lambda_x$, where Λ_x is the $\mathbb{Z}$-submodule of $\mathbb{C}^g$ generated by the rows of x_1 and x_3.
We have cycles $\delta_i \in H_1(A_x, \mathbb{Z})$, $i = 1, 2, \ldots, 2g$, which are defined by the property
$[\int_{\delta_i} dz_j] = \begin{pmatrix} x_1 \\ x_3 \end{pmatrix}$, where z_j, $j = 1, 2, \ldots, g$, are linear coordinates of $\mathbb{C}^g$. There is a
basis $\alpha_x = \{\alpha_1, \alpha_2, \ldots, \alpha_{2g}\}$ of $H^1_{\mathrm{dR}}(A_x)$ such that

$$[\int_{\delta_i} \alpha_j] = \begin{pmatrix} x_1 & x_2 \\ x_3 & x_4 \end{pmatrix}.$$

The polarization in $H_1(A_x, \mathbb{Z}) \cong \Lambda_x$ (which is defined by $[\langle \delta_i, \delta_j \rangle] = \Psi_0$) is an
element $\theta_x \in H^2(A_x, \mathbb{Z}) = \bigwedge^2_{i=1} \mathrm{Hom}(\Lambda_x, \mathbb{Z})$. It gives the following bilinear map

$$\langle \cdot, \cdot \rangle : H^1_{\mathrm{dR}}(A_x) \times H^1_{\mathrm{dR}}(A_x) \to \mathbb{C}, \ \langle \alpha, \beta \rangle = \frac{1}{2\pi i} \int_{A_x} \alpha \cup \beta \cup \theta_x^{g-1}$$

which satisfies $[\langle \alpha_i, \alpha_j \rangle] = \Psi_0$.

The triple $(A_x, \theta_x, \alpha_x)$ that we constructed in the previous paragraph does not depend on the action of $Sp(2g, \mathbb{Z})$ from the left on P; therefore, for each $x \in U$, we have constructed such a triple. In fact U is the moduli space of the triples (A, θ, α) such that A is a principally polarized abelian variety with a polarization θ and α is a basis of $H^1_{\mathrm{dR}}(A)$ compatible with the Hodge filtration $F^1 \subset F^0 = H^1_{\mathrm{dR}}(A)$ and such that $[\langle \alpha_i, \alpha_j \rangle] = \Psi_0$.

We constructed the moduli space U in the framework of complex geometry. In order to introduce Siegel quasi-modular forms, we have to study the same moduli space in the framework of algebraic geometry. We have to construct an algebraic variety T over $\mathbb{C}$ such that the points of T are in one-to-one correspondence with the equivalence classes of the triples (A, θ, α). We also expect that T is an affine variety and it lies inside another affine variety $\tilde{T}$ which describes the degeneration of varieties (as it is explained in Sect. 4.3). The pullback of the $\mathbb{C}$-algebra of regular functions on $\tilde{T}$ through the composition

$$\mathbb{H} \to P \to U \overset{\mathrm{pm}^{-1}}{\to} T \hookrightarrow \tilde{T}$$

is, by definition, the $\mathbb{C}$-algebra of Siegel quasi-modular forms. The first map is given by

$$z \to \begin{pmatrix} z & -I_g \\ I_g & 0 \end{pmatrix}$$

and the second is the canonical map. The period map in this case is a biholomorphism. If we impose a functional property for f regarding the action of G_0, then this will be translated into a functional property of a Siegel quasi-modular form with respect to the action of $Sp(2g, \mathbb{Z})$. In this way we can even define a Siegel quasi-modular form defined over $\overline{\mathbb{Q}}$ (recall that we expect $\tilde{T}$ to be defined over $\overline{\mathbb{Q}}$). It is left to the reader to verify that the $\mathbb{C}$-algebra of Siegel quasi-modular forms is closed under derivations with respect to z_{ij} with $z = [z_{ij}] \in \mathbb{H}$. For the realization of all these in the case of elliptic curves, $g = 1$; see the Introduction and [14]. See the books [4, 10, 12] for more information on Siegel modular forms.

5.2 Hodge Numbers, 1, 1, 1, 1

In this section we consider the case $m = 3$ and the Hodge numbers $h^{30} = h^{21} = h^{12} = h^{03} = 1$, $h = 4$. The polarization matrix written in an integral basis is given by

$$\Psi_0 = \begin{pmatrix} 0 & 0 & 1 & 0 \\ 0 & 0 & 0 & 1 \\ -1 & 0 & 0 & 0 \\ 0 & -1 & 0 & 0 \end{pmatrix}.$$

Let us fix a basis $\omega_1, \omega_2, \omega_3, \omega_4$ of V_0 compatible with the Hodge filtration $F_0^\bullet$, a basis $\delta_1, \delta_2, \delta_3, \delta_4 \in V_{\mathbb{Z}}(x)^\vee$ with the intersection matrix Ψ_0, and let us write the period matrix in the form $\mathrm{pm}(x) = [x_{ij}]_{i,j=1,2,\ldots,4}$. We assume that the polarization ψ_0 in the basis ω_i is given by the matrix

$$\Phi_0 := \begin{pmatrix} 0 & 0 & 0 & 1 \\ 0 & 0 & 1 & 0 \\ 0 & -1 & 0 & 0 \\ -1 & 0 & 0 & 0 \end{pmatrix}.$$

The algebraic group G_0 is defined to be

$$G_0 := \left\{ g = \begin{pmatrix} g_{11} & g_{12} & g_{13} & g_{14} \\ 0 & g_{22} & g_{23} & g_{24} \\ 0 & 0 & g_{33} & g_{34} \\ 0 & 0 & 0 & g_{44} \end{pmatrix}, \; g^t\Phi_0 g = \Phi_0, \; g_{ij} \in \mathbb{C} \right\}.$$

One can verify that it is generated by six one-dimensional subgroups, two of them isomorphic to the multiplicative group $\mathbb{C}^*$ and four of them isomorphic to the additive group $\mathbb{C}$. Therefore, G_0 is of dimension 6. We consider the subset $\tilde{\mathbb{H}}$ of P consisting of matrices

$$\tau = \begin{pmatrix} \tau_0 & 1 & 0 & 0 \\ 1 & 0 & 0 & 0 \\ \tau_1 & \tau_3 & 1 & 0 \\ \tau_2 & -\tau_0\tau_3 + \tau_1 & -\tau_0 & 1 \end{pmatrix}, \tag{20}$$

where τ_i, $i = 0, 1, 2, 3$, are some variables in $\mathbb{C}$ (they are coordinates of the corresponding moduli space of polarized Hodge structures and so this moduli space is of dimension four). The particular expressions for the $(4, 2)$ and $(4, 3)$ entries of the above matrix follow from the polynomial relations (15) between periods. The connection matrix A restricted to $\tilde{\mathbb{H}}$ is

$$d\tau^t \cdot \tau^{-t} = \begin{pmatrix} 0 & d\tau_0 & -\tau_3 d\tau_0 + d\tau_1 & -\tau_1 d\tau_0 + \tau_0 d\tau_1 + d\tau_2 \\ 0 & 0 & d\tau_3 & -\tau_3 d\tau_0 + d\tau_1 \\ 0 & 0 & 0 & -d\tau_0 \\ 0 & 0 & 0 & 0 \end{pmatrix}.$$

The Griffiths transversality distribution is given by

$$-\tau_3 d\tau_0 + d\tau_1 = 0, \quad -\tau_1 d\tau_0 + \tau_0 d\tau_1 + d\tau_2 = 0,$$

and so if we consider τ_0 as an independent parameter defined in a neighborhood of $+\sqrt{-1}\infty$, and all other quantities τ_i depending on τ_0, then we have

$$\tau_3 = \frac{\partial\tau_1}{\partial\tau_0}, \quad \frac{\partial\tau_2}{\partial\tau_0} = \tau_1 - \tau_0\frac{\partial\tau_1}{\partial\tau_0}. \tag{21}$$

In [15] we have checked the conjectures in Sect. 4.3 for the Calabi–Yau threefolds of mirror quintic type. In this case $\dim(T) = 7 = 1 + 6$, where 1 is the dimension of the moduli space of mirror quintic Calabi–Yau varieties and 6 is the dimension of the algebraic group G_0. Hence, we have constructed an algebra generated by seven functions in τ_0, which we call it the algebra of quasi-modular forms attached to mirror quintic Calabi–Yau varieties. The image of the geometric period map lies in $\mathbb{H}$ with

$$\tau_1 = -\frac{25}{12} + \frac{5}{2}\tau_0(\tau_0 + 1) + \frac{1}{(2\pi i)^2} \sum_{n=1}^{\infty} \left(\sum_{d|n} n_d d^3 \right) \frac{e^{2\pi i \tau_0 n}}{n^2}. \tag{22}$$

Here, n_d's are instanton numbers and the second derivative of τ_1 with respect to τ_0 is the Yukawa coupling. The Yukawa coupling itself turns out to be a quasi-modular form in our context but not its double primitive τ_1. The set $\mathbb{H}$ is a subset of $\tilde{\mathbb{H}}$ defined by (21) and (22). As far as I know this is the first case in which the Griffiths period domain is not Hermitian symmetric, and we have an attached algebra of quasi-modular forms and even the Global Torelli problem is true; that is, the period map is globally injective (see [6]). However, note that in [15] we have only used the local injectivity of the period map. In this case we can prove that the pullback map from the algebra of regular functions on $\tilde{T}$ to the algebra of holomorphic functions on $\mathbb{H}$ is injective. Our quasi-modular form theory in this example is attached to mirror quintic Calabi–Yau varieties and not the corresponding period domain. There are other functions τ_1 attached to one-dimensional families of varieties and the corresponding period maps. They may have their own quasi-modular forms algebra different from the one explained in this section.

Acknowledgements I would like to thank the organizers of the Fields workshop for inviting me to participate and give a talk. In particular, I would like to thank Charles F. Doran for his support and stimulating conversation.

References

1. W.L. Baily Jr., A. Borel, Compactification of arithmetic quotients of bounded symmetric domains. Ann. Math. (2) **84**, 442–528 (1966)
2. A. Clingher, C.F. Doran, Lattice polarized K3 surfaces and Siegel modular forms. Adv. Math. **231**(1), 172–212 (2012)
3. P. Deligne, J.S. Milne, A. Ogus, K.-y. Shih, Hodge cycles, motives, and Shimura varieties, in *Lecture Notes in Mathematics*, vol. 900. Philosophical Studies Series in Philosophy, vol. 20 (Springer, Berlin, 1982)
4. E. Freitag, in *Siegelsche Modulfunktionen*. Grundlehren der Mathematischen Wissenschaften, vol. 254 (Springer, Berlin, 1983)
5. S. Usui Periods of integrals on algebraic manifolds: summary of main results and discussion of open problems. Bull. Am. Math. Soc. **76**, 228–296 (1970)
6. P.A. Griffiths, Generic Torelli theorem for quintic-mirror family. Proc. Japan Acad. Ser. A Math. Sci. **84**(8), 143–146 (2008)
7. A. Grothendieck, On the de Rham cohomology of algebraic varieties. Inst. Hautes Études Sci. Publ. Math. **29**, 95–103 (1966)
8. K. Kato, S. Usui, Borel-Serre spaces and spaces of SL(2)-orbits, in *Algebraic Geometry 2000*, Azumino, Hotaka. Advanced Studies in Pure Mathematics, vol. 36 (Mathematical Society of Japan, Tokyo, 2002), pp. 321–382

9. K. Kato, S. Usui, Classifying spaces of degenerating polarized Hodge structures, in *Annals of Mathematics Studies*, vol. 169 (Princeton University Press, Princeton, 2009), pp. xii+336
10. H. Klingen, in *Introductory Lectures on Siegel Modular Forms*. Cambridge Studies in Advanced Mathematics, vol. 20 (Cambridge University Press, Cambridge, 1990)
11. M. Kontsevich, Homological algebra of mirror symmetry, in *Proceedings of the International Congress of Mathematicians*, Zürich, 1994, vol. 1, 2 (Birkhäuser, Basel, 1995), pp. 120–139
12. H. Maass, in *Siegel's Modular Forms and Dirichlet Series*. Lecture Notes in Mathematics, vol. 216 (Springer, Berlin, 1971)
13. H. Movasati, Moduli of polarized Hodge structures. Bull. Braz. Math. Soc. (N.S.) **39**(1), 81–107 (2008)
14. H. Movasati, On differential modular forms and some analytic relations between Eisenstein series. Ramanujan J. **17**(1), 53–76 (2008)
15. H. Movasati, Eisenstein type series for Calabi–Yau varieties. Nucl. Phys. B **847**, 460–484 (2011)
16. H. Movasati, Modular-type functions attached to mirror quintic Calabi–Yau varieties. Preprint (2011) [arXiv:1111.0357]

The Zero Locus of the Infinitesimal Invariant

G. Pearlstein and Ch. Schnell

Abstract Let v be a normal function on a complex manifold X. The infinitesimal invariant of v has a well-defined zero locus inside the tangent bundle TX. When X is quasi-projective, and v is admissible, we show that this zero locus is constructible in the Zariski topology.

Key words: Normal functions, Hodge classes, Algebraic cycles, Mixed Hodge modules

Mathematics Subject Classifications (2010): Primary 14D07; Secondary 32G20, 14K30

1 Introduction

Let $\mathcal{H}$ be a variation of Hodge structure of weight -1 on a Zariski open subset of a smooth complex projective variety X. We shall assume that $\mathcal{H}$ is polarizable and defined over $\mathbb{Z}$. We denote the Hodge filtration on the underlying flat vector bundle $\mathcal{H}_{\mathcal{O}}$ by the symbol $F^\bullet \mathcal{H}_{\mathcal{O}}$. Let v be a normal function, that is to say, a holomorphic and horizontal section of the family of intermediate Jacobians $J(\mathcal{H})$. For any local lifting $\tilde{v}$ to a holomorphic section of $\mathcal{H}_{\mathcal{O}}$, we have

$$\nabla \tilde{v} \in \Omega^1_X \otimes_{\mathcal{O}_X} F^{-1}\mathcal{H}_{\mathcal{O}},$$

G. Pearlstein (✉)
Department of Mathematics, Michigan State University, East Lansing, MI 48824, USA
e-mail: gpearl@math.msu.edu

C. Schnell
Department of Mathematics, Stony Brook University, Stony Brook NY 11794-3651
e-mail: cschnell@math.sunysb.edu

R. Laza et al. (eds.), *Arithmetic and Geometry of K3 Surfaces and Calabi–Yau Threefolds*,   589
Fields Institute Communications 67, DOI 10.1007/978-1-4614-6403-7_24,
© Springer Science+Business Media New York 2013

which is independent of the choice of lifting modulo $\nabla(F^0\mathcal{H}_\mathcal{O})$. We are interested in the subset of the tangent bundle TX defined by the condition $\nabla\tilde{v} \in \nabla(F^0\mathcal{H}_\mathcal{O})$. Concretely, this is the set

$$I(v) = \{\,(x,\xi) \in TX \mid \nabla_\xi(\tilde{v} - \sigma)(x) = 0 \text{ for some } \sigma \in F^0\mathcal{H}_\mathcal{O}\,\}.$$

The following theorem describes the structure of $I(v)$ for admissible v.

Theorem 1.1. *Suppose that v is an admissible normal function on a Zariski open subset of a smooth complex projective variety X. Then $I(v)$ is constructible with respect to the Zariski topology on TX.*

Recall that a subset of an algebraic variety is *constructible* if it is a finite union of subsets that are locally closed in the Zariski topology. At least locally, there is no reason for $I(v)$ to be closed, as the following example shows.

Example 1.2. On the unit disk Δ with coordinate t, the constant local system $\mathbb{Z}^3$ can be made into a variation of mixed Hodge structure $\mathcal{V}$, letting $W_{-1}\mathcal{V}_\mathcal{O}$ be spanned by the vectors $(1,0,0)$ and $(0,1,0)$, and letting $F^0\mathcal{V}_\mathcal{O}$ be spanned by the vectors $(1, i + t^2, 0)$ and $(0, t, 1)$. Since $\text{Im}(t^2) < 1$ for $t \in \Delta$, the variation of Hodge structure on $W_{-1}\mathcal{V}_\mathcal{O}$ is polarized by the standard alternating form on $\mathbb{Z}^2$. For the corresponding normal function v, one shows easily that the tangent vector $a\frac{d}{dt}$ at the point $t \in \Delta$ belongs to the set $I(v)$ if and only if there is a holomorphic function f such that

$$a \cdot f'(t) = 0 \quad \text{and} \quad a \cdot \left(t^2 f'(t) + 2t f(t) + 1\right) = 0$$

are satisfied at t. It follows that $I(v)$ is equal to the tangent bundle of Δ with all nonzero tangent vectors over $t = 0$ removed, and therefore not closed.

Remark 1.3. If one defines the infinitesimal invariant of v via the quotient sheaf construction

$$\delta v = [\nabla\tilde{v}] \in (\mathcal{F}^{-1} \otimes \Omega^1)/(\nabla\mathcal{F}^0)$$

then v is locally constant as soon as δv vanishes on any stalk. In contrast, as shown above, we are defining the vanishing of the infinitesimal invariant pointwise.

Remark 1.4. The previous example can be made geometric as follows: Let $Q \subset \mathbb{P}^3$ be the quadric defined by $x_0^2 + x_1^2 + x_2^2 + x_3^2 = 0$. Fix $a \in \mathbb{C} - \{1, 0, -1\}$. Then, for each $u \in S = \mathbb{C} - \{1, 0, -1, -a\}$ the quadric $Q_u \subset \mathbb{P}^3$ defined by $ax_0^2 + x_1^2 - x_2^2 - ux_3^2 = 0$ intersects Q in a smooth curve E_u of genus 1. The algebraic cycle Z on Q given by twice the difference of two non-parallel lines determines a normal function v_Z over S which vanishes at the point $u = a$ (see Example 3.7). Therefore, if we set $u = a + t^2$ we obtain a normal function v over Δ such that $v(0) = 0$ and the derivative of the local period map of the associated variation of Hodge structure also vanishes to order 1 at $t = 0$.

The proof of Theorem 1.1 is given in Sect. 2. In Sect. 3, we describe the relationship between this paper and the study of algebraic cycles via the approach to the Hodge conjecture by Green–Griffiths [7] using singularities of normal functions.

2 Proof of the Theorem

2.1 Algebraic Description of the Zero Locus

Since X is a projective algebraic variety, it is possible to describe the zero locus $I(v)$ of the infinitesimal invariant of v purely in terms of algebraic objects. In this section, we shall do this by a straightforward classical argument.

Using resolution of singularities, we may assume without loss of generality that v is an admissible normal function on $X - D$, where X is a smooth projective variety, and $D \subseteq X$ is a divisor with normal crossings. Let $\mathcal{V}$ be the admissible variation of mixed Hodge structure with $\mathbb{Z}$-coefficients corresponding to v; then $W_{-1}\mathcal{V} = \mathcal{H}$ and $W_0\mathcal{V}/W_{-1}\mathcal{V} \simeq \mathbb{Z}(0)$ by our choice of weights. The integrable connection $\nabla \colon \mathcal{V}_{\mathcal{O}} \to \Omega^1_{X-D} \otimes \mathcal{V}_{\mathcal{O}}$ on the underlying holomorphic vector bundle $\mathcal{V}_{\mathcal{O}}$ has regular singularities; because X is projective algebraic, it follows from [5] that $\mathcal{V}_{\mathcal{O}}$ and ∇ are algebraic. Admissibility implies that each Hodge bundle $F^p\mathcal{V}_{\mathcal{O}}$ is an algebraic subbundle of $\mathcal{V}_{\mathcal{O}}$; note that they satisfy $\nabla(F^p\mathcal{V}_{\mathcal{O}}) \subseteq \Omega^1_{X-D} \otimes F^{p-1}\mathcal{V}_{\mathcal{O}}$ because of Griffiths transversality.

To prove the constructibility of $I(v)$, our starting point is the exact sequence

$$0 \to F^0\mathcal{H}_{\mathcal{O}} \to F^0\mathcal{V}_{\mathcal{O}} \to \mathcal{O} \to 0 \tag{1}$$

of algebraic vector bundles on $X - D$. Let U be any affine Zariski open subset of $X - D$ with the following two properties: (1) both $F^0\mathcal{H}_{\mathcal{O}}$ and $F^{-1}\mathcal{H}_{\mathcal{O}}/F^0\mathcal{H}_{\mathcal{O}}$ restrict to trivial bundles on U; (2) there are coordinates $x_1, \ldots, x_n \in \Gamma(U, \mathcal{O}_U)$, where $\Gamma(U, -)$ always denotes the space of all algebraic sections of an algebraic coherent sheaf. Since $X - D$ can be covered by finitely many such open subsets, it is clearly sufficient to show that $I(v) \cap TU$ is a constructible subset of TU.

By our choice of U, the tangent bundle TU is trivial; let $\xi_1, \ldots, \xi_n \in \Gamma(TU, \mathcal{O}_{TU})$ be the coordinates in the fiber direction corresponding to the algebraic vector fields $\partial/\partial x_1, \ldots, \partial/\partial x_n$. Let $q = \mathrm{rk}F^0\mathcal{H}_{\mathcal{O}}$ and $p = \mathrm{rk}F^{-1}\mathcal{H}_{\mathcal{O}} \geq q$; we can then choose algebraic sections $e_1, \ldots, e_p \in \Gamma(U, F^{-1}\mathcal{H}_{\mathcal{O}})$ such that $e_1, \ldots, e_q \in \Gamma(U, F^0\mathcal{H}_{\mathcal{O}})$ are a frame for $F^0\mathcal{H}_{\mathcal{O}}$, and $e_1, \ldots, e_p$ are a frame for $F^{-1}\mathcal{H}_{\mathcal{O}}$. For $i = 1, \ldots, q$, we get

$$\nabla e_i = \sum_{k=1}^{n} \sum_{j=1}^{p} dx_k \otimes a^k_{i,j} e_j$$

with certain functions $a^k_{i,j} \in \Gamma(U, \mathcal{O}_U)$. Let $\tilde{v} \in \Gamma(U, F^0\mathcal{V}_{\mathcal{O}})$ be any lifting of the element $1 \in \Gamma(U, \mathcal{O}_X)$; then $\nabla \tilde{v} \in \Gamma(U, \Omega^1_U \otimes F^{-1}\mathcal{H}_{\mathcal{O}})$ can be written in the form

$$\nabla \tilde{v} = \sum_{k=1}^{n} \sum_{j=1}^{p} dx_k \otimes f^k_j e_j$$

for certain functions $f^k_j \in \Gamma(U, \mathcal{O}_U)$. By definition, a point $(x, \xi) \in TU$ lies in the zero locus $I(v)$ of the infinitesimal invariant if and only if there are holomorphic functions $\varphi_1, \ldots, \varphi_q$ defined in a small open ball around $x \in U$, such that

$$\nabla\left(\tilde{v} - \sum_{i=1}^{q} \varphi_i e_i\right)$$

vanishes at the point (x, ξ). When expanded, this translates into the condition that

$$\sum_{k=1}^{n} \sum_{j=1}^{p} \xi_k f_j^k(x) e_j = \sum_{i=1}^{q} \sum_{k=1}^{n} \xi_k \frac{\partial \varphi_i}{\partial x_k}(x) e_i + \sum_{i=1}^{q} \varphi_i(x) \sum_{k=1}^{n} \sum_{j=1}^{p} \xi_k a_{i,j}^k(x) e_j.$$

This is a system of p linear equations in the $q(n+1)$ complex numbers

$$\varphi_i(x) \quad \text{and} \quad \frac{\partial \varphi_i}{\partial x_k}(x) \qquad (\text{for } 1 \leq i \leq q \text{ and } 1 \leq k \leq n),$$

with coefficients in the ring $\Gamma(TU, \mathscr{O}_{TU})$ of regular functions on TU. The proof of Proposition 2.9 shows that the set of points $(x, \xi) \in TU$, where this system has a solution, is a constructible subset of TU. This completes the proof of Theorem 1.1.

Remark 2.1. In the case of an normal function ν arising from a family of cycles defined over a subfield k of $\mathbb{C}$, $I(\nu)$ is defined over a finite extension of k. More precisely, let $\nu : S \to J(\mathcal{H})$ be a k-motivated normal function as in Sect. 4.5 of [10]. Then, as in the proof of Theorem 89 of [10], we see that by virtue of the algebraicity of the Gauss–Manin connection, $\mathrm{Gal}(\mathbb{C}/k)$ permutes the components of $I(\nu)$.

2.2 A More Sophisticated Description

For some purposes, it is better to have a natural extension of $I(\nu)$ to the entire cotangent bundle TX, without modifying the ambient variety X. In this section, we indicate how such an extension can be constructed using the theory of mixed Hodge modules [11].

We begin by recalling how one associates a short exact sequence of the form

$$0 \to F_0 \mathcal{M} \to F_0 \mathcal{N} \to \mathscr{O}_X \to 0 \tag{2}$$

to the given admissible normal function; here $F_0 \mathcal{M}$ and $F_0 \mathcal{N}$ are algebraic coherent sheaves on X, and all three morphisms are morphisms of algebraic coherent sheaves.

The polarizable variation of Hodge structure $\mathcal{H}$ extends uniquely to a polarizable Hodge module with strict support equal to X. We denote by $\mathcal{M}$ the underlying regular holonomic $\mathscr{D}_X$-module; it is the minimal extension of the flat vector bundle $\mathcal{H}_{\mathscr{O}}$. It has a good filtration $F_\bullet \mathcal{M}$ by $\mathscr{O}_X$-coherent subsheaves, and $F_k \mathcal{M}$ is an extension of the Hodge bundle $F^{-k} \mathcal{H}_{\mathscr{O}}$. Since X is a complex projective variety, each $F_k \mathcal{M}$ is an algebraic coherent sheaf, and $\mathcal{M}$ is an algebraic $\mathscr{D}_X$-module.

Because the normal function v is admissible, the corresponding variation of mixed Hodge structure extends uniquely to a mixed Hodge module on X; in fact, this condition is equivalent to admissibility [12], p. 243. Let $\mathcal{N}$ denote the underlying regular holonomic $\mathscr{D}_X$-module, and $F_\bullet\mathcal{N}$ its Hodge filtration; as before, $\mathcal{N}$ is an algebraic $\mathscr{D}_X$-module, and each $F_k\mathcal{N}$ is an algebraic coherent sheaf. We have an exact sequence of regular holonomic $\mathscr{D}_X$-modules

$$0 \to \mathcal{M} \to \mathcal{N} \to \mathcal{O}_X \to 0,$$

in which all three morphisms are strict with respect to the Hodge filtration; in particular, (2) is an exact sequence of algebraic coherent sheaves on X. Because $\mathcal{N}$ is a filtered $\mathscr{D}_X$-module, we have $\mathbb{C}$-linear morphisms $\mathscr{T}_X \otimes F_k\mathcal{N} \to F_{k+1}\mathcal{N}$; note that they are not $\mathcal{O}_X$-linear.

We can use the exact sequence in (2) to construct an extension of the zero locus $I(v)$ to all of X. Inside the tangent bundle TX, we define a subset

$$\tilde{I}(v) = \{ (x,\xi) \in TX \mid (\xi \cdot \sigma)(x) = 0 \text{ for some } \sigma \in F_0\mathcal{N} \text{ with } \sigma \mapsto 1 \},$$

where the notation "$\sigma \in F_0\mathcal{N}$" means that σ is a holomorphic section of the sheaf $F_0\mathcal{N}$, defined in some open neighborhood of the point $x \in X$.

Lemma 2.2. *We have $\tilde{I}(v) = I(v)$ over the Zariski open subset of X where the variation of Hodge structure $\mathcal{H}$ is defined.*

Proof. This is obvious from the definitions.

Denote by $p\colon TX \to X$ the projection. The pullback p^*T_X of the tangent sheaf has a tautological global section θ; in local holomorphic coordinates $x_1, \ldots, x_n$ on X, and corresponding coordinates $(x_1, \ldots, x_n, \xi_1, \ldots, \xi_n)$ on TX, it is given by the formula

$$\theta(x_1, \ldots, x_n, \xi_1, \ldots, \xi_n) = \xi_1 \frac{\partial}{\partial x_1} + \cdots + \xi_n \frac{\partial}{\partial x_n}.$$

Let $\tilde{\mathcal{M}}$ denote the pullback of $\mathcal{M}$ to a filtered $\mathscr{D}$-module on the tangent bundle; because p is smooth, we have $\tilde{\mathcal{M}} = p^*\mathcal{M}$ and $F_k\tilde{\mathcal{M}} = p^*F_k\mathcal{M}$. Similarly define $\tilde{\mathcal{N}}$.

Lemma 2.3. *In the notation introduced above, we have*

$$\tilde{I}(v) = \{ (x,\xi) \in TX \mid (\theta \cdot \tilde{\sigma})(x,\xi) = 0 \text{ for some } \tilde{\sigma} \in F_0\tilde{\mathcal{N}} \text{ with } \tilde{\sigma} \mapsto 1 \}.$$

Proof. The set on the right-hand side clearly contains $\tilde{I}(v)$. To prove that the two sets are equal, suppose that we have $(\theta \cdot \tilde{\sigma})(x,\xi) = 0$ for some holomorphic section $\tilde{\sigma}$ of $F_0\tilde{\mathcal{N}}$, defined in a neighborhood of the point $(x,\xi) \in TX$. Since $F_0\tilde{\mathcal{N}} = p^*F_0\mathcal{N}$, we can write $\tilde{\sigma} = \sum_k f_k \cdot p^*\sigma_k$ for suitably chosen $f_k \in \mathcal{O}_{TX}$ and $\sigma_k \in F_0\mathcal{N}$. Define $\sigma = \sum_k f_k(-,\xi)\sigma_k$ by setting ξ constant; then $\sigma \in F_0\mathcal{N}$ and $\sigma \mapsto 1$. A brief calculation in local coordinates shows that

$$(\xi \cdot \sigma)(x) = (\theta \cdot \tilde{\sigma})(x,\xi) = 0,$$

and so we get $(x,\xi) \in \tilde{I}(v)$ as desired.

The next step is to show that $\tilde{I}(v)$ is the zero locus of a holomorphic section of an analytic coherent sheaf on TX. Let $\mathscr{F}$ denote the analytic coherent sheaf on TX obtained by taking the quotient of $F_1\tilde{\mathcal{M}}$ by the analytic coherent subsheaf generated by $\theta \cdot F_0\tilde{\mathcal{M}}$. For any local holomorphic section $\sigma \in F_0\tilde{\mathcal{N}}$ with $\sigma \mapsto 1$, we have $\theta \cdot \sigma \in F_1\tilde{\mathcal{M}}$, and the image of $\theta \cdot \sigma$ in the quotient sheaf $\mathscr{F}$ is independent of the choice of σ, due to the exactness of (2). In this manner, we obtain a global holomorphic section s of the sheaf $\mathscr{F}$.

Lemma 2.4. $\tilde{I}(v)$ *is the zero locus of the section s of the coherent sheaf $\mathscr{F}$.*

Proof. If $(x, \xi) \in \tilde{I}(v)$, then we have $(\theta \cdot \tilde{\sigma})(x, \xi) = 0$ for some choice of $\tilde{\sigma} \in F_0\tilde{\mathcal{N}}$ with $\sigma \mapsto 1$; in particular, $s(x, \xi) = 0$. Conversely, suppose that we have $s(x, \xi) = 0$ for some point $(x, \xi) \in TX$. By definition of $\mathscr{F}$, we can then find local sections $\tilde{\sigma}_k \in F_0\tilde{\mathcal{M}}$ and local holomorphic functions $f_k \in \mathscr{O}_{TX}$, such that

$$\theta \cdot \tilde{\sigma} - \sum_k f_k \theta \cdot \tilde{\sigma}_k$$

vanishes at the point (x, ξ). Set $a_k = f_k(x, \xi) \in \mathbb{C}$; then

$$\theta \cdot \left(\tilde{\sigma} - \sum_k a_i \tilde{\sigma}_k \right) = \theta \cdot \tilde{\sigma} - \sum_k a_k \theta \cdot \tilde{\sigma}_k$$

also vanishes at (x, ξ), and this shows that $(x, \xi) \in \tilde{I}(v)$.

Despite the analytic definition, both $\mathscr{F}$ and s are actually algebraic objects.

Lemma 2.5. $\mathscr{F}$ *is an algebraic coherent sheaf on TX, and $s \in \Gamma(TX, \mathscr{F})$ is an algebraic global section.*

Proof. Each $F_k\tilde{\mathcal{M}} = p^* F_k\mathcal{M}$ is an algebraic coherent sheaf on TX, and since the tautological section $\theta \in \Gamma(TX, p^*\mathscr{T}_X)$ is clearly algebraic, it follows that $\mathscr{F}$ is an algebraic coherent sheaf. To show that the global section $s \in \Gamma(TX, \mathscr{F})$ is algebraic, observe that we have an exact sequence of algebraic coherent sheaves

$$0 \to F_0\tilde{\mathcal{M}} \to F_0\tilde{\mathcal{N}} \to \mathscr{O}_{TX} \to 0;$$

indeed, (2) is exact, and $p: TX \to X$ is a smooth affine morphism. At every point $(x, \xi) \in TX$, we can therefore find an algebraic section $\sigma \in F_0\tilde{\mathcal{N}}$, defined in a Zariski open neighborhood of (x, ξ), such that $\sigma \mapsto 1$. This clearly implies that s, which is locally given by the image of $\theta \cdot \sigma$ in $\mathscr{F}$, is itself algebraic.

To prove Theorem 1.1, it is clearly sufficient to show that the set $\tilde{I}(v)$ is constructible in the Zariski topology on TX. Lemmas 2.4 and 2.5 reduce the problem to the following general result in abstract algebraic geometry: On any algebraic variety, the zero locus of a section of a coherent sheaf is constructible (but not, in general, Zariski closed). This fact is certainly well-known, but since it was surprising to us at first, we have decided to include a simple proof in the following section.

Remark 2.6. As outlined in the introduction of [3], higher Chow cycles give rise to higher normal functions which can be viewed as extensions of $\mathbb{Z}(0)$ by variations of Hodge structure negative weight. The formalism of mixed Hodge modules remains valid in this setting, and hence the corresponding infinitesimal vanishing locus $\tilde{I}(v)$ is the zero locus of a section of a coherent sheaf, and hence constructible.

2.3 Zero Loci of Sections of Coherent Sheaves

In this section, we carefully define the "zero locus" for sections of coherent sheaves, and show that it is always constructible in the Zariski topology. This is obviously a local problem, and so it suffices to consider the case of affine varieties. Let R be a commutative ring with unit; to avoid technical complications, we shall also assume that R is Noetherian. For any prime ideal $\mathfrak{p} \subseteq R$, we denote by the symbol

$$\kappa(\mathfrak{p}) = R_\mathfrak{p}/\mathfrak{p}R_\mathfrak{p}$$

the residue field at $\mathfrak{p}$; it is isomorphic to the field of fractions of the local ring $R_\mathfrak{p}$. Let $X = \operatorname{Spec} R$ be the set of prime ideals of the ring R, endowed with the Zariski topology. For any ideal $I \subseteq R$, the set

$$V(I) = \{\, \mathfrak{p} \in X \mid \mathfrak{p} \supseteq I \,\}$$

is closed in the Zariski topology on X, and any closed subset is of this form; likewise, for any element $f \in R$, the set

$$D(f) = \{\, \mathfrak{p} \in X \mid \mathfrak{p} \not\ni f \,\}$$

is an open subset, and these open sets form a basis for the Zariski topology.

Definition 2.7. A subset of X is called *constructible* if it is a finite union of subsets of the form $D(f) \cap V(I)$.

Here is how this algebraic definition is related to constructibility on complex algebraic varieties. Suppose that R is a $\mathbb{C}$-algebra of finite type. Let $X(\mathbb{C})$ be the set of all maximal ideals of R, endowed with the classical topology; it is an affine complex algebraic variety, and the inclusion mapping $X(\mathbb{C}) \hookrightarrow X$ is continuous.

Definition 2.8. A subset of $X(\mathbb{C})$ is called *constructible* (in the Zariski topology) if it is the set of maximal ideals in a constructible subset of X.

Any coherent sheaf on $X = \operatorname{Spec} R$ is uniquely determined by the finitely generated R-module of its global sections; conversely, any finitely generated R-module M defines a coherent sheaf on X, and hence by restriction to the subset $X(\mathbb{C})$ an algebraic coherent sheaf $\mathscr{F}_M$ on $X(\mathbb{C})$. Its fiber at the point corresponding to a maximal ideal $\mathfrak{m} \subseteq R$ is the finite-dimensional $\mathbb{C}$-vector space

$$M \otimes_R \kappa(\mathfrak{m}) = M_\mathfrak{m}/\mathfrak{m}M_\mathfrak{m}.$$

Similarly, any element $m \in M$ defines an algebraic global section s_m of the sheaf $\mathscr{F}_M$. Obviously, s_m vanishes at the point corresponding to a maximal ideal $\mathfrak{m} \subseteq R$ if and only if m goes to zero in $M \otimes_R \kappa(\mathfrak{m})$. Thus if we define

$$Z(M,m) = \{\, \mathfrak{p} \in X \mid m \text{ goes to zero in } M \otimes_R \kappa(\mathfrak{p}) \,\},$$

then the zero locus of s_m on $X(\mathbb{C})$ is precisely the set of maximal ideals in $Z(M,m)$. Thus the desired result about zero loci of sections of coherent sheaves is a consequence of the following general theorem in commutative algebra.

Proposition 2.9. *Let R be a commutative Noetherian ring with unit. Then for any finitely generated R-module M, and any $m \in M$, the set $Z(M,m)$ is constructible.*

Proof. We are going to construct a finite covering

$$\operatorname{Spec} R = \bigcup_{k=1}^{n} D(f_k) \cap V(I_k)$$

with $f_1, \ldots, f_n \in R$ and $I_1, \ldots, I_n \subseteq R$, such that for every $k = 1, \ldots, n$, one has

$$Z(M,m) \cap D(f_k) \cap V(I_k) = D(f_k) \cap V(I_k + J_k),$$

for a certain ideal $J_k \subseteq R$. This is sufficient, because it implies that

$$Z(M,m) = \bigcup_{k=1}^{n} D(f_k) \cap V(I_k + J_k)$$

is a constructible subset of $\operatorname{Spec} R$.

Since M is finitely generated and R is Noetherian, we may find a presentation

$$R^{\oplus q} \xrightarrow{A} R^{\oplus p} \twoheadrightarrow M, \tag{3}$$

in which A is a $p \times q$-matrix with entries in R. Let $y \in R^{\oplus p}$ be any vector mapping to $m \in M$. Then $Z(M,m)$ is the set of $\mathfrak{p} \in \operatorname{Spec} R$ such that the equation $y = Ax$ has a solution over the field $\kappa(\mathfrak{p})$.

We construct the desired covering of $\operatorname{Spec} R$ by looking at all possible minors of the matrix A. Fix an integer $0 \le \ell \le \min(p,q)$ and an $\ell \times \ell$-submatrix of A; to simplify the notation, let us assume that it is the $\ell \times \ell$-submatrix in the upper left corner of A. Let f be the determinant of the submatrix, and let I be the ideal generated by all minors of A of size $(\ell + 1) \times (\ell + 1)$; if $\ell = 0$, we set $f = 1$, and if $\ell = \min(p,q)$, we set $I = 0$. We can then make a coordinate change in $R^{\oplus q}$, invertible over the localization $R_f = R[f^{-1}]$, and arrange that

$$A = \begin{pmatrix} f & 0 & \cdots & 0 & 0 & \cdots & 0 \\ 0 & f & \cdots & 0 & 0 & \cdots & 0 \\ \vdots & \vdots & & \vdots & \vdots & & \vdots \\ 0 & 0 & \cdots & f & 0 & \cdots & 0 \\ a_{\ell+1,1} & a_{\ell+1,2} & \cdots & a_{\ell+1,\ell} & a_{\ell+1,\ell+1} & \cdots & a_{\ell+1,q} \\ \vdots & \vdots & & \vdots & \vdots & & \vdots \\ a_{p,1} & a_{p,2} & \cdots & a_{p,\ell} & a_{p,\ell+1} & \cdots & a_{p,q} \end{pmatrix} \in \mathrm{Mat}_{p \times q}(R).$$

Note that our change of coordinates involves multiplication by the adjugate of the submatrix, and hence the determinant of the submatrix is now f^ℓ. To continue, let $J \subseteq R$ be the ideal generated by the elements

$$f y_i - \sum_{j=1}^{\ell} a_{i,j} y_j$$

for $i = \ell + 1, \ldots, p$. Then we have

$$Z(M, m) \cap D(f) \cap V(I) = D(f) \cap V(I + J).$$

Indeed, suppose that $\mathfrak{p}$ is any prime ideal with $f \notin \mathfrak{p}$ and $I \subseteq \mathfrak{p}$. Since $a_{i,j} \in \mathfrak{p}$ for every $\ell + 1 \leq i \leq p$ and $\ell + 1 \leq j \leq q$, the equation $y = Ax$ reduces over the field $\kappa(\mathfrak{p})$ to the equations $y_i = f x_i$ for $i = 1, \ldots, \ell$, and

$$y_i = \sum_{j=1}^{\ell} a_{i,j} x_j$$

for $i = \ell + 1, \ldots, p$; they are obviously satisfied if and only if $J \subseteq \mathfrak{p}$.

We now obtain the assertion by applying the above construction of f, I, and J to all possible $\ell \times \ell$-submatrices of A.

Here is a simple example to show that, when the coherent sheaf is not locally free, the zero locus of a section need not be Zariski closed.

Example 2.10. Let $R = \mathbb{C}[x, y]$, let M be the ideal of R generated by x, y, and let $m = x$. Then M has a free resolution of the form $R \to R^{\oplus 2}$, and $\mathfrak{p} \in Z(M, m)$ if and only if the equations $1 + yf = 0$ and $xf = 0$ have a common solution $f \in \kappa(\mathfrak{p})$. A simple computation now shows that

$$Z(M, m) = \{ \mathfrak{p} \in \mathrm{Spec}\, R \mid x \in \mathfrak{p} \text{ and } y \notin \mathfrak{p} \}.$$

As a subset of $\mathbb{C}^2$, the zero locus consists of the y-axis minus the origin; it is constructible, but not Zariski closed.

3 Relation to Algebraic Cycles

3.1 Green-Griffiths Program

Our interest in the algebraicity of $I(v)$ is motivated in part by the program [7] of Green and Griffiths to study the Hodge conjecture via singularities of normal functions. More precisely, given a smooth complex projective variety X, a very ample line bundle $L \to X$ and a non-torsion, primitive Hodge class ζ of type (n, n) on X, Griffiths and Green construct an admissible normal function

$$v_\zeta : P - \hat{X} \to J(\mathcal{H})$$

on the complement of the dual variety $\hat{X}$ in $P = \mathbb{P}H^0(X, \mathcal{O}(L))$. At each point $\hat{x} \in \hat{X}$, the cohomology class of v_ζ localizes to an invariant

$$\operatorname{sing}_{\hat{x}}(v_\zeta) \in IH^1_{\hat{x}}(\mathcal{H})$$

called the *singularity* of v_ζ at $\hat{x}$. A normal function v_ζ is said to be *singular* if there is a point $\hat{x} \in \hat{X}$ at which $\operatorname{sing}_{\hat{x}}(v_\zeta)$ is non-torsion.

Conjecture 3.1. Let (X, L, ζ) be as above. Then, there exists an integer $k > 0$ such that after replacing L by L^k, the associated normal function v_ζ is singular.

Theorem 3.2 ([2, 4, 7]). *Conjecture* (3.1) *holds (for every even dimensional X and every non-torsion, primitive middle dimensional Hodge class ζ) if and only if the Hodge conjecture holds (for all smooth projective varieties).*

Now, as explained in part III of [7] one can also define a notion $\operatorname{sing}_{\hat{x}}(\delta v_\zeta)$ of the singularities of infinitesimal invariant δv_ζ of v_ζ. Moreover,

$$\operatorname{sing}_{\hat{x}}(\delta v_\zeta) = \operatorname{sing}_{\hat{x}}(v_\zeta)$$

for $L \gg 0$. As a first attempt at constructing points at which v_ζ is singular, observe that

$$Z(v_\zeta) = \{ p \in P - \hat{X} \mid v_\zeta(p) = 0 \}$$

is an analytic subset of $P - \hat{X}$, and hence it is natural to ask if its closure is an algebraic subvariety of P which intersects $\hat{X}$ at some point where v_ζ is singular. An affirmative answer is provided by the following two results:

Theorem 3.3 ([1, 9, 14]). *If S is a smooth complex algebraic variety and $v : S \to J(\mathcal{H})$ is an admissible normal function then $Z(v)$ is an algebraic subvariety of S.*

Proposition 3.4 ([13]). *Let v_ζ be the normal function on $P \setminus \hat{X}$, associated to a non-torsion primitive Hodge class $\zeta \in H^{2n}(X, \mathbb{Z}) \cap H^{n,n}(X)$. Assume that $Z(v_\zeta)$ contains an algebraic curve C, and that $P = |L^d|$ for L very ample and $d \geq 3$. Then v_ζ is singular at one of the points where the closure of C meets $\hat{X}$.*

The caveat here, which is illustrated in the example (3.7) below, is that there is no reason for $Z(v_\zeta)$ to contain a curve. The advantage of working with the infinitesimal invariant is that it is often easier to compute [8], and will vanish along directions tangent to $Z(v)$. Of course, $I(v)$ will also contain the directions tangent to any m-torsion locus of v, as well as potentially other components.

Question 3.5. Is there an analog of Proposition (3.4) for $I(v_\zeta)$?

Remark 3.6. The study of zero loci of normal function also arises in connection with the construction of the Bloch–Beilinson filtration on Chow groups. For a survey of results of this type, see [10].

The determination of a good notion of the expected dimension of the zero locus of a normal function is an important open problem in the study of algebraic cycles. In particular, in the Green–Griffiths setting, if a smooth projective variety has moduli, any reasonable expected dimension count is probably only valid at the generic point of the locus where the class ζ remains a Hodge class.

In the case of a smooth projective surface X, if $L = \mathcal{O}(D)$ is a very ample line bundle, then a Riemann–Roch calculation shows the expected dimension of the zero locus of the associated normal functions arising from the Green–Griffiths program (i.e., comparing the dimensions of the fiber and the base) is

$$-(D \cdot K_X) + \chi(\mathcal{O}_X) - 2$$

where K_X is the canonical bundle of X. For X of general type, on the basis of this calculation one would expect the zero locus to be empty for all sufficiently ample L. We close with a careful study of a simple example of normal function of Green–Griffiths type for which the naive expected dimension count is positive.

Example 3.7. Let $X = \mathbb{P}^1 \times \mathbb{P}^1$ viewed as the smooth quadric $Q = V(q) \subseteq \mathbb{P}^3$ defined by the vanishing of $q = x_0^2 + x_1^2 + x_2^2 + x_3^2$. Let L_α and L_β be the lines on Q defined by the equations

$$L_\alpha : t \mapsto [1, t, it, i], \qquad L_\beta : t \mapsto [1, t, -it, i]$$

Then, the difference $\zeta = [L_\alpha] - [L_\beta]$ is a primitive Hodge class on X. For future use, we also introduce the line

$$L_\gamma : t \mapsto [1, t, -it, -i]$$

which is parallel to L_α and intersects L_β at $t = \infty$.

Let $P = \mathbb{P}H^0(X, \mathcal{O}(2))$. Then, the associated normal function v_ζ assigns to each smooth section

$$X_\sigma = V(q) \cap V(\sigma)$$

the class of $(L_\alpha - L_\beta) \cap V(\sigma)$ in the Jacobian of X_σ. A naive expected dimension count for the zero locus of v_ζ can be obtained as follows: The dimension of P is $8 = 10 - 1 - 1$ since the space of quadratic forms on $\mathbb{C}^4$ has dimension 10, and we need mod out by Q and then projectivize. The adjunction formula shows the fibers X_σ to have genus 1. Accordingly, the graph of ν_ζ in the associated bundle of Jacobians $J \to P$ has codimension 1. Likewise, the zero section of J is also codimension 1, and so to first approximation the zero locus of ν_ζ in this case should have codimension 2 in J, which corresponds to a seven-dimensional subvariety of P.

To see that the zero locus of ν_ζ is in fact empty, let $Y \subset \mathbb{P}^3$ be a smooth quadric which intersects X in a smooth curve E. Let $\Lambda \subset X$ be a line of the form $\{z\} \times \mathbb{P}^1$ which intersects E in a pair of distinct points

$$e = (z, w), \qquad f = (z, w')$$

Let the line $\Upsilon = \mathbb{P}^1 \times \{w\}$ intersect E in the divisor $e + g$. Then, since every line on X is parallel to either Λ or Υ, it follows that $L_\alpha - L_\beta$ intersects E in a divisor which is linearly equivalent to

$$(e + f) - (e + g) \sim f - g$$

Accordingly, if ν_ζ vanishes at Y then $f \sim g$ and hence $\Lambda = \Upsilon$.

As a consequence of symmetries however, the 2-torsion locus of ν_ζ is non-zero. To be explicit, let $S = \mathbb{C} - \{-1, -i, 0, i, 1\}$ and $\mu : S \to P$ be the map which associates to a point $s \in S$ the quadric

$$Q_s = V(s^2 x_0^2 + x_1^2 - x_2^2 - s^2 x_3^2)$$

Then, for each $s \in S$, the associated curve $X_{\mu(s)}$ is smooth.

Let θ be the involution of $\mathbb{P}^3$ induced by the linear map

$$(c_0, c_1, c_2, c_3) \mapsto (-c_3, -c_2, c_1, c_0)$$

on $\mathbb{C}^4$. Then, the lines L_α and L_γ are the projectivizations of the $\pm i$-eigenspaces of this map, and hence are pointwise fixed under the action of θ. The involution θ also fixes the quadrics Q and Q_s, and hence the curve $X_{\mu(s)}$. Consequently, the fixed points of the action θ on $X_{\mu(s)}$ are exactly the four points

$$\alpha_1 = [1, is, -s, i], \qquad \alpha_2 = [1, -is, s, i]$$
$$\gamma_1 = [1, is, s, -i], \qquad \gamma_2 = [1, -is, -s, -i]$$

corresponding to the intersection of the lines L_α and L_γ with Q_s. The line L_β on the other hand intersects Q_s at the points

$$\beta_1 = [1, is, s, i], \qquad \beta_2 = [1, -is, -s, i]$$

which are interchanged under the action of θ.

Let

$$F_1 = sx_0 + ix_1 - x_2 + isx_3$$
$$F_2 = sx_0 - ix_1 + x_2 + isx_3$$
$$F_3 = ix_1 + x_2$$

Then, direct calculation shows that $V(F_1)$ is a plane passing through $\{\alpha_1, \alpha_2, \gamma_1\}$ which is also tangent to E_s at γ_1. Similarly, $V(F_2)$ is a plane passing through $\{\alpha_1, \alpha_2, \gamma_2\}$ which is tangent to E_s at γ_2. Finally, $V(F_3)$ is a plane passing through $\{\beta_1, \beta_2, \gamma_1, \gamma_2\}$. Moreover, one can easily check that these planes have no additional points of intersection or tangency other than the ones listed above. Therefore, the rational function

$$F = (F_1 F_2)/F_3^2$$

on $\mathbb{P}^3$ restricts to a meromorphic function on E_s with divisor

$$(\alpha_1 + \alpha_2 + 2\gamma_1) + (\alpha_1 + \alpha_2 + 2\gamma_2) - 2(\beta_1 + \beta_2 + \gamma_1 + \gamma_2) = 2(\alpha_1 + \alpha_2) - 2(\beta_1 + \beta_2)$$

and hence $2\nu_\zeta$ vanishes along the image of μ.

Finally, to get a 7-dimensional subvariety of P as predicted above, observe that the group $SO(4)$ has dimension 6 and acts on $\mathbb{P}^3$ preserving the quadric Q. This action also fixes the integral Hodge class ζ, and hence acts on the 2-torsion locus. The orbit of S under the action of $SO(4)$ therefore provides a 7-dimensional complex analytic subvariety of P on which $2\nu_\zeta$ vanishes.

Acknowledgements G.P. is partially supported by NSF grant DMS-1002625. C.S. is supported by the World Premier International Research Center Initiative (WPI Initiative), MEXT, Japan, and by NSF grant DMS-1100606. We thank Patrick Brosnan, Matt Kerr and James Lewis for helpful discussions. G.P. also thanks the IHES for partial support during the preparation of this manuscript.

References

1. P. Brosnan, G. Pearlstein, On the algebraicity of the zero locus of an admissible normal function, [arXiv:0910.0628]
2. P. Brosnan, H. Fang, Z. Nie, G. Pearlstein, Singularities of admissible normal functions. Invent. Math. **177**(3), 599–629 (2009)
3. P. Brosnan, G. Pearlstein, C. Schnell, The locus of Hodge classes in an admissible variation of mixed Hodge structure. C. R. Acad. Sci. **348**, 657–660 (2010)
4. M.A. de Cataldo, L. Migliorini, On singularities of primitive cohomology classes. Proc. Am. Math. Soc. **137**, 3593–3600 (2009)
5. P. Deligne, in *Equations différentielles à points singuliers réguliers*. Lecture Notes in Mathematics, vol. 163 (Springer, Berlin, 1970)
6. A. Dimca, M. Saito, Vanishing cycle sheaves of one-parameter smoothings and quasi-semistable degenerations. J. Algebr. Geom. **21**(2), 247–271 (2012)
7. M. Green, P. Griffiths, Algebraic cycles and singularities of normal functions. Lond. Math. Soc. Lect. Note Ser. **343**, 206–263 (2007)
8. P. Griffiths, Infinitesimal variations of Hodge structure (III): determinantal varieties and the infinitesimal invariant of normal functions. Compos. Math. **50**(2–3), 267–324 (1983)
9. K. Kato, C. Nakayama, S. Usui, Analyticity of the closures of some Hodge theoretic subspaces. Proc. Jpn. Acad. **87**, 167–172 (2011)
10. M. Kerr, G. Pearlstein, An exponential history of functions with logarithmic growth. MSRI Publ. **58**, 281–374 (2010)
11. M. Saito, Mixed Hodge modules. Publ. Res. Inst. Math. Sci. **26**(2), 221–333 (1990)

12. M. Saito, Admissible normal functions. J. Algebr. Geom. **5**(2), 235–276 (1996)
13. Ch. Schnell, Two observations about normal functions. Clay Math. Proc. **9**, 75–79 (2010)
14. Ch. Schnell, Complex-analytic Néron models for arbitrary families of intermediate Jacobians. Invent. Math. **188**, 1–81 (2012)

Printed by Publishers' Graphics LLC
LMO130627.15.14.14